DYNAMICS OF THE BERING SEA

Thomas R. Loughlin and Kiyotaka Ohtani, Editors

*A Summary of Physical, Chemical,
and Biological Characteristics,
and a Synopsis of Research on the Bering Sea*

North Pacific Marine Science Organization (PICES)

Published by University of Alaska Sea Grant, Fairbanks, Alaska

AK-SG-99-03 • 1999 • $40.00

Elmer E. Rasmuson Library Cataloging in Publication Data:

Dynamics of the Bering Sea / Thomas R. Loughlin and Kiyotaka Ohtani, editors. — Fairbanks : Alaska Sea Grant College Program, 1999.

xi, 838 p.   cm. — (University of Alaska Sea Grant ; AK-SG-99-03)

Includes bibliographical references and index.

ISBN 1-56612-062-4

1. Bering Sea. 2. Marine biology—Bering Sea. 3. Oceanography—Bering Sea. 4. Chemical oceanography—Bering Sea. 5. Marine productivity—Bering Sea. 6. Primary productivity (Biology)—Bering Sea. 7. Bering Sea—Research.   I. Title. II. Loughlin, Thomas R. III. Ohtani, Kiyotaka. IV. Series: Alaska Sea Grant College Program report ; AK-SG-99-03.

GC841.D96 1999

Citation: Loughlin, T.R., and K. Ohtani, eds. 1999. Dynamics of the Bering Sea. University of Alaska Sea Grant, AK-SG-99-03, Fairbanks. 838 pp.

Acknowledgments

This book is the result of a joint effort by the North Pacific Marine Science Organization (PICES), and the University of Alaska Sea Grant College Program.

PICES is an intergovernmental scientific organization established in 1992. PICES member-countries are Canada, Japan, the People's Republic of China, the Republic of Korea, the Russian Federation, and the United States of America. The purposes of PICES are (1) to promote and coordinate marine scientific research in the northern North Pacific and adjacent marginal seas, particularly northward from 30°north; (2) to advance scientific knowledge about the ocean environment, global weather and climate change, living resources and their ecosystems, and the impact of human activities on them; and (3) to promote the collection and rapid exchange of scientific information on these issues. The PICES Secretariat is in Sidney, B.C., Canada, phone (250) 363-6366.

Sea Grant is a unique partnership with public and private sectors combining research, education, and technology transfer for public service. This national network of universities meets changing environmental and economic needs of people in our coastal, ocean, and Great Lakes regions. University of Alaska Sea Grant is cooperatively supported by the U.S. Department of Commerce, NOAA National Sea Grant Office, under grant no. NA86RG-0050, project A/75-01; and by the University of Alaska with state funds. The University of Alaska is an affirmative action/equal opportunity institution.

University of Alaska Sea Grant
(Publisher)
P.O. Box 755040
205 O'Neill Bldg.
Fairbanks, Alaska 99775-5040
FYPUBS@uaf.edu
Phone (907) 474-6707
Fax (907) 474-6285
http://www.uaf.edu/seagrant/

Contents

Chemical Distributions and Dynamics

Biological Dynamics

Synopsis of Major Research Programs

Foreword

A Bering Sea Working Group was established at the first North Pacific Marine Science Organization (PICES) Annual Meeting, in 1992. The group is interdisciplinary as well as international, and its activities have included identification of key unanswered questions (see Appendix) and organization of a Bering Sea Symposium held in October 1995 during the Fourth Annual Meeting of PICES, in Qingdao, China. An early charge was to initiate preparation of a book reviewing present knowledge of the climatology, oceanography, and biology of the Bering Sea.

Recently, several substantial Bering Sea volumes have appeared. Russian papers edited by Kotenev and Sapozhnikov (1995) are concerned with the Bering Sea ecosystem, and Russian investigations are also the basis for Mathisen and Coyle's (1996) compilation of papers on ecology of the Bering Sea. A different sort of review is found in the Committee on the Bering Sea Ecosystem's (1996) analysis of the state of the Bering Sea ecosystem.

Each of these publications is specialized, in purpose, geography, nationality, or scientific discipline. Changes in the political relations among countries working in the Bering Sea, and the advent of a regional international scientific organization, PICES, have now provided the opportunity to bring the findings of scientists of all disciplines and countries together in a contemporary synthesis of the oceanography of this region.

Vast as the Bering Sea is (2.3 million km^2 according to Fairbridge [1966]), it is a convenient microcosm for such a synthesis. Although the basin is semi-enclosed, in a geographical sense, it actively exchanges with the Arctic Ocean and the North Pacific. Atmospheric forcing is on a large scale, with the Aleutian Low playing an important role. Biological conditions reflect the oceanic forcing, with tides being particularly important on the continental shelf. The ecosystems are complex like those of other high latitude regions, and are enormously productive of finfish and shellfish, along with marine mammals and seabirds. It seems likely that even before commercial harvesting of living resources began, the resident populations exhibited large fluctuations in distribution and abundance, resulting just from the physical forcing and the internal dynamics of the ecosystems.

One reason for the recent interest in the Bering Sea is the perception that commercial fisheries, among the largest in the world, have seriously damaged Bering Sea ecosystems. Evidence cited for this damage includes declines in several conspicuous populations, including king crab, Steller sea lions, and several species of seabirds. There is a popular sentiment that all human activities in favored ecosystems are inherently bad, and fishery biologists have traditionally assumed that essentially all changes, but especially decreases, in fish stocks have resulted from human activity.

Meanwhile, evidence is accumulating that climate variations on decadal to century scales have identifiable ecosystem effects and that interdecadal changes such as the 1976 regime shift have been associated with major changes in fish stocks in the North Pacific. For example, a recent study (Wyllie-Echeverria 1996) showed that the coverage of winter sea ice in the Bering Sea, which correlates with the distribution of pollock the following summer, was significantly different before and after the late 1970s. Variations in ocean circulation and mixing, driven by the atmosphere, may affect ecosystems either from the bottom up or the top down, and the linkages are as yet poorly understood. At the same time, it is obvious that ecosystems are also forced by human predation, and our skill at sorting out the relative importance of these kinds of forcing in any given case is limited.

Such questions led PICES in 1993 to initiate planning for a program called Climate Change and Carrying Capacity (CCCC), as part of an international program on Global Ocean Ecosystem Dynamics (GLOBEC). The goal of CCCC is to understand the effect of climate variations on marine ecosystems. A comparable effort is also needed to establish the effect of fishing on these ecosystems. Ultimately, predicting the response of climate and human forcing on any specific marine populations will depend on a much better understanding of these systems than we now have. For any given region, such as the Bering Sea, a synthesis of existing knowledge such as the present volume, is a key to the eventual achievement of that predictive capability.

Warren S. Wooster
School of Marine Affairs
University of Washington
Seattle, Washington

References

Committee on the Bering Sea Ecosystem. 1996. The Bering Sea ecosystem. National Academy Press, Washington, D.C. 305 pp.

Fairbridge, R.W. (ed.). 1966. The encyclopedia of oceanography. Reinhold, New York. 135 pp.

Kotenev, B.N., and V.V. Sapozhnikov (eds.). 1995. Complex studies of the Bering Sea ecosystem. VNIRO, Moscow. 412 pp. (In Russian.)

Mathisen, O.A., and K.O. Coyle (eds.). 1996. Ecology of the Bering Sea: A review of Russian literature. University of Alaska Sea Grant, AK-SG-96-01, Fairbanks. 306 pp.

Wyllie-Echeverria, T. 1996. The relationship between the distribution of one-year-old walleye pollock, *Theragra chalcogramma*, and sea-ice characteristics. In: R.D. Brodeur, P.A. Livingston, T.R. Loughlin, and A.B. Hollowed (eds.), Ecology of juvenile walleye pollock. NOAA Technical Report NMFS-126, pp. 47-56.

Preface

Dr. Warren Wooster mentioned in the foreword that this book was originally conceived in the PICES Bering Sea Working Group chaired by Dr. Al Tyler.
The last review (Hood and Calder 1981) was outdated, and because of the
current interests in Bering Sea resources the working group felt that a
timely review was needed. This book reviews past publications on the
Bering Sea, looks at the type and direction of present research in and
adjacent to the Bering Sea, and makes suggestions for future studies. The
intent is to summarize the current body of work on the Bering Sea as a
resource guide to scientists, managers, and others from all nations working in the Bering Sea and similar ecosystems.

The Bering Sea is one of the most productive ecosystems in the world.
It is at least a seasonal home to some of the largest marine invertebrate,
fish, bird, and mammal populations among the world's seas. The North
Pacific Ocean and Bering Sea are not only one of the most productive
ocean regions in the world, they also support many of the world's largest
populations of groundfish, salmon, and crab. For example, large-scale
commercial fisheries for groundfish in Alaska waters are now a major
industry with 1992 landings totaling almost 2 million metric tons generating ex-vessel revenues of over $600 million. When the landings and val-
ues generated by fisheries for Pacific halibut, crabs, and salmon are also
considered, the commercial fisheries off Alaska alone generate several
billion dollars each year, as well as millions of metric tons of protein for
the world.

The resources of the Bering Sea also provide for the subsistence needs
of residents of the Bering Sea region. Food, clothing, cultural and religious
artifacts, handicraft items, and economically valuable products have been
taken from the Bering Sea for thousands of years by inhabitants of the
Bering Sea coast. These subsistence resources are essential today to the
livelihood of American and Russian Natives and other residents of Bering
Sea coastal communities.

The U.S. National Research Council recently established the Committee on the Bering Sea Ecosystem and charged it with (1) evaluating the
environmental factors and ecological relationships that control the Bering
Sea food web, (2) evaluating the probable cause and effects of observed
population fluctuations, (3) estimating historical population dynamics of
marine mammals and birds, and commercially important species, (4) evaluating historical records of commercial fisheries, and (5) evaluating the
relationship between biological resources and subsistence cultures of in-
digenous people. Some of the recommendations for research and management within the NRC report are relevant to this book. They include
recommendations to (1) improve coordination of the many institutions
and groups conducting studies or utilizing the resources of the Bering
Sea, (2) adopt a broader ecosystem perspective for both scientific research
and management of Bering Sea resources, and (3) conduct research on the

structure of the ecosystem (with emphasis on the cause of pollock population dynamics and the role of sea ice in structuring the ecosystem). Many of these recommendations are included here.

Dynamics of the Bering Sea has four parts—the first part has chapters in particular disciplines, the second has chapters reviewing major programs formerly conducted in the Bering Sea, third is a concluding chapter summarizing the entire volume, and last is an appendix on research topics that need more study. Within each of the three discipline area sections (Physical Dynamics, Chemical Distribution and Dynamics, and Biological Dynamics), the first chapter is a review and the following assortment of chapters report specific research projects to provide a glimpse at the kind of studies currently under way in PICES member nations.

After receiving guidance on the general topics and framework for the chapters from us, the section editors selected the topics to be addressed in their sections, chose authors to write on the topics, solicited the papers, and arranged for peer review. Section editors for Physical Dynamics are Dr. S. Gladychev, Pacific Oceanographic Institute, Russian Academy of Sciences of the Far East, Vladivostok, Russia; and Dr. J. Schumaker, PMEL, Seattle, WA, U.S.A. Section editors for Chemical Distributions and Dynamics are Dr. T. Whitledge, University of Alaska Fairbanks, Fairbanks, AK, U.S.A. (formerly at the University of Texas); Dr. Y. Maita, Hokkaido University, Hakodate, Japan; and Dr. V. Sapozhnikov, VNIRO, Moscow, Russia. Section editors for Biological Dynamics are Dr. R. Francis, University of Washington, Seattle, WA, U.S.A.; and Dr. T. Ikeda, Hokkaido University, Hakodate, Japan.

Some highlights of the book follow:

1. The existence of interannual and decadal changes in biological resources and variability of oceanic conditions affected by meteorological and climatic changes.

2. The magnitude of the variations in biological resources caused by climate change is as large as, or even exceeds, those caused by human activities (fishing, etc.); however, our knowledge of the processes and mechanisms on those relationships are qualitative, fragmentary, and scarce.

3. An attempt is made to show a network of physical, chemical, and biological relations in the Bering Sea ecosystem in an effort to guide and enhance future investigations.

4. Cooperative research among all nations is needed to attain needed answers to complex ecosystem questions, which should be the ultimate goal of future efforts.

We wish to acknowledge the financial support and encouragement offered by PICES, particularly Drs. Warren Wooster and Doug McKone, in the publication of this volume. The National Marine Mammal Laboratory and Alaska Fisheries Science Center also provided financial and logistical assistance. We thank members of the Bering Sea Working Group for their involvement and guidance in this project. We gladly recognize Sue Keller and Carol Kaynor of the University of Alaska Sea Grant for their expert production of the book, Ron Dearborn for supporting the publishing effort, and Dave Brenner for designing the cover. We wish to give special recognition to Rebecca Paulson for her editorial support and considerable time and expertise in the management of the in-house preparation of the book. Lastly, we wish to thank the authors and their host institutions for their participation and enthusiasm which led to the completion of this project.

Thomas R. Loughlin
National Marine Fisheries Service
Seattle, WA, U.S.A.

Kiyotaka Ohtani
Faculty of Fisheries
Hokkaido University
Hakodate, Japan

CHAPTER 1

The Physical Oceanography of the Bering Sea

Phyllis J. Stabeno and James D. Schumacher
Pacific Marine Environmental Laboratory, Seattle, Washington

Kiyotaka Ohtani
Laboratory of Physical Oceanography, Hokkaido University, Hokkaido, Japan

Introduction

The Bering Sea is a semi-enclosed, high-latitude sea that is bounded on the north and west by Russia, on the east by Alaska, and on the south by the Aleutian Islands (Fig. 1). It is divided almost equally between a deep basin (maximum depth 3,500 m) and the continental shelves (<200 m). The broad (>500 km) shelf in the east contrasts with the narrow (<100 km) shelf in the west. Seasonal extremes occur in solar radiation, meteorological forcing, and ice cover. Large interannual fluctuations exist in climate, due both to the Southern Oscillation and the Pacific-North American atmospheric pressure patterns (Niebauer 1988; Niebauer et al., chapter 2, this volume). An amplification of global warming is predicted in the Bering Sea (Bryan and Spelman 1985). Basin scale climate variability profoundly impacts both the physical and biological environment (Schumacher and Alexander, chapter 6, this volume).

Interactions among ocean, ice, and atmosphere dominate the physics of the Bering Sea. Large-scale weather patterns in both the tropical South Pacific (El Niño-Southern Oscillation events) and the North Pacific (Pacific–North America patterns) have strong connections to the Bering Sea, mainly via the atmosphere (Niebauer 1988; Niebauer and Day 1989; Niebauer et al., chapter 2, this volume). The mode of teleconnection appears to perturb the passage of storms (areas of low sea level atmospheric pressure with closed isobars) along the Aleutian Island chain. The migration of storms results in a statistical feature known as the Aleutian Low, one of the two main low pressure systems in the high latitude Northern Hemisphere. During summer with its long periods of daylight and high insolation, the Aleutian Low is typically weak and weather benign. During winter, a marked change occurs in atmospheric pressure fields. High sea level

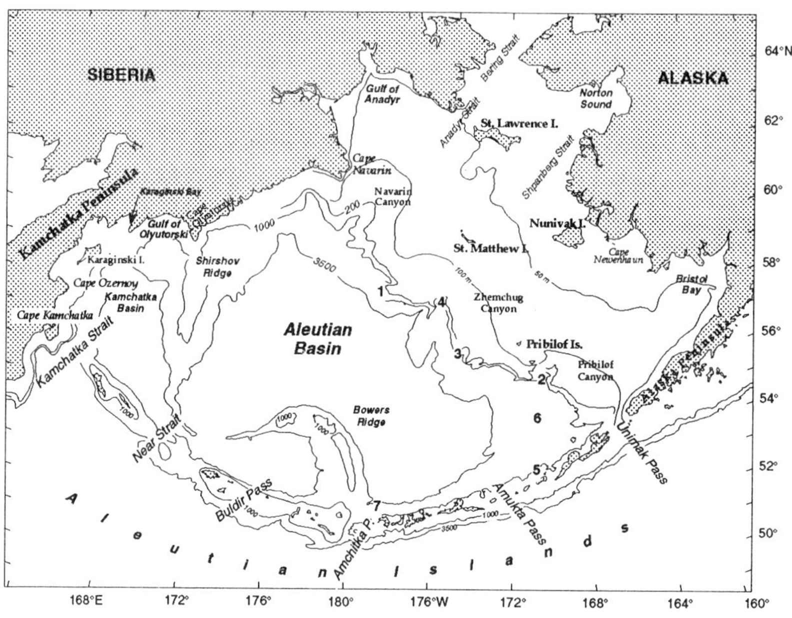

Figure 1.   *Geography and place names of the Bering Sea. The location of the seven mooring sites discussed in the text are indicated by bold numerals. Depth contours are in meters.*

pressure (Siberian High) dominates Asia, while the Aleutian Low intensifies and dominates weather over the North Pacific and Bering Sea. The juxtaposition of these features results in strong, frigid winds from the northeast. The frequency and intensity of storms in the southern Bering Sea decreases from winter to summer and frequency also decreases with increasing latitude (Overland 1981, Overland and Pease 1982). In the winter, an average of three to five storms per month move eastward along the Aleutian Chain forming the primary storm track, while less than two storms per month cross the northern Bering Sea. A secondary storm track curves northward along the Asian coast. A difference in the number of storms also occurs across the broad shelf, with more activity occurring over the outer shelf (Schumacher and Kinder 1983).

Due to the presence of the Aleutian Low, the wind torque over the Bering Sea is greater in winter than in summer by an order of magnitude. A climatology of the wind forcing (Bond et al. 1994) shows that eastward- and northward-propagating storm systems dominate the surface stress at short periods (<1 month), and their energy mixes the upper ocean. At longer periods (>1 month), the wind-driven (Sverdrup) transport accounts for roughly one-half of the observed transport in the Kamchatka Current.

Driven by changes in the location of the Aleutian Low, interannual variations in the Sverdrup transports occur that are ~25% of the mean.

Much of the physical oceanographic research in the Bering Sea has been coupled to fisheries research, since the combination of nutrient-rich slope waters and high summer solar radiation create one of the world's most productive ecosystems (Walsh et al. 1989). Seasonal primary production often begins with a bloom on the shelf associated with ice-edge melt; annual production varies from >200 g C/m^2 over the southeastern shelf to >800 g C/m^2 north of St. Lawrence Island. Over the western shelf maximum annual production (>400 g C/m^2) occurs over the continental slope (Arzhanova et al. 1995). These blooms often consume all available nutrients (Niebauer et al. 1995), so subsequent production depends on nutrients being supplied by mixing due to storms and/or advection. The blooms support strong higher trophic level production which in turn supports vast populations of marine mammals, birds, fish, and shellfish. During the mid-1980s, the pollock fishery was the world's largest single species fishery, and during the mid-1990s the salmon run along the Alaska Peninsula was the world's largest.

This paper both reviews and presents new data on the physical oceanography of the Bering Sea. It focuses on the basin, since a thorough review of the shelves has recently been completed (Schumacher and Stabeno 1998). Further, since this volume contains an extensive review of the water property characteristics (Luchin et al., chapter 3, this volume), the primary goal here is to review and update circulation over the basin. We begin with the known and estimated transports through each of the passes of the Aleutian Islands, followed by a discussion of the surface and deep currents of the Bering Sea basin and the flow on the Bering Sea shelves.

General Circulation

The circulation in the Bering Sea basin (Fig. 2) is often described as a cyclonic gyre, with the southward flowing Kamchatka Current forming the western boundary current and the northward flowing Bering Slope Current forming the eastern boundary current. Circulation in the Bering Sea is strongly influenced by the Alaskan Stream, which enters the Bering Sea through the many passes in the Aleutian Arc. The inflow into the Bering Sea is balanced by outflow through Kamchatka Strait, so that circulation in the Bering Sea basin may be more aptly described as a continuation of the North Pacific subarctic gyre. Circulation on the eastern Bering Sea shelf is generally northwestward. The net northward transport through the Bering Strait, while important to the Arctic Ocean, has virtually no effect on the circulation in the Bering Sea basin. It does, however, play the dominant role in determining the circulation on the northern shelf. The currents of the Bering Sea have been examined principally through inferred baroclinic flow and to a lesser extent by drifting buoys, current meter moorings, and models.

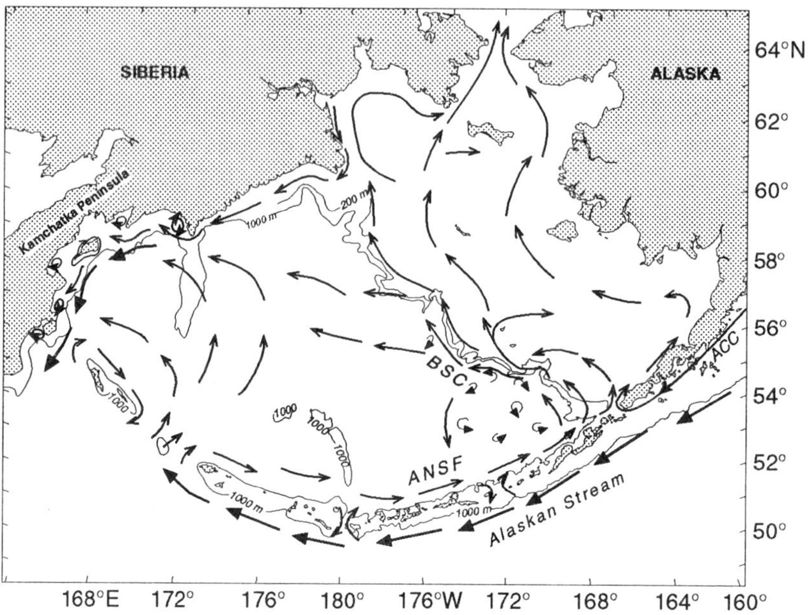

Figure 2.   Schematic of mean circulation in the upper 40 m of water column over the basin and shelf (after Stabeno and Reed 1994, Schumacher and Stabeno 1998). The arrows with solid heads represent currents with mean speeds typically >50 cm/s. The Alaskan Stream, Kamchatka Current, Bering Slope Current, and Aleutian North Slope Current are each indicated. The triangles with numbers indicate the location of the seven deep moorings (water depth >1,000 m) discussed in the paper. The 100-m flow and 1,000-m isobath are indicated.

Passes

The passes play a primary role in determining both circulation and distribution of water properties. The Aleutian Arc forms a porous boundary between the Bering Sea and the North Pacific (Fig. 3). Even though only three passes (Amchitka Pass, Near Strait, and Kamchatka Strait) extend deeper than 700 m, there is significant flow through many of the 14 main passes. Transport into the Bering Sea can vary by more than a factor of two and this affects the transport in the Kamchatka Current. The strong variability of flow through the passes has time scales which vary from weeks to years. The source of most of the flow into the Bering Sea basin is the Alaskan Stream, the northern boundary of the cyclonic North Pacific subarctic gyre. It provides relatively fresh surface waters and warm subsurface water.

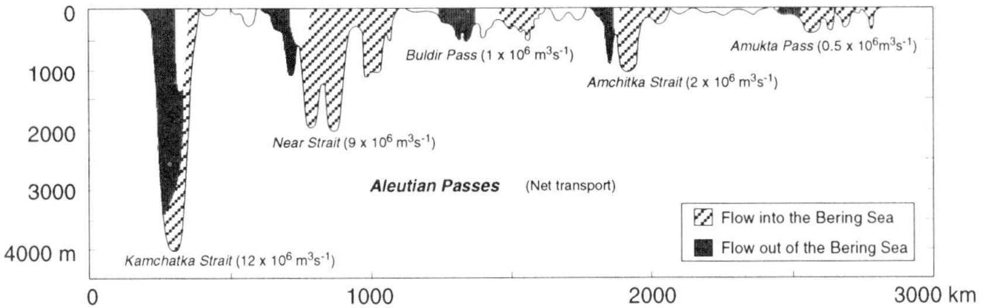

Figure 3.   A cross section of the passes of the Aleutians. Suspected or measured transports are given for each of the main passes. The hatched regions indicate northward flow and the shaded regions indicate southward flow. Unshaded regions indicate areas where either the flow is variable or unknown.

We first explore flow through the two passes on the eastern Bering Sea shelf, Unimak Pass (at the south), and Bering Strait (at the north). The influence of these passes is limited to the eastern Bering Sea shelf and plays virtually no role in modifying currents in the basin. We next examine flow through the major passes in the Aleutian Arc (starting on the eastern side and moving westward), which strongly influence the flow patterns and water properties of the basin.

Shelf Passes

Unimak Pass forms the only significant conduit between the shelves of the Gulf of Alaska and the eastern Bering Sea. This relatively shallow (<80 m) and narrow (~30 km) pass permits flow of a portion of the Alaska Coastal Current (ACC) into the Bering Sea. The ACC extends 1,000 km along the Gulf of Alaska coast from southern Alaska to Unimak Pass (Stabeno et al. 1995). A recent study of flow through Unimak Pass (Reed and Stabeno submitted) substantiates the earlier conjecture (Schumacher et al. 1982) that a strong seasonal signal in transport exists. The current appears to be uniform across the pass. Transport is predominantly baroclinic (>70%). Net baroclinic transport is northward with a maximum in the fall and winter (>0.50 × 10⁶ m³/s), a minimum in the late spring and summer (~0.10 × 10⁶ m³/s), and a mean transport of 0.23 × 10⁶ m³/s. Maximum daily-averaged barotropic flow is also northward and can exceed 100 cm/s in the winter (tidal currents can strengthen this, resulting in the flow exceeding 200 cm/s), although it is weak during the summer. The barotropic flow through this pass results from sea level changes induced by winds along the peninsula (Schumacher et al. 1982). While direct observations do not exist to establish the magnitude of the long-term variability, the portion of transport resulting from winds likely varies significantly.

Of all the passes that connect the surrounding oceans with the Bering Sea, only the physical oceanography of Bering Strait is well documented (Coachman et al. 1975, Aagaard et al. 1985b, Roach et al. 1995). An approximate 0.4 m mean sea level difference between the Bering Sea and Arctic Ocean drives a net northward transport ($\sim$0.8 $\times$ 10^6 m^3/s) through Bering Strait (Coachman 1993). Reversals of the northward flow occur during periods of southward winds. The strongest winds occur during the autumn/winter, resulting in a seasonal signal in the transport. The strongest monthly mean northward transport through Bering Strait occurs in July ($\sim$1.4 $\times$ 10^6 m^3/s), and the weakest in December (0.3 $\times$ 10^6 m^3/s). Thus the annual signal is out of phase with the inflow through Unimak Pass. There is also strong intra-annual variability, with mean annual transports ranging from 0.6 $\times$ 10^6 m^3/s to >1.0 $\times$ 10^6 m^3/s.

The northward flow through Bering Strait strongly influences the currents over much of the northern Bering Sea shelf. One consequence of the northward transport is that a supply of nutrient-rich water to the northern shelf upwells near St. Lawrence Island, thereby stimulating primary production (Nihoul et al. 1993). The northward flow also provides the only connection and exchange of water between the Pacific and Atlantic Oceans in the Northern Hemisphere. During ice formation, cold saline water (>34 psu and <$-$1.5°C) is produced over the northern shelf and flows northward through Bering Strait. Globally, this water plays a role both in maintaining the Arctic Ocean halocline and in ventilation of the deep waters (Aagaard et al. 1985a).

Deep Aleutian Passes

The easternmost major conduit between the North Pacific Ocean and the Bering Sea is Amukta Pass (Figs. 1 and 3). Recent measurements have shown that the net baroclinic transport through this pass is highly variable (Reed and Stabeno 1997), ranging from 1.4 $\times$ 10^6 m^3/s northward to a weak (<0.10 $\times$ 10^6 m^3/s) southward flow, with a mean of 0.6 $\times$ 10^6 m^3/s. Maximum northward geostrophic speeds vary from 25 to 73 cm/s, while the southward speeds are weaker (7-32 cm/s).

This transport, along with the flow through Amchitka Pass, is the source of the Aleutian North Slope Current (Reed and Stabeno, chapter 8, this volume) and ultimately the Bering Slope Current. The width of the pass is sufficient (greater than the internal Rossby radius) that there is generally northward flow on the east side and southward flow on the west side. The relative strength of these two branches determines the direction and magnitude of the net transport. The source of the flow through this pass is usually the Alaskan Stream; however, at times ($\sim$10%) the eastward flow on the north side of the Aleutian Islands turns southward through the western side of the pass and retroflects through the eastern side. A more common scenario, however, is that the Alaskan Stream enters on the eastern side of the pass with some retroflecting, supplying a portion of the southward branch.

Two conductivity temperature depth (CTD) sections across Amukta Pass, done a year apart during August (Fig. 4), show water properties during a period of strong (Fig. 4a,b) and of weak (Fig. 4c,d) inflow. The warm water in the upper 100 m during both years resulted from summer warming. The warm water at depth during 1994 resulted from Alaskan Stream inflow, while during 1995 there was little inflow of Alaskan Stream water. While variability on time scales of days exists, the data presented in Reed (1995) and Reed and Stabeno (1997) suggest that inflow of Alaskan Stream water through Amukta Pass varies mainly on the time scale of months to years.

In the Aleutian Basin, a subsurface temperature maximum occurs between 150 and 400 m. The sigma-*t* density of the maxima is between 26.6 and 26.9 and the temperature ranges between 3.5 and 4.6°C (Fig. 5a,b). The variability in this temperature results from the absence or presence of flow of Alaskan Stream water through Amukta Pass. When flow is present, a signature of warm (>4.0°C) subsurface water is evident in the eastern basin and slope region (Reed 1995). Inferring lack of inflow from the absence of the warm subsurface water indicates that for periods of months inflow of Alaskan Stream water through Amukta Pass can be absent.

Few measurements have been made of transport through Amchitka Pass, the third deepest pass in the Aleutian Arc, and only three hydrocast sections have been reported in the literature through the narrow sill region at 51°28′N (Reed and Stabeno, chapter 8, this volume). Measured baroclinic transports range from 2.8×10^6 m^3/s southward to 2.8×10^6 m^3/s northward, with a mean of 0.3×10^6 m^3/s northward. The flow pattern across the pass is similar to that observed through Amukta Pass, with inflow on the eastern side and outflow on the western side. The source of outflow is twofold. Some originates at Near Strait and Buldir Pass and flows eastward across Bowers Ridge, while the remainder consists of a retroflection of the inflow through the eastern part of Amchitka Pass.

A vertical section of temperature and salinity reveals the inflow of Alaskan Stream water on the eastern side of the pass where the depth of the 4°C contour dipped to ~400 m (Fig. 6a). As suggested by isopleths, the net baroclinic transport was weak (0.3×10^6 m^3/s northward). Two satellite-tracked drifters were deployed shortly after the CTD section in the vicinity of Amchitka Pass. The trajectories (Fig. 7) reveal the presence of two eddies that inhibited much of the flow through the pass. The northern drifter remained in the cyclonic eddy for 22 days, making 2.5 circuits around the northern eddy. The other drifter made four circuits of the southern eddy before malfunctioning. Typical speeds of rotation were the same for each eddy, 30 cm/s. Flow of significant amounts of Alaskan Stream water was thus blocked from the pass. During 1991, there was also evidence from drifter trajectories (not shown) of an eddy in the central portion of the pass, but these drifters did not remain in the eddy for more than one rotation. Evidence of an eddy in the Alaskan Stream south of Amchitka Pass which affected the transport through Amchitka Pass, was revealed in altimeter data (Okkonen 1996).

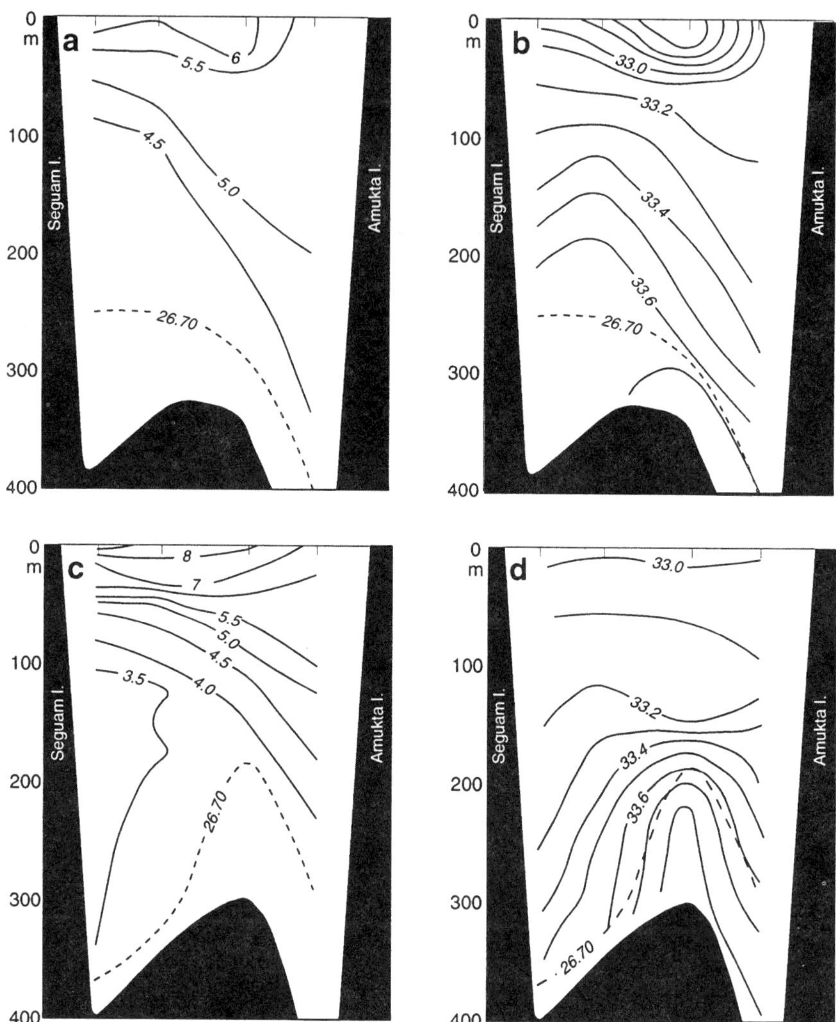

Figure 4.   *(a) Vertical sections of temperature (˚C) and (b) salinity (psu) in Amukta Pass in August 23, 1994, during a period of strong inflow (net 1.0 × 10⁶ m³/s) of Alaskan Stream water into the Bering Sea. (c) Vertical sections of temperature (˚C) and (d) salinity (psu) during a period of weak (net flow 0.1 × 10⁶ m³/s) outflow from the Bering Sea.*

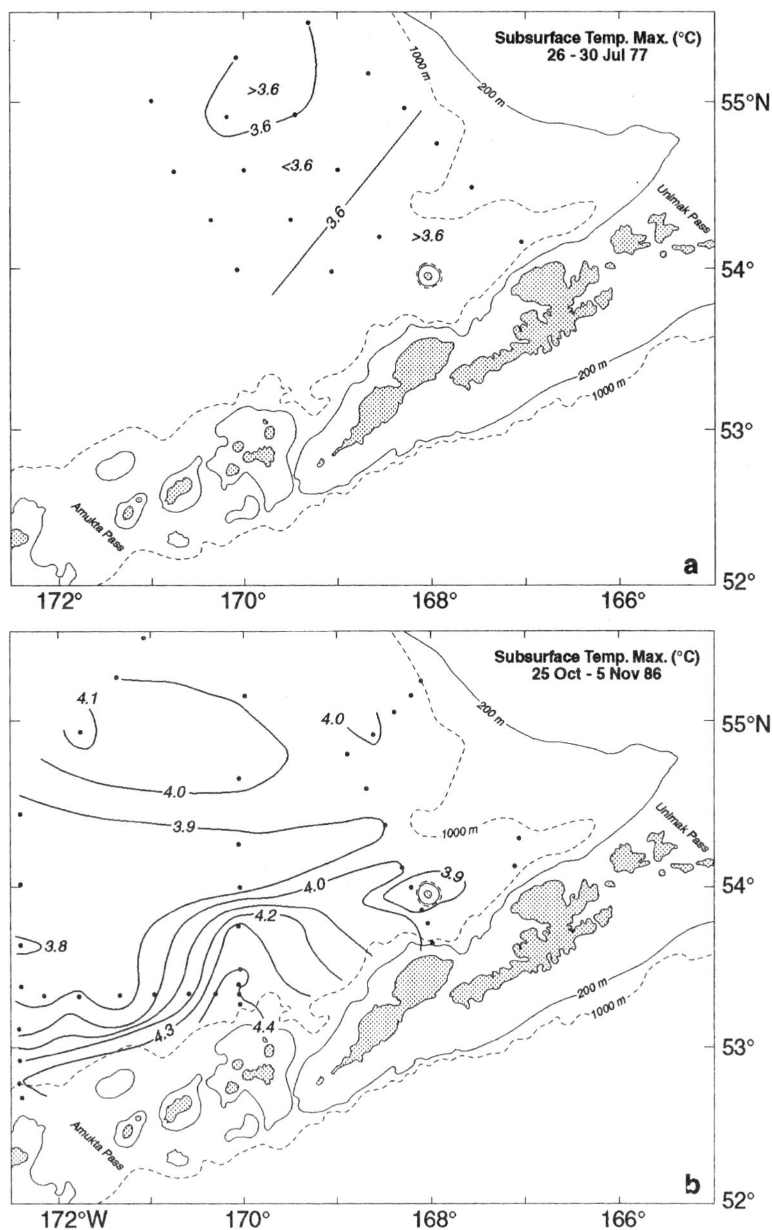

Figure 5. (a) Distribution of temperature (°C) at the subsurface maximum during a period when the temperature was at a minimum. (b) Distribution of the temperature of the subsurface maximum when substantial areas were >4.0°C. (This figure is adapted from Figs. 3 and 5 in Reed 1995).

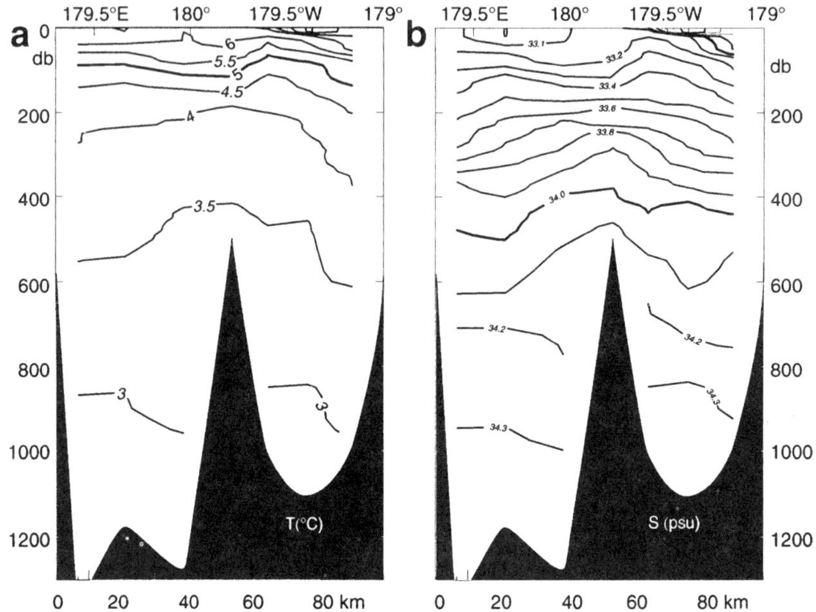

Figure 6.   Vertical sections of (a) temperature (°C) and (b) salinity (psu) along 51°30'N in Amchitka Pass on September 15, 1992.

Results from the one current meter mooring, which was positioned north of the pass, have been reported in the literature (Reed 1990). The net flow (14 cm/s) is to the northeast (Fig. 8). Interpretation of measurements from a single mooring within a bi-directional flow is ambiguous; however, results from this current meter integrated with hydrographic surveys suggest a northward flow of $2\text{-}3 \times 10^6$ m³/s through the eastern part of the pass. Through the western portion of the pass a southward flow occurs, resulting in a net transport northward of probably $1\text{-}2 \times 10^6$ m³/s.

Virtually no direct measurements of current have been made in Buldir Pass. Trajectories of satellite-tracked drifters indicate that northward flow can occur through the pass. Stabeno and Reed (1992) estimated an inflow of less than 1.0×10^6 m³/s.

Most of the transport into the Bering Sea occurs through Near Strait $(6\text{-}12 \times 10^6$ m³/s). Historically, transport of Alaskan Stream water through Near Strait was estimated at $\sim 10 \times 10^6$ m³/s and believed to be relatively constant (Favorite 1974). More recent research indicates that the flow through this pass, as through the other passes, is extremely variable. Occasionally, instabilities (eddies) occur in the Alaskan Stream that inhibit

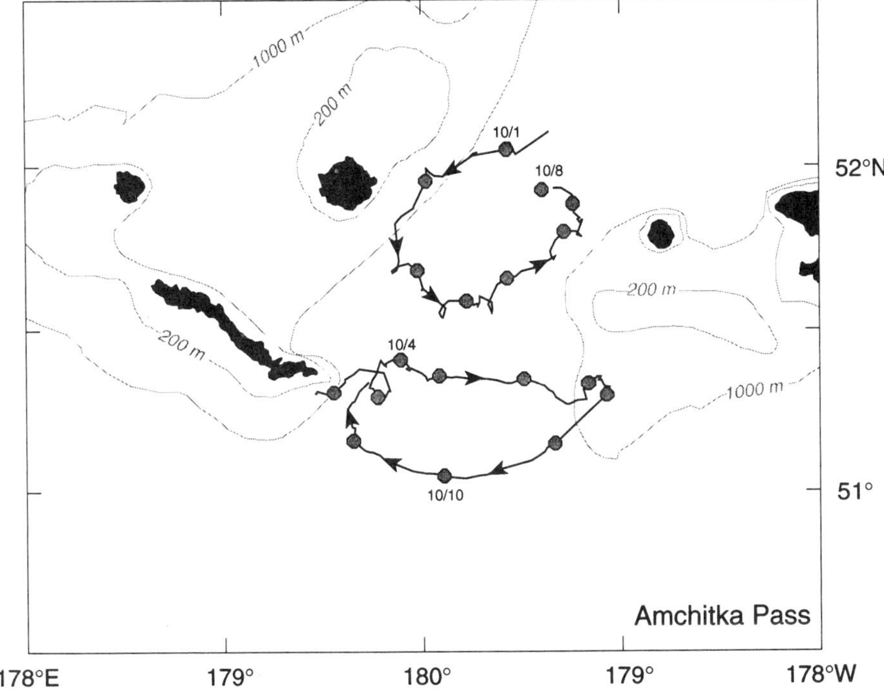

*Figure 7.	Trajectories from satellite-tracked drifting buoys deployed in the vi-
cinity of Amchitka Pass in September 1992. Each drifter had a holey
sock drogue at 40 m depth. Daily positions are indicated by circles
on the trajectories. The southern drifter remained in the eddy for 25
days, making 2.5 circuits. The northern drifter made four circuits of
the eddy. Mean rotational speed was 30 cm/s in both eddies.*

flow into the Bering Sea through Near Strait. Such an eddy in 1991 resulted in a net northward baroclinic transport of less than 3×10^6 m³/s (Stabeno and Reed 1992). Near Strait is the only deep pass that has been
monitored with current moorings (Reed and Stabeno 1993). A series of
moorings deployed in the eastern portion of the pass indicate that the
periods of reduced inflow vary on scales of months. Yearly average velocity at ~100 m was 6 cm/s into the Bering Sea in the eastern portion of the
pass, and decreased to 1-2 cm/s in the central part of the pass. It was
moderately surprising that the strongest, steadiest flow occurred near the
bottom of the narrow passage on the eastern side of the pass, a result of
tidal rectification.

Not only is Kamchatka Strait the location of the majority of outflow,
but it is the only conduit into the Bering Sea that is deep enough (>2,000 m)
to permit inflow of Deep Pacific Water (DPW). In the upper 1,500 m, flow is

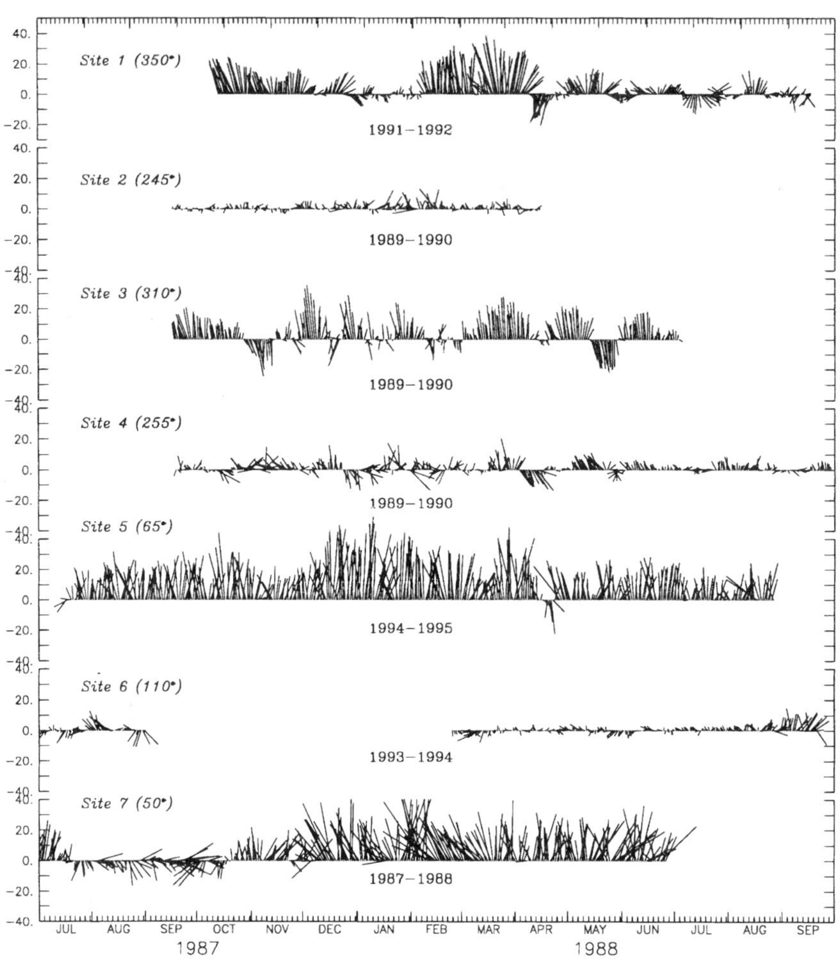

*Figure 8.	Daily currents measured at seven locations indicated in Fig. 2. A low pass filter (35-hour Lancoz squared) was applied to each series and the series were then rotated to their net direction (indicated in parentheses). The depth of each instrument was between 250 and 350 m. Site locations are shown in Fig. 1.*

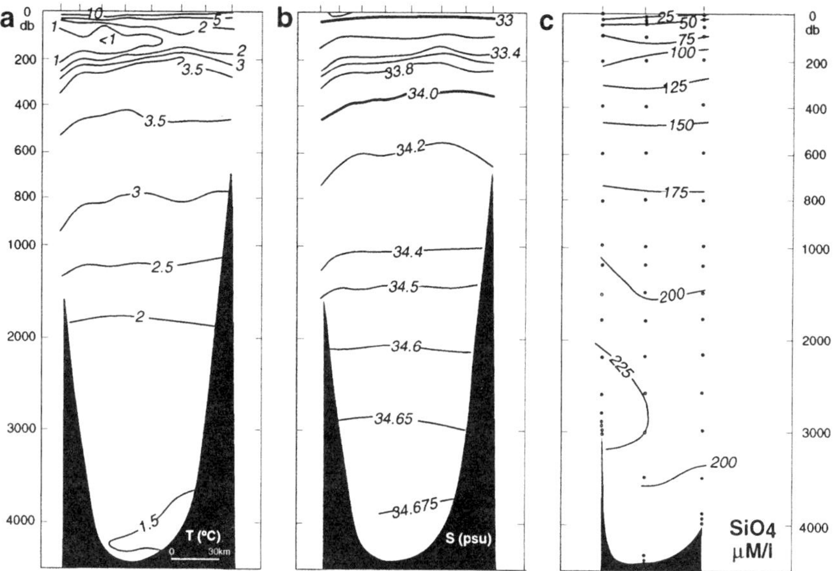

Figure 9.   *Vertical sections of (a) temperature, (b) salinity, and (c) silica (mM/L) across Kamchatka Strait on August 13-14, 1991. Modified from Reed et al. (1993).*

dominated by the southward-flowing Kamchatka Current on the western side of the strait and a weak inflow occurs on the eastern side (Stabeno and Reed 1994, Cokelet et al. 1996). The inflow of DPW occurs below 2,000 m predominately on the eastern side of the strait (Reed et al. 1993), and thus the highest salinities in the Bering Sea basin are observed here (Fig. 9b). Salinities decrease in the basin as the high salinity DPW mixes with surrounding water. The Bering Sea deep water contains the highest concentrations of silica in the world's ocean (Mantyla and Reid 1983). Between 2,000 and 3,000 m, high concentrations of silica (>225 m M/L; Fig. 9c) occur in the strait (Reed et al. 1993). Concentrations similar to these are observed in the Bering Sea basin farther to the northeast. This suggests a southward flow of deep Bering Sea water beneath the Kamchatka Current and above the inflow of DPW. Measurements of transport of the Kamchatka Current in the pass range from 5×10^6 m^3/s (Verkhunov and Tkachenko 1992) to 15×10^6 m^3/s (Ohtani 1970).

Mean Upper Ocean Circulation

General circulation over the deep basin is characterized by a cyclonic gyre, with three well-defined, distinct currents: the Kamchatka Current along the western boundary; the Bering Slope Current along the eastern boundary; and the Aleutian North Slope Current connecting the inflow through Amukta Pass and Amchitka Pass with the Bering Slope Current (Fig. 2). Transport within the gyre can vary by more than 50%. Modeling studies have simulated such large changes in transport and identified the causal mechanisms to be fluctuations of Alaskan Stream inflow (Overland et al. 1994) and/or changes in the wind-driven transport within the basin (Bond et al. 1994).

The main features of the upper-ocean circulation are shown in Fig. 3. This figure has been modified from those in Stabeno and Reed (1992, 1994) and Reed (1995). Much of the information is from the water property and geostrophic flow results from Favorite (1974), Hughes et al. (1974), and Sayles et al. (1979). Information from satellite-tracked drifting buoys is taken from Stabeno and Reed (1994) and from the limited number of current meter moorings which have been deployed in the basin. A more detailed flow field (Fig. 10) derived from satellite-tracked drifters (following Stabeno and Reed 1994) from 1984 through 1996, reveal each of the main current systems along with flow through the passes and weak westward flow across the basin.

The Alaskan Stream

While the Alaskan Stream (AS) is not in the Bering Sea, it is the main source water for much of the flow that occurs in the Bering Sea and so merits a brief discussion. The AS, which is the northern boundary current of the Pacific subarctic gyre, extends from the head of the Gulf of Alaska to the western Aleutians. It is characterized as a narrow, high speed current, with small eddy kinetic energy to mean kinetic energy (KE′/KE) ratios (KE′/KE<1.0) from $165°$W to $173°$E (Stabeno and Reed 1990). It turns northwestward at Amchitka Pass ($180°$) following the bathymetry, slowing and broadening. Maximum daily averaged buoy speeds east of $180°$ were 70-95 cm/s, while to the west of $180°$ maximum speeds were 40-65 cm/s. Although the position of the Alaskan Stream relative to the Aleutian Islands is relatively constant, there is some variability, which, coupled with eddies, results in changes in transport through the passes.

Aleutian North Slope Current

Although there is some indication of an eastward flow along the north side of the Aleutian Islands between Kamchatka Strait and Near Strait, and also between Near Strait and Amchitka Pass (Stabeno and Reed 1994), the only well-documented persistent eastward flow occurs east of Amchitka Pass (Reed and Stabeno, chapter 8, this volume). This current is strongly influenced by flow through the passes. East of Amchitka Pass, it is not completely continuous, but rather originates from inflow on the eastern side of one deep pass and continues eastward until it (or some portion) exits through the western side of the next deep pass.

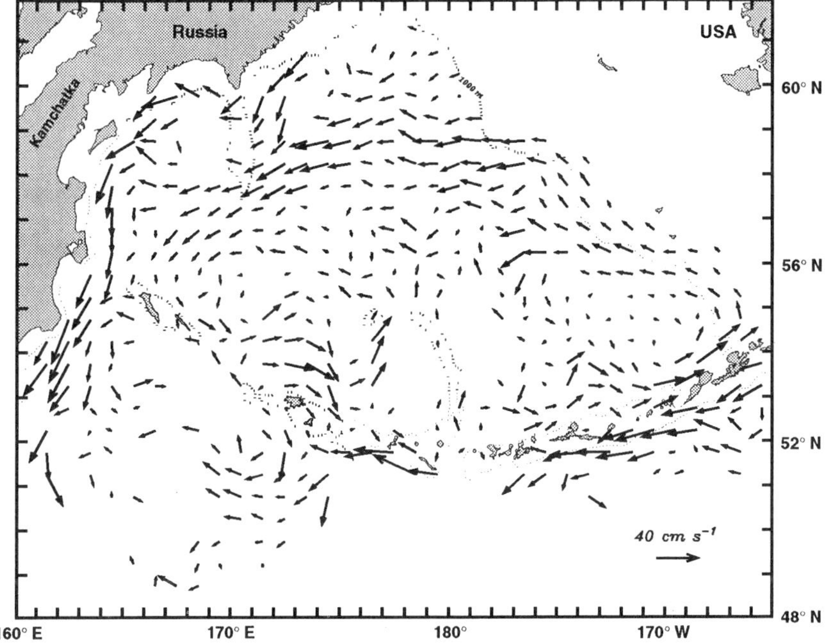

Figure 10. Velocity at 40 m derived from satellite-tracked drifters. The method used was the same as described in Stabeno and Reed (1994).

The Aleutian North Slope Current (ANSC) is best documented between 174°W and 167°W by hydrographic surveys (Reed and Stabeno 1997), satellite-tracked drifting buoys (Stabeno and Reed 1994), and more recently in current meter records (Reed and Stabeno 1997). It is here that the flow merits being called a current. The ANSC is a narrow (~20 km), stable current that appears to have an annual signal, as indicated by both the trajectories of satellite-tracked drifting buoys and time series from current meter moorings. Currents (measured at mooring site 5, Fig. 1) are northeastward following bathymetry. Few reversals are evident in these 13-month-long current records (Fig. 8). The strongest daily average currents (>40 cm/s at 100 m) occur during winter, and the mean velocity at 100 m exceeds 20 cm/s. The flow at 50 m above the bottom has no seasonal signal and is statistically significant at 4.5 cm/s northeastward. The KE′/KE ratio is very small (<0.5) in the upper ~300 m and near the bottom. This vertical structure, with decreasing speeds, is also evident in baroclinic estimates of flow from hydrographic surveys. Measurements of baroclinic transport range from 3×10^6 m³/s (Reed and Stabeno, chapter 8, this volume) to 5.5 $\times 10^6$ m³/s (unpublished data from February 1997).

Bering Slope Current

Eastward of 167°N, the ANSC turns northwestward, forming the Bering Slope Current (BSC), the eastern boundary current of the Bering Sea gyre (Schumacher and Reed 1992, Kinder et al. 1975). The characteristics of this flow differ markedly from the ANSC. This can easily be seen in comparing time series and statistics of two moorings, one in the ANSC and the other in the BSC (Fig. 1). Unlike the ANSC, the presence of an annual cycle is not clearly evident in the time series at any of the sites in the BSC (sites 1, 2, 3, and 4), although model simulations show an intensification in fall-winter (Overland et al. 1994). The reversals evident in the BSC time series are likely a result of eddies. Only weak flow occurs at the sites associated with the canyons (sites 2 and 4). At the northernmost site, the strong currents are associated with eddies and the background flow is weak. The eddy:mean kinetic energy ratios at site 3 best represent the BSC. Seaward of the 1000-m isobath the flow is more variable and the BSC is often less well defined. The mean velocity (site 3) at 100 m (11 cm/s) is approximately half that observed in the ANSC. The BSC is also shallower than the ANSC, with mean flow at 500 m ~1 cm/s. Total transport ($3\text{-}6 \times 10^6$ m^3/s) is of course similar to that observed in the ANSC. These are consistent with the BSC being wider than the ANSC.

The BSC is strongest and most persistent south of 56.5°N. Most of the flow separates from the slope at the latitude of Pribilof Canyon or Zymchug Canyon, forming the source of the weak westward flow across the basin (Stabeno and Reed 1994). To the north of 58°N the flow is still generally northwestward, but is weaker, more intermittent, and confined to shallower water depth.

Studies of the BSC suggest that two significantly different structures or modes exist. Many CTD surveys reveal an ill-defined, highly variable flow interspersed with eddies, meanders, and instabilities (Kinder et al. 1975, Reed 1991). Other surveys, however, have revealed a more regular northwestward flowing current (Reed and Stabeno 1989). The trajectories from the more than 50 satellite-tracked drifters deployed in the southeast Bering Sea support this dichotomy in the structure of the BSC. They reveal the strong variability in the flow patterns that occur along the shelf break. In some trajectories, the BSC appears as a well-behaved current flowing northwestward along the shelf break (Stabeno and Reed 1994). At other times, chaos dominates the dynamics of the system (Reed and Stabeno 1990). At still other times, eddies, ranging in size from ~40 km to 150 km, are imbedded in the flow (Schumacher and Stabeno 1994). The structure of the BSC directly influences both the advection and dynamics along the slope, and also the occurrence of across-shelf fluxes. This dichotomy of structure, of either a series of eddies or a uniform flow, is supported by numerical model simulations of the subarctic North Pacific portion of the NRL Pacific Ocean model.

The Kamchatka Current

The third distinct current system, the Kamchatka Current, forms the western boundary current of the Bering Sea gyre. It originates near 175°E (Shirshov Ridge), as the weak westward flow (Stabeno et al. 1994). The source of this water is a combination westward flowing water (from the BSC) and northward flowing water entering the Bering Sea through Near Strait (Stabeno and Reed 1994, Khen 1989). The bathymetric feature of Shirshov Ridge causes a southward deflection and reduction in speed of the current (Stabeno and Reed 1994). As the Kamchatka Current continues southward along the Russian coast it both strengthens and deepens. Near Shirshov Ridge flow (>5 cm/s) is evident only in the upper 1,000 m, while in Kamchatka Strait it has deepened to >1,500 m. The maximum daily averaged speeds observed in the Bering Sea occur in the Kamchatka Current (40-77 cm/s). Daily averaged velocities as large as 100 cm/s occur in Kamchatka Strait. Meanders and eddies are common in the Kamchatka Current (Solomon and Ahlnas 1978, Stabeno et al. 1994, Cokelet et al. 1996). These features cause the long-term mean velocities to be much smaller than observed in the more stable Alaskan Stream.

Favorite (1974) estimated the transport in the Kamchatka Current as $\sim10 \times 10^6$ m³/s, but recent results indicate that it can be quite variable ranging from 7 to 15×10^6 m³/s (Stabeno and Reed 1992, Reed and Stabeno 1993, Overland et al. 1994, Cokelet et al. 1996). Variations in wind-stress curl are substantial with a winter maximum and might affect the transport (Overland et al. 1994, Bond et al. 1994). At times flow in the Kamchatka Current recirculates in the Bering Sea, with only a portion flowing through Kamchatka Strait and the remainder flowing eastward along the north side of the Aleutian Islands (Reed et al. 1993).

Interior Flow

Resulting from the strong inflow through Near Strait, a weak northward flow is evident in buoy trajectories and model results over the basin (Figs. 2 and 10). The BSC is the source of the westward flow across the basin (Royer and Emery 1984), which originated at the two areas along the slope where the 1,000-m isobath is oriented east-west (Stabeno and Reed 1994). At the southern source (~56.5°N) the flow bifurcates with a significant portion forming a small sub-gyre in the southeast corner of the basin. At the northern source (~58°N) the flow continues eastward across the center of the basin.

There have been no long-term current moorings in the central basin, but a mooring (site 6) has been located in the southeast Bering Sea (2,200 m water depth) during spring and summer of 1992-1994 (Fig. 8). These records show generally weak flow interrupted by eddies (Cokelet and Stabeno 1997). The background flow contrasts sharply from the other two sites (one in the ANSC and the other in the BSC) discussed earlier. At 80 m the flow is weakly northward (~1 cm/s) with high eddy-mean kinetic energy ratios. Below 150 m the currents are negligible. The eddies that intersperse the weak background flow are far more energetic.

Eddies

As elsewhere in the world's oceans, eddies are ubiquitous in the Bering Sea (Solomon and Ahlnas 1978, Kinder et al. 1980, Paluszkiewcz and Niebauer 1984, Schumacher and Stabeno 1994). They occur on horizontal scales ranging from ~10 km to 200 km. Proposed mechanisms for the creation of these eddies include instabilities, wind forcing, strong flows through the eastern passes, and topographic interactions (Schumacher and Stabeno 1994).

The influence of eddies in straits have already been discussed, as has the presence of eddies in the BSC. One of the best time series, showing eddies in the basin has been collected during the spring and summer over a 3-year period at site 6. Eddies are common in these current records. These observations characterize the eddies as often anti-cyclonic, 20-100 km in diameter, extending to a depth of 400-1000 m, and with rotational speeds >20 cm/s (Schumacher and Stabeno 1994, Cokelet and Stabeno 1997).

On the western side of the basin, eddies result from instabilities in the Kamchatka Current. The eddies are often anti-cyclonic, 20-100 km in diameter, and have rotational speeds of >40 cm/s. They have been observed in satellite-tracked buoy trajectories, hydrographic surveys, and satellite images (Solomon and Ahlnas 1978). In three embayments on the Kamchatka Peninsula anti-cyclonic eddies were evident in the tracks of many of the buoys. They resulted from the interaction of the Kamchatka Current with topographic features and are likely semi-permanent, since they appeared in trajectories from more than 1 year (Schumacher and Stabeno 1998).

West of Bowers Ridge a large (~200 km), energetic (velocities ~30-40 cm/s) eddy was observed in 1991 (Reed et al. 1993, Cokelet et al. 1996). Buoys were deflected around the feature, with none entrained into the center. Although this feature persisted for several months, it was only observed during 1991. Thus its persistence is unknown.

Eddies are common at the eastern shelf break and shoreward to depths of ~150 m. A recent interpretation of hydrographic observations (Reed, in press) suggests anticyclonic eddies exist in the region between 100 and 122 m <20% of the time. Shoaler than 100 m, eddies are uncommon.

Deep Circulation

Little comprehensive information is available on deep circulation in the Bering Sea. However, a study that examines all of the available information on deepwater properties, and infers flow, is nearing completion (T.E. Whitledge, University of Texas at Austin, pers. comm., November 1995 [now at University of Alaska Fairbanks, School of Fisheries and Ocean Sciences, Fairbanks, AK 99775-7220]). Sayles et al. (1979) presented limited data to 2,500 m and inferred flow at 1,000 and 1,400 m, referred to 2,500 m. The only significant flow at these levels was the Kamchatka Current.

Reed et al. (1993) presented vertical sections of temperature salinity sigma-*t*, and silica across the Kamchatka Strait. Inflow of deep water occurred below 2,500 m on the eastern side of the strait. Silica data suggested a return flow of upper deep water (near 2,500 m) on the western side of the strait. In July 1993, a World Ocean Circulation Experiment section was occupied in the Bering Sea from the continental slope southwestward through Amchitka Pass (Roden 1995). The deep Bering Sea is warmer, less salty, less dense, less oxygenated, and with greater concentrations of silica than the open waters south of the Aleutians. Silica increases somewhat near the continental slope (Roden 1995, Tsugonai et al. 1979). Clearly more data are needed, but the path of deep water must be northward and eastward from Kamchatka Strait with some return flow above 3,000 m on the western side of the strait. The map of deep flow is consistent with hydrographic data and water properties (Fig. 11).

We are not aware of any long-term measurements of flow below 1,000 m. A year-long record at ~1,000 m depth along the eastern shelf break indicates that all significant flow occurred above 500 m (Schumacher and Reed 1992). Reed (1995) reviewed previous efforts to establish reference levels for geostrophic flow computations in the western Bering Sea near the Kamchatka Current. Deep levels (to 3,000 m) appeared most suitable, but levels of 1,500 m or above were suggested for the eastern Bering Sea. Cokelet et al. (1996), compared measurements between an acoustic Doppler current profiler and geostrophic shear; these results supported the regional pattern suggested above.

Shelf-Slope Exchange

A connection exists between the basin and shelf that is fed by shelf-slope exchange. The northward transport through Bering Strait (0.8×10^6 m^3/s) and through Unimak Pass requires a net onshelf transport of ~0.5×10^6 m^3/s. One school of thought is that a "river" of water flows onto the northern shelf. This is manifest in some schematics (e.g., Shuert and Walsh 1993) that show the BSC flowing northward along the slope and bifurcating south of Cape Navarin. One branch (sometimes called the Anadyr Current) then flows across isobaths along the Gulf of Anadyr where a canyon exists which influences model simulations (e.g., Overland and Roach 1987) through Anadyr Strait and thence to Bering Strait. Direct observations (Stabeno and Reed 1994), however, show that the BSC tends to leave the slope near 59°N and then flow eastward. While temperature, salinity, and nutrient concentrations in the waters that eventually go through Bering Strait are all of slope origin, their source is likely not a "river" which flows from

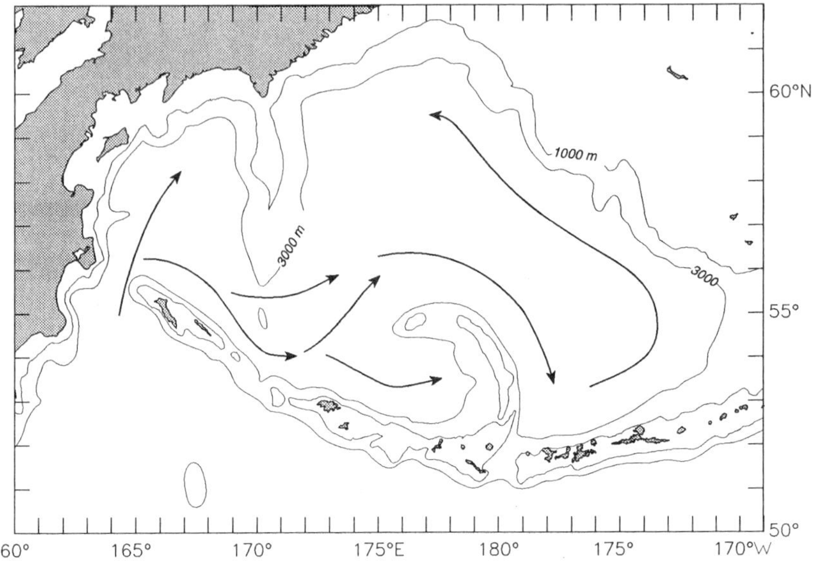

Figure 11. The deep flow in the Bering Sea basin below 3,000 m. This pattern is inferred from water properties as described in the text.

the basin onto the shelf south of Cape Navarin (Schumacher and Stabeno 1998).

Shelf-slope exchange can occur virtually anywhere along >1,200 km of shelfbreak north of Unimak Pass. Two regions exist where preferential transport onto the shelf has been observed. The first is Bering Canyon which lies along the Aleutian Islands near Unimak Pass. The enhanced concentration of nutrients observed near Unimak Pass likely originate from ANSC waters interacting with canyon topography and coming onto the shelf (Schumacher and Stabeno 1998). The second region occurs west of Pribilof Islands, where the narrowing of the shelf break accelerates the flow along the 100-m isobath, which then turns northward. The accelerated flow over the outer shelf between St. George Island and Pribilof Canyon results in water being entrained from the adjacent slope (Stabeno et al., chapter 9, this volume). Flow near the Pribilof Islands is evident in satellite-tracked drifter trajectories (drogue depth 40 m).

Episodic events of onshelf flow have been observed resulting from both eddy-topography interactions and as a result of instabilities in the BSC (van Meurs and Stabeno, in press). Along the slope of the central shelf (~56.7°N), current records reveal that eddies are common and estimates of salt and temperature fluxes indicate that significant onshore fluxes exist (Schumacher and Reed 1992). In 1992 an excellent set of data was collected when an eddy translated onto the shelf. (Schumacher and Stabeno 1994). Three satellite-tracked drifters had been deployed in the center of an eddy. The drifters remained in the eddy for weeks, until the eddy moved eastward onto the shelf. The drifters were deposited onto the shelf as the eddy spun down. Both of these episodic events of onshelf flow were related to the stability of the BSC.

The net effect of onshelf fluxes from the slope, either regionally or by intermittent processes, is the presence of slope waters over the outer shelf of the eastern Bering Sea. Using the generally accepted value of mean speed along both the southeastern and central outer shelf (0.05 m/s in the upper 100 m of the ~100 km wide domain), we compute a transport of 0.5×10^6 m^3/s. It is these waters that are westward intensified and become the strong steady current that flows through Anadyr Strait, providing nutrients for the high productivity observed there.

Shelf Flow

Low Frequency Currents

The schematic of flow on the eastern shelf (Fig. 2) is a synthesis of moorings, satellite-tracked drifting buoys, and hydrographic sections (Schumacher and Stabeno 1998). Flow through Unimak Pass results in a weak, although persistent, flow along the 50-m isobath. Measurements are not sufficient to determine if there is a seasonal signal in this flow, reflecting the variability of the flow through Unimak Pass. The flow along 50-m isobath continues through Shpanberg Pass and accounts for about a third of the flow through Bering Strait (0.35 m^3/s). Some flow through Unimak Pass follows along the 100-m isobath (Reed and Stabeno, submitted). This flow continues northwestward along the ~100 m isobath, although its position appears to vary on annual or longer scales (Reed, in press). It is augmented by the onshelf flow associated with Bering Canyon and episodic events. Additional flux of onshelf flow occurs south of the Pribilof Islands. North of the Pribilof Islands a portion of northward flow along the 100 m isobath separates and flows eastward across the shelf, augmenting the flow through Unimak Pass. The flow that continues along the 100-m isobath appears as a broad, slow northwestward flow.

This northward flow intensifies along the east coast of Siberia forming the only strong low-frequency currents on the shelf (the Anadyr Current). The flow turns eastward across the shelf along the southern coast of Siberia and eventually exits the Bering Sea through Bering Strait. The flow through Bering Strait dominates the current dynamics of the northern shelf.

The freezing and melting of sea ice can result in a distribution of mass that generates baroclinic flow. As with all high latitude seas, sea ice is one of the primary characteristics of the Bering Sea. For approximately half the year the Bering Sea is free of ice, but during November ice begins to form or be advected into the Bering Sea through Bering Strait. Although the ice is generally limited to the Bering Sea shelves, the extent of the seasonal advance is the largest observed in any arctic or subarctic region. The formation of the ice is best described by "conveyor belt" analogy (Pease 1980, Overland and Pease 1982). Ice forms along the leeward side of the coasts and islands in polynyas (open regions of water). The ice is then advected southward (or southwestward) where it is melted by the warmer water. At the polynyas the freezing sea water produces regions of high salinity (>34 psu) and at the melting edge the ice freshens the water (~30 psu). The amount of production and advection of ice depends upon which storm track dominates in a given winter, with greatest ice production occurring in years when the Aleutian Low is well developed and storms migrate along the primary storm track. By late March or April, the ice generally has begun to retreat, leaving behind a freshened water column and shelf with relatively cold water.

Cold saline water from the Gulf of Anadyr and Anadyr Strait polynyas provides a substantial fraction of the total salt advected into the Arctic Ocean (Cavalieri and Martin 1994). The ice melt produces a lens of fresh water that can assist in the formation of a strong two-layer system over the middle Bering Sea shelf. Associated with ice melt is a bloom of phytoplankton (during April) that accounts for 10-65% of the total annual primary production over the eastern shelf (Niebauer et al. 1990, Stabeno et al. 1998) with estimates of 45-70% for the western shelf (Mordasova 1995). During years when the ice is not present over the southeastern Bering Sea shelf, the spring bloom is delayed (May).

Far less is known about the flow along the western shelf. Satellite-tracked drifter trajectories reveal a narrow, westward flowing current (15-25 cm/s) along the south coast of Siberia. These observations support the inferred flow from water property distributions.

Tidal Currents

Tides enter the Bering Sea from the Pacific Ocean, and to a much lesser degree from the Arctic Ocean through Bering Strait (Pearson et al. 1981, Mofjeld 1986). Most of the kinetic energy south of 60°N is provided by tidal currents. North of this latitude the strength of the tides decreases and the mean flow strengthens (Schumacher and Stabeno 1998, Coachman 1986).

Tidal currents play a vital role in the physical oceanography over the shelves. They provide sufficient energy to mix the bottom ~40 m of water over the southeastern shelf, thus setting up a two-layer density structure in water depths of 50-100 m. Shoaler than this the water column is weakly stratified or well mixed (Schumacher and Stabeno 1998). Tidal currents may contribute to the cross-shelf flux of salt, nutrients and heat necessary to maintain the high levels of primary production over the eastern shelf. Tidal currents also provide a mechanism for the generation of subtidal flow resulting from rectification of tidal currents with topography (Kowalik, chapter 4, this volume).

Future Directions

While our knowledge of the physical oceanography of the Bering Sea has expanded greatly over the past few decades, many phenomena exist which are not understood primarily because observations are limited or do not exist. That a vast percentage of the Bering Sea lies within the domain of two different nations has not facilitated research programs that could provide the needed observations. Further, while the eastern continental shelf continues to have ongoing research programs and interest in the role of physical processes over the western shelf is growing, the deep basin remains largely unexamined.

The Bering Sea is a vital part of the general circulation of the North Pacific Ocean; fluxes of heat, salt, nutrients, and other dissolved constituents and planktonic material are exchanged through the passes. Primary questions, however, remain about the variability, magnitude and mechanisms influencing transport through the Aleutian Passes. In contrast, the flux northward through Bering Strait, which is important to conditions on the Arctic shelves and ocean, is relatively well described and understood.

While it is recognized that flow through the passes is a primary source of circulation within the basin, many questions remain regarding the current systems of the Bering Sea. What is the nature of the annual signal of the ANSC, and if there is a significant annual signal, does it occur in both speed and transport? The BSC apparently can be characterized by two modes, yet the phenomena that generate these are not known. While some studies have elucidated the nature of the Kamchatka Current, little is known of the temporal variability in transport and eddy kinetic energy. What is the magnitude and variability of inflow of DPW into the Bering Sea Basin? The flow patterns of the deep basin have been inferred from a limited number of hydrocasts, so we know neither the temporal nor spatial variability. The behavior of the source waters for deep circulation, the inflow of the DPW through Kamchatka Strait, is not known.

The processes that result in the exchange of slope and shelf waters have not yet been determined. Hence, we do not know the mechanisms that provide nutrients to the euphotic zone and are responsible for the region of prolonged biological production known as the "Green Belt" (Springer et al. 1996). While the processes are unknown, the results of their interactions are evident. The continental shelf of the Bering Sea exhibits extremely high productivity, and this richness applies throughout the food chain. Not only are there vast quantities of commercially valuable species, but the eastern shelf is the summer feeding ground for numerous marine bird and marine mammal populations of the North Pacific Ocean. The eastern Bering Sea provides an ideal location to examine exchange mechanisms between slope water of an eastern boundary current and a continental shelf. Because the coast and its inherent topographic and coastal convergence processes are far removed from the slope, the processes involved in shelf/slope exchange should provide a clear signal. The continental shelf of the western Bering Sea is bounded by a typical western boundary current, so that contrasts of processes between the eastern and western shelves should be fruitful.

Future studies that focus on how the extant physical phenomena affect marine populations offer the best opportunity to enhance our understanding of ecosystem dynamics. This, in turn, could lead to management strategies aimed at sustainable production to ensure a rich ecosystem for our future generations. The observational database for the Bering Sea is not adequate, in both spatial and temporal coverage, to answer most of the questions noted above. In addition to further observations, modeling efforts need to be improved. A primitive equation basin-shelf model coupled to both outflow through Bering Strait and exchange in the North Pacific Ocean is a likely starting place. Once the model provides accurate simulations of the physical features, then biophysical processes and rates can be incorporated. Some of the questions that must be addressed to understand the ecosystem are best investigated by modeling efforts.

Acknowledgments

We thank the numerous scientists who planned, conducted, and published their research. We also thank all the technical staff who assisted with the data acquisition, preparation, and analysis. This contribution was funded by the Coastal Ocean Programs Bering Sea FOCI and by the Fisheries Oceanography Coordinated Investigations of NOAA (#BS302), and is Pacific Marine Environmental Laboratory's contribution #1878.

References

Aagaard, K., L.K. Coachman, and E. Carmack. 1985a. On the halocline of the Arctic Ocean. Journal of Geophysical Research 90:4833-4846.

Aagaard, K., J.D. Schumacher, and A.T. Roach. 1985b. On the wind-driven flow through Bering Strait. Journal of Geophysical Research 90:7213-7221.

Arzhanova, N.V., V.L. Zubarevich, and V.V. Sapozhnikov. 1995. Seasonal variability of nutrient stocks in the euphotic zone and assessment of primary production in the Bering Sea. In: B.N. Kotenev and V.V. Sapozhnikov (eds.), Complex studies of the Bering Sea ecosystem. VNIRO, Moscow, pp. 162-179.

Bond, N.A, J.E. Overland, and P. Turet. 1994. Spatial and temporal characteristics of the wind forcing of the Bering Sea. Journal of Climate 7:1119-1130.

Bryan, K., and M.J. Spelman. 1985. The ocean's response to a CO_2-induced warming. Journal of Geophysical Research 90:11,679-11,688.

Cavalieri, D.J., and S. Martin. 1994. The contribution of Alaskan, Siberian, and Canadian coastal polynyas to the cold halocline layer of the Arctic Ocean. Journal of Geophysical Research 99:18,343-18,362.

Coachman, L.K. 1986. Circulation, water masses, and fluxes on the southeastern Bering Sea shelf. Continental Shelf Research 5:23-108.

Coachman, L.K. 1993. On the flow field in the Chirikov Basin. Continental Shelf Research 13:481-508.

Coachman, L.K., K. Aagaard, and R.B. Tripp. 1975. Bering Strait: The regional physical oceanography. University of Washington Press, Seattle. 172 pp.

Cokelet, E.D., and P.J. Stabeno. 1997. Mooring observations of the thermal structure, density stratification and currents in the southeast Bering Sea basin. Journal of Geophysical Research 102(C10):22947-22964.

Cokelet, E.D., M.L. Schall, and D. Dougherty. 1996. ADCP-referenced geostrophic circulation in the Bering Sea basin. Journal of Physical Oceanography 26:1113-1128.

Favorite, F. 1974. Flow into the Bering Sea through Aleutian island passes. In: D.W. Hood and E.J. Kelley (eds.), Oceanography of the Bering Sea with emphasis on renewable resources. Occasional Publication No. 2, Institute of Marine Science, University of Alaska, Fairbanks, pp. 3-37.

Hughes, F.W., L.K. Coachman, and K. Aagaard. 1974. Circulation, transport and water exchange in the western Bering Sea. In: D.W. Hood and E.J. Kelley (eds.), Oceanography of the Bering Sea with emphasis on renewable resources. Occasional Publication No. 2, Institute of Marine Science, University of Alaska, Fairbanks, pp. 59-98.

Khen, G.V. 1989. Oceanological conditions of the Bering Sea biological productivity. In: Proceeding of the International Scientific Symposium on Bering Sea Fisheries. NOAA Tech. Memo NMFS F/NWC-163, pp. 404-414. (Available through NTIS.)

Kinder, T.H., L.K. Coachman, and J.A. Galt. 1975. The Bering Slope Current System. Journal of Physical Oceanography 5:231-244.

Kinder, T.H., J.D. Schumacher, and D.V. Hansen. 1980. Observations of a baroclinic eddy: An example of mesoscale variability in the Bering Sea. Journal of Physical Oceanography 10:1228-1245.

Mantyla, A.W., and J.L. Reid. 1983. Abyssal characteristics of the world ocean waters. Deep-Sea Research 30:805-833.

Mofjeld, H.O. 1986. Observed tides on the Northeastern Bering Sea Shelf. Journal of Geophysical Research 91:2593-2606.

Mordasova, N.V. 1995. Chlorophyll in the western Bering Sea. Oceanology 34:503-509. (English translation.)

Niebauer, H.J. 1988. Effects of El Niño-Southern Oscillation and North Pacific weather patterns on interannual variability in the subarctic Bering Sea. Journal of Geophysical Research 93:5051-5068.

Niebauer, H.J., and R.H. Day. 1989. Causes of interannual variability in the sea ice cover of the eastern Bering Sea. Geology Journal 18:45-59.

Niebauer, H.J., V. Alexander, and S. Henrichs. 1990. Physical and biological oceanographic interaction in the spring bloom at the Bering Sea marginal ice edge zone. Journal of Geophysical Research 95:22229-22242.

Niebauer, H.J., V. Alexander, and S.M. Henricks. 1995. A time-series study of the spring bloom at the Bering Sea ice edge. I: Physical processes, chlorophyll and nutrient chemistry. Continental Shelf Research 15:1859-1878.

Nihoul, J.C.J., P. Adam, P. Brasseur, E. Deleersnijder, S. Djenidi, and J. Haus. 1993. Three-dimensional general circulation model of the Northern Bering Sea's summer ecohydrodynamics. Continental Shelf Research 13:509-542.

Ohtani, K. 1970. Relative transport in the Alaskan Stream in winter. Journal of the Oceanographic Society of Japan 26:271-282.

Okkonen, S.R. 1996. The influence of an Alaskan Stream eddy on flow through Amchitka Pass. Journal of Geophysical Research 101:8839-8852.

Overland, J.E. 1981. Marine climatology of the Bering Sea. In: D.W. Hood and J.A. Calder (eds.), The eastern Bering Sea shelf: Oceanography and resources, volume one. Published by the Office of Marine Pollution Assessment, National Oceanic and Atmospheric Administration and Bureau of Land Management, pp. 15-30. (Distributed by the University of Washington Press, Seattle, WA 98105.)

Overland, J.E., and C.H. Pease. 1982. Cyclone climatology of the Bering Sea and its relation to sea ice extent. Monthly Weather Review 110:5-13.

Overland, J.E., and A.T. Roach. 1987. Northward flow in the Bering and Chukchi seas. Journal of Geophysical Research 92:7097-7105.

Overland, J.E., M.C. Spillane, H.E. Hurlburt, and A.J. Wallcraft. 1994. A numerical study of the circulation of the Bering Sea basin and exchange with the North Pacific Ocean. Journal of Physical Oceanography 24:736-758.

Paluszkieicz, T., and H.J. Niebauer. 1984. Satellite observations of circulation in the eastern Bering Sea. Journal of Geophysical Research 89:3663-3678.

Pearson, C.A., H.O. Mofjeld, and R.B. Tripp. 1981. Tides of the Eastern Bering Sea Shelf. In: D.W. Hood and J.A. Calder (eds.), The eastern Bering Sea shelf: Oceanography and resources, volume two. Published by the Office of Marine Pollution Assessment, National Oceanic and Atmospheric Administration and Bureau of Land Management, pp. 111-130. (Distributed by the University of Washington Press, Seattle, WA 98105.)

Pease, C.H. 1980. Eastern Bering Sea ice processes. Monthly Weather Review 108:2015-2023.

Reed, R.K. 1990. A year-long observation of water exchange between the North Pacific and the Bering Sea. Limnology and Oceanography 35:1604-1609.

Reed, R.K. 1991. Circulation and water properties in the central Bering Sea during OCSEAP studies, Fall 1989-Fall 1990. NOAA Technical Report, ERL 446-PMEL 41, NTIS PB90-155847. 13 pp.

Reed, R.K. 1995. On the variable subsurface environment of fish stocks in the Bering Sea. Fisheries Oceanography 4:317-323.

Reed, R.K. In press. Confirmation of a convoluted flow over the southeastern Bering Sea shelf. Continental Shelf Research.

Reed, R.K., and P.J. Stabeno. 1989. Circulation and property distributions in the central Bering Sea, Spring 1988. NOAA Technical Report, ERL 439-PMEL 39, NTIS PB90-155847. 13 pp.

Reed, R.K., and P.J. Stabeno. 1990. Flow trajectories in the Bering Sea: Evidence for chaos. Geophysical Research Letters 17:2141-2144.

Reed, R.K., and P.J. Stabeno. 1993. The return of the Alaskan Stream to Near Strait. Journal of Marine Research 51:515-527.

Reed, R.K., and P.J. Stabeno. 1994. Flow along and across the Aleutian Ridge. Journal of Marine Research 52:639-648.

Reed, R.K., and P.J. Stabeno. 1996. On the climatological net circulation over the eastern Bering Sea shelf. Continental Shelf Research 16(10):1297-1305.

Reed, R.K., and P.J. Stabeno. 1997. Long-term measurements of flow near the Aleutian Islands. Journal of Marine Research 55:565-575.

Reed, R.K., and P.J. Stabeno. Submitted. Inflow through Unimak Pass, Alaska. Geophysical Research Letters.

Reed, R.K., G.V. Khen, P.J. Stabeno, and A.V. Verkhunov. 1993. Water properties and flow over the deep Bering Sea basin, summer 1991. Deep-Sea Research 40:2325-2334.

Roach, A.T., K. Aagaard, C.H. Pease, S.A. Salo, T. Weingartner, V. Pavlov, and M. Kulakov. 1995. Direct measurements of transport and water properties through Bering Strait. Journal of Geophysical Research 100:18,443-18,457.

Roden, G.I. 1995. Aleutian Basin of the Bering Sea: Thermohaline, oxygen, nutrient and current structure in July 1993. Journal of Geophysical Research 100:13,539-13,554.

Royer, T.C., and W.I. Emery. 1984. Circulation in the Bering Sea, 1982-1983, based on satellite-tracked drifter observations. Journal of Physical Oceanography 14:1914-1920.

Sayles, M.A., K. Aagaard, and L.K. Coachman. 1979. Oceanographic Atlas of the Bering Sea Basin. University of Washington Press, Seattle. 158 pp.

Schumacher, J.D., and T.H. Kinder. 1983. Low-frequency current regimes over the Bering Sea shelf. Journal of Physical Oceanography 13:607-623.

Schumacher, J.D., and R.K. Reed. 1992. Characteristics of currents over the continental slope of the eastern Bering Sea. Journal of Geophysical Research 97:9423-9433.

Schumacher, J.D., and P.J. Stabeno. 1994. Ubiquitous eddies of the eastern Bering Sea and their coincidence with concentrations of larval pollock. Fisheries Oceanography 3:182-190.

Schumacher, J.D., and P.J. Stabeno. 1998. The continental shelf of the Bering Sea. In: A.R. Robinson and K.H. Brink (eds.) The Sea: The global coastal ocean regional studies and synthesis, Vol. XI. John Wiley and Sons, New York, pp. 869-909.

Schumacher, J.D., C.A. Pearson, and J.E. Overland. 1982. On exchange of water between the Gulf of Alaska and the Bering Sea through Unimak Pass. Journal of Geophysical Research 87:5785-5795.

Shuert, P.G., and J.J. Walsh. 1993. A coupled physical-biological model of the Bering-Chukchi seas. Continental Shelf Research 13:543-573.

Solomon, H., and K. Ahlnas. 1978. Eddies in the Kamchatka Current. Deep-Sea Research 25:403-410.

Springer, A.M., C.P. McRoy, and M.V. Flint. 1996. The Bering Sea Green Belt: Shelf edge processes and ecosystem production. Fisheries Oceanography 5:205-223.

Stabeno, P.J., and R.K. Reed. 1990. Recent Lagrangian measurements along the Alaskan Stream. Deep-Sea Research 38:289-296.

Stabeno, P.J., and R.K. Reed. 1992. A major circulation anomaly in the western Bering Sea. Geophysical Research Letters 19:1671-1674.

Stabeno, P.J., and R.K. Reed. 1994. Circulation in the Bering Sea basin by satellite tracked drifters. Journal of Physical Oceanography 24:848-854.

Stabeno, P.J., R.K. Reed, and J.E. Overland. 1994. Lagrangian measurements in the Kamchatka Current and Oyashio. Journal of Oceanography 50:653-662.

Stabeno, P.J., R.K. Reed, and J.D. Schumacher. 1995. The Alaska Coastal Current: Continuity of transport and forcing. Journal of Geophysical Research 100:2477-2485.

Stabeno, P.J., J.D. Schumacher, R.F. Davis, and J.M. Napp. 1998. Under-ice observations of water column temperature, salinity and spring phytoplankton dynamics: Eastern Bering Sea shelf. Journal of Marine Research 56:239-255.

Tsugonai, S., M. Kusakabe, H. Iizumi, I. Koike, and A. Hattori. 1979. Hydrographic features of the deep water of the Bering Sea, the sea of silica. Deep-Sea Research, Part A 26:641-659.

van Meurs, P., and P.J. Stabeno. In press. Evidence of episodic on-shelf flow in the southeastern Bering Sea. Journal of Geophysical Research.

Verkhunov, A.V., and Y.Y. Tkachenko. 1992. Recent observations of variability in the Western Bering Sea Current system. Journal of Geophysical Research 97:14,369-14,378.

Walsh, J.J., C.P. McRoy, L.K. Coachman, J.J. Georing, J.J. Nihoul, T.E. Whitledge, T.H. Blackburn, P.L. Parker, C.D. Wirick, P.G. Shuert, J.M. Grebmeier, A.M. Springer, R.D. Tripp, D.A. Hansell, S. Djenidi, E. Deleersnijder, K. Henriksen, B.A. Lund, P. Andersen, F.E. Müller-Karger, and K. Dean. 1989. Carbon and nitrogen cycling with the Bering/Chukchi Seas: Source regions for organic matter effecting AOU demands of the Arctic Ocean. Progress in Oceanography 22:277-359.

CHAPTER 2

An Update on the Climatology and Sea Ice of the Bering Sea

Henry J. Niebauer
Institute of Marine Science, University of Alaska Fairbanks, Fairbanks, Alaska

Nicholas A. Bond
Pacific Marine Environmental Laboratory, Seattle, Washington

Lev P. Yakunin and Vladimir V. Plotnikov
Pacific Oceanographic Institute, Russian Academy of Sciences of the Far East, Vladivostok, Russia

Abstract

The Bering Sea is a region of extreme seasonal as well as substantial interannual variability in its air-ice-ocean environment. This chapter updates work on the weather/climate and ice environment of the Bering Sea and its surrounding environs. Topics with extra emphasis include polynyas, analyses of ~50-year long time series of the air-ice-ocean parameters, analysis of a climate "regime shift," and the connection of the Bering Sea environment with interannual variability in the Aleutian Low (i.e., indices of interannual atmospheric variation such as the North Pacific [NP] and western Pacific [WP] oscillations in the North Pacific as well as the Southern Oscillation Index [SOI] in the tropical Pacific).

Introduction

This chapter is an update and expansion upon, but does not replace, the reviews of synoptic climatology and the effects on sea ice of the Bering Sea by Overland (1981) and Niebauer (1981a,b). This chapter is also meant to compliment the Bering Sea atlases of Brower et al. (1977) and Brower et al. (1988), the satellite microwave sea ice atlases of Parkinson et al. (1987) and Gloersen et al. (1992), as well as a recent review of the physical environment of the Bering sea by Niebauer and Schell (1993). One of the topics

---

Current address for H.J. Niebauer is Dept. of Atmospheric and Oceanic Sciences, University of Wisconsin, Madison, WI.

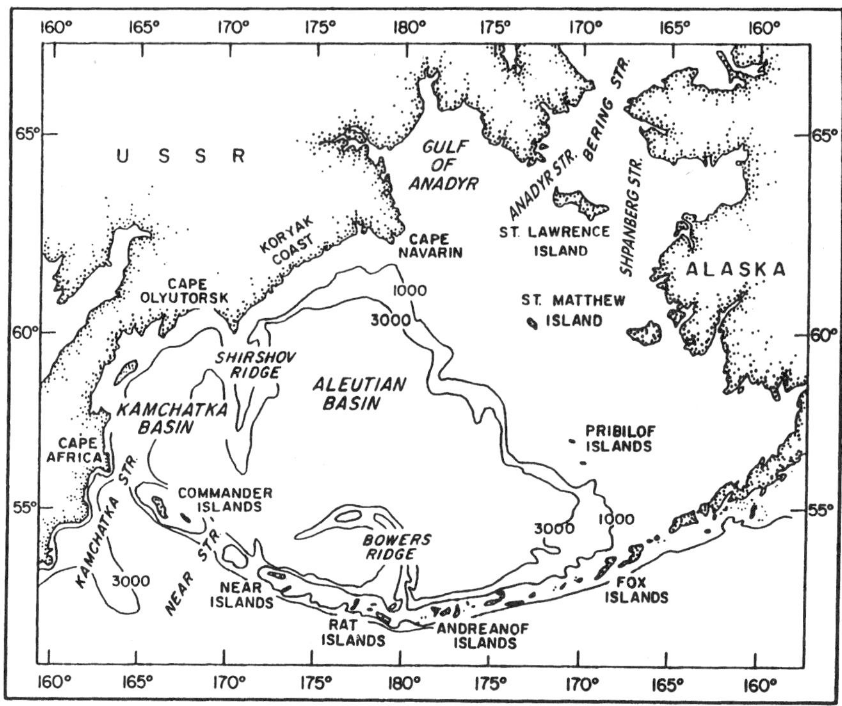

Figure 1.   *The Bering Sea is about half abyssal with depths greater than 3.5 km while the northeast half is a wide, shallow continental shelf with depths less than 200 m.*

that we emphasize is the update of time series, of ~50 years in length, of the air-ice-ocean environment of the Bering Sea as well as indices of the Aleutian Low (both North Pacific and western Pacific) and El Niño–Southern Oscillation (ENSO). In this time series analysis, we also update the quantitative relationships among the time series outlining the role of the Aleutian Low and its impact on the Bering Sea environment before and after a climatic "regime shift" (e.g., Ebbesmeyer et al. 1991) that occurred in the late 1970s. Our focus is primarily the southeastern Bering Sea shelf, but we do include some Russian material on the western Bering Sea.

Updated General Background

The Bering Sea is the semi-enclosed sea separating Alaska and Siberia and is the only direct ocean link between the Arctic and Pacific oceans (Fig. 1). The dimensions of the Bering Sea are ~1,500 km north-south by ~3,000 km along the Aleutian Chain. The northeast half of the basin is the widest

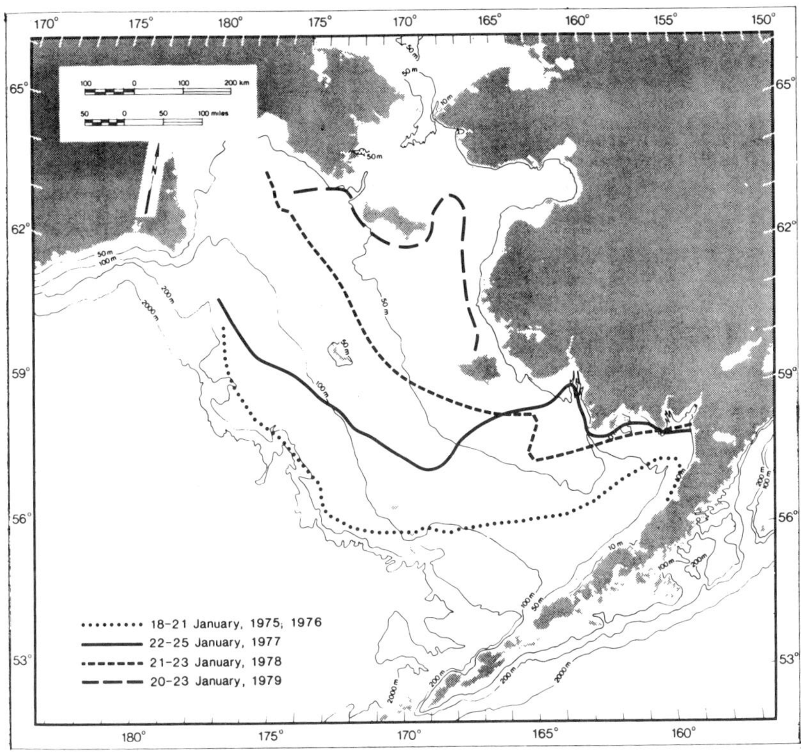

Figure 2. Eastern Bering Sea shelf region with southern ice limit for January 1975-1979 (from Niebauer 1981).

continental shelf outside the Arctic Ocean at ~500 km wide (Fig. 1). However, most of this shelf is shallow at <100 m deep with the shelf break ~170 m. The southwest half of the Bering Sea is deep ocean basin with maximum depths of ~4,000 m. The net ocean flow through the Bering Sea is from south to north, that is from Pacific to Arctic. This flow is somewhat restricted by the Aleutian Chain between the Pacific and the Bering Sea, but is more restricted by the Bering Strait (~85 km by ~45 m) between the Bering Sea and Arctic Ocean. The long-term mean transport through the Bering Strait is 1-2 Sv (Coachman and Aagaard 1988), or ~15% of the world ocean inflow to the Arctic Ocean. However, recent estimates are ~1.2 Sv (Roach et al. 1995) or even lower at ~0.7 Sv for the 5-year period 1990-95 (K. Aagaard, Applied Physics Lab, University of Washington, Seattle, WA; and T. Weingartner, SFOS, University of Alaska, Fairbanks, AK, pers. comm.).

The Bering Sea is subjected to large seasonal variations in daylight and insolation that result in large variations in weather and ice patterns

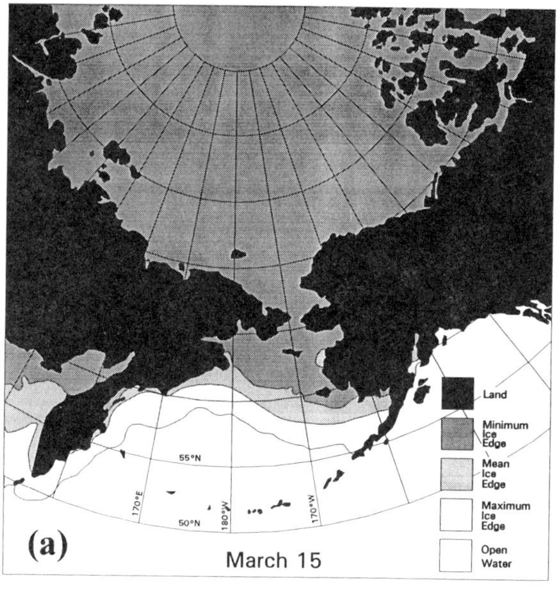

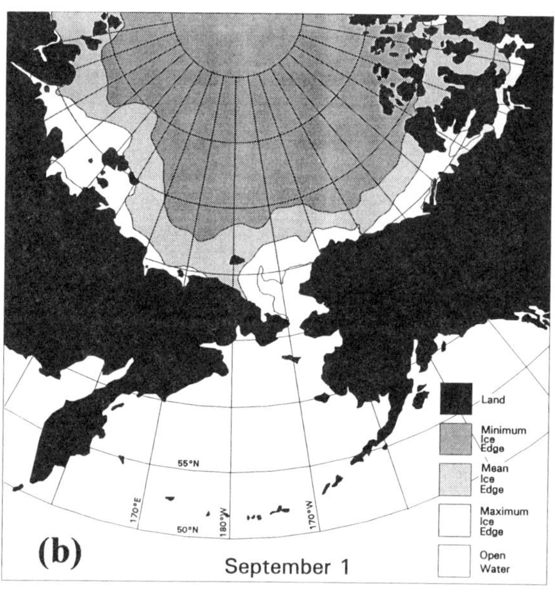

Figure 3.   *Maximum, mean, and minimum extent of the ice edge for: (a) March 15 (typically the maximum winter ice extent), and (b) September (typically the minimum summer ice extent) for the region of the Bering Sea. Ice data are means for 1973-1986. (From U.S. Naval Oceanography Command 1986, after Niebauer and Schell 1993.)*

(e.g., Figs. 2-5). Overland (1981) has suggested that the weather and climate of the Bering Sea are related to the presence and fluctuations in sea ice. Fig. 3 shows the region-wide distribution of sea ice in winter and summer. Approximately the northern 75% of the shelf region is ice covered in most winters and the entire Bering Sea is typically ice free in summer. This seasonal sea ice advance and retreat is the largest of any of the Arctic or subarctic regions, averaging ~1,700 km (Walsh and Johnson 1979), while interannual variability has been as great as ~400 km, or ~25% of the seasonal range (Niebauer 1983). Maximum ice cover (Fig. 3a) occurs in March or early April and lags minimum insolation (late December) by ~3 months mainly due to the heat capacity of the ocean. At this point, nearly all the Arctic Ocean is ice covered as well as ~ ⅓ to ½ of the Bering Sea. In any one year, heaviest ice is usually found in the western Bering Sea. In the mean, the maximum ice edge is ~900 km south to southeast of the Bering Strait but this can vary from ~700 km in light (in extent) ice years to ~1,100 km in heavy (in extent) ice years. In an average year, the maximum amount of the Bering Sea that is ice covered is ~37%, but in an icy year, this can go to ~56% while in a light year, ~20% (Plotnikov 1990). Duration of ice cover also varies from a minimum of ~3 months to a maximum of isolated instances of some ice actually lasting through the summer in isolated areas (Plotnikov 1990). Minimum sea ice occurs in September (Fig. 3b), again lagging maximum insolation by ~3 months, with the ice edge in the Chukchi Sea far to the north of the Bering Sea.

Winter ice formation in the Bering Sea has been described by a "conveyor belt" analogy (e.g., Muench and Ahlnas 1976, Pease 1980, Burns et al. 1981). Sea ice forms along the south facing coasts where polynyas are formed as the predominantly northerly winds carry sea ice southward away from the coasts (e.g., Fig. 4). Major polynyas occur south or downwind of the Chukchi Peninsula, St. Lawrence and St. Matthew islands, and the Seward Peninsula (Fig. 4). Sea ice is blown toward the south until it reaches warmer water and melts. The ice edge thus advances southward as the melt water cools the upper ocean layer, but seldom farther than the warmer deep water off the shelf break. Ice drift rates vary from 17 to 22 km/day to as fast as 28-32 km/day (Shapiro and Burns 1975, Muench and Ahlnas 1976, Weeks and Weller 1984, Glueck and Niebauer 1996). In the Bering Strait speeds as fast as 50 km/day have been observed, although there are reversals driven by wind events (Pease 1980). Undeformed ice thickness is ~0.5-1.0 m in the open Bering Sea but rafting does occur.

While most of the ice is formed in situ, there is transport of ice into and out of the Bering Sea through the Bering Strait in the north and through Kamchatka Strait in the far western Aleutians (Yakunin 1987). The average annual ice transport into the Bering Sea from the Arctic Ocean through the Bering Strait is ~3 km³/year for the period 1964-1987, but with a range of ~54 km³/year out of the Bering Sea into the Arctic Ocean (1964) to ~44 km³/year in from the Arctic Ocean (1985). For comparison, in a typical year at maximum ice in late March, the average ice content of the Bering

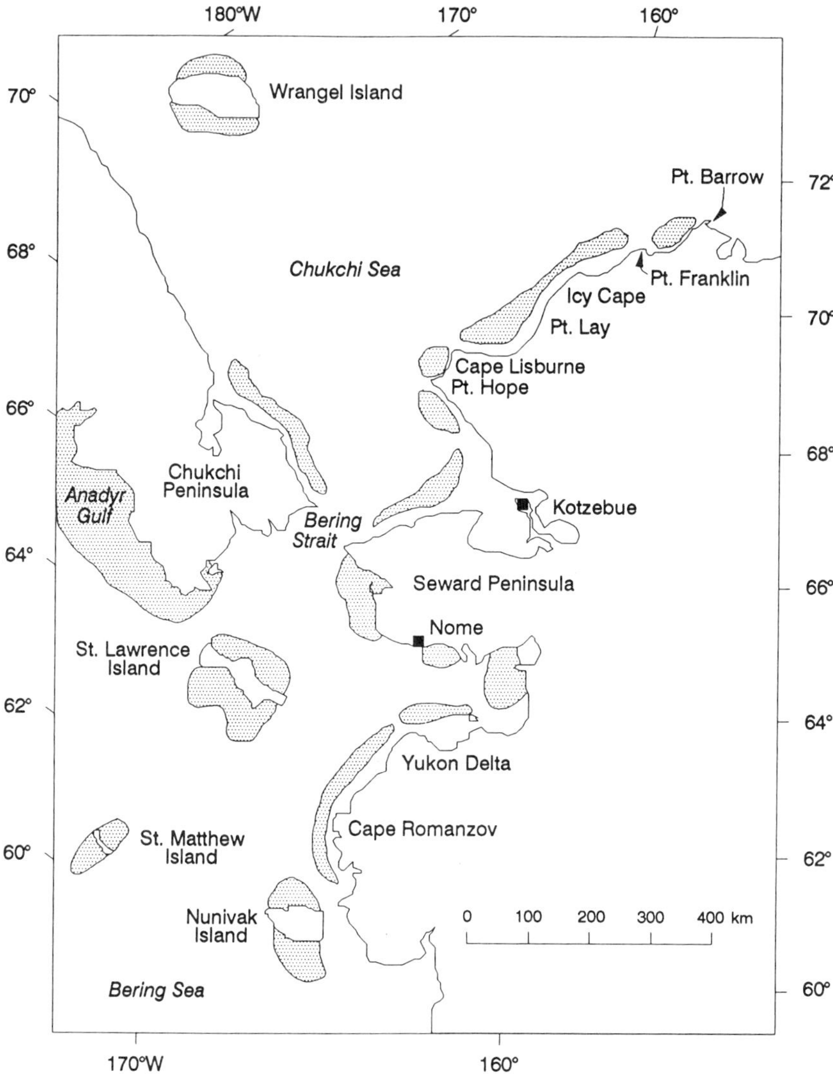

Figure 4.   Recurring polynyas in the eastern Bering and Chukchi seas. (From Niebauer and Schell 1993 as redrawn from Stringer et al. 1982.)

Sea is ~850 km³ (note that this is not the total amount of ice formed because of the "conveyor belt" ice formation analogy described above). In addition to interannual variability through the Bering Strait, there is also an intra-annual pattern superimposed on the ~3 km³/year net inflow through the Bering Strait, with mean ice transport into the Bering in ~January-March but mean transport of ice out ~April-May. Week to week variations are very much larger than this ~3 km³/year, ranging from 45 km³ (in) to 54 km³ (out). To the south, in Kamchatka Strait, the net flow of ice is out of the Bering Sea into the North Pacific in all years with a mean of ~36 km³/year and a range of ~15-90 km³/year (Yakunin 1987).

As mentioned above, polynyas are formed as winds carry sea ice away from the coasts (Fig. 4). Polynyas (defined here as areas of persistent and/ or recurring mesoscale size areas of ice-free water despite air temperatures well below the freezing point for sea water) are an important component in air-ice-ocean interaction in the Bering Sea. Polynyas are sites of high rates of heat exchange between atmosphere and ocean (Maykut 1978) resulting in zones of ice formation and brine production. This is true of ice covered seas in general. The Bering Sea polynyas are "latent" heat polynyas (Pease 1980) in which winds or currents cause open water resulting in formation of new ice (as opposed to "sensible" heat polynyas in which ice melts resulting in open water).

As an example of Bering Sea polynyas, the St. Lawrence Island Polynya (SLIP) as shown in Fig. 4 is one of the largest island polynyas in the Bering Sea. In a recent analysis of ERS-1 Synthetic Aperture Radar (SAR) imagery and SAR-derived ice vectors, Glueck and Niebauer (submitted) outlined several SLIP events (openings and closings of a week or two duration). While the SLIP does develop along the north coast of the island under southerly winds, SLIP events occurring off the south coast under the predominant northerly winds are more prevalent and dramatic. At maximum size, the southside SLIP extends tens to hundreds of kilometers offshore by >100 km in the long-shore direction. The SLIP outline has been observed at least 235 km downwind (Glueck and Niebauer, submitted). They also observed maximum ice speeds in the SLIP >30 km/day away from the island. Because they are episodically wind-driven, latent heat polynyas are often characterized by an outline or gradient of ice types becoming older with distance away from the land and open water.

Sea ice production rates are reported in the range of 10-12 cm/day in the SLIP from Cavalieri and Martin (1994) using SMMR, from Glueck and Niebauer (submitted) SAR observations, and from results of Pease's (1987) analytical model of polynya growth (Glueck and Niebauer, submitted). The calculations agree within ~20%. Associated calculations of brine production rates of ~0.016 Sv (range of 0.006-0.042 Sv) for the February 1992 southern SLIP event (Glueck and Niebauer, submitted) compare favorably with the results of Cavalieri and Martin (1994) of ~0.017 Sv (0.008-0.028 Sv). For comparison, this brine production is nearly identical in magnitude with the mass transport of the Mississippi River of ~0.018 Sv.

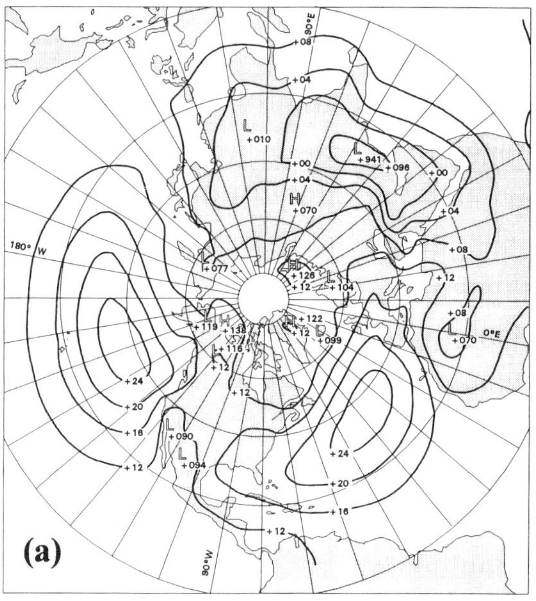

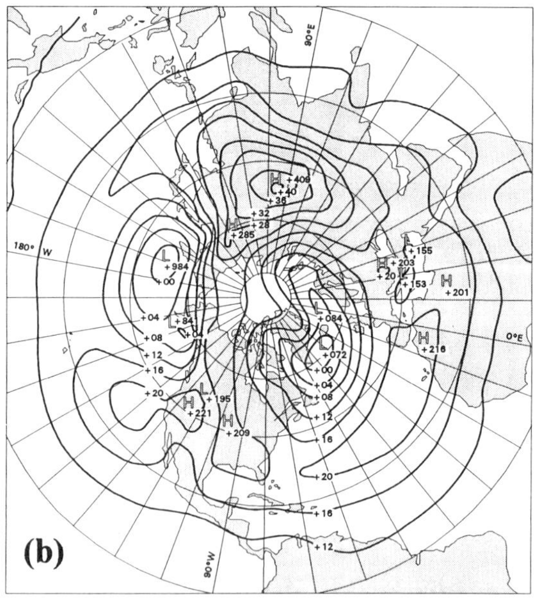

Figure 5.   Mean (1946-1983) atmospheric surface pressure chart for the Northern Hemisphere for (a) July and (b) January. Isobars are labeled as mb + 1,000 except for pressure below 1,000 mb, which are labeled as actual pressure. (U.S. Naval Fleet Numerical Weather Central in Monterey, CA.)

Dense brine formed as a result of salt rejection during ice formation in polynyas in the Bering Sea helps maintain the Arctic Ocean halocline (Aagaard et al. 1981, Cavalieri and Martin 1994). This brine becomes part of the northward flow across the shelf through the Bering Strait into the Arctic Ocean (Schumacher et al. 1983, Overland and Roach 1987). Aagaard et al. (1981) and Björk (1989, 1990) suggest that 1-2 Sv of brine are required to maintain the Arctic Ocean halocline with an associated renewal time of ~10-40 years (Aagaard et al. 1981, Wallace et al. 1987). Cavalieri and Martin (1994) estimate that the brine contribution from all the polynyas in the Arctic, including the Bering Sea, is ~0.9 Sv (0.7-1.2 Sv). For the western Arctic (i.e., Bering, Chukchi, Beaufort, and East Siberian seas), they estimate that average annual dense water contribution from polynyas is ~0.51 Sv of which the Bering Sea contribution is ~0.059 Sv. Altogether, these calculations suggest that the Bering Sea polynyas contribute ~6% of the brine to the Arctic halocline. SLIP brine production is ~30% of the Bering Sea contribution, ~3% of the western Arctic contribution, and ~1.8% of the total contribution to the maintenance of the Arctic Ocean halocline.

The weather of the Arctic and Bering Sea is dominated by major pressure systems that change seasonally as shown in Fig. 5. In summer (Fig. 5a), the Bering Sea is between the northern portions of a high pressure system over the North Pacific and low pressure over Asia. In the mean (and also generally in a day-to-day sense) the pressure gradients and winds over the Bering Sea in summer are relatively weak.

The change to winter is drastic (Fig. 5b); now the Asian continent is dominated by high pressure while the Aleutian and Icelandic lows dominate the North Pacific (including the Bering Sea) and North Atlantic respectively. (As a working definition, the Aleutian Low is a region of primarily wintertime low atmospheric pressure covering much of the North Pacific and Bering Sea and typically centered along the Aleutian Island chain as shown in Fig. 5b.) The winter pressure gradients are much stronger resulting in intensified winds, especially in storms. The Aleutian Low actually represents a composite of the migrating storms (e.g., Anderson and Gyakum 1989) that dominate the day-to-day weather, and so the Aleutian Low is statistical in nature. In the winter months, three to five storms per month (defined as areas of low sea level pressure [SLP] with closed isobars) move along the Aleutian Islands and into either the Gulf of Alaska or the Bering Sea–Bristol Bay. By contrast, less than two storms per month cross the northern Bering Sea (Overland 1981). This storm track is reflected in the mean wind stress curl (Bond et al. 1994), which exhibits a maximum extending from the southern tip of Kamchatka along or just south of the Aleutians into the Gulf of Alaska. The gradients in the root-mean squared surface stress are smaller, implying a more uniform distribution of wind mixing in the North Pacific and Bering Sea. The wind forcing of the Bering Sea is greater by an order of magnitude in winter than summer due to the seasonal modulation of the Aleutian Low.

The large annual variations in the Bering Sea are accompanied by substantial interannual variations in the air, ice, and ocean environment (e.g., Walsh and Johnson 1979; Niebauer 1980, 1983, 1988; Walsh and Slater 1981; Roger 1981; Overland and Pease 1982) especially during the winter. These interannual variations are driven largely by the atmosphere, specifically through its influence of the horizontal transports of heat, water vapor, and momentum, and the amount of energy radiated to space. The overall sense of the atmospheric circulation anomalies can be described using an index based on the intensity and position of the wintertime Aleutian Low. During the summer the atmospheric circulation tends to be weak and interannual variations are generally not prominent, with the summer of 1997 providing a notable exception. The interannual variability in the summer is expected to be controlled principally through radiative effects, in particular, via the fractional coverage of low cloud decks and their impact on insolation. Interannual fluctuations do not appear to be driven by the Bering Sea ocean circulation because ocean flow, at least on the shelf, is so sluggish (Reed 1978) and because flow from the Pacific through the Aleutian passes is restricted.

While Bering Sea variability appears to depend so much on the Aleutian Low, the Aleutian Low, in turn, responds to variations on a Pacific-wide and global scale (e.g., Barnstron and Livezey 1987, Wallace et al. 1990, Trenberth 1990, Bond et al. 1994, Latif and Barnett 1994, Trenberth and Hurrell 1994). During El Niño events, in the Northern Hemisphere winter, the Aleutian Low has been shown to deepen and move southeastward of normal (Bjerknes 1966, 1969, 1972). These El Niño patterns in the Aleutian Low have been related to warming in the Bering Sea (e.g., Niebauer 1988, Niebauer and Day 1989). In some winters, the atmospheric teleconnection mode called the Pacific–North American pattern (PNA, Wallace and Gutzler 1981) is prominent, with substantial effects on the Aleutian Low. Positive PNA events are associated with an intensified Aleutian Low along with ridging over western Canada and Alaska so that warmer air is driven northward. In connecting PNA and El Niño, Niebauer (1988) calculated a correlation coefficient of −0.27 (significant at a 0.01 level) with PNA lagging SOI by ~2 months, suggesting that at times positive PNA events are related to negative SOI, or El Niños. Conversely, during La Niña events, the Aleutian Low appears to become less intense and to move westward of normal and this pattern has been related to cooling in the Bering Sea (e.g., Niebauer 1988, Niebauer and Day 1989). In addition, atmospheric circulation over the North Pacific is also sensitive to local SST anomalies, as argued by Namias (1978) among others, and as suggested by recent numerical model results (e.g., Latif and Barnett 1994, see also Discussion below).

These physical weather-climate patterns have also been related to fluctuations in Bering Sea fisheries (e.g., Quinn and Niebauer 1995). For example, a major event, or regime shift, in environmental parameters in the North Pacific and Bering Sea in the late 1970s (e.g., Niebauer 1988, Trenberth 1990, Trenberth and Hurrell 1994, Ebbesmeyer et al. 1991) has had major implications for the entire biophysical environment, as found in zooplankton abundance (Brodeur and Ware 1992, Roemmich and McGowan 1995) and groundfish stocks (Hollowed and Wooster 1992).

Time Series

Data and Methods

The annual cycles for each of the environmental parameters described below and in the following sections are shown in Fig. 6. The actual time series of anomalies are shown in Fig. 7.

The ice data are for the eastern Bering Sea and Chukchi Sea and are a composite of monthly data for 1953-1977 from Walsh and Johnson (1979) and the weekly data for 1978-1994 from the Navy-NOAA Joint Ice Center in Suitland, Maryland. The weekly data are averaged to monthly. The monthly mean percent ice cover was calculated as the ratio of ice cover to the total area (see Niebauer 1981). This gives a semiquantitative estimate of ice cover as it considers all the ice enclosed by the southern limit of ice but without regard to thickness or concentration of ice. Note that the method of data collection, and thus the uniformity of the data collection, have varied over time. Satellite observations were not available until ~1973. The Electrically Scanning Microwave Radiometer (ESMR) only operated from 1973 to 1976 (Parkinson et al. 1987) but Scanning Multichannel Microwave Radiometer (SMMR) began in fall 1979 (e.g., Gloersen et al. 1992). While there was a shift in the way satellite data were collected in the period of the regime shift, the effects on the time series of anomalies appear minimal.

Monthly mean sea surface temperature (SST) for the period 1947-1994 were obtained from the Climate Research Group, Scripps Institution of Oceanography, UCSD. The SSTs used here are $5° \times 5°$ means centered south of the Pribilof Islands (55°N, 170°W) on the edge of the Bering Sea shelf but ~500 km from shore.

Air temperatures and surface winds for 1947-1994 and 1965-1994 respectively were obtained from the National Weather Service Local Climatological Data for St. Paul Island in the Pribilof Islands. The magnitude of the wind is resolved along due north-south but the wind anomalies still require caution in interpretation because, as plotted in Fig. 7, the actual direction of the anomaly is not indicated. In the Pribilofs, the monthly mean winds are from the north in all months except July and August (Fig. 6). Therefore, a negative anomaly can mean either less-strong winds from the north (i.e., less than the long-term mean for that month) or actual southerly flow.

The interannual variations in the weather patterns affecting the North Pacific and Bering Sea are described here using the North Pacific (NP) and West Pacific (WP) indices. These two indices characterize the teleconnection modes identified by Barnstron and Livezey (1987) that have large

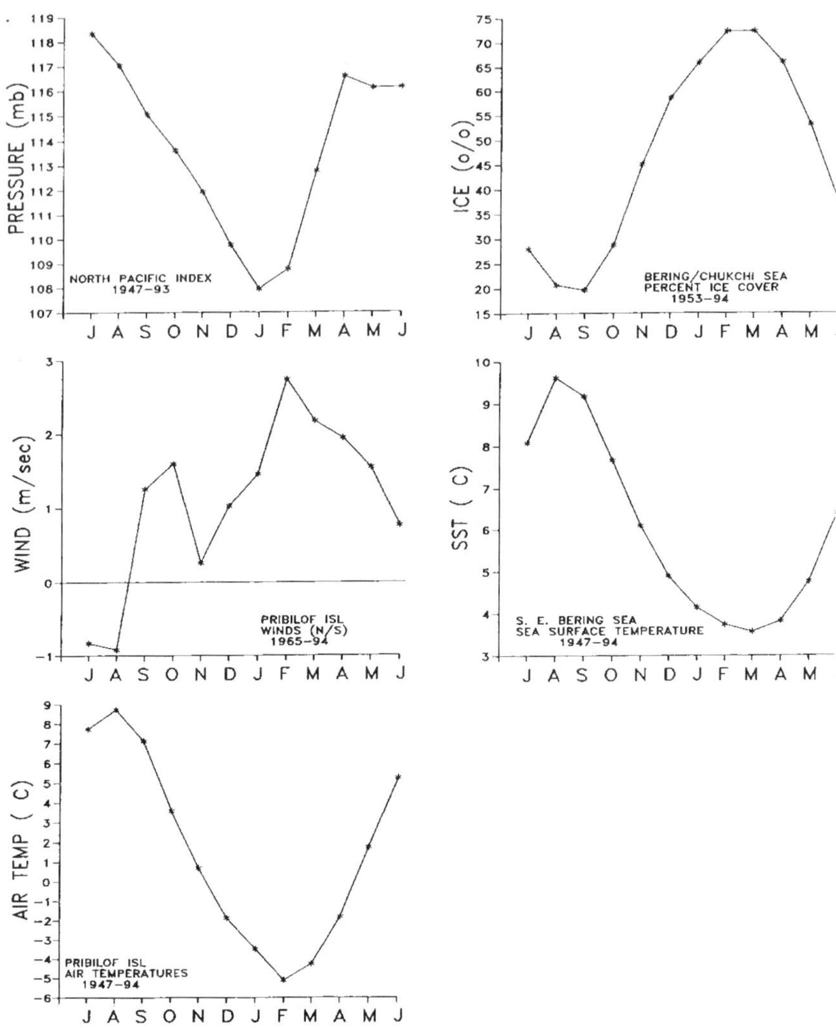

Figure 6.   The annual cycles for each of the Bering Sea air, ice, and ocean physical environmental parameters and the length of each data set.

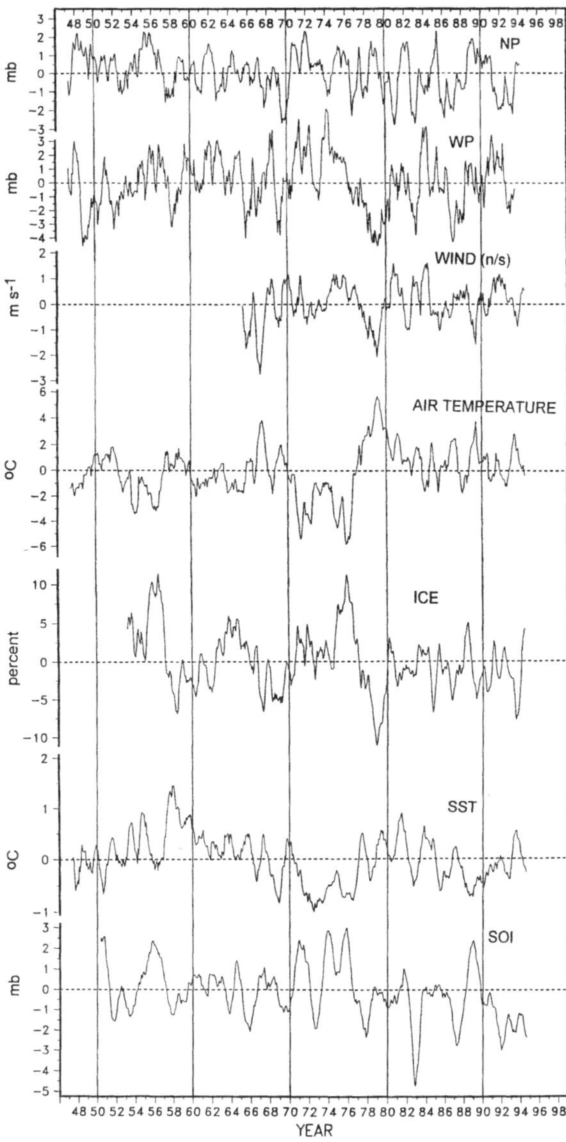

Figure 7. Deviations from monthly mean air, ice, and ocean physical environmental parameters from the Bering Sea, as well as the Southern Oscillation Index (SOI) from the south Pacific. NP = North Pacific index, WP = western Pacific index, SST = sea surface temperature, and WIND (n/s) = north-south component of the wind. For the wind, positive anomalies mean either more strongly from the north or less strongly from the south (see Fig. 6). All of these data are smoothed with a 9-month running mean.

amplitudes in the Bering Sea. The NP index is defined (after Trenberth and Hurrell 1994) as the areal averaged monthly mean surface air pressure for the mid- to North Pacific (27.5-60°N by 140°W-160°E). The actual sea level pressure used to calculate the NP index for 1947-1994 are from the Climate Research Group of Scripps Institution of Oceanography. Previously, in Niebauer (1988) and Niebauer and Day (1989), the Pacific-North American (PNA) index of Wallace and Gutzler (1981) was used but this index uses points from Hawaii, the North Pacific, western Canada, and the southeast United States. We wanted to concentrate in the North Pacific. The PNA and NP indices appear to measure the same phenomenon as their correlation coefficient = –0.82 (the 99% confidence level = –0.14) with no time lag, but falls off precipitously to –0.18 within about one month.

WP is a north-south dipole over the western North Pacific. The WP index is defined following Wallace and Gutzler (1981) as the difference in the normalized 500 mb height anomalies between two points (60°N, 155°E and 30°N, 155°E). Since the pattern associated with the WP index has an equivalent barotropic structure, it also characterizes the primary fluctuations in the sea-level pressure field in the western Pacific. A negative WP anomaly indicates a stronger Aleutian Low west of 180° with the stormtrack displaced northward, while a positive anomaly indicates a weaker Aleutian Low west of 180° with the stormtrack displaced southward.

The Southern Oscillation Index (SOI) is calculated as the difference between normalized sea level pressure (SLP) from Tahiti in the Society Islands in the South Pacific, and Darwin, Australia. These data were obtained from the Climate Diagnostics Bulletin of NOAA. El Niño events are negative anomalies (e.g., 1951-52, 57-58, 65-66, 72-73, 77-78, 82-83, 86-87, and 91-93) while La Niña events are positive anomalies (e.g., 1949-51, 55-56, 70-71, 73-74, 75-76, and 88-89).

In all of the time series in Fig. 7, we removed the mean annual cycles (Fig. 6) in an attempt to isolate the monthly deviations from the mean (except for the SOI which is already an anomaly series). We did this simply by averaging all the Januarys, all the Februarys, etc., and then subtracting the mean January, February, etc., from each individual January, February, etc. The resultant time series are shown in Fig. 7 smoothed with a 9-month running mean. Each of the unsmoothed time series were cross-correlated with all the other time series, leading and lagging one dataset with the other for up to 3 years (Fig. 8 and Table 1). Significance levels were calculated (Zar 1984) with an effective number of degrees of freedom (N_{eff}) calculated according to Trenberth (1984). N_{eff} depends upon autocorrelation such that the stronger the autocorrelation, the fewer the degrees of freedom.

Annual Means

The North Pacific index (NP) reaches its maximum in July, falling to a minimum in January (Fig. 6). This is indicative of the North Pacific High dominance in the Northern Hemisphere (NH) summer (Fig. 5a) followed by the

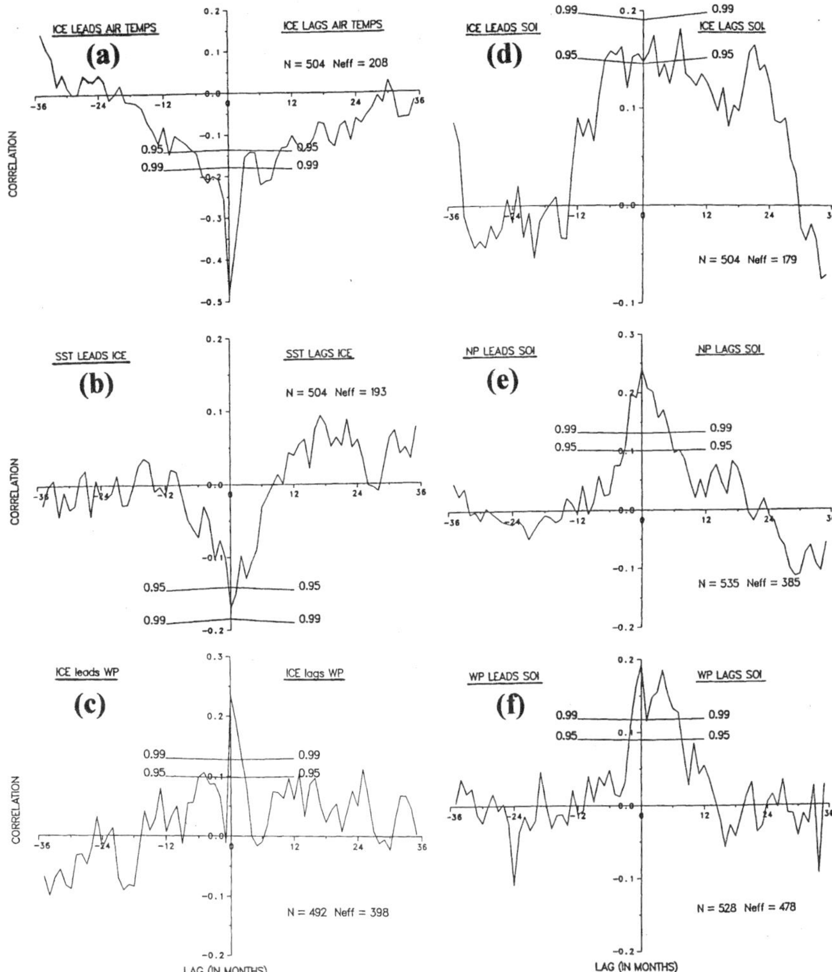

Figure 8. Cross-correlation plots of selected Bering Sea and north and south Pacific time series of Fig. 7: (a) ICE = sea ice vs. Bering Sea air temperatures, (b) SST = Bering Sea surface temperature vs. ICE, (c) ICE vs. WP = west Pacific index, (d) ICE vs. SOI = Southern Oscillation Index, (e) NP = north Pacific index vs. SOI, and (f) WP vs. SOI. The x-axis is the time lag between datasets. N is the number of data points while N_{eff} is the number of effective or independent data points available for use in calculating the confidence levels shown on each plot.

Table 1. Table of correlation coefficients matrix of monthly mean times series of deviations from mean percent ice cover (ICE), sea surface temperatures (SST), winds, air temperatures (AT), North Pacific (NP) and West Pacific (WP) indices from the Bering Sea, Niño 1+2 SST (N1+2) and the southern oscillation index (SOI) from the south Pacific as illustrated in Fig. 6.

	ICE	SST	WIND	AT	NP	WP	N1+2
SST	-0.17 (0-1, SST lags, 0.95) 2						
WIND	0.32 (0-1, ice lags, 0.99) 4	-0.12 (3, SST lags, 0.95) 1					
AT	-0.48 (0, no lag, 0.99) 16	0.26 (0, no lag, 0.99) 5	-0.51 (0, no lag, 0.99) 3				
NP	0.14 (7, ice lags, 0.95) 1 inconclusive	-0.1 (3&9, SST lags, 0.95) 3	-0.33 (0, no lag, 0.99) 3	-0.11 (4&5, AT lags, 0.95) 2			
WP	0.24 (0, no lag, 0.99) 3	-0.09-0.14 (3&12, SST lags, 0.95) 3	0.55 (0, no lag, 0.99) 1	-0.41 (0, no lag, 0.99) 3	0.13 (0, no lag, 0.99) 1		
N1+2	-0.17 (19, ice lags, 0.95) 1 inconclusive	0.17 (1-7, SST lags, 0.95) 7	0.11 (7&10, AT lags, 0.90) 0	0.15 (1, AT lags, 0.95) 2	-0.16 (-1 to 2, NP lags, 0.99) 8	-0.15 (0,6, WP lags, 0.99) 11	
SOI	0.17 (2, 7&21, ice lags, 0.95) 12	-0.12 (0-5, SST lags, 0.90) 0	INCONCLUSIVE	-0.21 (1&4, AT lags, 0.99) 8	0.24 (0, no lag, 0.99) 8	0.19 (0,4, WP lags, 0.99) 10	-0.48 (-1 to 3, N1+2 lags, 0.99) 14

The data in the first set of parentheses indicate the lag, in months, when maximum correlation occurred, which dataset lags, and the significance level. The numbers in the second set of parentheses indicate the number of months the correlations are greater than the 95% significance level. "INCONCLUSIVE" indicates that there are so many correlation peaks that the analyses are inconclusive.

dominance of the Aleutian Low dominance in the NH winter (Fig. 5b). The climb from the winter minimum to the summer high is not monotonic but rather plateaus for April-June before hitting the July high. The WP index is defined in terms of monthly anomalies and so its annual mean is identically zero.

Air temperatures in the Pribilof Islands lag the NP by one month with maximum temperatures of ~9°C occurring in August and minimum temperatures of ~ −5°C occurring in February. There is no plateau in temperature as in the pressure index.

The north-south component of the mean monthly surface wind from the Pribilof Islands is more complex than the other annual cycles (Fig. 6). The winds are southerly only in July and August and undergo a relatively rapid shift to northerly during early autumn. The minimum (maximum) in the northerly component observed in November (February) reflects a relatively high (low) incidence of cyclonic storms passing to the north and west.

Minimum sea ice occurs in September (Figs. 3, 6) with the ice edge in the Chukchi Sea far to the north of the Bering Sea. The Bering Sea is sea ice–free in summer. Maximum ice cover occurs in February-March (Figs. 3, 6). The seasonal latitudinal excursion of the ice is ~1,400 km.

Finally, sea surface temperature (SST) from the southeast Bering Sea (Fig. 6) peaks at just under 10°C in August before falling to the annual minimum of ~3.5°C in March.

Time Series Analyses

As remarked above, the time series shown in Fig. 7 have all been cross-correlated with one another within the Bering Sea, with indicators of ENSO activity in the South Pacific (SOI and Niño 1+2), and with time series of variability in the Aleutian Low (NP, WP). These cross-correlation statistics are shown in Table 1. The purpose of these analyses are to try to understand the extraordinary interannual variability, including the regime shift observed in the air, ice, and ocean environment of the Bering Sea as pointed out above. For comparison and update, a similar analysis was reported on these same datasets (except for the NP and WP indices) by Niebauer (1988) and Niebauer and Day (1989) when the datasets spanned 20-33 years (~1953-1985) as compared to the present >45 years (~1947-1995).

Most of the basic conclusions (summarized in this section) have not changed. However, a regime shift that occurred in the mid-late 1970s (e.g., air temperature and ice in Fig. 7), has affected the correlations shown in Fig. 8 and Table 1. For example, the magnitude of the correlation coefficients dropped an average of ~28% between the years 1974-1982 (Niebauer 1988) and this study. The reasons are related in the next section.

The interannual variability in the Pribilof air temperatures, surface winds and SST, and sea ice are nearly in phase with some lags (e.g., Fig. 8 and Table 1). The hierarchy of lags are that air temperatures, ice, and SST lag winds by 0-4 months, SST and ice lag air temperatures by 0-3 months, and SST lags ice by 0-1 month. This suggests that the interannual variability in the southern Bering Sea is primarily driven by the atmosphere rather than the ocean. This is consistent with the sluggish flow of 0.01-0.03 m/s or 0.9-2.6 km/day over this 500 km wide shelf and with the heat budget studies on the Bering Sea shelf by Reed (1978). On the other hand, stronger currents tend to occur in the western portion of the Bering Sea, and here oceanic heat transports may have a significant effect on ice extent.

The Bering Sea time-series are also significantly correlated with the El Niño-Southern Oscillation (ENSO) signals (Fig. 8 and Table 1) from the south Pacific as well as the NP and WP (Fig. 8 and Table 1) for the North Pacific. Because a significant amount of the Bering Sea air-ice-ocean climate depends upon the Aleutian Low weather system and because it has been known for at least 30 years that the Aleutian Low depends, at least in part, upon the Southern Oscillation (Bjerknes 1966, 1969, 1972), it is worthwhile to discuss briefly the El Niño-Southern Oscillation (ENSO).

The Southern Oscillation is a basin-scale seesaw in anomalous sea level pressure between the extreme western portion and the central and eastern portion of the tropical Pacific. Extremes in this dipole in anomalous pressure are associated with El Niño and La Niña events, with the former characterized by high pressure to the west and low pressure to the east, and vice versa. During El Niños, these pressure anomalies cause the normal easterlies along the equator to weaken, or with intense cases, to reverse to westerly. On the other hand, La Niñas feature an enhancement of the normal west to east pressure gradient, and hence stronger than normal easterlies. These anomalies in the surface winds affect the upper tropical Pacific in the following manner. Easterly surface winds tend to cause higher sea level, a deeper thermocline, and relatively warm SSTs in the western Pacific. These easterlies contribute to upwelling along the South American coast and the equator, causing a shallower thermocline and lower SSTs in the eastern Pacific. This SST pattern in the tropical Pacific, in turn, can positively reinforce the winds through its impact on atmospheric boundary layer temperatures and deep cumulus convection and hence sea level pressure (e.g., Wyrtki 1982). In essence, El Niños (La Niñas) represent a relaxation (intensification) of the east-west asymmetry in the coupled atmosphere-ocean system of the tropical Pacific. A typical evolution for the tropical Pacific (e.g., the canonical ENSO event as composited by Rasmusson and Carpenter 1982) includes a positive SOI anomaly, or La Niñas, followed by a negative SOI anomaly, or El Niño (e.g., 1970-1972 or 1975-1977 in Fig. 7). But the fluctuations of the tropical Pacific climate system are only quasi-periodic on time scales of 2-7 years; substantial decadal variations occur in the background state and in the frequency of individual El Niño and La Niña events (Fig 7).

The mechanism(s) by which ENSO anomalies affect the rest of the globe have been studied using observations (e.g., Horel and Wallace 1981, Lau and Wallace 1979, Mantua et al. 1997, Minobe 1997, Trenberth 1990,

Trenberth and Hurrell 1994) and numerical models (Graham et al. 1994, Graham 1995, Kumar et al. 1994, Meehl et al. 1993). These results suggest that it is SST anomalies along the equator that force anomalies in deep cumulus convection, which in turn are associated with anomalies in the upper-tropospheric flow gyres that straddle the equator. These perturbations at upper levels interact with the mid-latitude westerlies, bringing about the wide-ranging impacts of ENSO. An important point is that ENSO's effect on the higher-latitude circulation depends on the higher-latitude base state, which is subject to other competing factors. This means the forcing by ENSO of the global circulation is not highly deterministic.

Comparing the Bering Sea time series with the ENSO time series, the Bering Sea time series are significantly correlated with, but typically lag behind, the SOI (Fig. 8 and Table 1), with perhaps the exception of the winds. The lag ranges 0-12 months for correlation of CI > 0.95 (Table 1). The correlations with N1+2 tend to be poorer but are significant for Bering Sea SST and AT. The correlations suggest that warming in the Bering Sea (i.e., below normal ice and winds and above normal SST and AT) follows negative anomalies in the SOI, or El Niños. La Niñas, or positive anomalies, in the SOI tend to precede cooling in the Bering Sea as illustrated by above normal ice and northerly winds and below normal SST and AT.

The lag-correlations have been used to indicate which parameters are most strongly related, but additional information is generally necessary to identify the causal mechanisms for these relationships. We have suggested previously that a significant connection between the ENSO events in the Southern Hemisphere and interannual variability in the Bering Sea is atmospheric and is the wintertime intensity and position of the Aleutian Low. In the mean from December to March, the center of the Aleutian Low is located over the Aleutian Islands (Fig. 9a). During El Niño winters (Fig. 9b), the Aleutian Low is located slightly southeastward of its usual position, and is ~2 mb lower in central pressure. During La Niña winters (Fig. 9c), the Aleutian Low is displaced westward of its usual position, and is ~3 mb higher in central pressure. These differences in mean sea-level pressure, as also shown using anomaly maps (Figs. 10a,b), represent the cumulative effects of variations in the tracks and character of high-frequency weather disturbances. There is the tendency for more or stronger low-pressure cyclonic storms, and fewer or weaker high-pressure anticyclones, in the Bering Sea during El Niño winters, and vice versa. These transient disturbances are responsible for large meridional heat fluxes (e.g., Lau and Wallace 1979). A net poleward flux of warmer air tends to accompany cyclones, and a net equatorward flux of colder air tends to accompany anticyclones. Therefore, El Niño winters with their preponderance of cyclonic systems tend to be warm in the Bering Sea, especially in its southern portion, and La Niña winters favoring anticyclones tend to be cold. The heat fluxes due to the transient disturbances can be enhanced or suppressed by the temperature advection associated with anomalies in the mean meridional wind; as shown by Table 1 and Fig. 8, a systematic

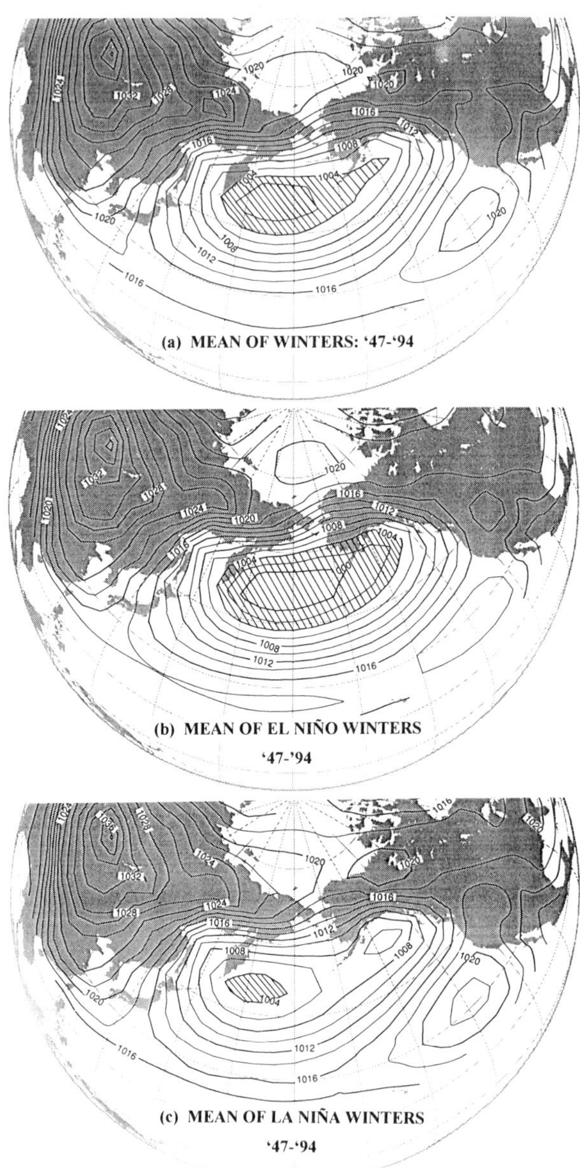

Figure 9.   Sea level pressure (SLP) patterns for the north Pacific and Bering Sea winters (December-March) for (a) all the winters for 1947-94, (b) all the El Niño winters (i.e., 1951-52, 57-58, 65-66, 72-73, 77-78, 82-83, 86-87 and 91-93), and (c) all the La Niña winters (i.e., 1949-51, 55-56, 70-71, 73-74, 75-76 and 88-89). Pressure less than 1,004 mb is shaded for comparison. (After Niebauer, submitted.)

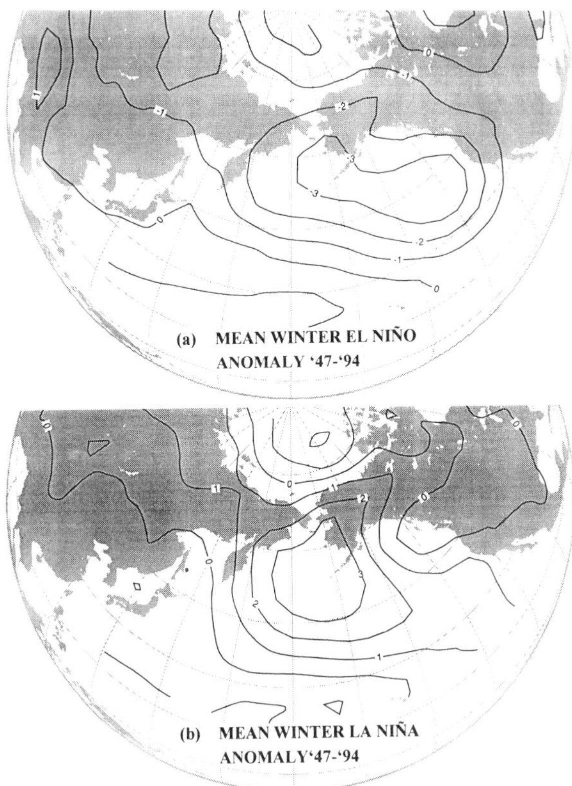

Figure 10. (a) El Niño winters minus mean winters for 1947-1994 (i.e., Fig. 9b minus Fig. 9a). Note the large area of drop in pressure (mb) over the Bering Sea and eastern North Pacific as the Aleutian Low intensifies and moves east and south of normal in teleconnection with El Niño events in the Southern Hemisphere. (b) La Niña winters minus mean winters for 1947-1994 (i.e., Fig. 9c minus Fig. 9a). Note the large area of rise in pressure (mb) over the southeastern Bering Sea and north Pacific as the Aleutian Low weakens and moves westward of normal in teleconnection with La Niña events in the Southern Hemisphere. (After Niebauer, submitted.)

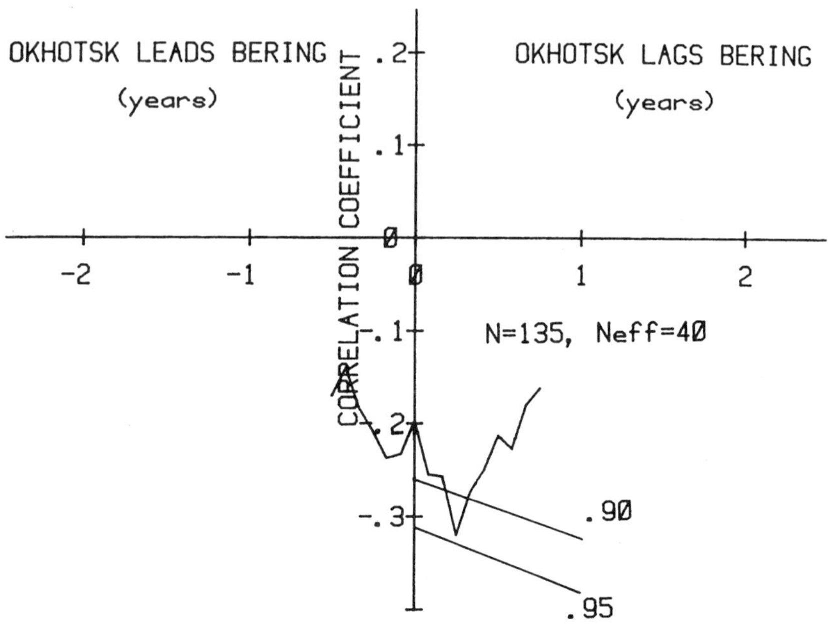

Figure 11. Same as Fig. 8 except the data are deviations from mean ice cover for the Sea of Okhotsk and the Bering Sea for January 1973-March 1984. (From Niebauer 1988.)

relationship between ENSO and the mean meridional wind is lacking, but during any particular winter the latter's (i.e., meridional wind) effects can be significant to the mean temperature.

There is additional evidence supporting this seesaw relationship between north-south interannual variability in the ice extent in the eastern Bering Sea, and the interannual variability in the position of the Aleutian Low (e.g., Niebauer 1988, Parkinson 1990, Fang and Wallace 1994). Sea ice extents in the Bering Sea and the Sea of Okhotsk are negatively correlated (Fig. 11), implying that the air temperatures in these regions are also negatively correlated. This result can be attributed to the typical horizontal scale (1,000-2,000 km) of the individual weather disturbances that account for most of the wintertime forcing of the Bering Sea. When storms are causing relatively warm conditions in the southeastern Bering, relatively cool air tends to be drawn down over the Sea of Okhotsk, and vice versa. The tendency for one state or the other will be reflected in the longitudinal position of the Aleutian Low. We do note that this is not the whole explanation as the correlation analysis explains only 10% of the variability. Further, Plotnikov (1990) points to a recent (~10 years) increase

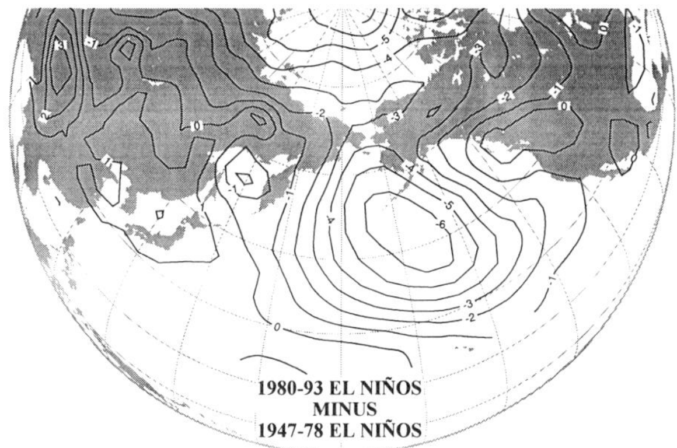

Figure 12. The El Niño winters of 1980-93 minus the El Niño winters of 1947-78. That is, the average of all the winter SLP (December-March) for the El Niño events in years 82-83, 86-87 and 91-93 minus the El Niño events in 1951-52, 57-58, 65-66, 72-73 and 77-78. Note the large area of drop in pressure (mb) over the eastern North Pacific as the Aleutian Low has become more intense and has moved farther east and somewhat south of normal in teleconnection with El Niño events in the Southern Hemisphere. (After Niebauer, submitted.)

in frequency of cases where this relationship does not hold very well, suggesting possible climatic changes in the hydrometeorology.

Interannual Variability for 1975-94

While a "typical" ENSO event has been described (e.g., Rasmusson and Carpenter 1982) as a coupled La Niña, or high intensity event, followed within a year or so by an El Niño, or low intensity event (e.g., 1956-1958 or 1970-1972 in Fig. 7), we note that there has not been a "typical" ENSO event since the mid-1970s (i.e., since the regime shift). This change in the ENSO patterns appears to have caused changes in the position and intensity of anomalies in the Aleutian Low associated with El Niños (Niebauer submitted; Fig. 12). To illustrate, there have been ~4 El Niños but only two La Niñas (see the SOI panel in Fig. 7) since 1976. The ENSO event of 1974-1978 actually had two La Niñas preceding an El Niño. This is also the ENSO event that occurred during extraordinarily strong shifts in the Bering Sea environment (e.g., especially air temperature in Fig. 7, and ice in Figs. 2, 7, and 13). Following that, the intense El Niño of 1982-1983, as well as the El Niño of 1986-1987, did not have a preceding La Niña. While there was a strong La Niña in 1988-1989 followed by an El Niño, the El Niño lasted

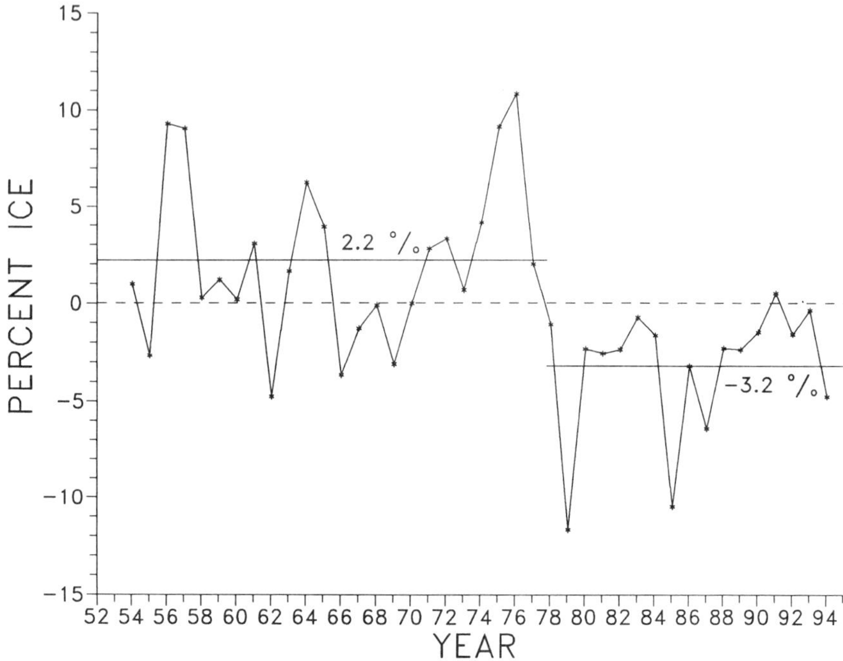

Figure 13. Winter averaged percent ice cover anomalies for the Bering-Chukchi seas for the period 1954-1994. Average anomalies for the periods 1954-1977 and 1978-1994 are also shown. (After Niebauer, submitted.)

about 4 years. (It is interesting to note that this 4-5 year-long El Niño coincides with the reduced flow of ~0.7 Sv through the Bering Strait over this same period, 1990-1995, as observed by Aagaard and Weingartner [pers. comm.]). Finally, the ratio of El Niño:La Niña has gone from 0.8 for 1947-1977 to 2.8 for 1978-1996 (Niebauer, submitted).

This shift has had a strong effect on the relationship between the Bering Sea ice cover and the SOI. The cross correlation between the SOI and Bering Sea ice for the entire period 1953-1995 is 0.17 (Fig. 8 and Table 1). However, for the period 1953-1978 (i.e., before the regime shift), the correlation coefficient was ~0.26, while after the regime shift, for the period 1978-1995, the correlation coefficient actually changes sign to ~ −0.18 (Niebauer, submitted). The change in sign of the correlation, combined with the lack of La Niña events, means that since the regime shift, *increases* in ice are now associated with El Niño events. However, this is consistent with El Niño-associated Aleutian Lows becoming even more intense and moving even farther east following the regime shift. These more easterly Aleutian Low pressure patterns are forcing southerly winds off the

Pacific into interior Yukon and Alaska and easterly winds carrying continental air from Alaska out over the Bering Sea. Before the regime shift, the Aleutian Low was farther west in association with a high frequency of storms and southerly winds, and hence warmth, over the Bering Sea.

Discussion

As mentioned earlier, in some winters, fluctuations associated with other teleconnection modes (e.g., the western Pacific oscillation [WPO or WP; Wallace and Gutzler 1981, Barnston and Livezey 1987] and/or the Pacific-North American pattern [PNA, Wallace and Gutzler 1981]) can have substantial effects on the Aleutian Low seemingly independent of ENSO events. For example, there is significant correlation between WP and SOI undoubtedly through association with the Aleutian Low. For the entire period 1947-1993, as well as for the period before the shift (1947-1977), the correlation between WP and SOI is ~0.2 with WP lagging SOI by 0 and 4 months (e.g., Fig. 8 and Table 1). After the shift, the correlation at 0 lag dropped a little to ~0.17 while the 4-month lag peak was greatly diminished. For WP and ice (1953-1996), the correlation is 0.23 (ice lagging 0-2 months) which is higher than between SOI and ice (0.18; Fig. 8 and Table 1). For the period after the regime shift (1978-1996), the correlation between WP and ice (0.30) is greater than between SOI and ice (–0.18), but of opposite sign. However, for the period before the regime shift (1953-1977), the correlation between SOI and ice (0.26) is slightly greater than the correlation between WP and ice (0.21 with no lag between WP and ice). (All of these cross correlations are significant at >0.1 level although none explain more than 10% of the variability.)

An important point here is that the significant correlation between SOI and ice lies over a broad period of lag (at least 12 months) as compared to a narrow range (0-2 months) for WP and ice. The narrow, sharp peak of WP and ice correlation (as well as WP vs. AT and wind in Table 1) suggests that WP and ice fluctuations are probably parts of the same or a local process (i.e., shift in location of the Aleutian Low) but different from the SO teleconnection; that ENSO events are a significant driving force for both WP and ice through teleconnection of the Aleutian Low with the Southern Oscillation. Niebauer (submitted) showed that average winter (Nov.-Mar.) ice was ~2.2% above normal for the 1950s to the regime shift while the average winter SOI was ~0.12 mb above normal. After the regime shift, both ice and SOI dropped to ~ –3.1% and ~ –0.7 respectively. This relationship between SOI and ice provides the possibility of predicting ice conditions further in the future because of this longer duration lag of statistically significant correlation, and because of (or perhaps in spite of) the change in sign of correlation between ice and SOI through the climate shift.

Against the larger picture of decadal time scales (i.e., Pacific-wide response and global warming), Trenberth and Hoar (1996) consider two questions: (1)"Is this pattern of change a manifestation of the global warming and related climate change associated with increases in greenhouse gases in the atmosphere?", or (2) "Is this pattern a natural decadal-timescale variation?" For the second question, Trenberth and Hoar (1996) analyze normalized SOI (negative Darwin sea level pressure) anomalies from the 1880s-1990s. From a statistical viewpoint, they suggest that the shift to more El Niños vs. La Niñas since 1976-1977, as well as the ~5 year long El Niño from 1990 to 1995, is unprecedented with a probability of occurrence about 1 in 2,000 years.

However, more recently, Mantua et al. (1997), Minobe (1997), and Zhang et al. (1997) show evidence of four reversals, or climatic regime shifts, around 1890, 1925, 1947, and 1977, in a recurring pattern of very large scale, interdecadal atmosphere-ocean climate variability centered over the midlatitude North Pacific. Mantua et al. (1997) call this the Pacific (inter) Decadal Oscillation or PDO. Minobe (1997), using reconstructed climate records (tree rings) for the last three centuries, presents evidence of a 50-70 year oscillation from the Eighteenth Century to the present. Minobe (1997) suggests that 50-70 year oscillation seen in ~300 year record is likely an oscillation of the ocean-atmosphere coupled system, although he suggests it may be modulated by solar radiation in the Twentieth Century.

The causes of decadal variability in the North Pacific is currently attracting a great deal of interest. Essentially, two different explanations have been offered. Certainly decadal or longer variations in ENSO are important, through their influence on the Northern Hemisphere weather via atmospheric teleconnections (e.g., Kumar et al. 1994, Graham et al. 1994, Graham 1995). The long-term variations in the systematic sense of ENSO, or the relative frequency of individual events, may be due both to ENSO's intrinsic variability (e.g., as modeled by Zebiak and Cane 1987) and to external forcing, such as increased concentrations of greenhouse gases (e.g., as modeled by Meehl et al. 1993). Alternatively, or in conjunction, long-term fluctuations in the North Pacific may be associated with local, positive feedbacks between the atmosphere and ocean. Anomalous SSTs may cause significant surface heat and moisture flux anomalies that in turn reinforce the atmospheric anomalies impacting the ocean (e.g., Latif and Barnett 1994, Trenberth and Hurrel 1994). The natural time scales for the North Pacific gyre circulations may then be responsible for the decadal time scales that have been observed. Regardless of their cause, these decadal variations of the North Pacific need to be recognized for their role in providing a varying background or "basic state" for the Bering Sea atmosphere-ice-ocean system.

Acknowledgments

This is contribution 1551 from the Institute of Marine Science, University of Alaska Fairbanks, contribution 1747 from the Pacific Marine Environmental Laboratory (PMEL), and contribution 367 from the Joint Institute for the Study of the Atmosphere and Ocean (JISAO). Support was provided by NOAA Coastal Ocean Program (Southeast Bering Sea Carrying Capacity) and is contribution S317. Support was also provided by the University of Alaska Fairbanks Cooperative Institute for Arctic Research (CIFAR/NOAA) grants CIFAR 96-14 (Southeast Bering Sea Carrying Capacity) and CIFAR 97-102R (Arctic Research Initiative). Support for HJN was also provided by the Office of Naval Research through grant N00014-94-1-00537. The support of the Department of Atmospheric and Ocean Science, University of Wisconsin-Madison is gratefully acknowledged. This contribution greatly profited from the comments of anonymous reviewers.

References

Aagaard, K., L.K. Coachman, and E. Carmack. 1981. On the halocline of the Arctic Ocean. Deep-Sea Research 28:529-545.

Anderson, J.R., and J.R. Gyakum. 1989. A diagnostic study of Pacific Basin circulation regimes as determined from extratropical cyclone tracks. Monthly Weather Review 117:2672-2686.

Barnston, A., and R.E. Livezey. 1987. Classification, seasonality and persistence of low-frequency circulation patterns. Monthly Weather Review 115:1083-1126.

Bjerknes, J. 1969. A possible response of the atmospheric Hadley circulation to equatorial anomalies of ocean temperature. Tellus 18:820-829.

Bjerknes, J. 1969. Atmospheric teleconnections from the equatorial Pacific. Monthly Weather Review 97:162-172.

Bjerknes, J. 1972. Large-scale atmospheric response to the 1964-65 Pacific equatorial warming. Journal of Physical Oceanography 2:212-217.

Björk, G. 1989. A one-dimensional time-dependent model for the vertical stratification of the upper Arctic Ocean. Journal of Physical Oceanography 19:52-67.

Björk, G. 1990. The vertical distribution of nutrients and oxygen 18 in the upper Arctic Ocean. Journal of Geophysical Research 95:16025-16036.

Bond, N.A., J.E. Overland, and P. Turet. 1994. Spatial and temporal characteristics of the wind forcing of the Bering Sea. Journal of Climate 7:1119-1130.

Brodeur, R.D., and D.M. Ware. 1992. Long-term variability in zooplankton biomass in the sub-Arctic Pacific Ocean. Fisheries Oceanography 1:32-38.

Brower, W.A., H. Diaz, A. Prechtel, H. Searby, and J. Wise. 1977. Climatic atlas of the outer continental shelf waters and coastal regions of Alaska: Vol. II, Bering Sea. NOAA/OCSEAP. Final Report Research Unit 347. 443 pp. (Available from AEIDC Library, 707 A Street, Anchorage, AK 99501.)

Brower, W.A., W.S. Baldwin, C.N. Williams, J.L. Wise, and L.D. Leslie. 1988. Climatic atlas of the outer continental shelf waters and coastal regions of Alaska: Vol. II, Bering Sea. NOAA. 519 pp. (Available from AEIDC Library, 707 A Street, Anchorage, AK 99501.)

Burns, J., L.H. Shapiro, and F.H. Fay. 1981. Ice as marine mammal habitat in the Bering Sea. In: D.W. Hood and J.A. Calder (eds.), The eastern Bering Sea shelf: Oceanography and resources, Vol. 2. NOAA. Distributed by the University of Washington Press, Seattle, WA 98105, pp. 781-797.

Cavalieri D.J., and S. Martin. 1994. The contribution of Alaskan, Siberian, and Canadian coastal polynyas to the cold halocline layer of the Arctic Ocean. Journal of Geophysical Research 99(C9):18343-18362.

Coachman, L.K., and K. Aagaard. 1988. Transport through Bering Strait: Annual and interannual variability. Journal of Geophysical Research 93:15535-15539.

Ebbesmeyer, C, D.R. Cayan, D.R. McLain, F.H. Nichols, D.H. Peterson, and K.T. Redmond. 1991. 1976 step in the Pacific climate: Forty environmental changes between (1968-1975 and 1977-1984). In: J.L. Betancourt and V.L. Sharp (eds.). 1990. Proceedings of the Seventh Annual Pacific Climate (PACLIM) Workshop, April 1990, California Department of Water Resources Interagency Ecological Studies Program Technical Report 26:129-141.

Fang, Z., and J.M. Wallace. 1994. Arctic sea ice variability on a timescale of weeks and its relation to atmospheric forcing. Journal of Climate 7:1897-1914.

Glueck, M., and H.J. Niebauer. Submitted. Ice circulation and brine production at a Bering Sea polynya using ERS-1 synthetic aperture radar imagery. Continental Shelf Research.

Gloersen, P., W.J. Campbell, D.J. Cavalieri, J.C. Comiso, C.L. Parkinson, and H.J. Zwally. 1992. Arctic and Antarctic sea ice, 1978-1987: Satellite passive-microwave observations and analysis, NASA SP-511. National Aeronautics and Space Administration, Washington, DC. 290 pp.

Graham, N.E. 1995. Simulation of recent global temperature trends. Science 267:666-671.

Graham, H.E., T.P. Barnett, R. Wilde, M. Porter, and S. Schubert. 1994. On the roles of tropical and midlatitude SSTs in forcing interannual to interdecadal variability in the winter Northern Hemisphere circulation. Journal of Climate 7:1416-1441.

Hollowed, A.B., and W.S. Wooster. 1992. Variability of winter ocean conditions and strong year classes of northeast Pacific groundfish. ICES Marine Science Symposium 195:433-444.

Horel, J.D., and J.M. Wallace. 1981. Planetary-scale phenomena associated with the Southern Oscillation. Monthly Weather Review 109:813-829.

Kumar, A., A. Leetmaa, and M. Ji. 1994. Simulations of atmospheric variability induced by sea surface temperatures and implications for global warming. Science 266:632-634.

Latif, M., and T.P. Barnett. 1994. Causes of decadal climate variability over the North Pacific and North America. Science 266:634-637.

Lau, N.-C., and J.M. Wallace. 1979. On the distribution of horizontal transports by transient eddies in the Northern Hemisphere wintertime circulation. Journal of Atmospheric Science 36:1844-1861.

Mantua, N.J., S.R. Hare, Y. Zhang, J.M. Wallace, and R.C. Francis. 1997. A Pacific interdecadal climate oscillation with impacts on salmon production. Bulletin of the American Meteorological Society 78:1069-1079.

Maykut, G.A. 1978. Energy exchange over young sea ice in the central Arctic. Journal of Geophysical Research 83:3646-3658.

Meehl, G.A., G.W. Branstator, and W.M. Washington. 1993. Tropical Pacific interannual variability and CO_2 climate change. Journal of Climate 6:42-63.

Minobe, S. 1997. A 50-70 year oscillation over the North Pacific and North America. Geophysical Research Letters 24:683-686.

Muench, R.D., and K. Ahlnas. 1976. Ice movement and distribution in the Bering Sea from March to June 1974. Journal of Geophysical Research 81:4467-4476.

Niebauer, H.J. 1980. Sea ice and temperature variability in the eastern Bering Sea and the relation to atmospheric fluctuations. Journal of Geophysical Research 85:7507-7515.

Niebauer, H.J. 1981a. Recent short-period wintertime climatic fluctuations and their effect on sea-surface temperatures in the eastern Bering Sea. In: D.W. Hood and J.A. Calder (eds.), The eastern Bering Sea shelf: Oceanography and resources, Vol. 1. NOAA. Distributed by University of Washington Press, Seattle, pp. 23-30.

Niebauer, H.J. 1981b. Recent fluctuations in sea ice distribution in the eastern Bering Sea. In: D.W. Hood and J.A. Calder (eds.), The eastern Bering Sea shelf: Oceanography and resources, Vol. 1. NOAA. Distributed by University of Washington Press, Seattle, pp. 133-140.

Niebauer, H.J. 1983. Multiyear sea ice variability in the eastern Bering Sea: An update. Journal of Geophysical Research 88:2733-2742.

Niebauer, H.J. 1988. Effects of El Niño-Southern Oscillation and North Pacific weather patterns on interannual variability in the subarctic Bering Sea. Journal of Geophysical Research 93:5051-5068.

Niebauer, H.J. Submitted. On the climatic "regime shift" in the North Pacific in the period 1947-96. Journal of Geophysical Research.

Niebauer, H.J., and R.H. Day. 1989. Causes of interannual variability in the sea ice cover of the eastern Bering Sea. GeoJournal 18:45-59.

Niebauer, H.J., and D.M. Schell. 1993. Physical environment of the Bering Sea population. In: J.J. Burns, J.J. Montague, and C.J. Cowles (eds.), The bowhead whale. The Society for Marine Mammalogy Special Pub. No. 2., Seattle, pp. 23-43.

Overland, J.E. 1981. Marine Climatology of the Bering Sea. In: D.W. Hood and J.A. Calder (eds.), The eastern Bering Sea shelf: Oceanography and resources, Vol. 1. NOAA. Distributed by the University of Washington Press, Seattle, pp. 15-22.

Overland, J.E., and C.H. Pease. 1982. Cyclone climatology of the Bering Sea and its relation to sea ice extent. Monthly Weather Review 110:354-384.

Overland, J.E., and A.T. Roach. 1987. Northward flow in the Bering and Chukchi seas. Journal of Geophysical Research 92:7097-7105.

Parkinson, C.L. 1990. The impact of the Siberian High and Aleutian Low on the sea-ice cover of the Sea of Okhotsk. Annals of Glaciology 14:226-229.

Parkinson, C.L., J.C. Comiso, H.J. Zwally, D.J. Cavalieri, P. Gloersen, and W.J. Campbell. 1987. Arctic sea ice, 1973-1976: Satellite passive-microwave observations, NASA SP-489, National Aeronautics and Space Administration, Washington, DC. 296 pp.

Pease, C.H. 1980. Eastern Bering Sea ice processes. Monthly Weather Review 108:2015-2023.

Pease, C.H. 1987. The size of wind-driven coastal polynyas. Journal of Geophysical Research 92:7049-7059.

Plotnikov, V.V. 1990. Seasonal and interannual variability of ice extent in the far-eastern seas. Transactions of the Far Eastern Hydrometeorological Institute, Valdivostok, Russia, Issue 40:65-75.

Quinn, T., and H.J. Niebauer. 1995. Relation of eastern Bering Sea walleye pollock *(Theragra chalcogramma)* to environmental and oceanographic variables. In: Climate change and northern fish populations. Canadian Special Publication of Fisheries and Aquatic Science 121:497-507.

Rasmusson, E.M., and T.H. Carpenter. 1982. Variations in tropical sea surface temperature and surface wind fields associated with the Southern Oscillation/El Niño. Monthly Weather Review 110:354-384.

Reed, R.K. 1978. The heat budget of a region in the eastern Bering Sea, summer of 1976. Journal of Geophysical Research 83:3635-3645.

Roach, A.T., K. Aagaard, C.H. Pease, and T.J. Weingartner. 1995. Direct measurements of transport and water properties through Bering Strait. Journal of Geophysical Research 100:18443-18459.

Roemmich, D., and J. McGowan. 1995. Climatic warming and the decline of zooplankton in the California Current. Science 267:1324-1326.

Schumacher, J.D., K. Aagaard, C.H. Pease, and R.B. Tripp. 1983. Effects of a shelf polynya on flow and water properties in the northern Bering Sea. Journal of Geophysical Research 88:2723-2732.

Shapiro, L.H., and J.J. Burns. 1975. Satellite observations of sea ice movement in the Bering Strait region. In: G. Weller and S.A. Bowling (eds.), Climate of the Arctic. Twenty-fourth Alaska Science Conference, Fairbanks, Alaska. August 15-17, 1973. Published by the Geophysical Institute, University of Alaska Fairbanks, pp. 379-386.

Stringer, W.J., J. Zender-Romick, and J.E. Groves. 1982. Width and persistence of the Chukchi Polynya. NOAA/OCSEAP Research Unit 267. 22 pp. (Available from NOAA, Ocean Assessments Division, Alaska Office, 222 West Eighth Avenue, Anchorage, AK 99513-7543.)

Trenberth, K.E. 1984. Signal versus noise in the southern oscillation. Monthly Weather Review 112:326-332.

Trenberth, K.E. 1990. Recent observed interdecadal climate changes in the Northern Hemisphere. Bulletin of the American Meteorological Society 71:988.

Trenberth, K.E., and J.W. Hurrell. 1994. Decadal atmosphere-ocean variations in the Pacific. Climate Dynamics 9:303-319.

Trenberth, K.E., and T.J. Hoar. 1996. The 1990-1995 El Niño-Southern Oscillation event: Longest on record. Geophysical Research Letters 23:57-60.

United States Naval Oceanography Command. 1986. Sea ice climatic atlas. Vol. I. 147 pp. (NTIS No. AD-A168-716.)

Wallace, D.W.R., R.M. Moore, and E.P. Jones. 1987. Ventilation of the Arctic Ocean cold halocline: Rates of diapycnal and isopycnal transport, oxygen utilization and primary production inferred using chlorofluoromethane distributions. Deep-Sea Research, Part A 34(12):1957-1979.

Wallace, J.M., and D.S. Gutzler. 1981. Teleconnections in the geopotential height field during the Northern Hemisphere winter. Monthly Weather Review 109:784-812.

Wallace, J.M., C. Smith, and Q. Jiang. 1990. Spatial patterns of atmosphere-ocean interaction in the northern winter. Journal of Climate 3:990-998.

Walsh, J.E., and C.M. Johnson. 1979. An analysis of Arctic sea ice fluctuations, 1953-77. Journal of Geophysical Research 9:580-591.

Walsh, J.E., and J.E. Slater. 1981. Monthly and seasonal variability in the ocean-ice-atmosphere systems of the North Pacific and North Atlantic. Journal of Geophysical Research 86:7425-7446.

Weeks, W.F., and G. Weller. 1984. Offshore oil in the Alaskan Arctic. Science 225:371-378.

Wyrtki, K. 1982. The Southern Oscillation, ocean-atmosphere interaction and El Niño. Marine Technology Society Journal 16:3-10.

Yakunin, L.P. 1987. An atlas of ice extent in the far-eastern seas. Vladivostok, POP PUGKS. 80 pp.

Zar, J.H. 1984. Biostatistical analysis, 2nd edn. Prentice-Hall, Englewood Cliffs, NJ. 718 pp.

Zebiak, S.E., and M.A. Cane. 1987. A model El Niño/Southern Oscillation. Monthly Weather Review 115:2262-2278.

Zhang, Y., J.M. Wallace, and D.S. Battisti. 1997. ENSO-like interdecadal variability: 1900-93. Journal of Climate 10:1004-1020.

CHAPTER 3

Thermohaline Structure and Water Masses in the Bering Sea

Vladimir A. Luchin, Vladimir A. Menovshchikov, and Vladimir M. Lavrentiev
Far Eastern Regional Hydrometeorological Research Institute (FERHI), Vladivostok, Russia

Ronald K. Reed
Pacific Marine Environmental Laboratory, Seattle, Washington

Abstract

Data from 35,700 hydrographic stations were grouped in areas of 1° latitude and 2° longitude and monthly, seasonal, and annual means calculated. We examined temperature, salinity, and water masses of the Bering Sea, studying their vertical structure, temporal variability, and features of their spatial and temporal distribution. We also considered data from coastal observations, information on the meteorological regime, and river outflow.

Previous Investigations

The first summaries of hydrographic research in the Bering Sea date from the previous century (Tanner 1890, Rathburn 1894). They contained general information about the temperature and salinity of the water. From 1906 to 1929, deepwater hydrographic work was not conducted; it was renewed only in the 1930s. Observations obtained during these years served as the basis for the work of Ratmanov (1937), who gave the first presentation of the distribution of temperature and salinity in the sea and the origins of the Anadyr and Olyutorsk cold zones. He also noted the interaction of waters from the northern part of the Pacific Ocean and the Bering Sea.

The shallow-water regime of the northern and eastern parts of the sea were studied by Goodman et al. (1942). They used data from expeditions in 1937 and 1938. Leonov (1947, 1960) extended the research in the Bering Sea to the start of the 1950s. He distinguished four water masses in the sea: the upper Bering Sea, intermediate Bering Sea, deep Pacific Ocean, and polar. He also studied the water balance, thermal regime, and distribution of salinity and density in the sea.

From 1950 to 1953, the Institute of Oceanology of the USSR carried out a set of studies in the Bering Sea; the studies continued in 1955 and 1956 in conjunction with the Pacific Ocean Institute of Fisheries Ecology and Oceanography (TINRO). Data from these studies formed the basis of work by Burkov (1958), Dobrovolskiy et al. (1959), Ivanenkov (1964), and Arsenev (1967). Summarizing observations from 1932 to 1955, Dobrovolskiy and Arsenev (1961) examined the influence of morphological and climatic factors on processes in the sea, described the vertical structure and origins of water masses, and summarized the distribution of parameters in different seasons. They distinguished three types of vertical water structure (arctic, subarctic, and temperate latitudes) and also examined in detail seasonal variability of temperature in the top 200-300 m layer. Ivanenkov (1964) also proposed a classification of water masses in the Bering Sea. In the deep basin, he distinguished four water masses: the surface Bering Sea, the subsurface, the intermediate, and the deep North Pacific. In the shallow water, he differentiated seven regions using the distribution of temperature, salinity, and chemical properties. In 1958, the Institute of Oceanology and TINRO began work on various expeditions. During three years, a total of 1,000 stations were occupied during the summer and winter. The results of this research are contained in the work of Natarov (1963). He concluded that the deep Bering Sea had the character of subarctic water, characterized by the presence in summer of four water masses: the warm surface layer from 0 to 50 m, the cold subsurface layer from 50 to 150-200 m, the warm intermediate layer from 200 to 600-800 m, and the deep water. In winter, this structure consisted of only three layers. Arsenev (1967) carried out a more complete study of the water masses of the Bering Sea. He used data from 1894 to 1959 and the results of earlier summaries. The author gave detailed characteristics of the main water masses and examined the factors causing their formation.

In recent years, Russian scientists have concentrated on the western part of the Bering Sea (Davydov and Lipetskiy 1970; Lazo 1971; Davydov 1972, 1984; Naumov and Khistyaev 1972). These authors noted the large role of the Kamchatka Current in the formation of the thermohaline regime of the shelf water near the east coast of Kamchatka. The distributions of chemical parameters also attest to the formation of eddies in Karagin and Olyutorsk bays (Davydov and Lipetskiy 1970; Davydov 1972, 1984).

An article by Poluektov and Khistyaev (1981) stands out from the more fragmentary work published in the 1970s and 1980s (Lazo 1971, Tiguntsev 1971, Naumov and Khistyaev 1972, Milyeiko 1973, Izrael' 1983, Yarichin 1984, Khen and Voronin 1986). Poluektov and Khistyaev systematized data of bathythermograph observations from 1971 to 1976 (more than 1,200 observations, taken from January to May). They studied the variation of the lower limit of the isothermal layer, distinguished types of temperature profiles, divided the area into five homogeneous regions, and calculated correlation coefficients between temperature at the surface and temperature in the lower layers.

Short-period variability of water temperature in Bering Strait was examined by Tiguntsev (1971). He found that water temperature varied by up to 6°C during the course of a day; this variation was a significant part of the total variability of temperature in this region. Khen and Voronin (1986) examined the influence of atmospheric processes and sea ice on the interannual oscillations of parameters in the eastern part of the Bering Sea. These authors distinguished three periods of warming and two periods of cooling in the shelf waters.

In the post-war years, Japanese organizations began to carry out expeditions in the Bering Sea. Most of the data were obtained from the ships *Oshoro Maru* and *Hokkusai Maru*. The results of these expeditions are reflected in the work of Mishima and Nishizawa (1955), Koto and Fuji (1958), Maeda et al. (1968), Kihara and Uda (1969), Kitano (1970), Ohtani (1969, 1973), and Takenouti and Ohtani (1974). Researchers examined the characteristics of the cold intermediate layer (Koto and Fuji 1958), the distribution of bottom temperature on the shelf, the penetration of warm water from the deep basin onto the shelf (Koto and Maeda 1965), and the distribution of temperature and salinity in the bottom layer of the shallow eastern part of the sea (Maeda et al. 1968). Maeda et al. (1968) also examined interannual variability of temperature and salinity in the bottom water on the shelf, showed the location of a cold zone southwest of St. Lawrence Island, and described the spreading of cold bottom water to the southeast, toward Bristol Bay. Kihara and Uda (1969) examined the formation of the bottom water mass in the eastern Bering Sea. They inferred that the sources were Alaskan coastal water, water from the northern part of the sea, and water from the Alaskan Stream.

Ohtani (1969, 1973) carried out the most complete summary of research conducted by Japanese and American expeditions from 1959 to 1966. He presented a detailed classification of the waters of the Bering Sea. He differentiated three water systems: oceanic, shelf, and continental. Ohtani noted links between systems of currents and the thermohaline structure of the water. He also gave a detailed characterization of shelf waters and examined the process of convective mixing in winter and the vertical distributions of temperature and salinity in the deep basin and on the shelf. More recently, work by Japanese researchers has dealt mainly with the eastern and southeastern regions of the sea. Results in various publications (Takenouti 1976, Kitani and Kawasaki 1979, Otobe et al. 1983) examined the structure and conditions for the formation of subarctic water, as well as features of the fine structure of water near the continental slope.

Early American work in the Bering Sea was by Barnes and Thompson (1938) and Saur et al. (1952). In the 1950s and 1960s, American and Canadian scientists carried out programs to study the northern part of the

Pacific, including the Bering Sea. Observations from expeditions of this period were used in monographs and summaries (Dodimead et al. 1963; Favorite et al. 1961, 1976; Favorite 1974). Dodimead et al. (1963) also examined the oceanographic conditions in Bristol Bay. They noted the existence of an inner front dividing nearshore and mid-shelf waters. They also indicated the existence of an analogous front along the outer edge of the shelf.

Until about 1975, most of the American and Japanese research dealt with the deep part of the sea and the region of Bering Strait. Work in this period was done to determine the mean conditions in the sea (Hood and Kelley 1974), the structure of currents (Dowling 1962, Hughes et al. 1974), and to characterize water masses. Sayles et al. (1979) compiled an atlas of oceanographic characteristics and examined the water masses of the deep basin as well as the distribution of temperature, salinity, and dynamic topography of the water in individual seasons. A paper by Swift and Aagaard (1976) examined upwelling and mixing near Samalga Pass. They confirmed the existence of upwelling, which could occur along the entire length of the Aleutian Islands.

Coachman et al. (1975) were the first to summarize information obtained by American and Japanese expeditions in the region of Bering Strait from 1922 to 1973. Research distinguished three water masses there (Anadyr, Bering Shelf, and Alaskan Coastal). The authors made the important conclusion that lateral mixing is extremely limited in Bering Strait.

After 1975, American scientists concentrated on the study of the continental shelf. Their interest shifted from the study of climatic variability to processes at smaller scales. Research included the region from the Beaufort Sea to the Gulf of Alaska. Observations were reflected in works by Muench (1976), Coachman and Charnell (1977), Kinder (1977), Kinder and Coachman (1978), Reed (1978), Coachman (1979), Schumacher et al. (1979), Coachman et al. (1980), and Ingraham (1981). These significantly expanded knowledge about processes at various time scales on the shelf of the Bering Sea.

In 1981, Hood and Calder edited a monograph containing preliminary results of OCSEAP (Outer Continental Shelf Environmental Assessment Program) on the eastern shelf. Results of this program (Ingraham 1981, Kinder and Schumacher 1981, Muench et al. 1981, Mofjeld 1986) covered in detail the structure of the water, the meteorological and sea ice regimes, and the dynamics of the water. Kinder and Schumacher (1981) analyzed the structure of shelf waters. They summarized previous research, and examined data observed during 1975-1978. They divided the shelf into three structural zones (coastal, middle, and outer). The borders of their zones roughly conform to the 50 m and 100 m isobaths. The authors noted that the different regions are more pronounced in the summer.

Coachman (1986) summarized the physical research of the PROBES (Processes and Resources of the Bering Sea Shelf) program. He arrived at the conclusion that the water mass of the southeast part of the shelf could be regarded as a system with three regimes of physical processes, separated by fronts. He stressed the importance of the vertical distribution of water properties. In the frontal zones, important changes are observed in the balance of factors influencing features of the water masses. He also examined the most important physical processes that determine the formation of water masses.

Materials and Methods

The data used are from deep-sea observations, carried out in the Bering Sea from 1932 to 1989. We processed data from 35,700 stations obtained during 588 cruises of three countries. Most of the data was collected by the U.S.S.R. (30,998 stations), but there is also material from Japanese (1,976 stations) and U.S. ships (2,726 stations). We examined short-period variability in temperature and salinity with data from 11 open-sea stations occupied many times from 1950 to 1989. All of the deep-ocean observations were subject to quality control. First, using analogous work (Belkin 1984, Gubenko and Tsipis 1984, Pereskokov 1984), we ascertained if our values were within the bounds of others' data. Then the data were divided by month. At standard depths, we calculated the mean and standard deviation of specific variables. Values greater than three standard deviations from the mean were rejected. Data were binned into trapezoids (1° of latitude by 2° of longitude). In each trapezoid, we calculated the mean, maximum, minimum, and standard deviation of each variable at standard depths. The statistics were assigned to the center of their respective trapezoids. The distribution of oceanographic stations in the region is presented in Fig. 1. It is obvious that few studies occupied the center of the deep part of the sea. In the deep basin, the number of observations decreases sharply with depth. The number of 500 m stations in individual quadrants sometimes reaches ~200, while the number of stations in each quadrant having data at 1,000 m is less than 100. In all of the quadrants, there were fewer than 25 stations at 2,000 m and deeper.

Of the total data, 59% are from the summer months (June–September). In the present work, we consider the active layer to be the upper layer of water, in which we observe an annual signal in the physical parameters. To determine its lower limit, we used profiles of the vertical distribution of monthly mean characteristics. To calculate the lower limit of the upper quasi-homogeneous layer, we took the point of intersection of the two tangents to the profile of vertical distribution (to the surface quasi-homogeneous layer and in the layer of seasonal gradients). To differentiate water

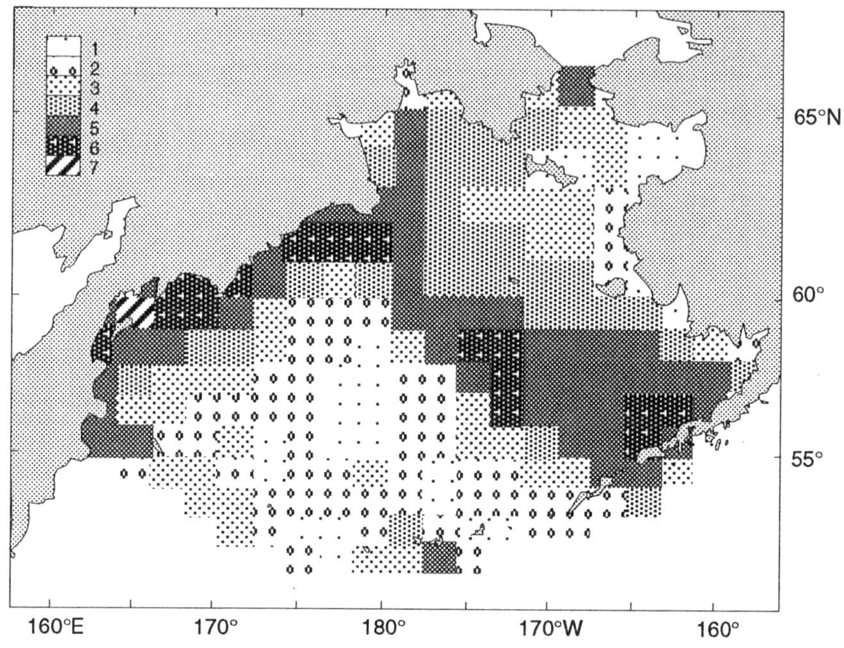

Figure 1.   Distribution of oceanographic stations in the Bering Sea: (1) 0-25 stations, (2) 26-50, (3) 51-100, (4) 101-200, (5) 201-500, (6) 501-1,000, and (7) more than 1,001 stations.

mass parameters, we used the method of temperature-salinity (TS) analysis.

Temperature

Vertical Distribution

The Bering Sea is located in the region of subarctic water structure, whose main feature is the presence of cold and warm intermediate layers (Burkov 1958, Arsenev 1967, Filyushkin 1968, Moiseev 1978). The subarctic structure is most marked in the deep western basin. Profiles of the vertical distribution of temperature indicate that the Bering Sea water column may be divided into several layers. The active layer of the sea is most susceptible to thermodynamic effects from the air-sea interface. The topography of its depth is shown in Fig. 2a. In the eastern part of the sea, the depth of the active layer is generally less than 150 m. In the western half of the sea, it increases to as much as 250 m, except near the coast (Fig. 2a). Its vertical extent depends on the intensity of the exchanges at the air-sea interface and on features in the large-scale circulation. An additional factor

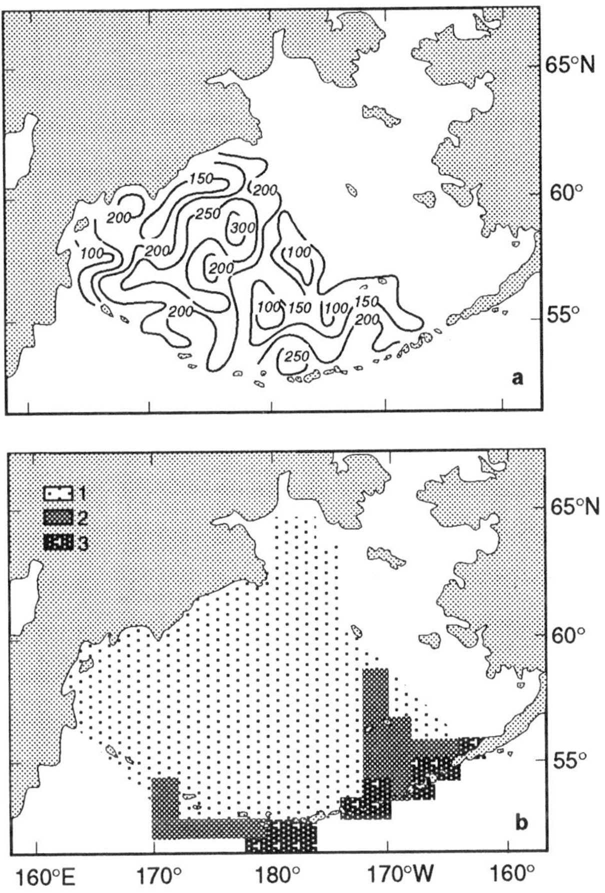

Figure 2. *(a) Depth (m) of the lower boundary of the active layer; (b) A measure of the development of the cold intermediate layer: (1) well-developed, (2) weakly developed, and (3) not apparent.*

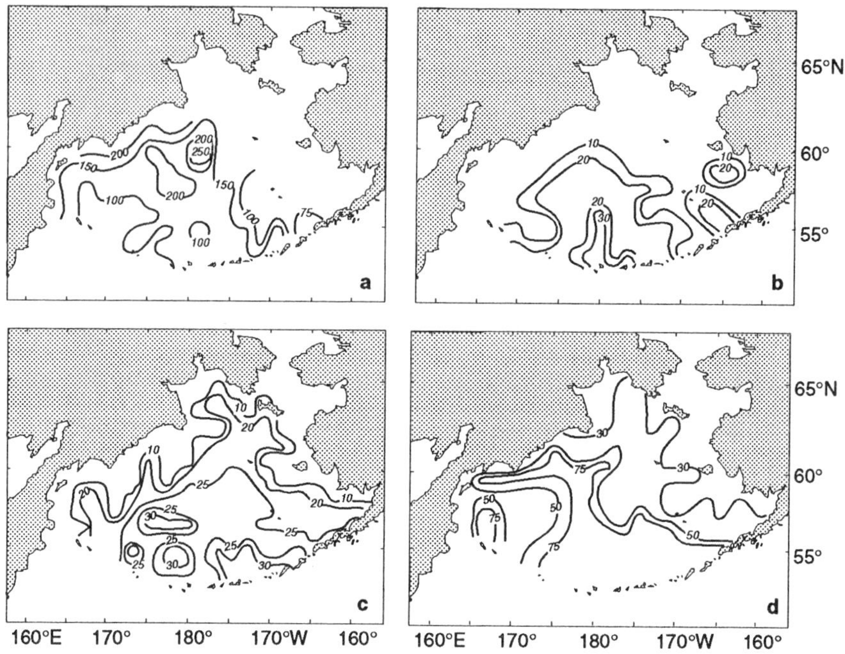

*Figure 3.   Depth of the lower boundary of the upper quasi-homogeneous layer: (a)
January-March, (b) June-July, (c) August-September, and (d) November.*

in determining its extent is the mixing of water in the passes of the Aleutian Islands.

The cold intermediate layer (CIL) is an important structural element of the active layer. It forms as a result of two processes: cooling of the water in autumn and winter and its warming in the spring and summer. Above and below the CIL, there are layers with increased temperature gradients. The CIL is not clearly obvious at some sites (Fig. 2b).

The vertical extent of the surface quasi-homogeneous layer (SQL) varies substantially during the course of the year and in different parts of the sea. Its maximum development is attained during the winter (Fig. 3a). In April and May, it is difficult to distinguish the SQL. During the warm period of the year, the minimum development of the SQL occurs with the greatest warming of the surface water in August-September (Fig. 3c). With the start of autumn (Fig. 3d), cooling of the water gradually increases the thickness of the SQL.

The formation of the CIL results from cooling of the sea in autumn and winter. Its vertical extent is limited to the depth attained by convection. Intensive cooling in winter leads to a homogeneous structure. The characteristic CIL appears only when a surface-warmed layer is formed. The magnitude of the minimum temperature and the depth of its occurrence vary within fairly wide limits in the region. This is a consequence of the variations in the intensity of mixing processes in different parts of the Bering Sea, the effect of the "winterizing" of Pacific Ocean water, and features of its distribution in the Bering Sea. The temperature of the core of the CIL is a striking indication of advective processes occurring in the subsurface layers of the Bering Sea. The most intense inflow of heat occurs in the Aleutian passes east of 172°W.

The lowest temperatures in the core of the CIL (as low as −1.2°C) are seen on the shelf south and southwest of St. Lawrence Island. In July and August, the core temperature of the CIL increases. Because of the advection of Pacific Ocean water, the temperature at the core of the CIL near the Kamchatka Peninsula reaches 1.8°-2.4°C. In September-October, wind speed increases, and thermal convection begins. The role of horizontal and vertical exchanges in the active layer of the sea increases. This leads to an increase in the magnitude of the core temperature of the CIL, especially on the shelf. At this time, the remaining cooled water exits the Bering Sea through Kamchatka Strait. The temperature and depth of the core of the warm intermediate layer (WIL) is determined by contact with the Pacific Ocean and by the extent of transformation of Pacific Ocean water in different parts of the sea. We can divide the deep part of the sea into two parts, separated by a line connecting Bering Island and St. Matthew Island. To the northwest of the line, the WIL is obvious, while to the southeast it is either absent or very weak. The core of the WIL is found at a depth of 250-500 m. Minimum values are seen in Near Strait and adjoining regions and in the southeastern part of the deep basin.

High-Frequency Variability

Large diurnal temperature fluctuations occur in areas where there is inflow of Pacific Ocean water into the sea (Fig. 4). In these regions, there are high current speeds, and the formation of eddies segregates waters with different characteristics. Maximum daily temperature variability (up to 4°C) occurs at the shelf break and near coasts. In these regions, currents are intensified and there is also an increase in the probability of eddy formation. In gulfs and bays, there may be periodic replacement of low-salinity surface waters with water from the open sea or underlying depths. On the eastern shelf, however, diurnal temperature fluctuations do not exceed ~0.7°C. In this region there are relatively small horizontal gradients of characteristics.

In winter, daily surface temperature fluctuations do not generally exceed 0.2°C. Below ~100 m, however, a layer with significant gradients in characteristics forms. This leads to well-developed internal waves at this depth and to short-period variations in water temperature. Temperature variability observed in the 150-200 m layer (Fig. 4a) is primarily from tidal

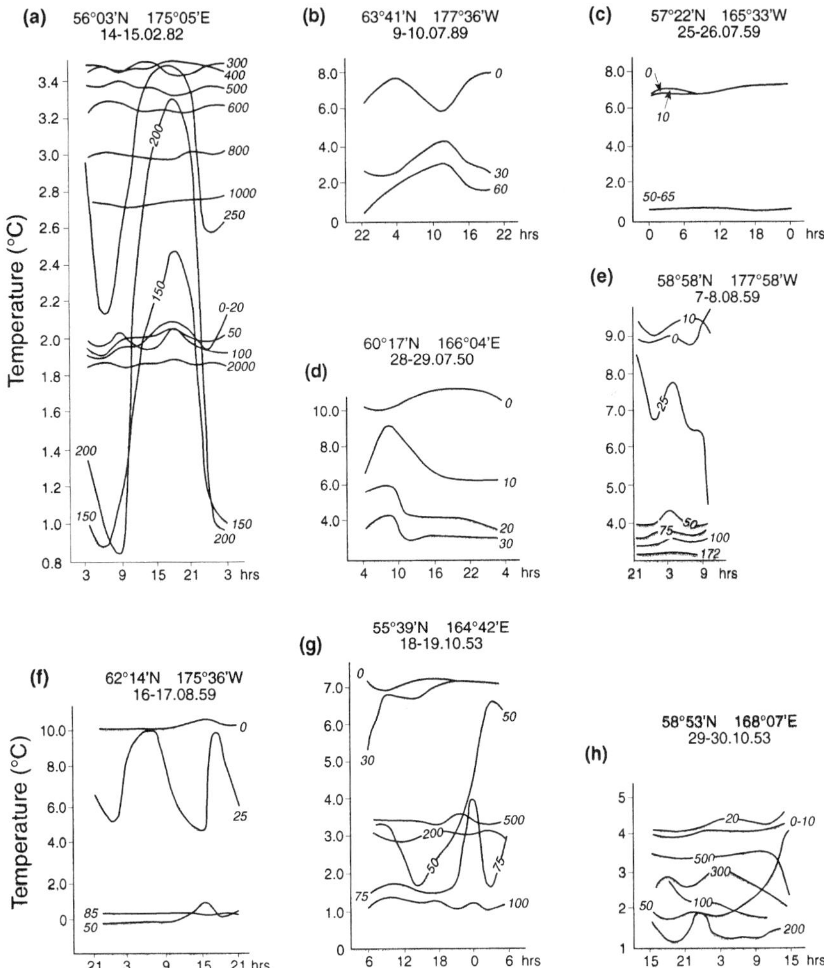

Figure 4. Daily variations of temperature at different times in various regions of the Bering Sea at different depths (m).

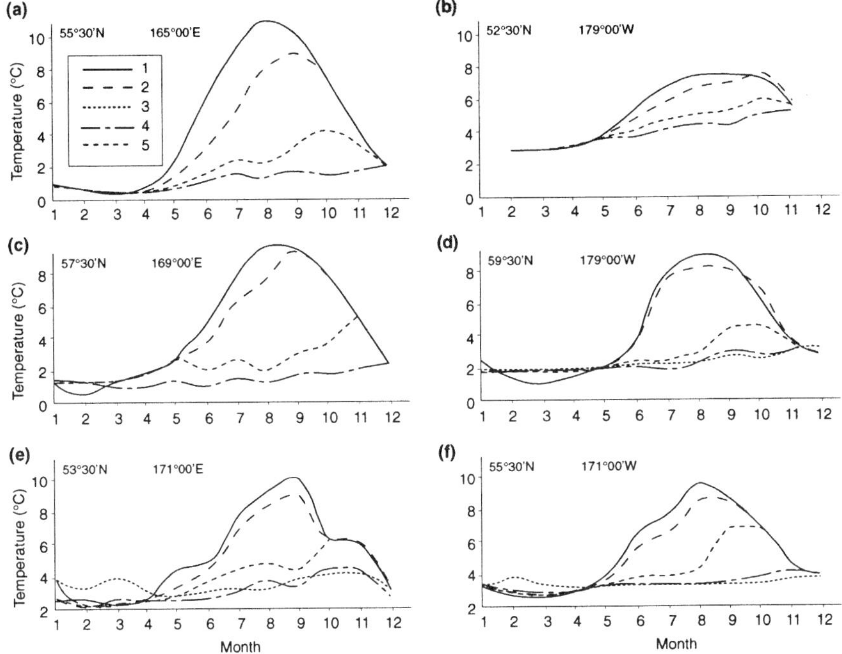

Figure 5. Intra-annual variability of water temperature in the Bering Sea at a depth of: (1) at the surface, (2) 20 m, (3) 50 m, (4) 100 m, and (5) 150 m, at the locations indicated.

effects, with some fortnightly variations being apparent. The range of daily temperature variation here is 1.0°-2.5°C.

In summer, subsurface layers may also have large temperature gradients and variations (Fig. 4). Temperatures may change by 2.0°-4.5°C during the day. In regions of the sea where dynamic processes are insignificant, temperature variability is minimum (Fig. 4c). At some stations, we observe the upwelling of underlying water to near the surface (Fig. 4e). In Pacific Ocean waters, two layers with high values of daily variability may form (Fig. 4g,h).

Seasonal Variability

Seasonal temperature variability depends on the variability of the components of the heat balance at the sea surface and the redistribution of heat in deeper layers due to horizontal and vertical movement of water. In most of the Bering Sea, the maximum temperature of the surface water occurs in August. In areas such as Near Strait and the central Aleutian passes, the maximum may be displaced respectively to September and

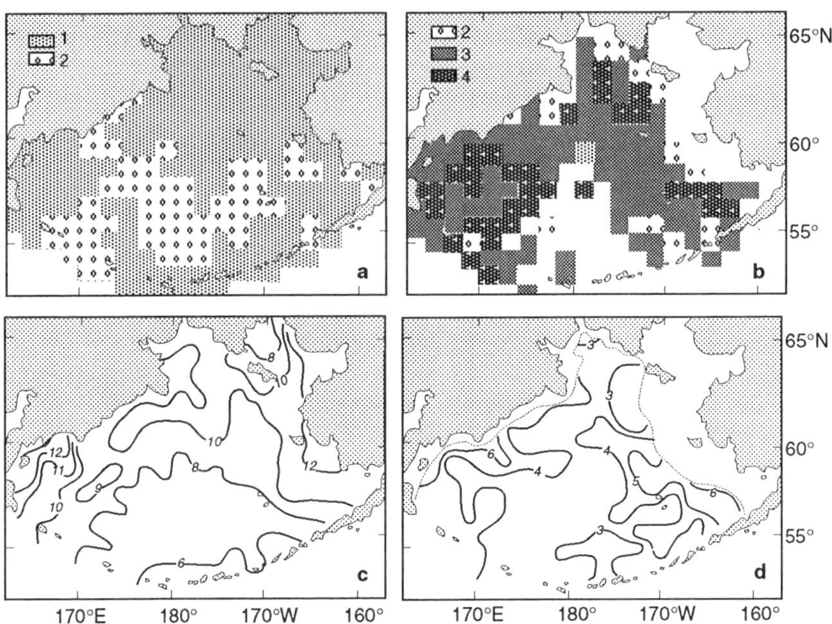

Figure 6.   *Timing of the occurrence of the maximum temperature: (a) at the surface, and (b) at 50 m: (1) August, (2) September, (3) October, and (4)
November. Also, variation of temperature (degrees C) (c) at the surface,
and (d) at 50 m.*

August-October. In regions in the vicinity of the straits, the maximum
temperature occurs in August-September (Figs. 5, 6a). Maximum temperatures at a depth of 20 m occur, as a rule, with the onset of thermal convection in September. In zones with intense vertical mixing, this shifts to
August-September (Fig. 6). With increased depth in the water column, there
is a further shift in the timing of the arrival of the maximum temperature.
This is connected to processes of thermal convection and the intensity of
horizontal and vertical exchange in the water column. The temperature at
50 m reaches its maximum value in October-November. In dynamically
active regions, the peak temperature occurs earlier (Fig. 6b).

The minimum surface temperatures in the deep part of the Bering Sea
occur in March-April. In areas of intense vertical mixing, they occur in
April. Seasonal variability in surface temperature is shown in Figure 6c.
Maximum magnitudes coincide with coastal regions of the sea, while the
lowest values are observed in and near the Aleutian passes.

Seasonal temperature variability at 50 m is a factor of 2-4 less than at
the sea surface (Fig. 6c,d). The greatest variability is seen in the gulfs and
along the shelf break. Minimum temperature variability occurs in the straits

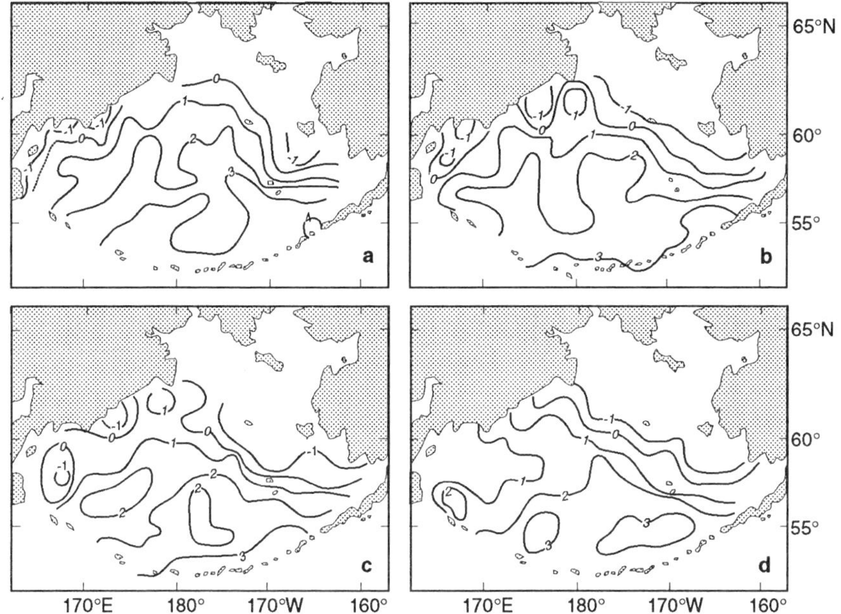

Figure 7.   Distribution of the sea surface temperature (°C) of the Bering Sea: (a) January, (b) February, (c) March, and (d) April.

of the Aleutian Islands, in the central part of the eastern Bering Sea shelf, and in the Gulf of Anadyr.

At a depth of 100-150 m, seasonal temperature variability is generally <3°C (Fig. 5). Maximum variations are seen in places where there is Pacific Ocean inflow, and also in regions of intense autumn-winter convection.

Spatial and Temporal Distribution of Temperature

In the Bering Sea, the year can be divided into two periods based on large-scale features of the distribution of sea surface temperature (Figs. 7-9). From November to June (winter type of distribution; Figs. 7-9), the Pacific Ocean water arriving in the Bering Sea is warmer than the Bering Sea water. This has two causes: first, Pacific Ocean water has a higher heat supply in its active layer, and second, vertical mixing in the Aleutian passes leads to a transfer of heat from underlying layers to the surface. In the warm part of the year (July-September; Figs. 8-9), water temperature near the Aleutian passes is significantly lower than in the Bering Sea. This is caused by the vertical mixing occurring in the passes. The temperature field in October is transitional between these two thermal conditions in the Bering Sea. The thermal regime in the Bering Sea depends to a large

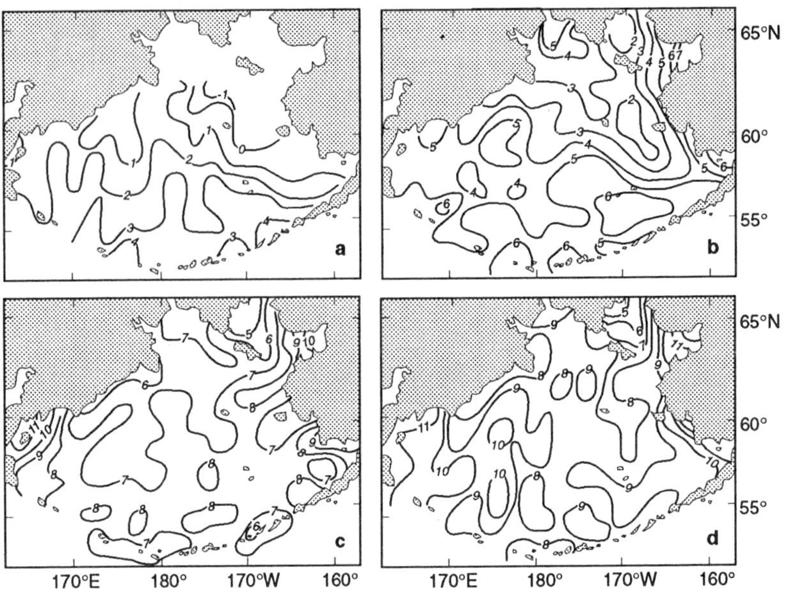

Figure 8.   Distribution of the sea surface temperature (°C) of the Bering Sea:
(a) May, (b) June, (c) July, and (d) August

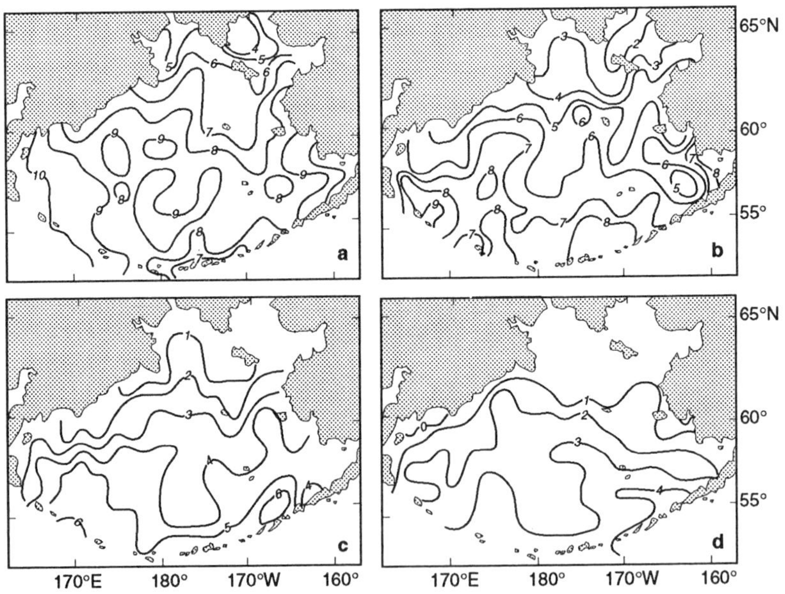

Figure 9.   Distribution of the sea surface temperature (°C) of the Bering Sea:
(a) September, (b) October, (c) November, and (d) December.

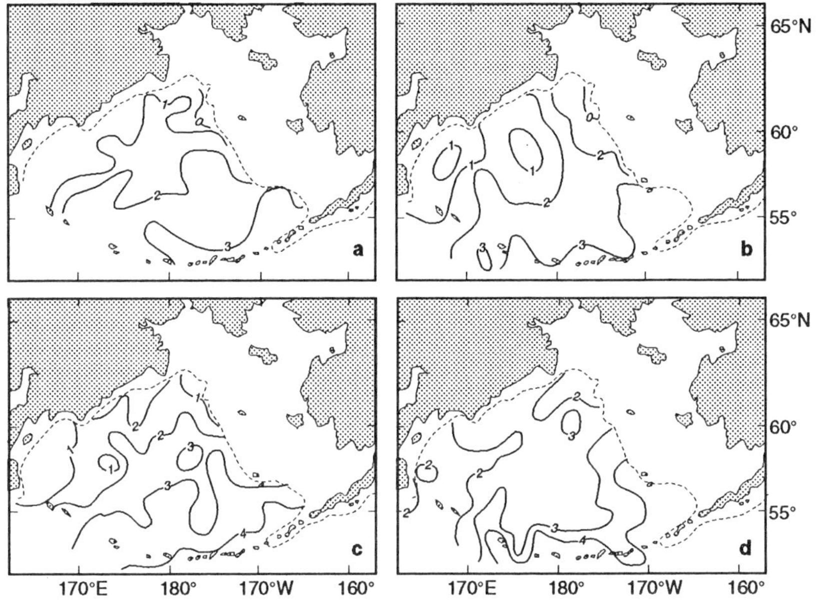

Figure 10. Distribution of the temperature (°C) of the Bering Sea at 100 m in: (a) March, (b) May, (c) August, and (d) October.

extent on water exchange with the Pacific Ocean and on further spreading of Pacific Ocean water within the Bering Sea. The configuration of the 2°C and 3°C isotherms in Figure 7 can illustrate the roles of different Aleutian passes in the exchange of water with the Pacific Ocean. From January to June (Figs. 7a,b, 8a,b), there is a hydrographic front between the shelf waters and the waters of the Bering Sea's deep basin. In December (Fig. 9d), two regions of inflow of warm Pacific Ocean water are seen: east of 167°W, and through the passes between 171° and 178°E.

Seasonal temperature variations at 100 m (Fig. 10) are small. Maximum temperatures (3°-5°C) are noted in the eastern Aleutian Island passes. A general warming of temperatures from May to October is a consequence of the displacement of cold water from winter convection by warmer Pacific Ocean water.

At a depth of 200 m (Fig. 11), there are still appreciable horizontal gradients. This results from the main inflow of Pacific Ocean water occurring through Near Strait and the central Aleutian passes. Maximum temperature at this depth is in autumn (Fig. 11d). At this time, convective processes begin to be seen, as well as an increase in vertical and lateral water exchange linked to an increase in wind speed.

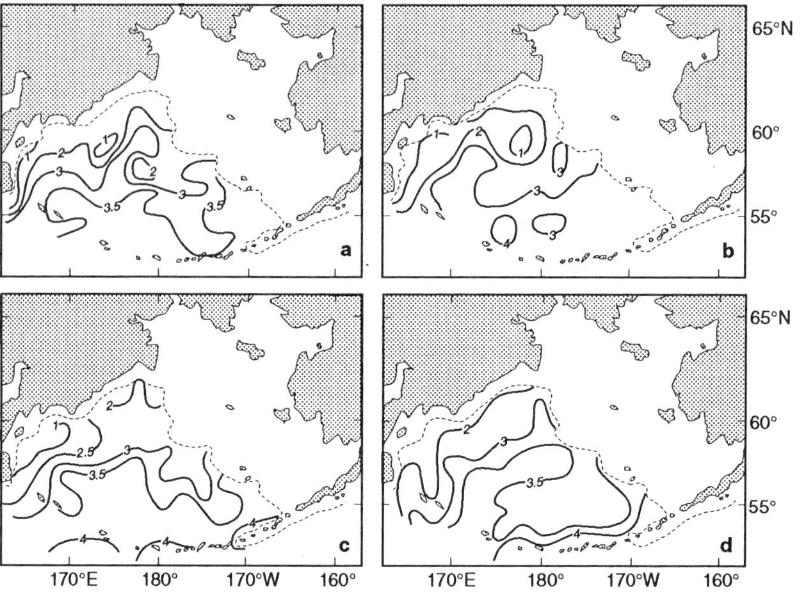

Figure 11. Distribution of the temperature (°C) of the Bering Sea at 200 m in: (a) March, (b) May, (c) August, and (d) October.

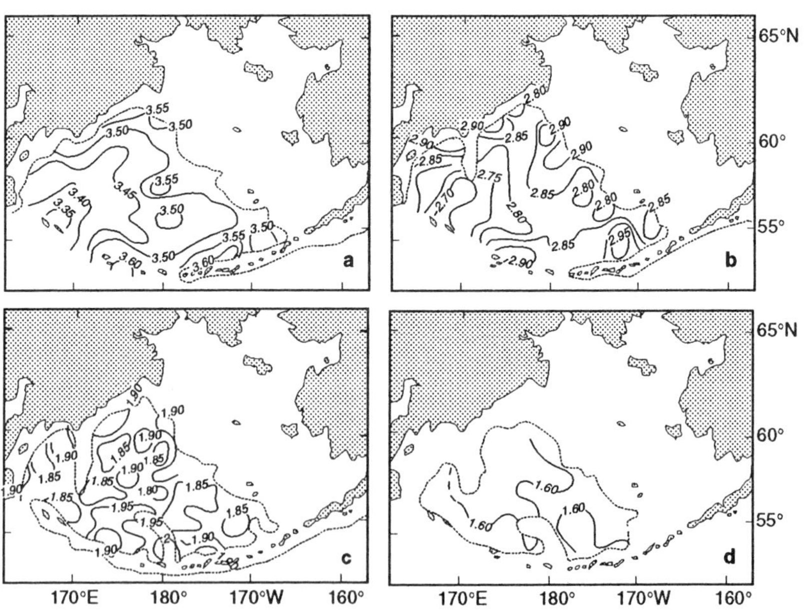

Figure 12. Temperature (°C) at a depth of: (a) 500 m, (b) 1,000 m, (c) 2,000 m, and (d) 3,000 m.

It is difficult to discern seasonal variability at depths greater than 200 m. At depths of 500 m and 1,000 m (Fig. 12), the main inflow of Pacific Ocean water occurs between the Commander Islands and 173°E. Temporal variations at these depths are less than 0.3°C. An increase in temperature is noted near the continental slope and near the Aleutian Islands east of 174°E. This may be associated with increased vertical mixing and the transfer of heat from overlying layers of water (Fig. 12a,b). At a depth of 2,000 m (Fig. 12c), water temperatures in the Bering Sea vary between 1.8° and 1.95°C, and at 3,000 m (Fig. 12d) they vary from 1.56°C to 1.70°C.

Salinity

Vertical Distribution

The salinity of the water in the upper layer of the Bering Sea depends on advection of Pacific Ocean water, the hydrological cycle between the surface layer and atmosphere, continental drainage, ice formation, and melting of ice. Currents and mixing of water only redistribute salt. At greater depths, the salinity depends on currents and water exchange with the Pacific Ocean. Salinity in the Bering Sea increases with depth; however, during the period of ice formation, there may be a slight salinity inversion in the surface layer.

The highest variability occurs in the upper layer of the sea, with the seasonal halocline being part of this layer. The layer of the main thermocline is limited at its lower level by the depth of the sharp change in vertical gradients. In this layer the salinity always increases with depth. The deepest structural feature extends from the lower limit of the main halocline to the bottom.

In the Bering Sea, seasonal variability in salinity does not penetrate beyond 150 m. Minimum thickness of the active layer is observed in the shelf regions of the sea, as well as near the shelf break. Maximum thicknesses of the active layer are seen in regions where there is inflow of Pacific Ocean water. Minimum values of the depth of the bottom of the main halocline are seen in Near Strait. In deep water, salinity increases slightly with depth. The magnitude of the salinity gradient in the layer of the main halocline is 3-5 times greater than in the underlying water column.

Short-Period Variability

The observed data show that daily salinity variability on the whole depends on dynamic processes. Oscillations up to 0.10 ppt are observed in regions far from shore on the eastern Bering Sea shelf. This value is typical for both the surface and underlying depths. One reason for the small salinity variability is probably the relatively weak currents. In regions affected by the Kamchatka Current and in zones of inflow of Pacific Ocean water (e.g., Near Strait), daily variability in salinity increases to 0.2-0.4

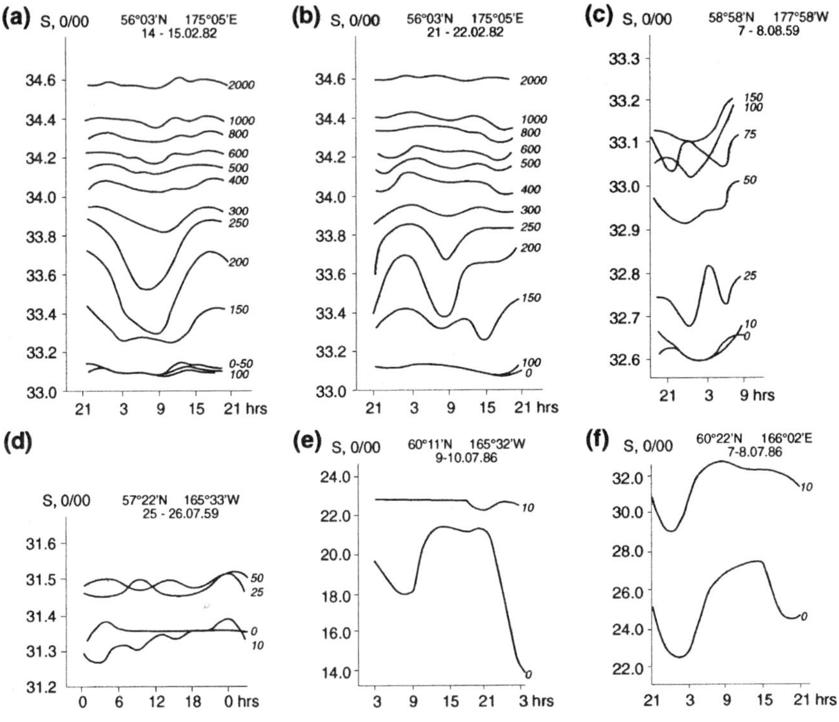

Figure 13. Hourly variations of salinity by date in various regions of the Bering Sea at different depths.

ppt. This increase is brought about by intense vertical and lateral exchange and by advection. In frontal zones, daily salinity variations increase to 1 ppt or more. Maximum values are observed in the surface layer.

During the cold part of the year, daily salinity variations in the upper layer practically disappear (Fig. 13). At the lower edge of the convective layer, however, salinity variability may reach 0.3-0.6 ppt (Fig. 13a). Temporal variation in salinity at the pycnocline indicates that the oscillations may be tidal. In the warm part of the year (Fig. 13c), two layers show increased magnitudes of daily variability: the lower boundary of the upper quasi-homogeneous layer and the layer affected by winter convection. In areas of the eastern Bering Sea shelf that are far from shore, daily salinity oscillations do not exceed ~0.2 ppt (Fig. 13c,d). Near the coast, the salinity oscillations strengthen, most notably in regions near the mouths of rivers (Fig. 13e,f). Greatest stability is seen in regions far from the coast. In zones where mixing of waters of different origin occurs, however, there is a substantial distortion of the vertical salinity distribution. In coastal

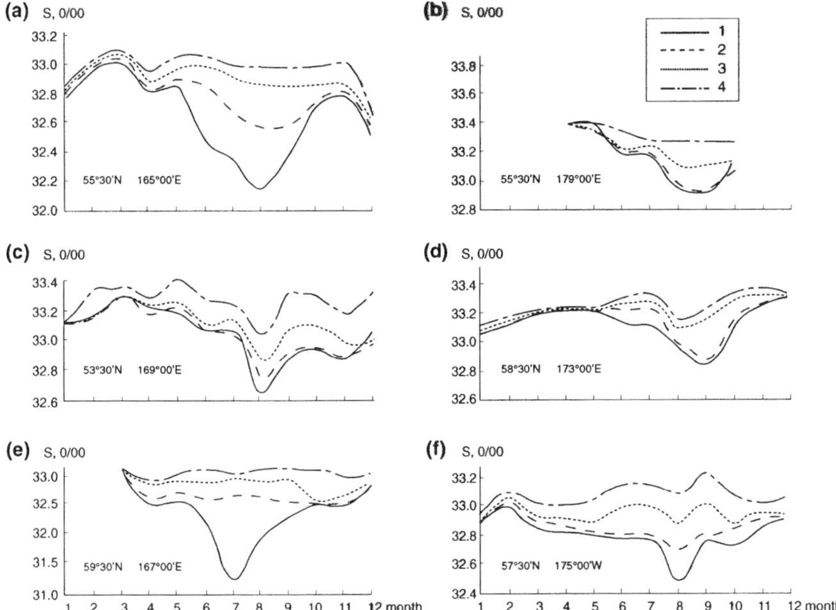

*Figure 14. Intra-annual variability of salinity (ppt) in the Bering Sea at a depth of:
(1) at the surface, (2) 20 m, (3) 50 m, and (4) 100 m.*

and estuarine regions, major transformation of the vertical profile does not occur.

Seasonal Variability

Seasonal variability in salinity is linked to oscillations in river outflow, the formation of ice cover, the balance between precipitation and evaporation from surface waters, and other factors. In the open part of the sea, annual salinity variation is substantially smaller than in nearshore areas. The highest variability in the deep sea occurs in the surface layer. Near shore, the density gradient between surface and subsurface waters increases, which impedes exchange between the two layers. Therefore the major salinity oscillations are limited to the upper 50 m (Fig. 14). Near the Aleutian Island passes and over the deep basin of the Bering Sea, the highest salinities in the active layer are often from March to May. Near the continental slope, the highest salinities occur in February-March. Minimum salinity in the deep part of the sea is often from July to September. Near the coast, salinity is minimum in July because of river flow. Over the deep basin in areas far from the coast, lowest salinities are often in August. Near the Aleutian

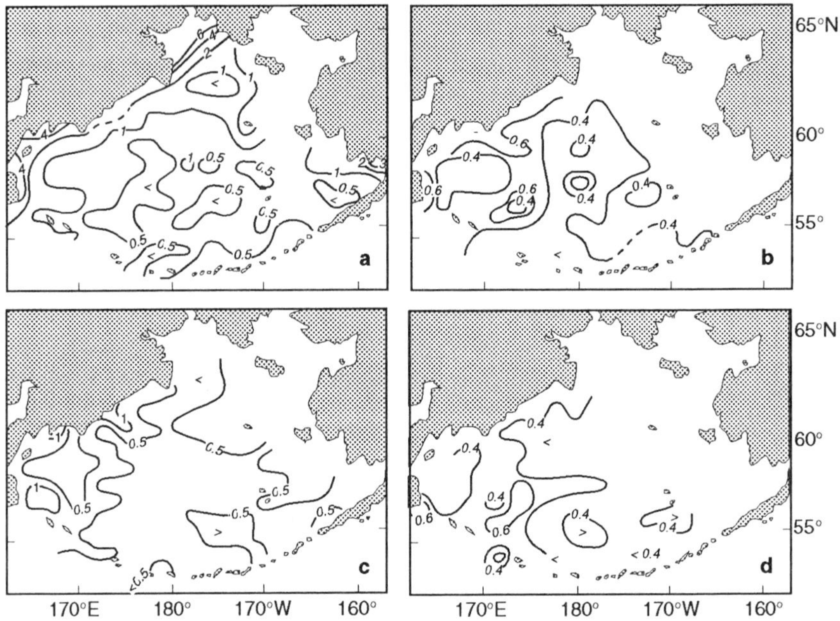

Figure 15. Variability of salinity (ppt) at a depth of: (a) at the surface, (b) 50 m, (c) 100 m, and (d) 150 m.

Island passes, minimum salinity is frequently in August-September (Fig. 14).

The greatest range of salinity variation is observed in the surface layer (Fig. 15). The largest values (up to 4-7 ppt) occur toward the periphery of the sea. In the deep basin, the annual salinity variation is less than 1 ppt. Minimum variability is observed near the Aleutian passes (Fig. 15). At a depth of 50 m (Fig. 15b), the main pattern in the annual salinity variation is similar to that at the surface, but the magnitude is less. At a depth of 100 m, there is a further decrease in the annual salinity variability (Fig. 15c). Even at the periphery of the sea, variability at this depth barely exceeds ~1 ppt. At a depth of 150 m, the difference in the magnitudes of salinity variability divides the sea into two regions (Fig. 15d). In the eastern section, annual salinity variability is small (0.2-0.4 ppt), while in the west variability is larger (0.4-0.6 ppt).

Spatial and Temporal Distribution of Salinity

The major large-scale features in the distribution of salinity in the surface layer are preserved throughout the year. Pacific Ocean water generally has the highest salinity. The salinity decreases approaching the coast, especially

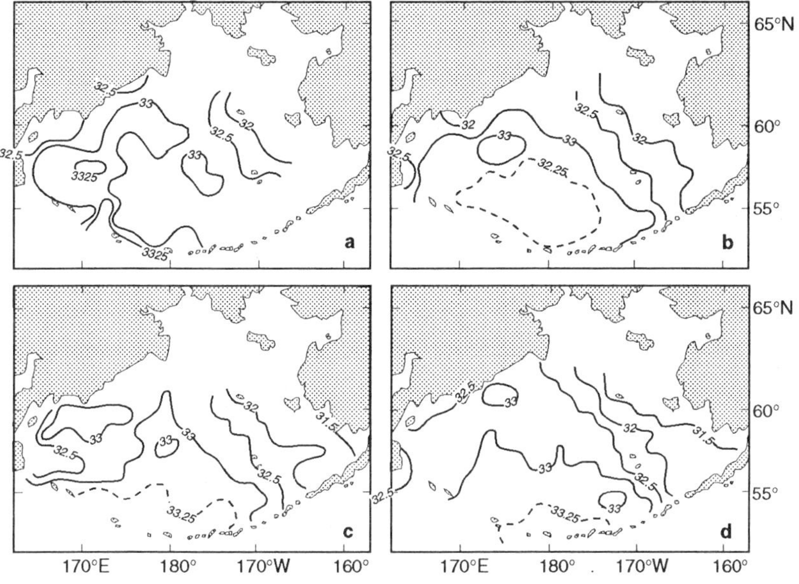

Figure 16. Salinity (ppt) distribution at the surface of the Bering Sea in: (a) January, (b) February, (c) March, and (d) April.

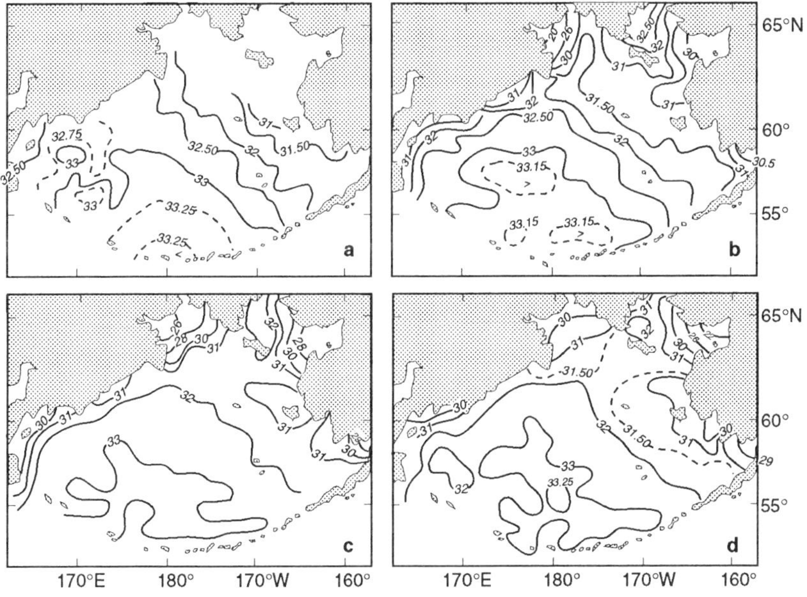

Figure 17. Salinity (ppt) distribution at the surface of the Bering Sea in: (a) May, (b) June, (c) July, and (d) August.

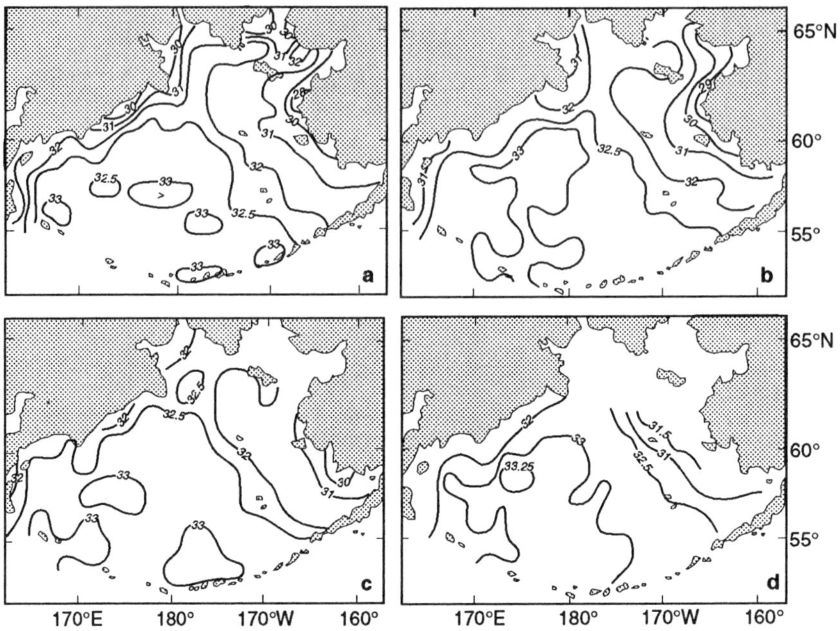

Figure 18. Salinity (ppt) distribution at the surface of the Bering Sea in: (a) September, (b) October, (c) November, and (d) December.

on the eastern Bering Sea shelf. A pronounced frontal zone divides the shelf waters from waters over the deep basin (Figs. 16-18). The lowest surface salinities are observed at the head of the Gulf of Anadyr (as low as 19-20 ppt), in Norton Sound (<29 ppt), and in Bristol Bay (as low as 30 ppt) in June-July (Fig. 17b,c). River input is less important along the Kamchatka coast, so that salinity in that region does not fall below ~30.5 ppt. The shapes of the 31.5 ppt and 32.0 ppt isohalines illustrate the flow of Pacific Ocean water onto the shelf east of Cape Navarin (Fig. 17b).

Near the Aleutian Islands (east of 176°E), a region of reduced salinity begins to form in June (Fig. 17b). Its formation is associated with continental runoff in the Gulf of Alaska as well as from the Aleutian Islands. The freshening of nearshore surface water is generally greatest in July (Fig. 17c). There is about a month lag between the peak river runoff in June and the minimum salinity of the shelf waters in July. In the deep basin of the Bering Sea, the change in salinity from June to July is <0.1 ppt. In October and November, the positions of the 32.0 and 32.5 ppt isohalines indicates the path of high-salinity water which flows along the outer edge of the Gulf of Anadyr toward Chirikov (Anadyr) Strait and also to the south of St. Lawrence Island (Fig. 18b).

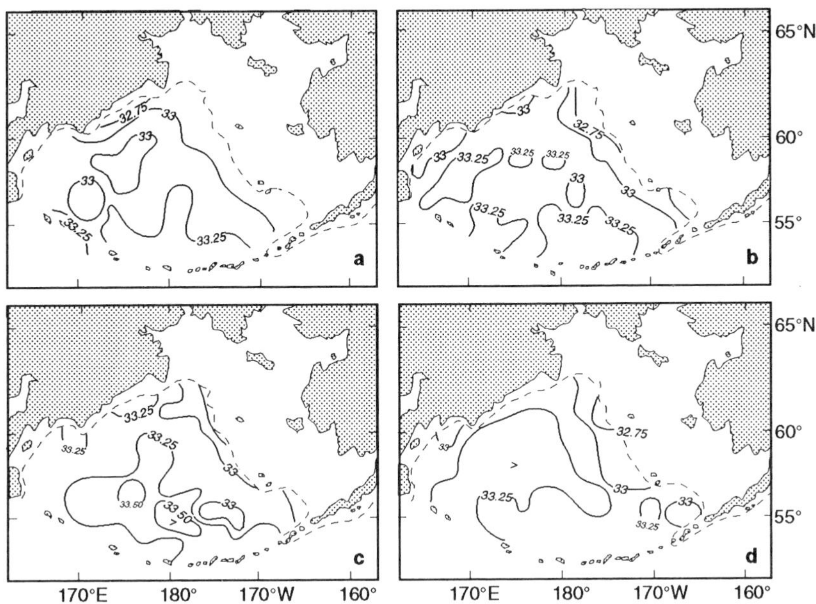

Figure 19. Salinity (ppt) distribution at 100 m in: (a) March, (b) May, (c) August, and (d) October.

At a depth of 50 m, all the main large-scale features of the surface salinity distribution are maintained. The characteristic features of the salinity field are a weak reflection of the surface distribution. The salinity at 50 m varies little during the course of the year. The smallest variability occurs on the eastern Bering Sea shelf. This is linked to the low current speeds and the sharp pycnocline between the surface and underlying waters. In the Gulf of Anadyr in October, at the start of autumn convection, salinities decrease to 32.8 ppt. The frontal zone dividing the shelf from the deep basin is observed all year long, although from September to November gradients in the zone decrease.

The salinity field at 100 m and 200 m (Figs. 19, 20) is subject to the influence of surface variations in salinity mainly in the Aleutian passes. This process is most clearly observed in October. In the other parts of the sea, salinity variability during the course of the year is linked to features in the large-scale circulation.

The distribution of salinity at 500 m and 1,000 m (Fig. 21) reflects the entry of Pacific Ocean water into the Bering Sea through the deep passes of the western part of the Aleutian Islands. Salinity is highest in areas adjoining these passes. As Pacific Ocean water spreads into the deep basin, its salinity is slightly and gradually decreased (Fig. 21a,b). At depths

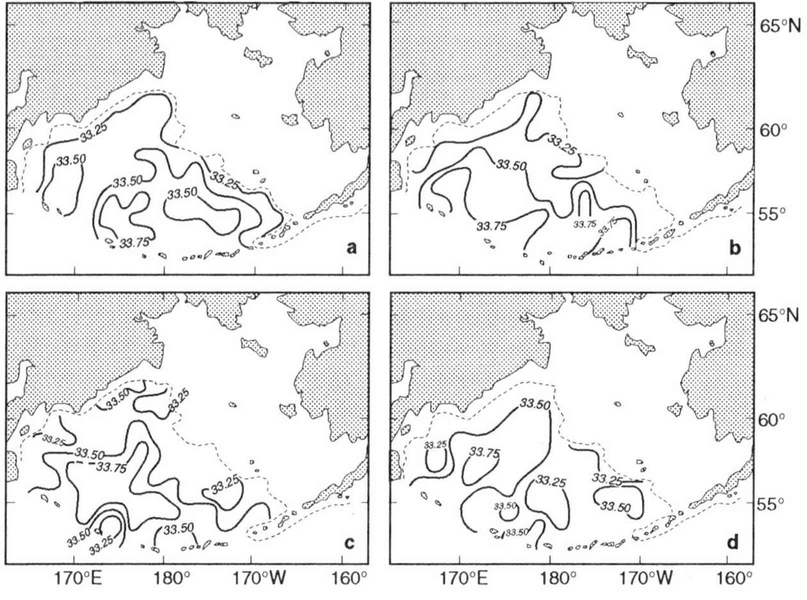

Figure 20. Salinity (ppt) distribution at 200 m in: (a) March, (b) May, (c) August, and (d) October.

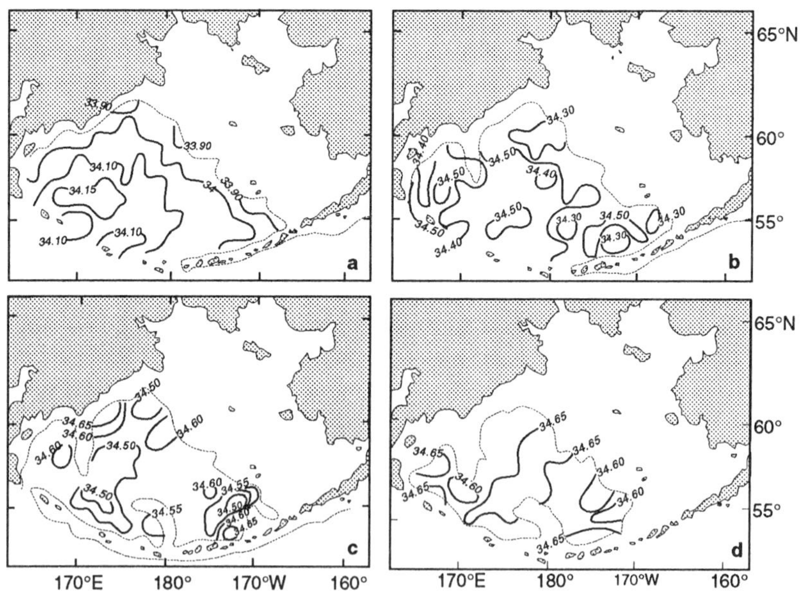

Figure 21. Salinity (ppt) distribution at a depth of: (a) 500 m, (b) 1,000 m, (c) 2,000 and (d) 3,000 m.

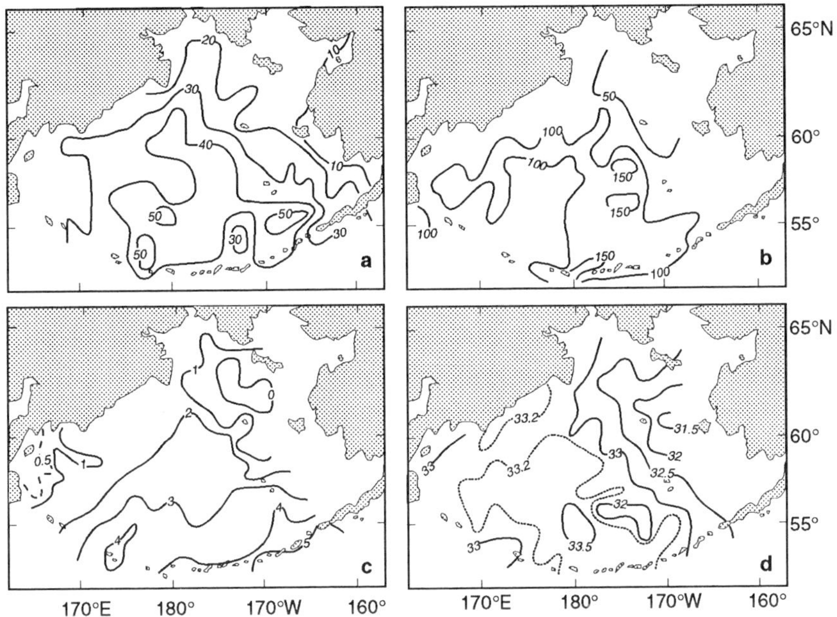

Figure 22. Characteristics of the intermediate Bering Sea water mass in summer: (a) location of the upper boundary in meters, (b) depth of the core of the water mass, in meters, (c) temperature (°C) of the core, and (d) salinity (ppt) of the core.

of 2,000 m and 3,000 m (Fig. 21c,d), it is difficult to distinguish a prevailing pattern.

Characteristics of Water Masses

Water masses in the upper layer of the sea are influenced by river runoff, advection of water from the Pacific Ocean, solar radiation, evaporation, precipitation, wind mixing, autumn and winter convection, the formation and destruction of ice, spreading of characteristics by currents, and turbulent mixing. At intermediate and deep depths, the dominant effects are water exchange with the Pacific Ocean and mixing of water by currents.

Warming and freshening of the surface water in the spring allow the formation of the upper boundary of the intermediate Bering Sea water mass (IBWM). The upper boundary of the IBWM in summer is shown in Figure 22a. The core of IBWM is deepest in regions of inflow of Pacific Ocean water and also in the region just west of the eastern shelf (Fig. 22b). Near the shelf break, there are large horizontal salinity gradients. However, the general large-scale pattern of its distribution is maintained. The

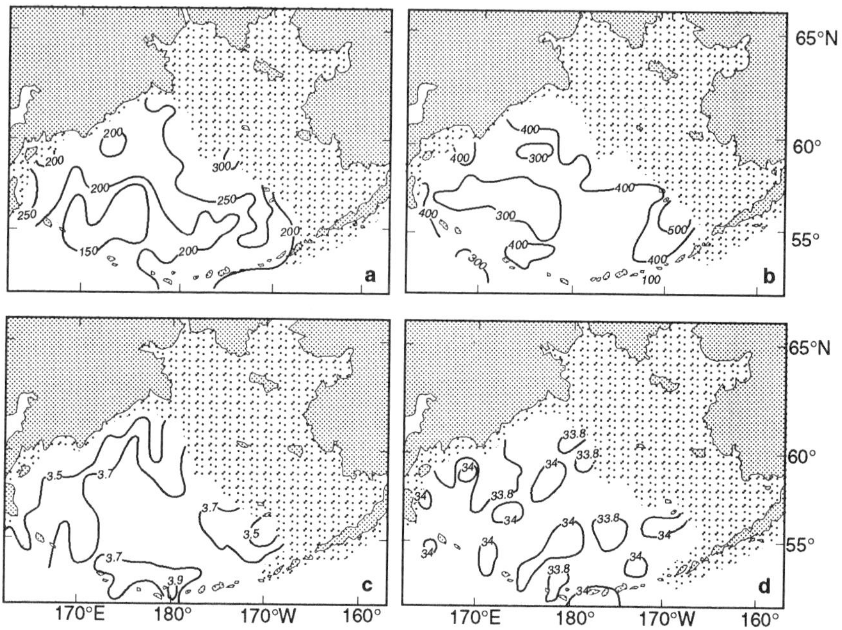

Figure 23. Characteristics of the intermediate Pacific Ocean water mass in August: (a) location of the upper boundary in meters, (b) depth of the core of the water mass, in meters, (c) temperature (°C) of the core, and (d) salinity (ppt) of the core. (The hatched area is shallower than the water mass).

only change is in its core temperature. The highest temperatures (up to 5°C) are noted in the straits in the central and eastern parts of the Aleutian Islands, while temperatures near the coast of Kamchatka are <1°C. The lowest core temperatures (<0°C) are distinguished in the anticyclonic eddy south of St. Lawrence Island (Fig. 22c). The salinity at the core of the IBWM in summer varies from 31.5 to 33.5 ppt (Fig. 22d).

The main feature of the intermediate Pacific Ocean water mass (IPWM) is its high temperature. There is little seasonal variation in its characteristics. Its upper boundary gradually deepens from March, the time of maximum cooling, through October, the start of the next cycle of cooling of the surface water. The core of the IPWM occurs at depths from 200 to 500 m, with maximum core depths occurring at the periphery of the deep basin (Fig. 23b). The temperature at the core of the IPWM varies within narrow limits (from 3.4° to 4.0°C; Fig. 23c). The highest temperatures occur where Pacific Ocean water enters the Bering Sea. The lowest values (3.4-3.5°C) are noted near the Kamchatka coast, associated with severe cooling of water there in the winter. The salinity at the core of the IPWM varies from 33.7-34.1 ppt (Fig. 23d). The maximum values in winter, spring, and summer are seen where Pacific Ocean water enters the Bering Sea, as well as near the coast of the Kamchatka Peninsula. In autumn, in dynamically active regions (the Aleutian Island passes and a section of the northeast slope) the core salinity of the IPWM is minimal because of mixing with fresher overlying water.

The deep water mass (DWM) is the most homogeneous in its characteristics. It has lower temperatures and higher salinities than the intermediate Pacific Ocean water mass. The upper boundary of the DWM is shallowest in regions of inflow of Pacific Ocean water and at the periphery of the deep part of the sea. The core of the deep water mass is near the bottom, in the deep part of the Bering Sea. Water temperature at depths below 3,000 m is 1.45-1.65°C, while salinity is 34.60-34.68 ppt.

Acknowledgments

We thank Sigrid Salo (NOAA/PMEL) for her many efforts in translation and preparation of this manuscript.

References

Arsenev, V.S. 1967. Currents and water masses of the Bering Sea. Nauka, Moscow. 135 pp.

Barnes, C.A., and T.G. Thompson. 1938. Physical and chemical investigations in the Bering Sea and portions of the North Pacific Ocean. University of Washington Publication in Oceanography 3(2):1-164.

Belkin, I.M. 1984. An assessment of editing oceanographic data (main principles). Trudy Vse Nauchno-Issledobatel'skogo Instituta Gydrometeorologicheskogo Informatsia-Mirny Danny Tsentr 113:108-113.

Burkov, V.A. 1958. The hydrography of the Komandoro-Kamchatka region of the Pacific Ocean in spring. Trudy Instituta Okeanologii Akademy Nauk SSSR 27:12-21.

Coachman, L.K. 1979. On lateral water mass interaction: A case study, Bristol Bay, Alaska. Journal of Physical Oceanography 9:278-297.

Coachman, L.K. 1986. Circulation, water masses, and fluxes on the southeastern Bering Sea shelf. Continental Shelf Research 5:23-108.

Coachman, L.K., and R.L. Charnell. 1977. Fine structure in outer Bristol Bay. Deep-Sea Research 24:869-889.

Coachman, L.K, K. Aagaard, and R.B. Tripp. 1975. Bering Strait: The Regional Physical Oceanography. University of Washington Press, Seattle. 172 pp.

Coachman, L.K, T.H. Kinder, J.D. Schumacher, and R.B. Tripp. 1980. Frontal systems of the southeastern Bering Sea Shelf. In: T. Casterns and T. McClimans (eds.), Stratified flows, the second IAHR symposium, Trondheim, June 1980. Tapir, Trondheim, pp. 917-933.

Davydov, I.V. 1972. On the question of oceanographic factors affecting the yield of individual generations of herring in the western Bering Sea. Izvestia TINRO 82:281-307.

Davydov, I.V. 1984. On the development of oceanographic conditions in the main fisheries regions of the Far Eastern Seas. Izvestia TINRO 109:3-17.

Davydov, I.V., and F.F. Lipetskiy. 1970. The hydrography of the Karaginskiy and Olyutorskiy-Karaginskiy fisheries regions in the Bering Sea. Izvestia TINRO 73:178-193.

Dobrovolskiy, A.D., and V.S. Arsenev. 1961. Hydrographic characteristics of the Bering Sea. Trudy Instituta Okeanologii Akademy Nauk SSSR 38:64-96.

Dobrovolskiy, A.D., A.S. Ionin, and G.B. Udintsev. 1959. The history of research in the Bering Sea. Trudy Instituta Okeanologii Akademy Nauk SSSR. 29:5-20.

Dodimead, A.J., F. Favorite, and T. Hirano. 1963. Salmon of the North Pacific Ocean. Part 2, Review of Oceanography of the Subarctic Pacific Region. Bulletin of the North Pacific Fisheries Commission 13. 195 pp.

Dowling, G. 1962. Subsurface horizontal temperature gradients and fluctuations in the Bering Sea. Journal of Geophysical Research 67:5-54.

Favorite, F. 1974. Flow into the Bering Sea through Aleutian Island passes. In: D.W. Hood and E.J. Kelley (eds.), Oceanography of the Bering Sea, Occasional Publication Number 2. Institute of Marine Science, University of Alaska, Fairbanks, pp. 3-37.

Favorite, F., A.J. Dodimead, and K. Nasu. 1976. Oceanography of the subarctic Pacific region, 1960-71. International North Pacific Fisheries Commission Bulletin, 33. 187 pp.

Favorite, F., J.W. Schantz, and C.R. Hebard. 1961. Oceanographic observations in Bristol Bay and the Bering Sea, 1939-1941 (USCGT Redwing). U.S. Fish and Wildlife Service Special Scientific Report, Fisheries 381. 323 pp.

Filyushkin, B.N. 1968. Thermal characteristics of the surface layer in the Northern Pacific Ocean. Mezhduvedomatvenniy Geofizicheskii Komitet Akademy Nauk SSSR/Okeanologicheskie Issledovaniya 19:22-69.

Goodman, J.R., J.H. Lincoln, T.G. Thompson, and F.A. Zeusler. 1942. Physical and chemical investigations: Bering Sea, Bering Strait, and Chukchi Sea during the summers of 1937 and 1938. University of Washington, Publication in Oceanography 3(3):81-103, plus app.

Gubenko, N.D., and Y.L. Tsipis. 1984. On the question of the formation of data bases in the computing system KOMPAS. Trudy Vse Nauchno-Issledobatel'skogo Instituta Gydrometeorologicheskogo Informatsia-Mirny Danny Tsentr 113:54-67.

Hood, D.W., and J.A. Calder. 1981. The eastern Bering Sea Shelf: Oceanography and resources. University of Washington Press, Seattle. 520 pp.

Hood, D.W., and E.J. Kelley. 1974. Oceanography of the Bering Sea. Occasional Publication 2, Institute of Marine Science, University of Alaska, Fairbanks. 623 pp.

Hughes, F.W., L.K. Coachman, and K. Aagaard. 1974. Circulation, transport, and water exchange in the western Bering Sea. In: D.W. Hood and E.J. Kelley (eds.), Oceanography of the Bering Sea. Occasional Publication 2, Institute of Marine Science, University of Alaska, Fairbanks, pp. 59-98.

Ingraham, W.J. 1981. Shelf environment. In: D.W. Hood and E.J. Calder (eds.), The eastern Bering Sea Shelf: Oceanography and resources. University of Washington Press, Seattle, pp. 31-52.

Ivanenkov, V.N. 1964. Hydrochemistry of the Bering Sea. Nauka, Moscow. 138 pp.

Izrael', Y.A. 1983. Research on the ecosystems of the Bering Sea. Gydrometeoizdat. 157 pp.

Khen, G.V., and V.F. Voronin. 1986. Interannual oscillations at the southern edge of shelf water in the eastern Bering Sea linked with large-scale variability. Trudy (DVNII) Dal'nye-Vostochnogo Nauchno-Issledobatel'skogo Gydro-Meteorologicheskogo Instituta 125:10-19.

Kihara, K., and M. Uda. 1969. Analytical studies of the mechanisms concerning the formation of demersal fishing grounds in relation to the bottom water masses of the eastern Bering Sea, Part 1. Tokyo University of Fisheries N55(2):83-90.

Kinder, T.H. 1977. The hydrographic structure over the continental shelf near Bristol Bay, Alaska, June 1976. Department of Oceanography Technical Report M773, University of Washington. 61 pp.

Kinder, T.H., and L.K. Coachman. 1978. The front overlying the continental slope in the eastern Bering Sea. Journal of Geophysical Research 83:4551-4559.

Kinder, T.H., and J.D. Schumacher. 1981. Hydrographic structure over the continental shelf of the southeastern Bering Sea. In: D.W. Hood and E.J. Calder (eds.), The eastern Bering Sea shelf: Oceanography and resources. University of Washington Press, Seattle, pp. 31-52.

Kitano, K. 1970. A note on the thermal structure of the eastern Bering Sea. Journal of Geophysical Research 75:1110-1115.

Kitani, K., and S. Kawasaki. 1979. Oceanographic structure in the region of the eastern Bering Sea. Part 1: The movement and physical characteristics of water in summer 1978. Bulletin of the Far Seas Fisheries Research Laboratory 17:1-12. (In Japanese.)

Koto, H., and T. Fuji. 1958. Structure of the waters in the Bering Sea and Aleutian region. Bulletin of the Fisheries Department, Hokkaido University 9:149-170.

Koto, H., and T. Maeda. 1965. On the movement of fish: Shoals and the change of bottom temperature on the trawl-fishing grounds of the eastern Bering Sea. Bulletin of the Japanese Society of Scientific Fisheries 1:263-268. (In Japanese.)

Lazo, A.V. 1971. Several features of the hydrographic regime of the coastal zones of Bering and Medniy Islands. Sbornik Rabot Petropavloskoi Gydro-Meteorologicheskogo Organizatsii 1:65-76.

Leonov, A.K. 1947. Water masses of the Bering Sea and currents at its surface. Meteorologiya i Gydrologiya 2:51-66.

Leonov, A.K. 1960. Regional oceanography. Part 1: The Bering Sea. Gidrometeoizdat, Leningrad. 765 pp.

Maeda, T, T. Fujii, and K. Masuda. 1968. On the annual fluctuation of oceanographic conditions in the summer. Part 2, Studies of the trawl-fishing grounds in the eastern Bering Sea. Bulletin of the Japanese Society of Scientific Fisheries 34:502-510. (In Japanese.)

Milyeiko, G.N. 1973. Forecasting water temperature distribution and the position of the ice edge in the Bering Sea during the cold part of the year. Trudy Gydro-Meteorologicheskogo Tsentra SSSR 127:100-104.

Mishima, S., and S. Nishizawa. 1955. Report on hydrographic investigations in Aleutian waters and the southern Bering Sea in the early summers of 1953 and 1954. Bulletin of the Faculty of Fisheries, Hokkaido University 6:85-124.

Mofjeld, H.O. 1986. Observed tides on the northeastern Bering Sea Shelf. Journal of Geophysical Research 91:2593-2606.

Moisyeev, L.K. 1978. Stratification of the temperature field. Trudy Vse Nauchno-Issledobatel'skogo Instituta Gydrometeorologicheskogo Informatsia-Mirny Danny Tsentr 45:36-62.

Muench, R.D. 1976. A note on eastern Bering Sea shelf hydrographic structure, August 1974. Deep-Sea Research 23:245-247.

Muench, R.D., R.B. Tripp, and J.D. Cline. 1981. Circulation and hydrography of Norton Sound. In: D.W. Hood and J.A. Calder (eds.), The eastern Bering Sea Shelf: Oceanography and resources. University of Washington Press, Seattle, pp. 77-93.

Natarov, V.V. 1963. On water masses and currents in the Bering Sea. Trudy Vladivostochnogo Nachno-Issledobatel'skogo Rybnogo Organizatsii 48:111-133.

Naumov, G.K., and Y.A. Khistyaev. 1972. Some features in the hydrography near the shelf break in the Bering Sea. In: Trudy All-Union Hydromet Service of the USSR, Oceanographic Forecasting and Calculations, Gidrometeoizdat, Leningrad, pp. 95-102.

Ohtani, K. 1969. On the oceanographic structure and ice formation on the continental shelf in the eastern Bering Sea. Bulletin of the Faculty of Fisheries, Hokkaido University 20:94-117.

Ohtani, K. 1973. Oceanographic structure in the Bering Sea. Memoirs of the Faculty of Fisheries, Hokkaido University 21:65-106.

Otobe, H., T. Nakai, and A. Hattori. 1983. Heat energy exchange across the sea surface of the Bering Sea and the North Pacific Ocean in summer: Estimates from direct measurements of radiation fluxes. Deep-Sea Research 30:1023-1031.

Pereskokov, A.I. 1984. Elimination of spikes during statistical processing of deep hydrometeorological data. Trudy Vse Nauchno-Issledobatel'skogo Instituta Gydrometeorologicheskogo Informatsia-Mirny Danny Tsentr 101:106-113.

Poluektov, S.V., and Y.A. Khistyaev. 1981. Thermal stratification of the active layer of the Bering Sea in winter. Trudy Vse Nauchno-Issledobatelskogo Instituta Gydrometeorologicheskogo Informatsia-Mirny Danny Tsentr 83:15-23.

Rathburn, R. 1894. Summary of the fishing investigations conducted in the North Pacific Ocean and Bering Sea from July 1, 1888 by the U.S. Fish Commission Steamer "Albatross." Bulletin of the U.S. Fisheries Commission 12:127-201.

Ratmanov, G.E. 1937. On the hydrology of the Bering and Chukchi seas. Issledobanniya Moryei 25:10-118.

Reed, R.K. 1978. The heat budget of a region in the eastern Bering Sea, summer 1976. Journal of Geophysical Research 33:3636-3645.

Saur, J.F.T, R.M. Lesser, A.J. Carsola, and W.M. Cameron. 1952. Oceanographic cruise to the Bering and Chukchi seas, summer 1949. U.S. Navy Electronics Laboratory Report 298. San Diego, CA. 38 pp.

Sayles, M.A, K. Aagaard, and L.K. Coachman. 1979. Oceanographic atlas of the Bering Sea Basin. University of Washington Press, Seattle. 158 pp.

Schumacher, J.D., T.H. Kinder, D.J. Pashinski, and R.L. Charnell. 1979. A structural front over the continental shelf of the eastern Bering Sea. Journal of Physical Oceanography 9: 79-87.

Swift, J.H., and K. Aagaard. 1976. Upwelling near Samalga Pass. Limnology and Oceanography 21:399-408.

Takenouti, A.Y. 1976. Recent Japanese physical oceanographic studies of the Bering Sea. Marine Science Communications 2:285-297.

Takenouti, A.Y., and K. Ohtani. 1974. Currents and water masses in the Bering Sea: A review of Japanese work. In: D.W. Hood and E.J. Kelley (eds.), Oceanography of the Bering Sea. Occasional Publication Number 2, Institute of Marine Science, University of Alaska, Fairbanks, pp. 39-57.

Tanner, Z.L. 1890. Explorations of the fishing grounds of Alaska, Washington Territory, and Oregon during 1888 by the U.S. Fish Commission Steamer "Albatross." Bulletin of the U.S. Fisheries Commission 8:1-92.

Tiguntsev, L.A. 1971. Variability of water temperature in the region of Bering Strait. Trudy Articheskogo i Antarticheskogo Nauchno-Issledobatel'skogo 302:50-57.

Yarichin, V.G. 1984. Water structure and water masses in the Bering Sea in the summer of 1982. Trudy (DVNII) Dal'nye-Vostochnogo Nauchno-Issledobatel'skogo Gydro-Meteorologicheskogo Instituta 111:83-97.

CHAPTER 4

Bering Sea Tides

Zygmunt Kowalik
Institute of Marine Science, University of Alaska Fairbanks, Fairbanks, Alaska

Abstract

This paper includes a review of research on tides in the Bering Sea and presents new results obtained from numerical models. The focus of investigations is a description of the enhanced tidal currents. Dynamics of the tidal currents are shown to be closely related to the tidal period. In both semidiurnal and diurnal bands of oscillations the tides travel from the North Pacific through the deep basin onto the shelf. The enhanced tidal currents for both bands of oscillations are generated due to topographic amplification. Examples of such amplification occur during propagation of tides from deep to shallow water over the continental shelf slope, in proximity to the land and islands and also in the narrow passages and straits. The most conspicuous topographic enhancement occurs in the triangular-shaped Bristol Bay where M_2 tide current increases to about 100 cm/s. In the diurnal band of oscillation, the enhanced currents are generated not only due to the topographic amplification but also due to the tidal shelf waves trapped against the continental shelf slope. Numerical computations and observations are applied to identify diurnal tidal currents related to the trapped shelf waves. Investigations reveal two major regions of diurnal tide enhancement: (a) off Cape Navarin, and (b) close to the Aleutian Islands between 170°W and 173°W. Regions of lesser diurnal tide enhancement are located along the shelf break where diurnal tides depict local maxima in the sea level distribution. Nonlinear interactions of the strong tidal currents also have been investigated. To study tidal current rectification over the bottom topography, a high-resolution numerical model was used in the region of the Pribilof Islands and Canyon. The existence of the model's residual currents is well corroborated by long-term time series obtained around St. Paul Island. Six current meters deployed for up to 1 year revealed a very stable clockwise flow in the range 4 to 18 cm/s.

Introduction

Tides provide the most important and consistent driving force in the Bering Sea. The tides and tidal currents on the Bering Sea shelf play an important role in such oceanographic processes as the maintenance of the density structure, sediment resuspension and transport, and the distribution of benthic and intertidal organisms. Observations in the Bering Sea described four hydrographic domains separated by three fronts (Kinder and Schumacher 1981; Schumacher and Stabeno 1998): coastal, middle shelf, outer shelf, and oceanic. Basic processes influencing fronts are tides, wind, and thermohaline circulations. The role of the tides in generating a coastal domain of uniform vertical distribution of salt and temperature is well understood. The role of tidal motion in sustaining middle shelf, outer shelf, and shelf break circulation needs further investigation. The kinetic energy of the tidal flow over middle and outer shelves is much greater than the kinetic energy of the mean flow. For example, Kinder and Schumacher (1981) showed that at the middle shelf in Bristol Bay the mean flow is about 2 cm/s while tidal currents attain 20 cm/s. Tidal currents can play an especially important role in the material transport between shelf edge and deep ocean. Depending on interaction with bathymetry, both upwelling and downwelling are feasible; therefore these processes either bring nutrients into the surface layer or transport particulate matter from the shelf into the deep ocean. This transport process was investigated by Yanagi et al. (1992) and is called "tidal pump." It can be important in explaining the shelf–deep water interaction in the Bering Sea.

The tides enter the Bering Sea as progressive waves from the North Pacific Ocean, principally through the central and western passages of the Aleutian-Commander Islands. The Arctic Ocean is a secondary source of tides. Therefore, tides in the Bering Sea are considered to be the result of co-oscillation with the Pacific and, to a lesser degree, with the Arctic Ocean. These tides are dominated by four constituents: M_2, N_2, K_1, and O_1. Observations show that over most of the shelf the tides are primarily semidiurnal. The diurnal tides are dominant, however, in proximity to the M_2 amphidromic points, in Norton Sound, at some locations along the shelf break, and near some Aleutian Islands. Pearson et al. (1981) constructed empirical charts for four major tidal constituents (M_2, N_2, K_1, and O_1) over the Bering Sea shelf. Mofjeld (1984, 1986) and Mofjeld et al. (1984) analyzed data for three diurnal components (O_1, P_1, K_1) and three semidiurnal components (M_2, S_2, N_2) over the northeastern Bering shelf and made extensive comparisons against numerical models and tidal theory.

Several numerical models have been applied to Bering Sea tides. They are of two basic types: vertically integrated models and three-dimensional models which simulate vertical structure of tidal currents. The vertically integrated models were applied to the Bering Sea shelf by Hastings (1976) and to the entire Bering Sea by Sunderman (1977). Three-dimensional models were built to simulate the tides in the Alaskan coastal waters by Liu and Leendertse (1979, 1981, 1982, 1990). A direct tidal analysis of the altimetry from the Geosat Exact Repeat Mission was carried out for the Bering Sea and other regional seas by Cartwright et al. (1991). These data have a spatial resolution of about 100 km. Recent world tide models by LeProvost et al. (1994) and Kantha (1995) include the Bering Sea domain with much finer spatial resolution. The model by Kantha (1995) has approximately 22 km resolution and specifically discusses M_2 and K_1 tides in the Bering Sea. The Russian investigations are summarized by Bogdanov et al. (1991).

The difference in the periods between diurnal and semidiurnal tidal waves results in the different dynamics of the tide propagation. Tidal waves can depict several different behaviors. Gravity, Coriolis force, and bathymetry play the major role in tidal dynamics. If gravity is the major restoring force, then the tide wave's local amplification may occur due to resonance in the local embayment adjacent to the open sea basin. The condition for resonance to occur in such a basin is that its length must be a quarter of the wavelength of the oscillation forced from outside. Quite different behavior of the tidal wave can occur under the influence of the Coriolis force, gravity, and bathymetry. Here the resonance condition may lead to so-called topographic waves. Tidal waves propagating into continental slope and shelf areas locally induce topographic waves where the local period of rotational-gravitational mode is equal to the tidal period. Tidally generated topographic waves occur only in the regions poleward from the critical latitude (i.e., the latitude where inertial period is equal to the tidal period). This latitude for the K_1 tide is 30, and for the M_2 tide is 74.5. Therefore, in the Bering Sea domain only diurnal topographic waves can be generated.

The important features of tidal dynamics I will examine and define here are:

(a) Distribution of enhanced currents in the semidiurnal band of periods. These often occur in shallow water, resulting in enhanced local mixing and tidal fronts.

(b) Distribution of the tidal currents and regions of near-resonant trapped tidal waves in the diurnal band of periods. Trapped tidal waves often occur at the edges of continental shelves. Therefore, they are important in the dynamical coupling and exchange of properties between shelf and deep ocean.

(c) Position and patterns of the residual currents. These currents occur permanently along horizontal and vertical directions. Often, along the horizontal direction, the residual motion occurs as clockwise and counterclockwise eddies. Residual currents play a considerable role in both small and mesoscale transport of nutrients and plankton in shallow water and between shelf and deep waters.

Important mechanisms of tidal current enhancement are internal waves and, especially, large amplitude internal wave packets. This field of research is under development in the Bering Sea; a few measurements and three-dimensional models by Liu and Leendertse (1990) indicate that tidal signal is of both barotropic and baroclinic origin.

I will also extend the investigations of Pearson et al. (1981), Mofjeld (1984, 1986), Mofjeld et al. (1984), and Liu and Leendertse (1979, 1981, 1982, 1990) on tides in the eastern part of the Bering Sea to the whole Bering Sea. Further, some new results on tidal current distribution will be given, and the geographic regions of enhanced currents will be delineated.

Tidal Equations and Parameters

Information on tides and tidal currents in the Bering Sea has been given in several papers which describe observations and numerical modeling. This information is mainly related to the Bering Sea shelf. The Bering Sea is too vast to be covered by a dense net of tidal observations and at present the amount of data is not sufficient to draw cotidal charts. On the other hand, the resolution given by the numerical model is sufficient to depict well-resolved tidal charts which can be compared against data. Measurements derived in the Bering Sea provide the possibility to test and validate model performance, but unfortunately, only at a few locations. A numerical model is used here as a unifying tool which will bring all investigations under one umbrella, although the numerical model also has inherent problems influencing its reliability. The main parameters which define a model's performance are resolution (defined by horizontal grid spacing) and bathymetry. I began computation in the entire Bering Sea with a resolution of about 10 km because this resolution is sufficient to describe tidally generated trapped shelf waves (Kowalik and Proshutinsky 1993). Even so, it does not resolve fully the nonlinear tidal interactions. Zimmerman (1978) and Robinson (1981) suggested that the tidal ellipse (tidal excursion) length ought to be resolved for the proper investigation of the nonlinear terms. Often tidal excursion of the water particle is on the order of several kilometers. This limits resolution to several hundred meters. In the Bering Sea high resolution bathymetry is available only around islands and close to the Alaska shoreline. In vast areas of the Bering shelf and in the deep basin, the bathymetry is poorly known.

The model domain includes the Bering Sea and a small portion of the North Pacific, from 48°N to 67°N and from 154°E to 142°W (Fig. 1). The Bering Sea bathymetry structure displays two large domains of importance for the tidal dynamics: deep domain (>3,000 m) located southwest and shallow domain located northwest. The depth distribution for the numerical model is based on an ETOPO5 file with many corrections compiled from the available charts. The compiled depth distribution and the major areas of interest are shown in Fig. 1. To study nonlinear effects, I chose a domain around the Pribilof Islands. Here the fine horizontal grid step will be introduced and afterward, around one of the islands (St. Paul Island), a super-fine grid will be used to obtain results for comparison against measurements (Kowalik and Stabeno submitted). Investigation of the high resolution tides began around St. Paul Island by analyzing the

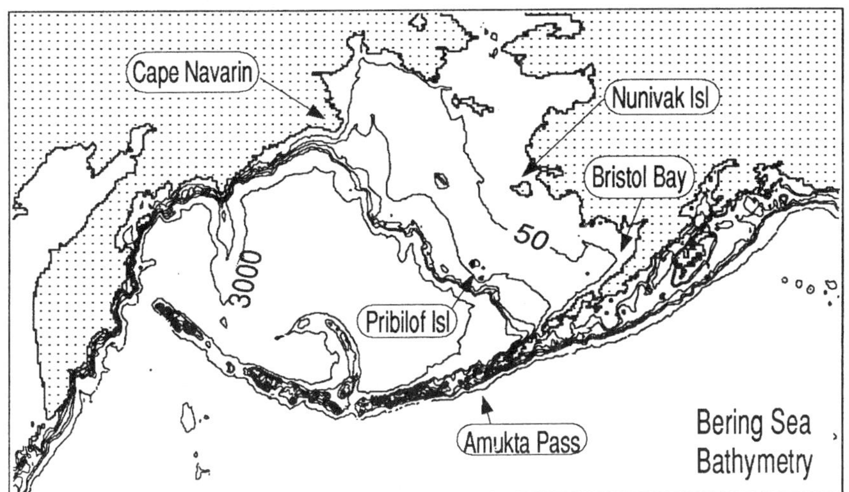

Figure 1. Bering Sea bathymetry.

large-scale model of the entire Bering Sea with resolution of 5 minutes along latitude and 10 minutes along longitude (Fig. 2). At the latitude of 60°N both steps are equal to 9.26 km; therefore I use this distance to describe the large-scale numerical lattice. In the entire Bering Sea a spherical system of coordinates is used. The results of computations from this domain serve for construction of the boundary conditions for the domain around the Pribilof Islands (Fig. 2). The resolution applied in the Pribilof Islands domain is close to 1.852 km and the rectangular system of coordinates is used. I have applied the results of computations from the latter domain to establish boundary conditions for the computations around St. Paul with the grid size of 617 m.

To obtain the distribution of tide amplitudes and phases, I used the vertically integrated equations of motion and continuity in the spherical coordinate system (Gill 1982):

$$\frac{\mathrm{D}u}{\mathrm{D}t} - fv - \frac{uv\sin\phi}{R\cos\phi} = -\frac{g}{R\cos\phi}\frac{\partial}{\partial\lambda}(\alpha\zeta - \beta\zeta_0) - \frac{\tau_\lambda^b}{\rho H} + Au \qquad (1)$$

$$\frac{\mathrm{D}v}{\mathrm{D}t} + fu + \frac{uu\sin\phi}{R\cos\phi} = -\frac{g}{R}\frac{\partial}{\partial\phi}(\alpha\zeta - \beta\zeta_0) - \frac{\tau_\phi^b}{\rho H} + Av \qquad (2)$$

$$\frac{\partial\zeta}{\partial t} + \frac{1}{R\cos\phi}\frac{\partial(Hu)}{\partial\lambda} + \frac{1}{R\cos\phi}\frac{\partial}{\partial\phi}(Hv\cos\phi) = 0 \qquad (3)$$

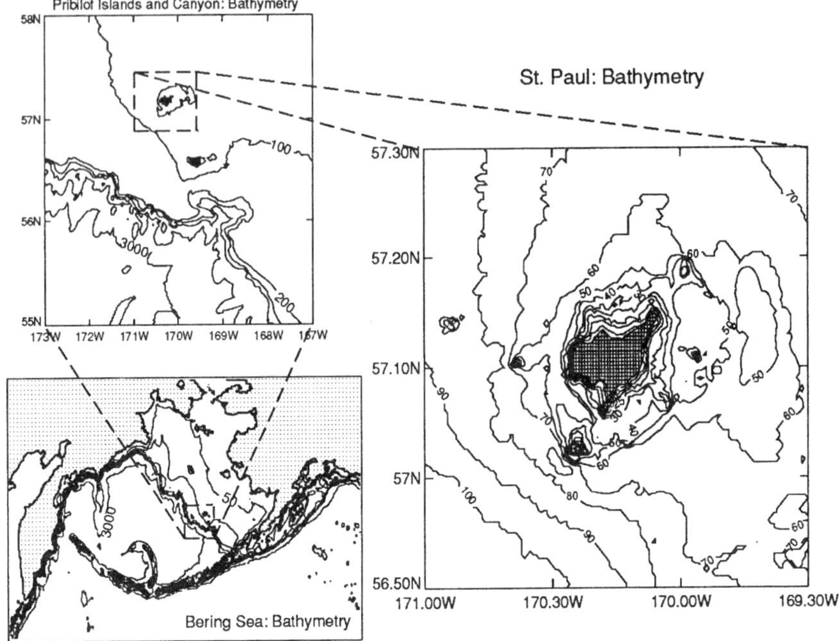

Figure 2.   Computational domains and bottom topography. Bering Sea: grid resolution 9.26 km. Pribilof Islands and Canyon: grid resolution 1.852 km. St. Paul Island: grid resolution 0.617 km.

Here

$$\frac{D}{Dt} = \frac{\partial}{\partial t} + \frac{u}{R\cos\phi}\frac{\partial}{\partial\lambda} + \frac{v}{R}\frac{\partial}{\partial\phi} \tag{4}$$

The advective nonlinear terms in equation (4) are about two orders of magnitude larger than the remaining nonlinear terms in equations (1) and (2); therefore only advective nonlinear terms are taken into account in the ensuing computations.

The operator in the horizontal friction term in equation (1) and (2) is

$$A = N_h\left[\frac{1}{R^2\cos^2\phi}\frac{\partial^2}{\partial\lambda^2} + \frac{1}{R^2\cos\phi}\frac{\partial}{\partial\phi}\left(\cos\phi\frac{\partial}{\partial\phi}\right)\right] \tag{5}$$

The bottom friction components are taken as

$$\tau_\lambda^b = \rho r u\sqrt{u^2+v^2}; \quad \tau_\phi^b = \rho r v\sqrt{u^2+v^2}. \tag{6}$$

In equation (6) r denotes the bottom drag coefficient; it will be taken as $r = 2.6 \times 10^{-3}$.

The following notations have been used in the above equations:

λ, φ = longitude and latitude
t　　= time
ζ　　= free surface elevation
u　　= velocity component along longitude
v　　= velocity component along latitude
ρ　　= water density
N_h　= horizontal eddy viscosity
H　= depth (does not include sea level)
f　　= Coriolis parameter
g　　= gravity acceleration
R　= radius of the earth
ζ_0　= the equilibrium tide
α, β = parameters accounting for tidal potential perturbations.

Seven tidal constituents, namely K_1, O_1, P_1, Q_1, M_2, N_2, and S_2, are included in the computations. Generation of the tidal constituents is done through the open boundaries of the computational domain and through astronomical forcing given in equations (1) and (2). The boundary condition at the coast is defined as no normal flow. This condition is easily implemented through the staggered C (Arakawa) grid. Along the open boundaries, amplitudes and phases for every tidal constituent are specified. This is accomplished through the results obtained by Kantha (1995) and Schwiderski (1979, 1981a-f). The tidal forcing is described in equations (1) and (2) through terms which are multiplied by coefficients α and β. These terms include the tide-generating potential with various corrections due to earth tide and ocean loading (Schwiderski 1979, 1981a-f). Coefficient α defines ocean loading; its value ranges from 0.940 to 0.953 according to Ray and Sanchez (1989). The term $\beta\zeta_0$ includes both the tide-generating potential and correction due to the earth tide. It is usually expressed as (Hendershott 1977)

$$\beta\zeta_0 = (1 + k - h)\zeta_0 \tag{7}$$

Here k and h denote Love numbers, which are equal to 0.302 and 0.602, respectively. These numbers are averaged over all tidal constituents. The equilibrium tide for the diurnal constituents is

$$\zeta_0 = H_n \sin 2\phi \cos(\sigma_n t + \lambda) \tag{8}$$

Here $H_n = 14.25$ cm for K_1 and $H_n = 10.06$ cm for O_1. For the semidiurnal constituents the equilibrium tide is

$$\zeta_0 = H_n \cos^2 \phi \cos(\sigma_n t + 2\lambda) \tag{9}$$

Here H_n =24.23 cm for M_2 and H_n = 4.639 cm for N_2. Hence the tide-generating terms in equation (1) and (2) can be calculated from equations (8) and (9). In the calculation process, along with the refining of the numerical grid, it is necessary to take into account the dependence of the horizontal exchange of momentum on the scale of numerical grid. The computations were carried out with $N_h = 2.0 \times 10^7$, $N_h = 5 \times 10^6$, and $N_h = 5 \times 10^5$ cm²/s for the space grids of 9.26 km, 1.852 km, and 617 m, respectively.

Tides in the Bering Sea

Generation of tides in the models is accomplished by both astronomical forcing and boundary conditions. The astronomical forcing is well defined in equations (1) and (2), but the boundary conditions ought to be interpolated from available data to the fine numerical grid used. At the southern and northern boundaries the amplitudes and phases are taken from Kantha's (1995) and Schwiderski's (1979, 1981a-f) computations and from Cartwright et al. (1991). In the Kantha model a resolution of about 22 km is used. In the Schwiderski model and in satellite-derived data a resolution of 1° is used. Missing values for my numerical grid are interpolated linearly from the Kantha and Schwiderski results. At the northern boundary, which is located close to the Bering Strait, a relatively variable tidal amplitude and phase are observed. Therefore the results derived from ground observations are used together with the model data to establish boundary conditions. The computations of mixed tides (seven waves) were carried out for a 2-month period. After one month the full energy of the system became stationary. During the second month the sea level and velocity were recorded every hour for the harmonic analysis (Foreman 1978). Below, only the four major constituents are discussed—M_2, N_2, K_1, and O_1.

Distribution of co-amplitude (continuous) lines and co-phase (dashed) lines for the principal semidiurnal lunar constituent (M_2) is depicted in Fig. 3. The calculated distribution of amplitudes and phases is similar to the distribution derived through numerical models by Sunderman (1977), Liu and Leendertse (1979, 1981, 1982, 1990), and Kantha (1995), as well as through measurements at the shelf domain by Pearson et al. (1981). Propagation of the M_2 constituent follows bathymetric division of the Bering Sea into the deep basin, where a small change in amplitude and phase occurs, and the shelf basin with many amphidromic points due to partial reflection of the tide wave in the semi-enclosed water bodies. The amplitudes increase in the narrow and shallow bays. The most conspicuous amplitude enhancement, up 150 cm, occurs in triangular-shaped Bristol Bay. All islands, small and large, introduce disturbance into otherwise smooth distribution of amplitude.

The M_2 current ellipses are given in Fig. 4. Ellipses are very suitable for studying behavior of the tidal wave and performing comparison against

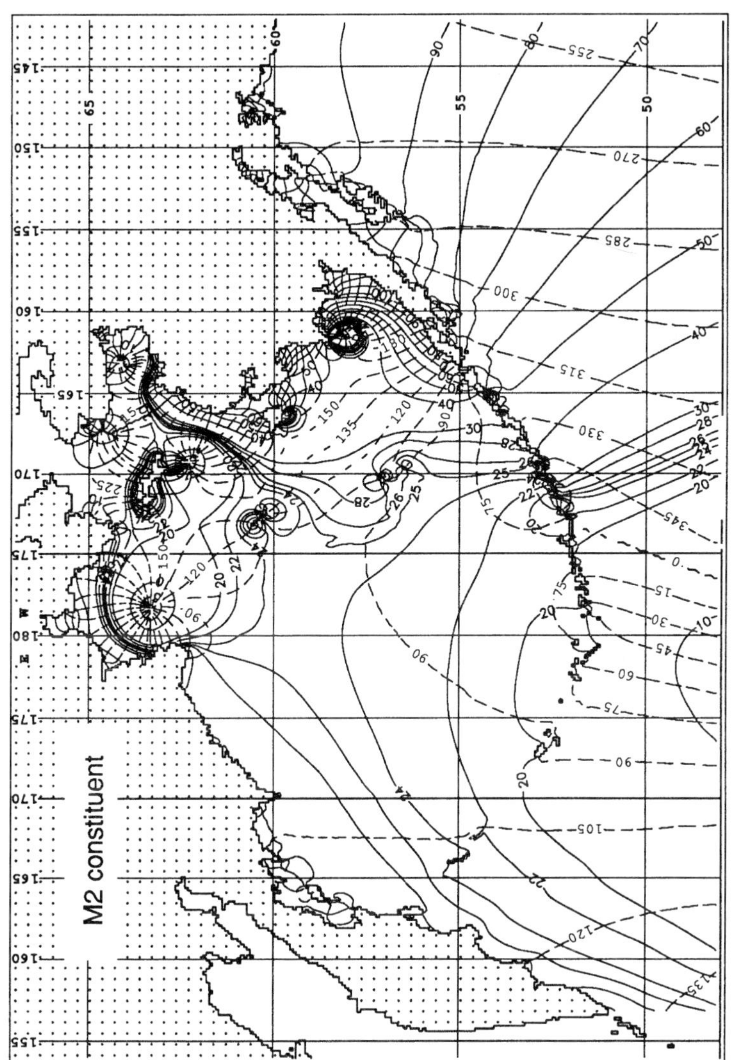

Figure 3. Computed amplitude (solid lines, in centimeters) and phase (dashed lines, in degrees) of surface elevation for the semidiurnal M_2 tide, grid spacing = 9.26 km. The contours for the phase are plotted every 15°. Phase is referred to Greenwich.

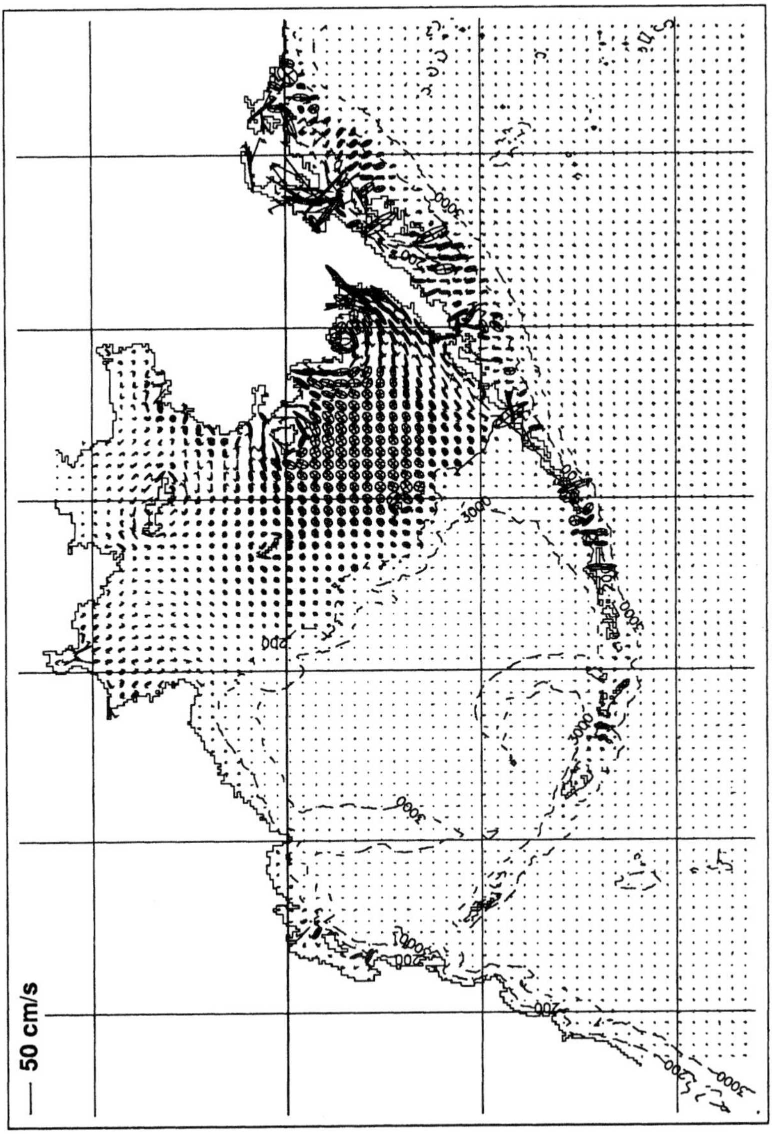

Figure 4.   Tidal current ellipses for M_2 wave in the Bering Sea.

data, but, because in Fig. 4 they are plotted in every fourth point of the numerical grid, it is difficult to follow the local velocity pattern. In the deep basin a velocity on the order of 1-2 cm/s occurs, while at the open shelf the velocity increases to 5-20 cm/s. The different velocity magnitude at the shelf and in the deep Bering Sea is obviously due to bathymetric enhancement of the tidal currents. Strong currents greater than 100 cm/s develop between Aleutian Islands due to narrow passages, and in Bristol Bay due to the elongated triangular shape of the bay. Also, high velocity on the order of 50 cm/s occurs close to the Pribilof Islands, St. Matthew Island, and in the strait dividing Nunivak Island from Alaska (Etolin Strait).

Distribution of co-amplitude and co-phase contours for the semidiurnal elliptical lunar constituent (N_2) is depicted in Fig. 5. This constituent generally repeats both the large and small scale pattern of the M_2 constituent with much smaller amplitudes. Again, the maximum of amplitude, about 70 cm, is located in Bristol Bay. The currents associated with this constituent (not shown) in the deep Bering Basin are very small. Over vast spaces of the shelf, currents are less than 5 cm/s. In shallow bays and in proximity to some islands they may reach 20 cm/s.

Distribution of co-amplitude and co-phase contours for the diurnal lunisolar declination constituent (K_1) is given in Fig. 6. The amplitudes are in centimeters; phase angle is referred to Greenwich and is expressed in degrees. Over the vast regions of the deep basin neither phase angle nor amplitude of the K_1 wave changes significantly. When impinging on a shelf slope region a diurnal tide develops maximum amplitude, approximately 40 cm high, with small local domes over canyons. An unusually broad local maximum of amplitude is located north of the Aleutian Islands between $170°$W and $173°$W (Amukta Pass). The largest amplitude, close to 90 cm, again occurs in the triangular Bristol Bay. As in a semidiurnal band of oscillations, the islands do modify a smooth run of isolines by introducing local perturbations. The calculated distribution of amplitudes and phases is similar to the distribution obtained by Kantha (1995) (the model's resolution was 22 km), and to the charts derived from satellite observations by Cartwright et al. (1991). Over shallow domains, models by Liu and Leendertse (1979, 1981, 1982, 1990) and measurements by Pearson et al. (1981) and Mofjeld (1986) agree well with the results depicted in Fig. 6.

In Fig. 7 the amplitude and phase for the O_1 wave are given. The general pattern of this constituent is close to the pattern of the dominant diurnal wave K_1. Again, across the shelf break from the deep basin toward the shelf the distribution of amplitude depicts the maximum. North of Amukta Pass in the Aleutian Islands the local maximum of the O_1 amplitude occurs in the region of K_1 maximum. The largest amplitudes, close to 60 cm, are located at the head of Bristol Bay. To describe the distribution of currents in the diurnal band of oscillations the maximum current velocity has been depicted in Fig. 8. The upper panel shows the distribution for K_1 tide, and the lower panel for O_1. The major constituent in the diurnal

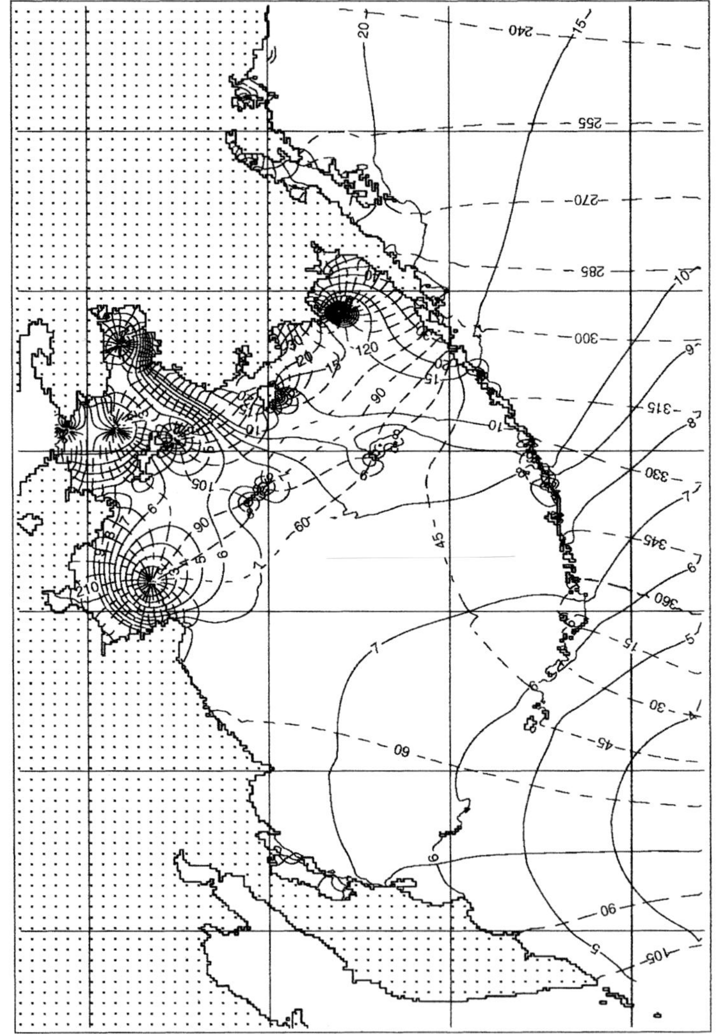

Figure 5. Computed amplitude (solid lines, in centimeters) and phase (dashed lines, in degrees) of surface elevation for the semidiurnal N_2 tide, grid spacing = 9.26 km. The contours for the phase are plotted every 15°. Phase is referred to Greenwich.

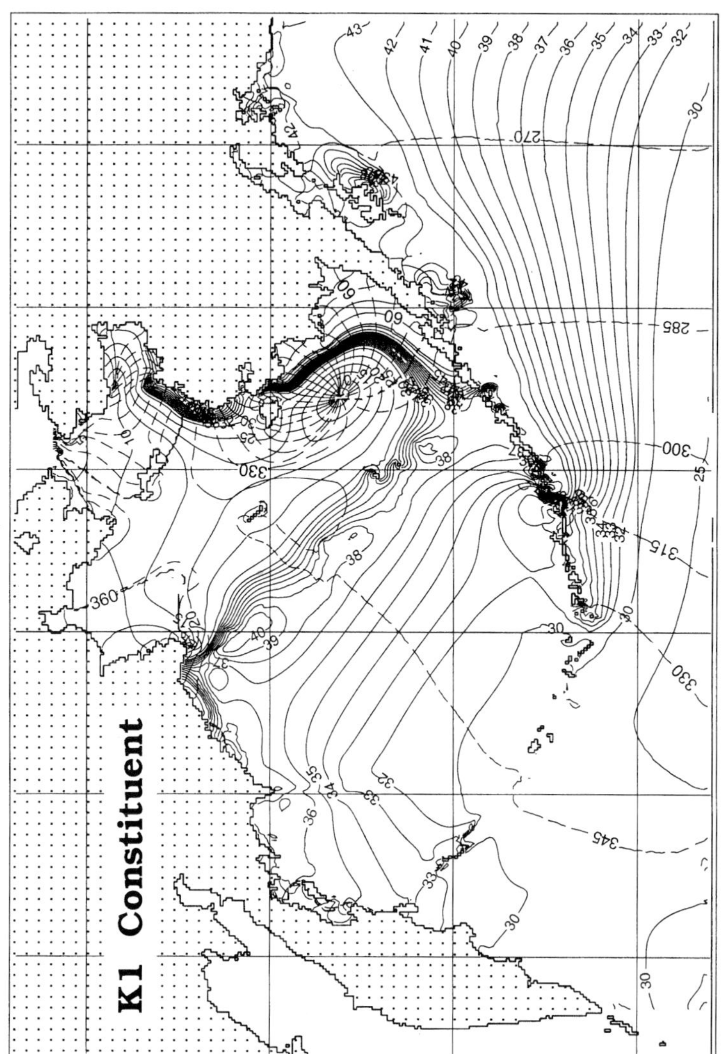

Figure 6. *Computed amplitude (solid lines, in centimeters) and phase (dashed lines, in degrees) of surface elevation for the diurnal K_1 tide, grid spacing = 9.26 km. The contours for the phase are plotted every 15°. Phase is referred to Greenwich.*

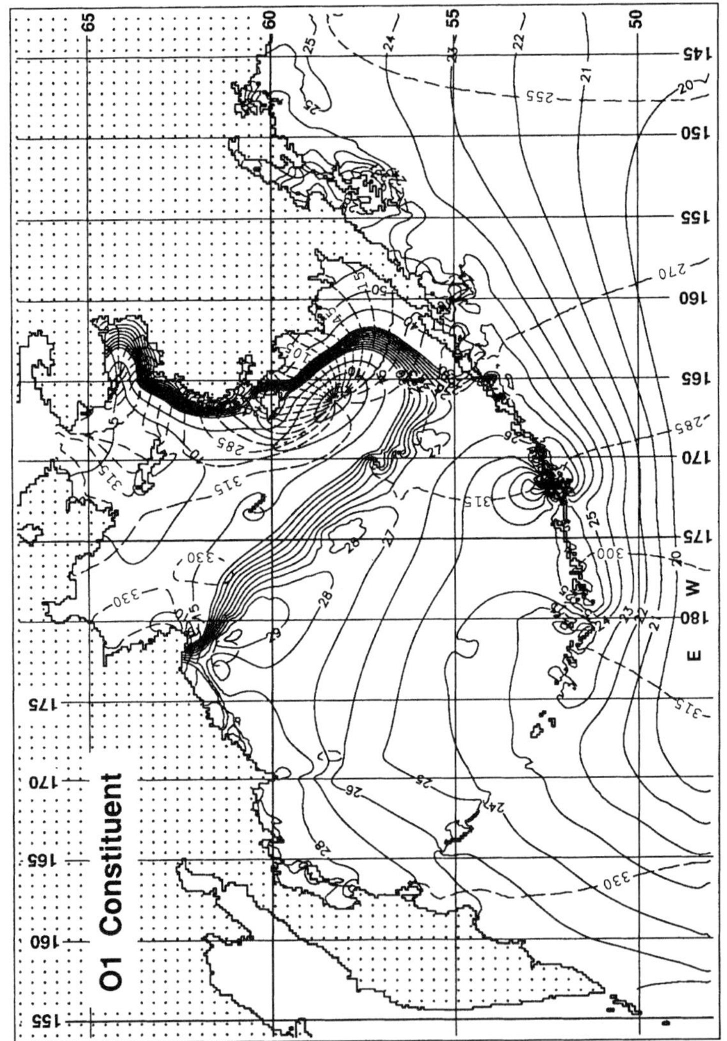

Figure 7. Computed amplitude (solid lines, in centimeters) and phase (dashed lines, in degrees) of surface elevation for the diurnal O_1 tide, grid spacing = 9.26 km. The contours for the phase are plotted every 15°. Phase is referred to Greenwich.

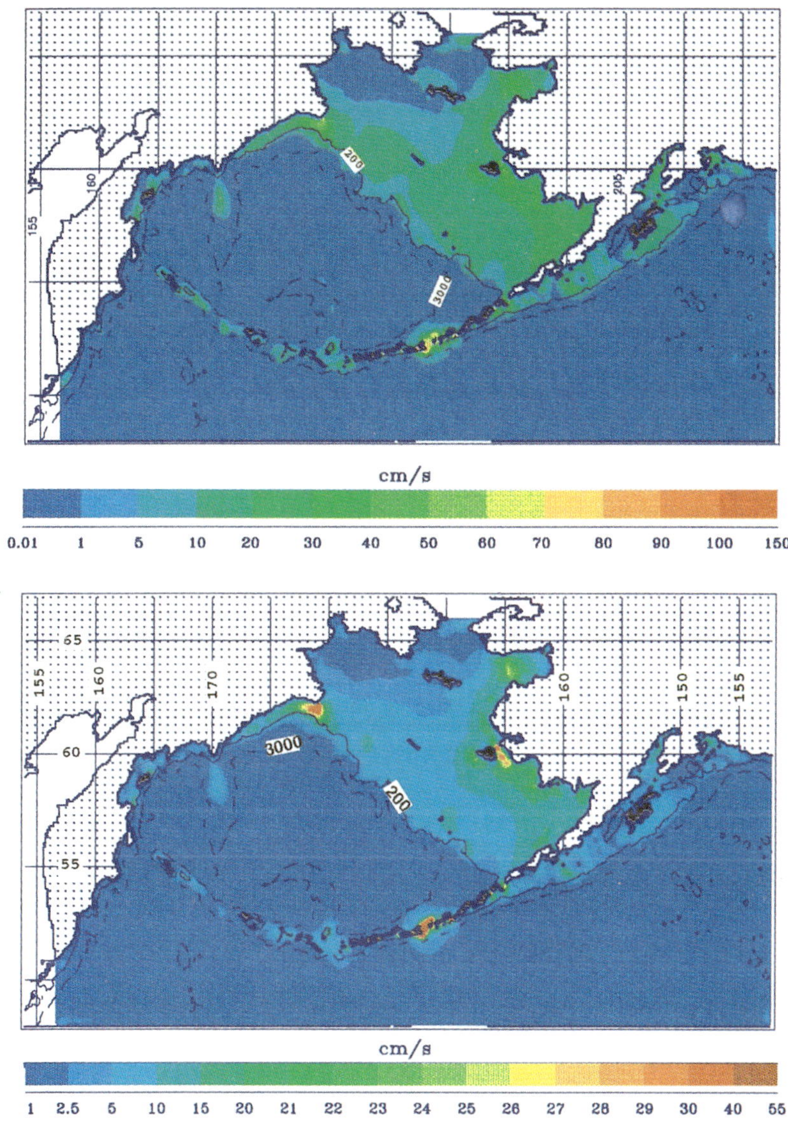

Figure 8. Maximum tidal currents, grid spacing = 9.26 km. Upper panel: K_1 wave, lower panel: O_1 wave.

band K_1, when compared against M_2, shows a stronger tendency for inter-action with the bathymetry. Over the deep Bering Sea basin at the Shir-shov Ridge, where the M_2 tide generated currents close to 5 cm/s, the K_1 current is enhanced up to 10 cm/s. The strongest currents on the order of 100 cm/s occur in Amukta Pass. Currents of about 50 cm/s are located south from Cape Navarin and in Etolin Strait. The maximum velocity for the O_1 tide is shown in Fig. 8 lower panel. This constituent depicts even more strongly the three dominant regions of current amplification: off Cape Navarin, Amukta Pass, and Etolin Strait. The largest currents are close to 50 cm/s. The resolution of 9.26 km used for construction of the above charts for maximum velocity is too crude to show all local regions of enhancements in proximity to the islands and in the narrow straits.

In the ensuing discussion I will refer to M_2 and K_1 as representatives of semidiurnal and diurnal bands of tidal oscillations. Upon comparing the model results for the amplitude of the M_2 (Fig. 3) and K_1 (Fig. 6) waves, one striking difference was noticed in the region of the shelf break in both the Bering Sea and in the North Pacific south of the Aleutian Islands. The am-plitude of the M_2 wave at the shelf break region remains practically un-changed, whereas the amplitude of the K_1 wave is enhanced over the same region. South of the Aleutian Islands the shelf is narrow so the amplifica-tion is less pronounced. In the Bering Sea the K_1 wave exhibits maximum amplitude over the shelf break with many local maxima. Occurrence of the local maxima in proximity to the shelf break in the diurnal band of oscillations suggests the presence of trapped or partially trapped diurnal shelf waves (Kowalik and Proshutinsky 1993). The above charts, as well as data presented by Mofjeld (1986) and computations of Liu and Leendertse (1990) over the shelf domain, and Kantha's (1995) world tidal model, con-firm existence of the maximum of the amplitude in the diurnal domain over the Bering Sea slope.

Velocity patterns of diurnal and semidiurnal waves are also quite dif-ferent. While the velocity of M_2 (Fig. 4) tends to be enhanced in shallow water and around islands, the velocity of K_1 (Fig. 8, upper panel) in addi-tion to the shallow water enhancement, is amplified at the shelf break depicting local maxima in proximity to the local maxima of sea level. Struc-tures of enhanced tidal currents for the semidiurnal and diurnal constitu-ents are often generated due to topographic amplification over shallow areas. An example of such amplification is shallow, triangular-shaped Bristol Bay (Figs. 4 and 8) and Etolin Strait. In the diurnal band of oscillations the maximum current is also associated with the shelf wave regions. Struc-tures of enhanced tidal currents are generated by near-resonant shelf waves of tidal origin which are trapped, or partially trapped, over the bottom topography. Resonant transfer of energy from tidal wave to shelf wave occurs when the period of rotational mode of oscillation is close to the tidal period. Maximum velocity for K_1, as well as for O_1 (Fig. 8), reveals two regions of enhanced currents in the Aleutian Islands and off Cape Navar-in. These regions can also be identified in the sea level charts as locations of the maxima in the sea level distribution. A number of smaller trapped waves can be identified along the Bering Sea shelf slope through the local maximum of the sea level (Fig. 6) and through the local maximum of the velocity (Fig. 8).

It is possible to delineate three modes of near-resonant interaction of diurnal tidal waves with bottom topography in the Bering Sea. The first mode exists in the Aleutian Islands region (Figs. 6 and 8) and most probably is caused by generation of resonance oscillation around islands (Longuet-Higgins 1969) or seamounts (Hunkins 1986, Chapman 1989, Haidvogel et al. 1993, Kowalik 1994). The second mode of oscillation occurs off Cape Navarin (Figs. 6 and 8) and is related to the abrupt changes in the shelf width along the coast. A shelf wave which enters such a domain is trapped or arrested and, if resonance conditions occur, local enhancement of the current follows (Kowalik and Proshutinsky 1993). The third mode of trapped oscillation occurs along the shelf break at the canyons. In ensuing investigations we shall study tides in the Pribilof Canyon, where measurements taken by Schumacher and Reed (1992) indicate strong diurnal currents. The horizontal scale of the trapped diurnal waves in the Bering Sea ranges from 20 km to 100 km. The tidal charts presented above are computed for the ice-free season. The pack ice usually covers the Bering Sea shelf with the ice boundary running along the shelf break from Cape Navarin to the Pribilof Islands. Along with the pack ice, some nearshore areas (especially in Norton Sound) are covered by shorefast ice. Mofjeld's (1986) investigations of the long series of currents and sea level data from the northeastern shelf of the Bering Sea shed some light on variation in the tidal harmonic constants. These variations are probably due to pack ice. One long series of sea level data from under the shorefast ice was analyzed by Johnson and Kowalik (1986). The physics of the ice-tide interaction is not well understood. The sea ice generally dumps shorter wave oscillations in the water column which results in less turbulence and smaller vertical and horizontal eddy viscosity coefficients. Due to ice cover an additional drag force occurs (similar to the bottom drag) between ice and water. In addition, fast ice can change the geometry of the flow by changing depth or channel width. Canadian observation from Tuktoyaktuk revealed strong dependence of the tide on the ice cover (Murty 1985). The M_2 constituent changed amplitude from 16 cm during summer to 11 cm during the winter. The nonlinear constituent M_4 increased up to 100% during the winter. The observations on ice drift and deformation in the Bering Sea indicate possible strong nonlinear ice-water coupling in the tidal band of oscillations since the M_4 tidal component of the ice velocity is relatively much stronger than the M_4 in the ocean currents (Pease and Turet 1989). Data described by Mofjeld (1986) from the northwestern shelf show that the tidal harmonic constants change seasonally. The diurnal amplitude increased and semidiurnal amplitude decreased during winter. The increase and decrease amounted to a few percent of the mean values. The maximum change in amplitude is close to 3 cm and the maximum change in phase is up to $15°$ for the M_2 tide in the Bering Strait. The relatively high changes in amplitude (37% of the mean amplitude) at this site possibly express damping of the tides by the sea ice in the Arctic Ocean (Kowalik 1981).

Johnson and Kowalik (1986) analyzed a series of sea level data taken at Stebbins ($\lambda = 162°20'W$, $\varphi = 63°30'N$), Norton Sound, in 1982, both under the fast ice and during summer. This region is dominated by diurnal tides (K_1 amplitude is 47.4 cm during summer). The harmonic analysis showed that the amplitudes of the main constituents, K_1, O_1, and M_2, increase from winter to summer by about 40%. The largest phase change ($75°$) occurred for the M_2 constituent.

Pribilof Islands and Canyon

The numerical lattice of 9.26 km, used for the whole Bering Sea, gives good resolution of the basic features of the tidal waves—including the identification of the major topographically trapped waves. A smaller scale is required for investigation of the tidal dynamics around islands, in the straits and at the shelf break, where interaction of the tide with the bottom slope changes tide dynamics. The region of the Pribilof Islands and Canyon (PIC) has been chosen for ensuing investigation because in this region the data are available for comparison against models. Generally, tides dominate the dynamics around the Pribilof Islands (Pearson et al. 1981). The interaction of tides and currents with bottom topography around islands, seamounts, and canyons often leads to current enhancement and, as a consequence, the nonlinear interactions become stronger in such regions (Haidvogel et al. 1993, Brink 1995). Enhanced tidal current causes stronger mixing along vertical directions; such mixing may erase vertical stratification and build a local tidal front. This process occurs in proximity to the Pribilof Islands. Both islands are surrounded by structure fronts that divide the well mixed inner shelf waters (0-40 m) from the two layers characteristic of the middle shelf (50-100 m). The tidal currents and the vertical mixing around islands can be important mechanisms in the transporting and concentration of nutrients, plankton, and zooplankton. Coyle et al. (1992) described such a mechanism through investigation of tidal currents around St. George Island and their effects on transport of zooplankton and adjustment of the feeding behavior of birds to the tidal cycle.

Satellite-tracked drifting buoys deployed on the Bering Sea southern shelf were advected northwestward to the vicinity of the Pribilof Islands (Kowalik and Stabeno, submitted; Stabeno et al., chapter 1, this volume). In this advective motion tidal loops are also evident. After reaching the Pribilof Islands, the buoys were trapped in a clockwise motion around the islands and then transported southward. A series of current meters deployed for several months around St. Paul Island revealed that the largest M_2 tide currents (greater than 70 cm/s) occur at the southern tip of St. Paul Island, where total depth is close to 25 m (Kowalik and Stabeno, submitted;

Stabeno et al., chapter 1, this volume). This circulation pattern, which results in a trapped flow around the islands, can play an important role as a dynamic pathway for the transport and trapping of nutrients, plankton, and fish larvae in the vicinity of the Pribilof Islands. How this circulation traps foreign material near the islands can affect the productivity of these waters.

A high-resolution numerical grid with a space step of 1 nautical mile (1.852 km) in the region of PIC (Fig. 2) is used to study tidal motion around the islands and at the canyon. The results from the Bering Sea model serve as a boundary condition for the fine-scale model in the PIC domain (Fig. 2). The sea level calculated in the Bering Sea model is used to define elevations along the open boundaries of the PIC. The fine grid in the PIC domain is connected to the coarse grid through the linear interpolation of the coarse grid data.

A tide wave enters the Bering Sea as a progressive wave from the North Pacific through Aleutian Island passages (Pearson et al. 1981). The PIC domain is located both in the deep Bering Basin and at the shelf. Here the tides propagate from the deep basin into the shallow domain. Over the latter domain both amplitude and currents are strongly enhanced. To identify differences in semidiurnal and diurnal tidal waves, results of calculations for the M_2 and K_1 waves are considered. In Fig. 9, upper panel, amplitude of surface elevation for the M_2 tide is given. Sea level amplitude slowly changes from 24 cm in the deep basin to 31 cm over the open shelf. An interesting pattern in the sea level occurs around the Pribilof Islands. The sea level depicts a dipole structure with minimum located at the southeastern shores and maximum at northern shores. Especially conspicuous is the 15 cm sea level change around St. Paul Island. (The rapid change of the sea level around this island refutes a general assumption that islands are good platforms for sea level measurements.)

Figure 9, lower panel, depicts the maximum of M_2 tidal currents. The deep basin's sluggish flow on the order of 1 cm/s is enhanced over the shelf break to 5-10 cm/s. The isolines of the current magnitudes actually tend to repeat bathymetry contours. This is well depicted in Pribilof Canyon. Around the Pribilof Islands, due to topographic amplification and nonlinear interactions in the shallow water, currents up to 50-70 cm/s are generated. The driving mechanism which generates this trapped motion around the islands is a dipole structure of the sea level shown in Fig. 9, upper panel. The strength of the tidal stream will be proportional to the sea level difference in the dipole structure.

The tidal dynamics in the diurnal band of oscillations is illustrated through the major K_1 constituent. Figure 10, upper panel, depicts a general pattern of slowly varying sea level from southwest to northeast with the maximum of amplitude, above 39 cm, along the shelf slope. A number of local sea level maxima can be discerned not only in the proximity of the islands but also along the shelf slope close to the 200 m depth contour. Especially pronounced is the maximum at the flank of the Pribilof Canyon.

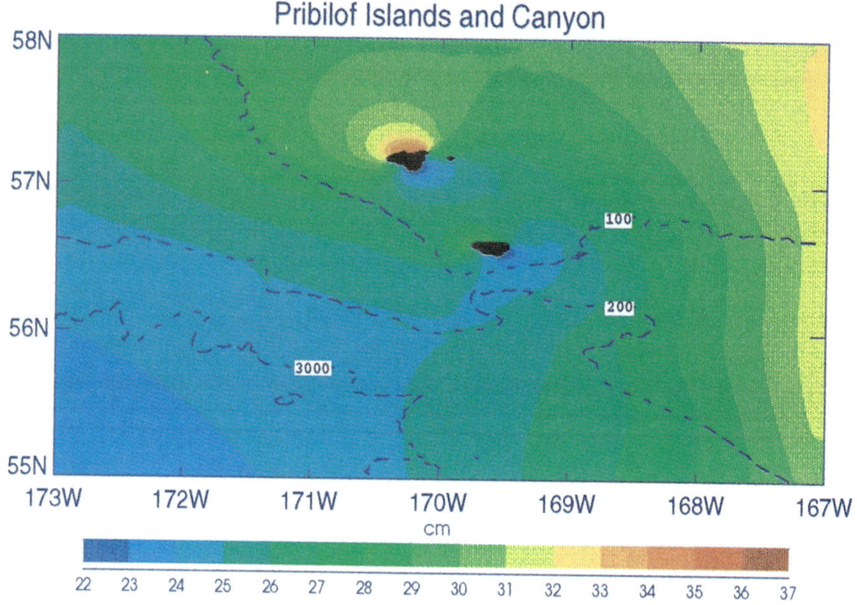

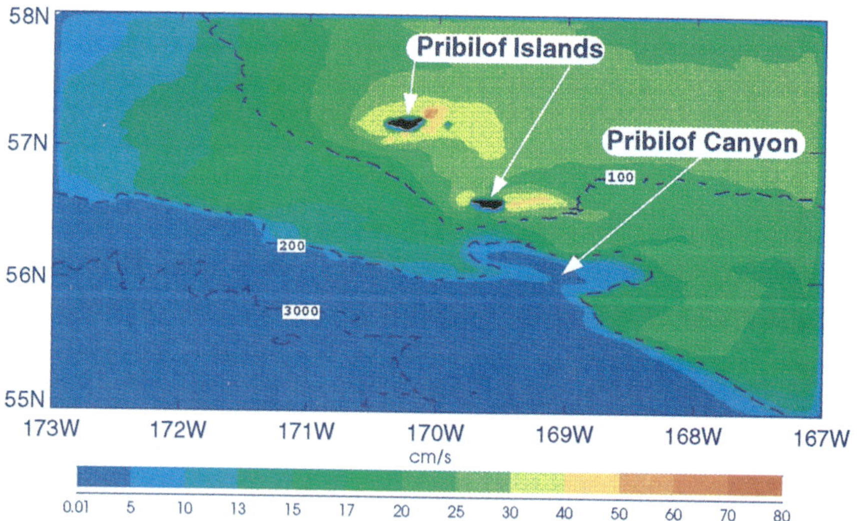

Figure 9.   Pribilof Islands and Canyon. Space resolution 1.852 km. Dashed lines denote bathymetry in meters. Upper panel: computed amplitude of surface elevation for the semidiurnal M_2 tide. Lower panel: maximum tidal currents of the M_2 tide.

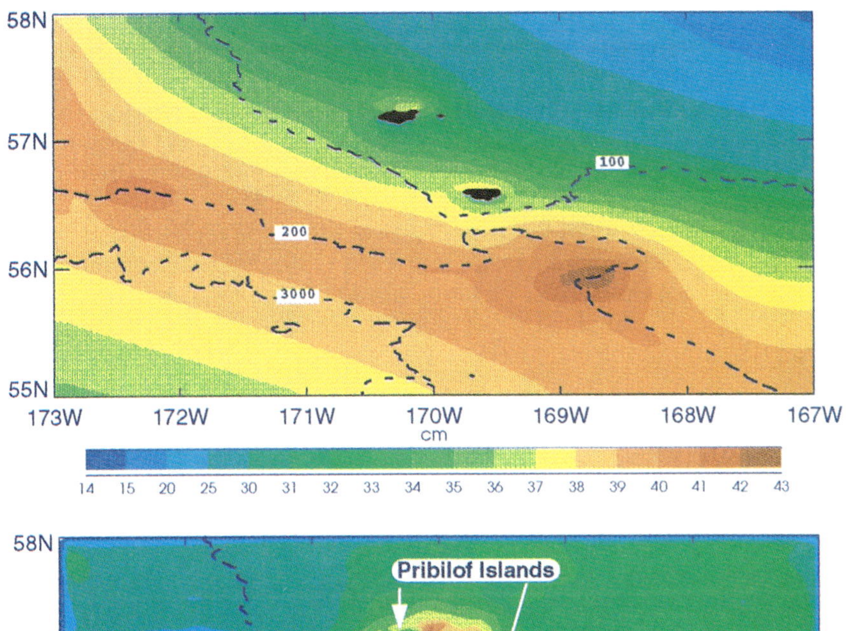

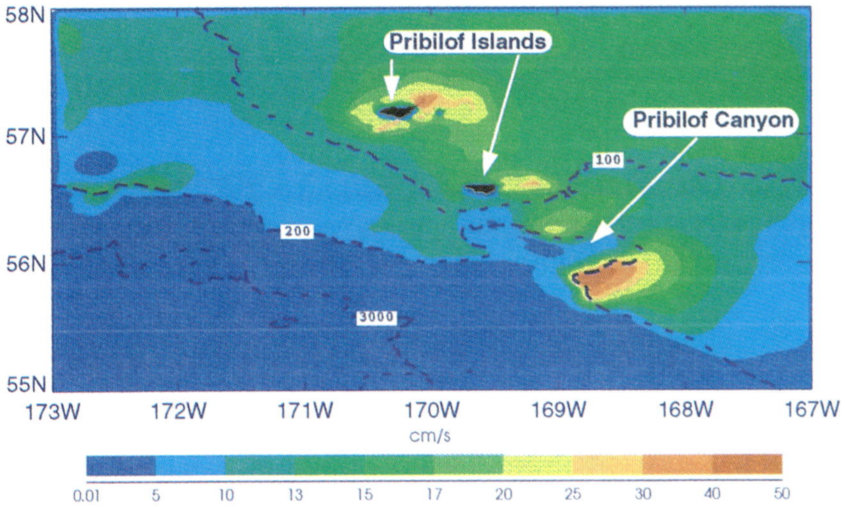

Figure 10. Pribilof Islands and Canyon. Space resolution 1.852 km. Dashed lines denote bathymetry in meters. Upper panel: computed amplitude of surface elevation for the diurnal K_1 tide. Lower panel: maximum tidal currents of the K_1 tide.

The K_1 maximum currents are shown in Fig. 10, lower panel. Although in the large scale pattern one can see the enhancement of current from the deep basin to the shelf and enhancements in the shallow waters in the islands' proximity, the complete picture is more complicated than for the M_2 tide. The enhancement of currents around islands is not the sole feature, because along the shelf slope in the domain of Pribilof Canyon, the new regions of the local enhancement have been generated. At the mid-slope location in Pribilof Canyon (272 m depth), long-term current meter records analyzed by Schumacher and Reed (1992) corroborate this local enhancement of the diurnal currents. The local regions of enhanced diurnal currents are related to the sea level distribution given in Fig. 10, upper panel. The size of the enhancement domain varies from about 20 km (Fig. 10, lower panel, northern flank of the Pribilof Canyon) to 50 km at the eastern flank of the canyon and to about 80 km at the Aleutian Islands (Fig. 8, upper panel).

The important conclusion is that the diurnal tide can generate enhanced currents not only in the shallow domains around islands but in the deeper domains in the shelf slope region as well. Again, this illustrates well the importance of the shelf waves in the cross-shelf exchange and suggests the possibility that along the shelf break north from the PIC region similar domains may exist.

The above computations were extended by Kowalik and Stabeno (submitted) into the local domain around St. Paul Island by application of a high-resolution 617 m numerical grid. The model data were used for comparison against current meter data. Several current meters were deployed in the vicinity of St. Paul Island from 13 September 1995 until 7 August 1996. The computed and observed currents were compared through the various ellipse parameters. The general pattern of the tidal motion deduced from measurements and model revealed that the largest tidal currents occurred at the southern tip of St. Paul Island, and that directions of the major axes of the tidal ellipses follow the equal bathymetry contours and are often parallel to the coastlines. The tidal current at the measuring site with the strongest flow ($\lambda = 170°17.49'W$, $\varphi = 57°5.36'N$) showed the following values for the major axes of the tidal ellipse: M_2—71.6 cm/s, N_2—21.3 cm/s, K_1—19.8 cm/s, O_1—11.7 cm/s. The M_2 tide dominates the tidal current field around the island. A glance at Figs. 9 and 10 reveals that the amplitude of diurnal and semidiurnal constituents is close to 30 cm. The difference is due to strength of the dipole in proximity to the island. The M_2 tide shows much stronger variations in the sea level distribution around the island.

Residual Tidal Currents

The nonlinear interactions of the tidal currents result in the new dynamics, which often differs from linear dynamics. The energy of the basic tidal constituents is transferred toward shorter and longer periods and the tidal currents averaged over tidal period generate steady (residual) current. Observations on generation of the new periods are scarce, but those that are available do confirm importance of the fortnightly periodicity in the Bering Sea dynamics (Schumacher and Reed 1992). Application of the three-dimensional model to the nonlinear interaction of the tides by Liu and Leendertse (1990) suggests that the generation of the new periods is more effective above the bottom boundary layer.

Residual currents can be an important mechanism contributing to the mean flow. These currents are generated as a result of rectification process as the oscillating velocity interacts either with the sea surface or with the bottom topography. The interaction of tidal currents with bottom topography usually dominates the rectification process. In the process of generation of residual flow a significant role is played by advective terms in the equation of motion. Several attempts have been made to define residual circulation in the Bering Sea. Mofjeld et al. (1984) presented estimates of residual currents by fitting simple tidal waves to the observed tidal currents from two stations at the Bering Sea shelf. Estimated residual currents were smaller than 1 cm/s. Schumacher and Kinder (1983) arrived at similar residual velocity over the open Bering Sea shelf (far from islands and coastline). Liu and Leendertse (1990) demonstrated, through various shelf models, the importance of residual currents in Bering Sea circulation. In their model a resolution of about 20 km was used to investigate residual tidal currents of both barotropic and baroclinic origin. Close to the Bering Sea slope the model depicted residual currents from 2 to 5 cm/s. Recent long series of measurements by Stabeno (Kowalik and Stabeno submitted) in the shallow water around Pribilof Islands revealed mean residual tidal current on the order of 15 cm/s.

In numerical computation the residual velocities for the several constituents interacting together have been obtained by averaging an hourly time series over a period of 29 days. Even computations carried out with a numerical lattice of 9.26 km in the entire Bering Sea show quite developed residual motion. Most interesting is residual current along the Bering Sea shelf break. The local residual currents generated in proximity to the shelf break are interconnected, resulting in a single flow-through along the Bering Sea shelf slope from southeast to northwest. The average velocity of this current is less than 1 cm/s. This result suggests that some of the residual currents investigated by Liu and Leendertse (1990) are barotropically driven. Strong tidal currents, through nonlinear interactions, generate quite strong residual currents at two regions where the wide Bering Sea shelf changes into the narrow shelf (i.e., close to Cape Navarin and in proximity to the Aleutian Islands). Residual tidal currents in the Aleutian and Pribilof Islands regions are shown in Fig. 11, lower panel. The maximum residual current in the passages between the Aleutian Islands is about 15 cm/s. The mean flow should play an important role in exchange of properties between the north Pacific and the Bering Sea. Around numerous Aleutian islands and around the Pribilof Islands, residual currents

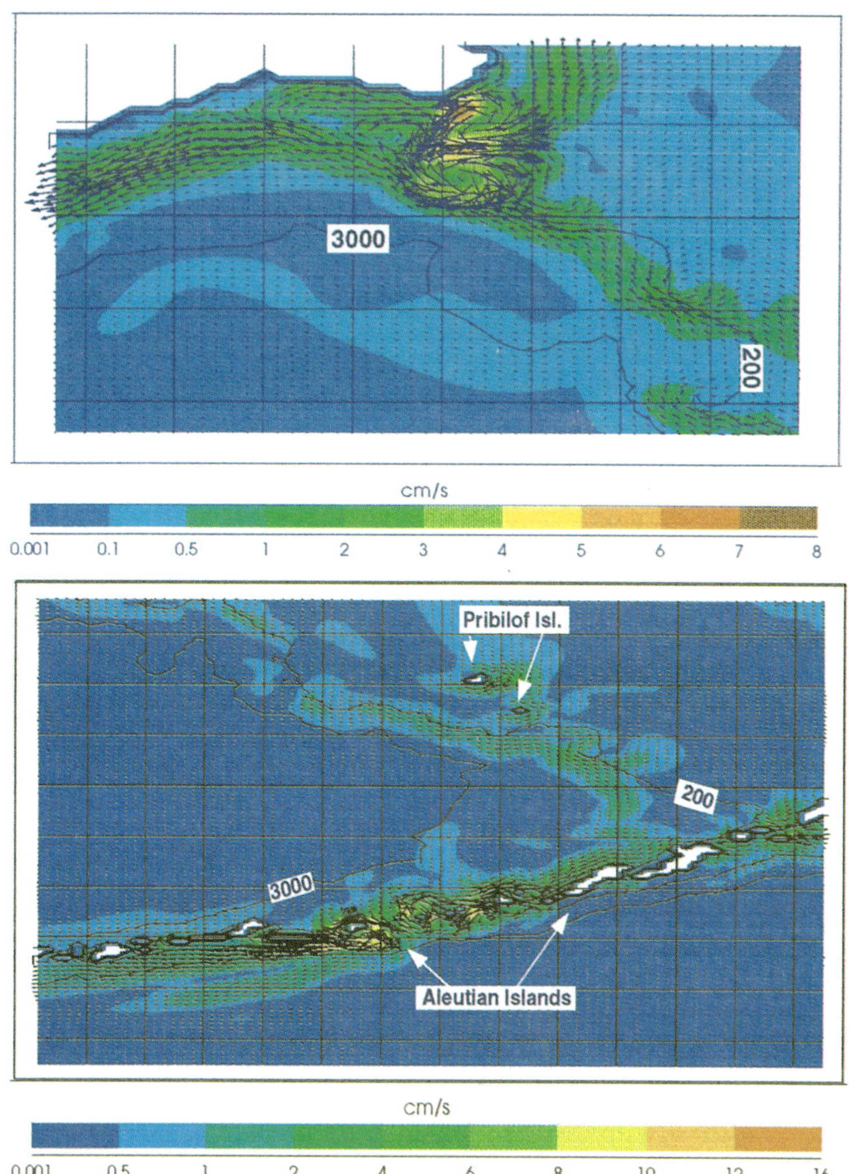

Figure 11. Residual tidal currents. Seven major constituents are considered, grid spacing = 9.26 km. Upper panel: off Cape Navarin. Lower panel: Aleutian Islands and Bering Sea slope.

render trapped clockwise circulation. In Amukta Pass such a circulation will result in an inflow to the Bering Sea on the eastern side of the pass and an outflow on the western side of the pass. This tidally driven circulation pattern agrees well with the general circulation pattern described in Amukta Pass by Reed and Stabeno (1997). Similar circulation was observed by Reed et al. (1993) in Near Strait and Amchitka Pass. Residual currents off Cape Navarin are given in Fig. 11, upper panel. A pair of eddies rotating in opposite directions, seen in this figure, is not confined to the shelf domain; this is probably an indication that these eddies play an important role in the exchange of properties between shelf and deep ocean.

Horizontal resolution of 9.26 km is too large to properly resolve all nonlinear interactions, but since the entire Bering Sea is being investigated the results quickly delineate major domains of interaction. A higher-resolution numerical grid with a space step of 1.852 km has been located in the PIC region to estimate residual flow both around the islands due to topographical enhancement of the tidal currents in the shallow water and at the canyon due to trapped shelf waves. Residual tidal motion calculated from a 9.26 km grid model for the whole Bering Sea depicts, in the shallow areas around the Pribilof Islands, quite a small velocity on the order of 1-2 cm/s. This small velocity exactly delineates the permanent clockwise eddy around the islands (see Fig. 11, lower panel). The mean currents depict about a 2.5-fold increase when grid spacing is diminished from 9.26 km to 1.852 km.

The horizontal resolution has been further increased around St. Paul Island to 617 m (see Fig. 2). This resolution is close to the grid distance recommended by Zimmerman (1978) for reproduction of the bottom topography with significant production of vorticity and nonlinearity. Calculations with this fine grid in the St. Paul subdomain confirm general clockwise circulation and the average and maximum residual velocities have grown considerably as can be seen in Fig. 12, upper panel. Maximum residual velocity has changed from approximately 10 cm/s in a 1.852 km grid up to 30 cm/s in a 617 m grid. In the island's vicinity the residual circulation is small and tends to be organized in local coastal eddies. At a distance of about 1-3 km from St. Paul Island the residual velocity attains greatest values and circles the island in continuous fashion. The average residual currents in this region are 10 to 15 cm/s, a considerable increase when this result is compared with 1-2 cm/s obtained in the 9.26 km grid. Additional results obtained for the M_2 constituent computed alone without interaction with the remaining tidal constituents show that this constituent is primarily responsible for the generation of residual currents around St. Paul Island.

All current meters deployed in the vicinity of St. Paul Island during 1995-1997, except for site one, were deployed close to the 25 m isobath, 5 m above the bottom. At mooring site 1, two current meters were deployed in a water depth of 65 m. The residual flow around the island for the whole measuring period depicts stable clockwise flow in the range 4

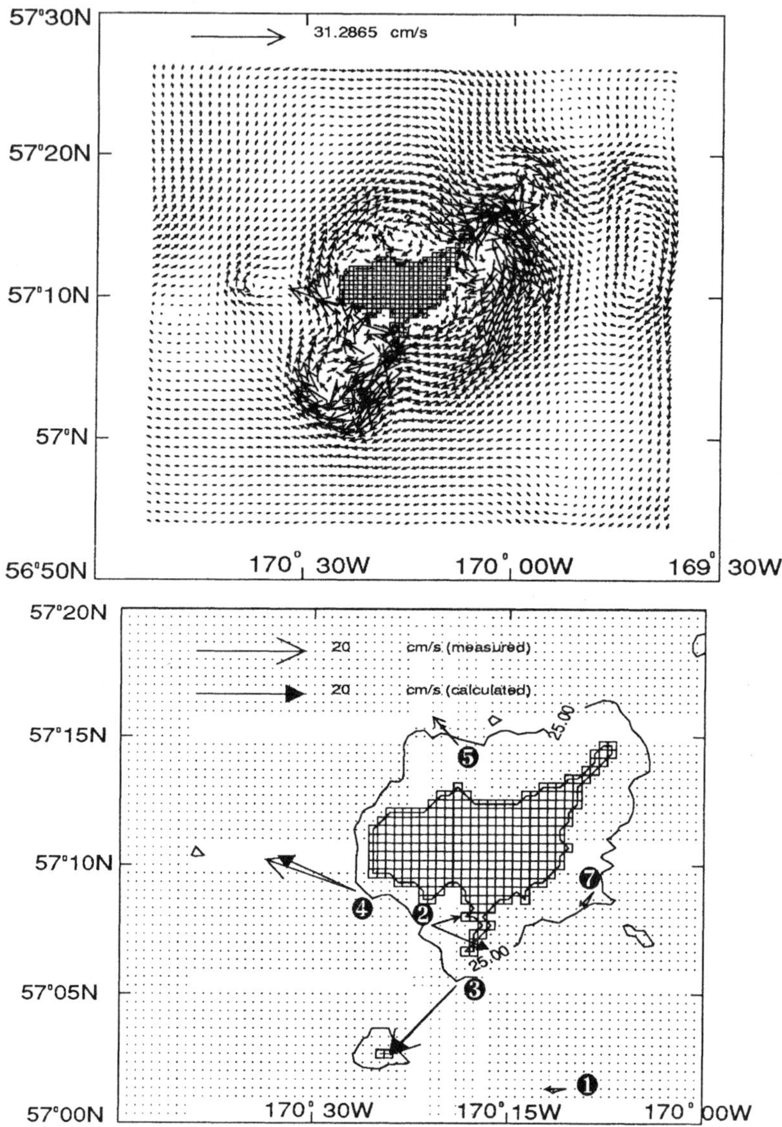

Figure 12. Residual tidal currents around St. Paul Island. Four major constituents are considered, grid spacing = 617 m. Upper panel: distribution derived from computation. Lower panel: measurements against model at the measuring sites.

to 18 cm/s. Stability in direction is quite amazing: flow reversal occurred only on 10 February. Circular currents around the island can be forced by density gradients or wind, but none of these mechanisms can depict stable behavior through the various seasons. The only permanent mechanism is rectifying the tidal currents to the low frequency currents.

A comparison of the measured residual currents with the currents obtained from the above computation is given in Fig. 12, lower panel. Computed current magnitude and direction turned out to be in good agreement with the current data obtained from the observations. The most interesting is current pattern in points 2 and 5. There, due to the local interaction with the bottom slope and coastline, the measured currents have been deflected from the general clockwise circulation. The local coastal eddies rendered through the fine resolution model (Fig. 12, upper panel) are responsible for this deflected flow pattern.

Phenomena of the tidally trapped motion and coastal eddies observed and measured around the Pribilof Islands probably occur in proximity to many islands in the Bering Sea. One confirmation of the tidally generated eddies is given by Ahlnas (1994) through satellite observations. The Synthetic Aperture Radar image showed multiple dipole eddies in the strait between St. Matthew Island (60°30′N, 172-173°W) and Hall Island. The eddies seem to be permanent features which concentrate nutrients, lead to considerable biological productivity, and sustain large bird populations in this area.

Model Versus Measurements

The model results can be verified against various observations. Sea level observations are available from more than 50 stations in the Bering Sea, including a few from Russian coastal observations at Provideniya, Anadyr, Petropavlovsk, Port Siber, and Bering Island. The set of data obtained by Pearson et al. (1981), Mofjeld (1984, 1986) and Mofjeld et al. (1984) covers the Bering Sea shelf. An additional source of data is Geosat altimetry measurements (Cartwright et al. 1991). The satellite data are quite smooth but very close to the quality of the data set derived from a numerical model by Schwiderski with resolution of 1° (R.D. Ray, NASA Goddard Space Fight Center, pers. comm., January 1998).

Current meter data from more than 30 locations are available. These data were acquired through the efforts of Pearson et al. (1981), Mofjeld (1986), Schumacher and Reed (1992), and Kowalik and Stabeno (submitted).

In a series of papers related to the Bering Sea shelf Mofjeld (1984, 1986) reported tidal harmonics for sea level and currents, together with comparisons with numerical models. The computed elevations were in satisfactory agreement with those obtained from observations. Especially good similarity was obtained by Liu and Leendertse (1979, 1990) through application of three-dimensional models. When these models were driven

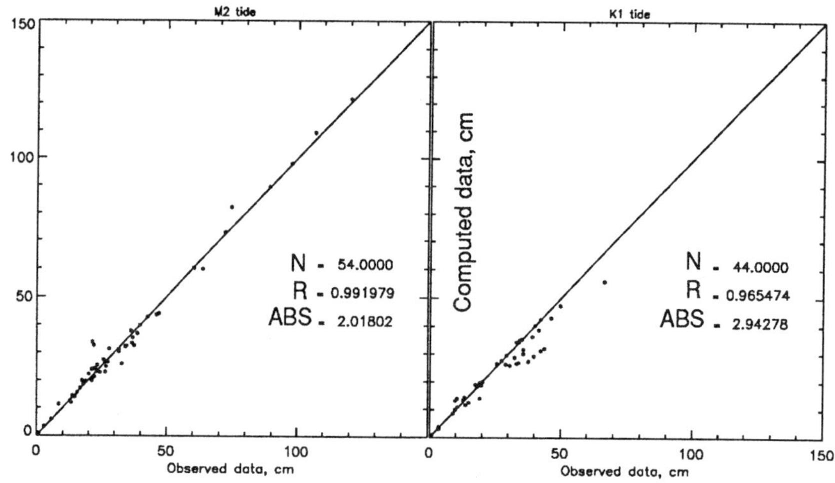

Figure 13. Computed versus observed sea level elevations in the Bering Sea, for the M_2 (left) and for the K_1 (right). N denotes number of stations, R coefficient of correlation, ABS absolute error between observed and calculated elevation in centimeters.

with the sea level forcing due to tides, the computed currents agreed well with the observed currents.

Here I extend the comparison between model and measurements to sea level over the entire Bering Sea domain. The comparison between measured and computed sea level for the M_2 and K_1 constituents is given in Fig. 13. A high correlation coefficient shows satisfactory agreement between measured and computed sea level. The few points with higher deviation from the line are attributed to the measurements taken in the river mouth and narrow bays where model resolution is inadequate for the reproduction of bathymetry and coastline. The general tidal charts obtained through computations are in satisfactory agreement with satellite-derived sea level. Enhancement of the diurnal sea level over the shelf break in the Bering Sea is especially well depicted.

For comparison of the computed versus measured tidal currents a data set from the shallow part of the Bering Sea is available. These data were thoroughly analyzed and compared with numerical models by Pearson et al. (1981), Mofjeld (1984, 1986), and Kowalik and Stabeno (submitted). A set of data for the Pribilof Canyon region has been acquired by Schumacher and Reed (1992). Here I limit my comparison to the data available in the region of the Pribilof Islands and Canyon where we applied a high resolution barotropic model (Figs. 9, 10). Data from six stations in the region are available for comparison against the model. Basic parame-

Table 1. Comparison of measured and computed tidal ellipse parameters.

| | | Depth of | | | K_1 | | | | M_2 | | | |
| | | | | | Major | | Minor | | Major | | Minor | |
Station	Total depth	current meter	Long	Lat	H	D	H	R	H	D	H	R
BC17	104	96	56 34	167 34	9.1	105	4.3	C	15.9	53	10.5	C
Calc					9.8	119	5.4	C	17.1	47	11.1	C
BBL2	69	64	57 37	167 45	11.7	296	4.5	C	21.3	42	16.6	C
Calc					12.6	305	4.6	C	20.1	48	16.2	C
BP1	140	50	56 16	169 48	8.5	176	3.4	C	9.0	43	6.5	C
BP1	140	125	56 16	169 48	15.7	168	2.9	C	12.0	108	9.3	C
Calc					8.5	138	5.6	C	9.2	59	6.0	C
BP2	287	62	56 14	169 42	4.6	167	1.1	C	12.7	36	10.6	C
BP2	287	137	56 14	169 42	10.8	155	5.8	C	9.8	10	5.8	C
BP2	287	272	56 14	169 42	19.8	137	4.7	C	6.5	122	4.5	C
Calc					5.1	107	1.6	C	7.9	54	5.1	C
BP2A	275	49	56 10	168 53	7.9	269	3.5	C	19.1	29	13.9	C
BP2A	275	124	56 10	168 53	6.8	272	2.6	C	11.1	52	7.1	C
BP2A	275	260	56 10	168 53	6.7	253	4.1	C	10.5	107	9.1	C
Calc					8.2	245	3.8	C	12.5	42	8.4	C
BP3	1,002	52	56 07	169 16	4.6	240	1.7	C	5.6	354	3.1	C
BP3	1,002	127	56 07	169 16	3.4	211	0.6	C	7.6	351	5.9	C
BP3	1,002	262	56 07	169 16	2.3	287	2.0	C	12.2	16	10.4	C
BP3	1,002	502	56 07	169 16	7.7	299	0.5	C	7.5	312	5.9	C
Calc					4.3	260	1.5	C	6.4	59	4.0	C

Depth is in meters, amplitudes H are in cm/s, D is direction of the major axis in degrees from the North, R refers to direction of rotation, and C denotes clockwise rotation.

ters of tidal ellipse, measured and calculated, are given in Table 1. Two relatively shallow water stations, BC17 (examined by Pearson et al. 1981) located east from Pribilof Canyon and BBL2 (examined by Mofjeld et al. 1984) located northeast from Pribilof Islands, show typical tidal velocities away from the shelf break. The computed tidal velocities are in satisfactory agreement with those obtained from the observations. Data from stations denoted as BP were collected by Schumacher and Reed (1992) at both the shelf break of the north-northwest side of Pribilof Canyon and the deep portion. Data at three stations show that the tidal current magnitude changes along the vertical direction, indicating that these tides are of baroclinic nature. The calculated tidal currents fall in the range of the lowest observed values along the vertical direction. In this geographical region the model predicts only slight enhancement of the diurnal tidal currents (Fig. 10). It is useful to note that at the three stations with baroclinic behavior the K_1 velocity increases toward the bottom. This pattern is consistent with baroclinic diurnal trapped waves described through models by Brink (1995) and Haidvogel et al. (1993) and through measurements by Eriksen (1991). The second set of data, which partly identifies the diurnal shelf waves at the shelf break, is a time series taken by Kitani and Kawasaki (1979). Two important conclusions can be drawn from their results. In all observations the ratio of the diurnal to semidiurnal tidal currents changes with depth and is much higher in the bottom layer than in the surface layer, thus confirming the general pattern observed in the trapped baroclinic waves. At two stations (St. 20 located at 60°N, 178°20′W and St. 61 located at 58°30′N, 173°15′W) diurnal currents dominate over semidiurnal currents in both the surface and bottom boundary layers. These two stations are situated close to the local maxima of velocity rendered in Fig. 8, upper panel.

It is difficult to find suitable measurements to confirm the general circulation pattern related to residual tidal currents along the Bering Sea slope. Fortunately, recent measurements in proximity to the Pribilof Islands confirm clockwise residual currents generated through the nonlinear interactions with quite high mean motion on the order of 15 cm/s (Kowalik and Stabeno, submitted). These observations compare well with the model simulations (Fig. 12).

Summary and Discussion

A tidal model with a resolution of about 10 km in the Bering Sea and submodels with resolution of 1.852 km and 0.617 km in the Pribilof Islands and Canyon have been constructed. The numerical solutions thus obtained are depicted through corange, cotidal, and tidal current charts. These numerical solutions together with all available information either from measurements or from models have been investigated. I focus especially on tidal currents and on origin and location of enhanced tidal currents. The high resolution used in computations allowed me to define details which were not attainable in previous computations and measurements. The numerical approach delineated regions of large tidal currents in the Bering Sea. Generally, computations confirmed enhancement of the tidal currents at the shallow water in bays and around islands as described by Mofjeld (1986), Liu and Leendertse (1990), and Kowalik and Stabeno (submitted). A more complicated picture is associated with the diurnal constituent along the shelf break. A homogeneous "classical" long wave tidal pattern along the continental slope is often broken by a smaller scale pattern related to trapped shelf waves of tidal origin. These regions require both special measuring and modeling approaches, because over a relatively short distance, the sea level and the currents vary considerably. In the Bering Sea two regions of enhanced diurnal tidal currents are found: in the Aleutian Islands (Amukta Pass) and off Cape Navarin. A number of smaller regions have been identified along the Bering Sea shelf break.

Do measurements provide a suitable test for shelf waves of tidal origin? Analysis of data by Mofjeld (1986) and three-dimensional models by

Liu and Leendertse (1990) shows clearly the maximum of the sea level in the diurnal band along the Bering Sea shelf break, and a decrease in amplitudes with distance shoreward. The same maximum is well depicted in the satellite data of Cartwright et al. (1991) and in Kantha (1995), as well as in the present model.

Only two sets of current observations point toward occurrence of the diurnal shelf waves. A time series taken by Kitani and Kawasaki (1979) shows enhanced diurnal tidal currents along the shelf break, and a time series obtained by Schumacher and Reed (1992) shows enhanced diurnal waves in the Pribilof Canyon. Future searches for new regions of enhanced currents should be facilitated by identification of the general regions of diurnal trapped waves made through the models and measurements.

Enhanced currents, through advective terms and bottom friction lead to stronger nonlinear interactions resulting in the new periodical oscillations and residual circulation. Mean residual flow can be a major driving mechanism for circulation around islands and in the passages between the Aleutian Islands. The presence of the strong oscillating and residual tidal currents in the Aleutian passages will certainly have a bearing on how tidal effects are to be incorporated into mechanisms that contribute to exchange of properties between the North Pacific and the Bering Sea. The homogenization of water properties in Amukta and Amchitka passes is believed to result from tidal motion (Reed et al. 1993, Reed and Stabeno 1994).

The occurrence of the enhanced currents in local regions along the shelf break should also modify our notion on the exchange mechanism between shelf and deep oceanic waters. Maximum production in the shallow northwest Bering Sea occurs, according to Springer and McRoy (1993), in pools of especially prolific growth. The investigations carried out during ISHTAR (Coachman and Hansell 1993) suggest an advective supply of nutrients to these pools via a north-flowing current that originates along the continental slope and bifurcates at Cape Navarin. This is a source region and the pathway for elevated primary production throughout spring, summer, and fall (Iverson et al. 1979). At present the pathways for the nutrients and phytoplankton in this region are not clear. Computations off Cape Navarin show strong tidal currents not only at the shelf but at the shelf break as well. This suggests the possibility of tidal pumping of nutrients from the deeper water to the shelf domain and subsequent transport of nutrients and plankton by residual tidal currents into the Gulf of Anadyr.

Acknowledgments

I would like to express my gratitude to L.H. Kantha from the University of Colorado, Boulder, for discussion of tides in the Pacific and for offering boundary data for the Bering Sea model; to A.Yu. Proshutinsky from the Institute of Marine Science, University of Alaska Fairbanks, for his help throughout the work; to R.D. Ray from NASA Goddard Space Flight Center for offering Geosat altimetry data and discussions; and to P. Stabeno from NOAA/PMEL for sharing her knowledge and data on tides around Pribilof Islands. I am grateful to J.D. Schumacher (Twocrow) from NOAA/PMEL and anonymous reviewers for comments and improvements that strengthened this paper. Support from the Office of Naval Research under grant N00014-95-1-0929 and from Cooperative Institute for Arctic Research, University of Alaska Fairbanks, is gratefully acknowledged.

References

Ahlnas, K. 1994. Tidally generated dipole eddies around St. Matthew Island, Bering Sea. In: T. Vihma (ed.), Report series in geophysics. University of Helsinki, Department of Geophysics. 28:11-17.

Bogdanov, K.T., V.V. Gorbachev, and V.V. Morozov. 1991. Atlas of tides of the Bering, Okhotsk and Japan seas. Okieanologicheskii Institute Akademi Nauk SSSR, Vladivostok. 39 pp.

Brink, K.H. 1995. Tidal and lower frequency currents above Fieberling Guyot. Journal Geophysical Research 100:10817-10832.

Cartwright, D.E., R.D. Ray, and B.V. Sanchez. 1991. Oceanic tide maps and spherical harmonic coefficients from Geosat altimetry. NASA Technical Memorandum 104544. 75 pp.

Chapman, D.C. 1989. Enhanced subinertial diurnal tides over isolated topographic features. Deep-Sea Research 36:815-824.

Coachman, L.K., and D.A. Hansell (eds.). 1993. ISHTAR: Inner shelf transfer and recycling in the Bering and Chukchi seas. Continental Shelf Research 13:473-704.

Coyle, K.O., G.L. Hunt Jr., M.B. Decker, and T.J. Weingartner. 1992. Murre foraging, epibenthic sound scattering and tidal advection over a shoal near St. George Island, Bering Sea. Marine Ecology Program Series 83:1-14.

Eriksen, C.C. 1991. Observations of amplified flows atop a large seamount. Journal Geophysical Research 96:15227-15236.

Foreman, M.G.G. 1978. Manual for tidal current analysis and prediction. Institute of Ocean Sciences, Pacific Marine Science Report 78-6. 70 pp.

Gill, A.E. 1982. Atmosphere-ocean dynamics. Academic Press, New York. 662 pp.

Haidvogel, D.B., A. Beckmann, D.C. Chapman, and R.-Q. Lin. 1993. Numerical simulation of flow around tall isolated seamount. Part II: Resonant generation of trapped waves. Journal Physical Oceanography 23:2373-2391.

Hastings, J.R. 1976. A single-layer hydrodynamical-numerical model of the eastern Bering Sea shelf. Marine Science Communication 2:335-356.

Hendershott, M.C. 1977. Numerical models of ocean tides. In: The sea, vol. 6. John Wiley & Sons, New York, pp. 47-96.

Hunkins, K. 1986. Anomalous diurnal tidal currents on the Yermak Plateau. Journal Marine Research 44:51-69.

Iverson, R.L., L.K. Coachman, R.T. Cooney, T.S. English, J.J. Goering, J.L.J. Hunt, M.C. Macaulay, C.P. McRoy, W.S. Reeburg, and T.E. Whitledge. 1979. Ecological significance of fronts in the southeastern Bering Sea. In: R.J. Livingston (ed.), Ecological processes in coastal and marine systems. Plenum Press, New York, pp. 437-466.

Johnson, W.R., and Z. Kowalik. 1986. Modeling of storm surges in the Bering Sea and Norton Sound. Journal Geophysical Research 91:5119-5128.

Kantha, L.H. 1995. Barotropic tides in the global oceans from a nonlinear tidal model assimilating altimetric tides. 1. Model description and results. Journal Geophysical Research 100:25283-25308.

Kinder, T.H., and J.D. Schumacher. 1981. Circulation over the continental shelf of the southeastern Bering Sea. In: D.W. Hood and J.A. Calder (eds.), The eastern Bering Sea Shelf: Oceanography and resources, vol. 1, pp. 53-76. Published by the Office of Marine Pollution Assessment, NOAA and BLM. Distributed by the University of Washington Press, Seattle, WA 98105.

Kitani, K., and S. Kawasaki. 1979. Oceanographic structure on the shelf edge region of the eastern Bering Sea: I. The movement and physical characteristics of water in summer, 1978. Bulletin Far Seas Fisheries Research Laboratory 17:1-12.

Kowalik, Z. 1981. A study of the M_2 tide in the ice-covered Arctic Ocean. Modeling, Identification, Control 2:201-223.

Kowalik, Z. 1994. Modeling of topographically-amplified diurnal tides in the Nordic Seas. Journal Physical Oceanography 24(8):1717-1731.

Kowalik, Z., and A.Y. Proshutinsky. 1993. Diurnal tides in the Arctic Ocean. Journal Geophysical Research 98:16449-16468.

Kowalik, Z., and P. Stabeno. Submitted. Trapped motion around the Pribilof Islands in the Bering Sea. Journal Geophysical Research.

LeProvost, C., M.L. Genco, and F. Lyard. 1994. Spectroscopy of the world ocean tides from a finite element hydrodynamic model. Journal Geophysical Research 99:24,777-24,797.

Liu, S.K., and J.J. Leendertse. 1979. Three-dimensional model for estuaries and coastal seas: VI: Bristol Bay simulations. The RAND Corp. R-2405-NOAA.

Liu, S.K., and J.J. Leendertse. 1981. A three-dimensional model of Norton Sound under ice cover. In: Proceedings of the Sixth International Conference Port. and Ocean Engineering under Arctic Conditions. POAC, Quebec, Canada, pp. 433-443.

Liu, S.K., and J.J. Leendertse. 1982. Three-dimensional model of Bering and Chukchi Sea. Coastal Engineering 18:598-616.

Liu, S.K., and J.J. Leendertse. 1990. Modeling of the Alaskan continental shelf waters. OCSEAP Final Reports 70:123-275.

Longuet-Higgins, M.S. 1969. On the trapping of long-period waves around islands. Journal Fluid Mechanics 37:773-784.

Mofjeld, H.O. 1984. Recent observations of tides and tidal currents from the northeastern Bering Sea shelf. NOAA Technical Memo ERL PMEL-57. 36 pp.

Mofjeld, H.O. 1986. Observed tides on the northeastern Bering Sea shelf. Journal Geophysical Research 91:2593-2606.

Mofjeld, H.O., J.D. Schumacher, and D.J. Pashinski. 1984. Theoretical and observed profiles of tidal currents at two sites of the southeastern Bering Sea shelf. NOAA Technical Memo ERL PMEL-62. 60 pp.

Murty, T.S. 1985. Modification of hydrographic characteristics, tides, and normal modes by ice cover. Marine Geodesy 9(4):451-468.

Pease, C.H., and P. Turet. 1989. Sea ice drift and deformation in the western Arctic. EEE Publ. N. 89CH2780/5, pp. 1276-1281.

Pearson, C.A., H.O. Mofjeld, and R.B. Tripp. 1981. Tides of the eastern Bering Sea shelf. In: D.W. Hood and J.A. Calder (eds.), The eastern Bering Sea Shelf: Oceanography and resources, University of Washington Press, Seattle, pp. 111-130.

Ray, R.D., and B.V. Sanchez. 1989. Radial deformation of the earth by oceanic tide loading. NASA Technical Memo 100743. 51 pp.

Reed, R.K., G.V. Khen, P.J. Stabeno, and A.V. Verkhunov. 1993. Water properties and flow over the deep Bering Sea basin, summer 1991. Deep-Sea Research 40(11/12):2325-2334.

Reed, R.K., and P.J. Stabeno. 1994. Flow along and across the Aleutian Ridge. Journal Marine Research 52:639-648.

Reed, R.K., and P.J. Stabeno. 1997. Long-term measurements of flow near the Aleutian Islands. Journal Marine Research 55:565-575.

Robinson, I.S. 1981. Tidal vorticity and residual circulation. Deep-Sea Research, Part A 28:195-212.

Schumacher, J.D., and T.H. Kinder. 1983. Low-frequency current regimes over the Bering Sea shelf. Journal Physical Oceanography 13:607-623.

Schumacher, J.D., and R.K. Reed. 1992. Characteristics of current over the continental slope of the eastern Bering Sea. Journal Geophysical Research 97:9423-9433.

Schumacher, J.D., and P.J. Stabeno. 1998. The continental shelf of the Bering Sea. In: A.R. Robinson and K.H. Brink (eds.), The sea: The global coastal ocean regional studies and synthesis, Vol. XI. John Wiley and Sons, New York, pp. 869-909.

Schwiderski, E.W. 1979. Global ocean tides, Part II: The semidiurnal principal lunar tide (M_2). Atlas of tidal charts and maps. Naval Surface Weapon Center, Dahlgren, VI 22248. 87 pp.

Schwiderski, E.W. 1981a. Global ocean tides, Part III: The semidiurnal principal solar tide (S_2), Atlas of tidal charts and maps. Naval Surface Weapon Center, Dahlgren, VA 22248. 96 pp.

Schwiderski, E.W. 1981b. Global ocean tides, Part IV: The diurnal lunisolar declination tide (K_1). Naval Surface Weapon Center, Dahlgren, VA 22248. 87 pp.

Schwiderski, E.W. 1981c. Global ocean tides, Part V: The diurnal principal lunar tide (O_1). Naval Surface Weapon Center, Dahlgren, VA 22248. 85 pp.

Schwiderski, E.W. 1981d. Global ocean tides, Part VI: The semidiurnal elliptical lunar tide (N_2). Naval Surface Weapon Center, Dahlgren, VA 22248. 86 pp.

Schwiderski, E.W. 1981e. Global ocean tides, Part VII: The diurnal principal solar tide (P_1). Naval Surface Weapon Center, Dahlgren, VA 22248. 86 pp.

Schwiderski, E.W. 1981f. Global ocean tides, Part IX: The diurnal elliptical lunar tide (Q_1). Naval Surface Weapon Center, Dahlgren, VA 22248. 86 pp.

Springer, A.M., and C.P. McRoy. 1993. The paradox of pelagic food webs in the northern Bering Sea—III. Patterns of primary production. Continental Shelf Research 13:575-599.

Sunderman, J. 1977. The semidiurnal principal lunar tide M_2 in the Bering Sea. Deutsche Hydrographie Zeitschrift 30:91-101.

Yanagi, T., M. Shimizu, T. Saino, and T. Ishimaru. 1992. Tidal pump at the shelf edge. Journal Oceanography 48:13-21.

Zimmerman, J.T.F. 1978. Topographic generation of residual circulation by oscillatory (tidal) currents. Geophysical Astrophysical Fluid Dynamics 11:35-47.

CHAPTER 5

Thermal Stratification and Mixing on the Bering Sea Shelf

James E. Overland and Sigrid A. Salo
Pacific Marine Environmental Laboratory, Seattle, Washington

Lakshmi H. Kantha
University of Colorado, Boulder, Colorado

Carol Ann Clayson
Purdue University, West Lafayette, Indiana

Abstract

A classical view of April-August stratification on the Bering Sea shelf is an isothermal inner domain, a two-layer middle domain, and a three-layer outer domain; domain boundaries are roughly 50 and 100 m. Based on a data set from 1974 to 1997, we show considerable spatial, intra-seasonal, and interannual variability in this picture. The primary physical balance is between solar heating and turbulent mixing by tidal currents and wind. The inner domain occasionally shows a warmer (1-3°C) surface layer, which is mixed out during subsequent storms. Greater depths in the middle domain have increased top to bottom temperature differences compared with shoaler regions. Transition zones between domains are deeper to the north of the Pribilof Islands, 50-60 m and 100-110 m, compared with the southeast Bering shelf, 40-50 m and 90-100 m. Such spatial variability is due to changes in turbulent kinetic energy, based on variations in tidal currents and water depth. Year-to-year summer temperature variability in the outer middle domain is 4°C in both the surface and bottom layers; this variability is primarily a result of initial temperatures in early spring, and secondarily of summer variations in solar heating.

Introduction

In this chapter we review the variability of the spring-summer thermal structure on the eastern and central Bering Sea shelf. We update the chapter on hydrographic structure in a previous Bering Sea review volume (Kinder and Schumacher 1981; hereafter referred to as KS). The data for their summary were collected between August 1975 and February 1978. We expand the KS picture by examining the spatial, seasonal, and interannual variability of stratification from 1974 to 1997 for April through August. We have attempted to collect all available conductivity, temperature, depth (CTD) data and have added a major collection of expendable bathythermograph (XBT) data from fisheries surveys. We review the data on vertical temperature profiles on the southeastern and central shelf from 30 to 110 m depth, which spans a horizontal extent of 250 km. The data set is three-dimensional: profiles vary with year, time of year, and depth interval. Averaging profiles over any of these three dimensions obscures interpretation of the physical processes involved. Dividing the data into 10-m depth increments and 2-week time increments, and plotting the mean of the profiles taken in each year, retains most of the available information on Bering Sea shelf variability for visual inspection. Our purpose is not to resolve every profile, but to show the envelope of observations as they vary across the shelf and across the years.

KS developed a conceptual model of summer mean hydrographic conditions; they define three domains on the Bering Sea shelf, based on stratification (Fig. 1). A coastal domain is well mixed or weakly stratified, while a middle domain is two-layered with a strong thermocline. An outer domain has surface and bottom mixed layers with an intermediate layer of more continuous stratification and lower levels of turbulence. The boundaries between these three domains were thought to be at approximately the 50 m and 100 m isobath (KS; Coachman 1986). The onset, location, and intensity of thermal stratification on the Bering Sea shelf is an important factor in regulation of the ecosystem. The Bering Sea shares this feature with other broad shelves such as Georges Bank and the Irish Sea, in that the three domains and the boundaries between them are important in determining the spatial variability of biological productivity.

We also investigate the evolution of stratification in the different domains with a high-resolution one-dimensional ocean model in the vertical. We show that this model is accurate in simulating the seasonal evolution of heating and stratification in the inner and middle domains, and adds verification that the primary processes controlling the thermal structure on the Bering Sea shelf are solar heating and tidal and wind mixing. Solar input has a daily and seasonal cycle, modified by cloudiness. Mixing is generated by shear production in top and bottom boundary layers. Because there is weak spatial variation of the mean wind forcing across the shelf (Bower et al. 1977), and tidal stress at the bottom varies in a regular manner, whether the stratification is well-mixed, two-layer, or continuous between an upper and lower boundary layer is roughly a function of water depth (KS). For tidal mixing in the inner domain the amount of energy available for mixing is proportional to U^3/h, where U is tidal velocity and h is water depth. The inner front, located between the inner and middle domains, should occur when the value of U^3/h drops below a given level

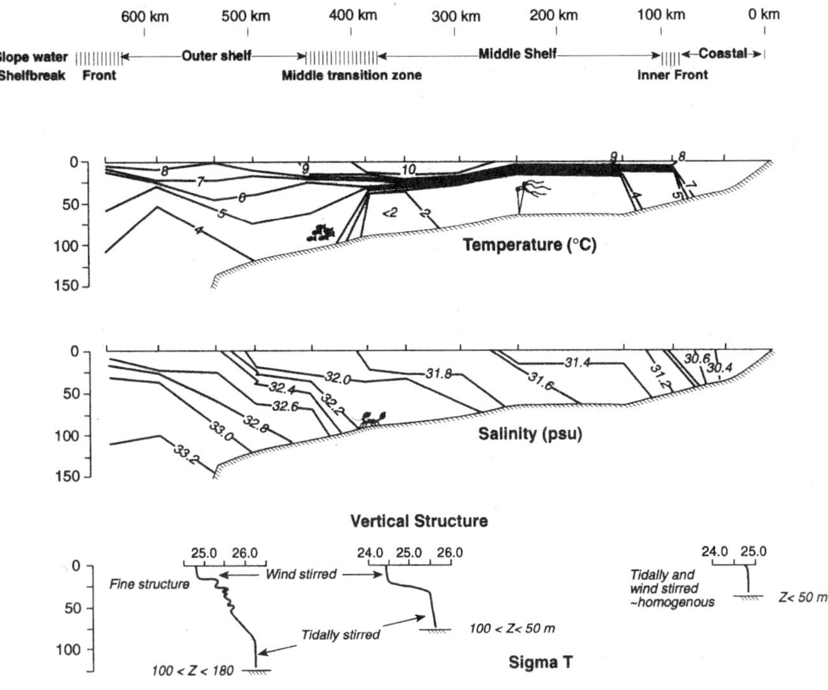

Figure 1.   *Vertical and horizontal sections of temperature and salinity across the southeast Bering Sea shelf. There is a one-layer domain inshore, a middle domain two-layer regime, and an outer shelf domain (after Kinder and Schumacher 1981).*

(Simpson and Hunter 1974, KS). A value of $-\log_{10}(U^3/h) = 3.5$ for the Bering Sea shelf was given by Schumacher et al. (1979). As the tidal velocity varies over a fortnightly cycle the position of the inner front should not, however, be fixed on an exact water depth.

Turbulent kinetic energy is also provided by the wind. Unlike the tides, this energy is provided on an episodic basis. Storms generally occur every 3-7 days but there can be periods of longer calms. As noted in other chapters of this volume, the mean circulation on the Bering Sea shelf is rather weak (<10 cm/s), so local events, solar input, winds, and tide predominate. However, the time dependence of storms or lack thereof, which regulates wind speed and clear skies, and the spatial dependence of tidal currents, have a major influence on the vertical structure in space and time. These complexities add considerable variability to the picture provided by KS and Coachman (1986).

In the outer domain, beyond 100 m, this simple picture is complicated by interactions with slope waters. There is a year-round halocline, generally below the seasonal thermocline. Salinity also plays an occasional role in the middle and inner domains.

Data Sources

We obtained the hydrographic data from four sources: Pacific Marine Environmental Laboratory (PMEL) archives of CTD data, National Oceanic Data Center (NODC) CTD archives, XBT data collected by the Resource Assessment and Conservation Engineering (RACE) Division of the National Marine Fisheries Service, and an archive provided by University of Alaska scientists which includes considerable Japanese data. Station data from the first source were evenly spaced at 1-m depth intervals, but the data from the latter three sources were unevenly spaced.

We grouped the hydrographic data into two areas, which we call the south-central and southeast, and excluded data from Aleutian and Pribilof regions. The two areas were then subdivided into bins by bottom depth. Figure 2 displays the spatial distribution of stations for all Julys. The row of stations running parallel to the southeast-Aleutian boundary and slightly to the northwest, is the Processes and Resources of the Bering Sea Shelf (PROBES) line where monthly hydrographic surveys were made from 1978 to 1981 (McRoy et al. 1986). Data were further divided by time into bimonthly periods in May and June, and into complete months for April, July, and August; this resulted in 56 bins. Figure 3 contains histograms showing the number of CTD and XBT casts from March through August in three depth ranges for the two regions: less than 50 m, from 50 to 100 m, and from 100 to 110 m. Although the station data are fairly well distributed spatially, there are yearly gaps, notably in the late 1980s and early 1990s, when little hydrographic information is available. The XBT data are an important resource, providing samples from the mid-1980s. Another problem is that few sites were occupied more than once, and the location of many sites, especially in early spring, was determined by the position of the ice edge, so that stations are farther south in heavy ice years.

We considered plotting only one profile for each year to show actual profiles. In some cases averaging different profiles within a 2-week, 10-m bin for a particular year with different mixed layer depths obscures the strength of the thermocline; however, profiles within a year could have temperatures that vary as much as 0.5-1.5°C. We have chosen to plot the average profiles within years to obtain more representative temperatures for each year. Some composite profiles have a multiple step profile and less thermocline strength that does not appear on individual profiles. However, in most 2-week, 10-m bins there is only one profile per year or small differences between profiles.

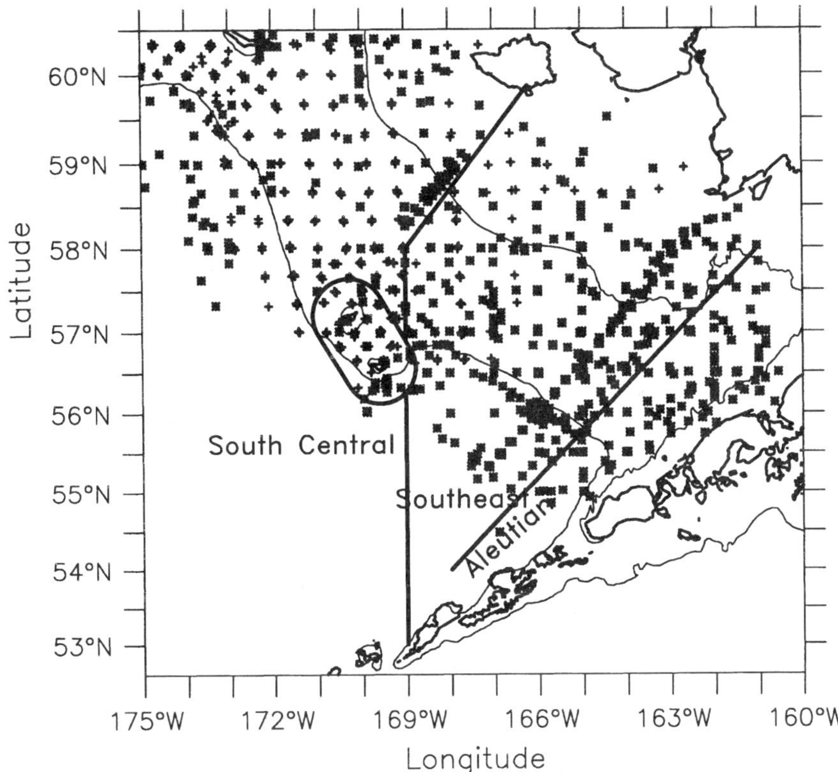

Figure 2. Spatial coverage of all vertical profile data used in this study for July in all years. The region is divided into four sectors: Aleutian, i.e., next to the peninsula; the Pribilof Island sector; and the southeastern and south-central sectors. Data are presented only for the southeast and south-central sectors.

Thermal Structure over the Southeast and South-Central Shelf

Temporal-Spatial Variability over the Southeast Shelf

The southeast shelf data set is larger than the south central (Fig. 3), especially from 1978 to 1981 through the PROBES project period. Only profiles for August are shown, divided into eight depth ranges (Fig. 4). For a water depth of 30-40 m all temperature profiles essentially showed vertical uniformity through May. The profiles for 1977 and 1980 began in the spring with temperatures below 0°C and warmed by nearly 2°C by the second

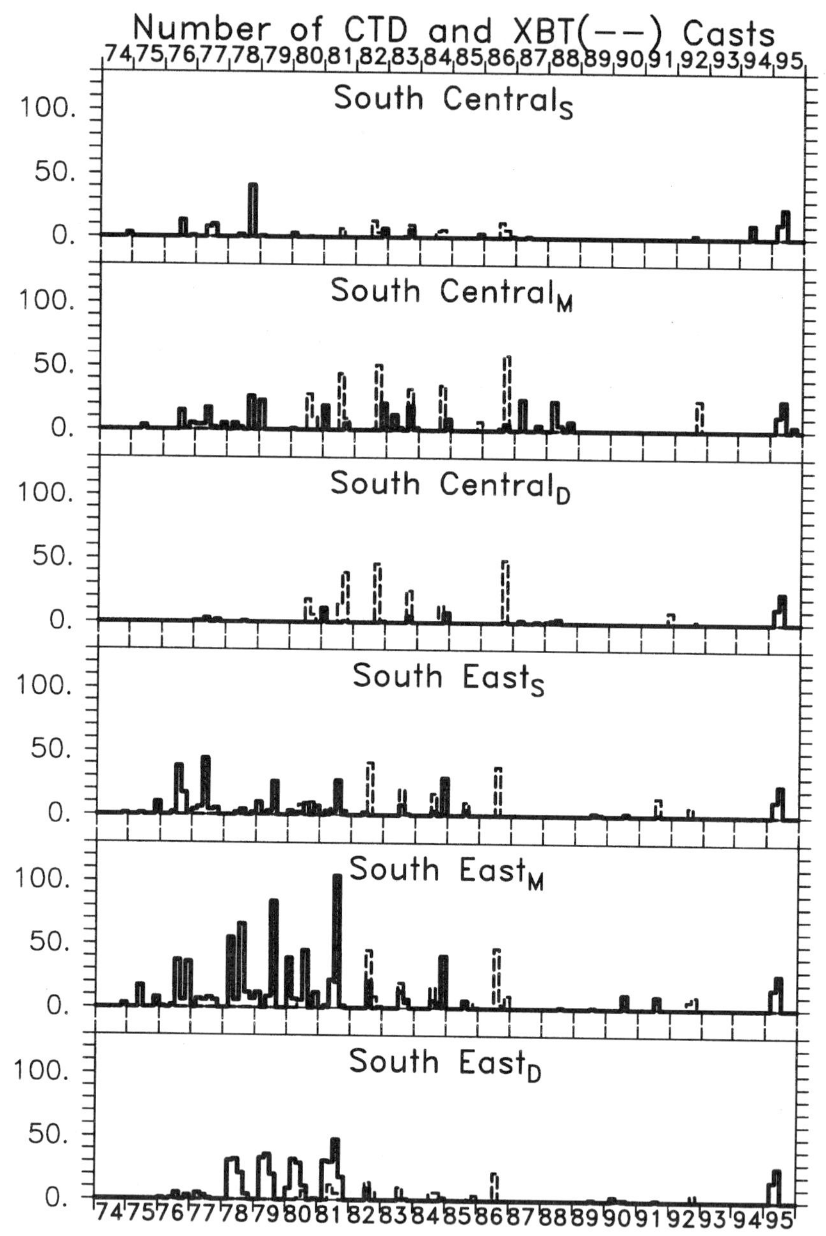

Figure 3. Temporal frequency of vertical profile data for the southeast and south-central sectors and further subdivided into shallow (<50 m), middle (50 m-100 m), and deep (>100 m). Most of the data are from 1977 to 1981 in southeast, but there are considerable interannual data for the middle domain in south-central.

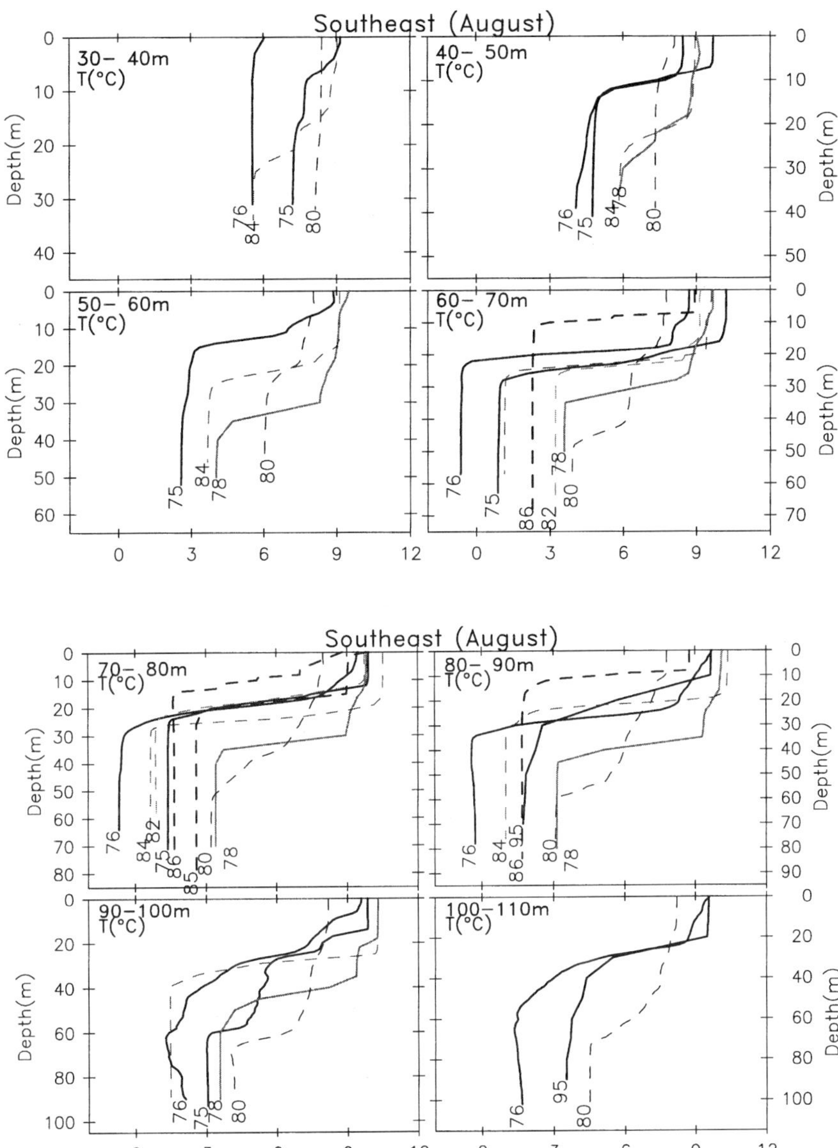

Figure 4. Vertical profile of temperature for the southeast region for August. It is stratified by 10 m depth bin. Each plot shows the average of all profiles for a given year in each 10-m bin.

half of May. Throughout the summer, year 1981 stayed warmer than 1980 and 1980 stayed warmer than 1977. June temperatures in 1981 and 1983 were near 6°C compared to 1976, which was near 2°C. 1976 remains the coldest year in the data set throughout the summer months. Temperatures for 1980 warm about 9°C from April until August. Profiles for 1984 in August and 1977 and 1981 for July showed a surface layer which is 2-3°C warmer than the lower layer with the transition centered at 25 m. Thus the vertical temperature structure is not always uniform in 30-40 m water depths on the southeast shelf. However, the top-to-bottom temperature difference below 10 m for three of the four profiles in August is less than 1°; the top-to-bottom temperature difference in 1984 was less than 1° in July and near 3° in August.

Figure 4 also shows August temperature profiles for 40-50 m water depth. Through June the profiles and year-to-year comparison were similar to the 30-40 m depth range. There is a tendency to form shallow warmer layers 1-2°C greater than the bottom temperatures, except for 1991, which had an increase of 5°C in early June. In July and August there is a mixture of profiles for different years; for most profiles there is a range of upper layer temperature increases of 2-5°C above those of the lower layer. The range of year-to-year August bottom temperatures is 4-7°C. 1980 was the only year that did not form a 2-layer system.

August temperature profiles for the middle domain are shown in Fig. 4 for water depths of 50-90 m. Most temperature profiles were uniform through early May; profiles in 1997, 1982, and 1995 for April and early May showed colder temperatures near the surface compared to depth. In late May and the first half of June there was a well-mixed layer below 35 m, independent of water depth. The upper layers showed temperatures decreasing from the surface. By the second half of June in most years and certainly by July, a strong two-layer system was established for water depths greater than 60 m. By August surface layer temperatures reached 8-10°C and bottom layer temperatures remained in the range of 0-4°C, except for 1980 at 50-60 m, which had 6°C, because of increased vertical mixing. For 50-60 m depths there are both strong and weak two-layer profiles in different years.

The profiles for 90-100 m (Fig. 4) appeared transitional between the middle and outer domains. The different depth ranges for the outer domain (Fig. 4b, 100-110 m; 110-150 m not shown) had several features in common. Except for 1976 all temperatures started warmer than 2°C and there was an indication of warmer water in the spring below 80 m. By the end of May there was only slightly warmer (2°C) water at the surface than at depth and an almost linear decrease in temperature down to a depth of 60 m. By the end of June there was a variety of profile shapes. In several years there was a shallow (20 m) wind mixed layer. Another typical temperature profile was an exponential shape decreasing from the surface to

80 m. Most summer surface temperatures were 8-10°C. In the outer domain, bottom temperatures started and remained near 4°C throughout the season, except in 1976 and 1984, which were colder in August. There were smaller top-to-bottom temperature differences than in the middle domain.

Temporal and Spatial Variability over the South-Central Shelf

Data for the south-central region (Fig. 3) does not have the same seasonal coverage as the southeast shelf during the PROBES years, 1978-1981, but has better interannual coverage during the mid-1980s.

August temperature profiles are shown in Fig. 5. Virtually all temperatures for the 30-40 depth range remained homogeneous throughout the spring-summer, except for an occasional 1-2°C warmer shallow (<10 m) surface layer. For 40-50 m there were three years with nearly homogeneous profiles throughout the spring and summer, 1978, 1982, and 1985; two profiles, 1984 and 1995, had 5° and 6°C top-to-bottom temperature differences. The bottom temperatures in south-central were mostly colder than southeast for the same years.

For water depths of 50-60 m, middle domain characteristics were seen. However, the top-to-bottom temperature differences were less than 5°C; this contrasts with the southeast, which had stronger thermoclines in this depth range. The profiles for April and early May for 1987 in the central middle domain, 60-70 m and 70-80 m, showed colder temperatures at the surface. Step structures were a feature of late May. Near bottom temperature profiles in July 1980, 1984, 1986, and 1992 had temperatures at or below 0°C, again in contrast with southeast. This lower layer is termed the cold pool. At 70-80 m in August, 1977 had warm bottom temperatures and the top-to-bottom temperature difference is less than other years: $\Delta T \sim 5°C$. The 1977 data are interesting because it began as a cold year. For the outer middle domain the presence of near-freezing temperatures in late spring was seen out past 90 m in 1977 and 1988. The depth range 90-100 m (Fig. 5) had characteristic middle-domain, two-layer structure in July and August.

In the outer domain (>100 m) April 1977 had surface temperatures below 0°C. The 100-110 m depth range showed a two-layer structure in July and August. Summer surface layer temperatures were in excess of 8°C.

In summary, the southeast and south-central regions have similar conditions. South-central in general starts out colder than southeast in agreement with a cold pool concept of colder temperatures toward the northwest. The transition zones between domains tend to occur in deeper water in south-central—50-60 m and 100-110 m compared with southeast—40-50 m and 90-100 m.

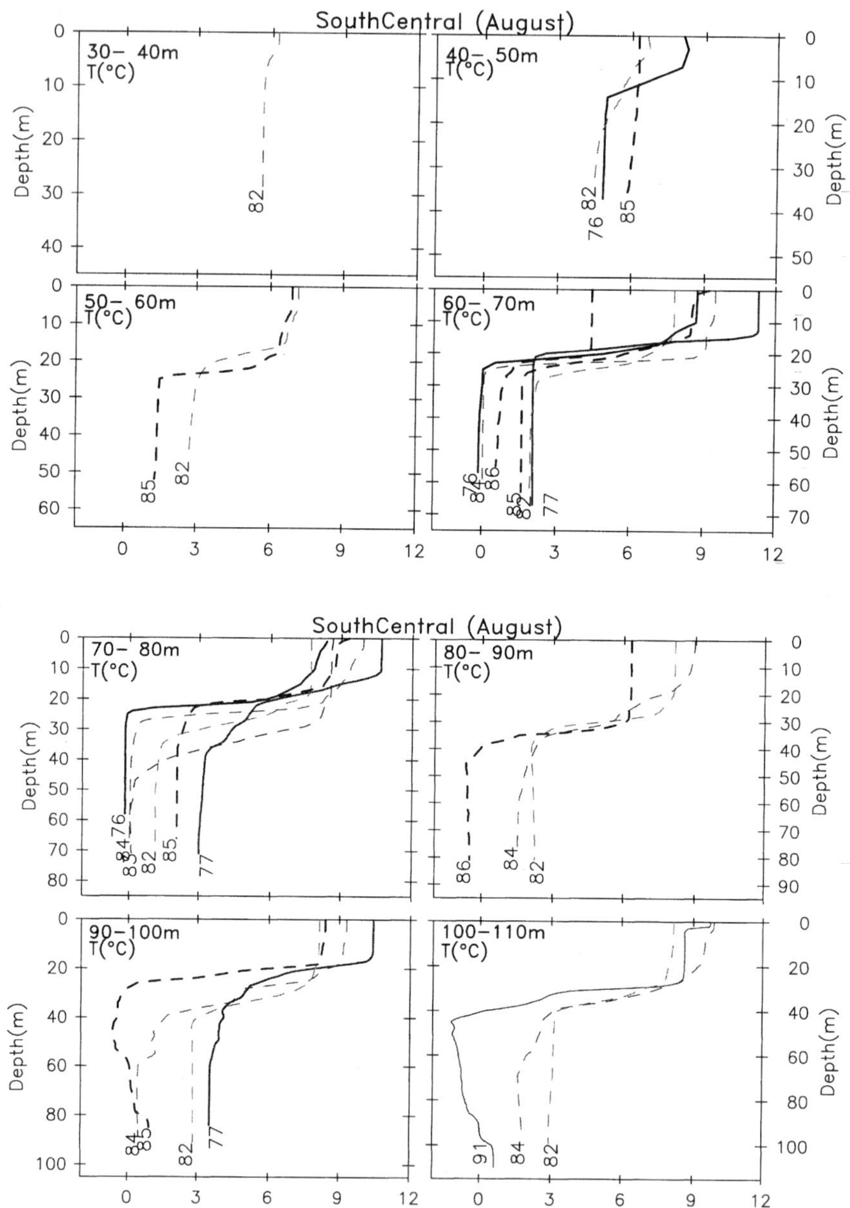

Figure 5.　Same as Fig. 4 for the south-central region.

Salinity

Salinity also contributes to density stratification. A 1.0 sigma-*t* value roughly corresponds to an 8°C top to bottom temperature difference, as often seen in the middle domain, which is equivalent to a 0.55 ppt salinity difference. Twenty-four of 196 (12%) representative salinity profiles for the southeast shelf showed vertical salinity gradients of 0.3-0.5 ppt for water depths less than 80 m. These were: 1977 in June and 1980 in July for 30-40 m, 1978 in May for 40-50 m, 1975 and 1982 in May and several years in July for 50-60 m, 1976 and 1982 for 60-70 m, and 1976 and 1982 for 70-80 m. Note that at most only 5 or 6 years are represented out of a 20-year record. For water depths of 80-90 m and 90-100 m, there was an increase in salinity below 70 m in most but not all profiles. By 100 m and deeper there was an increasing salinity profile with depth with top to bottom differences on the order of 0.7 ppt. These haloclines occur below the seasonal thermoclines.

Physical Processes and Model Simulation

The Model

The vertical transfer process of solar heating and turbulent mixing are fairly well understood on a quantitative basis. To obtain an understanding of the physics of the Bering Sea shelf we model the spring-summer thermal transition in the Bering Sea as a function of water depth. To carry out this task, we make use of a one-dimensional mixing model (Kantha and Clayson 1994). This model belongs to the Mellor-Yamada second moment turbulent closure hierarchy, but includes several improvements. The model has prognostic equations for momentum, heat, salt, turbulent kinetic energy (q^2) and its first spatial moment ($q^2\ell$). The latter includes a term to approximate logarithmic behavior near a boundary. Kantha and Clayson (1994) have included a parameterization for intermittent and episodic turbulence in strongly stratified but highly sheared regions away from boundaries. This feature is important for the transition to deeper water near the outer domain. Shortwave solar radiation penetration into the upper water column is included. The model has been validated against many oceanic mixed-layer data sets. We proceed with a simulation of the onset of thermal stratification during 1980, when time series were available for comparison at different depth locations.

Hindcast of 1980

We have chosen to hindcast a year in which there were monthly or bimonthly hydrocasts for a range of water depths in the southeastern Bering Sea. The model requires water depth and tidal current information for the rotary components of the M_2 and K_1 constituents.

The meteorological driving data are assigned from St. Paul Island. The wind, air temperature, and relative humidity were obtained from the hourly observations. The anemometer was well-exposed. We have increased the wind speed by a factor of 1.15 to account for the decreased friction over water. The solar insolation is modeled from the 12-hourly upper air profiles and the hourly surface weather observations (NREL 1992). We have decreased the solar input by a factor of 0.93 to account for surface reflection (Reed 1978). Time series of solar input and wind speed cubed for 1980 are shown in Fig. 6. Bottom roughness for the southeast Bering Sea shelf is taken as 10^{-4} m (Overland et al. 1984), corresponding to summer conditions. The M_2 tidal components are set at (0.13, 0.08) m/s and K_1 are set at (0.10, 0.01) m/s (Schumacher and Stabeno 1997). Temperature and salinity profiles are initialized at 15 April.

Data for comparison is shown in Fig. 7 with temperature profiles taken at four depths. In general, there are data for mid-April, May, and the end of August. All salinity profiles (not shown) start and remain vertical except for 113 m water depth; here there is a 0.8 ppt top to bottom difference with a halocline between 50 and 70 m.

Figure 8a shows the model results for 40-m water depth. The water temperature was initialized at $-0.8°$C and the plot shows temperatures for a depth/time history. Day 46 is the end of May and day 138 is the end of August. One sees near-vertical mixing with an August temperature near $8°$C similar to the observations. Based on wind variability one sees vertical mixing near days 62 and 120, or a slightly stable layer in the upper 10-20 m, days 90 and 135. The latter is eroded by day 150. Such a feature is seen in the observations for 1 June.

Figure 8a also shows the model results for 55-m water depth initialized at $-0.6°$C. Unfortunately, there are no August observations. The observed temperature of $2°$C occurs on 1 June while in the model it occurs 15 days later. The formation of the shallow weakly stable layer is also slightly delayed in the model. Note that the top-to-bottom temperature differences are all less than $1°$C. The model temperature in August is $6°$C, almost $2°$C cooler than in the 40-m case. Since both models have the same atmospheric forcing, the absorbed heat in the 55-m case must be spread over a greater depth interval.

Figure 8b shows the model results for 80 m, initialized at $0.3°$C. The end of May surface temperature is near $2°$C and a weak two-layer system is well simulated. The two-layer, middle-domain structure is established in June and continues to develop throughout the summer. August bottom temperatures in both the model and observations are near $3.7°$C. The model develops a surface layer of greater than $7°$C.

Figure 8b also shows the model results for 113 m, initialized at $2.7°$C and a top-bottom salinity difference of 0.8 ppt. One sees the establishment of a strongly stratified thermocline near 40 m in the model. A similar structure was seen in a model run without a salinity difference.

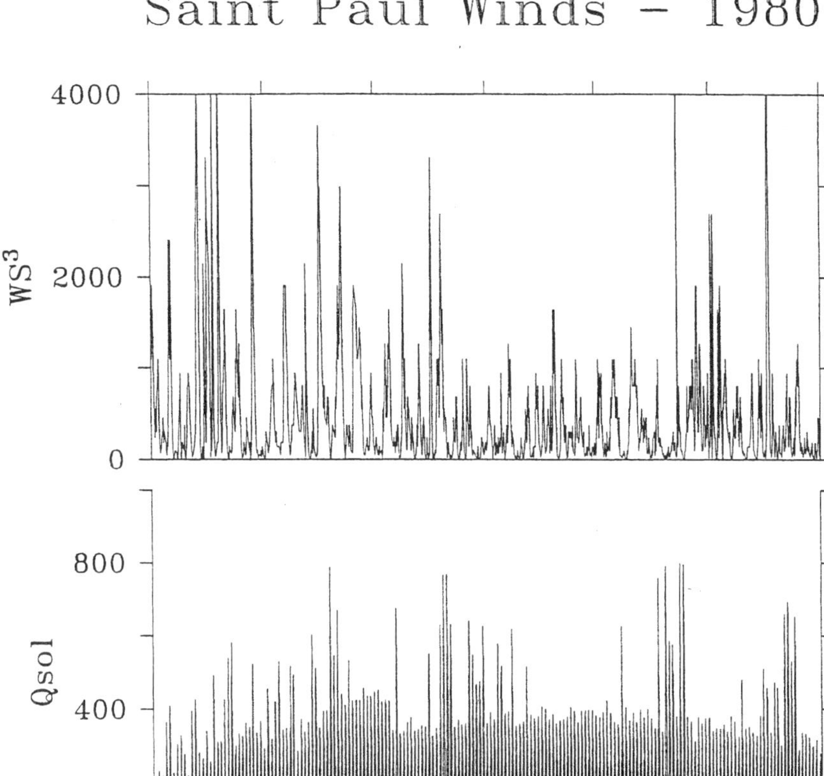

Figure 6. *Daily averaged solar input (W/m²) and wind speed cubed (m³/s³) at St. Paul Island for 1980.*

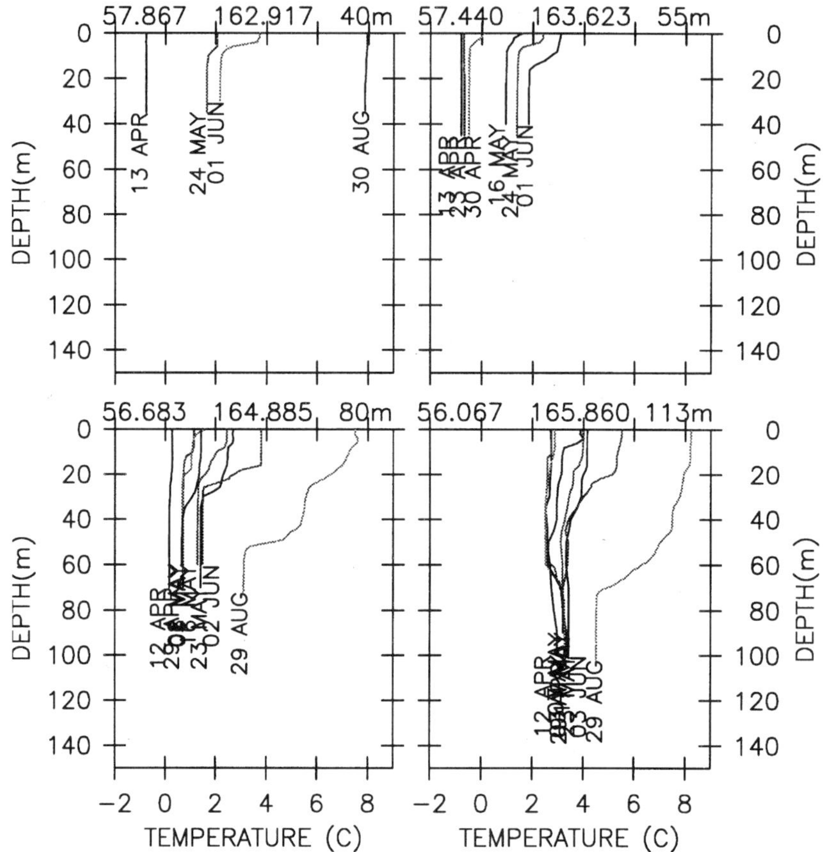

Figure 7. Spring-summer time history of vertical temperature profiles taken at four depths along the PROBES line in 1980; 40, 55, 80, and 113 m. These data are compared to the model calculations.

Observations show a deeper mixed layer. Both the model and observation only warm at the bottom by 1.5°C over the spring-summer period. The surface temperatures in the model of greater than 10°C do not match the 8°C observations. We conclude that the outer-domain temperature profiles at 113 m water depth are influenced by processes other than simple vertical mixing and heating.

Overall, the 1-D model contrasts the features of the inner and middle domains. Bottom mixing is important in the inner and middle domains. Variability in wind input creates temporary structure in the upper layers. Solar input can simulate the ~8°C warming through the spring-summer cycle. Stratification, once set up in the middle domain, reduces transfer of

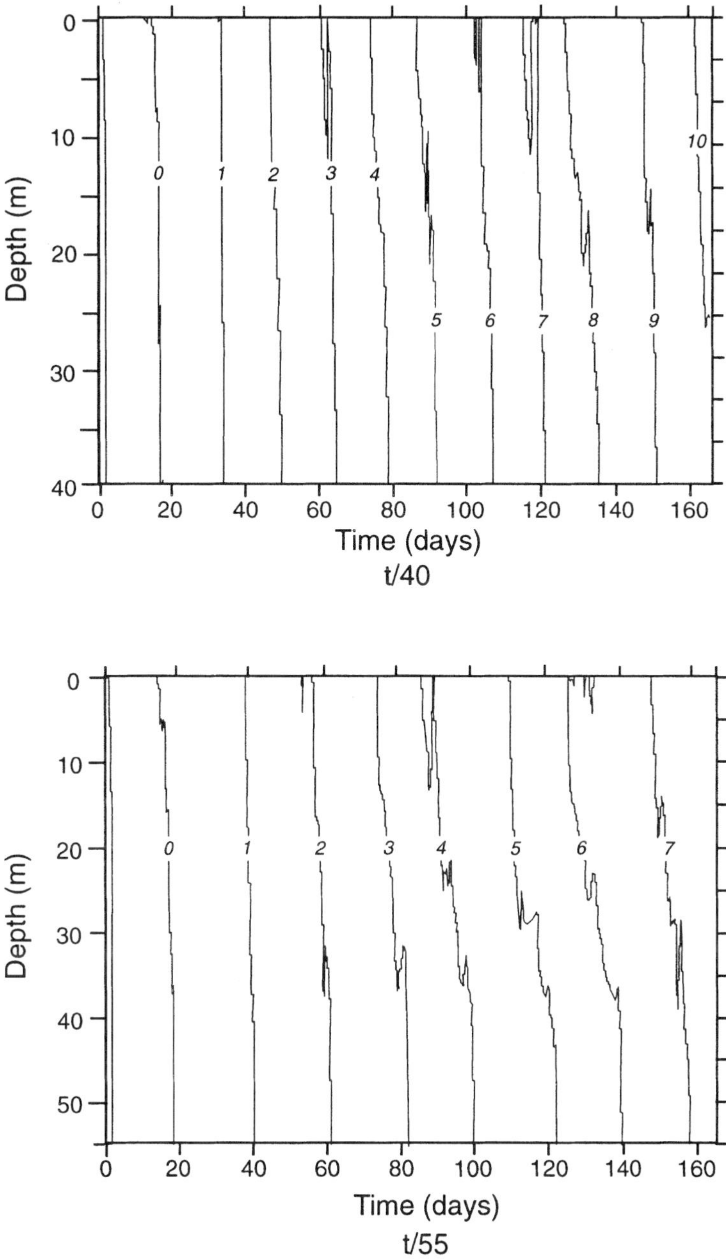

Figure 8a. One-dimensional vertical model calculations for two water depths, 40 m and 55 m. The plots show temperature plotted versus depth and time in days. All model runs are initialized for 15 April, so 1 June is day 47 and 30 August is day 138.

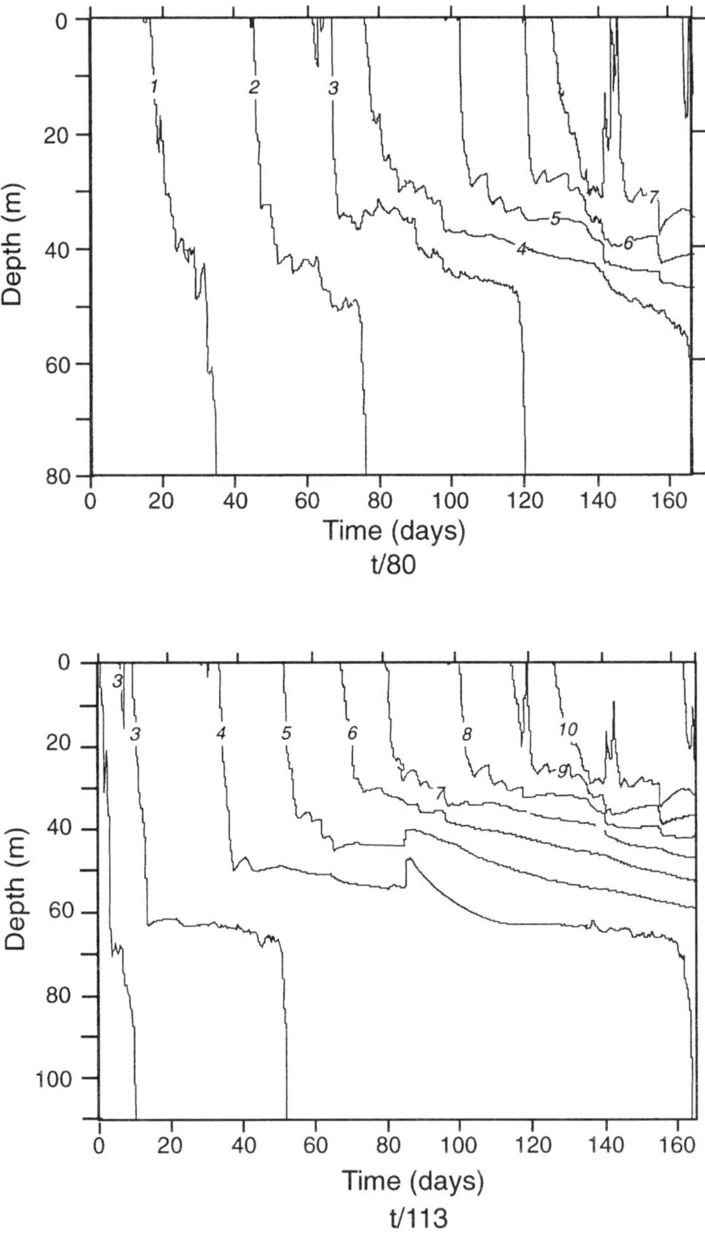

Figure 8b. (Continued.) One-dimensional vertical model calculations for two water depths, 80 m, and 113 m. The plots show temperature plotted versus depth and time in days. All model runs are initialized for 15 April, so 1 June is day 47 and 30 August is day 138.

heat to the lower layers. This is particularly evident in the deeper middle domains where the bottom and top turbulent boundary layers overlap only weakly. The 1-D model produced too strong a thermocline and too warm a surface temperature in the outer domain, which indicates the probable importance of horizontal processes.

Discussion and Summary

Our data from 1974 to 1997 confirm the existence of three thermal regimes on the Bering Sea shelf in summer. As seen from the model results these are governed by solar heating, tidal mixing, and wind mixing. There is a well-mixed inner domain, which occasionally shows a warmer (~2°C) surface layer which is mixed out during subsequent storms. Based on the model results the outer part of the inner domain will be colder than the inner part because the same amount of absorbed heat is spread over a greater depth. One can find both the well-mixed and vertical two-layer systems from 40-70 m. Because the top and bottom boundary layers overlap, water depths of 50-70 m will have less of a temperature difference than for deeper water. Thus the temperature structure in the middle domain cannot be thought of as uniform, but varies with depth of water. The transition zones between domains are deeper in south-central compared to southeast; this is due to increased tidal current amplitude. The seasonal cycle appears similar in different years. The temperature increase of 8-9° in the upper layer of the outer middle domain from April to August varied only a small amount, except for 1977. Detailed differences in vertical structure for a given water depth mostly depend on the initial temperature, and are modified fortnightly tidal cycle, wind history, and cloud history. We note local interannual variability in (1) summer mixed layer depths, due to differing wind events, and (2) absolute water temperatures throughout the season, particularly in the lower layers, based on differing water temperatures at the end of the previous winter season. For example, bottom temperatures in 1976 were particularly cold in April, and this anomalous feature remained through the summer. Our conclusions are similar to those reached by investigators of the North Sea (Bowers and Simpson 1990). The variability in vertical physical structure, due to variability in meteorological forcing, is driven by short- term events. These events are expected to influence not only the variance of physical and biological structures but can also influence their seasonal means in selected years (Ridderinkhof 1992).

Acknowledgments

This research was sponsored by NOAA's Coastal Ocean Program. We thank R. Reed for discussions, and R. Whitney for word processing. This is Fisheries-Oceanography Coordinated Investigations contribution FOCI-B259 and PMEL contribution 1642.

References

Bower, D.G., and J.H. Simpson. 1990. Geographical variations in the seasonal heating cycle in northwest European shelf seas. Continental Shelf Research 10:185-199.

Bower, W.A., H.F. Diaz, and S. Prechtel. 1977. Climate atlas of the outer continental shelf waters and coastal regions of Alaska: Vol. 2, Bering Sea. Arctic Environmental Information and Data Center (AEIDC) publication B-77. 441 pp.

Coachman, L.K. 1986. Circulation, water masses, and fluxes on the southeastern Bering Sea Shelf. Continental Shelf Research 5:23-108.

Kantha, L.H., and C.A. Clayson. 1994. An improved mixed layer model for geophysical applications. Journal of Geophysical Research 99:25,235-25,266.

Kinder, T.H., and J.D. Schumacher. 1981. Hydrographic structure over the continental shelf of the southeastern Bering Sea. In: D.W. Hood and J.A. Calder (eds.), The eastern Bering Sea Shelf: Oceanography and resources. University of Washington Press, Seattle, pp. 31-52.

McRoy, C.P., D.W. Hood, L.K. Coachman, J.J. Walsh, and J.J. Goering. 1986. Processes and resources of the Bering Sea shelf (PROBES): Development and accomplishments of the project. Continental Shelf Research 5:5-22.

National Renewable Energy Laboratory (NREL). 1992. National solar radiation data base, NSRDB: Volume 1. National Climate Data Center, Ashville, NC. 93 pp.

Overland, J.E., H.O. Mofjeld, and C.H. Pease. 1984. Wind-driven ice drift in a shallow sea. Journal of Geophysical Research 89:6525-6531.

Reed, R.K. 1977. On estimating insolation over the ocean. Journal of Physical Oceanography 7:482-485.

Reed, R.K. 1978. Heat budget of a region in the eastern Bering Sea. Journal of Geophysical Research 83:3635-3645.

Ridderinkhof, H. 1992. On the effects of variability in meteorological forcing on the vertical structure of a stratified water column. Continental Shelf Research 12:25-36.

Schumacher, J.D., and P.J. Stabeno. 1998. The continental shelf of the Bering Sea. In: A.R. Robinson and K.H. Brink (eds.), The sea: The global coastal ocean regional studies and synthesis, Vol. XI. John Wiley and Sons, New York, pp. 869-909.

Schumacher, J.D., T.H. Kinder, D.J. Pashinski, and R.L. Charnell. 1979. A structural front over the continental shelf of the eastern Bering Sea. Journal of Physical Oceanography 9:79-87.

Simpson, J.H., and J.R. Hunter. 1974. Fronts in the Irish Sea. Nature 250:404-406.

CHAPTER **6**

Variability and Role of the Physical Environment in the Bering Sea Ecosystem

James D. Schumacher
Pacific Marine Environmental Laboratory, Seattle, Washington

Vera Alexander
School of Fisheries and Ocean Sciences, University of Alaska Fairbanks, Fairbanks, Alaska

Abstract

Characteristics of the physical environment which vary over time scales from annual to decadal are examined in terms of their role in the ecosystem of the Bering Sea. The features examined are: solar activity, the lunar-nodal cycle of the moon, atmospheric circulation, ice cover, transport from the Pacific Ocean, and shelf circulation. The characteristics of the physical environment that are expected to respond to greenhouse gas-induced climate change are presented and their potential influence on the ecosystem discussed. We conclude by presenting a set of central issues and questions regarding how the Bering Sea functions.

Introduction

Features of the physical environment of the Bering Sea (Fig. 1), including solar radiation, atmospheric conditions, ice cover, and water column structure and temperature, change on a wide range of scales. These include a dramatic annual signal and smaller but potentially important longer-period fluctuations. This environmental variability, along with the flux of nutrient-rich slope waters onto the shelf, shapes one of the world's most prolific ecosystems (Walsh et al. 1989). Seasonal primary production begins with a bloom initiated by ice-edge melt; annual production varies from >200 g C/m^2 over the southeastern shelf to >800 g C/m^2 north of St. Lawrence Island. Over the western shelf, maximum annual production (>400 g C/m^2) occurs over the continental slope (Arzhanova et al. 1995). Subsequent to this ice edge bloom, summer production over the shelf depends on circulation patterns and the northward advection of nutrients.

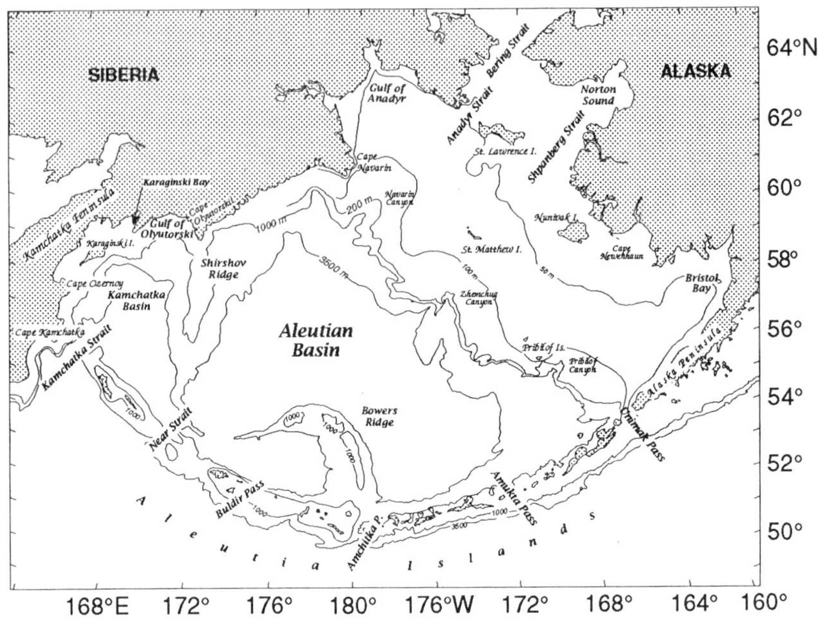

Figure 1. Geography of the Bering Sea, including a schematic representation of the major features of circulation.

Primary production supports vast populations at higher trophic levels, including marine mammals, birds, fish, and shellfish. At present, the walleye pollock *(Theragra chalcogramma)* fishery constitutes the largest single species fishery in the world, and the salmon run along the Alaska Peninsula is also the world's largest. In the past, these populations have exhibited marked variations in abundance (Fritz et al. 1993, Wespestad et al. 1994) that are likely related to interannual and/or longer-period variations in the physical environment.

The complex pathways that weave the living components of the Bering Sea into an ecosystem are not all known, nor are all the interactions among the various components well understood. We employ a conceptual model (Fig. 2; Steele 1995) to help define the time-space scales of interaction. In this model, the direct influence of the physical environment is a function of trophic level; the arrows between levels represent both biological interactions and the communication of physical influences. Spectral analysis of water temperatures in the Gulf of Alaska (Royer 1993) indicates that the annual signal has the greatest variance, with less energy at 18.6- and 22-year periods. Other analyses of physical and biological time series from the northeast Pacific (Ware 1995) also show that significant

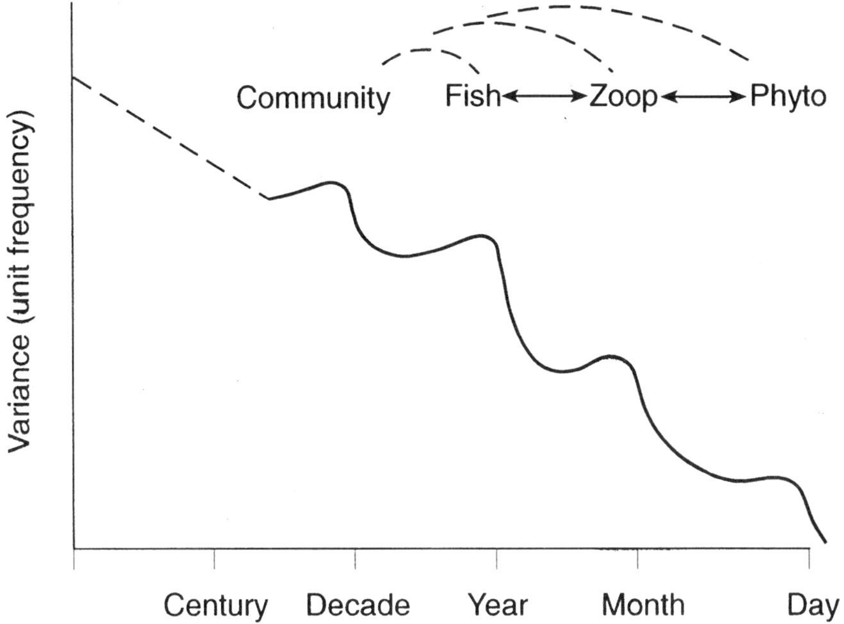

Figure 2.   An idealized representation of ocean power spectra as displayed in temperature observations, shown on a log-log scale. The time scales for trophic levels are indicated, and that for community is suggested. Added to the original (Steele 1995) is a spectral maximum at ~20 years to account for nodal tide and/or sunspot cycles.

peaks occur in spectral densities at a bidecadal period. Given these results, we have added a peak (period ~20 years) to Steele's conceptual model. We confine our discussion here primarily to periods greater than annual to decadal.

Understanding the influence of physical variations on the ecosystem is confounded by many factors. Physical conditions favorable for one life history stage may be detrimental for another. In addition, marine populations likely respond to change in a nonlinear fashion (Cury et al. 1995). For example, it is hypothesized that during cold years in the Bering Sea the spatial domain of age-1 pollock over the shelf is minimal (a behavioral response to temperature) and that increased densities of these fish result in enhanced take by predators (Ohtani and Asumaya 1995). Cold winters with extensive ice cover, however, provide conditions that can lead to more primary production over a shorter period in ice edge blooms (Niebauer et al. 1995), which could enhance the success of first-feeding pollock larvae and lead to a strong year class. In addition, it has been demonstrated that nitrate uptake on the southeastern Bering Sea shelf has a nonlinear relationship to wind-induced mixing, and timing of storms relative to the phase of the production system (i.e., respiration or nutrient limited period) is critical (Sambrotto et al. 1986).

We first review the major factors in variability of the physical environment on time scales from annual to decadal, followed by a discussion of how such variation, regardless of the exact nature of forcing mechanisms, may influence the ecosystem. We then present the features expected to characterize greenhouse gas-induced climate change (Hollowed 1996), and discuss their potential impacts on the Bering Sea ecosystem. Finally, we propose a set of central issues and questions whose answers will enhance understanding of both the physical factors and their impacts on the ecosystem.

Variable Features of the Physical Environment

The mechanisms that might be responsible for decadal period variations have not been examined for the Bering Sea. For the northeast Pacific, however, a recent review of climate variability (Ware 1995) presents the four leading candidates proposed to explain the origin of the bidecadal oscillation (BDO): (1) changes in solar activity; (2) the lunar-nodal cycle of the moon (Royer 1993); (3) atmospheric interaction between the Southern Oscillation and the Aleutian Low (Wooster and Hollowed 1995); and (4) an unstable air-sea interaction between the North Pacific subtropical gyre circulation and the Aleutian Low (Latif and Barnett 1994). Although their influence has not been directly examined, all of these mechanisms potentially could have a profound influence upon conditions in the Bering Sea.

We begin our discussion with the well known fluctuations in solar activity, which are nonstationary, with variable amplitude and phase (Wyatt et al. 1994). Such fluctuations in solar activity comprise a controversial forcing mechanism for many meteorological and biological time series. Solar activity is indexed by the number of sun spots (Wolf number; e.g., Bulatov 1989), or by solar emissions at a wavelength of 10.7 cm, which provides a more quantitative measure (Hill and Jones 1990). Dominant short periods for solar flux include a band that ranges from 7 to 17 years (average 11.1 years) and the 22-year Hale cycle.

Solar activity has been correlated with temperature and pressure in the atmosphere over the North Pole (Labitzke and van Loon 1988) and with ice extent off the east coast of Canada (Hill and Jones 1990), and has been suggested as a potential forcing mechanism for decadal oscillations in the coupled air-ice-sea system in the Northern Hemisphere (Ikeda 1990). While correlations exist, defining plausible mechanisms is problematic. Opposition to the hypothesis of forcing by solar activity centers on the small magnitude of its variation (~0.1%). For solar activity to be responsible for the observed phenomena requires either positive feedback from the sea (Ikeda 1990) or mechanisms not yet known.

The moon also exerts a pronounced influence on the earth, seen in tidal constituents with periods ranging from less than a day to interdecadal. Nodal tides (18.6-year) manifested through variations in mixing have been suggested to force decadal changes in coastal water temperatures off the east and west coast of North America (Loder and Garrett 1978). An approximate 17-year spectral peak exists in air temperature records from southeast Alaska and other high latitude locations (Royer 1993,1989). None of the records, however, were from stations adjacent to the Bering Sea. Royer showed that the low-frequency ocean temperature signal off Seward, Alaska, varies in concert with air temperatures, and suggested that forcing is provided by the nodal tide.

The primary control on atmospheric circulation, which in turn influences the Bering Sea, is the passage of low-pressure centers along the Aleutian Island chain. This results in a statistical feature known as the Aleutian Low. We have reliable time series of sea-level atmospheric pressure since 1924. Analysis of these observations shows that decadal and longer-period fluctuations are prominent, including a regime shift between 1976 and 1988, and that they are mirrored by sea surface temperature (SST) anomalies (Trenberth and Hurrell 1995). Regional atmospheric forcing reflects the intensity and location of the Aleutian Low, and some of the longer-period fluctuations are likely governed by the Southern Oscillation (Niebauer 1988, Wooster and Hollowed 1995). During the regime shift, the center of the Aleutian Low was farther east and deeper (Trenberth 1991). Patterns of SST in the eastern North Pacific (ENP) Ocean exhibit alternating warm and cool eras (average period ~17 years), and those in the Bering Sea appear to reflect these major shifts, although not necessarily in phase with those in the ENP (Wooster and Hollowed 1995). Although much attention has recently been focused on long-period changes, with their link to higher trophic levels and community structure (Fig. 2), the impact of individual storms on lower trophic levels remains important.

In the Bering Sea, interannual variations in overall ice coverage, time of advection over the open shelf, and subsequent melt-back are among the most striking features of the physical environment. Generation and advection of ice depends highly on atmospheric conditions. In winter, high pressure over the Arctic (the Siberian High) juxtaposed with the Aleutian Low results in cold dry northerly winds. Shelf waters are cooled to the freezing point by latent and sensible heat fluxes to the atmosphere. Winds then create and maintain open water regions, or polynyas, by downwind advection of ice. Observations of ice cover over the eastern Bering Sea shelf show a variation of nearly 40% about the mean (Niebauer 1988). Other characteristics of ice cover that show great variability include the duration of ice at its southern extent (3-15 weeks), time of retreat from the southernmost extent (between weeks 11 and 21), and number of weeks that ice lies over the middle shelf domain (3-28 weeks, mean of 20 weeks) (Wyllie-Echeverria 1995). While less definitive statistics are published on ice cover over the western shelf, observations indicate that similar general patterns obtain. Also, some ice can be exported to the Pacific Ocean via transport through Kamchatka Strait (Niebauer et al., chapter 2, this volume).

Volume transport to and from the Pacific Ocean through the passes in the Aleutian Island chain, as well as between the Bering and Chukchi seas through the Bering Strait, varies markedly (Stabeno et al., chapter 1, this volume). The Alaskan Stream, the dominant source of water flowing into the Bering Sea, largely determines transport in the Kamchatka Current and in the flow along the north side of the eastern Aleutian Island chain that later becomes the Bering Slope Current (Stabeno and Reed 1994; Reed and Stabeno, chapter 8, this volume). Recent results (Reed 1995) demonstrate how variations in inflow through the eastern passes influence subsurface water properties and can generate eddies (Schumacher and Stabeno 1994). Observations do not exist, however, to establish whether there are long-period signals in inflow of Alaskan Stream water. A climatology of wind forcing (Bond et al. 1994) shows that wind-driven (Sverdrup) transport accounts for roughly one-half of the observed transport in the Kamchatka Current and interannual variations in the Sverdrup transports are ~25% of the mean.

Over the eastern continental shelf there are seasonal variations in subtidal flow (Schumacher and Stabeno 1998), although again there is no indication of longer-period oscillation. Currents generated by residual tidal flow (Kowalik, chapter 4, this volume) will exhibit both spring-neap, annual, and nodal signals. Currents associated with melt at the ice edge (Muench and Schumacher 1985) vary in both magnitude and location in accordance with ice cover. On the western shelf, processes related to ice formation and melt and the impingement of the Kamchatka Current on shelf flow (Khen, chapter 7, this volume) likely account for much of the variability of shelf currents.

Water properties in the upper 300-400 m over the basin result from the combination of inflowing Alaskan Stream water and ambient Bering Sea water, which is tempered by winter cooling. A similar condition exists over the broad eastern shelf. Waters less than approximately 100 m deep generally are isolated from the outer shelf and slope; their temperatures result from the intensity of cooling and ice cover in the previous winter.

Influence of Physical Variation on the Ecosystem

One school of thought is that water temperature is the most important factor that determines year-class strength of the pollock stock in the eastern Bering Sea (Bulatov 1989, Khen 1989). Fluctuations in water temperature closely follow the dominant signal in solar activity, which has an 11-year period (Bulatov 1989). It is suggested (Khen 1989) that solar activity forces these changes through variations in extent of ice cover. Analyses of ice cover (Niebauer and Day 1989, Wyllie-Echeverria 1995) reveal different regimes (i.e., cold, mixed, and warm) but no marked periodicity. The length of the series, considered accurate only since 1972, precludes delineating an 11-year periodicity. Since ice extent results from atmospheric forcing, perhaps an atmospheric index such as the North Pacific (NP) index displays periodicity of 11 years? Spectra of the NP index have peaks (none of statistical significance) at periods of 2 and 6 years, associated with the Southern Oscillation, and a broad peak at periods >20 years; however, there is an absence of energy at 7- to 12-year periods (Trenberth and Hurrell 1994). Pressure gradients in the atmosphere over the North Pacific, required for the transfer of solar energy via the atmosphere to the ocean, show no variation at the necessary periodicity. The winter high-pressure pattern associated with the Eurasian continent may be a candidate for a mechanism which links solar activity through variations in the atmosphere to the forcing of sea ice in the Bering Sea.

Nodal tides and their influence upon biota have been discussed for other high-latitude seas (e.g., the Gulf of Alaska [Parker et al. 1995]) and the waters adjacent to Norway (Wyatt et al. 1994). In the Gulf of Alaska, the nodal tide correlates with variations in recruitment abundance of Pacific halibut *(Hippoglossus stenolepis),* possibly through enhanced nutrient availability and primary production due to tidal modulation of mixing and/or tidally driven advection (Parker et al. 1995). Tidal currents dominate horizontal kinetic energy over much of the Bering Sea shelf, supplying energy for mixing and generating residual flow. Although no references exist regarding nodal tides in the Bering Sea, let alone their impact on the ecosystem, they could influence the ecosystem.

Ice cover and the accompanying formation of cold bottom water exerts an important influence on the distribution of fish over both the western (Radchenko and Sobolevskiy 1993) and eastern shelves (Bulatov 1989, Ohtani and Azumaya 1995, Wyllie-Echeverria 1995). Analysis of ice climatology has demonstrated a connection via atmospheric forcing to Southern Oscillation periodicity (Niebauer et al., chapter 2, this volume; Niebauer 1988; Niebauer and Day 1989); however, this relationship accounts for only a small fraction of the total variation in ice extent. Initiation of phytoplankton blooms is related to ice melt. The ice-edge melt bloom and ensuing carbon flux provide about 10-65% of the total annual primary production over the southeastern shelf (Niebauer et al. 1990).

Potential Influence of Global Warming

A group of scientists well versed in atmospheric and oceanic phenomena of the North Pacific and Bering Sea recently outlined the most likely impacts there of global warming (19-20 April 1995; Hollowed 1996). They based their thoughts on projections from Global Climate Model simulations

Table 1.   Hypothesized changes in the physical environment of the Bering Sea due to global warming (after Hollowed 1996).

Physical feature	Change
Atmospheric features	
Storm intensity	Decrease
Storm frequency	Increase
Surface air temperature and humidity	Increase
Sea-level pressure	Lower in N. Bering
Southerly wind component	Increase
Wind stress curl	Decrease
Precipitation	Increase/shift north
Fresh water runoff	Increase
Oceanic features	
Sea ice extent, thickness, and brine flux	Decrease
Volume of Alaskan Stream inflow	Decrease
Bering Slope and Kamchatka Current flow	Decrease
Sea surface temperature	Increase
Cold pool extent/temperature	Decrease/increase
Nutrient flux onto shelf	Decrease
Sea level	Increase

(e.g., Hall et al. 1994) that indicate secular warming of the atmosphere over the North Pacific, especially at higher latitudes. We summarize those results, presenting only those processes likely to undergo substantial change (Table 1). The shifts associated with global warming could be exceeded by the extant long-period variations, which will likely continue. While changes in atmospheric features exert some influence directly on the ecosystem, their most profound effect is through modification of oceanic features. For example, in 1979, mean winds from the south increased by 1.8 m/s, while SST and air temperature increased by ~1.5° and 6°C, respectively, resulting in ice cover 15% below normal (Niebauer and Hollowed 1993). Atmospheric and ice variability lead to interannual changes in the cold pool.

As ice extent and thickness decrease, both lower and higher trophic levels would be affected. In addition to reducing the amount of primary production, changes in timing and spatial patterns of sea ice could influence the timing of the spring bloom (Niebauer et al. 1995) and have a direct impact on habitat and the phasing of biological events. Marine mammals associated with ice cover (e.g., walrus, *Odobenus rosmarus*, bowhead whale, *Balaena mysticetus*, and ringed seal, *Phoca hispida*) would experience a decrease in habitat.

Under the proposed secular trend, the cold pool would be reduced, affecting pelagic habitat. Arctic cod would find their southern limit moved northward, while pollock could experience a much larger spawning region as their habitat increased northward (Wyllie-Echeverria 1995). Since pollock constitute a large biomass on the eastern Bering Sea shelf, any change in their status likely would have a pronounced ripple effect throughout the ecosystem. The benthic environment would also experience marked changes.

As the inflow through the passes decreases, the attendant reduction in the strength of the major basin currents would affect shelf ecosystems. Eddies that impinge from deeper water onto both the western and eastern shelves would likely be reduced by a decrease in transport. Over the eastern shelf, the resultant flux of slope water provides larval pollock (Schumacher and Stabeno 1994) and nutrients to the shelf. Water over the western shelf shows the influence of the Kamchatka Current as well as eddies. On both shelves, slope waters provide the essential nutrients that fuel the ecosystem. Under the suggested climate change, this flux would decrease. The physical mechanisms involved are not understood well enough to suggest anything more than a reduction in overall production.

Directions for Future Research

As recent scientific interest by PICES, GLOBEC, and NOAA's Coastal Ocean Program (COP) focuses on the Bering Sea, some basic questions have arisen regarding the physical environment and its influence on the ecosystem. These include: Is there evidence for a change in the species mix in the Bering Sea; is there substantial evidence of a regime shift; and, if either question can be answered in the affirmative, how were the changes related to features of the physical environment?

The quest to delineate and understand long-period fluctuations in the physical environment and the response of the ecosystem to such fluctuations is limited by a lack of knowledge of both the biology and the physics of the Bering Sea. A primary missing element is time series of adequate length. While patterns can be elucidated in series as short as 60-70 years, longer time series are required to resolve the difference between 22- and 18.6-year periodicity. In addition, new methods may be valuable. One approach that has been fruitful in other regions (Baumgartner et al. 1992) would be to examine anoxic sediments in which fish scales are preserved. Another possibility is that information on long-period changes can be obtained from investigation of glaciers located near the Bering Sea.

Several key physical processes that must have ramifications for ecosystem dynamics require elucidation. In terms of circulation and transport, the importance of the Alaskan Stream to the Bering Sea has been identified (Stabeno et al., chapter 1, this volume). We have not, however, clearly determined the causes of fluctuations or their temporal behavior. Strikingly, the chemical and biological characteristics of the inflow are largely unknown.

The flux of nutrients onto the shelf is crucial, but the forcing mechanisms and their time-space variations are not well understood. Conventional wisdom proposes different mechanisms for flow onto the northern and southern portions of the eastern shelf. It is suggested that the Bering Slope Current flows northward to the vicinity of Cape Navarin, where some of the water crosses the shelf and flows through Anadyr Strait, providing nutrients (Shuert and Walsh 1993). Recent model (Overland et al. 1994) and observational (Stabeno and Reed 1994) results, however, indicate that the Bering Slope Current may flow westward south of 59°N. In the case of the southern shelf, tidally driven diffusion has been suggested as the mechanism responsible for flux onto the shelf south of the Pribilof Islands (Coachman 1986). More recent results indicate that the coefficients required for tidally driven diffusion are larger than those that exist (P.J. Stabeno, Pacific Marine Environmental Laboratory, Seattle, pers. comm.). It is also now recognized that eddies can play an important role in supplying slope water to the shelf (Stabeno et al., chapter 1, this volume). The generally seaward motion of ice may suggest offshelf flow in an upper layer, which would require onshelf flow at depth. This mechanism may serve to replenish nutrients over the southeastern shelf.

The cold pool of bottom water over the eastern shelf, a marked feature of the physical environment, has many biological implications, yet the dynamics that shape and determine the boundaries of the cold pool have not been explored. We have some understanding of the coupled bio-physical processes of ice edge–melt and initiation of a phytoplankton bloom; however, the time-space nature of melt-back and its influence on the conventional spring bloom are not known. Recent observations show that even over the middle shelf advective events can dominate changes in water properties (Stabeno et al. 1998). Perhaps such events also replenish nutrients over the winter.

Acknowledgments

The authors acknowledge PICES and its Bering Sea working group, who had the foresight to recognize the value of updating our knowledge and understanding of the Bering Sea. The first author acknowledges P.J. Stabeno, R.K. Reed, and S. Salo for valuable discussions and comments. This publication is contribution FOCI-B284 to NOAA's Fisheries Oceanography Coordinated Investigations, and contribution 1774 from NOAA's Pacific Marine Environmental Laboratory. Some of the research presented here was funded by Bering Sea FOCI of NOAA's Coastal Ocean Program.

References

Arzhanova, N.V., V.L. Zubarevich, and V.V. Sapozhnikov. 1995. Seasonal variability of nutrient stocks in the euphotic zone and assessment of primary production in the Bering Sea. In: B.N. Kotenev and V.V. Sapozhnikov (eds.), Complex studies of the Bering Sea ecosystem. VINRO, Moscow, pp. 162-179. (In Russian with English abstract.)

Baumgartner, T.R., A. Soutar, and V. Ferreira-Bartrina. 1992. Reconstruction of the history of Pacific sardine and northern anchovy populations over the past two millennia from sediments of the Santa Barbara basin, California. CalCOFI Report 33:24-40.

Bond, N.A., J.E. Overland, and P. Turet. 1994. Spatial and temporal characteristics of the wind forcing of the Bering Sea. Journal of Climate 7:1119-1130.

Bulatov, O.A. 1989. The role of environmental factors in fluctuations of stocks of walleye pollock *(Theragra chalcogramma)* in the eastern Bering Sea. Canadian Special Publication of Fisheries and Aquatic Sciences 108:353-357.

Coachman, L.K. 1986. Circulation, water masses, and fluxes on the southeastern Bering Sea shelf. Continental Shelf Research 5:23-108.

Cury, P., C. Roy, R. Mendelssohn, A. Bakun, D.M. Husby, and R.H. Parrish. 1995. Moderate is better: Exploring nonlinear climate effects on the Californian northern anchovy *(Engraulis modax)*. Canadian Special Publication of Fisheries and Aquatic Sciences 108:417-424.

Fritz, L.W., V.G. Wespestad, and J.S. Collie. 1993. Distribution and abundance trends of forage fishes in the Bering Sea and Gulf of Alaska. In: Is it food? Addressing marine mammal and seabird declines: Workshop summary. University of Alaska Sea Grant, AK-SG-93-01, Fairbanks, pp. 30-44.

Hall, N.M.J., B.J. Hoskins, P.J. Valdes, and C.A. Senior. 1994. Storm tracks in a high-resolution GCM with doubled carbon dioxide. Quarterly Journal of the Royal Meteorological Society 120:1209-1230.

Hill, B.T., and S.J. Jones. 1990. The Newfoundland ice extent and the solar cycle from 1860 to 1988. Journal of Geophysical Research 95:5385-5394.

Hollowed, A. (ed.) 1996. Report on climate change and carrying capacity of the north Pacific ecosystem. Scientific Steering Committee Coordination Office, Dept. of Integrative Biology, University of California, Berkeley, CA. U.S. Global Ocean Ecosystems Dynamics (GLOBEC) Report No. 15. 95 pp.

Ikeda, M. 1990. Decadal oscillations of the air-ice-sea system in the Northern Hemisphere. Atmosphere-Ocean 28:106-139.

Khen, G.V. 1989. Oceanographic conditions and Bering Sea biological productivity. In: Proceedings of the International Symposium on the Biology and Management of Walleye Pollock. University of Alaska Sea Grant, AK-SG-89-01, Fairbanks, pp. 79-89.

Labitzke, M., and H. van Loon. 1988. Associations between the 11-year solar cycle, the QBO and the atmosphere. I: The troposphere and stratosphere of the Northern Hemisphere in winter. Journal of Atmospheric and Terrestrial Physics 50:197-206.

Latif, M., and T. Barnett. 1994. Causes of decadal climate variability over the North Pacific and North America. Science 266:634-637.

Loder, J.W., and C. Garrett. 1978. The 18.6-year cycle of sea surface temperature in shallow seas due to tidal mixing. Journal of Geophysical Research 83:1967-1970.

Muench, R.D., and J.D. Schumacher. 1985. On the Bering Sea ice edge front. Journal of Geophysical Research 90:3185-3197.

Niebauer, H.J. 1988. Effects of El Niño–Southern Oscillation and North Pacific weather patterns on interannual variability in the subarctic Bering Sea. Journal of Geophysical Research 93:5051-5068.

Niebauer, H.J., and R.H. Day. 1989. Causes of interannual variability in the sea ice cover of the eastern Bering Sea. Geophysical Journal 18(1):45-59.

Niebauer, H.J., and A.B. Hollowed. 1993. Speculations on the connection of atmospheric and oceanic variability to recruitment of marine fish stocks in Alaska waters. In: Is it food? Addressing marine mammal and seabird declines: Workshop summary. University of Alaska Sea Grant, AK-SG-93-01, Fairbanks, pp. 45-53.

Niebauer, H.J., V. Alexander, and S.M. Henrichs. 1990. Physical and biological oceanographic interaction in the spring bloom at the Bering Sea marginal ice edge zone. Journal of Geophysical Research 95:22229-22242.

Niebauer, H.J., V. Alexander, and S.M. Henrichs. 1995. A time-series study of the spring bloom at the Bering Sea ice edge. I: Physical processes, chlorophyll and nutrient chemistry. Continental Shelf Research 15:1859-1878.

Ohtani, K., and T. Azumaya. 1995. Influence of interannual changes in ocean conditions on the abundance of walleye pollock *(Theragra chalcogramma)* in the eastern Bering Sea. Canadian Special Publication of Fisheries and Aquatic Sciences 121:87-95.

Overland, J.E., M.C. Spillane, H.E. Hurlburt, and A.J. Wallcraft. 1994. A numerical study of the circulation of the Bering Sea basin and exchange with the North Pacific Ocean. Journal of Physical Oceanography 24:736-758.

Parker, K.S., T.C. Royer, and R.B. Deriso. 1995. High-latitude climate forcing and tidal mixing by the 18.6-year lunar nodal cycle and low-frequency recruitment trends in Pacific halibut *(Hippoglossus stenolepis)*. Canadian Special Publication of Fisheries and Aquatic Sciences 121:447-459.

Radchenko, V.I., and Y.I. Sobolevskiy. 1993. Seasonal spatial distribution dynamics of walleye pollock *Theragra chalcogramma*, in the Bering Sea. Journal of Ichthyology 33:63-76.

Reed, R.K. 1995. On the variable subsurface environment of fish stocks in the Bering Sea. Fisheries Oceanography 4:317-323.

Royer, T.C. 1989. Upper ocean temperature variability in the northeast Pacific: Is it an indicator of global climate warming? Journal of Geophysical Research 94:18175-18183.

Royer, T.C. 1993. High-latitude oceanic variability associated with the 18.6-year nodal tide. Journal of Geophysical Research 98:4639-4644.

Sambrotto, R.N., H.J. Niebauer, J.J. Goering, and R.L. Iverson. 1986. Relationships among vertical mixing, nitrate uptake, and phytoplankton growth during the spring bloom in the southeast Bering Sea middle shelf. Continental Shelf Research 5:161-198.

Schumacher, J.D., and P.J. Stabeno. 1994. Ubiquitous eddies of the eastern Bering Sea and their coincidence with concentrations of larval pollock. Fisheries Oceanography 3:182-190.

Schumacher, J.D., and P.J. Stabeno. 1998. The continental shelf of the Bering Sea. In: A.R. Robinson and K.H. Brink (eds.), The sea: The global coastal ocean regional studies and synthesis, Vol. XI. John Wiley and Sons, New York, pp. 869-909.

Shuert, P.G., and J.J. Walsh. 1993. A coupled physical-biological model of the Bering-Chukchi Seas. Continental Shelf Research 13:543-574.

Stabeno, P.J., and R.K. Reed. 1994. Circulation in the Bering Sea basin observed by satellite-tracked drifters: 1986-1993. Journal of Physical Oceanography 24:848-854.

Stabeno, P.J., J.D. Schumacher, P.F. Davis, and J.M. Napp. 1998. Under-ice observations of water column temperature, salinity, and spring phytoplankton dynamics: Eastern Bering Sea shelf. Journal of Marine Research 56:239-255.

Steele, J.H. 1995. Climate change and community structure. Canadian Special Publication of Fisheries and Aquatic Sciences 121:5-9.

Trenberth, K.E. 1991. Recent climate changes in the Northern Hemisphere. In: M. Schlesinger (ed.), A critical appraisal of simulations and observations, Greenhouse Gas–Induced Climate Change, DOE Workshop 8-12 May 1989, Amherst, MA. Elsevier, New York, pp. 377-390.

Trenberth, K.E., and J.W. Hurrell. 1994. Decadal atmospheric-ocean variations in the Pacific. Climate Dynamics 9:303-319.

Trenberth, K.E., and J.W. Hurrell. 1995. Decadal coupled atmosphere-ocean variations in the North Pacific Ocean. U.S. GLOBEC CCCC Planning Meeting Summary 1996. Canadian Special Publication of Fisheries and Aquatic Sciences 121:15-24.

Walsh, J.J., C.P. McRoy, L.K. Coachman, J.J. Goering, J.J. Nihoul, T.E. Whitledge, T.H. Blackburn, P.L. Parker, C.D. Wirick, P.G. Shuert, J.M. Grebmeier, A.M. Springer, R.D. Tripp, D.A. Hansell, S. Djenidi, E. Deleersnijder, K. Henriksen, B.A. Lund, P. Andersen, F.E. Muller-Karger, and K. Dean. 1989. Carbon and nitrogen recycling with the Bering/Chukchi seas: Source regions for organic matter effection AOU demands of the Arctic Ocean. Progress in Oceanography 22:277-359.

Ware, D.M. 1995. A century and a half of change in the climate of the NE Pacific. Fisheries Oceanography 4:267-277.

Wespestad, V.G., P.A. Livingston, and J.E. Reeves. 1994. Juvenile sockeye salmon (*Oncorhynchus nerka*) predation on Bering Sea red king crab (*Paralithodes camtschaticus*) larvae as a cause of recruitment variation. ICES C.M. 1994/R:10. 18 pp.

Wooster, W.S., and A.B. Hollowed. 1995. Decadal-scale variations in the eastern subarctic Pacific. I. Winter ocean conditions. Canadian Special Publication of Fisheries and Aquatic Sciences 121:81-85.

Wyatt, T., R.G. Currie, and F. Saborido-Rey. 1994. Deterministic signals in Norwegian cod records. ICES Marine Science Symposium 198:49-55.

Wyllie-Echeverria, T. 1995. Seasonal sea ice, the cold pool and gadid distribution on the Bering Sea shelf. Ph.D. thesis, University of Alaska, Fairbanks. 281 pp.

CHAPTER 7

Hydrography of Western Bering Sea Shelf Water

Gennadiy V. Khen
Pacific Research Institute of Fisheries and Oceanography, Vladivostok, Russia

Abstract

This chapter uses data from two surveys in 1991 and 1992 to examine hydrography on the shelf of the western part of the Bering Sea. The shelf waters are divided into coastal, transitional, and oceanic zones. The borders of the oceanographic regions are not stationary but vary in relation to position of the Kamchatka Current. When the Kamchatka Current is far from the continental slope, there are only two regions on the shelf (coastal and transitional), and the oceanic zone is displaced beyond the shelf.

Winter conditions greatly affect water properties in summer. After a harsh winter, relatively cold water appears in summer; conversely, a mild winter precedes a summer with warm water.

Introduction

As Verkhunov (1994) noted, despite their economic importance, shelf waters of the western part of the Bering Sea have been studied little and are poorly understood even now. Only three publications (Davydov and Lipetskiy 1970, Davydov 1972, Verkhunov 1994) touch on some of the problems of zonation of the shelf region and on development of fronts in the western and northwestern parts of the Bering Sea. Temperature-salinity (TS) diagrams indicate that there are three regions of the shelf: coastal, transitional, and oceanic (Davydov and Lipetskiy 1970), which correspond to descriptions in the literature of hydrographic zones in the eastern part of the sea (Coachman and Charnell 1979, Kinder and Schumacher 1981, Coachman 1986). The similarity between the eastern and western shelves is strong. For example, frontal zones in both regions have similar nature and structure (Verkhunov 1994), and the isobaths associated with the fronts also coincide (i.e., the inner front occurs over a depth of 50 m, the middle front at 75-100 m, and the outer front near 170 m). Nonetheless, there are several differences between the areas. First, the western shelf is significantly narrower than the eastern shelf. Second, the coastline of the western

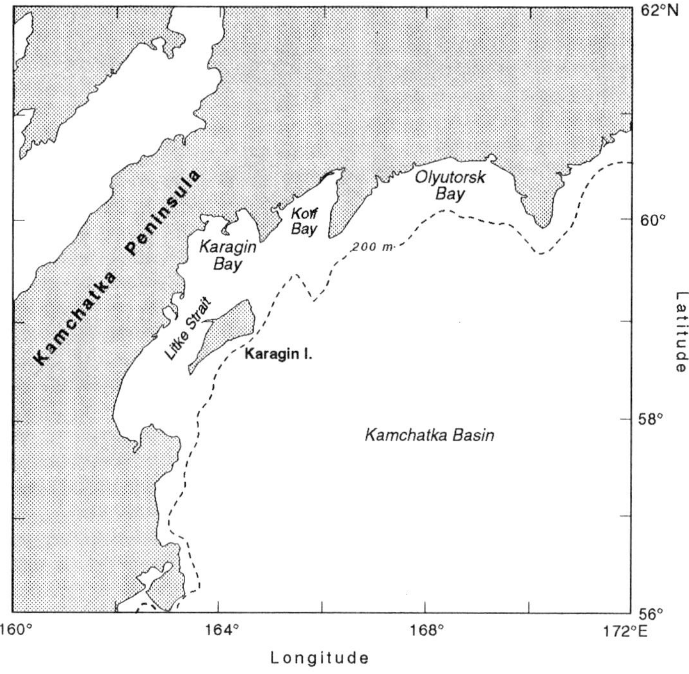

Figure 1.   The continental shelf of the western Bering Sea.

side is rough, with peninsulas that create half-isolated bays, such as Olyutorsk, Korf, and Karagin (Fig. 1). Third, the western shelf contains one of
the largest islands of the Bering Sea (Karagin Island).

The Siberian high–Aleutian low atmospheric system causes severe
conditions on the western Bering Sea shelf, accompanied by the processes
of ice formation. The ice begins to form in November, and in January the
entire shelf zone is covered by ice fields and becomes impassable to nav-
igation. The amount of ice gradually begins to decrease in the last half of
April; toward the end of May, ice survives only in the southern part of
Karagin Bay, the part most isolated from the influence of the relatively
warm water of the open basin. According to V.P. Pavlichev (TINRO [Pacific
Research Institute of Fisheries and Oceanography], pers. comm.), in 1995
ice was still present in this region at the end of June. In summer in the
western Bering Sea, as over the eastern Bering Sea shelf, a subsurface layer of residual water forms from winter cooling (Khen 1989). I assume that
these waters primarily form in shallow water and spread over the deep
Kamchatka basin (Figurkin 1992).

The nonstationarity of atmospheric processes, leading to coordinated
shifts in the dominant winds over the western shelf of the sea, and strong

meanders in the Kamchatka Current, are the principal causes of variability of the coastal currents. Thus, in some years the main flow of water through Litke Strait (Fig. 1) is from south to north, while in other years it is in the opposite direction (Davydov 1972). That is, the circulation of water around Karagin Island is sometimes anticyclonic and sometimes cyclonic. This is corroborated also by data from recent observations. In November 1990, the flow was clockwise (Verkhunov 1994), while in 1994, according to unpublished data from V.A. Abramov (TINRO), it was counterclockwise.

Distinctive features must be manifestations of the formation of oceanographic conditions on the shelf of the western part of the Bering Sea and their variability on seasonal or multiyear scales. In the present work, I will present characteristics of the temperature and salinity regimes in the summer on the basis of detailed observations.

Material and Methods

There have been no comprehensive oceanographic observations on the shelf of the western part of the Bering Sea in recent years. Therefore, a complete characterization of the entire area is not possible. A series of annual studies, including oceanographic observations, was carried out in the Russian economic zone under the leadership of V.P. Shuntov starting in 1989. I participated in only two of the ten cruises of this effort: a combined survey aboard the R/V *Professor Levanidov* and the *Professor Kaganovskiy* in June-July 1991, and a survey on the R/V *Professor Levanidov* in June-July 1992. The CTD (conductivity, temperature, depth) stations occupied during these cruises are shown in Fig. 2. Data from these surveys allowed me to study the spatial structure of the water and the interannual variability of the shelf waters of the western Bering Sea. Observations of temperature and salinity on the shelf were made to the bottom using a Neil Brown CTD system. The data were used to prepare temperature-salinity diagrams whose features were subsequently used to define hydrographic regions. Maps of the spatial distribution of temperature and salinity at the surface and the bottom were also prepared.

Results and Discussion

As expected, the analysis of TS curves (Figs. 3, 4) confirms the existence of three hydrographic regions, discussed earlier by Davydov and Lipetskiy (1970). The TS curves show a 2-layer structure in the coastal and transitional regions, composed of a warmer, less saline upper layer plus a lower, saltier layer formed from water cooled during the previous winter. The coastal region is characterized by sea surface salinities generally less than 31 practical salinity units (psu) and as low as 27-29 psu at some stations. Salinity increases rapidly to a depth of about 10 meters. The lower layer has salinities greater than 32 psu and temperatures less than 2°C, except at C1 in 1992, in both coastal regions. Surface salinity is generally 31-32

CTD Station Locations

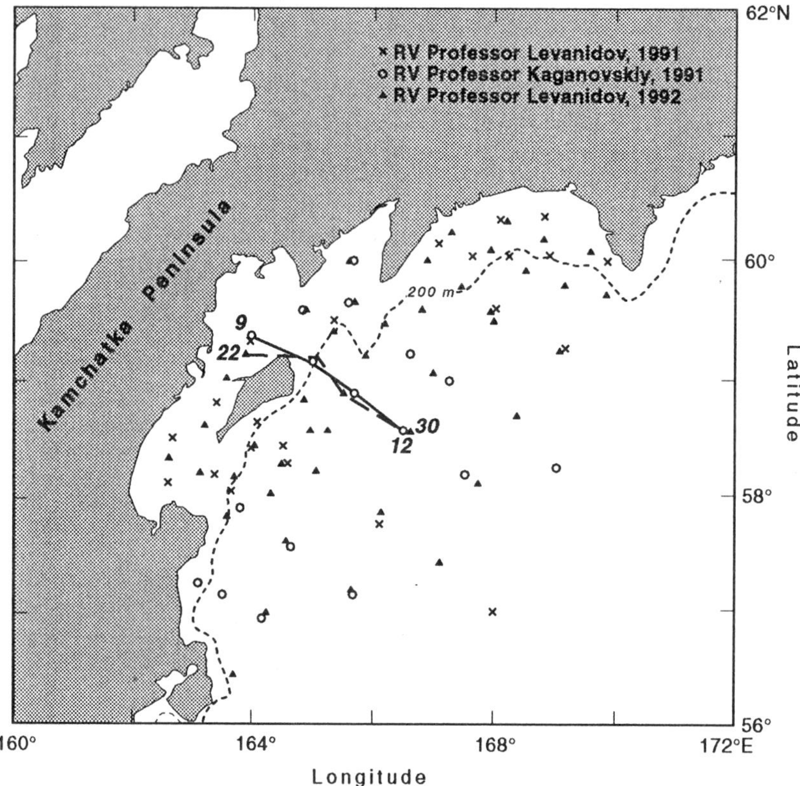

Figure 2.　Location of CTD stations discussed in this chapter. The solid and dashed lines indicate the location of stations used for vertical sections (Figs. 7, 8; 1991 and 1992, respectively).

psu in the transitional region. In the oceanic region, a mixed upper layer overlies a dichothermal layer characterized by temperatures less than 0.7°C. In the lower layer of the oceanic region, temperature increases slightly with depth to the bottom.

Unlike the eastern shelf, where frontal divisions are oriented along isobaths (Kinder and Schumacher 1981), the location of hydrographic regions is not always closely linked to the bathymetry over the western shelf. Here, a strong influence is exerted by dynamic outbreaks (baroclinic instability) originating at the periphery of the powerful Kamchatka Current and partly linked to changes in the bathymetry. Depending on the specific dynamic situation, one can expect powerful inflow of water from

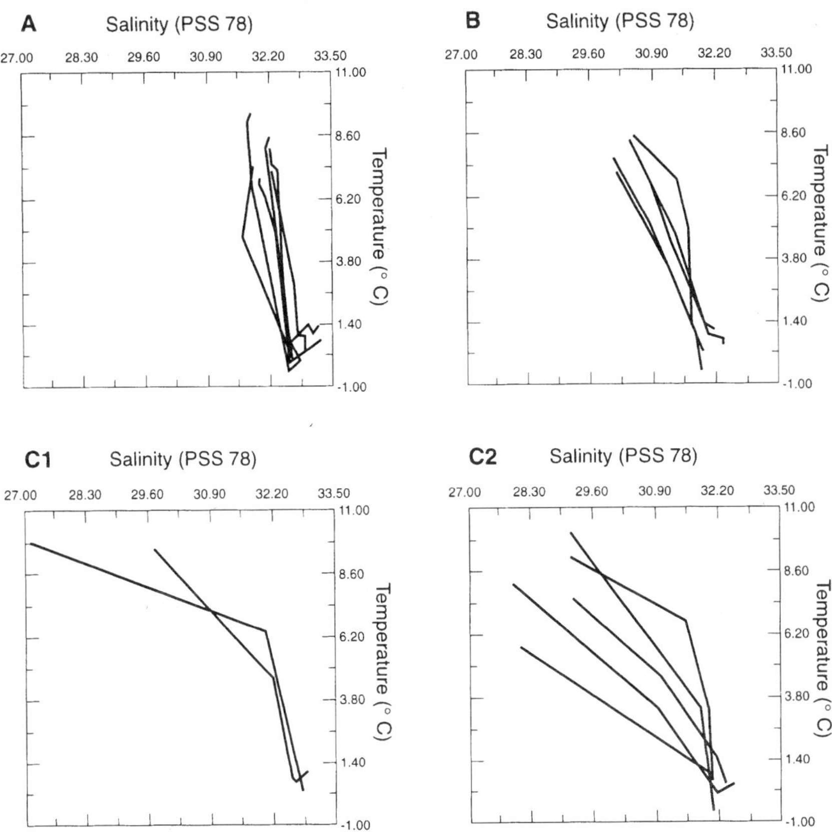

Figure 3. *Temperature-salinity (TS) curves of shelf water in the western Bering Sea in summer 1991. The labeled regions are: (A) Oceanic Region; (B) Transitional Region; (C1) Coastal Region in Olyutorsk Bay; (C2) Coastal Region in Karagin Bay. The label "salinity (PSS78)" refers to salinity units on the "practical salinity scale of 1978"; "psu," practical salinity units, is used in the text.*

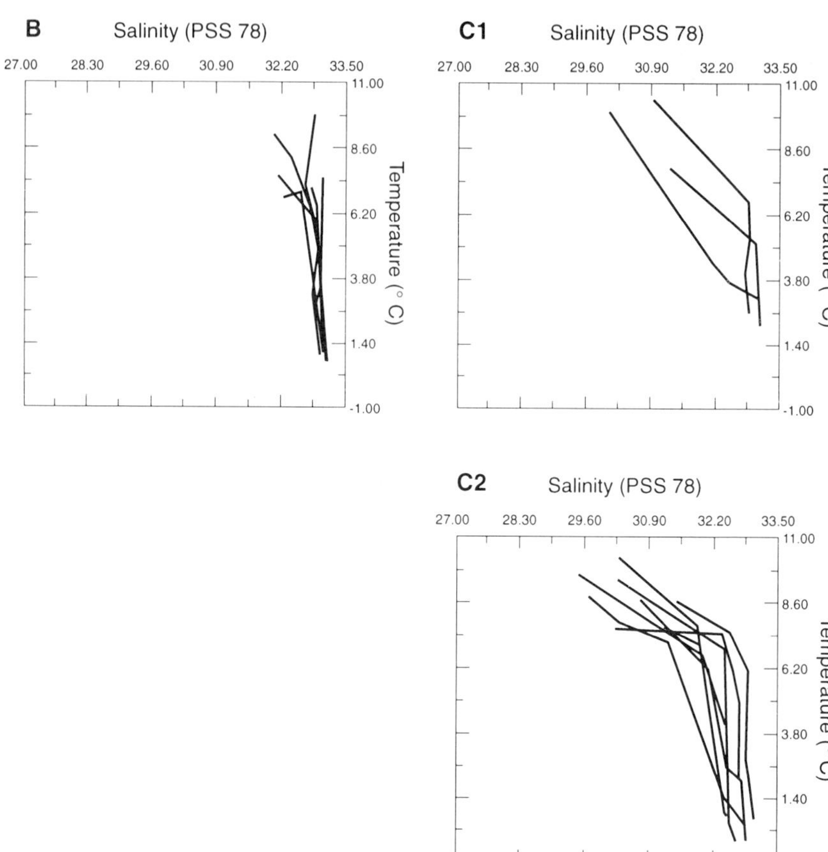

Figure 4.   Temperature-salinity (TS) curves of shelf water in the western Bering Sea in summer 1992. The labeled regions are: (A) Oceanic Region; (B) Transitional Region; (C1) Coastal Region in Olyutorsk Bay; (C2) Coastal Region in Karagin Bay. The label "salinity (PSS78)" refers to salinity units on the "practical salinity scale of 1978"; "psu," practical salinity units, is used in the text.

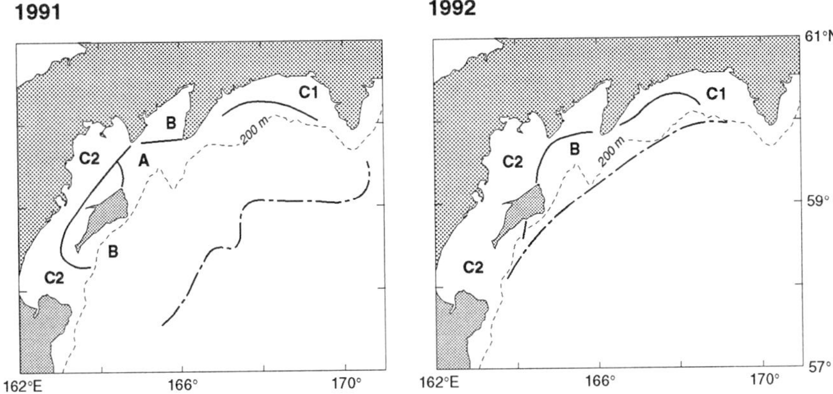

Figure 5. Hydrographic regions on the shelf of the western Bering Sea in the summers of 1991 and 1992. Regions A, B, C1, and C2 are as listed for Figures 3 and 4. The solid lines indicate the borders of these regions, and the dashed lines show the seaward extent of ice in April.

near the slope onto the shelf, or in contrast, spreading of cold low-salinity water from the gulfs. Therefore, the locations of the borders of the hydrographic regions are not stationary, but change from year to year (Fig. 5). Especially strong fronts may occur in northeastern Karagin Bay because the Kamchatka Current can easily penetrate the region. In 1991, I found a 3-layered water structure with a distinctive subsurface cold layer in this area and in the western part of Olyutorsk Bay. This is typical for the open sea part of the shelf (Davydov and Lipetskiy 1970). This structure is distinguished in Fig. 3 by the characteristic "hook" at the end of the TS curve. In 1992, there was a radical reorientation of hydrographic regions (Fig. 5). In place of the oceanic region, there was a transitional 2-layer structure consisting of a warm surface layer and cold lower layer. A 3-layer structure was seen only at stations in water deeper than 300 m, beyond the limit of the continental shelf.

The existence of an oceanic region on the shelf varies and depends on the dynamic situation, which, as I noted above, is subject to strong temporal variability. That is, the location of the Kamchatka Current is a major factor in causing the variability in coastal hydrography. In 1991, the current hugged the northwest slope of the Kamchatka Basin (Fig. 6), whereas in 1992 it had moved seaward. I did not succeed in fixing the exact position of the Kamchatka Current in 1992, since its main flow occurred beyond the limits of the hydrographic survey, despite the fact that the outer CTD stations were located 96-128 km beyond the continental slope. In such a year the influence of the Kamchatka Current on the hydrography of the western Bering Sea shelf must be minimal.

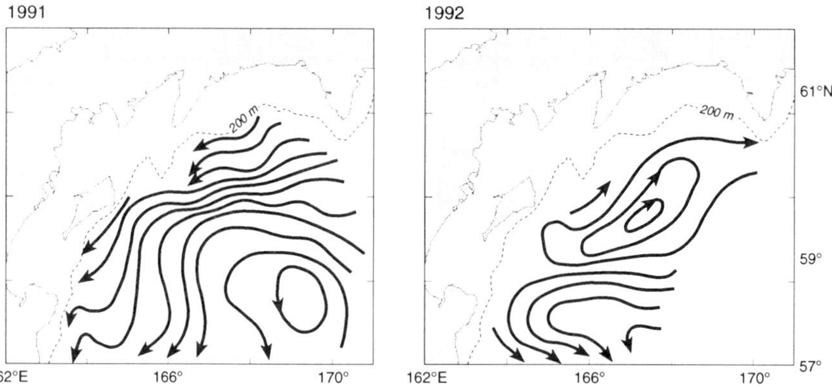

Figure 6.   Geostrophic currents in the summers of 1991 and 1992 in the Kamchatka Basin, using a reference level of 1,000 db. The 1991 data were taken from cruise reports of the combined expeditions of the R/V Professor Kaganovskiy and the R/V Professor Levanidov in 1991. The chief scientist of that expedition was V.P. Shuntov. The 1992 data were taken from the cruise report of the 1992 experiment aboard the R/V Professor Levanidov, headed by A.F. Volkov.

When the Kamchatka Current passes near the continental slope (Fig. 7), its waters are in direct contact with shelf waters, and the hydrographic conditions on the shelf have mixed characteristics. In this case, a 3-layer vertical structure is established over the outer shelf, with a relatively warm and saline near-bottom layer (>1°C and >33.2 psu), formed from near-slope water. Such a structure is typical in the deep basin (Arsenev 1967). In contrast, on the shallower part of the shelf, where access to near-slope water is restricted, nearshore factors play the dominant role in the formation of the hydrographic structure. These factors include coastal drainage, melting ice, and tidal strength, which, combined with solar radiation and wind mixing, contribute to the formation of a 2-layer structure: a warm dilute surface layer over a relatively cold and saline lower layer.

When the Kamchatka Current retreats from the continental slope, as in 1992 (Fig. 8), the intrusion of near-slope water into shallow water ceases, and the near-bottom layer of relatively warm and high-salinity water erodes. Over the outer shelf of Karagin Bay, a 2-layer structure typical of the transitional region is established: warm at the surface and cold below. A shift of all the hydrographic regimes toward the outer edge of the shelf occurs, and the oceanic region is displaced beyond the limits of the shelf. In such a year, the coastal region occupies practically the entire shelf except for the northeast part of Karagin Bay.

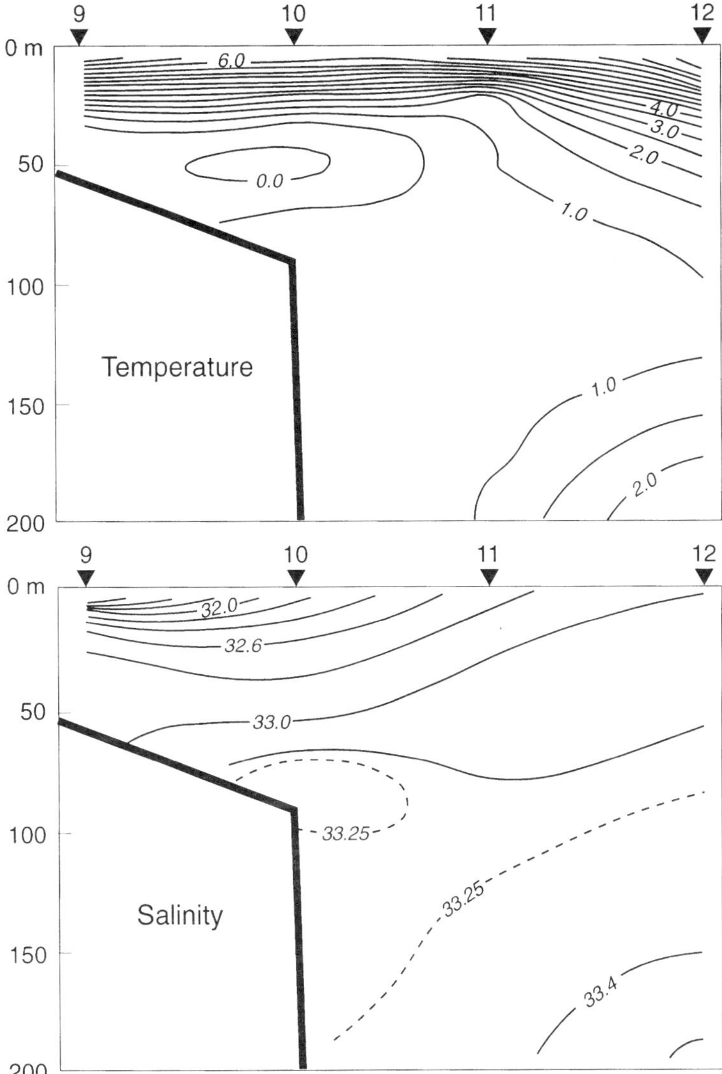

Figure 7.   Vertical sections of temperature (˚C) and salinity (psu), from the shelf to offshore waters (see Fig. 2), summer 1991.

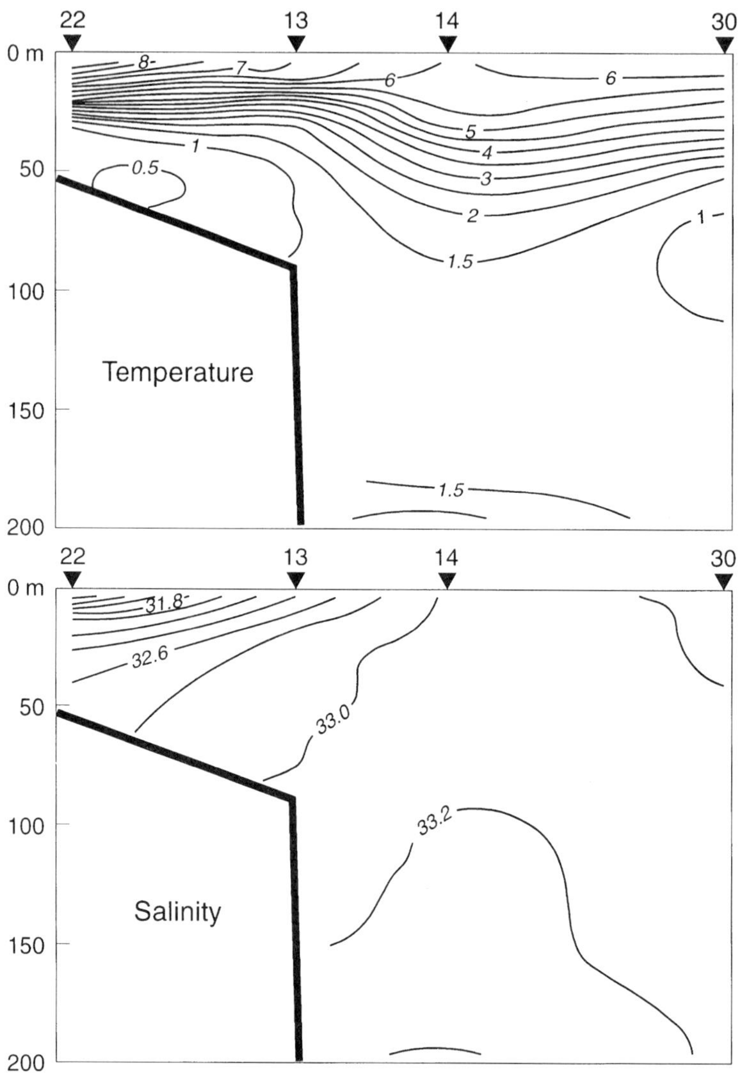

Figure 8.　Vertical sections of temperature (°C) and salinity (psu), from the shelf to offshore waters (see Fig. 2), summer 1992.

Table 1. Water structure on the shelf of the western Bering Sea in summer 1991.

Layer	Depth (m)	Temperature (°C) Range	Mean	Salinity (psu) Range	Mean
Oceanic Region (A)					
Surface	0-10	6.0-9.1	6.7	31.8-32.4	32.2
Pycnocline	10-30	0.5-8.0	3.9	32.2-32.6	32.4
Cold	30-75	−0.4-0.6	0.2	32.6-32.9	32.9
Lower	>75	0.3-0.8	0.7	32.9-33.3	33.2
Transitional Region (B)					
Surface	0-10	7.0-8.3	7.6	30.9-31.9	31.5
Pycnocline	10-20	0.9-7.5	5.1	31.6-32.7	32.1
Lower	>20	0.0-2.0	0.9	32.5-33.2	32.7
Coastal Region C1 (Olyutorsk Bay)					
Surface	0-5	9.0-9.6	9.3	27.1-30.0	29.2
Pycnocline	5-15	1.0-9.0	5.2	28.0-32.2	31.9
Lower	>15	0.3-1.0	0.6	32.0-32.8	32.7
Coastal Region C2 (Karagin Bay)					
Surface	0-5	7.0-9.0	7.9	28.8-31.7	30.2
Pycnocline	5-20	1.5-7.5	5.1	30.0-32.6	31.2
Lower	>20	−0.6-1.5	0.6	32.2-33.1	32.6

The coastal region is distinguished by its Brunt-Vaisala frequency, which is 0.012-0.03/s in its maximum layer. Increased vertical stability of water in the coastal region is caused by the seasonal halocline, which forms during the period of maximum freshwater input in May-June. The layer of maximum Brunt-Vaisala frequency coincides with the seasonal halocline, located in summer at 5-20 m (Table 1). The contribution of the seasonal thermocline to stabilizing vertical layers is noticeably smaller. Thus in the transitional region where the halocline is weak, the Brunt-Vaisala frequency does not exceed 0.01/s and is of the same order as in the oceanic region.

Apart from dynamic factors, synoptic processes that regulate heat exchange between the atmosphere and hydrosphere also have a strong influence on the locations of thermohaline regimes over the shelf. Specific atmospheric conditions change the rate and intensity of formation of an ice field and the period of its melting. These features are quite important in the formation of thermohaline conditions in the warm half of the year, since the entire shelf of the western part of the Bering Sea is covered with ice in winter.

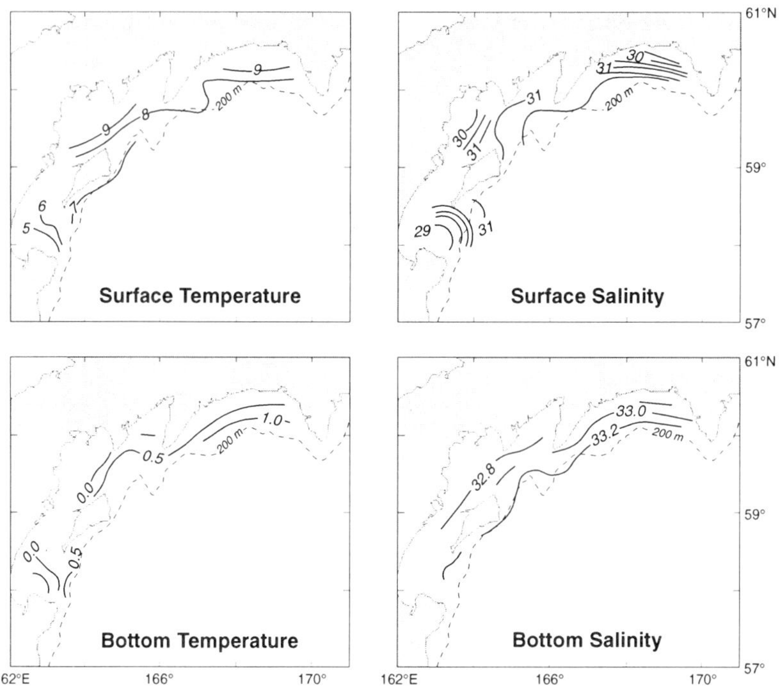

Figure 9. Distribution of temperature (°C) and salinity (psu) on the shelf of the western Bering Sea in summer 1991. The distributions in the surface and bottom layers are shown.

Temperature and salinity are generally aligned with bathymetry on the shelf (Figs. 9, 10). Both at the surface and near the bottom, the salinity decreases from the edge of the shelf toward the shore. In contrast, surface temperature increases shoreward. This latter feature is linked to an increase in the stability of coastal waters, which impedes the penetration of solar radiation below the pycnocline. Heat builds up in a thin surface layer about 5 m thick (Tables 1, 2). This leads to a larger increase in temperature of the surface coastal water than in the transitional region, where heat penetrates to 10-15 m.

A salinity front, separating the coastal region from the transitional region, forms at the surface near the source of the coastal flow. In 1991, clearly marked fronts were seen in Olyutorsk Bay and also in the south and north of Karagin Bay. Salinity was greater than 32 psu at the outer edge of the front, and less than 30 psu at its inner edge. The salinity gradient was 0.05-0.07 psu/km. In 1992, surface salinity at the inner edge of the front was greater than 30 psu, and the gradient was 0.05 psu/km or

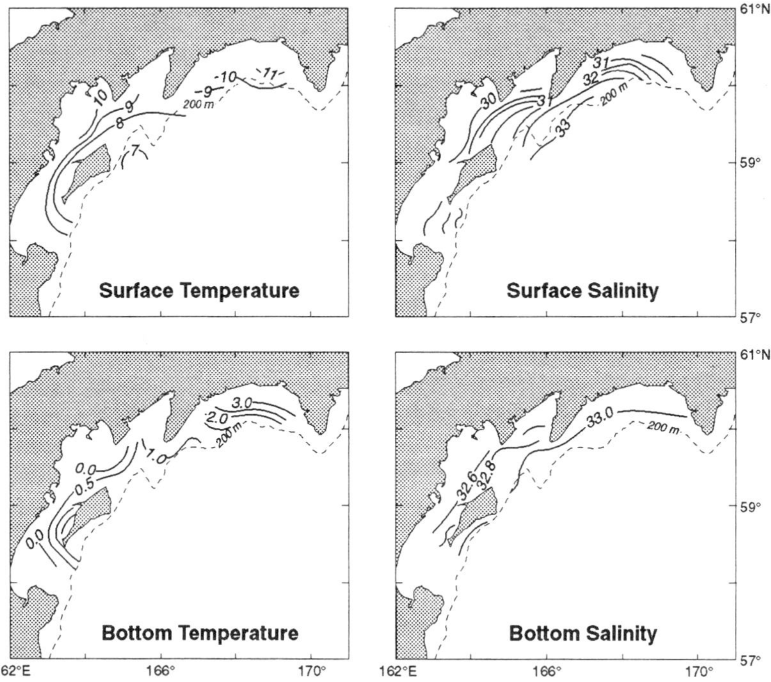

Figure 10. Distribution of temperature (°C) and salinity (psu) on the shelf of the western Bering Sea in summer 1992. The distributions in the surface and bottom layers are shown.

less. Interannual variability of salinity in the interior of the gulfs is largely determined by the volume of melting ice. Thus, the reduction of salinity in the summer of 1991, compared to 1992, is associated with the increased ice cover during the previous winter, as attested by the location of the ice edge in April.

The oceanographic conditions in the cold part of the year on the western Bering Sea shelf affect hydrographic conditions during the following summer. This was discussed by Davydov and Kutsikh (1968) for the case of the entire eastern coast of Kamchatka. If a winter has high ice concentrations, water the next summer will be cool, and its salinity will be low. If little ice is produced during a winter, a regime of water with higher temperature and salinity will be established the next summer. Ice conditions during their respective winters caused both temperature and salinity to be higher in 1992 than 1991 in all the hydrographic regions from the surface to the bottom (Tables 1, 2).

The quantitative characteristics cited in Tables 1 and 2 reflect conditions of the environment only in their respective years and do not necessarily

Table 2. Water structure on the shelf of the western Bering Sea in the summer of 1992.

Layer	Depth (m)	Temperature (°C)		Salinity (psu)	
		Range	Mean	Range	Mean
Transitional Region (B)					
Surface	0-15	6.5-9.7	7.4	32.0-33.0	32.3
Pycnocline	15-30	1.8-6.8	5.6	32.7-33.0	32.8
Lower	>30	0.7-3.5	0.9	32.9-33.1	33.0
Coastal Region C1 (Olyutorsk Bay)					
Surface	0-5	8.0-11.0	10.3	30.5-31.0	30.3
Pycnocline	5-15	4.1-8.0	6.6	30.5-32.5	32.1
Lower	>15	2.0-4.1	3.9	32.0-33.0	32.8
Coastal Region C2 (Karagin Bay)					
Surface	0-5	7.5-10.2	8.9	29.5-31.5	30.6
Pycnocline	5-20	0.6-8.8	5.8	31.0-32.9	31.9
Lower	>20	–0.2-1.5	1.3	32.5-33.0	32.7

apply to other years. Even a comparison of these two consecutive years points out the fairly wide range of possible interannual fluctuations in temperature and salinity of shelf waters. Hydrographic regions with characteristic TS curves will always be preserved, however, while their positions to some extent will reflect thermal dynamic conditions of the western part of the Bering Sea.

Any attempt to give a universal thermohaline assessment of each region is difficult because of their annual repositioning on the shelf. To do the assessment, I would have to define regions that vary over time, shifting from one year to the next. For example, the transitional region in 1992 would occupy the site of the 1991 oceanic region, while the 1991 transitional region was at the location of 1992's coastal region. Although the salinity in the transitional region in 1992 corresponds more to the salinity of the oceanic region from the previous year (Tables 1, 2), the vertical profiles of temperature form the basis of the decision to assign this part of the shelf in one case to the transitional and in the other to the oceanic region. Nevertheless, there is one feature in the distribution of salinity that distinguishes one part of the shelf from another. As noted above, in 1991 the salinity was on the whole lower than in 1992. This was true over the entire shelf except the northeast part of Karagin Bay, where a different hydrographic region was imposed. Here higher salinity, but only in the near-bottom layer, occurred in 1991. This layer occurred because the strong influence of the Kamchatka Current water in the summer of 1991 resulted in the formation of a relatively warm and salty near-bottom layer.

Conclusions

The hydrography of the western Bering Sea shelf is subject to strong influences from atmospheric processes that regulate the heat balance in the ocean. In winter, the entire shelf zone is covered by ice fields, and springtime processes largely depend on the volume of ice present. Cool summers such as 1991 follow heavy ice winters; water in these years is relatively fresh. In 1992, a relatively warm summer was preceded by a mild winter with little ice cover.

In the shelf zone of the western part of the sea, there are three hydrological regions, easily distinguished with TS diagrams: the coastal, transitional, and oceanic. The coastal region contains a 2-layer water structure with lowered salinity (<31 psu) in the surface layer. In the transitional region, the water structure is also 2-layered, but with a less dilute surface layer. A strong seasonal thermocline develops in both regions, separating a warm surface layer from a cold lower layer. One can distinguish the coastal region from the transitional region by their different Brunt-Vaisala frequencies in the layer of the seasonal maximum. In the coastal region the frequency is always greater than 0.012/s, with values up to 0.025-0.03/s near the source of fresher water. In the transitional region, where the halocline is weakly developed, the frequency does not exceed 0.01/s.

Unlike the eastern part of the Bering Sea, on the western Bering Sea shelf the positions of the hydrographic regions are not stationary; they depend on the position of the Kamchatka Current. When the Kamchatka Current flows near the continental slope, all the hydrological regions and their frontal positions are shifted toward the coast. When the current withdraws to the east, however, the regions move out toward the outer zone of the shelf. Another striking difference between waters over the eastern and western shelves is that waters inshore of 50 m in the southeastern Bering Sea are well mixed vertically (Kinder and Schumacher 1981, Coachman 1986), whereas they are not in the region examined here. This appears to result from the relatively strong tidal flow there, which greatly enhances mixing. In fact, semi-daily tidal flow ellipses are roughly ten times greater over the eastern shelf than over the western shelf (Sunderman 1977).

Acknowledgments

The author thanks V.P. Pavlichev and V.A. Abramov for valuable communications, V.I. Radchenko for allowing us to use data from the cruise reports, and S.V. Gladychev for comments and advice. [Translation from Russian by S. Salo and editorial assistance by R.K. Reed, PMEL/NOAA.]

References

Arsenev, V.S. 1967. Currents and water masses in the Bering Sea. Nauka, Moscow. 135 pp. (In Russian.)

Coachman, L.K. 1986. Circulation, water masses, and fluxes on the southeastern Bering Sea Shelf. Continental Shelf Research 5(1/2):23-108.

Coachman, L.K., and R.L. Charnell. 1979. On lateral water mass interaction: A case study in Bristol Bay, Alaska. Journal of Physical Oceanography 9:278-297.

Davydov, I.V. 1972. On the question of oceanographic bases for the formation of the productivity of individual generations of herring in the western part of the Bering Sea. Izvestiya TINRO 82:281-307. (In Russian.)

Davydov, I.V., and A.G. Kutsikh. 1968. Temperature of the core of the cold intermediate layer as a prognostic index of the thermal condition of waters adjoining Kamchatka. Izvestiya TINRO 64:301-308. (In Russian.)

Davydov, I.V., and F.F. Lipetskiy. 1970. On the hydrology of the Kamchatka and Olyutorskiy-Navarinskiy fisheries regions. Izvestiya TINRO 73:178-193. (In Russian.)

Figurkin, A.L. 1992. Some features in the formation and spread of CIL (Cold Intermediate Layer) water in the western part of the Bering Sea. In: I.F. Moroz (ed.), Oceanographic basis for biological production in the northwest part of the Pacific Ocean. TINRO Press, Vladivostok, pp. 20-29. (In Russian.)

Khen, G.V. 1989. Oceanographic conditions and Bering Sea biological productivity. In: Proceedings of the International Symposium on the Biology and Management of Walleye Pollock. University of Alaska Sea Grant, AK-SG-89-01, Fairbanks, pp. 79-89.

Kinder, T.H., and J.D. Schumacher. 1981. Hydrographic structure over the continental shelf of the southeastern Bering Sea. In: D.W. Hood and J.F. Calder (eds.), The eastern Bering Sea shelf: Oceanography and resources, University of Washington Press, Seattle, pp. 31-52.

Sunderman, J. 1977. The semidiurnal principal lunar tide M_2 in the Bering Sea. Deutsche Hydrographische Zeitschrift 30:91-101.

Verkhunov, A.V. 1994. Thermohaline characteristics of the shelf front in the western part of the Bering Sea. Okeanologia 34:356-369. (In Russian.)

CHAPTER 8

The Aleutian North Slope Current

Ronald K. Reed and Phyllis J. Stabeno
Pacific Marine Environmental Laboratory, Seattle, Washington

Abstract

Northward inflows of the Alaskan Stream through Near Strait, Amchitka Pass, and Amukta Pass help produce eastward flow along the northern side of the Aleutian Islands. We call this feature the Aleutian North Slope Current (ANSC). The ANSC is often fairly narrow and shallow with peak speeds sometimes >40 cm/s. Eddylike features also occur in this region. At times, an eastward velocity maximum is present below the near-surface ANSC; this appears to result from the structure of flow in Amchitka Pass. The ANSC transports $2\text{-}4 \times 10^6$ m^3/s and is the main source of the north-westward flowing Bering Slope Current, which in turn provides inflow to the Kamchatka Current on the western side of the Bering Sea basin.

Introduction

A synopsis of upper-ocean circulation over the deep Bering Sea basin, modified from results in Stabeno and Reed (1992, 1994) and Reed (1995a), is shown in Fig. 1. Circulation over the Bering Sea basin is cyclonic, and the mass is provided by inflow of the Alaskan Stream through various passes, but especially Near Strait (Fig. 1; Favorite 1974, Hughes et al. 1974). The Kamchatka Current provides the outflow on the western side of the basin (Dodimead et al. 1963, Ohtani 1970, Stabeno et al. 1994). The Bering Slope Current (Kinder et al. 1975) is the boundary current along the eastern continental slope. It is much weaker than the Kamchatka Current and is often rife with eddies.

The inflow through the Aleutian Island passes creates an eastward flow along the north side of the Aleutians which provides the source for the Bering Slope Current. The characteristics of this eastward flow are distinctive, and we feel it should be treated as a separate flow. It is here termed the Aleutian North Slope Current (ANSC) and is the subject of this contribution. Before discussing the ANSC, however, we examine flow through major Aleutian passes.

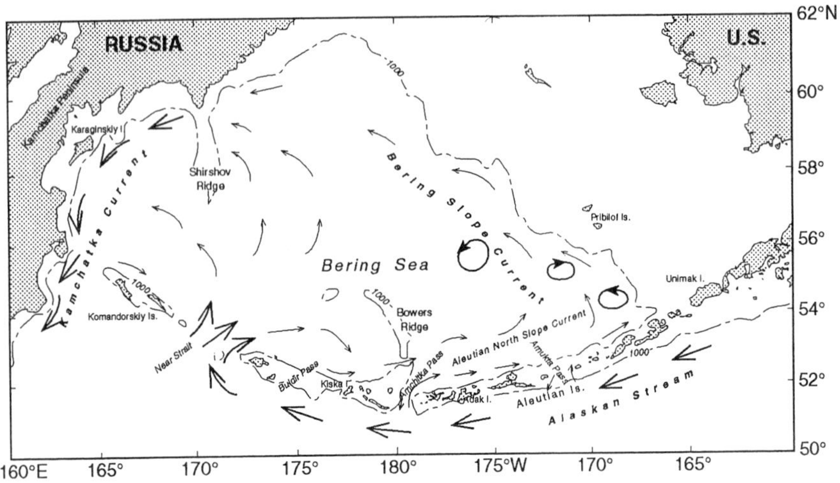

Figure 1.   Schematic upper-ocean circulation over and near the Bering Sea basin. Various geographic features and the 1,000-m isobath are shown.

Inflow Through Passes

Kamchatka Strait

A southward outflow through Kamchatka Strait (Kamchatka Current) is well known (e.g., Dodimead et al. 1963, Ohtani 1970) and appears to vary from ~5 to perhaps 15 Sv (10^6 m^3/s; Reed 1995a); however, upper waters on the eastern side of the strait sometimes flow northward and then eastward along the north side of the Commander Islands (Fig. 1; Reed et al. 1993, Stabeno and Reed 1994). Inflow of North Pacific deep water occurs on the eastern side of Kamchatka Strait below 3,500 m (Reed et al. 1993). This water is present throughout the deep Bering Sea, but the water properties are modified by the long residence time (Roden 1995).

Near Strait

Although inflow of the Alaskan Stream through Near Strait (sill depth ~2,000 m) must provide most of the mass needed for upper-ocean circulation in the western Bering Sea (Fig. 1), the actual long-term transport is not well known. Part of the difficulty results from the boundary between U.S. and Russian territorial waters being in the western strait; hence several cruises have only operated on one side of the boundary. Favorite (1974), however, estimated an inflow of ~10 Sv, based on continuity of Alaskan Stream transport to the south. Ohtani (1970) reported a similar value based on a survey in winter 1966. Stabeno and Reed (1992), however, reported an anomaly in 1991, from drifter and hydrocast data, whereby the Alaskan

Table 1. **Computed northward, southward, and net geostrophic volume transports from hydrocast sections across Amchitka Pass.**

Date	Northward transport (Sv)	Southward transport (Sv)	Net transport (Sv)
5-6 Aug 1991	1.6	4.4	2.8 south
14 Sep 1992	2.6	1.8	0.8 north
7 Sep 1993	4.1	1.3	2.8 north

Stream flowed to the west and south rather than through the strait. Three current moorings were placed in the strait in August 1991 and were retrieved in September 1992 (Reed and Stabeno 1993). These data showed stream inflow returning to Near Strait in fall 1991 and continuing until the end of the record. Overland et al. (1994) discussed model results which showed a disruption of inflow every few years. Hence inflow appears to be quite variable.

Buldir Pass

Buldir Pass (Fig. 1) has a sill depth of ~600 m and may permit significant inflow. Reed and Stabeno (1993) estimated an inflow of ~1 Sv in September 1992. Nothing is known about structure of flow in the pass or how typical this value may be.

Amchitka Pass

Although Favorite (1974) and Reed (1984) reported results from two hydrocast sections each, they were not well situated to derive transports. Reed (1990) also reported results from a year-long current mooring which gave a crude estimate of northward flow of 3 Sv. Again, however, the results were north of the pass and are not really definitive. During the summers of 1991, 1992, and 1993, we took hydrocast sections across the narrow sill region of the pass at 51°28′N (Fig. 1), using the same station spacing (~15 km; similar to the internal radius of deformation). Results were reported in Stabeno and Reed (1992) and Reed and Stabeno (1993, 1994), and they are given in Table 1. All of the results (geostrophic calculations, water properties, and some drifter paths) clearly show bidirectional flow; that is, northward flow (inflow) on the eastern side of the pass and southward flow (outflow) on the western side. The magnitudes of flow are quite variable, however. Although net flows are of interest, the northward transport is most relevant to transports along the Aleutians to the east and in the Bering Slope Current. (The southward branch of transport generally appears to be flows from the western Bering Sea that cross

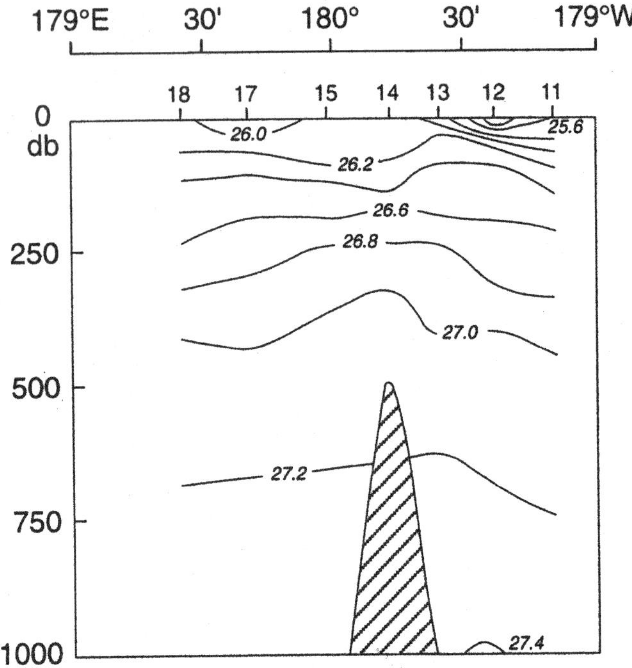

*Figure 2.   Vertical section of sigma-t density across Amchit-
ka Pass (at 51.5°N; stations 11-18, top), September
14, 1992. The stippled area below 500 m at station 14 indicates a shoal area or seamount.*

Bowers Ridge west of the pass.) Thus Table 1 indicates that northward transports varied from 2 to 4 Sv. These flows are similar to those farther east (Reed 1995a), except where flow is intensified in well-developed eddies. The typical density structure in Amchitka Pass associated with these geostrophic flows is shown in Fig. 2.

Data from the year-long current mooring at 51°46′N, 179°31′W in 1,189 m depth (Reed 1990) were used to estimate tidal currents in the deeper part of the pass. The combined M_2 and K_1 constituents, the major semidaily and daily tidal components, were approximately as follows at the three meter depths: at 39 m, 30 cm/s; at 199 m, 20 cm/s; and at 399 m, 15 cm/s. The current hourly vectors, during a tidal cycle, were elliptical and clockwise rotating. The above values are likely typical and fairly consistent in 1,000-1,200 m depths, where the geostrophic transports were derived. Only in shoaler areas such as at station 14 (Fig. 2) is tidal mixing appreciably enhanced as indicated by the relatively small vertical gradients of density.

It does not appear that wind stress or wind-stress curl influences the northward inflow of Alaskan Stream water through Amchitka Pass (Reed 1990). It was suggested instead that the offshore-trending "spur" in the bottom topography just east of the pass (see the 1,000-m contour in Fig. 1) is important. If the Alaskan Stream were in its typical inshore position, it would "feel" the shoal topography and turn to the right as a result of vorticity conservation. Conversely, if it were farther offshore, the bathymetric effect, and inflow, would be reduced.

Amukta Pass

The first hydrocast section across Amukta Pass (Fig. 1), which is 300-400 m deep, was taken in September 1993 and was described by Reed and Stabeno (1994). It should be stressed that this pass is only about one-half the width and one-third the depth of Amchitka Pass. Its circulation, though, was also bidirectional, with northward flow on the eastern side and southward flow on the western side. Subsequent sections (Reed and Stabeno 1997) show a continuation of the bidirectional pattern. The southward geostrophic flow (referred to a reference level at the bottom), however, often results from retroflection of the northward flow (Reed and Stabeno 1997). There was no apparent relation between computed flow and predicted tidal flow; effects of tidal mixing are apparent, but they appear to be fairly constant across the pass. Reed (1995b) examined subsurface temperatures north and east of the pass and found considerable variability. Hence evidence for Alaskan Stream water (subsurface temperature >4.0°C) was found at some times but not at others. Although net transport appears to generally be <1 Sv, large variations in subsurface temperature occur.

Near-Surface Flow North of the Aleutians

We now examine evidence for a distinct, narrow flow near the northern slope of the Aleutian Islands, which we call the Aleutian North Slope Current (ANSC; Fig. 1). We envision this flow as distinct from the Kamchatka Current and the Bering Slope Current, although it is the source of much of the latter flow.

Direct Measurements

The study by Stabeno and Reed (1994), based on data from 86 satellite-tracked drifting buoys during 1986-1993, clearly shows eastward flow all along the north side of the Aleutian Islands. The flow is quite narrow and strong (up to 40 cm/s), especially to the east of ~171°W. This is notable because the flow vectors were derived by spatially averaging over 0.5° of latitude and 1.0° of longitude, which would tend to reduce peak speeds. Stabeno and Reed (1994) termed this feature the "slope flow"; we now, however, use the term Aleutian North Slope Current (ANSC).

Table 2. **Information on the seven buoys shown in Fig. 3, with the date each entered the region, the number of days each was in the slope flow, and the mean velocity.**

Buoy number	Start date	Length (days)	Mean velocity (cm/s)
5611	3-18-86	12.0	37
2329	4-3-86	8.6	21
7241	8-27-87	25.1	17
7160	12-4-89	7.8	40
7167a	2-25-90	7.5	35
7167b	2-25-94	7.2	53
7237	7-30-94	10.3	35

The paths of seven satellite-tracked drifters, mainly between 172° and 167°W, are shown in Fig. 3. In general, ten or more fixes, with an accuracy of 0.2 km, were received each day. Some of the paths lack smoothness, especially of drifters 5611 and 7241, due to poor satellite reception. All of the drifters moved northeastward over a narrow zone, with some suggestion of bathymetric steering. The mean velocities varied between 17 and 53 cm/s, with an overall mean of 34 cm/s (Table 2). There is a suggestion of enhanced flow in winter, but the sample is small. The flow shown in Fig. 3 is stronger than the ANSC farther west (Stabeno and Reed 1994). We suggest that the strong flow to the east results from northward inflow through Amukta Pass. Although the volume transport of this inflow is not large, its impulse might increase velocities in the ANSC by impulse-momentum conservation ($Fdt = mdv$, where F, t, m, and v are force, time, mass, and velocity; see, for example, Collins and Pattullo 1970).

Two drifters to the west of Amukta Pass moved eastward (Fig. 4), but over a wider offshore zone than the drifters shown in Fig. 3. Both drifters moved southward across the Aleutian ridge through Amlia Canyon, the semicircular feature in the 1,000-m isobath near 173°W. One of the drifters entered the Alaskan Stream south of the islands, and the other remained on the shelf until it ceased transmission. Although we have stressed northward inflow through the deep passes, flow can occur in either direction through shallow passes between islands. Such flows, however, would have insignificant transport.

We have thus presented evidence from direct measurements that the ANSC is a coherent feature along the Aleutian Islands. To the east of ~171°W, the flow is narrow (perhaps only ~20 km), and nontidal speeds often exceed 40 cm/s.

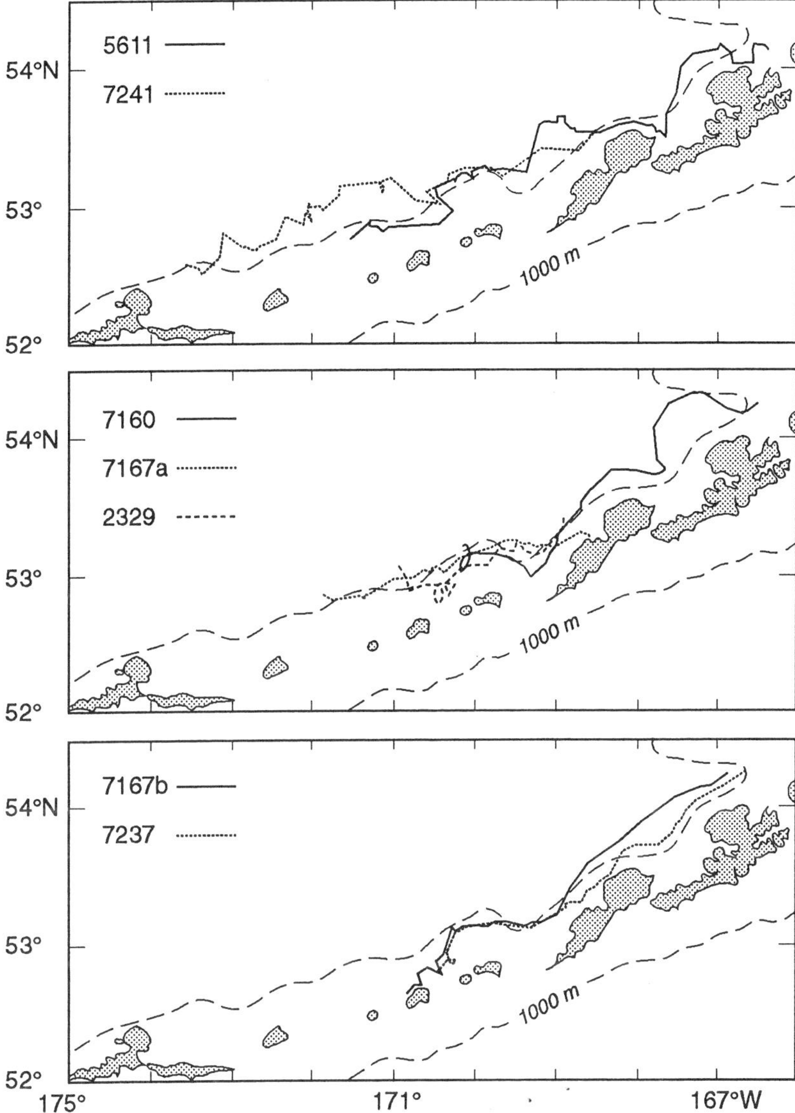

Figure 3. *Paths of seven satellite-tracked drifting buoys, drogued at 40 m, near the 1,000 m isobath north of the Aleutian Islands. The dates of the trajectories and the mean velocities are given in Table 2.*

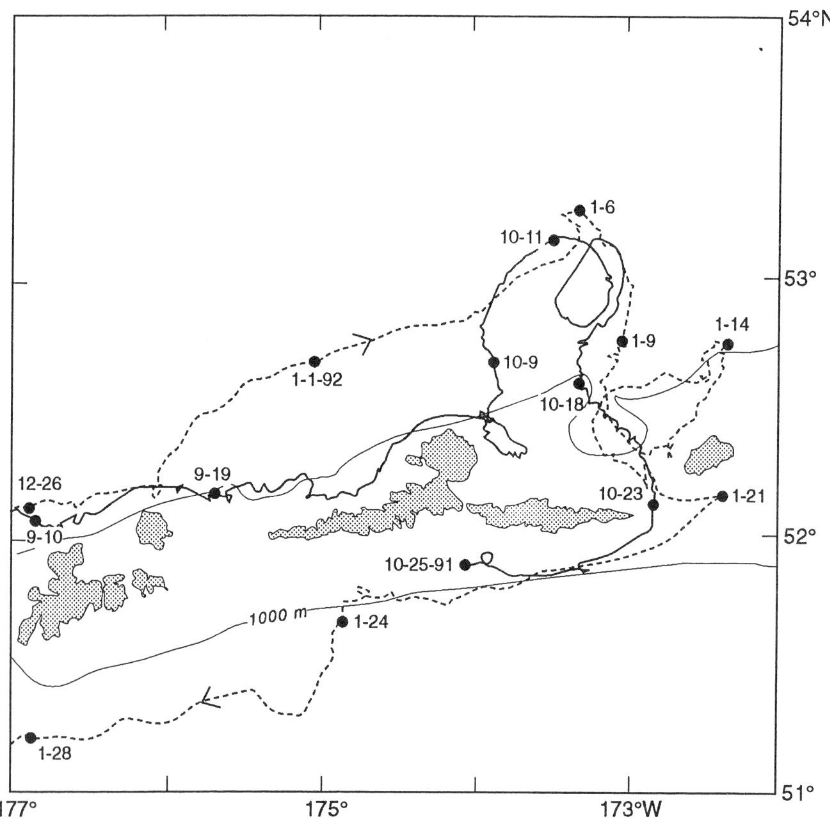

Figure 4.   *Paths of two satellite-tracked drifting buoys, drogued at 40 m, that moved southward through Amlia Canyon (the semicircular feature in the 1,000 m isobath near 173°W). Selected dates (month-day-year) are given.*

Geostrophic Flow

Figure 5 shows surface geostrophic flow, referred to 500 db, from data collected during a synoptic survey in September 1993 (Reed 1995a). The northward inflow of Alaskan Stream water in Amchitka Pass turned to the right and flowed eastward offshore of the Aleutian Ridge. At that time, the surface flow was not especially narrow or intense. Note, also, the well-developed cyclonic circulation to the east of 171°W. There was, however, an easily recognizable ANSC north of the Aleutian Ridge.

The general lack of synoptic data and determination of a proper reference level for computation of flow are major problems in using the geostrophic relation in the Bering Sea. In the Alaskan Stream and the eastward

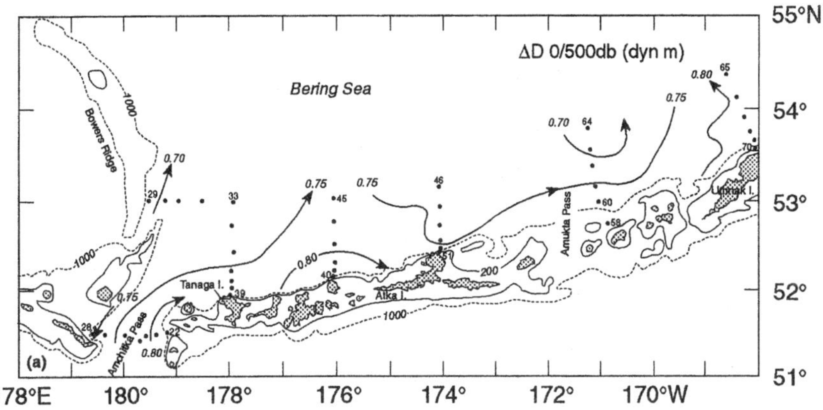

Figure 5.   *Geopotential topography of the sea surface, referred to 500 db, September 6–12, 1993.*

flow just south of it, baroclinic flow is quite deep; below the upper 200 m or so, however, it decreases continuously downward (Warren and Owens 1988). Hence one needs to use near-bottom reference levels to determine realistic total volume transports, but near-surface speeds can be derived (with errors of ~10-15%) by use of intermediate levels (1,500-1,000 m). Reed (1995a) examined this problem in the Bering Sea. Because direct current measurements were generally lacking during hydrocast surveys, some other method had to be used. The relation,

$$\frac{d(\Delta D_2 - \Delta D_1)}{dz} \cong 0 \tag{1}$$

where z is depth downward and $\Delta D_2, \Delta D_1$ are geopotential anomalies at stations 2 and 1, summed from the sea surface downward, was used (McLellan 1965). The method was found to give results that did not violate mass conservation and that were in agreement with physical property distributions. According to Reed (1995a), reference levels above 1,500 m occur perhaps half the time, especially near the ANSC, and improper choice of reference levels may change transport by a factor of two.

Figure 6 shows the vertical structure of geopotential anomaly differences from conductivity, temperature, depth (CTD) sections just north of the Aleutian Ridge, near 176-177°W, during three recent NOAA/PMEL cruises. The 1992 data indicate geopotential shear increasing downward to at least 1,000 db. Conversely, the 1993 and 1995 data have shear that suggests minimum velocity levels, using a standard error of ±0.002 dyn m, of 500-700 and 400-600 db, respectively. Figure 7 shows vertical sections of geostrophic flow along 176.1°W during 1993 and 1995, using reference

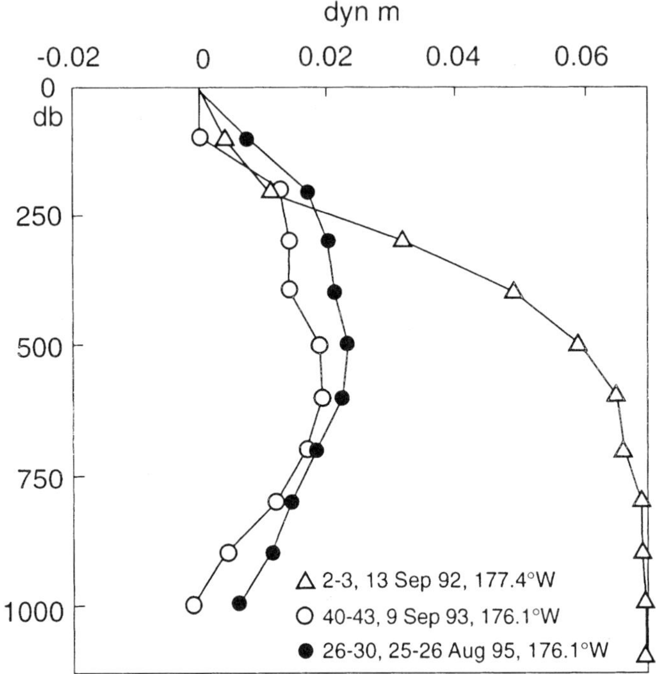

*Figure 6.   Vertical structure of geopotential anomaly differences
(in dyn m), between stations indicated, north of the
Aleutian Islands during three years.*

levels as inferred from the relation above. (The levels used for individual
station pairs depart somewhat from those across the entire flow as in Fig.
6). In September 1993, geostrophic speeds above 700 db were quite weak,
with some zones of westward as well as eastward flow. Between stations
40 and 41, however, there was a deep, narrow geostrophic flow with speeds
at 1,000 db of ~40 cm/s. The occupation of this section in August 1995
showed a surface flow >60 cm/s just offshore of 1,000 m but no appreciable deep flow. Hence considerable temporal variability occurs, especially
to the west of Amukta Pass. Most data sets also show considerable spatial
variability during synoptic surveys, as in Fig. 5. It should be stressed that
we have been dealing only with the density-driven or baroclinic flow, which
varies greatly in the vertical as in Fig. 7, and not with motion caused by
constant slopes, with constant velocity, in the vertical (barotropic flow).
This latter type of motion might well be significant near the margin in a
semi-enclosed sea, but one such estimate (Reed and Stabeno 1997) was
only 2 cm/s.

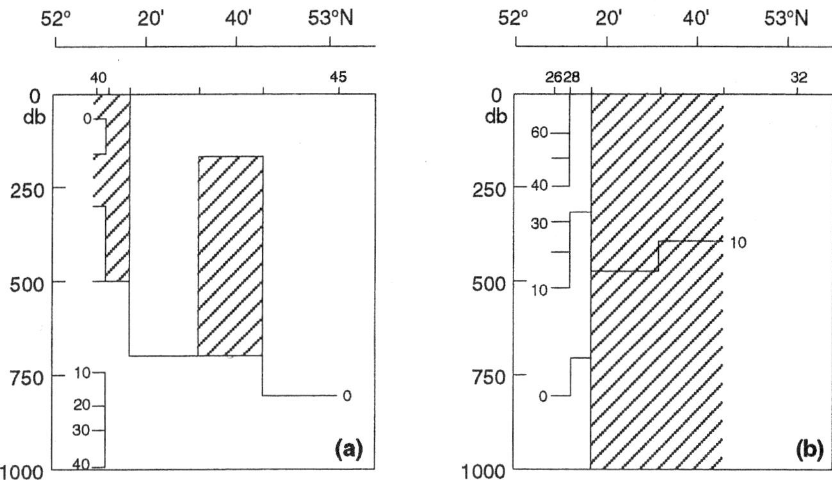

Figure 7. *Vertical sections of computed geostrophic flow (cm/s) across 176.1°W during (a) September 9, 1993 and (b) August 25–26, 1995. Flow is eastward, except where shaded (westward).*

What causes the variable reference levels, which are sometimes at intermediate depths near the continental margin to the west of Amukta Pass? Reed (1995a) suggested that the vertical structure of flow in Amchitka Pass and rising water near the bottom over the inner slope might be factors. Figure 8 shows geopotential shear between two sites in the inflow region of Amchitka Pass during three cruises (see also Table 1). In September 1992, shear increased downward to at least 800 db; to the east in the ANSC, shear increased downward to near-bottom in September 1992 (Fig. 6; Reed and Stabeno 1993). Conversely, in September 1993, Fig. 8 suggests a reference level at 300-400 db with northward flow below, which became quite weak near the bottom. To the northeast in the ANSC (Fig. 6), an intermediate reference level occurred. We suspect that the mechanism for this variability may be the inshore location of the Alaskan Stream. As discussed by Reed (1990) and above, the inflow of the stream may be affected by a bathymetric "spur" near 179°W (see Fig. 9). If the stream were inshore, its deeper part would be partially blocked by this feature, and flow in eastern Amchitka Pass and in the ANSC just downstream would be shallow.

Figure 9 shows sigma-*t* density at 800 db during the cruise in September 1993. Note the absence of water denser than 27.30 in the Bering Sea except at the most shoreward stations at 176.1° and 178°W. The increase there was 0.06 and 0.07 sigma-*t* toward shore on these sections, but the gradient was much larger at 176.1°W. Although the exact mechanism is

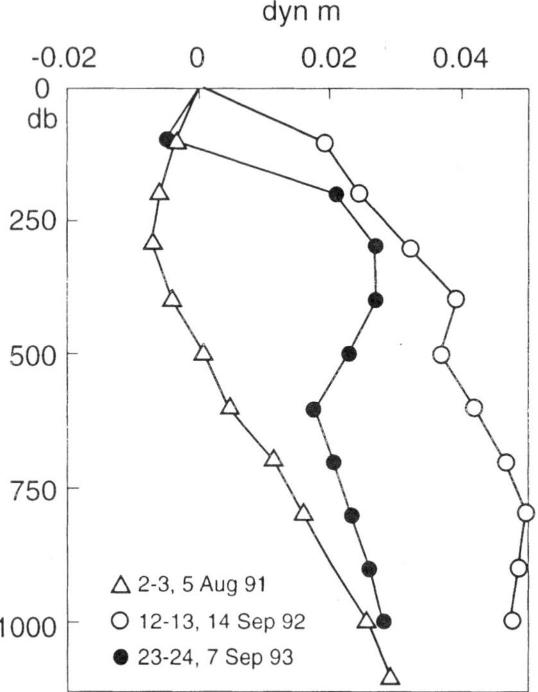

Figure 8.   *Vertical structure of geopotential anoma-*
ly differences (in dyn m) between stations
at the same locations in Amchitka Pass.

not known, it implies "rising" water near the bottom, and this feature seems to be crucial to the existence of the deep, strong eastward subsurface velocity maximum at 176.1°W in Fig. 7.

Existence of Subsurface Velocity Maxima

During a World Ocean Circulation Experiment cruise in July 1993 (two months before our cruise in September 1993), a subsurface velocity maximum (eastward flow) of 21 cm/s at 1,350 m was found just north of Amchitka Pass (Roden 1995). It was below a near-surface eastward flow with a maximum of 32 cm/s. It is not clear if Roden's feature is linked to our 40 cm/s subsurface flow (Fig. 7) because our section at 178°W (Fig. 9) only had a near-bottom eastward flow of 8 cm/s. Even if there was a linkage, it would appear that our flow at 176.1°W was enhanced locally. We do not claim that well-developed subsurface maxima are ubiquitous in the ANSC. First, geopotential shear in the ANSF often increases downward to the bottom, precluding such a feature. Second, even with an intermediate

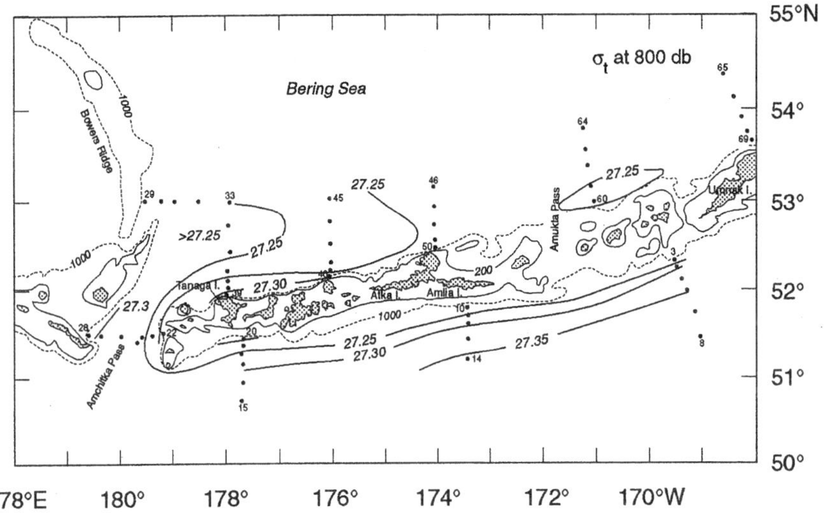

Figure 9. Sigma-t density at 800 db, September 4–12, 1993.

reference level, the near-bottom flow may be weak, as at 178°W in September 1993. It would appear that special conditions are required to create a strong subsurface velocity maximum.

Conclusions

The Aleutian North Slope Current (ANSC) has distinctive characteristics that set it apart from the Bering Slope Current to the north. The ANSC is narrow (~30 km), often shallow (<500 m), and near the north side of the Aleutian Ridge. Peak speeds in the ANSC are quite variable, but there is a tendency for higher speeds (often >40 cm/s, especially in winter; Stabeno and Reed 1994) to the east of Amukta Pass, perhaps as a result of the impulse of northward flow through the pass. Well north of the ANSC, there are sometimes significant flows or eddy-like features that may result from flow across Bowers Ridge; its source is presumably eastward flow north of the western Aleutians, which has not been discussed in detail here.

Transport of the ANSC is generally 2-4 Sv, which is the magnitude of the northward Alaskan Stream inflow through Amchitka Pass. The variation in this inflow, however, especially its vertical structure, appears to be crucial to the detailed structure of the downstream ANSC. Some evidence was also presented for local variations in density structure, and flow, near the ridge slope that may result from upward motion near the bottom. The ANSC is, of course, the source of the northwestward-flowing Bering Slope Current, whose mean transport is not well known. This latter feature, however, is unlike the ANSC in that it is broad, slow, convoluted, and rife with eddies. The ANSC is nearest the Alaskan Stream inflow that is the source of Bering Sea circulation; it is also more like the organized, stable stream flow than the other currents in the Bering Sea.

Acknowledgments

We thank the officers and crews of the NOAA ships *Miller Freeman* and *Surveyor.* We also thank W. Parker, C. DeWitt, C. Hadden, and L. Lawrence. This is contribution FOCI-B272 to the Fisheries Oceanography Coordinated Investigations and is part of the Coastal Ocean Program of NOAA; this is contribution 1722 from NOAA/Pacific Marine Environmental Laboratory.

References

Collins, C.A., and J.G. Pattullo. 1970. Ocean currents above the continental shelf off Oregon with a single array of current meters. Journal of Marine Research 28:51-68.

Dodimead, A.J., F. Favorite, and T. Hirano. 1963. Review of oceanography of the subarctic Pacific region. International North Pacific Fisheries Commission Bulletin 13. 195 pp.

Favorite, F. 1974. Flow into the Bering Sea through Aleutian Island passes. In: D.W. Hood and E.J. Kelley (eds.), Oceanography of the Bering Sea with emphasis on renewable resources. Occasional Publication No. 2, Institute of Marine Science, University of Alaska, Fairbanks, pp. 3-37.

Hughes, F.W., L.K. Coachman, and K. Aagaard. 1974. Circulation, transport and water exchange in the western Bering Sea. In: D.W. Hood and E.J. Kelley (eds.), Oceanography of the Bering Sea with emphasis on renewable resources. Occasional Publication No. 2, Institute of Marine Science, University of Alaska, Fairbanks, pp. 59-98.

Kinder, T.H., L.K. Coachman, and J.A. Galt. 1975. The Bering Slope Current system. Journal of Physical Oceanography 5:231-244.

McLellan, H.J. 1965. Elements of physical oceanography. Pergamon Press, Oxford. 151 pp.

Ohtani, K. 1970. Relative transport of the Alaskan Stream in winter. Journal of the Oceanographical Society of Japan 26:271-282.

Overland, J.E., M.C. Spillane, H.E. Hurlburt, and A.V. Wallcraft. 1994. A numerical study of the circulation of the Bering Sea basin and exchange with the North Pacific Ocean. Journal of Physical Oceanography 24:736-758.

Reed, R.K. 1984. Flow of the Alaskan Stream and its variations. Deep-Sea Research 31:369-386.

Reed, R.K. 1990. A year-long observation of water exchange between the North Pacific and the Bering Sea. Limnology and Oceanography 35:1604-1609.

Reed, R. K. 1995a. On geostrophic reference levels in the Bering Sea basin. Journal of Oceanography 51:489-498.

Reed, R.K. 1995b. On the variable subsurface environment of fish stocks in the Bering Sea. Fisheries Oceanography 4:317-323.

Reed, R.K., and P.J. Stabeno. 1993. The recent return of the Alaskan Stream to Near Strait. Journal of Marine Research 51:515-527.

Reed, R.K., and P.J. Stabeno. 1994. Flow along and across the Aleutian Ridge. Journal of Marine Research 52:639-648.

Reed, R.K., and P.J. Stabeno. 1997. Long-term measurements of flow near the Aleutian Islands. Journal of Marine Research 55:565-575.

Reed, R.K., G.V. Khen, P.J. Stabeno, and A.V. Verkhunov. 1993. Water properties and flow over the deep Bering Sea basin, summer 1991. Deep-Sea Research 40:2325-2334.

Roden, G.I. 1995. Aleutian basin of the Bering Sea: Thermohaline, oxygen, nutrient, and current structure in July 1993. Journal of Geophysical Research 100:13539-13554.

Stabeno, P.J., and R.K. Reed. 1992. A major circulation anomaly in the western Bering Sea. Geophysical Research Letters 19:1671-1674.

Stabeno, P.J., and R.K. Reed. 1994. Circulation in the Bering Sea basin observed by satellite-tracked drifters: 1986-1993. Journal of Physical Oceanography 24:848-854.

Stabeno, P.J., R.K. Reed, and J.E. Overland. 1994. Lagrangian measurements in the Kamchatka Current and Oyashio. Journal of Oceanography 50:653-662.

Warren, B.A., and W.B. Owens. 1988. Deep currents in the central subarctic Pacific Ocean. Journal of Physical Oceanography 18:529-551.

Physical Environment Around the Pribilof Islands

Phyllis J. Stabeno, James D. Schumacher, and Sigrid A. Salo
Pacific Marine Environmental Laboratory, Seattle, Washington

George L. Hunt Jr.
University of California, Irvine, California

Mikhail Flint
Shirshov Institute of Oceanology, Russian Academy of Sciences, Moscow, Russia

Abstract

The Pribilof Islands are located on the southeast Bering Sea shelf, near the shelf break. They form a natural laboratory to study changes in climate. Mean wind speeds are weakest during July and typically out of the south. From September through June winds are typically out of the north, with strongest winds occurring in winter. Northerly winds result in the advection of ice toward the Pribilof Islands. During most winters (~75%), ice reaches St. Paul Island. The currents flow clockwise around the islands. Sea surface temperatures are determined by the depth of the mixed layer. North of the islands bottom temperature is determined by the previous winter's ice extent, while to the south bottom temperature is determined by advective processes.

Introduction

The Pribilof Islands (Fig. 1) comprise an archipelago of five islands, of which the two largest (St. Paul and St. George islands) comprise over 97% of the total area. Located on the seaward edge of the continental shelf approximately 475 km from the Alaska mainland and 370 km from Unimak Pass and the Alaska Peninsula, they are the only islands on the eastern Bering Sea shelf in proximity to the shelf break. Each island is surrounded by a band of weakly stratified or well-mixed water, stirred by tidal currents and wind, which forms a coastal domain analogous to that found around the perimeter of the southeastern Bering Sea. A structure

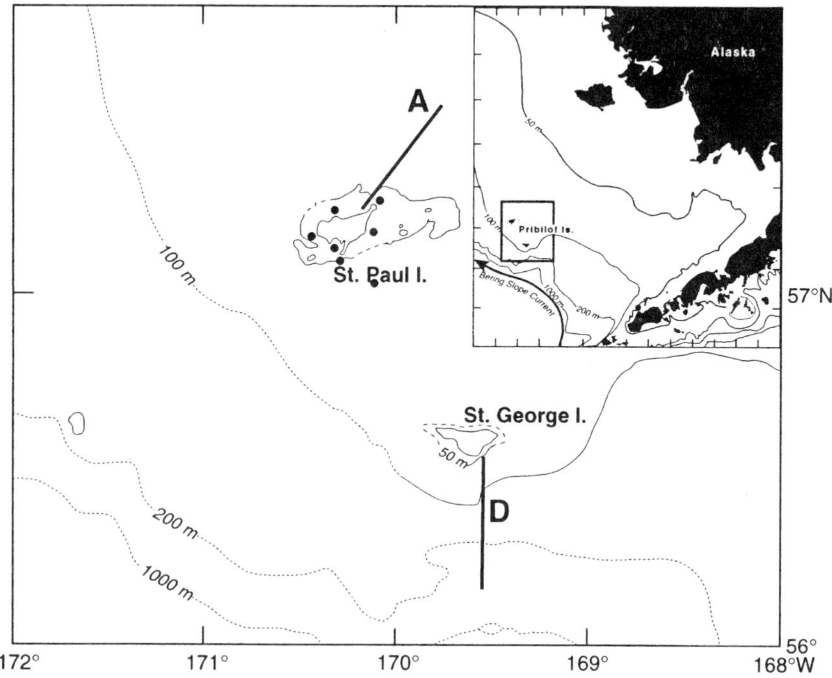

Figure 1.   The Pribilof Islands. The hydrographic lines discussed in this manuscript
are indicated by the dark lines and mooring locations are indicated as
dots.

front separates this domain from the surrounding two-layered middle shelf water (Schumacher et al. 1979; Coachman 1986; Schumacher and Stabeno 1998). Simulations from a tidal current model (Kowalik, chapter 4, this volume) indicate that the interaction of tides with island bathymetry results in a residual flow around the islands. In addition, the proximity of St. George Island to the shelf break restricts the broad flow over the outer shelf into a narrow (~25 km wide) band. With this marked change in the cross-sectional area comes an increase in the mean speed (Coachman 1986; Schumacher and Stabeno 1998).

Owing to their location near the shelf break, these islands are home for large populations of marine mammals, seabirds, and fish. The system of fronts around the islands influences the distribution of phytoplankton found in the vicinity of the islands. This, in turn, influences where zooplankton and their predators concentrate (Coyle and Cooney 1993, Hunt et al. 1996, Brodeur et al. 1997). Interaction of tidal currents with the shelf break, acceleration of the along-shelf current (which could draw up deeper water from Pribilof Canyon), and interaction between eddies and the shelf break are all possible mechanisms for providing nutrients vital to sustain this productive ecosystem. It is thought that this region, also known as the Green Belt (Springer et al. 1996), experiences prolonged production. The Pribilof Islands are positioned on the shoreward edge of the most productive part of the Green Belt.

Although the Pribilof Islands occupy an important place in the ecosystem, there is no research publication, which can be referenced, presenting the nature of the physical environment. To fill this void, we present a compilation of meteorological, sea ice, and physical oceanographic observations of the Pribilof Island area. The analysis of these data permit a description of the salient characteristics of the physical environment, providing insight into annual and longer period fluctuations, variations about mean conditions, and needs for further research. We begin our examination of the Pribilof Island environment with a description of atmosphere and ice features, followed by direct observations of circulation. We conclude our presentation of results by describing the hydrographic features.

Data Sources and Preparation

Atmospheric

To examine the climate at St. Paul Island, we obtained time series (spanning the years 1945-1996) of winds, temperature, air pressure, and precipitation from the National Climatic Data Center. The data from January 1950 to August 1997 are relatively complete, excluding a gap from September to December 1955. The frequency of observations varies, but except for the period from 1976 to 1981, gaps greater than 6 hours are uncommon. During this period, data were often collected (every 3 hours) from 0100 to 1300 local time. During the rest of the record, observations were made hourly or every 3 hours.

The wind, temperature, and pressure data were fitted using a spline and then subsampled at 3-hour intervals. We then averaged the resampled data to provide daily-mean and monthly-mean values. An annual signal was constructed by averaging each day's data over all the existing years. In non-leap years, we calculated the daily averages for day 366 by averaging the values from its adjoining days. The 4-month gap in 1955 was not filled.

Precipitation data were originally qualitatively coded, for example, as "light rain" or "light snow showers." We quantified precipitation by assigning a number to each type of precipitation: 0 for no precipitation, 1 for light drizzle, 2 for light rain or moderate drizzle, and up to 6 for heavy rain or snow showers. Using these codes, average daily liquid precipitation, frozen precipitation, and total precipitation were calculated by summing the codes from each observation in a day and then dividing by the number of observations.

Oceanic

Location of the ice edge in the Bering Sea was obtained from the compact disc (CD) produced by the National Ice Center, the Fleet Numerical Meteorology and Oceanography Center, and the National Climatic Data Center. This data set contains information on the ice concentration and ice thickness from weekly satellite images from 1972 through 1994.

Conductivity, temperature, and depth (CTD) transects around the Pribilof Islands were conducted during eight summers. During 1987 and 1988, data along sixteen transects (eight radiated out from each island) were collected. In 1989 and 1990 transects south of St. George Island were made by scientists from NOAA's Fisheries Oceanography Coordinated Investigations (FOCI) as part of Outer Continental Shelf Environmental Assessment Program (OCSEAP). In 1993, eight transects were occupied, and in 1994-1997 FOCI and South East Bering Sea Carrying Capacity (a Coastal Ocean Program of NOAA) occupied three to four lines each year. Unfortunately, none of the transects from 1987, 1988, and 1993 were synoptic enough to permit the presentation of contours of water properties and calculation of geostrophic flow around the islands. In addition, during 1994-1997, too few transects were completed to define the continuous water properties around the islands. Therefore, we have chosen to compare the two most densely sampled areas, one northeast of St. Paul Island and the other south of St. George Island (Fig. 1).

On September 13, 1995, seven taut-wire moorings with eight current meters (Aanderaa model RCM 7) and three miniature temperature recorders (MTR) were deployed in the vicinity of St. Paul Island (Fig. 1) as part of a study funded by the Environmental Protection Agency (EPA). On August 7, 1996, six moorings were recovered; the seventh mooring was recovered on September 17, 1996. Current speed and direction, temperature, and conductivity were sampled at hourly intervals. Rotor fouling eventually occurred on each of the current meters, but useful data were obtained for at least 2 months at each site. Temperature on all current meters and the MTRs was obtained for the full term of deployment. The time series presented in this paper were low pass filtered with a cosine-squared, tapered Lanczos filter (half amplitude 35 hours, half power 42 hours). This filter is designed to remove the daily and semi-daily tidal current signals and other less energetic high frequency variability. The time series were then resampled at 6-hour intervals.

Satellite-tracked drifters were deployed on the southeast Bering Sea shelf. All drifters were equipped with holey sock drogues centered at a depth of ~40 m. Positions were obtained through Service Argos. In this region an average of twelve fixes were obtained each day. Each buoy had a tilt switch to determine when the drogue was lost. Only trajectories of drifters with drogues are presented.

Observations

Atmosphere-Ice

The climatological annual signal for sea level pressure (SLP), air temperature, and wind vectors were calculated over the 47-year time series (Fig. 2). An annual signal is evident in each of the time series, with the greatest variability in each time series occurring during the winter months. Maximum SLP occurs during July and minimum during December. Warmest air temperature is in August, when the daily mean temperature often exceeds 10°C. In November, the mean temperature drops below 0°C and stays below freezing until the end of April. The lowest mean temperatures (–5°C), and the lowest minimum temperature ever recorded (–20°C) occurred in February. From daily time series (not shown), it is apparent that the coldest temperatures were "cold snaps" associated with northerly winds that lasted a few days. These usually occurred when sea ice was in the vicinity of the islands. During such times, the temperature typically dropped to below –10°C. Excluding these cold snaps, the air temperature (strongly influenced by the surrounding water) was near or slightly above 0°C for much of the winter. The warmest days occurred during periods of weak or southerly winds. Under these wind conditions, temperatures above 0°C occurred even when the ice edge was actually at or south of St. Paul Island.

Mean wind speed (Figs. 2d, 3) is weakest during July. Typically winds are below 10 m/s, with calm days (<2 m/s) occurring 7% of the time (Fig. 3). Strongest winds occur in winter, particularly in February. From September through June, winds are typically out of the north, but during early July the winds shift and come out of the south (Fig. 2d). Northerly winds result when high atmospheric pressure (caused by cold dry air) occurs over the continents to the north, while a low pressure overlies the western Aleutian Islands. Typically, the southerly winds last through August even as the wind speeds increase. The seasons observed at St. Paul Island are comparable to those based on atmospheric circulation patterns (Overland 1981).

Ice plays an important role in modifying the climate of the Pribilof Islands; however, its maximum extent is highly variable. Sea ice forms in polynyas located on the leeward side of the islands in the northern Bering Sea and of the mainland, and is advected southward (Overland and Pease 1982). The leading edge continues to melt, cooling the water to near freezing and reducing the ambient salinity. When the sea ice extends into the warmer slope and basin water, it is unable to sufficiently cool the deeper basin water which is warmed by the northward flowing Bering Slope Current (BSC).

The position of the ice edge varies week to week by tens of kilometers dependent largely upon the direction and magnitude of the winds. In 1976

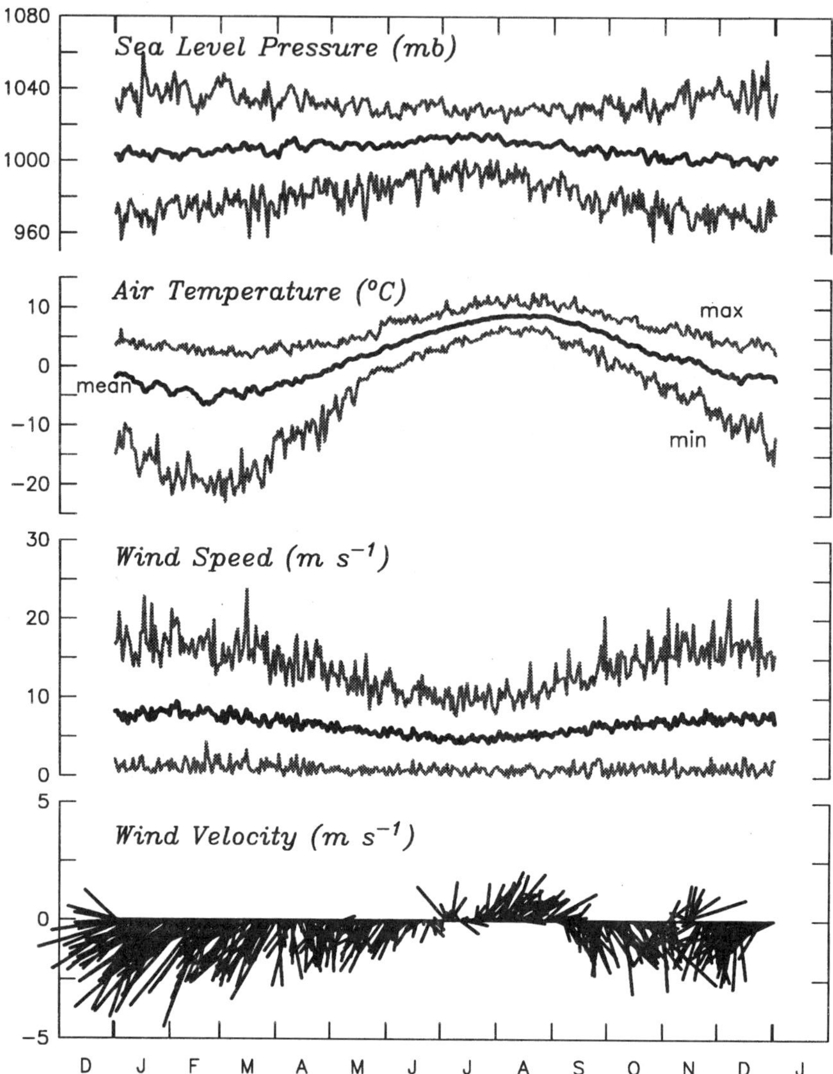

Figure 2.   Daily mean, minima and maxima of: (a) sea level pressure, (b) air temperature, (c) wind speed, and (d) mean daily wind velocity. All daily averages were computed for the period 1950-1954 and 1956-1996.

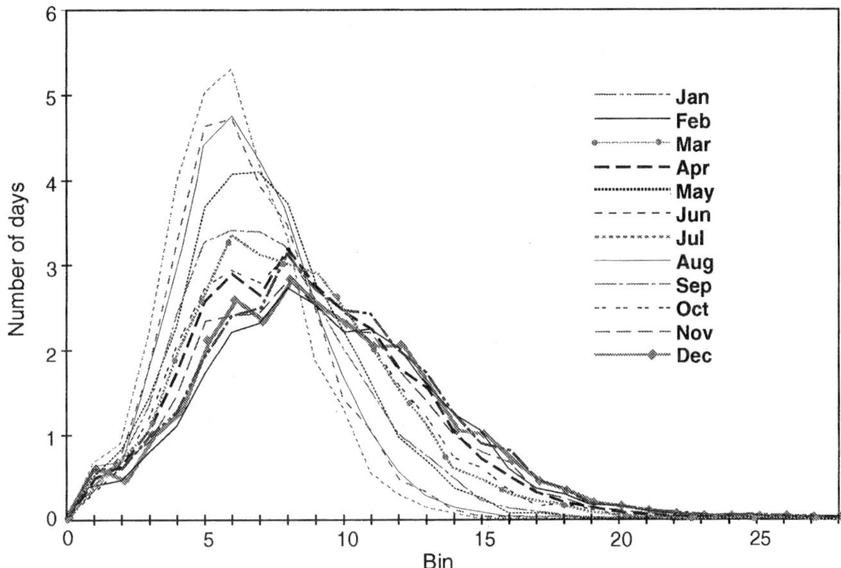

Figure 3.   Wind speed (1947-1997) displayed as a number of days in an average month that would have speeds in bin range indicated. The first bin represents wind speed of 0-1 knots; bin 2 is when the winds are 2-3 knots; etc. Knots are used since this is the unit in which the original data were recorded.

(Fig. 4), the most extensive ice year since satellite coverage began, the ice edge was at or south of St. Paul Island from early February through early May (except for a brief period in March). In 1979, a year with minimal ice, the sea ice never advanced south of ~59°N. During the last decade, 1995 had one of the more extensive ice covers, with sea ice arriving in early February and not retreating until late April (Fig. 5). This resulted from persistent winds out of the north starting in January. In contrast, during 1987 the winds were variable, resulting in the islands remaining ice free. Sea ice reached the Pribilof Islands for at least 1 week in 18 of the 25 years from 1972 to 1997, though in many of those years it quickly retreated northward. Even though the sea ice usually reached the Pribilof Islands, the median ice extent along 170°W (Fig. 5f) remained north of the islands. This resulted from the variable arrival time and the short duration that sea ice was near the islands. From the end of May through December, sea ice is absent from the water around the Pribilof Islands.

Total precipitation had little annual variability, although liquid precipitation predominated in the summer, and frozen precipitation was higher in the winter. Plots (not shown) of simple daily-averaged total precipitation

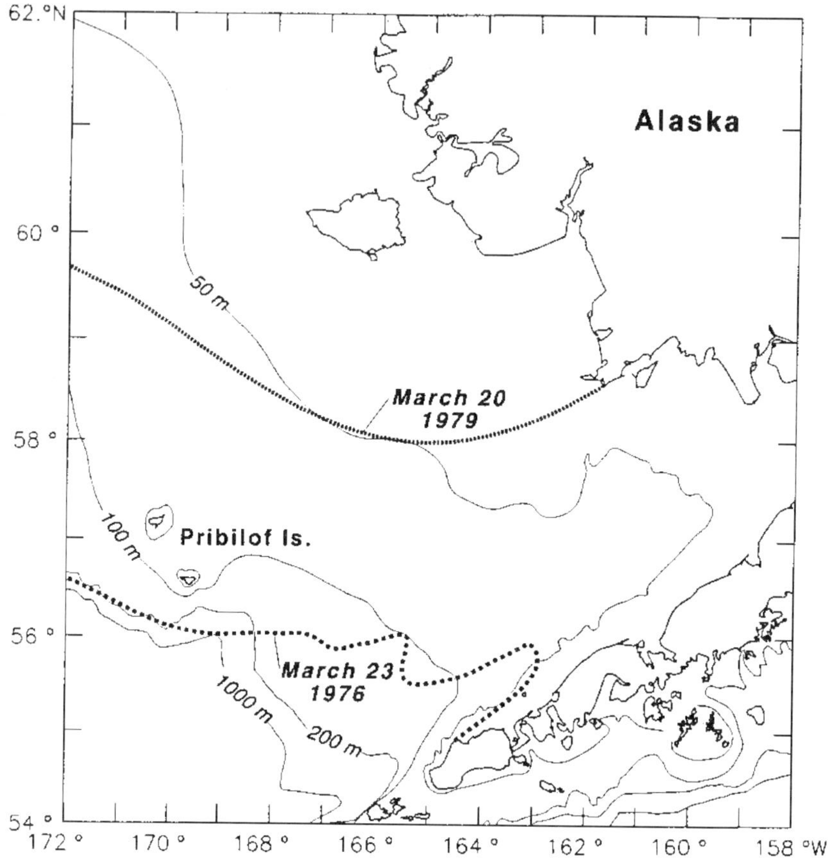

Figure 4.   Approximate location of the edge of sea ice during maximum (1976) and minimum (1979) conditions.

revealed that the weather in the Pribilof Islands consisted of many days of "code 1" rain, but these days were interspersed with days without rain and "code 4" days. There does not appear to be a rainy or dry season at St. Paul Island.

Currents

Between 1987 and 1997, approximately 80 satellite-tracked drifters were deployed in the southeast Bering Sea. A common trajectory of these drifters was northwestward along the shelf break or along the 100 m isobath. Drifters deployed in April or May on the shelf near Unimak Pass were advected northwestward to the Pribilof Islands, arriving by September (Fig. 6a,b). In addition, some drifters were advected from the slope onto the

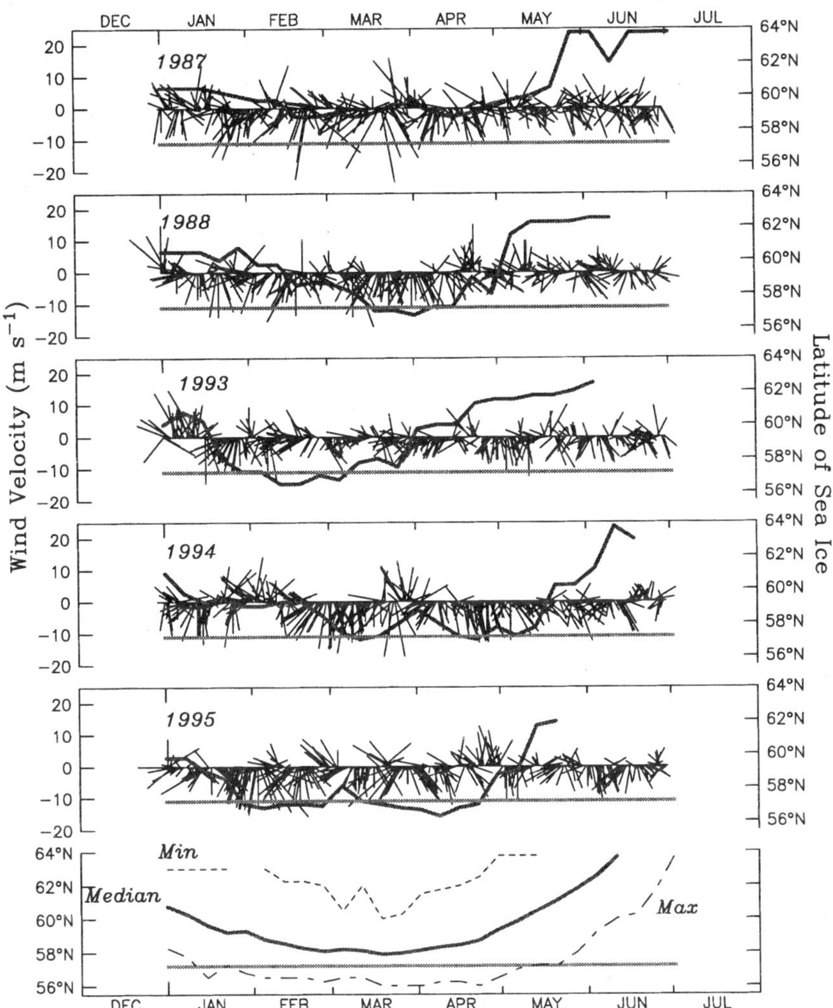

Figure 5. *Daily mean wind and latitude of the ice edge along 170°W for selected years discussed in the text. The bottom panel shows the median ice extent bracketed by the minimum and maximum extent. In each panel the latitude (~57°20'N) of St. Paul Island is indicated.*

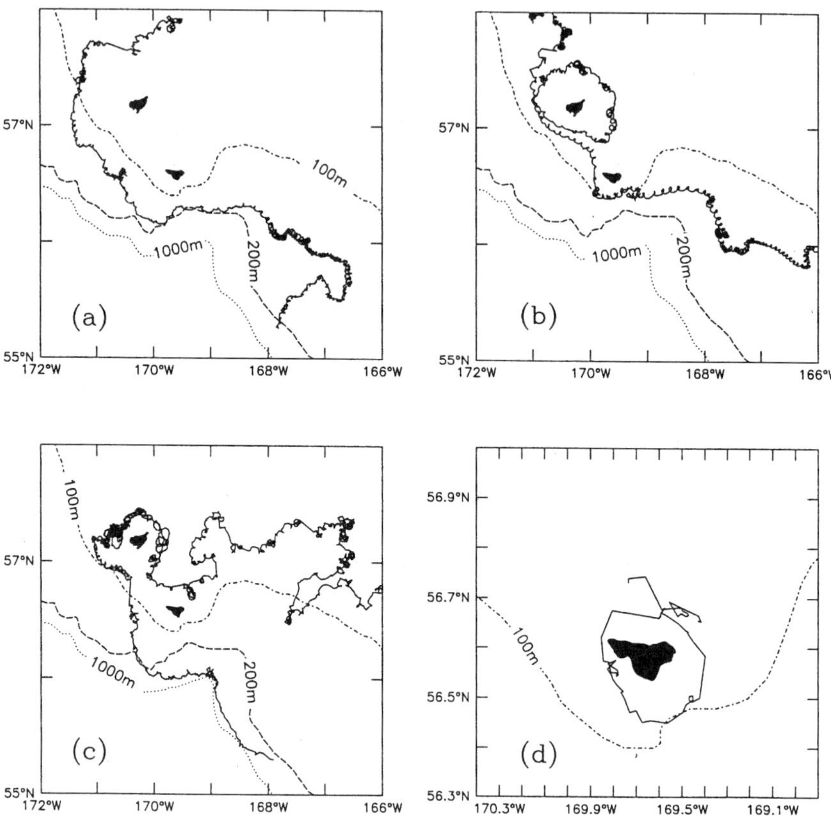

Figure 6.   *Trajectories of satellite-tracked drifters with a drogue depth of 40 m. Drifters were deployed on: (a) 1 May 1990, (b) 27 April 1993, (c) 19 April 1990, and (d) 7 March 1993. The drifter deployed on 7 March 1993 (d) entered clockwise circulation around St. George Island on 20 March and circled the island five times (only one period of rotation around the island is shown).*

shelf south of St. George Island (e.g., Fig. 6c). Upon reaching the Pribilof Islands, drifters continued northward along isobaths or joined the eastward flow across the shelf (Stabeno et al., chapter 1, this volume; Schumacher and Stabeno 1998). Other drifters became entrained in a clockwise circulation around the islands. Some drifters circled an island repeatedly (Fig. 6d), suggesting a "trapped" circulation around St. Paul or St. George islands. Although the flow evident from drifter trajectories are persistent, they are typically weak (<5 cm/s) except for south of St. George Island. Here the currents are usually 20-25 cm/s. The northward flow on the west side of the islands broadens and weakens to ~5 cm/s and the continuation northward along the 100 m isobath is weaker still (~3 cm/s). The eastward flow across the shelf is also weak at ~1 cm/s. These flow patterns are also evident in baroclinic flow calculated from hydrographic data (Reed and Stabeno 1996).

Moored current observations were made in the vicinity of St. Paul Island (Fig. 1) during 1995-1996. These moorings were in the well-mixed inner shelf or coastal domain in ~20 m water depth. As elsewhere on the southeast Bering Sea shelf, the tidal currents dominate the kinetic energy of water motion (Schumacher and Stabeno 1998). The dominant semidiurnal component is M_2 and the dominant diurnal component is K_1 (Kowalik and Stabeno, submitted). The tidal ellipses (Fig. 7c,d) are highly eccentric and are approximately parallel to the bathymetry. The strongest tidal currents occur between St. Paul and St. George Islands, likely a result of flow being constricted between these islands. These tidal currents provide more mixing energy than is found elsewhere on the shelf, in similar water depth, thus affecting the hydrographic structure. This is discussed in more detail in the section on hydrography.

The low frequency currents were rectilinear, with the axis of variability directed along bathymetry (Fig. 7a). The net flow indicated a generally clockwise circulation around the island. There were markedly few events when wind reversed or even modified the low frequency flow (Fig. 8) in spite of the regular occurrence of storms, particularly in the winter. The strongest currents were observed south of St. Paul Island at sites 3 and 4. The mean flow weakened at site 5, where it was directed northwestward along a bathymetric feature. The weakest flow occurred on the east side of the island at site 7. This may have resulted from either a reduction of transport around the island (i.e., some of the transport continues eastward across the shelf and away from the islands as shown in a climatology of hydrographic observations; Reed and Stabeno 1996) or from the increase in cross-sectional area of the around-island transport. The flow at site 2 in English Bay revealed weak mean and tidal currents.

Two possible mechanisms that result in flow around St. Paul Island are tidal rectification and continuation of the flow south of St. George Island. Model results reveal a strong flow due to rectification of tidal currents around St. Paul Island (Kowalik, chapter 4, this volume). Tidal rectification was most evident in observations from site 4. A demodulation of

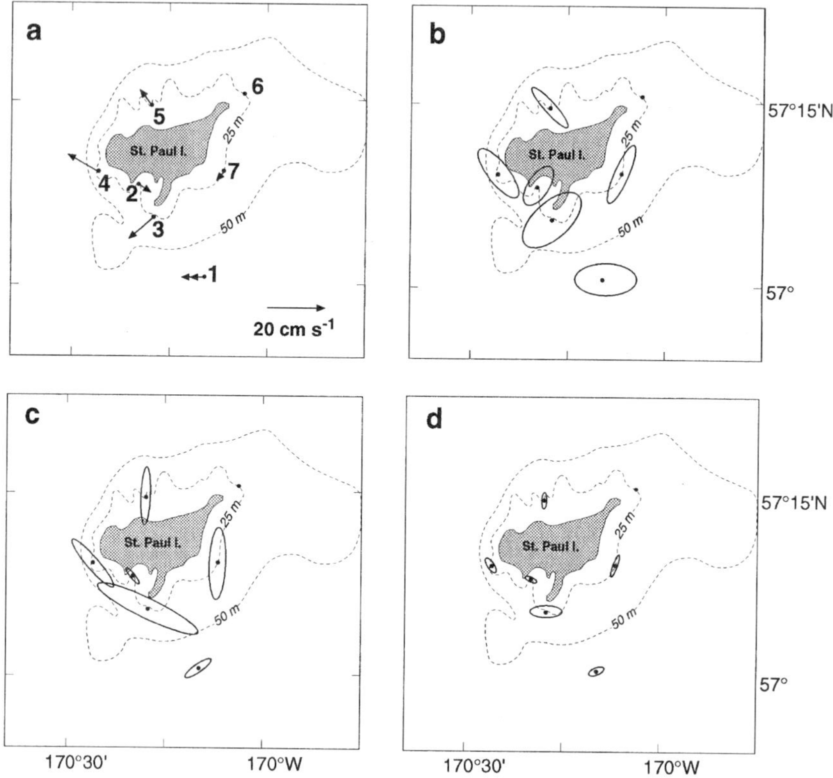

Figure 7.   Results from the six current meter records in the vicinity of St. Paul Island: (a) mean velocity, (b) subtidal current ellipses oriented on the axis of greatest variance, (c) the tidal ellipse for the M$_2$ tides, and (d) the tidal ellipse for K$_1$ tides.

the currents provides a time series of the amplitude of the tidal currents. The demodulated time series and the low-frequency, along-isobath flow are well correlated. Both had a strong fortnightly signal (Fig. 9). The lack of a seasonal signal in the flow around the island indicates that the direct wind forcing is not a candidate for driving the mean circulation. Although the winds may not drive the mean flow around the island, they do introduce variability into it. The most striking event of direct wind forcing occurred on February 10. The response to the strongest storm of that winter (Fig. 10a) is clearly evident in each of the current records. The magnitude of the wind stress is best shown in Fig. 10a. It is remarkable, however, that there are so few events in these records that clearly resulted from the wind.

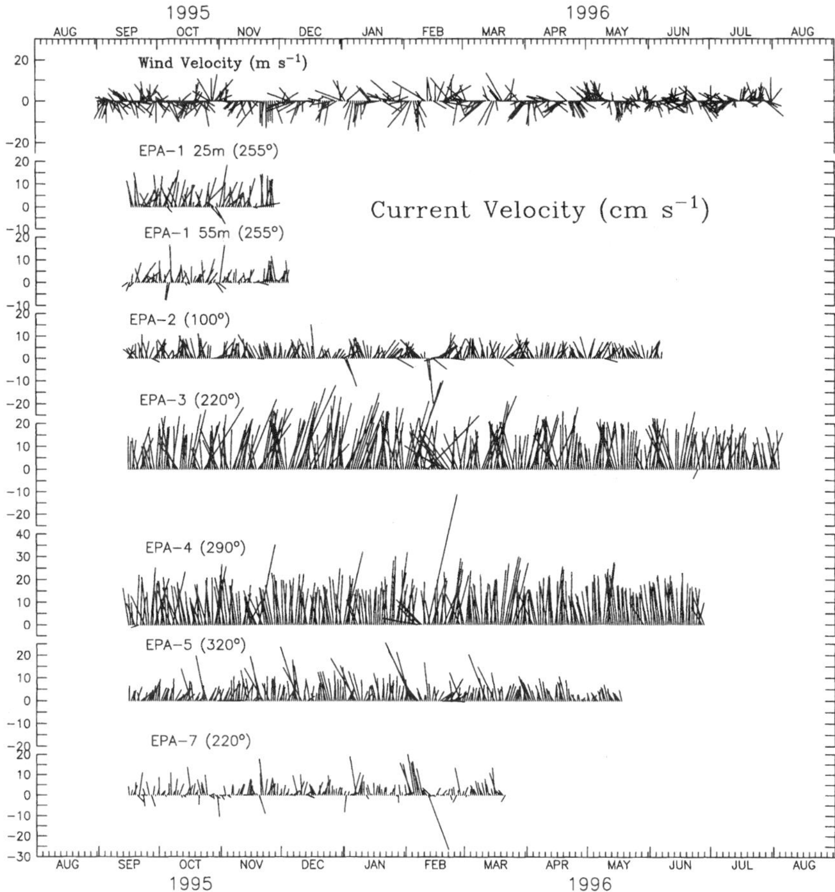

Figure 8. *Low-pass filtered currents from the six moorings in the vicinity of St. Paul Island (Fig. 1) and the winds as measured at St. Paul Island. The currents are presented along their major axis given in °T.*

Hydrographic Features

The only time series of temperature in the vicinity of the islands was collected in 1995-1996 (Fig. 1) at the mooring sites discussed in the previous section. The changes in temperature revealed a very coherent signal around the island, indicating that the changes in temperature are a result of exchange of heat with the atmosphere rather than advection of water (Fig. 10b). This was not a typical year, however, since sea ice was not advected into the vicinity of the Pribilof Islands. This is reflected in the relatively warm temperatures measured around the island. The minimum temperatures at each mooring site were greater than –0.5°C, whereas temperatures less than –1.5°C are common over the middle shelf during cold years (Schumacher and Stabeno 1998).

Site 1 was situated to the southwest of St. Paul Island in deeper water (60 m) in what was expected to be the two-layer middle shelf domain. The vertical structure was weakly stratified during the late summer 1995 with a temperature difference of less than 3°C between the top and bottom of the water column. This difference compares to a mean of ~4.5°C observed north of St. Paul Island (Fig. 11). The weakness and/or absence of a two-layer vertical structure during the summer of 1996, however, is striking. While the temperatures were not always uniform with depth, the lower part of the water column responded quickly to the seasonal warming (Fig. 10b,c) and even weak storms resulted in a well-mixed column. The lack of sea ice in the vicinity of the Pribilof Islands during the previous winter (1995-1996) provides one explanation for the weak stratification. The melting of the ice would have enhanced the vertical density structure, with fresher water in the upper part of the water column and more saline water near the bottom. This, combined with summer warming of the upper layer, would have inhibited vertical mixing and resulted in a more marked two-layer system. Alternately, however, this is a region of strong tidal flow (mixing energy) which would resist the establishment of the vertical density structure that occurs elsewhere on the shelf at this water depth. The lack of a clearly two-layer system permits greater horizontal exchange of material between the inner and middle shelf and between the two main islands.

The focus of our analysis of hydrographic observations is from two transects, each of which has been repeated many times during late summer in the last 11 years. Line A extends northeast from St. Paul Island and line D extends due south from St. George Island (Fig. 1). The dominant processes which determine the hydrographic characteristics differ markedly in the two regions. The temporal variability at line A can be largely explained by in situ processes (heat exchange with the atmosphere and the addition of fresh water through the melting of sea ice). The variability at line D is largely determined by advection and the influence of slope dynamics on the narrow shelf south of St. George Island.

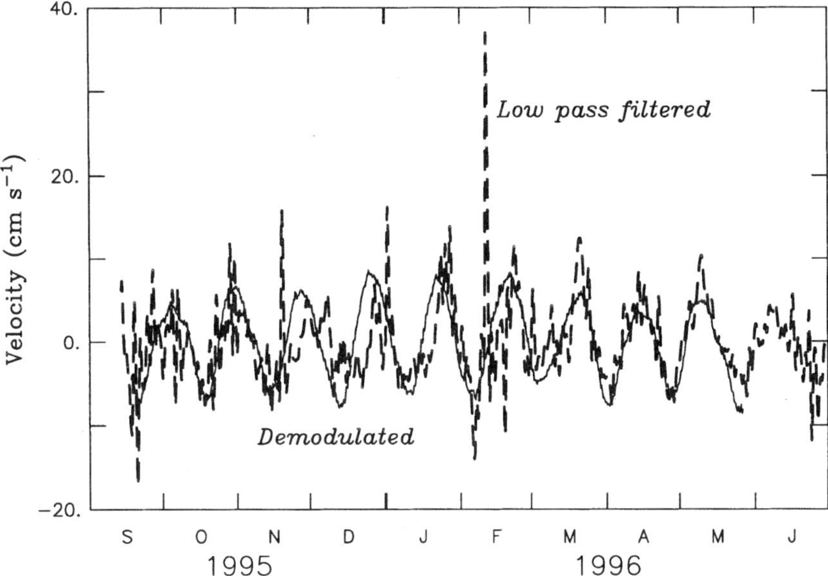

Figure 9. Time series of low-pass filtered current data resolved along the net current direction and the demodulated time series at site 4.

Line A extends 60 km to the northeast, although often only the 40 km closest to St. Paul Island were occupied. It starts in the shallow mixed coastal domain and extends through the inner front into the middle shelf. As already noted, unlike the water column between St. Paul and St. George islands, there was a strong two-layer structure in a water depth of 60 m along this line (Figs. 11 and 12). Tides provide much of the mixing energy, but do not vary on annual time scales. The large variations evident in the salinity and temperature are primarily a result of the amount of melting sea ice and wind mixing, respectively. The on-shelf flow from Pribilof Canyon, which wraps around St. Paul Island, plays a secondary effect. The transport of this more oceanic water plays an important role of increasing the ambient salinity, which is reduced as a result of melting of sea ice.

The lowest salinity (<31.4 psu) was observed in 1995, a year with extensive ice coverage. Ice reached St. Paul Island in late January and did not retreat permanently until late April. In contrast, 1987, 1988, and 1996 were all weak ice years. The lack of ice melt and resulting fresh water input permitted higher salinities to persist northeast of the islands. While ice melt can cool the water column to below –1.7°C, these cold temperatures quickly disappear in the upper mixed layer during summer heating.

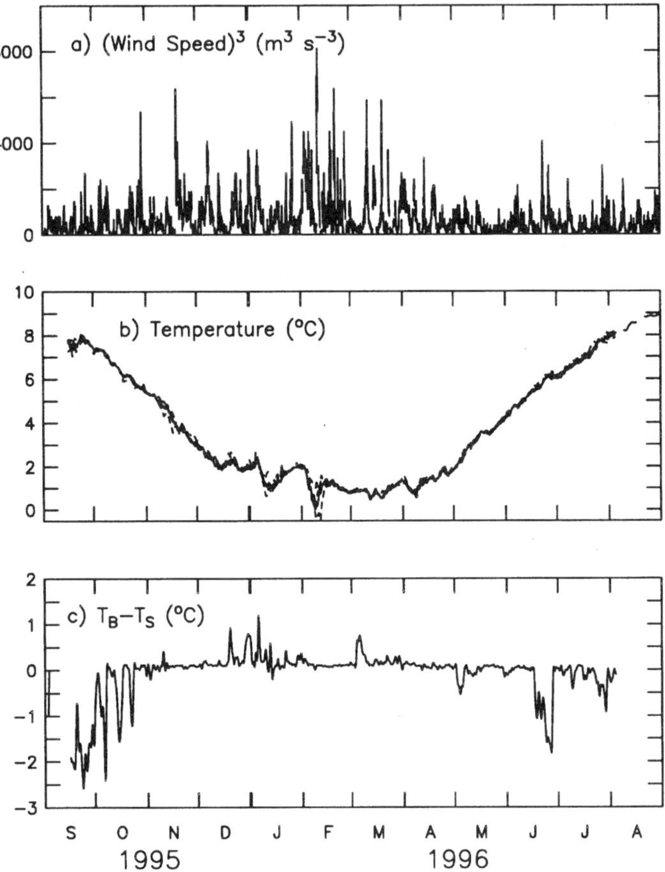

Figure 10. Environmental conditions between September 1995 and August 1996: (a) wind speed cubed (proportional to mixing energy) measured at St. Paul Island, (b) time series of temperature of the moorings located in 20 m of water around the island, and (c) time series of the temperature difference between 55 m (near bottom) and 25 m (the upper layer) at site 1.

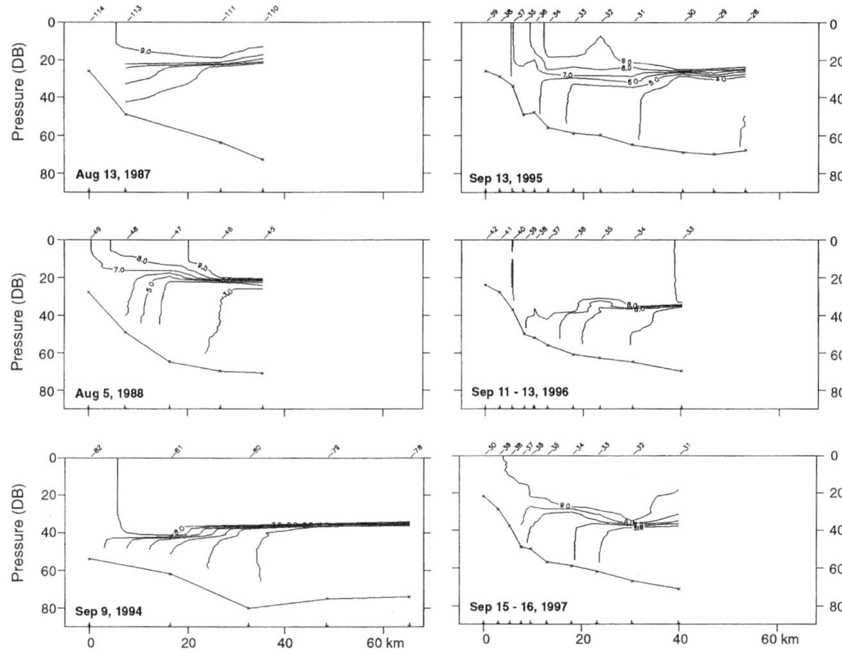

Figure 11. Temperature observed along line A during the periods indicated. Locations of each cast are shown at the top of each panel and distance from the station nearest to shore is given in kilometers across the bottom.

In contrast, near bottom temperatures over the middle shelf domain can persist throughout the summer and thus reflect the extent of ice coverage the previous winter. The amount of mixing (and hence heat transfer) between the upper and lower layers over the middle shelf is dependent upon the strength of the density gradient between the two layers, and of magnitude and duration of storms. It is not surprising that 1996, with its minimal ice extent, had relatively warm near-bottom temperatures, and 1995, with its large ice extent, had colder temperatures, although by September bottom temperatures were modified by summer heating.

Surface temperature is determined by wind mixing energy and the resulting thickness of the upper layer. The warmest surface temperatures were related to the shallowest wind mixed layers (1987, 1988, and 1995). During years when the upper layer is deeper (>40 m in 1994) the surface heating is distributed over a greater depth and thus sea surface temperature is cooler. It should be noted that winds during June, July, and August 1997 were particularly weak, but in September strong winds mixed the water column. The strong winds in September are responsible for the relatively deep mixed layer observed during September of that year at line A.

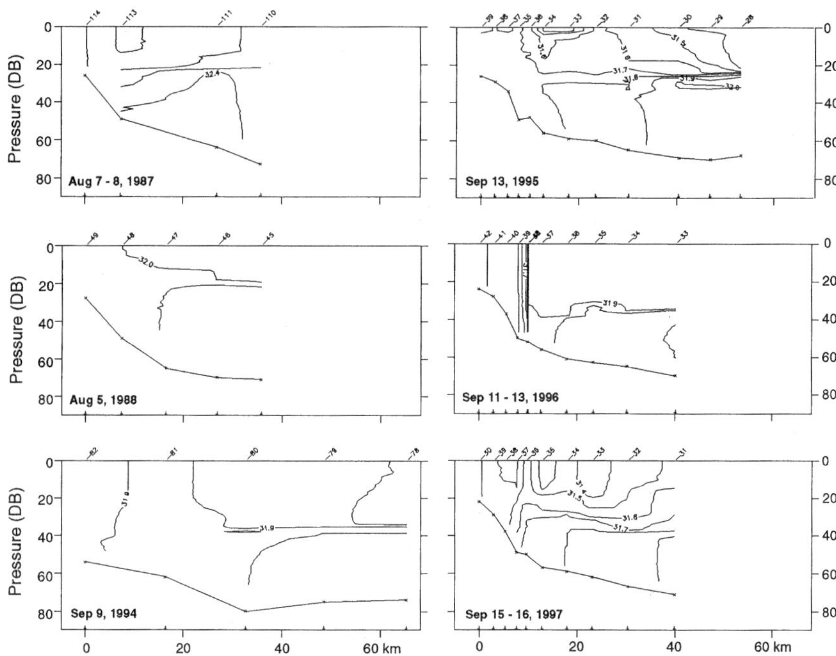

Figure 12. Salinity observed along line A during the periods indicated. Locations of each cast are shown at the top of each panel and distance from the station nearest to shore is given in kilometers across the bottom.

The position of the inner front is defined at its nearshore end as the first station that is well mixed (or weakly stratified). Identifying where the transition from weak stratification to two layers occurs is more challenging. We use changes in temperature to define frontal characteristics because temperature shows greater contrast than salinity. Further, station separation allows definition only in 1995, 1996, and 1997. Previously (Schumacher et al. 1979), the width of the front was defined using closely spaced (~1 km) XBTs to measure the horizontal gradient of thermal stratification. This parameter has low values in either the inner shelf or the middle shelf domains, but relatively large values within the transition zone itself. This definition provides inconsistent results when applied to data with greater spatial separation of stations. Instead, we define the front as being bounded by the deepest mixed (or weakly stratified) hydrographic station and by the station where the vertical extent of the thermocline between the upper and lower layer broadens by a factor of two over its width in the clearly two-layered middle shelf. Using this definition, in 1995 the width was ~24 km, in 1996 it was 12 km, and in 1997 it was >30 km. Thus with

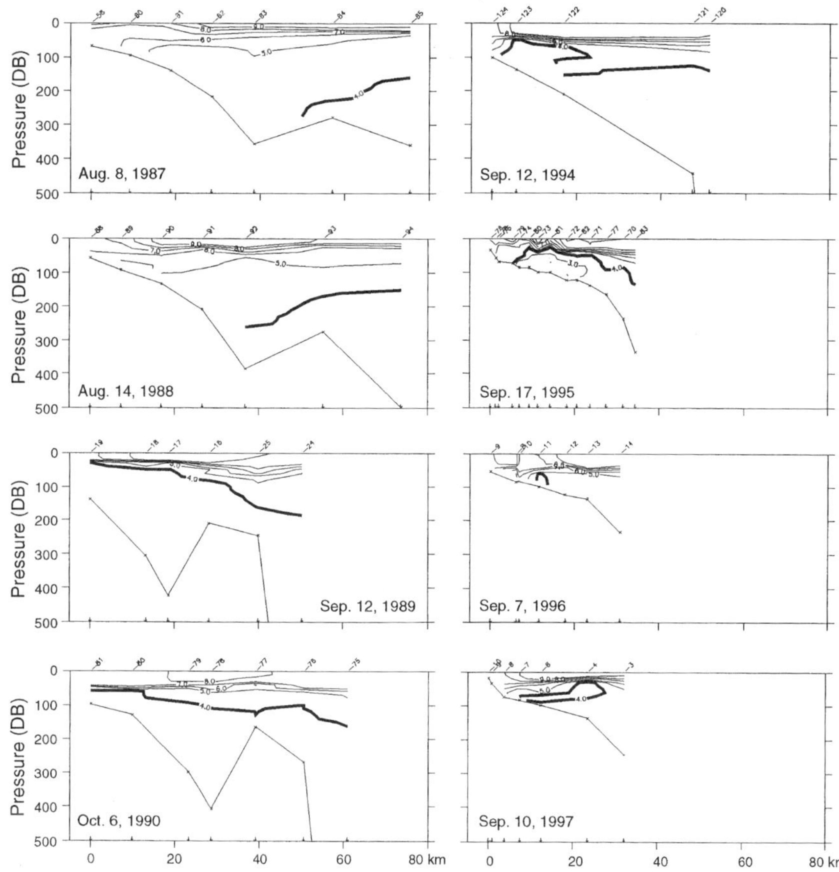

Figure 13. Temperature observed along line D during the periods indicated. Locations of each cast are shown at the top of each panel and distance from the station nearest to shore is given in kilometers across the bottom.

the new definition the widths in 1995 and 1997 are broader than widths obtained using the older definition, whereas the width in 1996 is consistent with earlier estimates.

In contrast to line A, changes in water properties at line D (Figs. 13 and 14) are mainly a result of advection. Variability in temperature is evident, but difficult to attribute to specific mechanisms. On the shelf (shoaler than 200 m) the near surface temperature ranges from 8°C to slightly greater than 9°C. These surface temperatures are similar to those observed on line A and, as there, strongly related to the depth of wind-mixed layer. Near bottom temperatures generally ranged from <4°C to 5°C; however, in 1995 a lens of <3°C bottom temperatures existed over the outer shelf

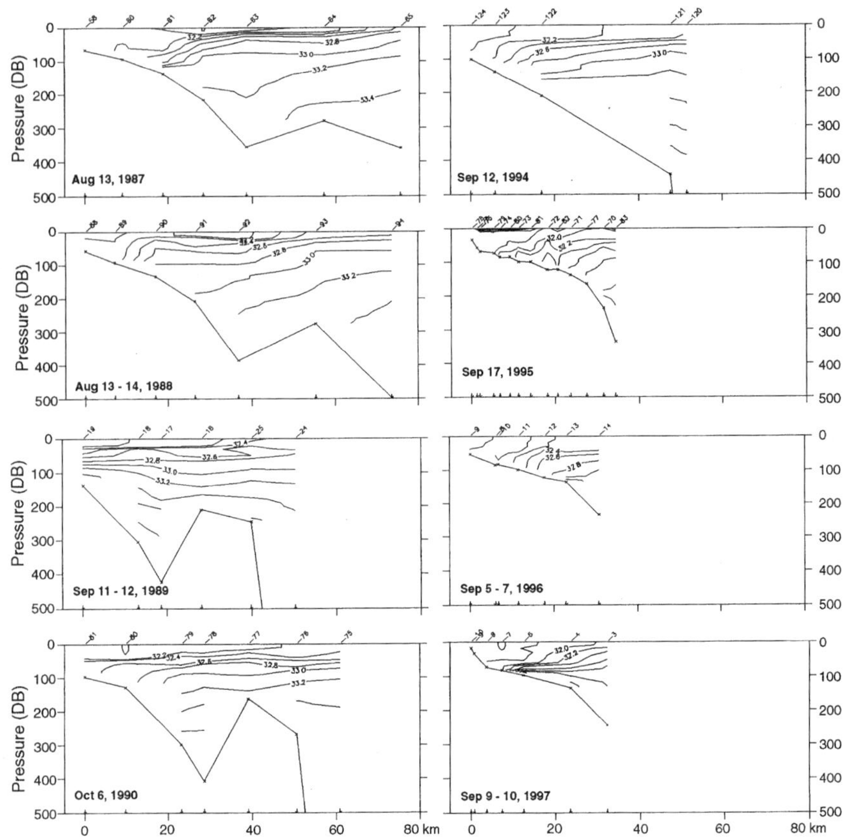

Figure 14. Salinity observed along line D during the periods indicated. Locations of each cast are shown at the top of each panel and distance from the station nearest to shore is given in kilometers across the bottom.

between 100 and 150 m isobaths. It is not known whether these cooler temperatures were related to the extensive ice coverage of the previous winter.

The 33 psu isohaline is an indicator of on-shelf flux, since these salinities do not occur on the shelf except in polynyas. Typically this isohaline is found only in water depths greater than 100 m (1987, 1988, 1990, 1994, 1995, and 1996) and often only in water greater than 200 m deep. Only in 1997 (and possibly in 1989 although there was no cast in water less than 100 m deep), does the 33 psu isohaline impinge upon the shelf. Since deeper basin water is associated with higher nutrients, this is an indication of on-shelf flux of nutrients.

Discussion

The Pribilof Islands form a natural laboratory to study changes in climate. Time series extending back to the 1940s exist of weather, and shorter, but still significant, records of ice and water properties permit us to study the initial impact of interannual and longer period climate variability on the Bering Sea ecosystem. Using the available data, we have noted several striking features that represent conditions typical of the last decades. The presence of sea ice and duration around the islands varies greatly year to year. Its presence influences water properties and vertical structure of the water column. A structure front circles each island from spring through early fall. These fronts are the center of biological activity (Hunt et al. 1996, Brodeur et al. 1997). Such a front likely limits the exchange of materials from the coastal domain with the waters over the middle shelf. The magnitude of horizontal and vertical gradients of water properties, as well as the width of the front, vary on interannual time scales. A mean current flows clockwise around each island that is at least partially due to the interaction of topography with the tides. Lagrangian and water properties observations reveal an on-shelf flow south of St. George Island.

While research has made the first steps toward understanding this complex ecosystem, numerous questions still must be addressed. A few will be mentioned here. While we know that the fronts exist and that they are important to biological activity (Brodeur et al. 1997), it is not clear what mechanisms promote the high productivity observed. It could be simply that the weaker vertical stratification facilitates the entry of nutrients into the upper mixed layer where they can be used to prolong primary productivity, or it may be a more complex interaction of mechanisms. The mechanisms of dispersion across the structure front around each island are also not well understood.

The causes of the often observed on-shelf flux in the vicinity of Pribilof Canyon is not well understood. This on-shelf flow is likely an important source of nutrients for the region around the Pribilof Islands. Interannual variations are likely to occur in on-shelf transport, since the processes controlling it are associated with the Bering Slope Current (transport varies from 2 to 7×10^6 m^3/s on interannual time scales [Stabeno et al., chapter 1, this volume]). Further longer period changes in climate will force changes in this current and will likely reduce the flux of nutrients onto the shelf (Schumacher and Alexander, chapter 6, this volume). While we know that the southward flow on the east side of St. Paul Island is weaker than flow observed on the west side, it is not known if this is due to a reduction in transport. Likely a portion of the north flowing current on the west side of the Pribilof Islands flows eastward across the shelf. Such a flow is likely important for supplying nutrients to the middle and coastal domains from Nunivak Island northward.

The high productivity of the eastern Bering Sea shelf is directly related to high concentrations of nutrients observed on the shelf. Approximately half of the nutrients on the shelf are regenerated (Whitledge et al. 1986), but the remainder must be introduced from the nutrient reservoir at the shelf break. It is evident from satellite-tracked drifters that the region south of the Pribilof Islands is a location where cross shelf fluxes often occur. Higher concentrations of nutrients are thus introduced at St. George Island and advected northward along the west side of the islands. An unknown portion of the nutrients remain in the vicinity of the islands, enhancing primary production, with the remainder advected either eastward across the shelf or northward.

The effect of changes in buoyancy flux, winds, and stratification as a result of changes in climate are difficult to predict. The winds, temperature, and sea-ice extent on the Bering Sea shelf, however, have been linked to the El Niño-Southern Oscillation (ENSO) and to parameters called the Pacific–North American (PNA) pattern index and the Aleutian low index (Niebauer 1988). Niebauer et al. (chapter 2, this volume) found warm temperatures in the Bering Sea are only weakly correlated with ENSO time series; thus much of the variability of the environment in the Bering Sea results from local forcing. While a decrease in sea-ice extent and storm intensity is expected to accompany warming, these factors have competing implications for changes in the Pribilof Island environment. An understanding of the long-term effects of climate change on this rich ecosystem must be a priority for future investigations.

Acknowledgments

Contribution 1941 from NOAA/Pacific Marine Environmental Laboratory and B326 from NOAA's Fisheries Oceanography Coordinated Investigations. G. Hunt's research supported by grants from Office of Polar Programs and NSF.

References

Brodeur, R.D., M.T. Wilson, J.M. Napp, P.J. Stabeno, and S. Salo. 1997. Distribution of juvenile pollock relative to frontal structure near the Pribilof Islands, Bering Sea. In: Forage fishes in marine ecosystems. Proceedings of the International Symposium on the Role of Forage Fishes in Marine Ecosystems. University of Alaska Sea Grant, AK-SG-97-01, Fairbanks, pp. 573-589.

Coachman, L.K. 1986. Circulation, water masses, and fluxes on the southeastern Bering Sea shelf. Continental Shelf Research 5:23-108.

Coyle, K.O., and R.T. Cooney. 1993. Water column sound scattering and hydrography around the Pribilof Islands. Continental Shelf Research 13:803-827.

Hunt, G.L., Jr., K.O. Coyle, S. Hoffman, M.B. Decker, and E.N. Flint. 1996. Foraging ecology of short-tailed shearwaters near the Pribilof Islands, Bering Sea. Marine Ecology Progress Series 141:1-11.

Kowalik, Z., and P.J. Stabeno. Submitted. Trapped motion around Pribilof Islands, Bering Sea. Journal of Geophysical Research.

Niebauer, H.J. 1988. Effects of El Niño-Southern Oscillation and North Pacific weather patterns on interannual variability in the subarctic Bering Sea. Journal of Geophysical Research 93:5051-5068.

Overland, J.E. 1981. Marine climatology of the Bering Sea. In: D.W. Hood and J.A. Calder (eds.), The eastern Bering Sea shelf: Oceanography and resources, Vol. 2. Published by the Office of Marine Pollution Assessment, NOAA and BLM. Distributed by the University of Washington Press, Seattle, WA, pp. 15-22.

Overland, J.E., and C.H. Pease. 1982. Cyclone climatology of the Bering Sea and its relation to sea ice extent. Monthly Weather Review 110:5-13.

Reed, R.K., and P.J. Stabeno. 1996. On the climatological mean circulation over the eastern Bering Sea shelf. Continental Shelf Research 16(10):1297-1305.

Schumacher, J.D., and P.J. Stabeno. 1998. The continental shelf of the Bering Sea. In: A.R. Robinson and K.H. Brink (eds.), The sea: The global coastal ocean regional studies and synthesis, Vol. XI. John Wiley and Sons, New York, pp. 869-909.

Schumacher, J.D., T.H. Kinder, D.J. Pashinski, and R.L. Charnell. 1979. A structural front over the continental shelf of the eastern Bering Sea. Journal of Physical Oceanography 9:79-87.

Springer, A.M., P. McRoy, and M.V. Flint. 1996. The Bering Sea Green Belt: Shelf-edge processes and ecosystem production. Fisheries Oceanography 5:205-223.

Whitledge, T.E., W.S. Reeburgh, and J.J. Walsh. 1986. Seasonal inorganic nitrogen distributions and dynamics in the southeastern Bering Sea. Continental Shelf Research 5:109-132.

CHAPTER **10**

Summary of Chemical Distributions and Dynamics in the Bering Sea

Terry E. Whitledge
Institute of Marine Science, University of Alaska Fairbanks, Fairbanks, Alaska

Vladimir A. Luchin
Far Eastern Regional Hydrometeorological Research Institute (FERHI), Vladivostok, Russia

Introduction

For the first 40 years of this century oceanographic studies in the Bering Sea were not systematic and were conducted in shallow water during the warm seasons. Chemical investigations were started in connection with the Second International Polar Year and were restricted to measuring basic salinity and dissolved oxygen content in the western and northern areas of the Bering Sea (Ratmanov 1937a,b). American oceanographic observations of Barnes and Thompson (1938) and Goodman and Lincoln (1942) described dissolved oxygen data for the eastern and southern areas of the Bering Sea in spring and summer 1937-1938. They verified Ratmanov's conclusion that there was a close connection between the southern part of the Bering Sea and the North Pacific Ocean.

Mokievskaya (1956, 1958, 1959) separated four layers in the deepwater area of the Bering Sea based on the distribution of oxygen, phosphate, silicate, and nitrites: the surface layer; a layer of variable concentrations which in winter was located at 150-200 m and in the summer at 200-250 m; a layer of low oxygen and maximal phosphate concentration (300-1,500 m); and deep water in which oxygen content increases and phosphates concentration decreases (from 1,500 m to the near bottom layers). She examined the seasonal variation of the concentration of biogenic material in the upper 500 m and suggested that the decreased phosphate concentration near the continental slope was due to adsorption by suspended matter (Mokievskaya 1958) and was later corroborated by Lisitsyn's hypotheses (1955).

Davidovich (1963) determined that phosphate and dissolved silica concentrations in western Bering Sea waters were larger than in eastern shelf waters. In the transition from the deepwater zone to the continental slope, phosphate concentration decreases in the photic layer. It was also observed that phosphate concentration in the Bering Sea was nearly twice as high as in the northwestern Pacific Ocean. Moiseev (1964) noted that near the slope of the Bering Sea shelf there was an upwelling of water with high nutrients from 300-800 m to the surface layers. This contributed to an intensive growth of phyto- and zooplankton.

Ivanenkov (1964) critically analyzed all available hydrochemical data of observations conducted in the Bering Sea since 1961. He examined the vertical and horizontal distribution and the seasonal variation of hydrochemical parameters. Using daily station data, Ivanenkov calculated total oxygen production in the photosynthetic layer, the production of organic carbon and organic matter in units of volume, and daily rates for each season and for all bloom periods. He noticed the high seasonal variability of nutrient elements in the euphotic layer (maximum in the winter and minimum in the bloom period). He also noticed that in shallow water areas nitrites were found from the surface to the bottom, but in deep water they were only present in the upper 250 m. Maximum nitrite values were found near the pycnocline and were dependent on the biomass of zooplankton rather than phytoplankton.

McRoy et al. (1972) determined primary production and nitrogen productivity rates. They estimated that nitrate uptake comprised 5-40% of the total near the Aleutian Islands and 18-89% in Bering Strait in summer. These rates, combined with nitrate concentrations ranging from 3.7 to 14.5 µM, would support productivity for two weeks or longer. Park et al. (1975) described the vertical distribution of hydrographic variables and nutrients on the 1973 GEOSECS station and discovered an unusual decrease in the concentration of silicate near the bottom while profiles of nitrate and phosphate were typical. Hattori and Wada (1974) constructed a nitrogen budget for the epicontinental eastern Bering Sea which indicated that 51% of the nitrogen on the shelf was regenerated. Hattori and Goering (1981) described the nitrogen distributions on the southeast Bering Sea shelf and described the importance of wind mixing to the replenishment of nitrate in the euphotic zone.

Salinity

The basic salt composition of seawater of the northwestern Pacific Ocean was determined by Miyake (1939) and Fukai and Shiokava (1955). Ratmanov (1937a) described the basic salt composition of Bering Sea waters. He used data from two shallow-water stations and one deepwater station (nine samples were taken). He observed that the chemical composition of Bering

Sea water depended on the origin of water replacement by the Pacific and northern Arctic oceans, and on the biological and chemical processes of the continental slope. Hydrological processes and their seasonal variability appeared to have great influence on the concentration of the major elements.

Dissolved Oxygen

Vertical Exchange of Dissolved Oxygen

The presence of oxygen in seawater is the result of transfer from the atmosphere at the surface and biological activity. The concentration of dissolved oxygen below the surface layer is governed by primary production and regenerative processes. Profiles of the vertical distribution of dissolved oxygen show several layers: an upper layer, a layer of maximal gradients, a layer of minimal oxygen, and a layer of constant or slight increase in concentration with depth.

The overall oxygen content in the upper layer indicates that the lowest values were observed in Pacific waters. Near the slope the accumulation of oxygen reaches maximum values, which is probably due to the increased solubility of oxygen related to a decrease in temperatures, salinity, and physical processes. Intensification of currents at the periphery of deepwater basins and the generation of anticyclonic eddies near the slope lead to an intensification of vertical mixing and the transfer of oxygen to more significant depths. Winter convection reaches the near-bottom layers in shelf areas so the accumulation of oxygen in these waters is transferred from the surface to the underlying layers.

The formation of an oxygen-minimum layer is probably a result of negligible mixing with overlying layers, and the biochemical utilization of oxygen. The position of the upper and lower boundaries of the oxygen-minimum layer can be represented by the 1 mg per L isopleth. The depth of the position of the upper boundary of this layer clearly coincides with the pattern of water circulation. It deepens in the area of penetration and subsequent movement of Pacific waters. The core of the oxygen-minimum layer in the Bering Sea is located at 600-1,000 m and deepens slightly at the center of the deepwater basins. This is probably connected with cyclonic circulation which leads to upwelling. Near island chains and the continental slope there is an intensification of the vertical exchange and a deepening of the core of the minimum. The oxygen content in the core of the oxygen-minimum layer fluctuates from 0.43 to 1.57 mg per L. Its minimal concentration occurs in the center of the deepwater troughs and at Near Strait and is associated with the penetration of intermediate Pacific water with a lower oxygen content into the Bering Sea. There is a slight increase in oxygen concentration in deep water and near-bottom waters of the Bering Sea. It reaches 3.20-3.43 mg per L at 3,000-3,500 m.

Seasonal Variability of Oxygen

The seasonal variation of dissolved oxygen concentration is basically determined by the annual variability of temperatures and salinity, and the intensity of primary production processes. During the course of a year a high concentration of dissolved oxygen is maintained in the upper layer. However, the scope of the seasonal fluctuations in different regions varies considerably.

Any increase in oxygen content in winter at the surface is primarily an increase in its solubility. In winter, surface waters are in a state of slight undersaturation. Over a large part of the Bering Sea the deficit does not exceed 1-5% and only rises to 10-13% near the coast of Kamchatka and in the shallow-water passes of the Aleutian Islands. Near Kamchatka this low level is associated with the formation of ice cover and an increase of salinity at surface waters, which reduces oxygen solubility. In deep basins, water exchange with the Pacific Ocean maintains relatively high water temperatures which decreases oxygen solubility. The large heat reserve inhibits convective mixing at greater depths. The intensive mixing of water in Aleutian Islands passes leads to a transfer of oxygen from the surface to the underlying layers, which increases the concentration of oxygen in subsurface waters and decreases its content at the surface.

The maximal content of oxygen in Bering Sea surface waters is produced in spring (Fig. 1) and is connected with low water temperature and intensive photosynthesis of phytoplankton which can produce a supersaturation of oxygen at the surface. In large parts of the Bering Sea during summer a slight supersaturation (105-110%) occurs. The supersaturation reaches 115-120% at the shelf near the slope and in Karagin Bay, and is most likely associated with relatively large amounts of primary production in subsurface layers.

The annual fluctuation of oxygen at the surface ranges from 0.3 to 2.3 mg per L. The highest changes are noted in the passes and minimal fluctuations are characteristic for Pacific water that has a relatively stable rate of primary production. In the southern region of the open sea the profiles are distorted either because photosynthesis is very weakly expressed or because conditions for phytoplankton growth are not favorable. In spring the subsurface maximum of dissolved oxygen occurs in the upper 30 m of the water column. In the eastern Bering Sea shelf the maximum is located at 10-20 m and rises to the surface near the coast. The core of subsurface oxygen maximum shifts in the summer to 20-30 m for most areas of the Bering Sea.

Space-Time Distribution of Oxygen

Dissolved oxygen in surface waters during winter varies from 9.5 to 13.0 mg per L but shelf areas, where water temperatures are lowest, have concentrations that increase to 11-13 mg per L. In spring, oxygen content increases because of photosynthesis so the slope and shelf produce an

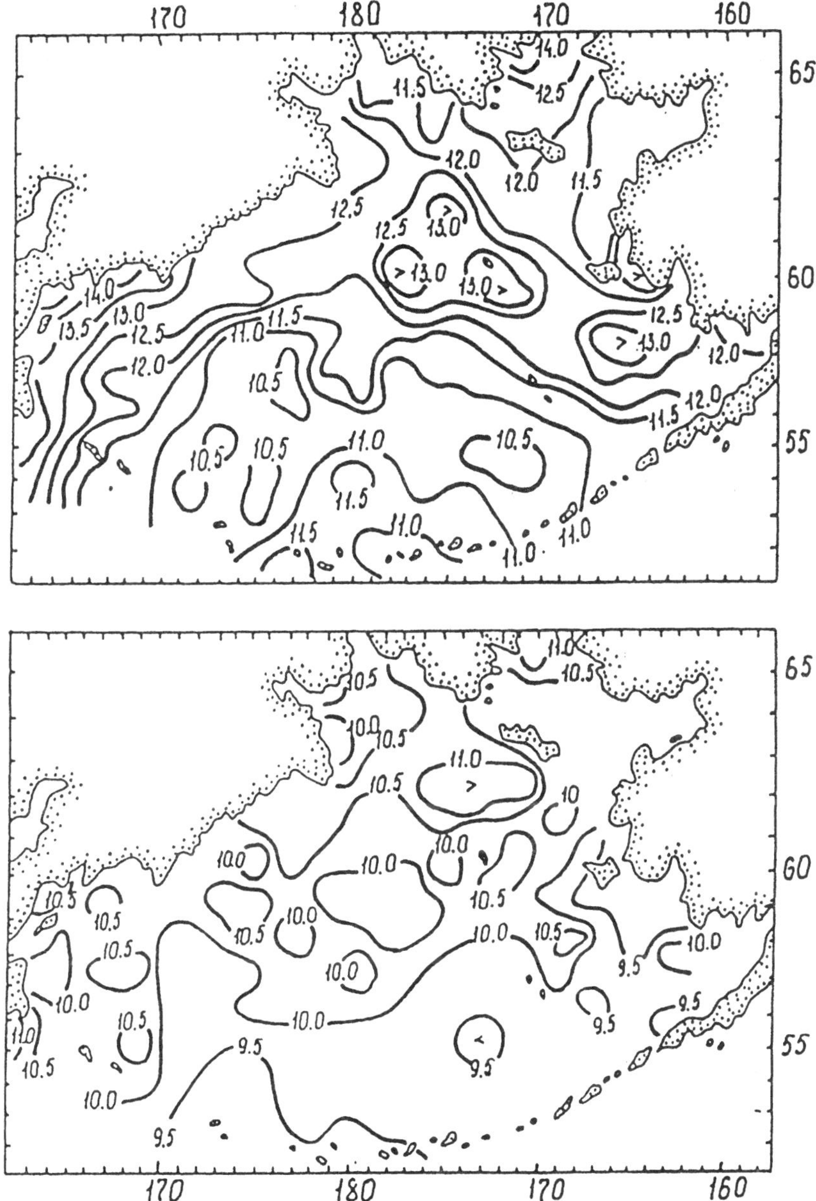

Figure 1. The distribution of dissolved oxygen (mg per L) in the surface layer in spring (top) and summer (bottom).

explosive growth of phytoplankton. Consequently, oxygen content there is 13.0-14.5 mg per L. In summer there is a smoothing of the spatial gradients of oxygen between the separate areas (Fig. 1). Maximum oxygen (over 11.0 mg per L) is recorded in the northern area near the Kamchatka coast. In autumn oxygen content at the surface, as a rule, ranges from 10.0 to 10.5 mg per L. The maximum value (over 11.0 mg per L) is located between St. Matthew Island and St. Lawrence Island due to a fall mixing of the subsurface oxygen maximum. In winter oxygen saturation ranges from 85 to 95% and minimal values are recorded in regions with the lowest water temperatures (for example, near the coasts of the Kamchatka Peninsula). Pacific waters are close to a saturated condition and Bristol Bay water is supersaturated. In spring surface waters are generally supersaturated with oxygen in the range of 110-140%. A minimum amount is found in Aleutian Islands passes probably as a result of the intensive vertical exchange. In summer the relative oxygen content at the surface is reduced but the saturation level remains rather high (100-110%). In shallow water regions in autumn there is an intensive cooling which produces a low level of saturation (up to 95%), but in the remaining parts surface waters are close to saturation.

In winter oxygen distribution in the upper 50 m is similar to the surface. In spring the content of dissolved oxygen is somewhat higher than in winter. It is possible that this is a result of the predominance of observations after severe winters. In summer, due to the rise of water temperatures, oxygen content is reduced. The minimum concentration (9.0-9.5 mg per L) is recorded in Pacific waters. In autumn the effect of convection begins to develop down to 50 m. Consequently, in Anadyr Bay the oxygen concentration rises to 2.0-2.5 mg per L which is similar to the spring concentration.

The oxygen content below 100 m is dependent on dynamic factors. In all seasons minimal values are recorded in the passes of the Aleutian chain and also at the approaches to the continental slope. This is explained by increased intensity of metabolic processes in water from underlying layers. Maximum oxygen content is recorded at 100 m in spring and summer. At 500 and 1,000 m minimal concentrations (0.7-0.8 mg per L) are recorded in the central part of the deepwater basins where an upwelling of deep water is observed (Fig. 2). Maximum values for oxygen (from 1.5 to 2.0 mg per L at 500 m and 1.0 to 1.1 mg per L at 1,000 m) are observed close to the continental slope. They form there due to an intensification of metabolic processes.

pH and Alkalinity

Vertical Distribution of pH

The general large-scale patterns of the vertical distribution of pH allow division of the water column into several layers with values for the vertical

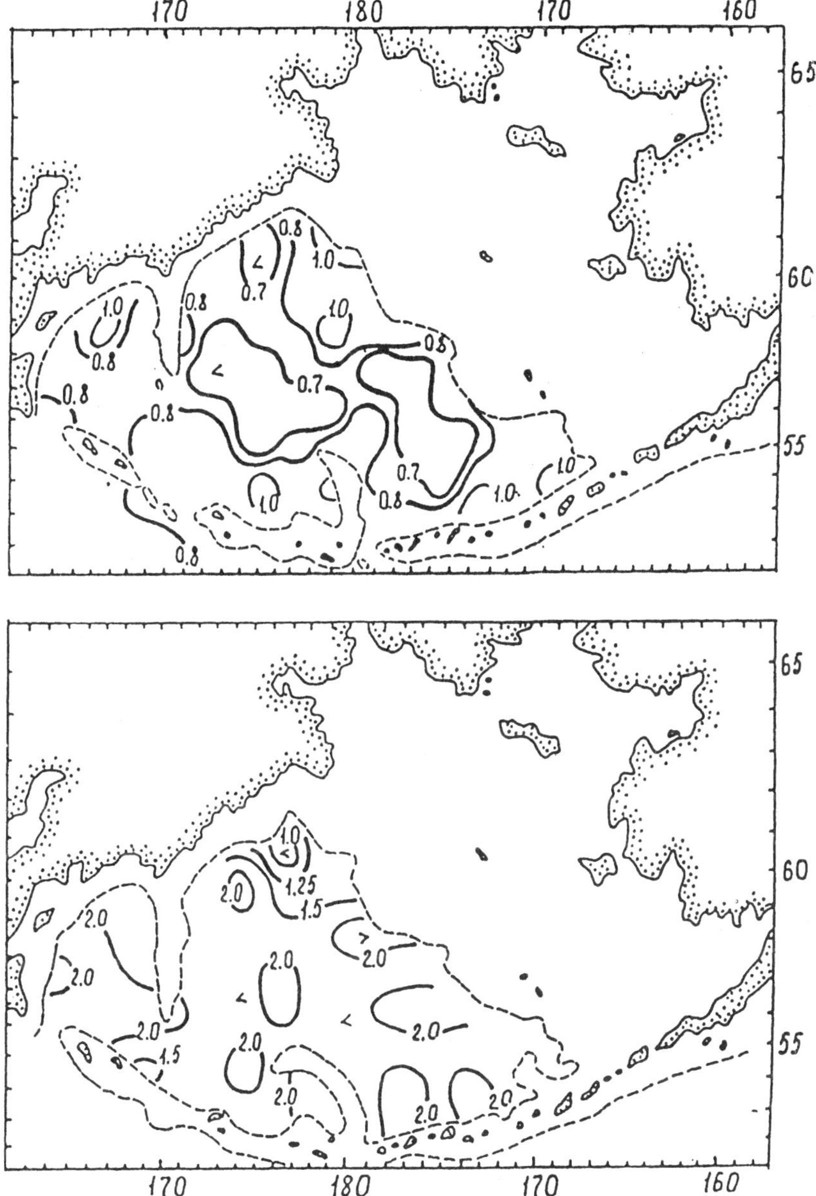

Figure 2.   The distribution of dissolved oxygen (mg per L) at 1,000 m (top) and 2,000 m (bottom).

gradients as the criteria. The upper layer, ignoring the transformation of the profiles in the photic layer in spring and summer, extends to 100-200 m. In these layers the vertical gradients are either minimal or generally absent. The layer of maximal gradients extends to depths of 300-500 m. The vertical gradients of pH in this layer are approximately two times higher than in the overlying layer. At the edge of deep basins the vertical gradients in these two layers are nearly identical. Next follows a layer of minimum pH. There is an increase of pH values in deep water and near-bottom waters of the Bering Sea.

The lower boundary of the upper layer of Pacific waters is located closest to the surface. The vertical gradients of pH in this layer vary from 0.0-0.05 to 0.25 pH per 100 m. Lower gradient values (up to 0.1 pH per 100 m.) are recorded for the deepwater area. The vertical gradients rise near the continental slope and the region of the eastern Bering Sea shelf and are located within the limits of the upper layer. At the present time it is difficult to find other structural elements in this vast area.

The depth of the position of the core of minimum pH is about 300-400 m at Near Strait and in the Aleutian Basin. With the proximity of Pacific water to the continental slope its position deepens to 700-800 m and pH values did not exceed 7.65-7.70. Maximum values of 7.80 pH were found near the central Aleutian Islands. In depths over 3,000 m minimal values of pH were observed in the central part of the deepwater basins of the Bering Sea, but at its periphery increased values of pH were recorded.

Seasonal Variability of pH

Seasonal variation of pH was tracked within the euphotic zone of the Bering Sea. Maximum changes of pH in this layer were found in spring or spring-summer, but only in summer for the Chirikov Basin and in Bering Strait (Fig. 3). Here ice cover breaks up late and the intensive growth of phytoplankton is delayed. Profiles of the vertical distribution of pH indicate that a pronounced seasonal variation only develops at 0-30 m. During the warm season a subsurface maximum of pH is distinguishable in these layers. It is clearly expressed at the shelf areas, at the periphery of deepwater basins, and in the centers of large-scale gyres. Near the slope and in the central part of gyres a dense stratification forms which provides favorable factors for phytoplankton growth.

Space-Time Distribution of pH

At 10 m the basic patterns of pH spatial distribution are practically indistinguishable from the surface. Only an insignificant decrease in values is observed (not more than 0.5 pH). The photosynthetic activity of phytoplankton has practically no effect at 50 m but the spatial distribution of pH is distinct from the surface. In winter the pH value at 50 m varies from 8.04 to 8.10. Here, as in the surface layer, there is no connection between the distribution of pH with water currents. In spring the pH at 50 m is somewhat

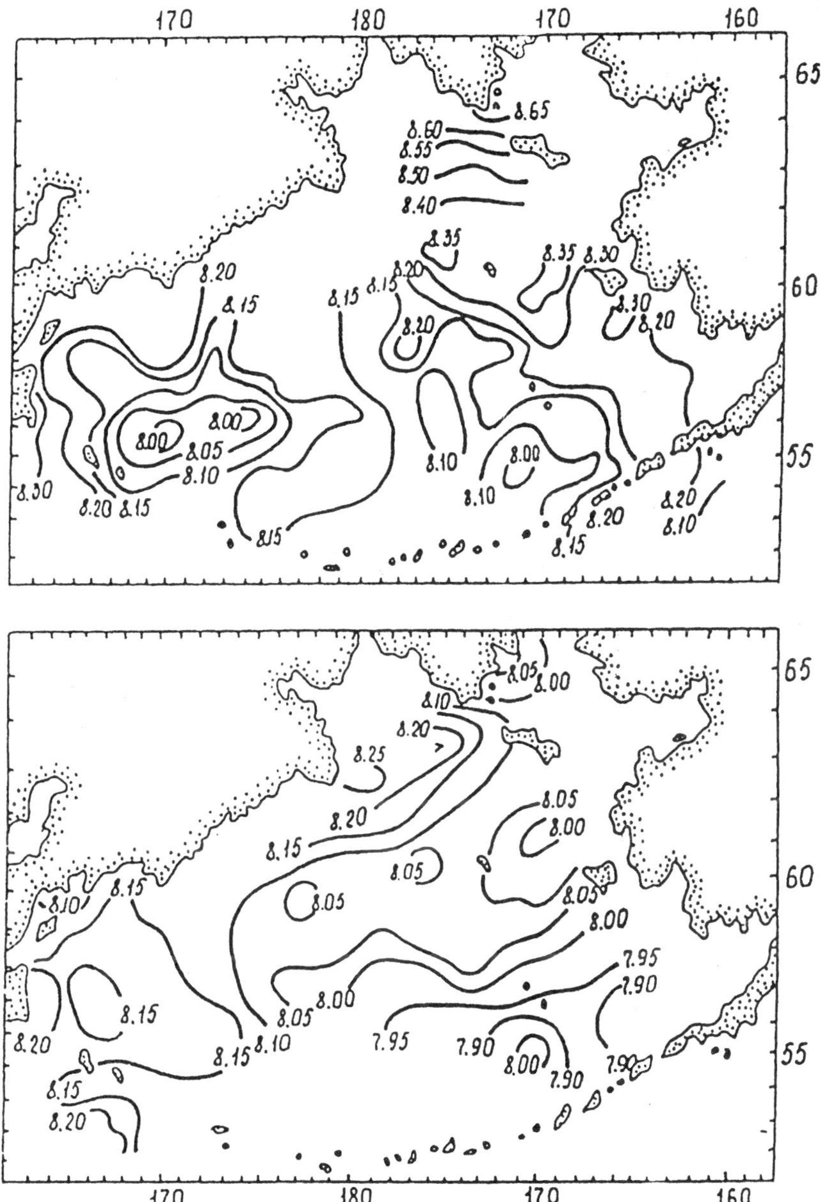

Figure 3. *The distribution of pH in the surface layer in spring (top) and autumn (bottom).*

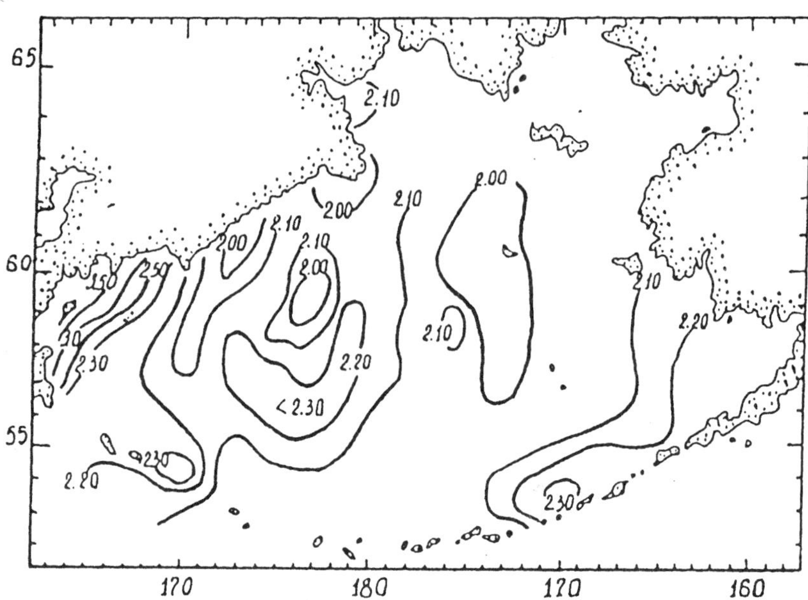

Figure 4.   The distribution of alkalinity (mmole per L) in summer.

higher and varies from 8.05 to 8.15. The increase is most likely a result of photosynthesis in the overlying layers. In deepwater basins where photosynthetic activity is not high, pH values from the surface to 50 m vary insignificantly. In summer at 50 m pH values of 8.05-8.10 are predominant. Magnitudes over 8.10 pH are characteristic for Pacific waters. Minimal values were seen near the coasts (from Cape Olyutorsk to Cape Navarin), and also between St. Matthew Island and St. Lawrence Island. These properties develop due to circulation factors. In autumn a convection begins to appear at 50 m. The distribution of pH in this layer is similar to the surface. However, the magnitudes of pH from the surface to 50 m are lower by 0.5-0.10 pH and at 100-200 m area a further reduction in the values of pH occurs. Minimum pH values are recorded at 100-200 m in Pacific waters and are most clearly developed in winter and autumn. An increase in pH values occurs near the continental slope which is most noticeable at 200 m. This feature can be explained by the increase in intensity of metabolic processes near the slope. At 500 and 1,000 m there is a distinctly visible penetration of Pacific waters which have higher pH values through Near Strait and in the passes of the central parts of the chain. Lower pH values were recorded at 500 and 1,000 m near the continental slope to the east of the Cape Navarin meridian. At 1,500 and 2,000 m maximal pH values were found in Pacific waters. At 1,500 m the basic

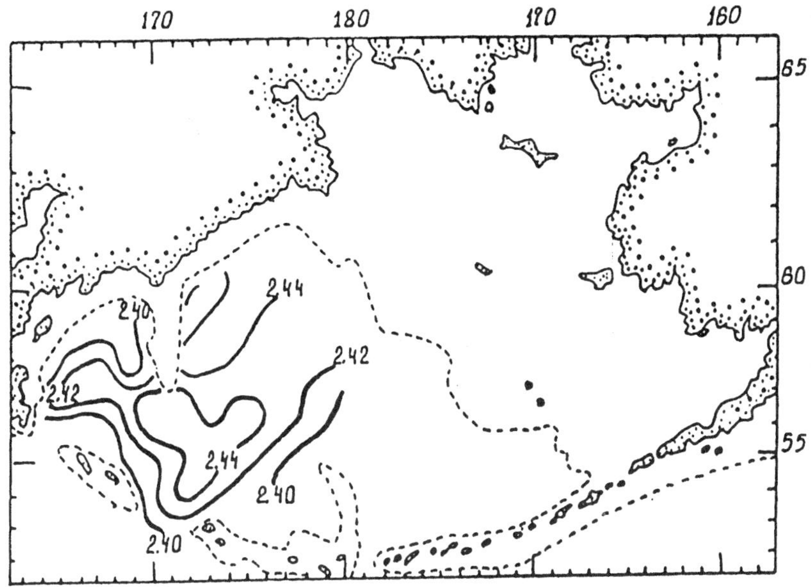

Figure 5.   The distribution of alkalinity (mmole per L) at 1,000 m.

penetration of these waters occurs through Near Strait. At 2,000 m Pacific waters were followed through Kamchatka Strait. Minimum pH values at 1,500 and 2,000 m were evident near the continental slope east of Shirshov Ridge.

Seasonal Variation of Alkalinity

There is a seasonal variation of alkalinity in the upper 200 m but it is most significant in thin surface layers (10-20 m). Seasonal variation of alkalinity in the upper 200 m occurs synoptically. Maximal seasonal variations were recorded in surface waters of Karagin Bay and also at the boundaries of Pacific water (over 0.1 mmol per L). In Pacific waters the seasonal variability of alkalinity is minimal (approximately 0.01-0.02 mmol per L). At 50 m the spatial properties of seasonal alkalinity variation are similar to the surface. Maximum alkalinity values in the active layer were found in spring. Near the coasts of the Kamchatka Peninsula there could be a shifting of maximum alkalinity in summer due to river discharge. Minimal contents of alkalinity in the active layer were seen in the cold season.

Space-Time Distribution of Alkalinity

In winter alkalinity values at the surface varied from 2.27 to 2.29 mmol per L. Minimal values are characteristic for Pacific waters, but the largest values (from 2.28 to 2.29 mmol per L) are found near the coasts of the Kamchatka Peninsula where upwelled waters have higher values of alkalinity. In spring maximum values in surface waters (up to 2.31 mmol per L) are characteristic for Pacific waters. The values significantly decreased near the coasts of the Kamchatka Peninsula. The distribution of alkalinity in surface waters in summer is determined by replacement with Pacific water. Consequently, high alkalinity values (more than 2.2 mmol per L) are recorded at sites of penetration of Pacific water into Bristol Bay and in the western part of deepwater basins (Fig. 4). Maximum alkalinity values in surface waters in summer (up to 3.5 mmol per L) are found in Karagin Bay, probably the result of river discharge. In large parts of the shelf, alkalinity is at a minimum which is a result of phytoplankton growth and the consumption of carbonates. In autumn, maximum alkalinity values of surface water (from 2.28-2.32 mmol per L) are seen close to Near Strait. Minimum alkalinity values (approximately 1.9-2.0 mmol per L) are recorded in autumn near the coasts of the Kamchatka Peninsula. Here the effect of the river discharge is practically negligible, and the hydrological conditions still are conducive to the growth of phytoplankton. In winter and spring alkalinity values at 50 m are slightly distinct from the surface. In summer alkalinity values at 50 m are higher than at the surface by 0.2-0.5 mmol per L. In autumn the alkalinity content at 50 m is distinct from the surface only in waters near the coasts of the Kamchatka Peninsula. Here alkalinity values exceed the surface by 0.2-0.3 mmol per L).

The distribution of alkalinity at 100-200 m is determined by replacement with Pacific water. In summer the alkalinity content in Pacific waters is lower than in Bering Sea waters. In winter, spring and autumn the distribution of alkalinity changes to the opposite. At 500-2,000 m the effects of the intra-water movement and the advection of Pacific waters are predominant (Fig. 5). Maximal values of alkalinity in this layer are evident in areas which are distant from coasts. At the periphery of basins the values are minimal, possibly because of the cyclonic motion of water in deepwater basins.

Nutrients

Dissolved nutrients provide the raw materials needed to synthesize protein and other organic matter by autotrophic populations. Nitrogen along with carbon provides the main constituents for amino acid and protein construction and to a significant degree determines all life in the sea. Phosphorus is encountered in all cellular formations and it regulates their life processes: photosynthesis, respiration, and metabolism (Bruevich et al. 1960). Dissolved silicon enters into the composition of the skeletal components of marine organisms. The main source of phosphorus and silicate in the ocean is continental runoff. Consequently relatively large concentrations of phosphorus and silica can be evidence of the discharge of river water into the sea. There is also evidence that bottom sediments are significant sources of remineralized phosphorus and silica.

Nitrate

Nitrate is normally considered the biogenic nutrient that limits the quantity of new primary production that occurs in the Bering Sea since surface concentrations are often depleted (Hattori and Goering 1981, Whitledge et al. 1986). Unfortunately, basin scale spatial and temporal distributions of nitrate have not been measured so only mesoscale variations will be discussed in the following sections.

Nitrate Cross-Shelf Distribution

The cross-shelf distributions of nitrate throughout an annual cycle demonstrate the interaction of deep water in the open basin with enrichment processes on the shelves (Whitledge et al. 1988). The near-bottom onshore fluxes of nitrate-rich water often originate as deep as 300-400 m and therefore as much as 35 µM vitrate were frequently observed near the bottom at the shelf break; surface values were observed as high as 25 µM (Fig. 6) (Whitledge et al. 1986). Nitrate concentrations as high as 25 µM were observed near the shelf break while shelf concentrations were 15-20 µM as the spring phytoplankton bloom started. The largest nitrate concentrations were observed in the bottom layer during late summer and autumn when deeper open basin waters were being carried upward from depths of 300 m or greater. Surface layer concentrations of nitrate declined rapidly during the spring bloom to values between undetectable and 1 µM. The inner portion of the shelf landward of the 50 m isobath was depleted of nitrate throughout the water column

Seasonal Variability of Nitrate

The largest concentrations of nitrate observed in surface waters over the eastern Bering Sea shelf occur in late winter and early spring. Winter fluxes of nitrate onto the shelf transported as much as 20 µM in surface waters of the outer shelf at the 130 m isobath before the spring phytoplankton bloom commenced (Fig. 7), while the middle shelf region at the 75 m isobath only reached about 15 µM (Whitledge et al. 1986). As the pycnocline developed, nitrate was quickly depleted to <1 µM in the middle shelf. The outer shelf declined at a much slower rate probably due to stronger vertical mixing which decreased light availability and maintained nutrient fluxes. The relatively strong pycnocline produced a two-layered system with depleted nitrate in the upper layer and uniform but higher concentrations in the bottom layer. The autumn period produced stronger mixing which weakened the pycnocline and enriched the surface with about 10 µM of nitrate.

The interannual variability of seasonal nitrate concentrations was observed to be surprisingly similar over nearly a 4-year period. Integrated concentrations over the upper 40 m of the water column showed that nearly 900 mg-at per m^2 occurred over the outer shelf before the spring bloom occurred (Fig. 8). The middle shelf was also very regular over the 4 years with a maximum each year of about 650 mg-at per m^2. The rapid

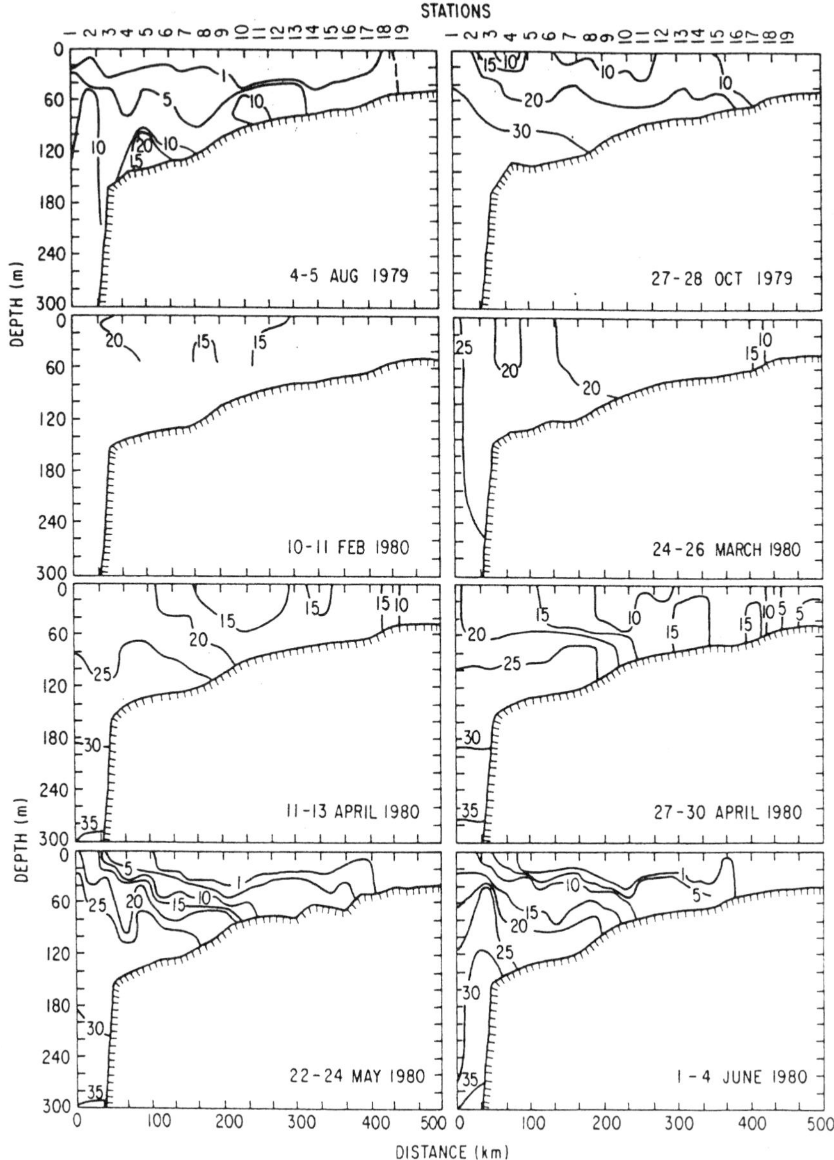

Figure 6.   The distribution of nitrate (µM) across the southeastern Bering Sea shelf for August 1979 to June 1980.

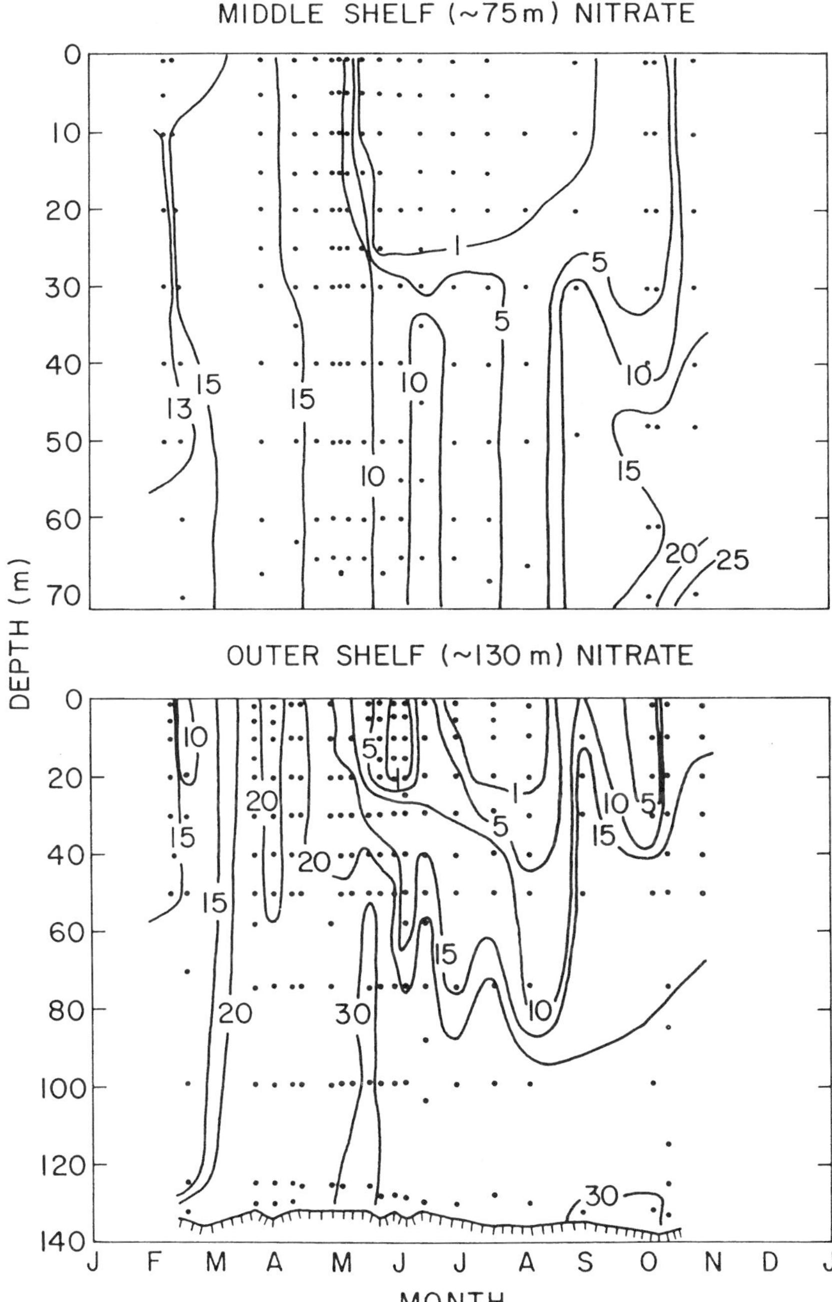

Figure 7. *The temporal distributions of nitrate (μM) over an annual cycle at the 75 m and 130 m isobaths in the southeastern Bering Sea.*

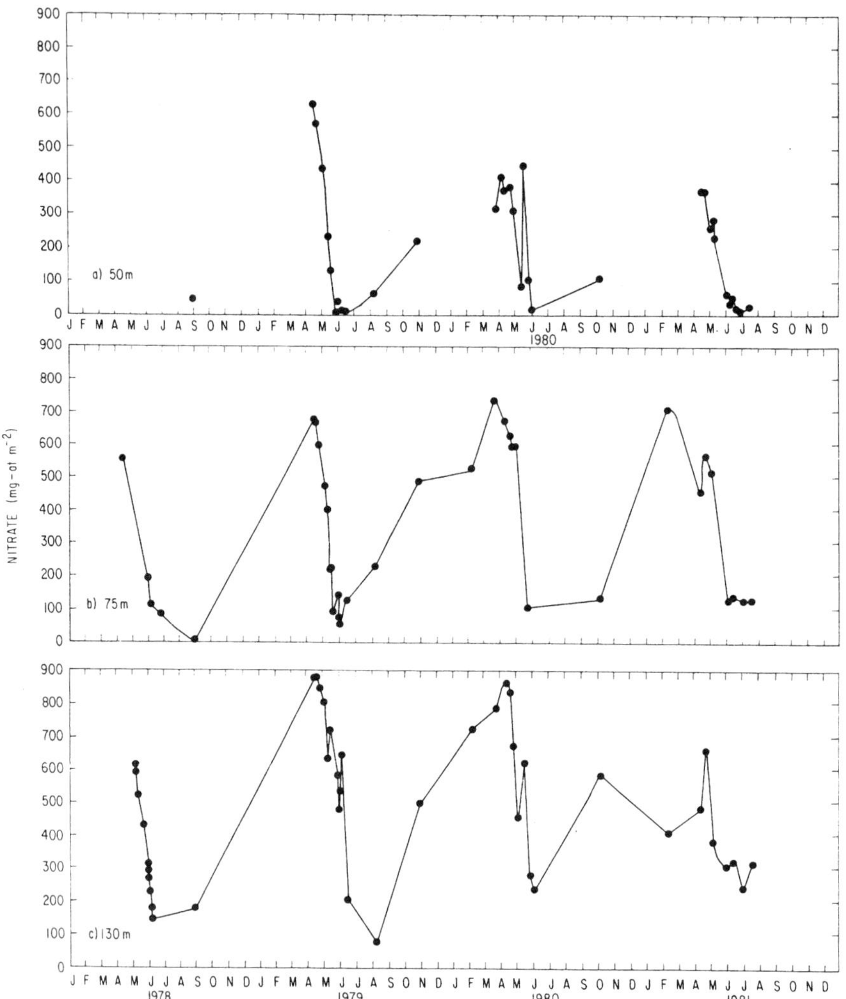

Figure 8.   Integrated concentrations of nitrate (mg-at per m²) over the upper 40 m at the 50 m, 75 m, and 130 m isobaths in the southeastern Bering Sea.

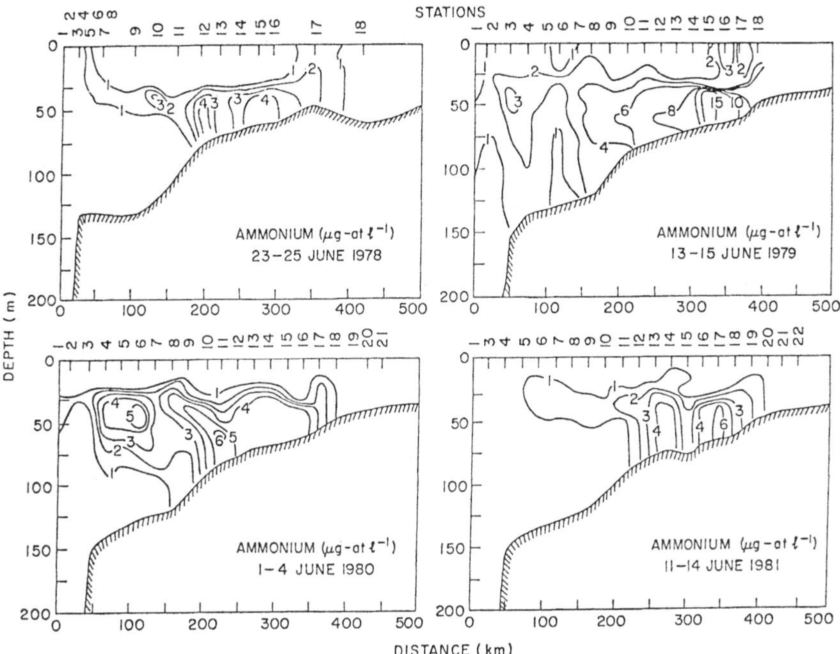

Figure 9.   The distributions of ammonium (µM) across the southeastern Bering Sea shelf for the years 1979-1981.

decline in nitrate occurred in both areas during approximately a month. Variability in the sharp decline in nitrate concentrations was attributable to short duration wind mixing from storm events (Whitledge et al. 1986).

Ammonium

Ammonium is not observed in open ocean waters of the Bering Sea Basin except at very low concentrations. As a result, the only significant ammonium concentrations are observed on the shelves. Ammonium is the principal product of remineralization of organic nitrogen and its utilization by phytoplankton populations sustains primary production when nitrate is depleted from the water column. The production of ammonium occurs predominately in the water column as a result of microbial processes such as ammonification. Some ammonium can be generated in the sediments, especially if dissolved oxygen concentrations are low enough to support denitrification processes. Most Bering Sea shelves sustain high dissolved oxygen concentrations so denitrification processes are minimal and not conducive to ammonium production. Zooplankton and fish populations also produce ammonium as excretory products which can produce sig-

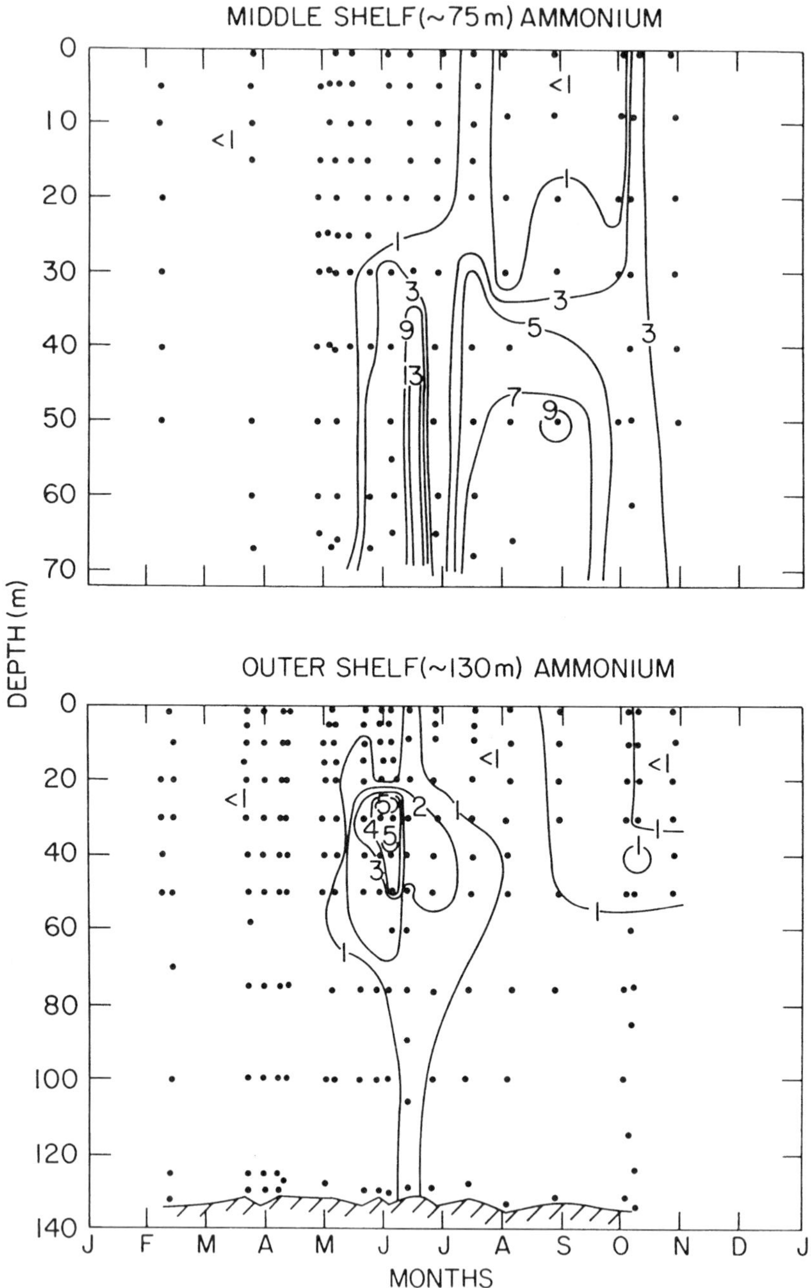

Figure 10. *The temporal distributions of ammonium (µM) over an annual cycle at the 75 m and 130 m isobaths in the southeastern Bering Sea.*

nificant concentrations of ammonium over a few days (Dagg et al. 1982, Vidal and Whitledge 1982).

Ammonium Cross-Shelf Distribution

Ammonium was observed on the eastern Bering Sea shelf mainly in summer following the spring phytoplankton bloom. The outer shelf near the shelf break sustained low ammonium concentrations at the surface and bottom but often an enhanced concentration was observed near 50 m which probably had a source on the middle shelf (Whitledge et al. 1986). The low ammonium concentrations near the bottom on the outer shelf are indicative of areas with water fluxes from the open ocean which have extremely low concentrations. Since the pycnocline and vertical stratification is well established, very small amounts of ammonium were observed in the upper layer, but concentrations of 1-6 µM were measured in the bottom layer of the middle shelf (Fig.9). During unusual years of increased primary production and vertical sinking of the phytoplankton, ammonium concentrations exceed 15 µM near the bottom. However, each of four years produced similar ammonium distributions which confirm that similar physical and biological processes were present even though concentrations were not constant.

Seasonal Variability of Ammonium

Winter concentrations of ammonium were as low as 0.1 µM over the entire eastern Bering Sea shelf. After the spring phytoplankton bloom declines the initial enhancements of ammonium were observed in the lower layer near the bottom at the 75 m isobath (Fig. 10). These middle shelf concentrations were larger since maximal ammonium regeneration has been observed at the sediment interface. The outer shelf also contained high concentrations of ammonium at depths between 25 and 40 m which were either produced locally or advected from inshore ammonium sources (Whitledge et al. 1986). The decline of ammonium in bottom waters of the middle shelf in September-October probably represents in situ nitrification which is verified by increases of nitrate at the site. It was estimated that about 20-30% of the nitrate resulted from this nitrification process (Whitledge et al. 1986).

Nitrite

Spatial-Temporal Variability of Nitrite

In winter and autumn in the active layer of the Bering Sea one observes a very small concentration of nitrites, likely a result of convective processes and low water temperatures. During spring in littoral areas (10-30 m), minimal nitrite contents are observed in the surface layer. Minimal nitrite concentrations in the thin surface layer in shallow waters are associated with high dissolved oxygen content. In the middle part of the shelf the vertical distribution of nitrites is more complex. Two maximums are re-

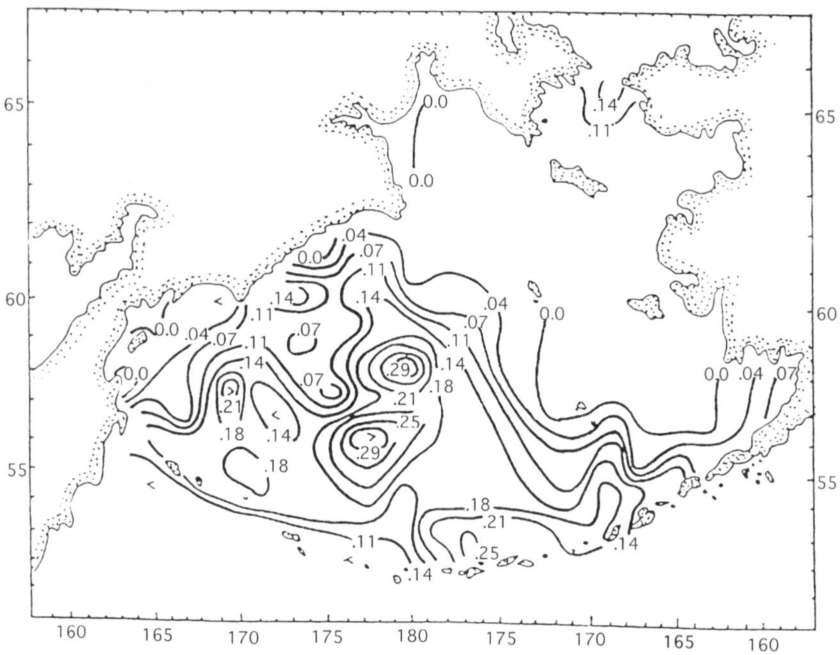

Figure 11. The distribution of nitrite (μM) in summer.

corded here: in the subsurface layers and near the bottom. Minimum values were recorded in surface waters in outer areas of the shelf and maximum nitrite content occurred in the subsurface layers (Bruevich 1954). A subsurface nitrite maximum was also observed in deepwater areas in spring and summer (Fig. 11).

There is a possible formation of two subsurface maxima near the slope and in the southern part of the shelf. This formation is associated with the mixing of Pacific and shelf waters. The subsurface nitrite maximum during spring in deepwater basins does not exceed 0.07-0.11 μM. In shelf areas it increases to 0.14-0.17 μM and is maximal near the continental slope (0.36-0.57 μM) where a complex alternation of both high and low values was observed. The subsurface nitrite maximum in summer varies from 0.1 to 0.4 μM.

In intermediate and deep waters there are two types of vertical distribution of nitrites. In the western and northern areas of deepwater basins, beginning with the lower boundaries of the seasonal pycnocline, the contents of nitrites decrease to undetectable values. In the central part of deepwater basins and close to the eastern Bering Sea slope there is a minimum of nitrites located at 400-1,000 m which clearly coincides with the position of minimum oxygen. In the central part of deepwater basins the values do not exceed 0.01-0.02 µM, but close to the slope they rise to 0.07-0.13 µM.

Dissolved Silicon (Silicate and Silica)

Seasonal Variability of Dissolved Silicon

In the littoral part of the Bering Sea the seasonal variability in silica content is influenced by fluvial discharge. In spring and summer (in comparison with winter) the volume of river discharge can increase up to 10-50 times. Phytoplankton blooms, especially diatoms (the main consumers of silica) also have seasonal peaks of activity which can greatly affect silicate distributions. The maximal content of silica at the surface in winter is due to the small amount of phytoplankton growth and intensive mixing of the water column. Its maximum concentration (up to 55 µM) is present in Pacific surface waters which fill the southern part of the deep basins. At the periphery of the basins it decreases to 30-35 µM. In the outer parts of the shelf (with depths over 100 m), silica content does not exceed 10-30 µM (Fig. 12).

The increase of continental discharge in spring and summer leads to an enrichment of silica in littoral waters. However, this is accompanied by the active growth of diatoms that consume silicate. Consequently, in littoral areas, it is difficult to partition the uptake and regeneration parts of the silica cycle. The silicate content in the euphotic layer is minimal in spring and summer in deep water of the shelf.

Spatial-Temporal Distribution of Dissolved Silicon

In the Bering Sea there are few records of silica in winter but concentrations exceed 40 µM in both middle and outer shelf areas (Fig. 13). In summer minimal contents of silica are observed at the surface in the upper mixed layer where the photosynthetic activity of phytoplankton decreases concentrations markedly. The middle shelf region, with the stronger pycnocline, depletes silicate <1 µM while the bottom layer maintains 20-30 µM. The outer shelf region at the 130 m isobath had dissolved silicon concentrations <10 µM but were not as strongly depleted as the middle shelf, probably due to larger vertical mixing. Bottom concentrations on the outer shelf were as high as 60 µM during late summer and early fall periods.

In deepwater areas maximal silica (up to 35-55 µM) is observed in Pacific waters. In summer the values are reduced in the basins, particularly in slope areas. At the shelf, summer distribution of dissolved silicon is significantly distinct from spring (Fig. 14). Minimum values (from 0-3 µM) are seen in Anadyr Bay. In the southeastern part of the shelf summer values at 3-5 µM are higher than in spring which are caused by the deepening of the layer of the seasonal pycnocline and the inflow of water from the Pacific. In autumn on the shelf the effect of convective mixing leads to

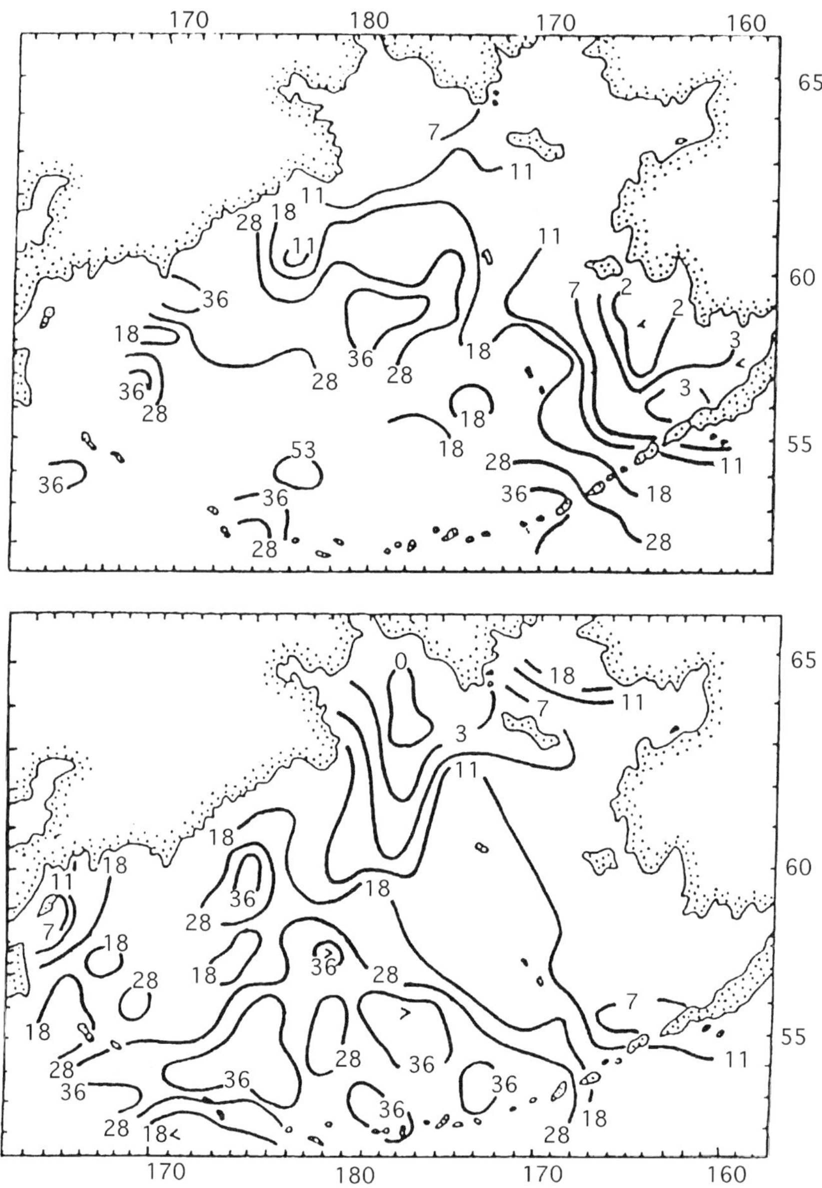

Figure 12. The distribution of silica (µM) in the surface layer in spring (top) and summer (bottom).

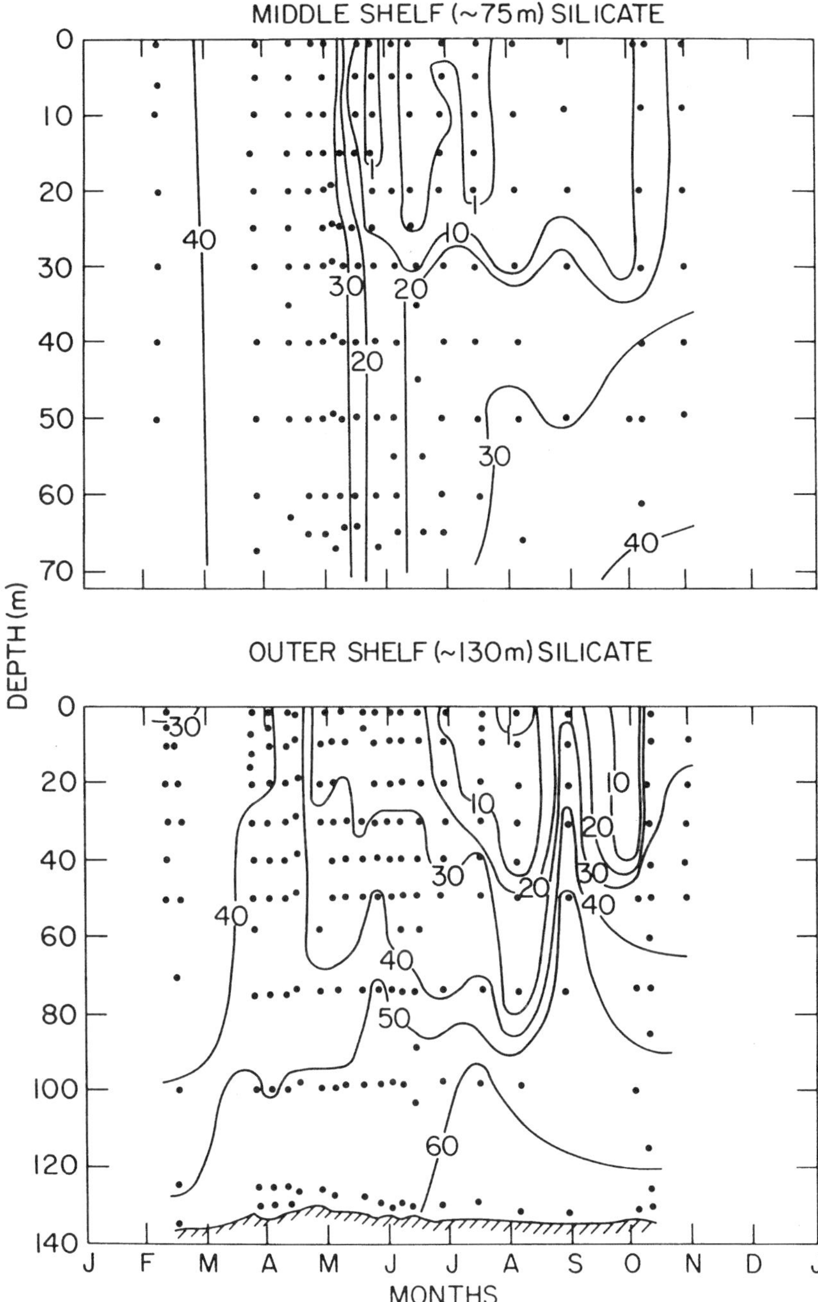

Figure 13. The temporal distributions of silicate (μM) over an annual cycle at the 75 m and 130 m isobaths in the southeastern Bering Sea.

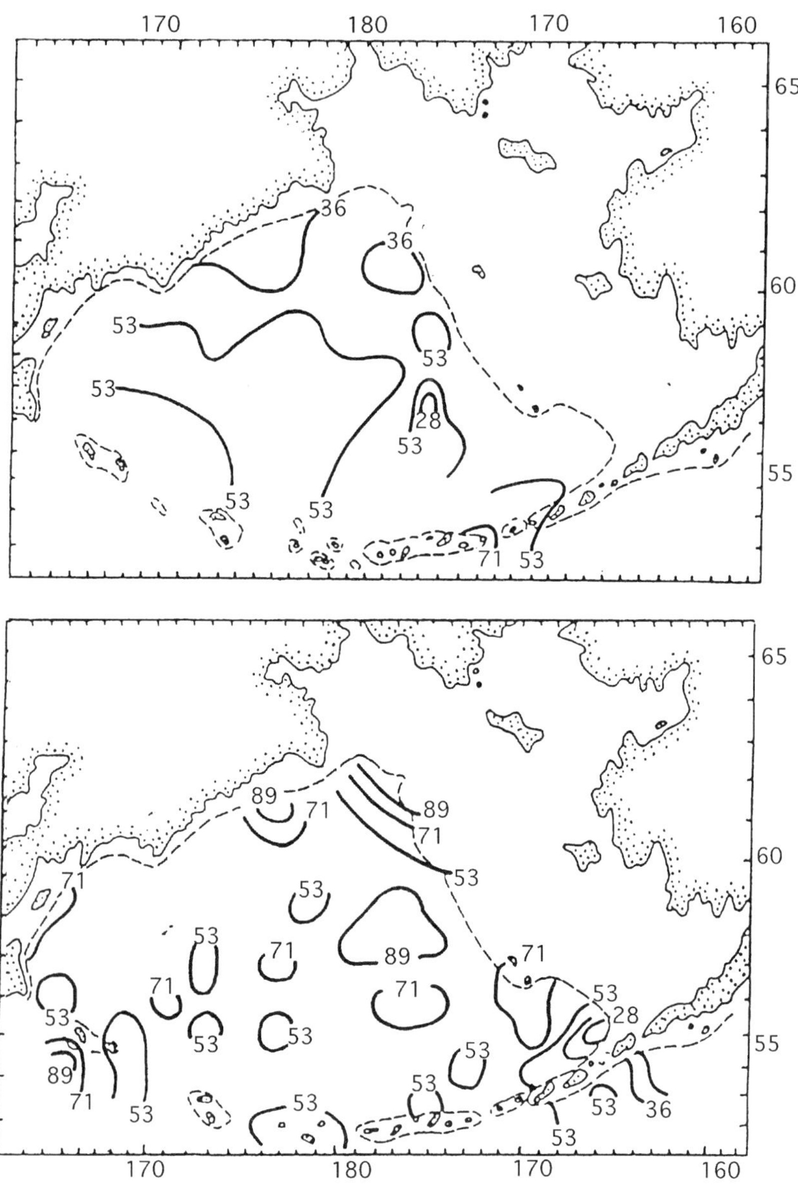

Figure 14. The distribution of silica (µM) at 100 m in spring (top) and summer (bottom).

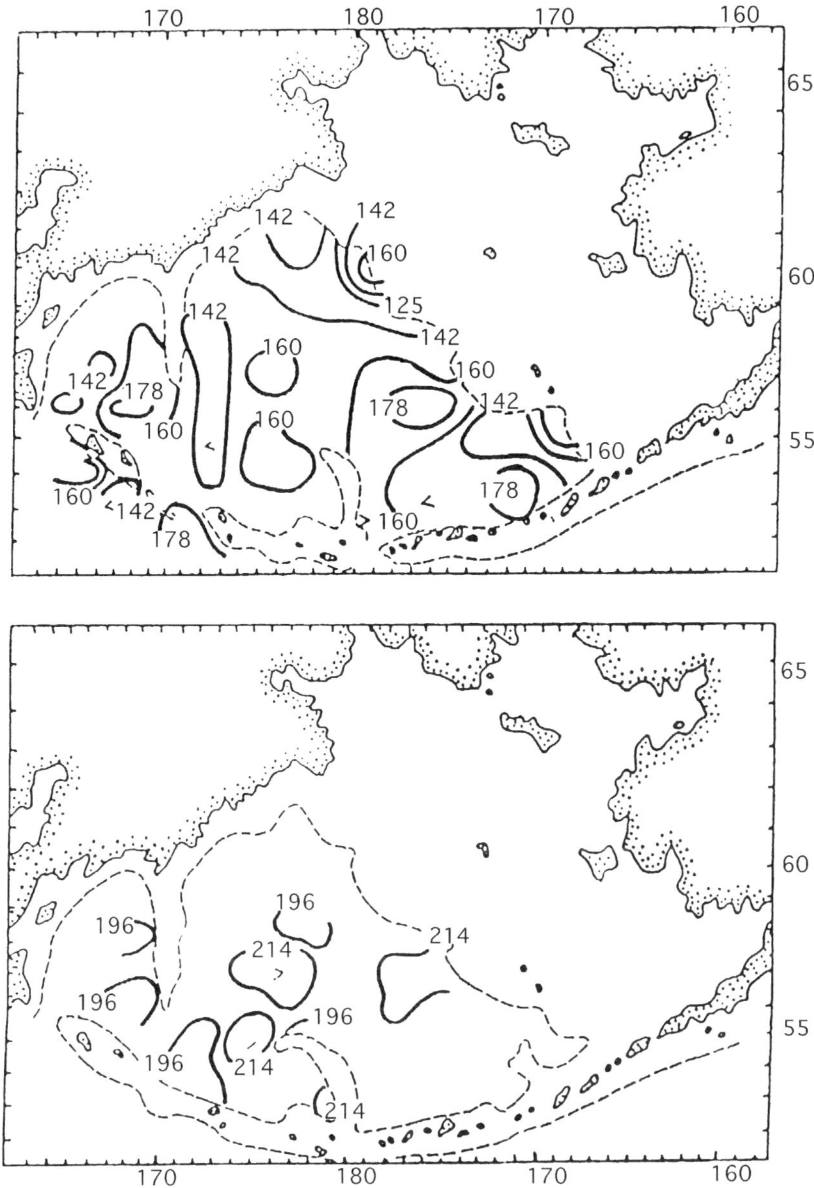

Figure 15. *The distribution of silica (µM) at 1,000 m (top) and 2,000 m (bottom).*

an increase of the values. The high values (>35 μM) in the southeastern part of the shelf occur as a result of the advection of Pacific waters.

At 50 m the photosynthetic activity of phytoplankton only weakly affects silicate; consequently, seasonal differences are weakly exhibited in the deepwater areas. Only near the continental slope, due to the intensification of metabolic processes, was a decrease observed in the concentration of silica in the warm time of the year. The distribution of silica at 100 m in spring is not distinct on a chart of the overlying layers. Only in the southeastern part of the basins was an increase in values recorded up to 50 μM. In summer the concentration of silicate is significantly higher. Near the continental slope, due to the intensification of currents and the formation of eddies, silicate content increases in summer to 70-90 μM. In summer and autumn, due to the downwelling into anticyclonic eddies near the eastern part of the slope, localized areas can contain decreased contents of silica dioxide (up to 28.6-35.7 μM).

At 200 m silica distribution is determined by dynamic factors. In spring maximum concentrations (up to 70-90 μM) are recorded in Pacific waters. Their values decrease near the slope to 35-55 μM. In summer maximum values can be observed at both sites where penetration of Pacific waters into the Bering Sea occurs. In autumn minimal silica contents are recorded in Pacific waters but the maxima are near the continental slope. It is possible that seasonal variation of silica at 200 m is associated with variations in silica in the subsurface layers in waters of the Alaska current.

There is no connection between the distribution of silica at 500 and 1,000 m with the large-scale circulation of water (Fig. 15). In deep water there was a further increase in silica concentration. Moreover, at 1,500 m, as in intermediate waters, we observed an alteration of enclosed regions with high and low values (from 160 to 200 μM).

Phosphate

Seasonal Variability of Phosphate

The vertical distribution of phosphates is separated into three layers: surface waters where the productive processes predominate and minimal concentrations of phosphates are observed; intermediate waters where phosphate concentration increases; and deep waters in which one sees either an unchangeable phosphate content or a slight concentration reduction.

Minimal phosphate concentrations were recorded in the euphotic zone. In winter phosphate concentration there changes slightly. The vigorous growth of phytoplankton in spring leads to a reduction in phosphate concentration in the photic layer. In spring and summer, the vertical distribution curve shows a surface layer with minimal concentrations (up to 0.2-0.3 μM) (Fig. 16). The lower boundary of this layer coincides with the upper boundary of the cold, intermediate layer of water.

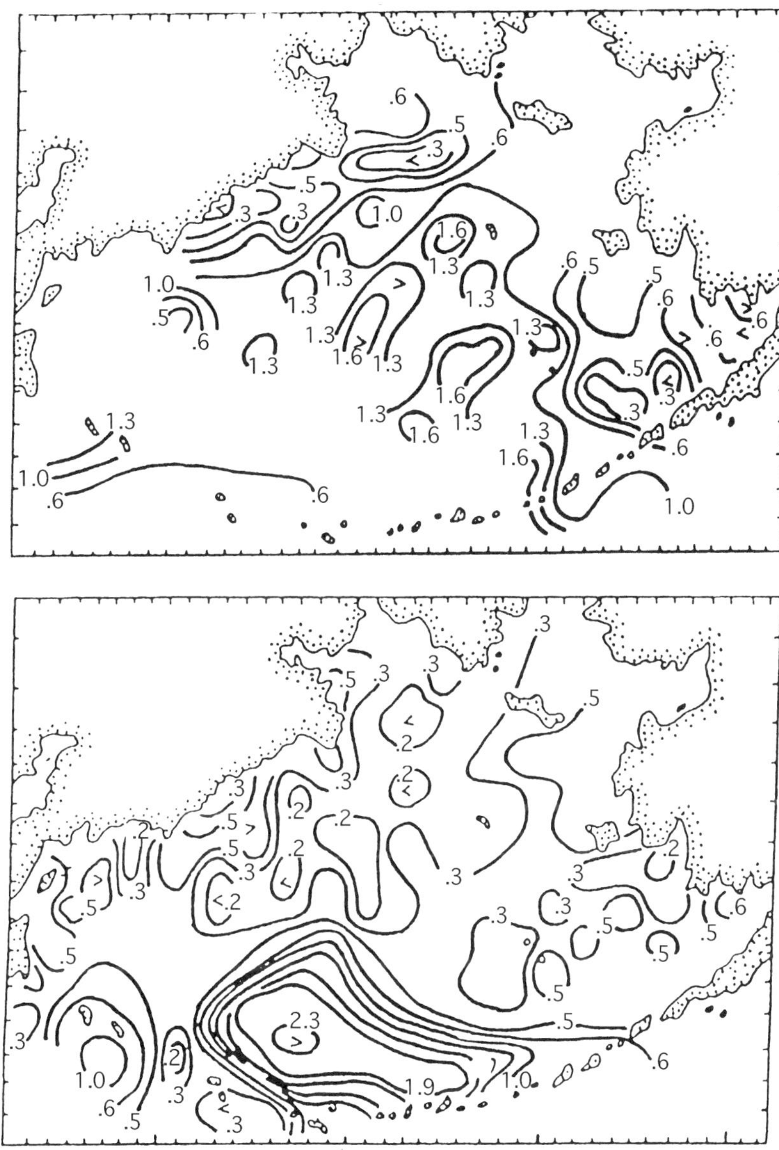

Figure 16. *The distribution of phosphate (μM) in the surface layer in spring (top) and summer (bottom).*

Phosphate dramatically increases at the lower boundary of the euphotic zone. Beneath the layer of the seasonal pycnocline there was a comparatively homogeneous distribution of phosphates, likely a result of the autumn-winter convective mixing. In shallow areas a nearly complete depletion of phosphates occurred in the warm period of the year down to the near bottom layer. In the expanses of water where there was considerable advection of water from adjacent areas, phosphate content was somewhat higher. For the deepwater photic layer there was a characteristic occurrence of a clearly expressed phosphate minimum at 10-20 m. In the Aleutian Islands passes and near the continental slope phosphate in the upper 100-200 m layers was either invariable along the vertical or increased slightly.

An increase in the concentration of phosphates was observed in intermediate waters. The core of intermediate phosphate maximum was located from 300-1,000 m. It deepens minimally in Pacific waters. Near the continental slope a deepening of the core of maximum phosphates occurred. Phosphate contents in this layer vary from 2.25 to 3.9 µM. In Pacific waters the concentration was minimal but increased to 3.9 µM near the slope. In deep waters phosphate concentration either remained invariable or increased slightly with depth.

Space-Time Distribution of Phosphate

Maximum phosphate was recorded at the surface in winter on the shelf and in the passes of the central Aleutian Islands and was due to the intensity of vertical exchange. In spring a map of phosphate distribution corresponded closely to the growth of phytoplankton. Consequently, in a large part of the shelf there was a low concentration (from 0.15 to 0.6 µM). In deep water phosphates reached 1.3-1.6 µM; in shallow waters in summer typical phosphate values never exceeded 0.15-0.3 µM. Areas of reduced values also extended to the northern and western parts of deepwater basins. Maximal values (2 to 2.25 µM) were seen to the north of central Aleutian Islands passes associated with intensive mixing of water in the passes. In autumn phosphate content increased at the surface. This was connected with a reduction in intensity of metabolic processes of phytoplankton and an increase in autumn convective mixing. This development was particularly noticeable in shallow water.

Phosphate distribution in winter at 50 m was not distinct from the surface. In spring at 50 m maximum phosphate was located in shallow water passes of the eastern Aleutian chain resulting from water mixing in the passes. In summer at 50 m maximum magnitudes (2-2.5 µM) were observed in Pacific waters which flowed through the Aleutian Islands passes. Near the slope there was a high value for the spatial gradients. This zone separated the productive shelf waters from the water of oceanic origin. The high phosphate concentration at 50 m in summer in the shelf area (0.6-1.3 µM) was connected with the density of the stratification of

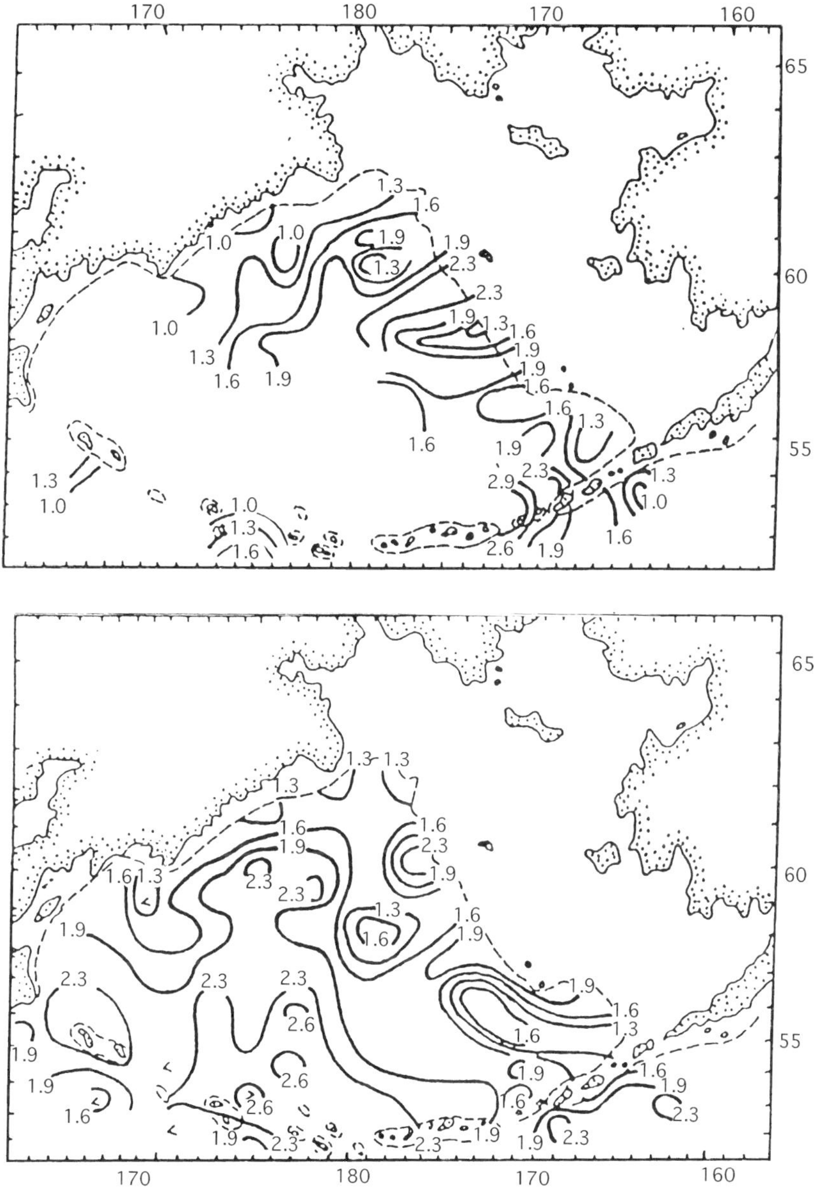

Figure 17. The distribution of phosphate (µM) at 100 m in spring (top) and summer (bottom).

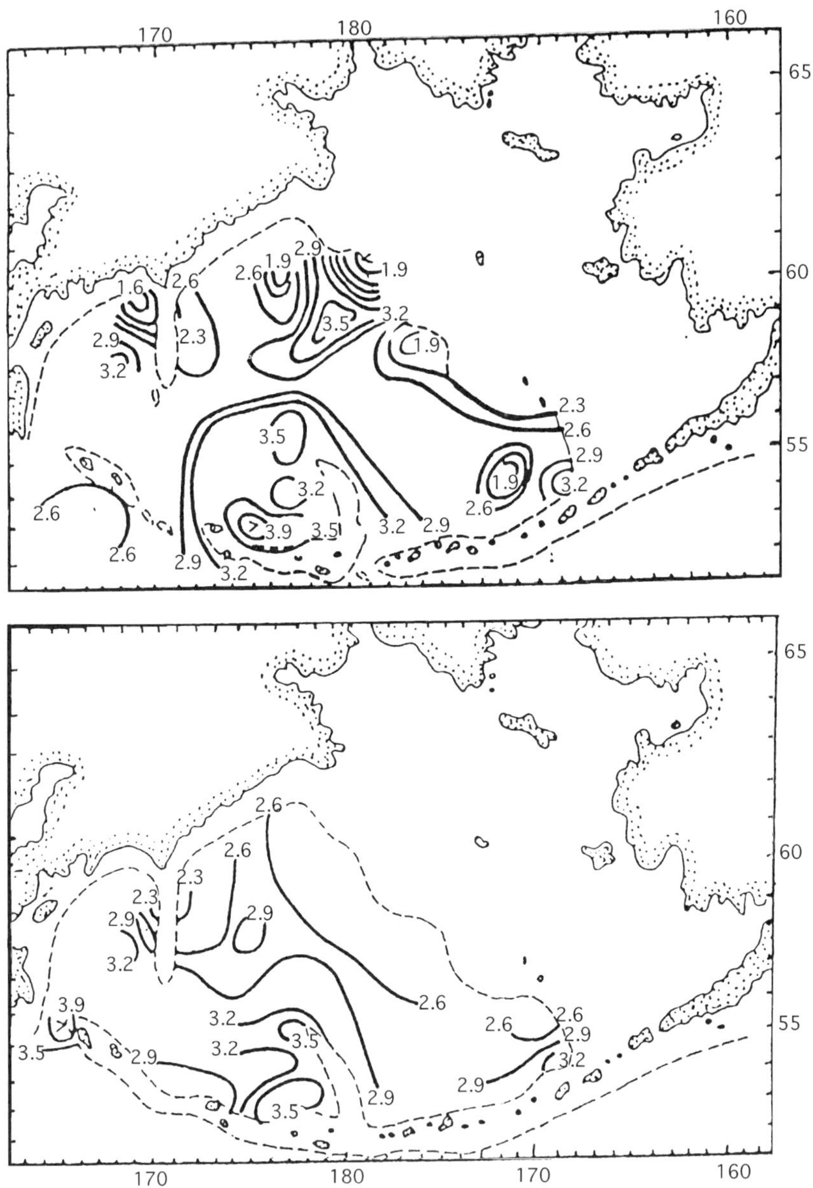

Figure 18. The distribution of phosphate (µM) at 1,000 m (top) and 1,500 m (bottom).

water in the overlying layers. The distribution of phosphate at 50 m in autumn attests to the large roles of autumn convection and the advection of Pacific waters.

The spatial variation of phosphate values at l00 and 200 m was dependent on the intensity of internal water exchange. Their contents at 100-200 m were maximal in winter and in spring maximal phosphate magnitudes (2.6-2.9 μM) were recorded in the eastern passes of the Aleutian chain (Fig. 17). At the periphery of the Bering Sea they were reduced to 1.3-2.25 μM. In summer maximal phosphate concentrations (2-3 μM) were seen in Pacific waters. This decreased significantly near the continental slope. In autumn at the periphery of the deepwater basins the magnitudes increased (in comparison with summer) by 0.3-0.4 μM. This was connected with the intensification of the metabolic processes near the slope due to the start of the autumn convection.

There was a deep maximum of phosphate at 500 and 1,000 m for which values reached 3.2-3.9 μM. The distribution of the magnitudes attests to its Pacific water origination (Fig. 18). Phosphates were reduced to 1.6-2.3 μM near the continental slope. The phosphate content at 1,500-2,000 m (in relation to the layers of the deep maximum) was reduced approximately by 0.3 μM. This was associated with the decrease in water replacement through Near Strait; water replacement basically occurred through Kamchatka Strait. Consequently at 2,000 m there were maximal values (3.6-3.9 μM) in the southwestern area of the sea.

Conclusions

The hydrochemistry of Bering Sea waters significantly depends on the advection of Pacific waters, bottom relief, and the biological productivity of various water masses. In the euphotic zone and in intermediate waters there is a significant accumulation of dissolved oxygen. It is basically derived from convective processes in the autumn-winter period, the photosynthetic activity of phytoplankton in warm seasons, and intensification of metabolic processes in the passes of the Commander-Aleutian Islands chains.

In intermediate waters there is a distinct decrease in the concentration of dissolved oxygen, nitrites, and magnitudes of pH with a simultaneous increase of alkalinity. The lower boundary of this layer is located from 200-300 m to 500-600 m, but in separate instances there can be a deepening to 1,000 m.

Additional expeditionary observations are needed (particularly in winter) for obtaining data on seasonal variability of water parameters of the Bering Sea. In the active layer observations are needed in increments of no more than a month in order to investigate the consequences of photosynthetic activity of phytoplankton. It is critical to obtain precise and accurate measurements of trace metal concentrations, especially those that are utilized as micronutrients.

Acknowledgments

Partial support for this work was provided by NSF/Arctic Natural Sciences grant number OPP-9617287 and the NOAA Coastal Ocean Program through the Southeast Bering Sea Carrying Capacity project.

References

Barnes, C.A., and T.G. Thompson. 1938. Physical and chemical investigations in Bering Sea and adjacent portion of the North Pacific Ocean. University of Washington, Publication in Oceanography 3(2). 140 pp.

Bruevich, S.V. 1954. Nitrites and nitrification in the sea. Trudy Instituta Okeanologii Akademy Nauk SSSR 8:3-17. (In Russian.)

Bruevich, S.V., A.N. Bogoyavlenskiy, and V.V. Mokievskaya. 1960. The hydrochemical characteristics of the Okhotsk Sea. Trudy Instituta Okeanologii Akademy Nauk SSSR 42:125-198. (In Russian.)

Dagg, M.J., J. Vidal, T.E. Whitledge, R.L. Iverson, and J.J. Goering. 1982. The feeding, respiration, and excretion of zooplankton in the Bering Sea during a spring bloom. Deep-Sea Research 29:45-63.

Davidovich, R.A. 1963. The hydrochemical properties of the southern and southeastern parts of the Bering Sea. Trudy VNIRO-TINRO 50:83-87. (In Russian.)

Fukai, R., and F. Shiokava. 1955. On the main chemical components dissolved in the adjacent waters to the Aleutian Islands in the North Pacific. Bulletin of the Chemical Society of Japan No. 28.

Goodman, J.R., J.H. Lincoln, T.G. Thompson, and F.A. Zeusler. 1942. Physical and chemical investigations: Bering Sea, Bering Strait, and Chukchi Sea during the summers of 1937 and 1938. University of Washington, Publication in Oceanography 3(3). 130 pp.

Hattori, A., and J.J. Goering. 1981. Nutrient distributions and dynamics in the Eastern Bering Sea. In: D.W. Hood and J.A. Calder (eds.), The eastern Bering Sea Shelf: Oceanography and resources. University of Washington Press, Seattle, pp. 975-992.

Hattori, A., and E. Wada. 1974. Assimilation and oxidation-reduction of inorganic nitrogen in the North Pacific Ocean. In: D.W. Hood and E.J. Kelley (eds.), Oceanography of the Bering Sea with emphasis on renewable resources. Occasional Publication No. 2, Institute of Marine Science, University of Alaska, Fairbanks, pp.149-162.

Ivanenkov, V.N. 1964. Hydrochemistry of the Bering Sea. Nauka, Moscow. 137 pp. (In Russian.)

Lisitsyn, A.P. 1955. Some data about the distribution of suspended particles in the waters of the Kuril-Kamchatka troughs. Trudy Instituta Okeanologii Akademy Nauk SSSR 12:62-96. (In Russian.)

McRoy, C.P., J.J. Goering, and W.E. Shiels. 1972. Studies of primary production in the eastern Bering Sea. In: A.Y. Takenouti et al. (eds.), Biological oceanography of the northern North Pacific Ocean. Idenitsu Shoten, Tokyo, pp. 199-216.

Miyake, Y. 1939. Chemical studies of the western Pacific Ocean: The chemical composition of the oceanic salts. Bulletin of the Chemical Society of Japan 14(2).

Moiseev, P.A. 1964. Some results of the investigation of the Bering Sea scientific-commercial expedition. Trudy VNIRO 52:7-31. (In Russian.)

Mokievskaya, V.V. 1956. Some data on the chemistry of biogenic elements of the Bering Sea. Trudy Instituta Okeanologii Akademy Nauk SSSR 17:176-191. (In Russian.)

Mokievskaya, V.V. 1958. The distribution of forms of phosphorus in the Far Eastern seas. Trudy Instituta Okeanologii Akademy Nauk SSSR 26:16-25. (In Russian.)

Mokievskaya, V.V. 1959. Biogenic elements in the upper water layers of the Bering Sea. Trudy Instituta Okeanologii Akademy Nauk SSSR. 3:150-165. (In Russian.)

Park, P.K., W.S. Broecker, T. Takahashi, and W.S. Reeburgh. 1975. GEOSECS Bering Sea station, a brief hydrographic report. In: Bering Sea oceanography: An update 1972-1974. Results of a seminar and workshop on Bering Sea oceanography, Fairbanks, Alaska, 7-11 October 1974. Institute of Marine Science Report 75-2, pp. 207-238.

Ratmanov, G.E. 1937a. The hydrology of the Bering and Chukchi seas. In: Investigations of the USSR seas. Gidrometeoizdat 25:10-118. (In Russian.)

Ratmanov, G.E. 1937b. The distribution of hydrochemical elements in the northwestern section of the Bering Sea and Chukchi Sea. Gidrometeoizdat 25:119-135. (In Russian.)

Vidal, J., and T.E. Whitledge. 1982. Rates of metabolism of planktonic crustaceans as related to body weight and temperature of habitat. Journal of Plankton Research 4:77-84.

Whitledge, T.E., W.S. Reeburgh, and J.J. Walsh. 1986. Seasonal inorganic nitrogen distributions and dynamics in the southeastern Bering Sea. Continental Shelf Research 5:109-132.

Whitledge, T.E., R.E. Bidigare, S.I. Zeeman, R.N. Sambrotto, P.F. Roscigno, P.R. Jensen, J.M. Brooks, C. Trees, and D.M. Veidt. 1988. Biological measurements and related chemical features in Soviet and United States regions of the Bering Sea. Continental Shelf Research 8:1299-1319.

CHAPTER 11

Mesoscale Anticyclonic Eddies at the Shelf Break and Their Impact on the Formation of Hydrochemical Structures of the Bering Sea

Victor V. Sapozhnikov
Russian Federal Research Institute of Fisheries and Oceanography (VNIRO), Moscow, Russia

Abstract

The impact of mesoscale eddies on Bering Sea hydrochemical structures is shown in the distribution of phosphate, nitrate, ammonium, dissolved organic phosphorus, and dissolved organic nitrogen. The specific role of powerful anticyclonic eddies is discussed. Water sinking along the continental slope to 600-1,000 m depths reaches the layer of minimum oxygen concentrations. Renewal of these layers and upwelling of deep water occurs in adjacent eddies. Thus, the renewal of deep water takes place considerably faster than was expected.

Introduction

The effects of mesoscale eddies on hydrochemical structures were first reported during the POLYGON-70 studies (Ivanenkov et al. 1972). There have been numerous reports of changes in these hydrochemical features caused by topographic eddies over sea mountains (e.g., Darnitski and Staritsin 1978, Chernyakov et al. 1988). Still, mesoscale eddies at the shelf break remain the most interesting of these phenomena (Fedorov 1983, Ovchinnikov 1990). Comprehensive hydrochemical studies of mesoscale anticyclonic eddies at the shelf break were first undertaken in the Black Sea during 1988 (Sapozhnikov 1990, 1991). At that time, sinking water was observed in zones of local convergence at centers of anticyclonic eddies. Studies of these effects were continued in the Bering Sea during 10 October 1990 to 10 January 1991.

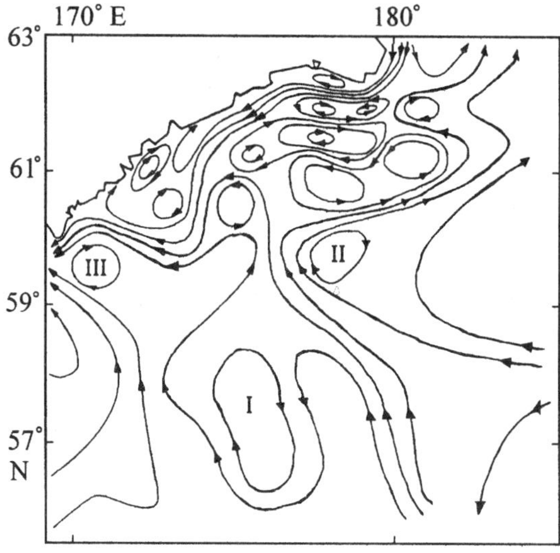

Figure 1.   Surface currents of the western Bering Sea.

Materials and Methods

The studies comprised 17 transects perpendicular to the western Bering Sea coast. Determination of the vertical distribution of temperature, salinity, oxygen, and phytopigments was performed with a Neil Brown (CTD) profiler at each station. Water samples were collected with sets of Go-Flo plastic sampling bottles. Phosphates, nitrites, nitrates, silicates, ammonium nitrogen, as well as dissolved organic forms of phosphorus and nitrogen (after combustion), were analyzed with an RFA-300 ALPKEM flow autoanalyzer. Currents were recorded with an acoustic Doppler current profiler (ADCP). Data obtained during the 1990-1991 winter-spring studies in the Bering Sea were also used.

Results and Discussion

Formation of the Hydrochemical Structure

Along with the general cyclonic circulation of the Bering Sea, there are numerous vortices above the continental slope (Fig. 1). These characteristics are shown by a peculiar pattern of hydrochemical features (Fig. 2). The cyclonic gyres were fairly well marked by maximal concentrations of silicates at the surface (>45 µM), phosphates (>1.80 µM), and nitrates (>24 µM) in the off-slope areas of Olyutorsk Bay and southwestward off

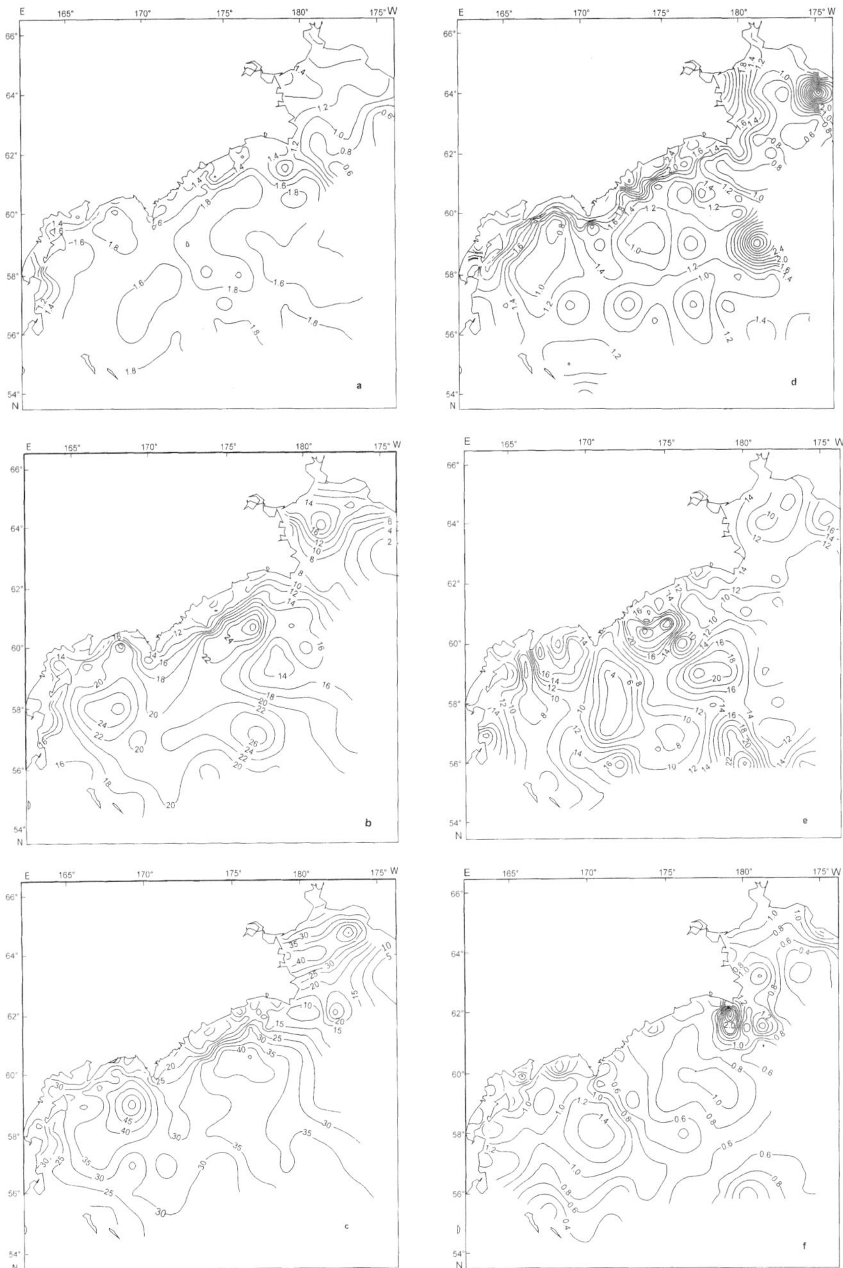

Figure 2.   Distribution of nutrients (µM) in the surface waters of the western Bering Sea: (a) phosphate; (b) nitrate; (c) silicate; (d) ammonium nitrogen; (e) organic nitrogen; (f) organic phosphorus.

Cape Navarin. These cyclonic effects were also characterized by maximal values of the Si/P ratio which exceeded 24, indicating upwelling of intermediate waters.

Cyclonic eddies transport water with low concentrations of ammonium nitrogen (1.00-1.20 µM) up to the surface. These values were lower than those in coastal waters, or in adjacent deepwater anticyclonic eddies.

Surface waters in the core of the quasi-stationary eddy were depleted of silicates, nitrates, and phosphates (8-10 µM, 6-8 µM and 1.0-1.2 µM, respectively), but were rich in ammonium nitrogen (2.0-2.4 µM), dissolved organic phosphorus (0.8-2.0 µM) and dissolved organic nitrogen (12-16 µM). A sharp increase of dissolved oxygen (up to 8.6-8.8 ml/L) and O_2 saturation (up to 116-120%) was observed in the center of the eddy. These features indicate fairly intensive processes of photosynthesis in the surface layer.

The shelf break, characterized by the convergence of the Bering Sea Central Current (BCC) and coastal currents, is a region of intensive turbulence with a complex system of eddies of different orientation. There is a sequence of six vortices along 178°E southward off Cape Navarin, starting with an anticyclonic eddy in shallow coastal waters and finishing with a cyclonic gyre, part of the Aleutian circulation.

The pattern of eddies in coastal and slope waters recorded with ADCP was confirmed by data on the distribution of nutrients. Thus, the hydrochemical charts, the geostrophic calculations, and direct measurements using current meters contribute to a comprehensive and precise picture of the currents.

The system of anticyclonic eddies developing over the shelf break seaward of the BCC is a dynamic effect peculiar to Northern Hemisphere seas with cyclonic circulation. These eddies influence the formation of hydrochemical structures, both at the shelf break and in the deep Bering Sea. It is in these zones of convergence of coastal and BCC waters that the centers of eddies provide favorable conditions for the development of phytoplankton, bacteria, and protozoa. The anticyclonic eddies were marked by maximum concentrations of ammonium and organic compounds of nitrogen and phosphorus. High values of dissolved oxygen (up to 9 ml/L) and O_2 saturation (up to 115%) were also observed.

Surface waters spread over upper cold intermediate water (CIW), or even form it, as these waters sink at the convergence zone along the shelf break, especially at anticyclonic eddy centers. Measurements in the upper CIW showed maximum values of oxygen and minimal values of salinity in the central part of the Aleutian cyclonic gyre, even far seaward from the shelf break. During winter cooling the CIW forms in the zone of convergence above the slope with further spreading over the rest of the basin. This process is biologically important as it drives a considerable aeration of intermediate waters. Turbulent movements over the slope also cause accumulation and downward transport of warmer waters with low salinity and particulate matter (i.e., ice, algae, phytoplankton, eggs, and larvae). In addition, cold waters oversaturated with oxygen carry a great amount of new production to the deep Bering Sea. New production was estimated from O_2 oversaturation. According to assessments conducted in the central Aleutian basin, new production ("autumn yield") in the upper 50 m equaled 23.1 gC/m^2 and the input of allochthonous organic matter from the shelf break amounted to 13.1 gC/m^2 in the upper CIW. Approximately one-third of the organic matter transfers to central deep waters which originate from the shelf break area. Thus, total abundance of organic matter in the central Bering Sea is 30-40 gC/m^2, which provides a good organic base for higher trophic level consumers.

Theories on oceanic turbulence have been advanced recently by our understanding of the role of eddies: the formation of fronts drives the eddies, which in turn are responsible for developing more complex frontal zones. The latter could be characterized by a vigorous vertical and horizontal mixing of waters which drives intensive transport of heat and salts through the stable pycnocline. Frontal zones provide areas of high productivity, which are important for commercial fisheries. Assuming that the fronts are not as rare as previously believed, the possibility of commercial fisheries in open waters is quite realistic (Fedorov 1983). This idea gains further support from the role of eddies and fronts as sources of organic matter fluxes to open seas and oceans.

Assessment of Mineral and Organic Nutrients

High accuracy and reliability of the ALPKEM flow autoanalyzer was of great importance, not only for quantitative calculations of primary production, but also for the detailed comparison of hydrochemical data with those on water dynamics. The distribution of phosphates, nitrates, silicates, and ammonium, and organic compounds of nitrogen and phosphorus, repeats the pattern of quasi-stationary geostrophic currents and eddies.

The distribution of values for phosphorus and nitrogen sometimes proved to be the best indication of vertical water movements. Being somewhat conservative, these features repeat the pattern of dynamic processes developing in the water column.

The absolute maxima of phosphates (3.8-4.2 µM), nitrates (55-62 µM), and silicates (190-200 µM) were observed in the minimum oxygen layer (800-1,500 m). These Bering Sea values are the highest in the entire World Ocean. One should bear in mind that the Bering Sea is the terminus in the global chain of nutrient deposit and oxygen decline (Bruevich et al. 1966). The observed maximum concentrations are similar to the Redfield-Richards stoichiometric ratio for waters with low values of oxygen (<0.5 ml/L). This phenomenon could be caused by the extremely high values of new production and low temperatures in the Bering Sea which hamper mineralization of organic matter and regeneration of nutrients. Thus, organic matter from new production reaches 1,000-1,500 m. High concentrations of dissolved organic phosphorus (0.4-1.2 µM) and dissolved organic nitrogen (4.0-40 µM) were observed in the deep Bering Sea.

During the autumn-winter cruise, data on the distribution of ammonium nitrogen were first obtained. The fairly high concentrations of ammonium (0.8-1.2 µM) in deep waters were double those observed during the winter-spring studies (0.4-0.6 µM). Moreover, at several transects perpendicular to the western coast, some stations reported water "sinking" with high concentrations of ammonium along the continental slope to 500-700 m (i.e., minimum oxygen layer) and spreading farther seaward. The "sinking" waters were also marked by high concentrations of dissolved organic nitrogen and phosphorus. The effect suggests another possibility for development of minimum oxygen and the rates of nutrient exchange.

Data from both cruises suggest the following picture of nutrient exchange. In spring, the phytoplankton bloom and its grazing cause nutrient accumulation in tissues on all trophic levels resulting in the depletion of nutrient stocks, not only in surface waters, but also in deeper layers due to intensive vertical mixing. Surface concentrations of phosphates, nitrates, and silicates were less than 0.2, 2.0, and 5.0 µM, respectively. Deep waters were also characterized by a decrease of nutrient concentrations: nitrates declined to 40-45 µM and phosphates down to 3.2 µM. Ammonium decreased more than two times.

It should be noted that the decrease in nutrients in the deep Bering Sea during the year suggests a rather high velocity of vertical mixing, which is uncommon. But the same is true with direct transport of surface waters via "sinking" along the slope to 500-800 m; the direct aeration of the minimum oxygen layer, and the input of a large amount of organic matter and high values of ammonium nitrogen occurs during autumn cooling. Obviously, we should revise our traditional views on the velocity of turbulence in the water column.

In autumn, the phytoplankton bloom and the consequent nutrient assimilation were characterized by phyto- and microzooplankton mixing together with organic nutrient compounds at the subsurface layer as a result of vertical mixing. In shallow waters, vertical turbulence usually reached the bottom and the resultant waters, with high values of organic nitrogen, phosphorus, silica, ammonium, and phytoplankton flow, down to the quasi-stationary zone of convergence at the shelf break where they sank into the CIW.

A vigorous anticyclonic eddy at the shelf break could drive the water flux along the continental slope to 500-800 m and cause partial aeration of the minimum oxygen layer with an increase of O_2 values by 1.5-2.0 ml/L (Fig. 3). Concentrations of ammonium nitrogen, organic phosphorus, and organic nitrogen attained 1.0-1.2; 1.2-1.4, and 6 µM, respectively. Meanwhile, concentrations of nitrates, phosphates, and silicates decreased slightly due to water turbulence: the flux of surface waters, even after mixing

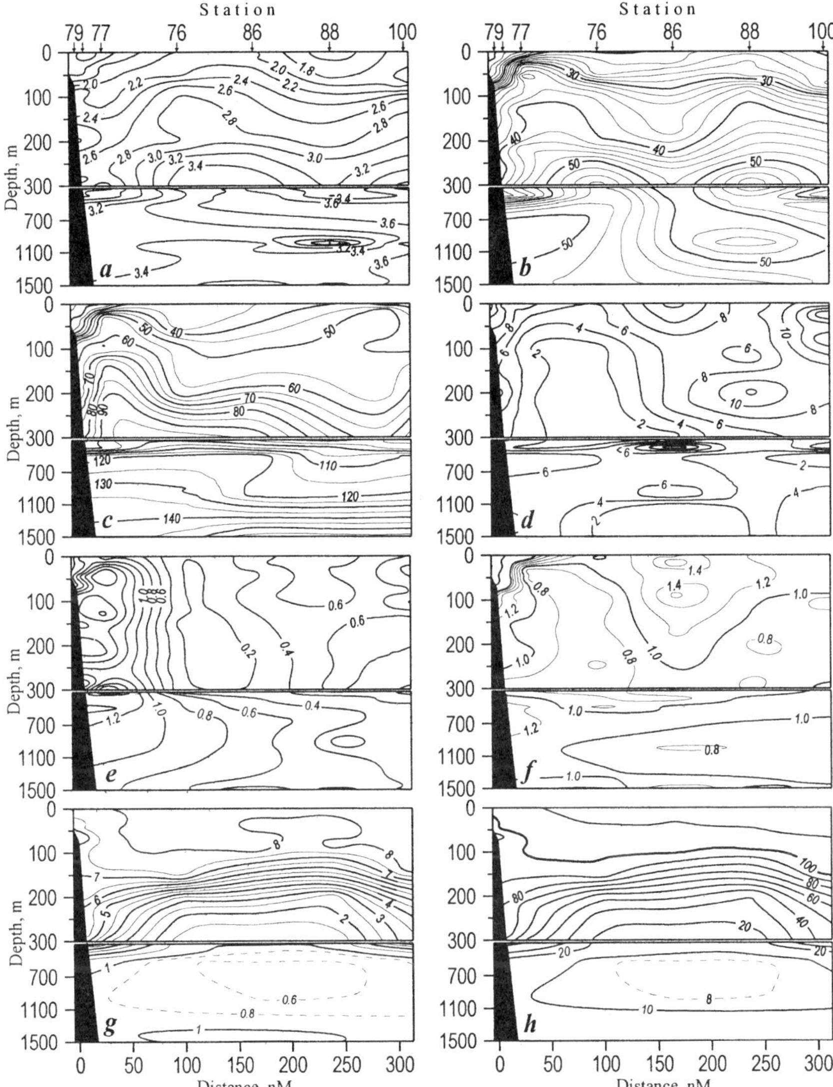

Figure 3. Distribution of nutrients along a transect from the Koryak coast to the central Bering Sea: (a) phosphate (µM); (b) nitrate (µM); (c) silicate (µM); (d) organic nitrogen (µM); (e) organic phosphorus (µM); (f) ammonium nitrogen (µM); (g) oxygen (ml/L); (h) oxygen (%).

with intermediate water, carried nutrients of lower values than the usual property of this layer.

Slope flux was generally compensated by upwelling with phosphates (2.0-2.2 μM), nitrates (24-25 μM), and silicates (>35 μM) in adjacent cyclonic eddies. According to the surface distribution of oxygen in the centers of the two main cyclonic eddies, upwelling velocity and depth could be different. The Karagin eddy showed a decrease of oxygen to 7.4-7.6 ml/L (100-102%) driving up waters from 100-150 m, while the increase of oxygen to 8.4 ml/L (116%) observed in the Olyutor eddy was caused by water input from the subsurface maximum. Simultaneously, the increase of silicate to 150-160 μM and maximal values of nitrates (58-62 μM) and phosphates (3.8-4.0 μM) were observed in deep waters under the "cyclonic dome."

Thus, the vertical circulation of nutrients within large cyclonic eddies is described. We could assume that the rates of water turnover should be much longer than 12-15 years (estimations by Fedosov and Davidovich 1963). The partial exchange in the water column down to 800-1,000 m obviously takes a year, and the vigorous turbulence over the slope influences not only autumn primary production, but also the abundance of nutrients in the surface layer resulting from compensatory upwelling in cyclonic eddies.

References

Bruevich, S.V., V.N. Ivanenkov, and V.V. Sapozhnikov. 1966. Conclusions: The Pacific Ocean. In: The chemical features of the Pacific Ocean. Nauka, Moscow, pp. 335-342. (In Russian.)

Chernyakov, A.M., Y.F. Lukashev, and S.O. Borodkin. 1988. Hydrochemical features of the mesoscale anticyclonic eddy in the deep waters. In: Mesopolygon studies: Hydrophysical studies. Nauka, Moscow. 145 pp.

Darnitski, V.B., and D.K. Staritsin. 1978. Oceanologic changes and eddies at the Wanganell Bank. In: Studies on ichthyology and fisheries. TINRO 9:26-34. (In Russian.)

Fedorov, K.N. 1983. The physical nature and structure of the oceanic fronts. Hydrometeoizdat, Leningrad. 296 pp. (In Russian.)

Fedosov, M.V., and R.L. Davidovich. 1963. Peculiarities of the hydrochemical regime of the Bering Sea. In: Trudy, VNIRO 48:51-59. (In Russian.)

Ivanenkov, V.N., V.V. Sapozhnikov, A.M. Chernyakova, and A.N. Gusarova. 1972. Rates of the chemical processes in the photosynthesis layer in the tropical Atlantic. Oceanology 12(2):243-251. (In Russian.)

Ovchinnikov, I.M. 1990. The main features of the hydrometeorological regime and hydrology of the region. In: The Black Sea: The practical ecology of the marine regions. Naukova Dumka, Kiev, pp. 11-34. (In Russian.)

Sapozhnikov, V.V. 1990. Hydrochemical studies of the continental shelf of the Black Sea. In: The Black Sea: The practical ecology of the marine regions. Naukova Dumka, Kiev, pp. 46-53. (In Russian.)

Sapozhnikov, V.V. 1991. New understanding of the hydrochemical structure of the Black Sea. In: Variability of the ecosystem of the Black Sea: Natural and anthropogenic factors. Nauka, Moscow, pp. 34-46. (In Russian.)

CHAPTER 12

Organic Matter of the Bering Sea

Alina Ivanovna Agatova, Natalya Vladimirovna Arzhanova, and Nadezhda Ivanovna Torgunova
Russian Federal Research Institute of Fisheries and Oceanography (VNIRO), Moscow, Russia

Abstract

The results of studies are presented on the horizontal and vertical distribution of particulate and dissolved organic matter (OM) in the western Bering Sea and Anadyr Bay in June 1992 and July 1993. The concentrations of dissolved organic forms of carbon, nitrogen, and phosphorus were in the range of 1.5-6.0 mg/L, 0.017-0.504 mg/L, and 0-0.056 mg/L, respectively. Particulate forms of the same elements were 0.3-2.1 mg/L, 0.005-0.136 mg/L, and 0.0003-0.0100 mg/L, respectively. The variability of concentrations of dissolved and particulate elements of OM in the photic layer were determined by the status of primary productivity in the area. Particulate and dissolved forms of organic carbon amounted to 2-58%. The high percentage content of dissolved organic matter (DOM) characterized shelf waters and areas of high primary productivity. Investigations in Anadyr Bay showed two main processes which drove the quantitative and qualitative changes of DOM and particulate organic matter (POM): the first was a physical chemical process of flocculation; the second was development of a biological barrier, i.e., an increase of heterotrophs in the microplankton population which caused high rates of uptake of OM carried by river discharge.

Introduction

All investigators of the Bering Sea ecosystem have noted high levels of primary and secondary production as its characteristic feature. Bering Sea productivity could be compared with that of upwelling zones of the Pacific and Atlantic oceans (Walsh and McRoy 1986, Walsh 1988, Izrael and Tsiban 1990). Therefore, studies of organic matter (OM) and its spatial and temporal variability become important in efforts to evaluate the energy supply for development of heterotrophic life, to identify changes of OM during its utilization by the community, and changes caused by various abiotic effects on this ecosystem.

Unfortunately, measurements of dissolved and particulate organic carbon (C_{org}), which closely reflect OM concentrations in the Bering Sea, have not been thoroughly studied. There are few data on distribution of dissolved and particulate C_{org} (POC) in the northeastern Bering Sea and Chukchi Sea (Kinney et al. 1971, Loder 1971), and recent data on the western Bering Sea (Lyutsarev et al. 1988, Pashkova et al. 1988). Joint Soviet-American investigations in 1984 and 1988 were aimed at a more comprehensive study of particulate organic matter (POM) in all major ecological subsystems of the Bering Sea (Whitledge et al. 1988, Hansell et al. 1989, Izrael and Tsiban 1990).

This paper discusses horizontal and vertical distribution of POM and dissolved organic matter (DOM) in shelf and open waters of the western Bering Sea, based on observations during the 21st and 23rd cruises of the R/V *Akademik Alexander Nesmeyanov* in June 1992 and July 1993, respectively.

Methods

Samples were collected with a Neil Brown CTD profiler, equipped with plastic Niskin bottles. To remove mesoplankton, water samples from two or three bottles closed at the same depth were passed through 60 μm gauze into a plastic bucket. To collect particulate matter, 1-5 L of water was filtered onto a fiberglass filter GF/F pretempered at 450°C. The filtered water was used in measurements of dissolved C_{org}. The filter with the collected particulate matter was homogenized in a glass homogenator and the resultant substance was used to prepare a suspension with distilled water. As a rule, the volume of the suspension was 10 ml.

The concentration of dissolved C_{org} was determined with UV and potassium persulfate (Collins and Williams 1977) on a Technicon autoanalyzer, or a high temperature catalytic oxidation (Sugimura and Suzuki 1988) on a Shimadzu TOC-500.

POC was analyzed in an aliquot of filtrate suspension with a method of wet oxidation with potassium bichromate under 100°C (Spravochnik 1991) and on a CHNS-analyzer (Karlo Erba, Italy).

Mineralization of organic nitrogen (N_{org}) and phosphorus (P_{org}) was performed with a combustion of unfiltered samples, in accordance with the Koroleff method modified by Volderrama. Determination of total phosphorus was made by the Murphy-Riley method and total nitrogen by the Bendschneider-Robinson method, respectively, on an AA-IIC autoanalyzer (Spravochnik 1991). Concentrations of N_{org} and P_{org} were calculated on the difference between their total and mineral values.

It should be noted that DOM of the Bering Sea is difficult to mineralize completely. Determination of dissolved C_{org} concentrations showed that results could vary by an order of magnitude, depending on the conditions resulting in oxidation to CO_2. Thus, analysis of water samples from the Bering Sea on the Technicon nutrient analyzer without a preliminary hy-

Table 1. Variability in dissolved C_{org} (mg/L).

| Sampling depths (m) | Autoanalyzers | | TOC-500 |
| | Technicon | | |
	Without hydrolysis	With hydrolysis	
0	2.10/0.105	2.39/0.358	2.62/0.131
40	1.61/0.080	2.23/0.034	2.46/0.123
200	1.29/0.064	2.06/0.309	2.39/0.120
800	0.89/0.044	2.31/0.346	2.29/0.114

Note: The denominator is the RMS deviation

drolysis resulted in 0.3-2.5 mg/L, but a more rigorous preliminary hydrolysis (boiling the sample with 2N H_2SO_4 or 1N NaOH for 30 min) results in a value of 2-6 mg/L. The latter results were analogous to those obtained on the TOC-500. DOM from shelf waters and pelagic waters (photic layer and deep layers) showed resistance to oxidation on the Technicon. Table 1 presents data on concentrations of dissolved C_{org} obtained from the different methods on two devices at one station (44°3′N, 150°4′E). Note that measurements of dissolved C_{org} concentrations need rigorous oxidation conditions, preferably high-temperature catalytic oxidation as discussed by Sugimura and Suzuki (1988).

Results

Figure 1 presents the location of stations where dissolved and particulate C_{org} were measured. The stations covered most of Anadyr Bay in 1992 and these data are discussed individually.

Dissolved C_{org}

Spatial distribution of dissolved C_{org} was not uniform vertically or horizontally. Concentrations of dissolved C_{org} varied in the photic zone in the range of 1.5-6.0 mg/L (Fig. 2a).

In 1992, maximum concentrations of dissolved C_{org} in the photic layer (about 6 mg/L) were observed in Kamchatka Basin (Stations 2106-2111), while minimum values (below 1.5 mg/L) occurred in Kamchatka Strait (Station 2097). No consistencies were found in C_{org} distribution in shelf or open waters. At one location in shelf waters of Kamchatka Strait (Station 2097), concentrations of dissolved C_{org} were half those in waters of the Karagin shelf. An increase of DOM concentrations in warm intermediate waters (WIW) in the seaward direction was observed in Kamchatka Strait (Station 2097—1.6 mg/L; Station 2099—2.7 mg/L). Alternatively, Kamchatka basin showed a decrease of C_{org} concentrations in the seaward direction. However, both a decrease and increase of dissolved C_{org} in open waters

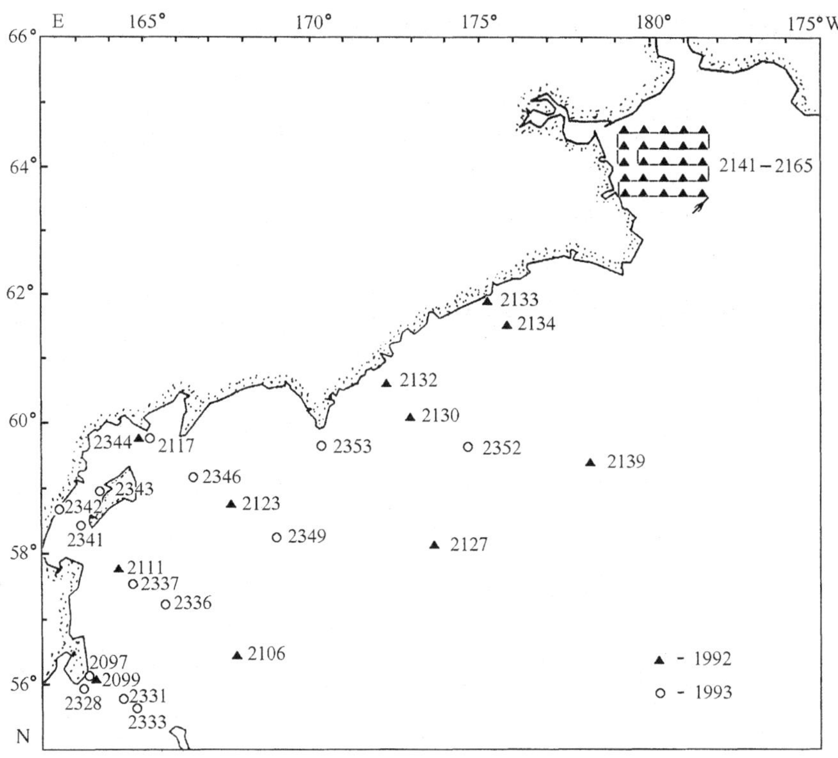

Figure 1.   The location of sampling stations in the Bering Sea.

was recorded in the northern Aleutian basin (Stations 2133-2134, 2127-2132, respectively).

Observations in 1993 differed from those in 1992 by indicating high concentrations of C_{org} (3-4 mg/L) in shelf waters of Kamchatka Strait. Values were 1.5-2 times higher than those obtained on the Karagin shelf. It is interesting that shelf waters north of Kamchatka Strait showed a gradual increase of C_{org} content to the northeast, almost to Shirshov Ridge. A similar tendency of increase of dissolved C_{org} was observed in deep waters in the photic layer, the Cold Intermediate Waters (CIW), and in WIW. This tendency was associated with the intensity of primary productivity and results in an increase of zooplankton biomass.

The vertical distribution of DOM in coastal waters of both Karagin and Koryak shelves were similar and were characterized by relative uniformity; C_{org} concentrations ranged from 3.0 to 3.4 mg/L from the surface to the bottom (Fig. 3a). In Kamchatka Strait C_{org} concentrations increase from 1.5 to 2.5 mg/L in the 25 m layer. Stations on the continental slope showed a

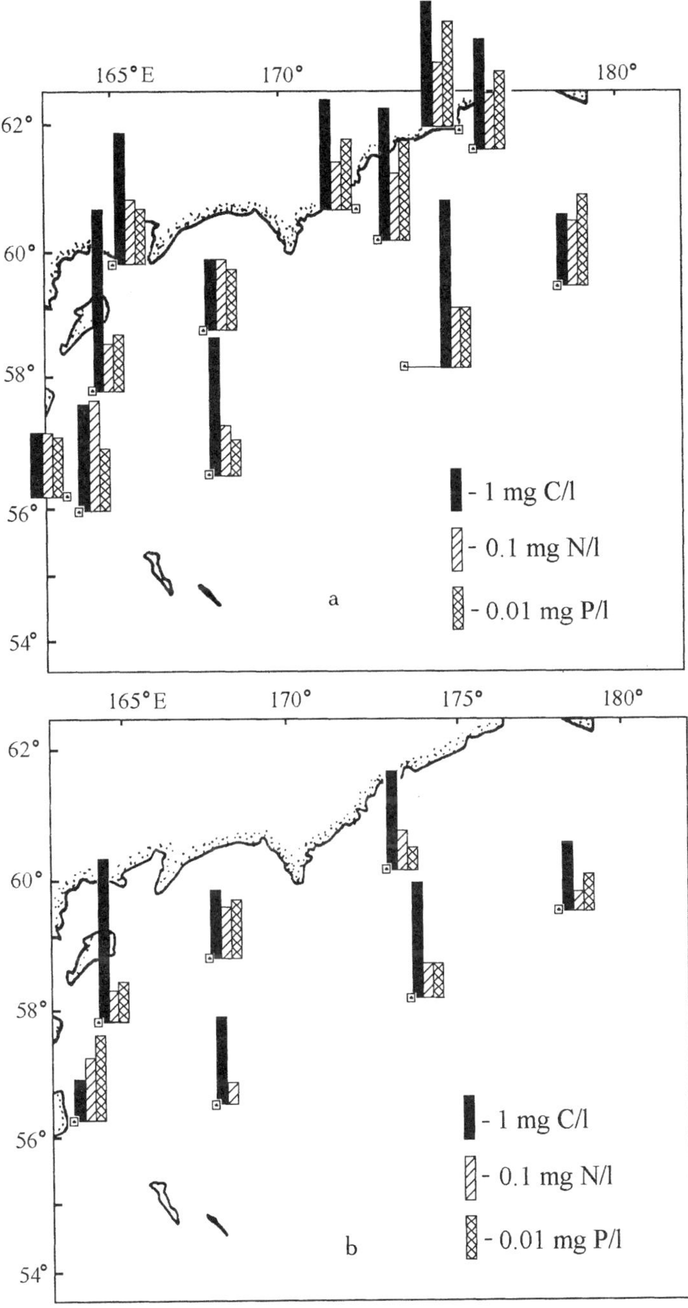

Figure 2.   Mean concentrations of dissolved C_{org}, N_{org}, and P_{org} in the Bering Sea: (a) the 0-30 m layer; (b) the 200-600 m layer.

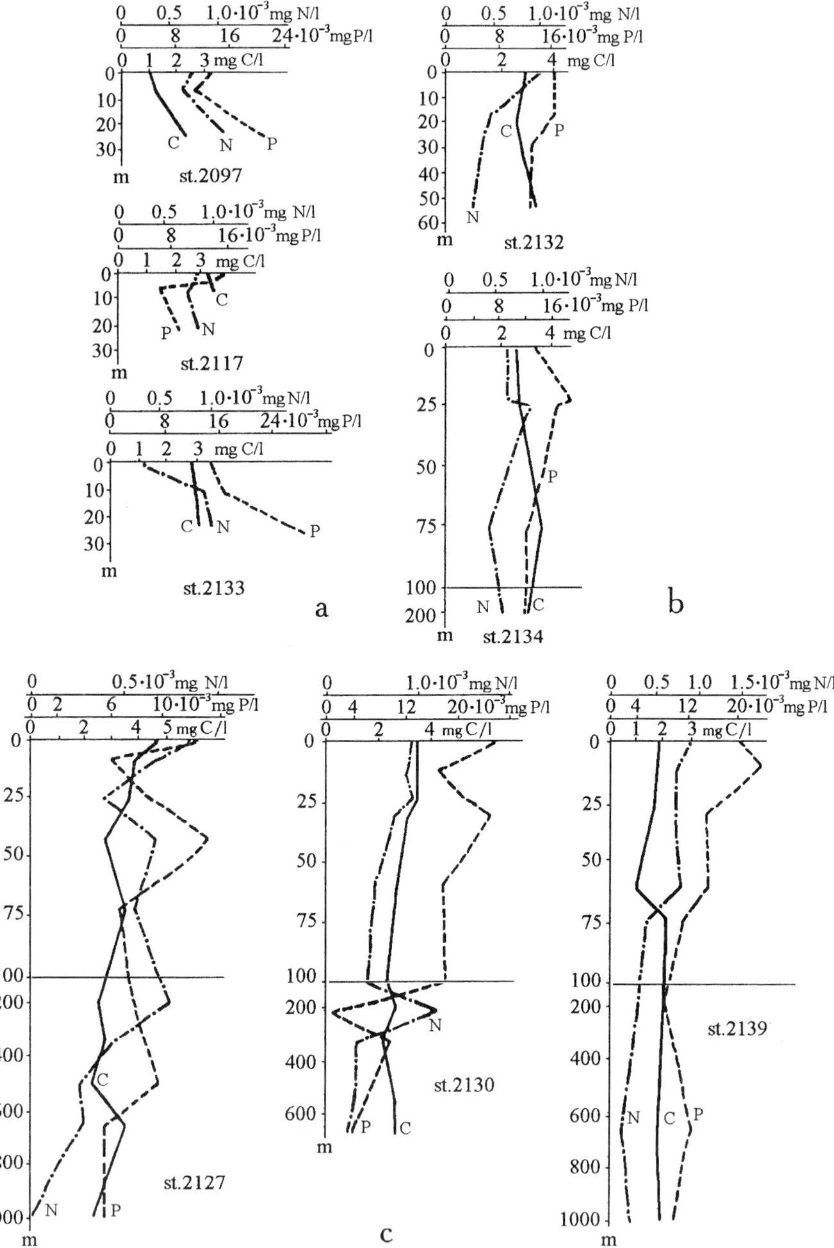

Figure 3.   *The vertical distribution of dissolved C_{org}, N_{org}, and P_{org} in the Bering Sea: (a) shelf waters; (b) slope waters; (c) pelagic waters.*

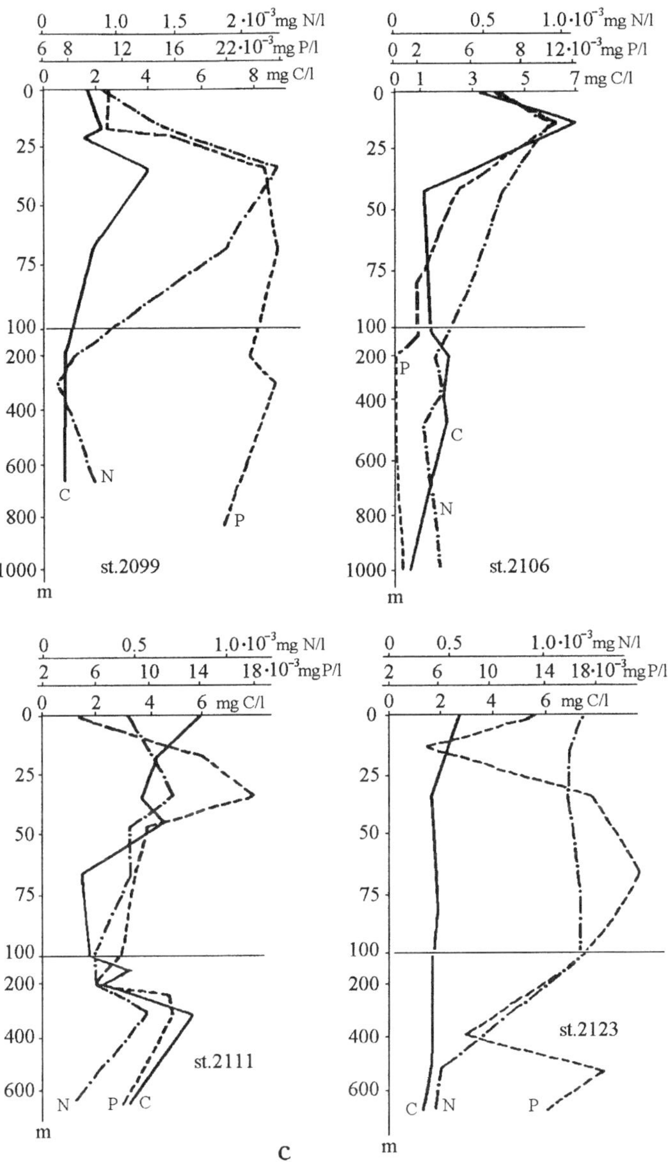

Figure 3.   *(Continued.)*

slight increase in concentration with depth. A tendency for a decline of DOM values with depth was observed at pelagic stations (Fig. 3c). The difference between values in the photic layer and the underlying waters can either be significant or negligible (ΔC_{org} was in the range of 0.0-1.7 mg/L).

The analysis of vertical profiles of C_{org} in the 0-1,000 m water column (there were no measurements below this level) suggested two patterns (Fig. 3c). The first pattern was characterized by a uniform distribution of values with positive or negative deviations (no more than 1 mg/L) related to surface concentrations. The second type had a zigzag pattern of sharp steps with C_{org} attaining 5 mg/L. The second pattern was characterized by two maxima in the upper layers; one in the subsurface layer (10-30 m) and the other in the boundary of the photic layer (40-70 m). There were also one or two maxima in the 200-600 m layer. Concentrations in the upper and lower maxima were generally similar. Both patterns were observed in the Kamchatka and Aleutian basins. The zigzag pattern in the 0-1,000 m water column was probably associated with a change in the intensity of biological processes at different depths.

Dissolved N_{org}

The spatial distribution of N_{org} was also diverse (Fig. 2). In 1992, dissolved N_{org} concentrations in the photic layer and in the entire water column were higher than in 1993. In 1992 and 1993, the highest values (0.28 mg/L and 0.30 mg/L) were observed in Kamchatka Strait (Stations 2099 and 2328, respectively). In open waters a significant decline of dissolved N_{org} was observed in the deep layers, compared to the photic layer (Fig. 2a,b). In 1992, the maximum difference between the photic layer and the WIW changed by almost a factor of 9 (Station 2130), attaining 0.13 mg/L (Stations 2127, 2130). In 1993, these areas approximated similar rates of variation in dissolved N_{org} concentrations (0.07-0.11 mg/L) in the same layers, but the ratio was double, on average (Stations 2331, 2336, 2346, 2352).

In 1992 and 1993, the coastal zone generally showed a slight increase in N_{org} concentrations with depth. Maximum values were observed at 10-30 m. The maximum step was found at Station 2097 in Kamchatka Strait: 0.06 mg/L on the surface, and 0.15 mg/L in the 24-m layer.

Stations located on the continental slope had a decrease in N_{org} concentration with depth, opposite to the pattern for C_{org} (Fig. 3b). The vertical distribution of N_{org} in pelagic waters had the same zigzag pattern as the C_{org} profile. Thus, we expected a relatively consistent C/N ratio in the water column. However, variations in the values were high both vertically and horizontally. The C/N ratio changed from 5 to 628 in 1992, and from 7 to 144 in 1993. Minimal C/N values were found in Kamchatka Strait (Stations 2099, 2328, 2333). It is interesting that these stations showed a vertical change in C/N within the range of 5-14, which approximates the value of 6.6 characteristic of live cells of phyto- and bacterioplankton (Redfield et al. 1963). The remainder of stations on the shelf and in open water had a C/N ratio higher than 6.6 indicating local DOM depletion of nitrogen compounds.

Several features were noted in the vertical variations in the C/N ratio of DOM within the study area. The pelagic stations in Kamchatka basin (Stations 2106, 2111, 2123, 2336, 2337) regularly had minimal C/N values for the upper layer (0-200 m), and maximum values for the underlying layers. Alternatively, the Aleutian basin (Stations 2130, 2139, 2346) showed a near uniform increase in the C/N ratio.

The C/N ratio in the Karagin shelf coastal zone varied little (in the range of 20-23 in 1992, and 13-28 in 1993) from the surface to the bottom; pelagic stations changed vertically by as much as six times. For example, the C/N ratio changed from 10 at the boundary of the photic layer to 61 in the core of the CIW (Station 2106, 1992); in 1993 it equaled 13 on the surface and 111 at 53 m, the layer of lowest temperatures (Station 2353).

Dissolved P_{org}

The spatial distribution of P_{org} was as diverse as that of N_{org}. Concentrations varied over a broad range (0-0.030 mg/L in 1992, 0-0.020 mg/L in 1993) along the entire survey area, except in Anadyr Bay.

In 1992, mean P_{org} concentrations in the photic layer southwest of the Shirshov Ridge were 1.5-2 times lower than in waters to the northeast (Fig. 2a). The opposite was found in the distribution of the mean P_{org} values in the WIW (Fig. 2b). Maximum concentrations of dissolved P_{org} were found in bottom waters (23 m) of the Koryak shelf (Station 2133). This station showed a uniform increase of P_{org} concentrations from 0.015 at the surface to 0.030 mg/L at the bottom. In coastal waters of the Karagin shelf, the P_{org} concentration at the surface was double that in bottom waters. The shallow waters of the Karagin shelf showed an increase of the P_{org} concentrations in 1993. Values were in the range of 0.004 mg/L (9 m, Station 2341) to 0.017 mg/L (5 m, Station 2343). A slight decrease of P_{org} with depth was observed at the continental slope both in 1992 and 1993. The values were in the range of 0.013-0.019 mg/L (1992) and 0.003-0.013 mg/L (1993).

In 1992-1993, the most significant vertical variations in P_{org} were observed in pelagic waters. As a rule, one could find two P_{org} maxima on the vertical distribution curves in the Kamchatka and Aleutian basins in 30-60 m and 300-600 m layers (Fig. 3). These maxima often coupled with the vertical distribution of the maxima of dissolved C_{org} and N_{org} (e.g., at Stations 2111, 2336, 2337); however, sometimes these patterns were reversed (Station 2127).

The C_{org}/P_{org} ratio varied over a broad range both horizontally and vertically (106-2,000 in 1992 and 106-2,530 in 1993). This variation differs from the molar ratio of C:P (106:1) which characterizes live cells (Redfield et al. 1963). Minimal values of C/P were observed in Kamchatka Strait (Stations 2097, 2099, 2328), in the Aleutian basin (Station 2139), and off Goven Peninsula (Stations 2341, 2342, 2343). It was difficult to see any regularity in the vertical distribution of the C/P ratio. Thus, values of C/P steadily increased (79-120) with depth in shallow waters off Cape Africa (west side of Kamchatka Strait), while at the Karagin shelf the distribution curve was different with higher values characterized by a greater amplitude of variations (190-570). The value of C/P ratio steadily declined from the surface to the bottom on the Koryak shelf (Fig. 3a). At pelagic stations, as a rule, the C/P values were maximal in the 0-10 m layer and minimal in the 30-200 m layer. However, several stations reported the minimal C/P in the 300-600 m layer (Fig. 3c).

Particulate C_{org}

The spatial distribution of POC is as diverse as that of dissolved C_{org}. Concentrations vary in a broad range of 0.038-1.040 mg/L. Similar variations of C_{org} concentrations were observed previously (Glebov et al. 1988, Lyutsarev et al. 1988).

No obvious trends were found in the distribution of POC in the photic layer. Maximum values were observed in Kamchatka Strait and on the Koryak shelf (Fig. 4). It is interesting that in 1992 the mean concentration of POM in the photic layer of pelagic waters was one-half to two-thirds that in the 200-600 m layer; in 1993 we found no such relationship. Generally, the amount of POM in 1993 was significantly lower than in 1992; the mean concentrations in the photic layer varied from 0.087 to 0.336 mg/L, respectively (Stations 2346 and 2353), a relative decrease of POC (on the average 2-2.5 times) was observed in the 200-600 m layer.

The vertical distribution of POC was diverse (Fig. 5). Kamchatka Strait and the Karagin shelf had a general decrease of POM concentrations with depth, while the Koryak shelf values increased with depth (Fig. 5a). In 1992, the continental slope stations had a tendency for increase of POC concentrations with depth; however, this change was not steady.

In pelagic waters one could define two major distribution patterns; an almost uniform decrease of the concentrations with depth (Station 2139) and a zigzag pattern with considerable deviations (up to 0.7-0.8 mg/L; Fig. 5c). Maximum values in deep water were higher than the values in the photic layer (Stations 2106, 2111, 2123, 2127, and 2130). Station 2099 (Kamchatka Strait) was the exception and had the highest subsurface maximum. The first distribution pattern was found only in the Aleutian basin, while the second one was observed both in the Kamchatka and the Aleutian basins. These patterns differ from those of dissolved C_{org}. In 1993, despite the diversity of POC distribution, vertical variations were considerably less (no more than 0.2 mg/L).

It should be noted that in 1992 POM content, compared to DOM, was 2-4 times higher than in 1993 and averaged 10-20% in the study area. The maximum amount and total contribution of POM was observed in Kamchatka Strait at Stations 2097-2098 (0.7-1.0 mg/L and 34-58%, respectively). High values of POM/DOM (21-25%) were recorded in shelf waters off Karagin Island (Station 2133), and in the photic layer in the Aleutian basin (Station 2139; 33-38%). Minimal values (below 10%) were found at Stations

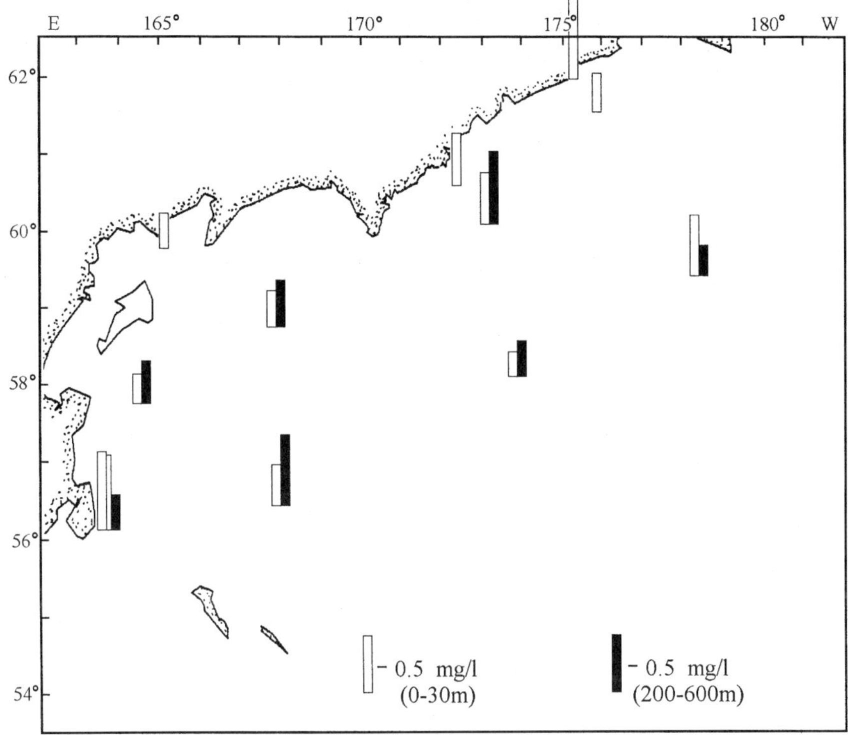

*Figure 4.   The mean concentrations of particulate C_{org} in the Bering Sea: in the 0-
30 m and 200- 600 m layers.*

2111 and 2127. In 1993, when the total contribution of POM was low (4-6 %), Station 2353 had higher concentrations of POM (0.28-0.34 mg/L) and a POM contribution of 17%.

In 1992, the C/N ratio in POM varied in a narrower range than in DOM (5-48). Minimal values (5-9) generally characterize the photic layer of pelagic waters. The Karagin and Koryak shelves showed the C/N values close to 8. As a rule, below 100 m the C/N value increases up to 20-50. However, several stations had values even lower at deeper layers than in the photic layer (Stations 2099 and 2106), probably due to the increase of the amount of nitrogen compounds at these depths (Fig. 5). In 1993, the ratio of C/N in POM was still more uniform and was generally 4-5. Only stations 2336 and 2337 located in the Kamchatka basin showed values of 10-17 mg/L in the CIW.

POM was characterized by a higher contribution of phosphorus, compared to DOM. Therefore, the C/P ratio was lower than in DOM and varied in the range of 70-850 (1992), and 61-277 (1993). The Karagin and Koryak

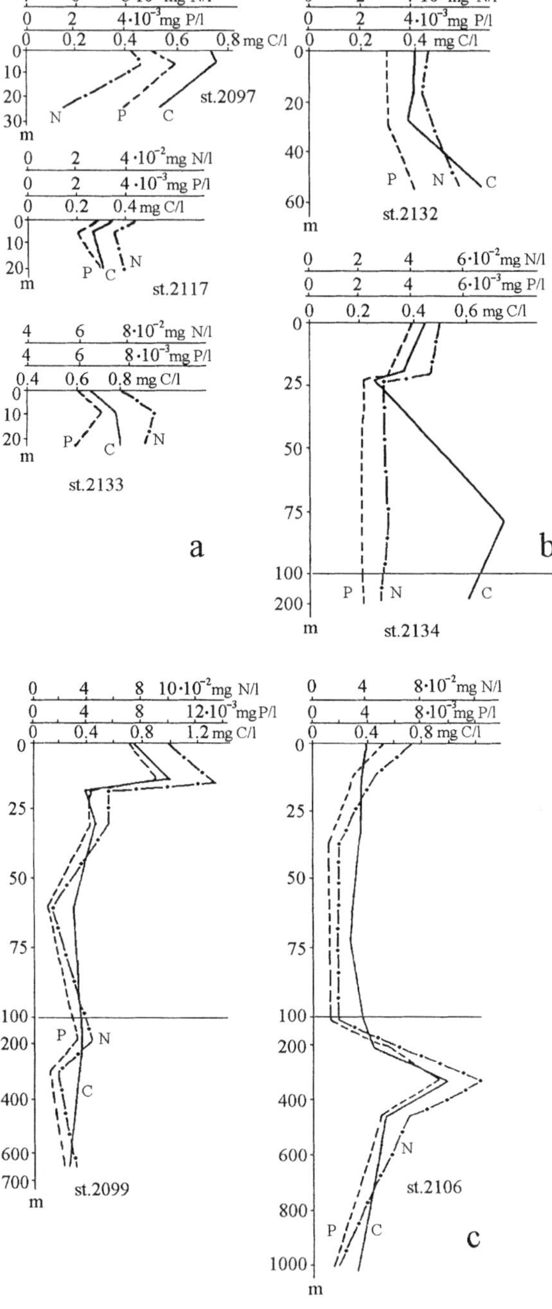

Figure 5.   *The vertical distribution of particulate C_{org}, N_{org}, and P_{org} in the Bering Sea: (a) shelf waters; (b) slope waters; (c) pelagic waters.*

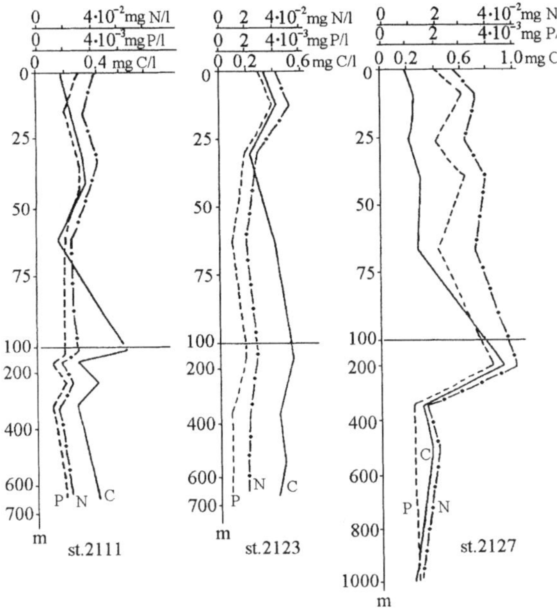

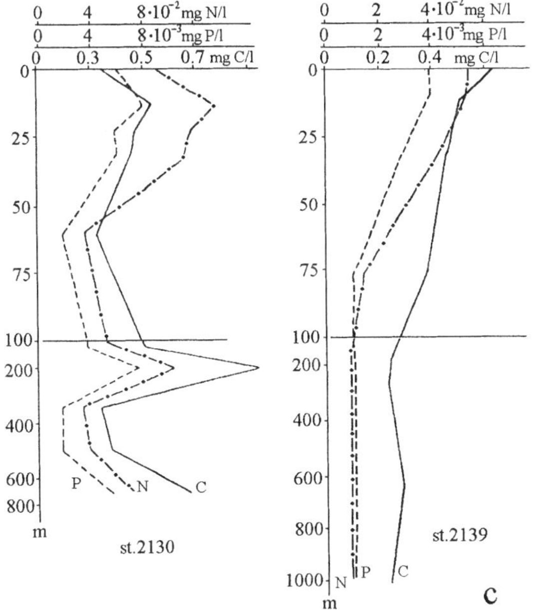

Figure 5.   (Continued.)

Table 2.   The concentrations of dissolved C_{org}, N_{org}, and P_{org} (mg/L) in the Bering Sea during 1993 at two sampling depths.

Station	Surface microlayer			0-0.5 m		
	C_{org}	N_{org}	P_{org}	C_{org}	N_{org}	P_{org}
2327	6.7	0.226	0.02	3.4	0.141	0.005
2333	3.6	0.646	0.009	2.3	0.232	0.013
2336	5.2	–	–	2.1	0.188	0.013
2341	4.2	0.43	0.003	2.4	0.089	0.007
2346	4.3	–	–	3.8	0.119	0.006
2349	7.6	–	–	5.2	0.126	0.008

Table 3.   The concentrations of particulate C_{org}, N_{org}, and P_{org} (mg/L) in the Bering Sea during 1993 at two sampling depths.

Station	Surface microlayer			0-0.5 m		
	C_{org}	N_{org}	P_{org}	C_{org}	N_{org}	P_{org}
2327	0.53	0.082	0.005	0.14	0.029	0.007
2333	0.493	0.078	0.005	0.131	0.03	0.002
2336	0.582	0.058	0.004	0.147	0.026	0.002
2341	0.92	0.129	0.009	0.204	0.036	0.002
2342	0.954	0.145	0.01	0.089	0.014	0.001
2346	0.38	0.055	0.004	0.259	0.043	0.003
2349	0.624	0.095	0.006	0.169	0.036	0.002

shelves had C/P values in the range of 85-145. Minimal C/P ratios in POM (61-100) were observed in the photic layer of the Kamchatka basin (Stations 2123, 2127, 2331, 2336, 2337).

The vertical distribution of particulate P_{org} is as diverse as that of C_{org} and N_{org}. However, there was a tendency for POM depletion in phosphorus in deep layers, compared to the photic layer (Fig. 5).

In 1993, the first measurements of dissolved and particulate C_{org}, N_{org}, and P_{org} in the surface microlayer (SML) were obtained. These values were significantly higher in the SML than in the surface layer (Tables 2 and 3). Concentrations of particulate P_{org} were lower in the SML than in the surface layer only at Station 2327. The accumulation of dissolved and particulate C_{org}, N_{org}, and P_{org} in the SML was likely due to the physical/chemical properties of the layer which facilitates retention of polar molecules.

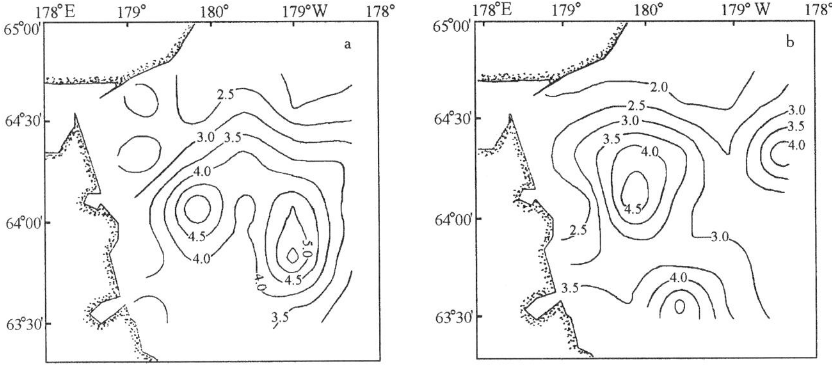

Figure 6.   Distribution of dissolved C_{org} in Anadyr Bay: (a) surface layer; (b) bottom layer.

Organic Matter in Anadyr Bay

River discharges in Anadyr Bay had minimal effect on the distribution of dissolved C_{org} in the surface and bottom layers there (Fig. 6). Concentrations of C_{org} were 3.4-4.3 mg/L in the surface layer and 3.4-5.2 mg/L in the bottom layer in waters far from the direct influence of the river (Stations 2141-2145). Minimal concentrations in the surface and bottom layers were observed at Stations 2159-2162 (2.0-2.9 mg/L and 1.9-2.9 mg/L, respectively). These stations also reported low concentrations of N_{org} (0.13-0.15 in the surface layer and 0.11-0.14 mg/L in the bottom layer). The highest C_{org} was measured in the central part of the Anadyr study area (3.7-5.8 mg/L in the surface and 3.2-5.1 mg/L in the bottom layer).

Concentrations of dissolved N_{org} and P_{org} were average (0.14-0.21 mg/L, 0.011-0.021 mg/L in the surface and 0.08-0.14 mg/L, 0.010-0.0022 mg/L in the bottom layer, respectively) (Figs. 7 and 8). Maximum concentrations of dissolved N_{org} (0.21-0.38 mg/L) and P_{org} (0.026-0.67 mg/L) characterized the stations which were located far from river discharge and closer to open waters (Stations 2142-2145, 2148, 2149, 2151).

The values of C/N were in the range of 7-15, and C/P varied within 90-330. It is interesting that OM rich in nitrogen generally characterized surface waters, while the P-rich OM was found in bottom waters. DOM in the center of the polygon was poor both in nitrogen and phosphorus. Values of C/N and C/P were high here and attained maximum values of 48 and 1,160, respectively. Areas under the strongest influence of Anadyr River discharge had lower C/N and C/P ratios in DOM than in DOM at the stations of the central study area. But on average the ratios were higher at stations in the central study area than in OM at stations farthest from the effect of the river.

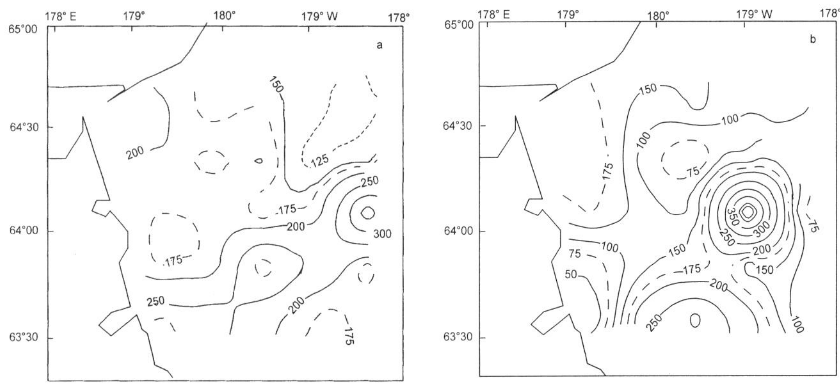

Figure 7.   Distribution of dissolved N$_{org}$ in Anadyr Bay: (a) surface layer; (b) bottom layer.

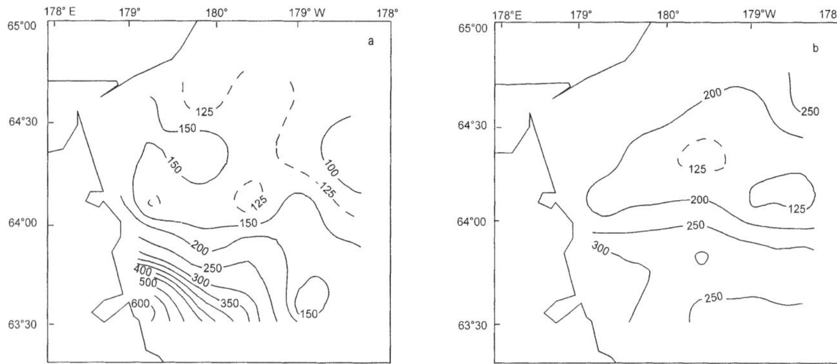

Figure 8.   Distribution of dissolved P$_{org}$ in Anadyr Bay: (a) surface layer; (b) bottom layer.

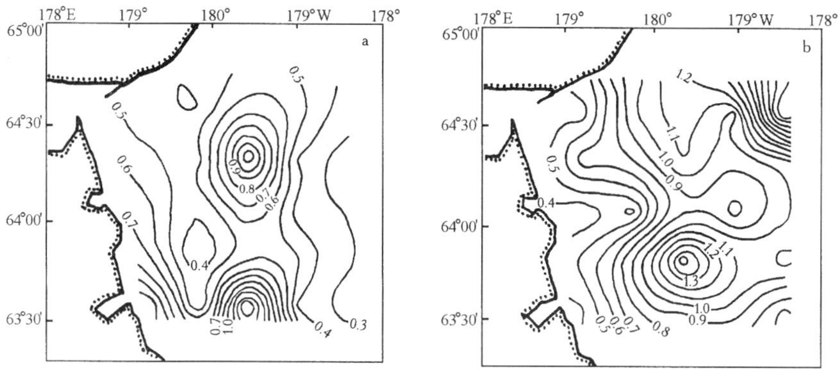

Figure 9. Distribution of particulate C_{org} in Anadyr Bay: (a) surface layer; (b) bottom layer.

POC was distributed unevenly along the entire polygon (Fig. 9) with concentrations varying 6× in surface waters, and 8× in bottom layers (0.27-1.47 mg/L and 0.33-2.74 mg/L, respectively). The amount of POC in the surface layer was 2-5 times lower than in bottom water. Coastal stations in the bay and Station 2143 were the exception: here the POM on the surface was 2-3 times higher, or equal to that of the bottom layer. These stations (except Station 2143) had minimum concentrations of POC in the bottom layer.

In the Anadyr study area, the contribution of particulate matter in the OM was high for both surface layers (7-38%) and bottom water (10-105%).

Generally, POM was richer in nitrogen and phosphorus than DOM in the Anadyr polygon. The range of variations for C/N was 8-28, and for C/P was 100-400. POM in the surface layer held more nitrogen and phosphorus than the bottom layer. Minimal C/N and C/P values were measured in POM in areas under the direct influence of the Anadyr River (Stations 2161-2165) and coastal Stations 2144-2146.

Discussion

Mean concentrations of dissolved C_{org} in the Bering Sea were 1.5 times higher than earlier studies and 3.5 times higher than those reported by Loder (1971) and Kinney et al. (1971). This could be explained by a difference in sampling methods, filtration techniques, and the duration of sample storage, but principally by different methods of OM combustion. The combustion of DOM with $K_2S_2O_8$ at 100°C (Loder 1971) allows one to oxidize only a small part of the C_{org} to CO_2. The combustion of OM, which passed a preliminary fractionation under more rigorous conditions with $K_2Cr_2O_7$ at 130°C using an Ag_2SO_4 catalyst, allowed Lyutsarev et al. (1988) to obtain a complete oxidation of a considerable part of dissolved C_{org}. We noticed that carbon of high-polymer OM with nitrogen compounds (proteins and nucleic acids) was the most resistant to complete oxidation. The same was true with carbon with branched polymers like starch and cellulose. Suzuki et al. (1985) noticed that nitrogen, of the natural high-polymeric compounds that are dissolved in seawater, is also resistant to oxidation, and therefore the wet combustion with potassium persulfate results in N_{org} content one-tenth to one-fourth times that from the method of high temperature catalytic oxidation. The wet combustion does not always result in complete oxidation of phosphates (Ridal and Moore 1990).

The determination of P_{org} based on differences in concentrations of total and mineral phosphorus ($P_{tot} - P_{min}$) can also result in values which are 10-20% lower. This difference may result from the Murphy-Riley method of P_{min} analysis which allows an easily hydrolyzed phosphate to split from phospho-organic compounds, which is then determined to be a mineral. Thus, our values of C_{org}, N_{org}, and P_{org} are not presented as values corresponding to the absolute concentrations of the substances, but only as relative parameters which reflect the amount and quality of OM. From this perspective, it is possible that the maximum ratio of C/N and C/P in several layers were underestimations of N_{org} and P_{org} and not the actual content of N and phosphorus in OM.

The observed diversity in the horizontal and vertical distribution of DOM and POM can be caused by a whole range of hydrological, physical/chemical, and biological factors (Lee and Wakeham 1989). Here we discuss the most important of these that control OM distribution in the study area.

We note that a reverse dependency between concentrations of particulate and dissolved C_{org} was observed on the shelf and the photic layer of pelagic waters, while below 100 m no dependency between these two factors was found (Fig. 10). It is difficult to explain this feature by a simple absorption of DOM on particulate suspensions, which could extract 10-30% of OM from solution (Handa and Tanoue 1981, Abdel-Moati 1990). If so, a reverse dependency between the concentrations of particulate and dissolved C_{org} should be observed below the photic layer at pelagic stations. Perhaps several processes are involved. First, the processes of sorption are important for the photic layer because the DOM of the photic layer contains more high-polymeric compounds These are more easily adsorbed on suspended particles (Sugimura and Suzuki 1988). Second, on the shelf and in the surface layer of pelagic waters, up to 43% of DOM can transfer to POM by coagulation resulting from wind mixing (Kepkay and Johnson 1988). As a rule, these waters also contain a greater number and biomass of heterotrophic microplankton which consume DOM (Sorokin 1995). And third, the contribution of small picoform plankton increases with depth (Sherr and Sherr 1991). These small forms can pass through the GF/F filter (Cho and Azam 1988) resulting in our recording this fraction as DOM.

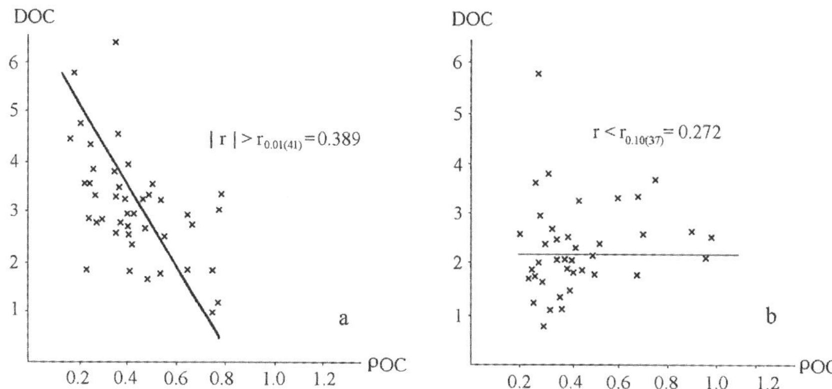

Figure 10. The dependency between the concentrations of dissolved C_{org} (DOC) and particulate C_{org} (POC) in the Bering Sea: (a) 0-30 m; (b) below 100 m.

The variety of OM in horizontal and vertical distributions was associated with hydrological conditions in the study area. For example, the intrusion of bottom water from the Karagin shelf into Kamchatka Strait caused an increase in dissolved and particulate C_{org}, N_{org}, and P_{org} at 30-60 m (Station 2099). Concentrations of particulate forms of elements increased in the mixing layers of Bering and Pacific waters (300-600 m) and in the Kamchatka basin (Stations 2106 and 2111; Figs. 3, 5). Here both the quantitative and qualitative composition of OM change; the share of nitrogen and phosphorus increased and the biochemical composition changed (Agatova et al. 1995). Japanese researchers noted an analogous increase of particulate C_{org} and N_{org} in the deep layers at stations in the Aleutian basin (Handa and Tanoue 1981). They associated the increase of particulate C_{org} and N_{org} in deep layers with the transfer of particulate matter from the continental shelf along the continental slope. However, during our investigations flux from the eastern shelf to the Aleutian basin was almost insignificant in the distribution of dissolved and particulate C_{org}, N_{org}, and P_{org} (Station 2139). Some investigators estimate about 51% of primary production is oxidized on the shelf and about 10-20% of OM is exported to deep waters. Walsh and McRoy (1986) estimate the amount of OM input from shelf to deep waters as no less than 48%.

The mixing of different types of water often changes the vertical structure of C_{org}, N_{org}, and P_{org} distribution. However, this is not a result of passive transfer of the element, but a result of the activation of a whole range of biological processes of microplankton. Thus, the interaction of Pacific Ocean and Bering Sea waters causes an increase of ETS enzyme activity and often of alkaline phosphatase in mixed waters (Agatova and Lapina 1995). It is likely that the primary mechanisms controlling variability of carbon, nitrogen, and phosphorus concentrations are biological processes. Obviously, even during periods of active productivity, concentrations of organic forms of nitrogen and phosphorus in the photic layer are not always determined by phytoplanktonic biomass. For example, Hansell et al. (1989) showed that the concentration of particulate N_{org} on the northeastern Bering Sea shelf could be determined using chlorophyll *a* (Chl *a*) concentration with the regression equation:

$$N_{org} = 0.542 \text{ Chl } a + 1.338$$

However, the calculated values of particulate N_{org} on the Karagin and Koryak shelves, and in the Kamchatka and Aleutian basins, were usually 1.5-3 times higher than the values calculated with the equation. This discrepancy suggests that bacterioplankton play a significant role, along with phytoplankton, in the process of N_{org} formation in the photic layer.

The diverse vertical distribution of DOM and POM is likely due to bacterioplankton. In mesopelagic waters the biomass of bacterioplankton, especially small, free-moving forms (0.1 μm^3 size), determines maximum concentrations of dissolved and particulate forms of the major elements of OM. Cho and Azam (1988) found picoplankton aggregations at 250 and 600 m in the North Pacific and suggested that picoplankton regulate the equilibrium of DOM and POM there.

When discussing such a basin rich in ichthyofauna, one should also take into account the contribution of phytoplankton that are grazed by mesoplankton and ichthyofauna, and the organic pellets of ichthyofauna in OM flux from phytoplankton to bacterioplankton (Banse 1990). Large amounts of DOM occur, not only in the photic layer, but also in deep layers. It is quite possible that in Kamchatka Strait (Station 2099) and in Kamchatka basin (Station 2123), this mechanism of DOM formation is dominant; the ratio of C/N and C/P are minimal there, even below the photic layer, and the bacterial formation of DOM from POM leads mainly to C_{org} extraction, but not N_{org} and P_{org} (Karl et al. 1991).

In the Gulf of Anadyr ecosystem, there are two major processes which determine the quantitative and qualitative variations of POM and DOM; the physical/chemical process of flocculation and the development of a biological barrier. The latter occurs when the number of heterotrophic organisms increases in the microplanktonic population facilitating a rapid utilization of OM introduced by river discharge. For example, a very high percentage of POM by DOM can be associated with microplanktonic development. This assertion is supported by high concentrations of particulate protein and nucleic acids in surface and bottom layers (Agatova et al. 1995). Conversely, one should not exclude the influence of OM conglomeration development at the change in ion power. Low molecular OM can also participate in this process (flocculation) which leads to a sharp decrease of DOM and an increase in POM concentrations. It is not by chance that there are almost no high polymeric compounds (e.g., nucleic acids) in solution in the bottom layers; they are also absent in the surface layers.

An abundance of microorganisms develop in areas of river and seawater interaction. These organisms rapidly utilize most of the OM introduced by river discharge (Sigleo at al. 1982). This uptake by microorganisms may result in the low C/N and C/P ratios in POM and rather low values in DOM in water under the direct influence of the Anadyr River (Stations 2161-2165).

The intensity of primary productivity also influences the characteristics and distribution of OM in Anadyr Bay. Thus, DOM and POM in areas of high concentrations of phytopigments (Stations 2144-2146) (Mordasova et al. 1995) are rich in nitrogen and phosphorus. In the eastern part of the bay, the characteristics and distribution of OM are influenced by flocculation. There the concentrations of POM in bottom layers are 2-7 times higher than in the surface layers and in bottom layers the contribution of POC is maximal (51-105%).

Conclusions

DOM matter in the Bering Sea is resistant to complete mineralization. The easily hydrolyzed DOM fraction makes up no more than 30% of the total, and often less. Concentrations of dissolved C_{org} are in the range of 1.5-6.0 mg/L, N_{org} is 0.017-0.504 mg/L, and P_{org} is 0.000-0.056 mg/L. POC concentrations vary in the range of 0.3-2.1 mg/L; N_{org} is 0.009-0.136 mg/L, and P_{org} is 0.0005-0.010 mg/L.

Variations in the concentrations of dissolved and particulate forms of major OM elements are determined by the status of primary productivity in the area. POC totals 2-58%. The high percentage of POM by DOM is characteristic of shelf and zones of high primary productivity. The spatial distributions of OM are diverse both horizontally and vertically. It is difficult to find any definite difference between shelf and open waters.

Two major types of vertical distribution of carbon, nitrogen, and phosphorus in pelagic waters were noted: (a) a steady decrease in concentrations with depth in the 0-1,000 m layer; and (b) a zigzag distribution pattern with considerable deviations. The second distribution pattern is characterized by the presence of two maxima in the upper layers (10-30 m and 40-70 m) and one or two maxima in the 200-600 m layer. The upper maximum coincides with the chlorophyll maxima, and the deep ones are possibly caused by the presence of small-size bacterioplankton.

In Anadyr Bay, the quantitative and qualitative variations in DOM and POM are determined mainly by two processes: the physical/chemical process of flocculation and development of a biological barrier. The latter occurs when the number of heterotrophic organisms increases in microplanktonic populations and causes a rapid utilization of OM introduced by river discharge.

References

Abdel-Moati, A.R. 1990. Adsorption of dissolved organic carbon (DOC) on glass fibre filters during particulate organic carbon (POC) determination. Water Research 24(6):763-764.

Agatova, A.I., and N.M. Lapina. 1995. The rates of organic matter transformation and nutrient regeneration in the Bering Sea. In: Komplexnie issledovanya ekosistemi Beringova morya, Sbornik nauchnikh trudov, Moscow, VNIRO, pp. 226-240. (In Russian.)

Agatova, A.I., N.M. Lapina, and N.I. Torgunova. 1995. The biochemical composition of dissolved and particulate organic matter. In: Komplexnie issledovanya ekosistemi Beringova morya, Sbornik nauchnikh trudov, Moscow, VNIRO, pp. 204-226. (In Russian.)

Banse, K. 1990. New views on the degradation of organic particles as collected by sediment traps in the open sea. Deep-Sea Research 37:1177-1195.

Cho, B.C., and F. Azam. 1988. Major role of bacteria in biogeochemical fluxes in the ocean's interior. Nature 332:441-443.

Collins, K.J., and P.J. Le B. Williams. 1977. An automated photochemical method for the determination of dissolved organic carbon in sea and estuarine waters. Marine Chemistry 5:123-141.

Glebov, B.V., V.I. Medinets, and V.G. Solovyev. 1988. Intensity of biochemical processes. In: Results of the 3rd U.S.-U.S.S.R. Bering Chukchi Seas expedition, pp. 224-230. (In Russian.)

Handa, N., and E. Tanoue. 1981. Organic matter in the Bering Sea and adjacent areas. In: D.W. Hood and J.A. Calder (eds.), The eastern Bering Sea shelf: Oceanography and resources, Vol. 1. Published by the Office of Marine Pollution Assessment, NOAA and BLM. Distributed by the University of Washington Press, Seattle, WA 98105, pp. 359-382.

Hansell, D.A., J.J. Goering, J.J. Walsh, C.P. McRoy, L.K. Coachman, and T.E. Whitledge. 1989. Summer phytoplankton production and transport along the shelf break in the Bering Sea. Continental Shelf Research 9:1085-1104.

Izrael, Y.A., and A.V. Tsiban (eds.). 1990. Issledovanie ekosistemi Beringova morya (2) 1990. Leningrad, Gidrometeoizdat, p. 344. (In Russian.)

Karl, O.M., B.D. Tilbrook, and I. Tien. 1991. Seasonal coupling of organic matter production and particle flux in the Bransfield Strait, Antarctica. Deep-Sea Research 38:1097-1126.

Kepkay, P.E., and B.D. Johnson. 1988. Microbial response to organic particle generation by surface coagulation in seawater. Marine Ecology Progressive Series 48(2):193-198.

Kinney, P.G., T.C. Loder, and J. Groves. 1971. Particulate and dissolved organic matter in Amerasian basin of the Arctic Ocean. Limnology and Oceanography 16:132-137.

Lee, C., and S.G. Wakeham. 1989. Organic matter in seawater: Biogeochemical processes. Chemical Oceanography 9:1-49.

Loder, T.C. 1971. Distribution of dissolved and particulate organic carbon in Alaskan polar, subpolar, and estuarine waters. Ph.D. dissertation, University of Alaska, Fairbanks.

Lyutsarev, S.V., Y.V. Konnova, and V.A. Konnov. 1988. The composition and distribution of the particulate organic matter in the Bering Sea. Oceanology 1(28):72-77. (In Russian.)

Mordasova, N.V., M.P. Metreveli, and M.V. Ventsel. 1995. The phytoplankton pigments in the western Bering Sea. In: Komplexnie issledovanya ekosistemi Beringova morya, Sbornik nauchnikh trudov, Moscow. VNIRO, pp. 256-264. (In Russian.)

Pashkova, E.A., N.I. Gul'ko, S.V. Lutsarev, and S.I. Tsekhonya. 1988. Colloidal and dissolved forms of gold and organic carbon in the seawater of the Bering Sea and the North Pacific. Oceanology 3(28):393-398. (In Russian.)

Redfield, A.C., B.H. Ketchum, and F.A. Richards. 1963. The influence of organisms on the composition of sea water. In: M.N. Hill (ed.), The sea. Interscience, New York, pp. 26-49.

Ridal, I.I., and R.M. Moore. 1990. A re-examination of the measurement of dissolved organic phosphorus in seawater. Marine Chemistry 29:19-31.

Sherr, E.B., and B.F. Sherr. 1991. Planktonic microbes: Tiny cells at the base of the ocean's food webs. Trends in Ecology and Evolution 6(2):50-54.

Sigleo, A.C., T.C. Hoering, and I.R. Helz. 1982. Composition of estuarine colloidal material: Organic components. Geochimica et Cosmochimica Acta 46:1619-1626.

Sorokin, Y.I. 1995. The primary production in the Bering Sea. In: Complex research on the Bering Sea, collected studies. VNIRO, pp. 264-276. (In Russian.)

Spravochnik. 1991. Reference book of hydrodynamics: Fisheries. Agropromizdat, Moscow. 224 pp. (In Russian.)

Sugimura, Y., and Y. Suzuki. 1988. A high temperature catalytic oxidation method of non-volatile dissolved organic in seawater by direct injection of liquid samples. Marine Chemistry 20:113-119.

Suzuki, Y., T. Sugimura, and T. Itoh. 1985. A catalytic oxidation method for the determination of total nitrogen dissolved in seawater. Marine Chemistry 16:83-97.

Walsh, J.J. 1988. On the nature of continental shelves. Academic Press, New York. 520 pp.

Walsh, J.J., and C.P. McRoy. 1986. Ecosystem analysis in the southeastern Bering Sea. Continental Shelf Research 5:259-288.

Whitledge, T.E., R.R. Bidigare, S.I. Zeeman, R.F. Sambrotto, P.F. Roschigno, P.R. Jensen, J.M. Brooks, C. Trees, and D.M. Veidt. 1988. Biological measurements and related chemical features in Soviet and United States regions of the Bering Sea. Continental Shelf Research 8:1299-1319.

CHAPTER 13

Silica in Bering Sea Deep and Bottom Water

Lawrence K. Coachman
School of Oceanography, University of Washington, Seattle, Washington

Terry E. Whitledge
Marine Science Institute, University of Texas, Port Aransas, Texas

John J. Goering
Institute of Marine Science, University of Alaska Fairbanks, Fairbanks, Alaska

Abstract

Neither the enormously high concentrations of silicic acid, higher than in any other ocean basin, nor the source, rate of supply, and flushing of the deep and bottom waters of the Bering Sea basins have been adequately explained. In this paper these questions are examined using the few available data, and a model describing the silicate distribution is proposed. The source for Bering Sea bottom water is North Pacific water from ~3,500-4,000 m depths, which enters through the westernmost pass in the Aleutian-Commander island arc (Kamchatka Strait) with high silicate concentrations, and then circulates into the other basins. The bottom water slowly displaces the deep water upward; at the same time silicic acid concentrations are increased by regeneration both within the water columns and from the bottom. Model results suggest bottom regeneration rates are about 4-5 times faster than those within the water columns, and that total residence times for the deep water are about 250-300 years. The deep Bering Sea acts like an "appendix" to the North Pacific Ocean—it may be an important location to monitor certain aspects of both climate change and anthropogenic pollution.

Introduction

Silicate concentrations of deep Bering Sea basin water exceed 230 µM, values higher than in any other basin of the World Ocean and nearly 40%

T.E. Whitledge is currently at Institute of Marine Science, University of Alaska Fairbanks, Fairbanks, AK.

greater than at the same depths in the adjacent North Pacific. The Bering Sea has aptly been dubbed "the sea of silica" (Tsunogai et al. 1979). Whence these uniquely high values? There appear three viable possibilities: (1) production over the deep basins of siliceous phytoplankton is extraordinary, such that the high concentrations of dissolved silicate are produced by dissolution at depth of an extra-large "rain" of diatoms and radiolaria; (2) flushing of deep basin waters is inordinately slow, so that high concentrations are long-time accumulations of dissolution products from more ordinary production levels (could deep Bering Sea water be the oldest water in the World Ocean?); and (3) regeneration rates are extraordinarily high, particularly at the sediment-water interface. Tsunogai et al. (1979) estimated that the regeneration rate in the deep Bering Sea is 4-5 times greater than that in the deep North Pacific. Perhaps some combination of these is likely. The rate of renewal of the deep water in the Bering Sea must be an important factor, but the source of Bering Sea bottom water and its rate of supply and basin flushing have never been described. Likewise, no measurements have ever been made of particulate silica, biogenic silica production, or dissolution rates of silica in the water columns of the deep basin. It is the purpose of this paper to examine, using the few available data, the question of bottom water formation—sources, rates of supply, and flushing—and then to present a preliminary model explaining the enormous concentrations of silica in the deep and bottom water of the Bering Sea which can serve as the basis for future research.

The deep Bering Sea (depths of 3,900-4,000 m) is not "at the greatest depths" considered as one basin, but three (Fig. 1). The westernmost Kamchatka Basin is connected with the North Pacific at depths to 4,000 m via Kamchatka Strait. Shirshov Ridge extends south from Cape Olyutorsk almost to the Aleutian Islands separating Kamchatka Basin from those to the east. There is a small connection with an approximate 25 km-wide gap northwest of Attu Island with depths to about 3,700 m. Geologically, Shirshov Ridge connects with Bowers Ridge which arcs northeastward and then south to the Aleutian Islands, nearly isolating small Bowers Basin from the large and deep Central Basin. The best bathymetric chart (Heezen and Tharp 1975) suggests that at abyssal depths, 3,900-4,000 m, only the deepest 200 m of these three basins is actually isolated by sills, but the direct connection for water exchange between them below ~3,000 m is narrow and tortuous.

The deep Bowers and Southeast basins (it is sometimes convenient to consider the large Central Basin in two parts, North Central and Southeast, though there is no topographic separation) are almost completely isolated from the North Pacific by the Aleutian-Commander island arc. The five major passages for water through the arc are listed in Table 1 and their locations indicated in Fig. 1.

There is no sill in Kamchatka Strait separating Kamchatka Basin from the North Pacific in the far west. Sill depths in Near Strait are close to 2,000 m in two or three narrow (20-30 km wide) gaps due south of the

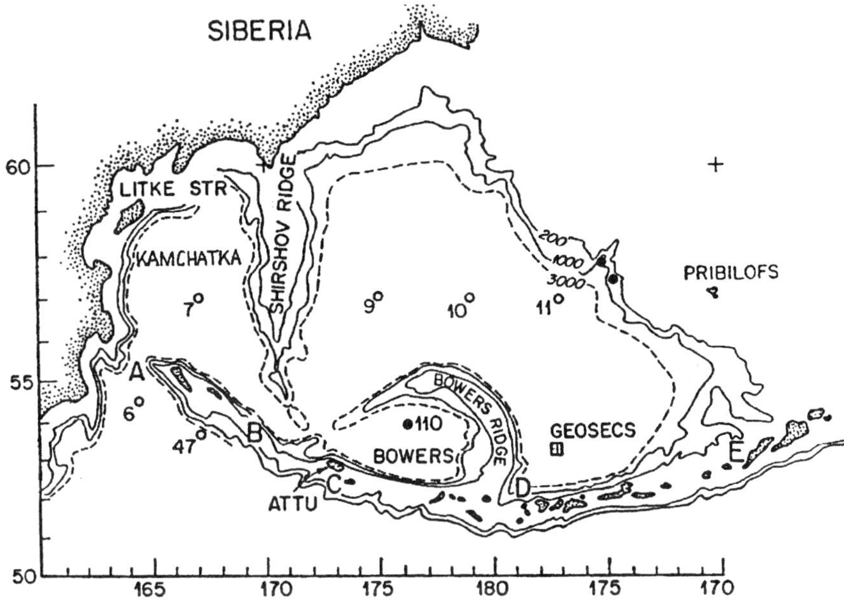

Figure 1. *The basins and separating ridges of the deep Bering Sea. Sills dividing Bowers Basin from Karagin and the Central basins rise only about 200 m above the deeps; no sill divides the North Central from the Southeast Basin. Letters indicate locations of the major passes through the Aleutian-Commander island arc (Table 1). Shown also are locations of deep-reaching hydrochemical stations from Argo (47; 1966), GEOSECS (1973), Hakuho Maru (6,7,9-11; 1975), and R/V Korolëv (110, GEOSECS; 1988).*

Shirshov Ridge; these actually connect directly with the southeast corner of Kamchatka Basin rather than with the basins to the east. Elsewhere in Near Strait, sill depths are about 1,100 m. Bowers Basin is open to the North Pacific via Buldir Pass with sill depth <700 m. Amchitka and Amukta passes connect with the Southeast Basin with sill depths <1,200 m in Amchitka Pass.

In the only study to date discussing bottom water, Tsunogai et al. (1979) hypothesized the source to be deep water from the North Pacific. They proposed, very generally, that the deep North Pacific water enters through Kamchatka Strait and from there spreads eastward over the whole basin. They argued that the vertical distribution of oxygen supported their hypothesis (without stating how), and that increasing values of the deep silicate maxima from the Kamchatka Basin through the Central Basin represent an upstream-downstream progression of deepwater flow. These arguments are not convincing, and there are other discrepancies in their

Table 1.   Sill depth and section area of the five largest passes through the Aleutian-Commander island arc.

Pass		Sill depth (m, approx.)	Area (km^2)
Kamchatka Strait	(A)	4,000	335.3
Near Strait	(B)	<2,000	239.0
Amchitka	(C)	1,155	45.7
Buldir	(D)	640	28.0
Amukta	(E)	430	19.3

From Favorite 1974. Letters show locations in Fig. 1.

discussion which require confirmation. For example, they state that the primary site of silicate regeneration is the sediment-water interface even though most deep silicate profiles actually show maxima some hundreds of meters above bottom.

There are few data from the deep Bering Sea. The total of NODC hydrographic stations from all basins with observations deeper than 3,000 m number less than 100. This paper uses the GEOSECS station of 1973 from the Southeast Basin (Park et al. 1975), 4 stations from *Hakuho Maru* KH-75-4 in 1975 (Tsunogai et al. 1979), and 8 stations from the *Akademik Korolëv* in 1988 which provide deep-reaching high-quality hydrographic and hydrochemical data.

Overview of Deep Hydrography

Interbasin Differences

Fewer than 100 stations extending >3,000 m have ever been taken in the Bering Sea. The data span many years (back to 1933) and were taken by many expeditions. Bottom water characteristics among the basins are very similar. In these data, salinities range between 34.62 and 34.68 ppt, and coldest potential temperatures are slightly below 1.3°C. To assess interbasin differences, the NODC data were organized into groups by location, as shown in Fig. 2. θ/S envelopes for the stations of each group were compared in Fig. 3 and encompass the deepest observations at each station (all those >3,300 m). Though there is no guarantee that the maximum salinity or minimum temperature is recorded for each location, there are sufficient stations in each group that we believe the envelopes show values close to the temperature extremum.

The only difference among groups that can be inferred for salinity is that water in Kamchatka Strait is slightly more saline than anywhere within the deep Bering Sea basins, perhaps by 0.01-0.02 ppt. Otherwise salin-

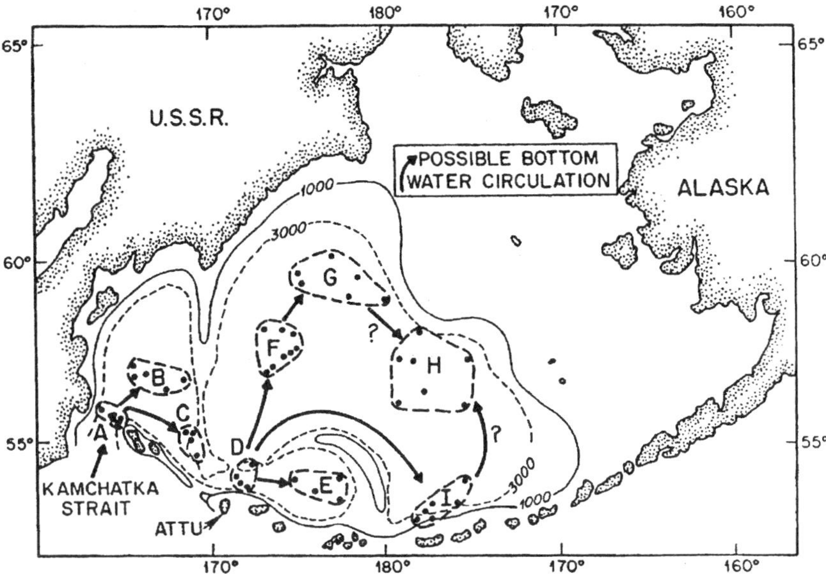

Figure 2.   Groupings of deep-reaching NODC stations for which the θ/S envelopes are shown in Fig. 3. A possible circulation of bottom water interpreted from potential temperature is indicated by arrows.

ity values of the historical data are too inaccurate to allow meaningful interpretation.

In minimum temperature, however, there appear to be definite differences between basins, and these differences have an upstream-downstream trend. Coldest potential temperatures are in Kamchatka Strait (Fig. 3; A). Next warmest shown in Fig. 3 are groups B and C, about equidistant to the north and to the southeast from Kamchatka Strait. Then comes group D in the constricted region northwest from Attu, and the minimum θs of the Bowers Basin group are even warmer. The four groups from the Central and Southeast Basins (Fig. 3; B) all show potential temperature minima warmer than the group at the entrance to these basins north of Attu (group D), but the differences among them are small. Possibly group H, on the east side of the Central Basin, shows the highest minimum temperatures.

The R/V *Akademik Korolëv* during the Third Soviet-American Bilateral Research Expedition to the Bering and Chukchi Seas, summer 1988, took five deep stations in Bowers Basin, two on the east side of the North Central Basin, and repeated the GEOSECS station of 1973 in the Southeast Basin (Fig. 1). Two plots showing within and between basin differences in the deep water suggested by these data are presented in Figs. 4 and 5. (Though the data were obtained from a Soviet oceanographic vessel, the

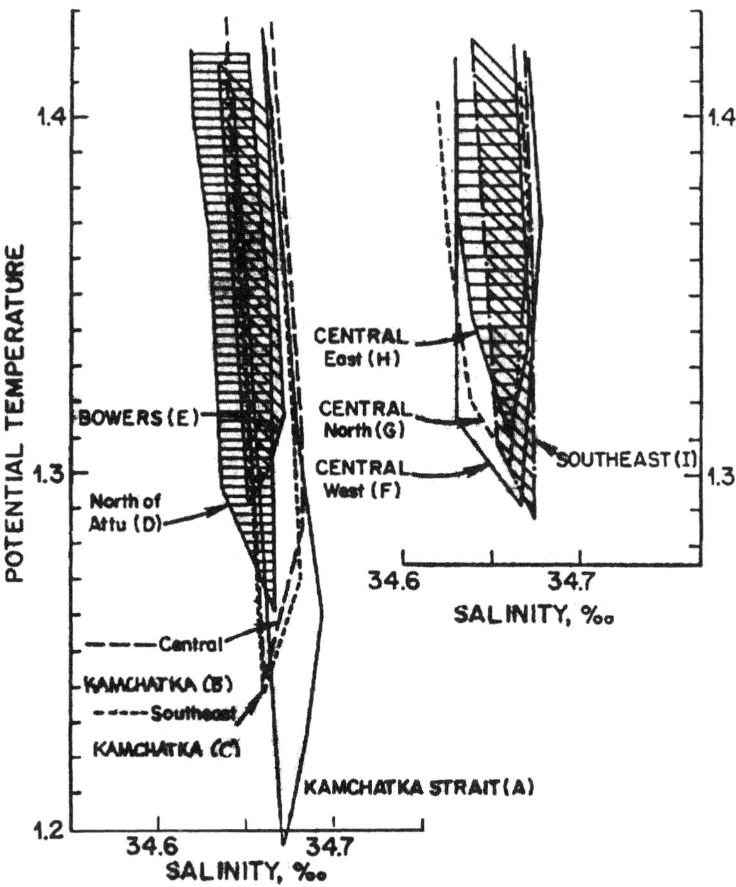

Figure 3.   θ/S envelopes for the station groupings of Fig. 2. A progressive warming from Kamchatka Strait to Bowers Basin can be interpreted from the envelopes on the left. On the right, all groups in the Central and Southeast basins appear to be downstream from Attu Island, but the direction of circulation is uncertain.

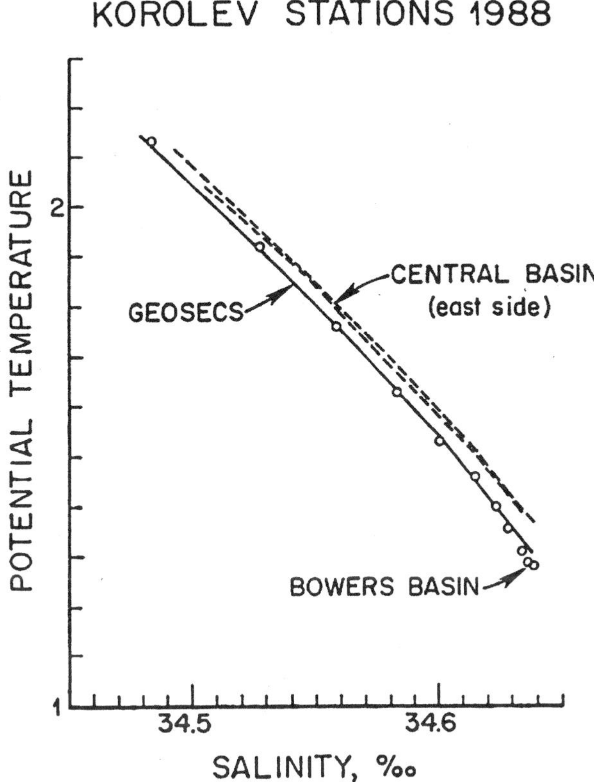

Figure 4.   θ/S curves from the R/V Korolëv for two stations from the Central Basin and the GEOSECS station (Fig. 1) and a composite of five Bowers Basin stations. Water columns in the Southeast and Bowers basins are indistinguishable, while in the eastern Central Basin the curves are displace upward to warmer temperatures, ~0.02-0.03°C.

CTD and chemical measurements were made by American scientists.) Figure 4 is the θ/S comparison between the two Central Basin stations, the GEOSECS station, and a composite Bowers Basin station (there are no significant differences among the five Bowers Basin stations in θ/S); Fig. 5 compares vertical distributions of dissolved nitrate and silicate, and their ratio, for water columns below 1,000 m for the same stations. In these data, Bowers Basin cannot be distinguished from the GEOSECS station in the Southeast Basin. This agrees with the interpretation from Fig. 3 where the minimum potential temperatures of these groups in these locations (E and 1) are essentially the same. On the other hand, the Bowers Basin and GEOSECS stations are distinguishable from the Central Basin water near the eastern continental shelf. The θ/S curves for the latter are displaced up to warmer temperatures for given values of salinity. This also agrees with Fig. 3 (group H shows them as warmer than E and I).

Total dissolved inorganic nitrogen and dissolved silicate show similar differences (Fig. 5). The Bowers Basin and GEOSECS stations both show higher silicates and lower nitrogen, and hence much lower differences in nitrogen/silicate ratio than do the stations from the east side of the Central Basin. Also, data from the *Hakuho Maru* (1975) show significant differences between the Kamchatka and Central basins in deep silicate maxima (Kamchatka is lower by ~20 μM).

In summary, the few available data from the Bering Sea deep basins show that there were subtle differences in bottom water properties within and between basins. Systematic increases in bottom potential temperatures and silicates were consistent with bottom water formed from North Pacific deep water advected into Kamchatka Basin through Kamchatka Strait and then into the other basins downstream sequentially.

Differences with the North Pacific

The Bering Sea water mass is compared with North Pacific waters in Fig. 6 which plots potential θ/S curves for four deep-reaching North Pacific stations from along the south side of the Aleutian Islands together with the GEOSECS station taken as representative of the Bering Sea (Fig. 4). Only small differences were obtained between the Bering Sea and North Pacific water masses. These are: (1) in the upper half of the water column (<2,000 m) at any given depth or salinity value the Bering Sea column was a little colder than the North Pacific mass, as much as 0.1°C; (2) below ~2,000 the situation reverses, so that at ~500 m Bering Sea water is warmer by about the same amount; and (3) at the bottom Bering Sea water is slightly fresher (~0.02 ppt) than at the same depths in the North Pacific.

Possible explanations for these subtle differences for (1) are that intermediate water in the Bering Sea, which is relatively shallow, is a little colder than in the North Pacific—thus, vertical heat flux warms the Bering deep water to a lesser degree than in the North Pacific. For (2) and (3), the reason is the isolation of Bering Sea bottom water after it enters from the North Pacific. Bottom water enters as a specific water type (i.e., 0-1.15°C;

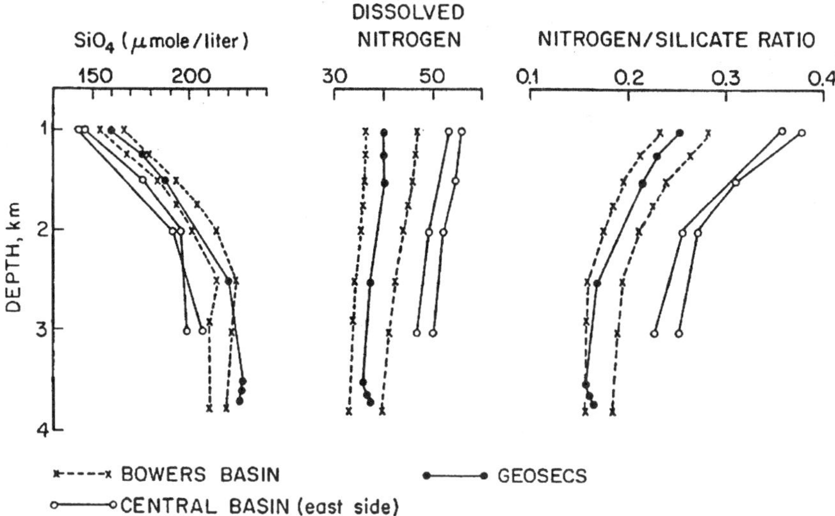

Figure 5.   *Vertical profiles of total dissolved inorganic nitrogen (NO_2 + NO_3 + NH_4) and SiO_4, and their ratio, from R/V Korolëv data summer 1988; the two Bowers Basin nitrogen profiles represent the range of values observed at that basin. The Central Basin stations show higher nitrogen and lower silicate and hence a much higher ratio than in Southeast and Bowers basins.*

salinity ~34.68 ppt) and then undergoes small warming and freshening in its travels around the basins (i.e., to 0-1.3°C; salinity ~34.66-34.67 ppt) (Fig. 3).

A small fraction of the warming of the bottom water might be due to heat flux from the bottom. The distance from Kamchatka Strait to the east side of the Central Basin is about 1,500 km; at 1 cm per s requires a travel time of 3.5 years. The heat flux of 1.6 μcal per cm² per s (Lee 1970) could warm a 100 m thick water layer ~0.015°C in this time. But heat flux from the earth cannot be a very significant factor in bottom water temperature change—at all stations θs decrease steadily to the bottom; that is, the values are not constant as they would be if there were significant warming from underneath.

Also shown in Fig. 6, and important in the discussion of bottom water formation, are estimates of the water mass property values occurring at sill depths of the major passes aside from Kamchatka Strait, which has no sill. Sill depth in Near Strait is ~2,000 m, and salinities at sill depth are never as high as 34.6 ppt. In Amchitka Pass, the third deepest passage (Table 1), salinity is never as high as 34.5 ppt. A search of historical data from these passes confirms these findings.

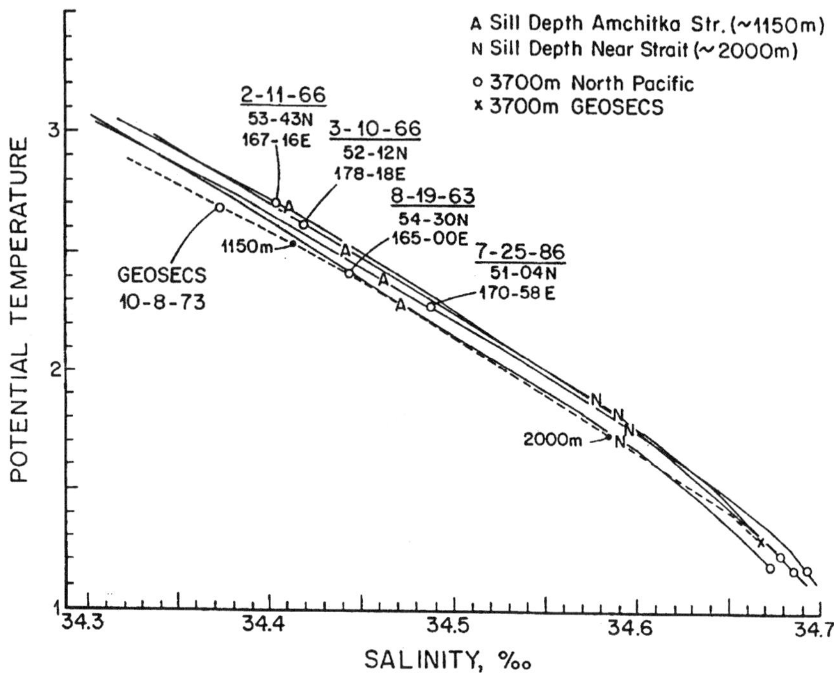

Figure 6.   *θ/S curves for the four deep-reaching North Pacific stations south to the Aleutian Islands compared with the 1973 GEOSECS station. Above ~2,000 m the Bering Sea water mass is slightly colder for a given salinity than in the North Pacific, while the reverse is true at depths below 2,000 m. Values at sill depths for Near and Kamchatka straits are indicated for each station.*

Bottom Water Formation—A Hypothesis

A primary constraint on the formation process is that the resulting water must have a salinity of at least 34.67 ppt. There are three hypothetical mechanisms by which bottom water might possibly be formed: (1) modification of surface (upper layer) water within the confines of the sea by cooling and salinity enhancement through ice formation, creating water sufficiently dense to sink to the bottom; (2) subsurface mixing of North Pacific water with appropriate Bering Sea waters as it crosses the sills in the Aleutian-Commander island arc passages; and (3) direct advection of deep North Pacific water in through Kamchatka Strait and then sequentially through the gaps into the other basins.

We can eliminate (1) and (2) as viable possibilities. For (1), generally over the Bering Sea salinities as high as 34 ppt in the surface layers are never observed (Sayles et al. 1979). Only in local areas in the shallow northern shelf regions does the formation of ice in polynyas occasionally generate salinities sufficiently high to meet the bottom water requirement (e.g., Schumacher et al. 1983). However, the water created in these areas (Gulf of Anadyr, near St. Lawrence Island, and in Norton Sound) is trapped in the regional northward advection and carried out of the Bering Sea into the Arctic Ocean.

One area on the west side of Kamchatka Basin at Litke Strait, which separates Karagin Island from Kamchatka (Fig. 1), is a possible location where this process might create sufficiently saline water. The strait is large with depths <100 m. In winter, with north and northwest winds from Siberia blowing off Kamchatka, the conditions would seem ripe for possible polynya formation and vigorous ice growth, and hence brine enhancement. There are, however, no oceanographic data to assess this possibility. We discount it, though, as a possible significant source of bottom water for two reasons: (a) This mechanism would produce water of varying salinity values, while the bottom water salinities exhibit a very narrow range and are practically the same in all basins; and (b) the average rate of production would be minuscule compared with the apparent required rate of bottom water supply (see below).

For (2), subsurface mixing in and near the major passes cannot be a source for bottom water. The highest salinities found near sill depths in the North Pacific are never greater than 34.60 ppt (Fig. 6). In order for this water to be turned into bottom water it must mix with water with salinity much greater than 34.7 ppt, and such water does not exist in the Bering Sea.

By deduction, the source of bottom water in the Bering Sea is deep North Pacific water at the approximate depths of the sill in Kamchatka Strait, viz. 3,500-4,000 m. This water enters the Bering Sea as a water type, with very narrow ranges in temperature and salinity (Fig. 3): θ = 1.15-1.20°C; S = 34.67-34.69 ppt. This water flows as a bottom layer through the basins and is progressively warmed and freshened by the downward fluxes of heat and freshwater. The general magnitude of the modification is shown by the comparison of bottom temperature and salinity values in Fig. 6; overall warming is about 0.1°C and freshening about 0.02 ppt.

General Circulation

The general path of bottom water renewal can be inferred from the progressive warming of deep potential temperatures. The minimum θs of each envelope of Fig. 3 showed a systematic increase from Kamchatka Strait (A) to locations B and C in the Kamchatka Basin (Fig. 2, i.e., Δ0-0.04°C). In the interbasin gap north of Attu from C to D, θ_{min} increases about 0.02°C more. Group D was a precursor to both Bowers Basin (group E) and the groups in the Central Basin; from D to these groups θs increase another ~0.02°C. Within the Central Basin temperature data were too inaccurate to infer the circulation unambiguously; θ increases were very small. The data do suggest, however, that groups F and I were about equidistant

Table 2.   Distance and estimated travel times of water masses moving within the Bering Sea.

From Kamchatka Strait to:		Distance (km)	Speeds (m/s)		
			0.01	0.001	0.0001
Central Kamchatka Basin	B	280	0.9 yr	8.9 yr	88.9 yr
Interbasin gap	D	450	1.4 yr	14.3 yr	143 yr
Central Bowers Basin	E	775	2.5 yr	24.6 yr	246 yr
Center of Central Basin		1,100	3.5 yr	34.9 yr	349 yr
GEOSECS Station	I	1,300	4.1 yr	41.2 yr	413 yr
Eastern Central Basin	H	1,550	4.9 yr	49.2 yr	492 yr

from D, while H, along the eastern side of the basin, was the farthest from the source.

The inferred bottom water movement is shown by arrows in Fig. 2. A summary of the distances to various points in the deep basins and possible bottom water travel times for a range of average speed values are given in Table 2.

Bottom Water Renewal

Bottom water enters the Bering Sea basins and circulates around as a bottom layer, and this layer more-or-less continuously displaces the deep-water layers above. Figure 7 compares vertical profiles of silicate from the three Bering basins with profiles typical for the North Pacific. In the North Pacific there is a silicate maximum between 1,500 and 2,000 m. Maximum values are 175-185 µM, while near-bottom values are <170 µM. In Kamchatka and Bowers basins and at some (but not all) Central Basin stations there is a deeper maximum, ~3,000 m, with values 40% greater—up to ~240 µM. These silicates are the highest observed in any deep waters of the World Ocean.

A silicate maximum in the deep water but not at the bottom can only be a result of the renewal process. In the North Pacific, water from the Antarctic floods the bottom, displacing the deep layers upward. The newer water to the deep basin, lower in silicate, is at the bottom, while the deepwater layers above have been in the basin longer and thereby collected much greater amounts of dissolved silicate from dissolution of the diatom "rain" from the surface layers. Thus it appears that in general the process which "flushes" the deep basins of the North Pacific and Bering Sea is similar, but, judging from the differences in depth and degree of silicate maxima, the rates of regeneration of silica in the water columns and at the sediment-water interface may be different.

Evidence for basin water renewal as a layer flooding in along the bottom is observed in the GEOSECS station data (Park et al. 1975). At this

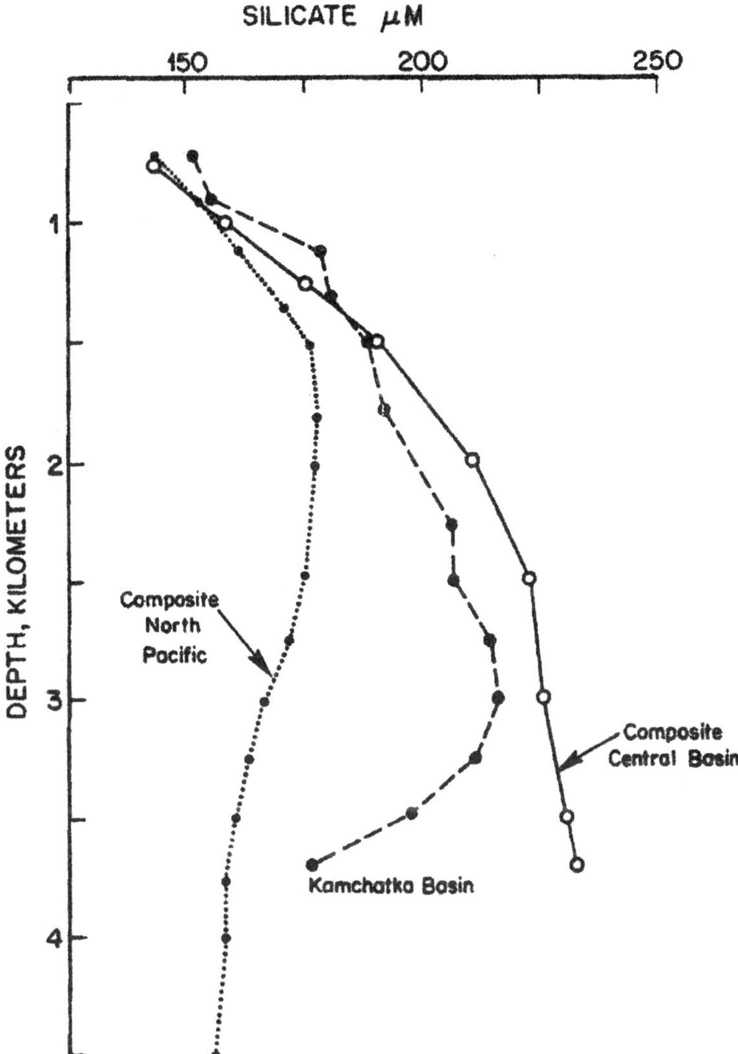

Figure 7.   *Silicate profiles for the North Pacific source water to the Bering Sea (composite of stations 6 and 47, Fig. 1), the Kamchatka Basin (station 7, Fig. 1), and the Central Basin (composite of stations 9-11, 110, GEOSECS, Fig. 1).*

station detailed measurements were made near-bottom of a number of properties. Figure 8 shows the profiles of silicate, dissolved oxygen, and turbidity. The bottom ~120 m layer was clearly a much "newer" water. The silicates were lower and oxygen higher, both indicating that this water was more recently exposed to the surface than the water above. Furthermore, the much increased turbidities of the layer suggest it may have been in significant motion.

Bottom water renewal may not be as a steady flow, but more episodic. The interface in Fig. 8 between the bottom layer and the deep water above is relatively sharp, suggesting such a recent addition that diffusion has not had time to smooth out the transition between them. Also, the *Korolëv* silicates, shown for comparison, suggest the presence of multiple, thinner layers rather than just one thicker one. Thus, it is probable that there are smaller temporal and spatial variations in the renewal process than can be resolved by the available data.

Preliminary estimates of vertical velocities in the deepwater columns and renewal rates can be estimated from the available data. Two time comparisons of θ/S curves are shown in Fig. 9: one curve compares the GEOSECS station with its reoccupation 15 years later, and the other compares the five *Korolëv* stations from Bowers Basin with two NODC stations made 18 and 20 years earlier. A slight freshening of the water column over 15-20 years is suggested by the curve comparisons. The curves begin diverging at about 700 m depth. If the freshening is real, then approximately the lower 2,000 m of the column has been affected. The indicated vertical displacement is 100-133 m/yr, or vertical velocities of $3.2\text{-}4.2 \times 10^{-6}$ m/s. The average area of the deep basins beneath 1,000 m is about 0.48×10^6 km^2, so the volume of the bottom 2,000 m layer is about 0.96×10^{15} m^3. To replace this volume in 15-20 years would require a continuous inward transport of 1.5-2 Sv of bottom water. The width of Kamchatka Strait at 3,500 m depth is about 65 km; with these transports a 100 m thick bottom layer would have to flow in continuously at ~0.25 m/s, which would be an upper bound—if the renewal layer were 500 m thick inflow speeds need to be only ~5 cm/s. However, to us these estimates seem very high. As both time comparisons depend on salinity values from only one cruise (*Korolëv*), confirmation of this result is required.

Vertical velocities estimated for open ocean basins are typically less than that calculated above by as much as an order of magnitude. For example Munk (1966) and Veronis (1981) cite values of $1\text{-}2 \times 10^{-7}$ m/s. Wyrtki (1961) adopted a value of 2×10^{-7} m/s in his study of the thermohaline circulation and Tsunogai et al. (1979) used an upward velocity estimate of 1.5×10^{-7} m/s, which led them to a very old age estimate for deep water (350-400 years).

Vertical velocities calculated to balance wind stress curl seem to be larger than those needed to balance property distributions. Stommel (1976) calculated $W = 9.5 \times 10^{-7}$ m/s (30 m/yr) to balance the wind stress transport in the North Atlantic, where the curl of wind stress is typically fairly

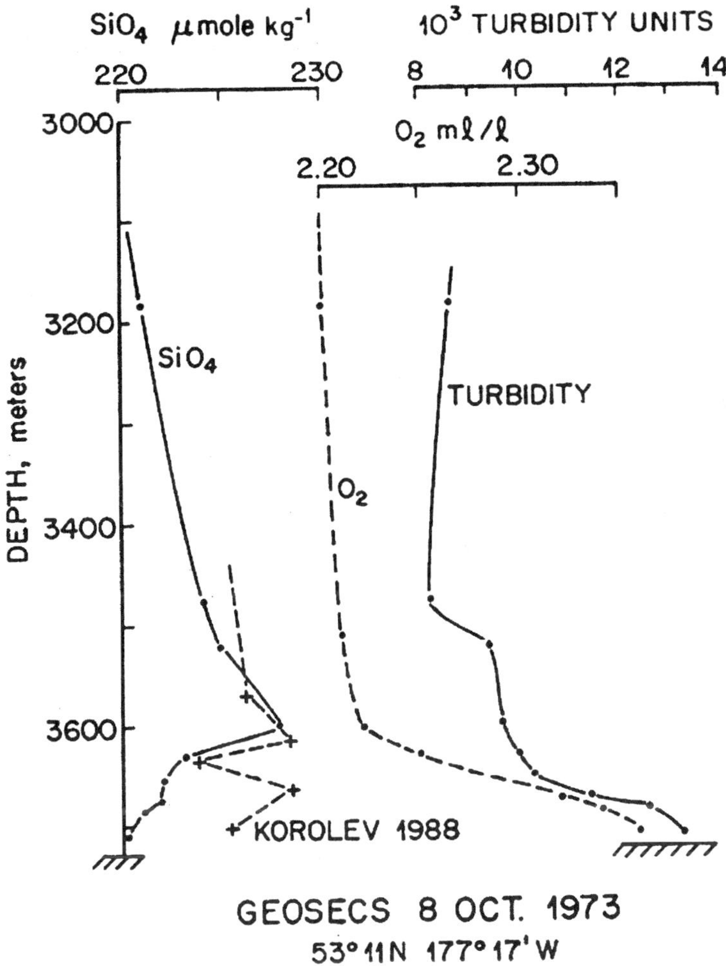

Figure 8.   Detailed near-bottom profiles from the 1973 GEOSECS station of silicate, dissolved oxygen, and turbidity. The bottom ~2,120 m of the column is a layer of much "newer" water with higher O₂ and lower silicate. It may also be in measurable motion (high turbidity). Silicates from the 1988 R/V Korolëv reoccupation are also shown.

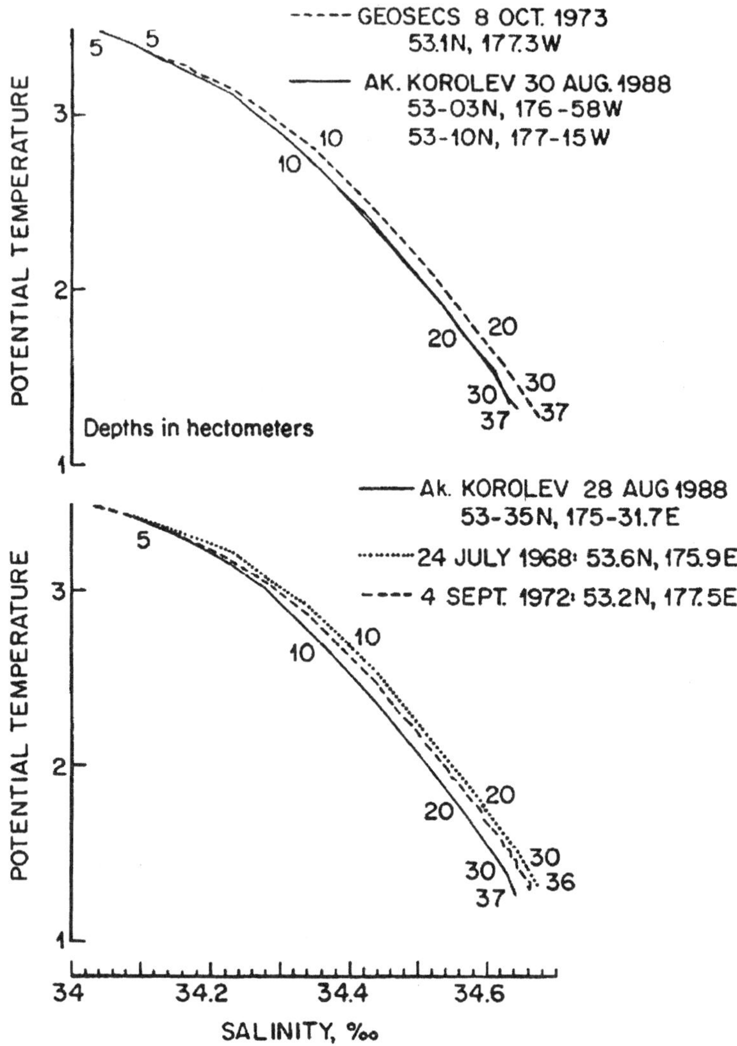

Figure 9.   θ/S *curves comparing: (upper) the GEOSECS station of 1973 with its reoccupation by the R/V Korolëv in 1988, and (lower) a composite of Bowers Basin stations from the Korolëv with two NODC stations in the same area made 18 and 20 years earlier. The comparison suggests a possible θ/S change over 15-20 years to salinities lower by 0.02-0.03 ppt.*

strong. The small, enclosed Bering basin has a strong wind stress curl. Hughes et al. (1974) calculated the mean monthly curl from climatological data, and reported the values as transports required to close the Sverdrup circulation. There is an order of magnitude annual variation (<1 Sv in summer to 13 Sv in winter), with a mean of ~5 Sv. A mean wind stress transport of 5 Sv would require a mean upward velocity of ~8×10^{-6} m/s, twice as great as estimated from the curve comparison. This value seems extraordinarily high.

Model of Silica

Combining the hypothesis for bottom water renewal in the Bering Sea with the silicate profiles of Fig. 7 leads us to propose the following model of silicate distribution. The bottom water is derived from North Pacific water at depths of 3,500-4,000 m with silicate values of ~170 μM. Progressing first into the Kamchatka Basin, then Bowers Basin, and lastly the Central Basin the water is progressively warmed. Ignoring possible but minor bottom heating, temperature is conservative so the controlling processes for steady temperature distribution are horizontal advection, vertical advection, and vertical diffusion, thus:

$$U\frac{\Delta T}{\Delta X} + W\frac{\Delta T}{\Delta Z} = K_z\frac{\Delta^2 T}{\Delta Z^2} \tag{1}$$

where U = speed of flow; $\Delta T/\Delta X$ = horizontal T gradient between groups (difference in near-bottom Ts divided by the approximate distance between the groups); K_z = vertical eddy conductivity; $\Delta^2 T/\Delta Z^2$ = curvature of the T profile; W = vertical velocity; and $\Delta T/\Delta Z$ the mean T gradient of the bottom layer (>3,000 m).

Credence for the simplicity of the temperature model is offered by the vertical T profiles themselves. Vertical T profiles for each station grouping of Fig. 2 are presented in Fig. 10. The profiles from all stations are very smooth, and fit very well ($r = 0.99$-1.00) by second-order polynomials of the form $T = a - bZ + cZ^2$. This simplicity of structure suggests that the bottom water progressing around the basins as a layer is warmed almost entirely by the net downward vertical diffusion of heat from the warmer surface layers, which would be modified by the upward vertical velocity in the negative temperature gradient.

The silicate in the bottom water, which increases regularly from 170 μM on entering to ~180-200 μM in the Kamchatka Basin and to ~230 μM in the Central Basin (Fig. 7), will be governed by these processes plus increases by two non-conservative processes: regeneration from the sediment-water interface (R_b) and regeneration within the water column from the diatom "rain" (R_w), thus:

$$U\frac{\Delta Si}{\Delta X} + W\frac{\Delta Si}{\Delta Z} = K_z\frac{\Delta^2 Si}{\Delta Z^2} + R_b + R_w \tag{2}$$

Table 3.   Parameters of temperature and silicate profiles within the deep and bottom layers.

	Bottom (3,000-4,000 m)	Deep (2,000-3,000 m)
$\Delta T/\Delta X$, C/m	0.09×10^{-6}	
$\Delta T/\Delta Z$, C/m	-0.15×10^{-3}	
$\Delta^2 T/\Delta Z^2$, C/m^2	0.10×10^{-6}	
W/K_Z, /m	2.4×10^{-3}	1.2×10^{-3}
$\Delta Si/\Delta X$, µM/m	60×10^{-6}	
$\Delta Si/\Delta Z$, µM/m	-20×10^{-3}	7×10^{-3}
$\Delta^2 Si/\Delta Z^2$, µM/m	28×10^{-6}	-20×10^{-6}

where Si denotes dissolved silicate.

The deepwater layer, approximately 2,000-3,000 m, which is being slowly raised upward by the bottom water intruding underneath, is presumed not to contain significant horizontal advection—the primary circulation in the sea which removes the water advected in as the bottom layer is in shallower layers (<1,500 m) closer to the source of wind stress. The temperature model of the deep layer is simply a 1-dimensional vertical balance of advection with diffusion, thus:

$$W \frac{\Delta T}{\Delta X} = K_Z \frac{\Delta^2 T}{\Delta Z^2} \tag{3}$$

The deep layer contains the silicate maximum within the Bering Sea. It will lose silicate upward through both advection and diffusion, while increases in silicate observed in this layer must be due to dissolution within the water column, thus:

$$W \frac{\Delta Si}{\Delta Z} = K_Z \frac{\Delta^2 Si}{\Delta Z^2} + R_w \tag{4}$$

Parameters for the model (Table 3) were calculated from the temperature profiles (Fig. 10) and silicate profiles (Fig. 7). The ratio W/K_Z was estimated from the simple balance of vertical advection to vertical diffusion for the temperature profiles (equation [3]), by the method of Wyrtki (1961).

With only the temperature and silicate fields mapped, and these not very well with high-quality data, accurate estimates of the process parameters U, K_Z, and W are not possible. We can, however, make useful first estimates by assuming a value for one of these. Typical values from deep basins of other oceans are: $U = 0.001$ m/s; $W = 10^{-7}$ to 10^{-6} m/s; and $K_Z = 10^{-4}$ m^2/s (Wyrtki 1961, Munk 1966, Stommel 1976, Veronis 1981, Pond and Pickard 1983, FRAM Group 1991). We choose $W = 2 \times 10^{-7}$ m/s (see

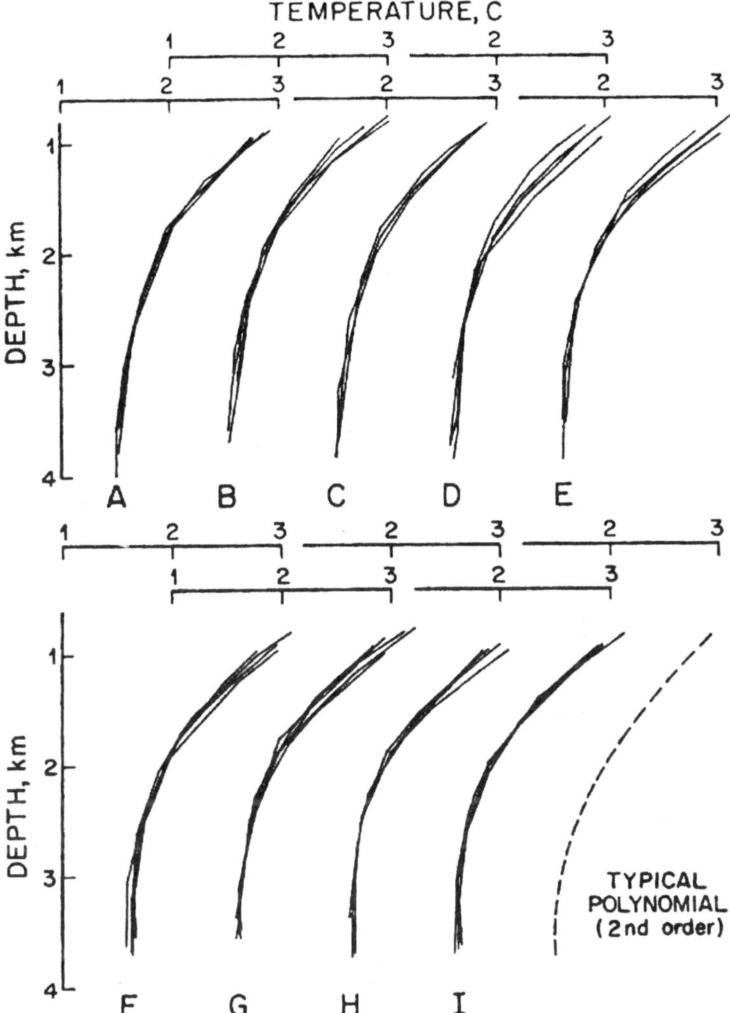

Figure 10. Vertical temperature profiles of the station groups (Fig. 2). The curves are all very smooth, and well fit (r = 0.99-1.00) by second order polynomials of the form $T = a - bZ + cZ^2$.

Table 4.　First estimates of results from model, with $W = -2 \times 10^{-7}$ m/s.

	Bottom	Deep
K_z, m^2/s	6.8×10^{-4}	1.7×10^{-4}
U, m/s	4.2×10^{-4}	
R_W, µM/s	20×10^{-10}	20×10^{-10}
R_b, µM/s	82×10^{-10}	

earlier discussion), because it is probably the most certainly known of the three. From equation (3), K_z is obtained which, when used in equation (4) gives the rate of regeneration of silicate (R_W) within the deep layer water column. In the bottom layer use of W/K_z allows an estimate of U through equation 1. Finally, by assuming, not unreasonably, that R_W is the same in the bottom layer as in the deep water, equation (2) allows estimation of the regeneration rate of silicate at the sediment-water interface (R_b). The results are presented in Table 4.

Discussion and Conclusions

The three deep (~3,900 m) Bering Sea basins are open only to the North Pacific without sill constraint via Kamchatka Strait, the westernmost passage through the Aleutian-Commander island arc. The western Kamchatka Basin connects with small Bowers and large Central basins to the east via small gaps in the underwater ridge topography, with a sill depth of ~3,500 m. Analyses of existing North Pacific and Bering Sea temperatures and salinities show that bottom water can neither be formed within the Bering Sea nor formed by mixing of Bering Sea water with North Pacific water over the sills in Aleutian Island passes. Thus, bottom water for the Bering Sea basins is derived from deep North Pacific water, between about 3,500 and 4,000 m depth. This layer enters through sill-less Kamchatka Strait and flows sequentially through Kamchatka Basin into Bowers and the Central basins. The bottom water renewal displaces the deepwater layers upward; the water is ultimately removed from the sea by circulation at depths <2,000 m. In transit through the basins downward fluxes of heat and freshwater warm the bottom water a maximum of ~0.1°C and reduce its salinity by –0.02 ppt.

　　Silica plays a special role in the Bering Sea in comparison with other world ocean basins. In the deep water silicates are as much as 40% greater than in the deep North Pacific, which suggests the possibility of great age of the deep waters, or inordinately high production of biogenic silica, or both. Profiles of silicate in North Pacific source water show maxima of ~180 µM at ~2,000 m and values in inflowing bottom water <170 µM. Within the basin, as the bottom water progresses sequentially from basin to basin, silicate values increase to ~230 µM. The cause of the increase is regeneration both in the water columns and at the sediment-water interface. In deep water the increase is less (water column regeneration only), such that the silicate maximum moves deeper, and sometimes is even erased from water columns in the Central Basin farthest removed from Kamchatka Strait by greater increases in the bottom layer from bottom regeneration.

A model describing the silicate distribution in deep and bottom water is proposed. In the bottom layer, the steady temperature distribution, being conservative, is a balance among horizontal advection, vertical advection, and vertical diffusion. Silicate in the layer is governed by these mechanisms plus regeneration both within the water column from the diatom "rain" and from the sediment-water interface. The deepwater layer above (2,000-3,000 m), being displaced upward in the water renewal process, has no significant horizontal advection. The temperature profile is the result of upward advection in the negative temperature gradient balancing downward heat diffusion, while the source for silicate in the layer is from water column regeneration. The gradients and curvatures in temperature and Si were estimated from the data. By assuming an upward advective velocity of 2×10^{-7} m/s, comparable with values from other deep ocean basins, the equations were solved to provide first estimates of mean bottom water velocity, eddy coefficients in the deep and bottom layers, and the regeneration rates of silicate in the water column and at the sediment-water interface.

The model results suggest that regeneration of silicic acid from bottom sediments is much more important than that taking place within the water column, by about one-half order of magnitude. This supports the results of Tsunogai et al. (1979) who reported that bottom regeneration in the Bering Sea is 4-5 times that occurring in the deep water of the North Pacific. Increasing the value of upward velocity in the model to $W = 5 \times 10^{-7}$ m/s leads to an even greater role for bottom regeneration in the silicate distribution, to values ~8 times greater than water column regeneration.

The low mean velocities in the bottom layer estimated from the model suggest that the transport of inflowing bottom water is small. Kamchatka Strait >3,000 m is ~65 km wide, so the transport of a 500 m thick layer would be ~0.01 Sv. It seems likely that velocities of bottom water within the Bering Sea are below detection by conventional meters, unless the renewal is strongly episodic (Fig. 8).

The oldest water in the Bering Sea will be the upper deep water in the Central Basin, at the downstream end of the water movement before it becomes entrained in the shallower circulation and removed from the sea. An estimate for the time required for a parcel of water to make this trip, approximately the flushing time for the Central Basin, is mean horizontal speed from model results in 4×10^{-4} m/s. To cover 1,500 km requires about 120 years (Table 2). For the parcel to move upward at least 1,000 m across the deep layer, at 2×10^{-7} m/s, requires another 150 years, for a total of perhaps 250-300 years. This estimate is less than that of Tsunogai et al. (1979), who estimated the residence time to be 350-400 years; nevertheless, it is still old. And, when it is considered that this is only the residence time within the Bering Sea proper, to be added to the age of the North Pacific deep source water when it enters the sea, Bering Sea deep water may well be the oldest water in the world ocean.

The large increase in silicate concentrations throughout the water column below 1,000 m is not accompanied by similar increases in other nutrients, notably nitrate, which leaves unanswered important questions regarding the fate of organic matter in the deep Bering Sea. Over the range where silicate increases (>1,000 m) dissolved inorganic nitrogen (predominantly nitrate) is constant or decreases slightly. This effect is shown in the increase with depth of SiO_4/DIN (dissolved inorganic nitrogen) ratios for the deep stations from the *Korolëv* cruise (Fig. 11). The dissolution of biogenic silica from the "diatom rain" should be accompanied by a similar production of nitrate. This is what occurs in most deep waters of the Atlantic and Pacific oceans where SiO_4/DIN values are relatively constant with depth; if differential remineralization were occurring a maximum of nitrate in the water column would be expected. The distributions in Fig. 11 show a slight (5-8 μmol/L) decrease in DIN below 1,000 m, but, given an increase in silicate of about 80 μmol/L between 1,000 and 3,600 m, one expects a concurrent release of at least 20 μmol/L of DIN (assuming the normal SiO_4/DIN in deep Pacific water to be about 4:1).

The explanation for the vertical decrease in DIN accompanying an increase in silicate probably involves very low oxygen concentrations (<2 ml; Whitledge et al. 1988) and losses to denitrification at depths below 1,000-1,500 m (Fig. 11). The rate of denitrification could be extremely slow given the projected age of the deep waters. The lowest dissolved oxygen concentrations occur between 1-3 km, so the decomposition of organic matter must have occurred within that depth range and nitrate should have been produced; in actuality, nitrate decreases slightly over that depth range. The production of nitrate from organic matter requires oxygen, but when concentrations of both these quantities decrease some additional processes must become important. Obviously, there is need for further study of the processes contributing to the decomposition of organic matter and their relationship to these unusual N-Si-O distributions.

The deep Bering Sea, acting like an "appendix" to the North Pacific Ocean, may be an important location to monitor certain aspects of both possible climate change and anthropogenic pollution. There are two modes of egress for materials into the deep water from the surface: by advection dissolved with the water, and as part of the "rain" of biogenic material. The time scales for these processes to deliver material into the deep water differ by orders of magnitude—for properties obtained when the water was near the surface in the Antarctic will be >1,000 years, but associated

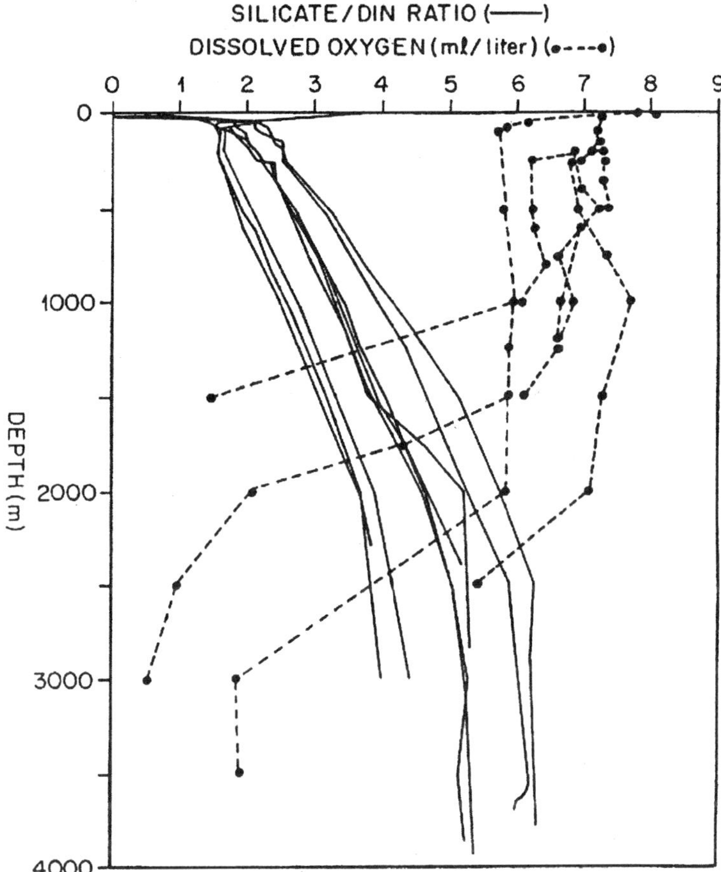

Figure 11. Vertical profiles of SiO_4/DIN ratio (solid lines) and dissolved O_2 (dashed lines) from the 1988 R/V Korolëv cruise (see Fig. 1 for locations). Dissolved O_2 typically displays a minimum of <2 ml/L between 1 and 3 km. The increasing SiO_4 ratio indicates that nitrogen is not increasing concurrently with silica, as is usual in other deep basins.

with fecal pellets and dead plankton only 1-10 years. We conclude that the deep Bering Sea may be an important location to look for those signals of climate change and pollution which have a preferential mode of transport; however, studies to better define the circulation and flushing are required in order to properly interpret the results. The model presented here can serve as the basis; measurement of a number of hydrochemical variables simultaneously should allow unique solutions for the process parameters.

The two junior authors would like to note that this chapter was the last professional work in this field by Professor Coachman.

Acknowledgments

This paper is funded in part by a grant to the University of Texas Austin by the National Science Foundation (NSF) and by grants to the University of Texas Austin and the University of Alaska Fairbanks from the National Oceanic and Atmospheric Administration (NOAA). The views expressed herein are those of the authors and do not necessarily reflect the views of NSF or NOAA or any of its subagencies.

References

Favorite, F. 1974. Flow into the Bering Sea through Aleutian Island passes. In: D.W. Hood and E.J. Kelley (eds.), Oceanography of the Bering Sea with emphasis on renewable resources. Occasional Publication No. 2, Institute of Marine Science, University of Alaska, Fairbanks, pp. 3-37.

FRAM Group. 1991. An eddy-resolving model of the Southern Ocean. Eos, Transactions of the American Geophysical Union 72(15):169,174-175.

Heezen, B.C., and M. Tharp. 1975. Bathymetry of map of the Arctic Region. The American Geographical Society, New York.

Hughes, F.W., L.K. Coachman, and K. Aagaard. 1974. Circulation, transport and water exchange in the western Bering Sea. In: D.W. Hood and E.J. Kelley (eds.), Oceanography of the Bering Sea with emphasis on renewable resources. Occasional Publication No. 2, Institute of Marine Science, University of Alaska, Fairbanks, pp. 59-98.

Lee, W.H.K. 1970. On the global variation of terrestrial heat flow. Physics of Earth and Planet Interiors 2:332-341.

Munk, W. 1966. Abyssal recipes. Deep-Sea Research 13:707-730.

Park, P.K., W.S. Broecker, T. Takahashi, and W.S. Reeburgh. 1975. GEOSECS Bering Sea station, a brief hydrographic report. In: Bering Sea Oceanography: An Update 1972-1974. Results of a seminar and workshop on Bering Sea oceanography, Fairbanks, AK, 7-11 October 1974. Institute of Marine Science Report 75-2, pp. 207-238.

Pond, S., and G.L. Pickard. 1983. Introductory Dynamical Oceanography, 2nd edn. Pergamon, Oxford.

Sayles, M.A., K. Aagaard, and L.K. Coachman. 1979. Oceanographic atlas of the Bering Sea basin. University of Washington Press, Seattle, WA. 158 pp.

Schumacher, J.D., K. Aagaard, C.H. Pease, and R.B. Tripp. 1983. Effects of a shelf polynya on flow and water properties in the northern Bering Sea. Journal of Geophysical Research 88:2723-2732.

Stommel, H. 1976. The Gulf Stream, 2nd edn. University of California Press, Berkeley, CA.

Tsunogai, S., M. Kusakabe, H. Lizumi, L. Koike, and A. Hattori. 1979. Hydrographic features of the deep water of the Bering Sea: The sea of silica. Deep-Sea Research 26:641-659.

Veronis, G. 1981. Dynamics of large-scale ocean circulation. In: B.A. Warren and C. Wunsch (eds.), Evolution of physical oceanography. MIT Press, Cambridge, MA, pp. 140-183.

Whitledge, T.E., R.R. Bidigare, S.I. Zeeman, R.N. Sambrotto, P.F. Roscigno, P.R.Jensen, J.M. Brooks, C. Trees, and D.M. Veldt. 1988. Biological measurements and related chemical features in the Soviet and United States regions of the Bering Sea. Continental Shelf Research 8:1299-1319.

Wyrtki, K. 1961. The thermohaline circulation in relation to the general circulation in the oceans. Deep-Sea Research 8:39-64.

CHAPTER 14

Variability in the Components of the Carbonate System and Dynamics of Inorganic Carbon in the Western Bering Sea in Summer

Alexander P. Nedashkovskiy
Pacific Oceanographic Institute, Russian Academy of Sciences of the Far East, Vladivostok, Russia

Victor V. Sapozhnikov
Russian Federal Research Institute of Fisheries and Oceanography (VNIRO), Moscow, Russia

Abstract

The distribution of components of the carbonate system in the western Bering Sea were mapped during 15-25 June 1992. The results were based on an assessment of total alkalinity and pH collected during the twenty-first cruise of the R/V *Akademik Alexander Nesmeyanov*. Variability of total inorganic carbon down to minimum oxygen depths was analyzed and found to be a result of production/destruction processes and gas exchange. Maximum vertical changes of inorganic carbon were recorded at the boundary between the upper quasi-homogeneous and cold intermediate water and deeper at all stations in the study area.

Introduction

Great attention has been paid to studies of variability of carbonate system components, particularly inorganic carbon in the ocean associated with problems of anthropogenic CO_2 consumption (Ivanenkov 1985, Poisson and Chen 1987, Bordovskiy and Makkaveev 1991). Investigations conducted in high-productivity areas are of interest because carbon dioxide exchange between the ocean and atmosphere is dependent on the intensity of production/destruction processes (Ivanenkov 1985).

Methods

The twenty-first cruise of the R/V *Akademik Alexander Nesmeyanov* occurred in the western Bering Sea, one of the most productive waters of the Pacific, during 15-25 June 1992 (Fig. 1). Measurements of pH and total alkalinity (Alk) were collected during the cruise and were the basis for calculations of seawater saturation with carbon dioxide (CO_2, %), levels of inorganic carbon (C_{tot}), specific inorganic carbon (C_{tot}/Cl), and specific alkalinity (Alk/Cl).

Measurements of pH were made in a constant temperature cell which prevented gas exchange with the atmosphere. Measurements were made below $20°C$ with a glass Orion 91-01 electrode and an Ionanalyser-901. The electrode was calibrated with a borate and phosphate buffer. The error in pH measurements did not exceed 0.01 during the cruise. Total alkalinity was determined by a reverse potentiometric titration with an automatic determination of the point of equivalence (BAT-15). Each sample was titrated twice. Relative error in alkalinity measurements was calculated on the results of parallel determinations and totaled 0.3%. A sodium carbonate solution was prepared from commercial Russian crystals and was used as a standard solution.

Carbonate system elements were calculated with the Henry constant, according to the equation in Weiss (1974). Carbonic acid dissociation constants were calculated by Mehrbach et al. (1973), and the boric acid dissociation constant by Lyman (1956). Millero's (1979, 1983) methods were applied to express pH values as in situ and to consider the pressure influence on the apparent constants of the carbonate equilibrium.

Results and Discussion

Analysis of the vertical distribution of carbonate system components revealed a number of features which were characteristic of the western Bering Sea. The surface layer (0-35 m) was marked by high pH values (8.1-8.4) and, consequently, low saturation values for carbon dioxide (45-95%). The relatively uniform layer between 35-180 m became oversaturated with CO_2 (up to 145%) with a pH value of 8.0. This relatively uniform layer was underlain by waters which showed a significant decrease in pH (to 7.6 at 700-1,000 m). Water saturated with CO_2 increased to 350-400% there. Figure 2 presents the characteristic vertical profiles of CO_2 (%), C_{tot}/Cl, concentrations of silicates, nitrates, phosphates, oxygen saturation (%), Alk/Cl, and temperature. Comparison of profiles indicated that the variability mentioned above also characterized the distribution of hydrochemical parameters and temperature, and was possibly determined by water mass structure typical for the western Bering Sea in summer (Ivanenkov 1964, Arsenev 1967). This water mass structure consisted of the upper quasi-uniform layer (UQL), which resulted from warming of surface water in summer, cold intermediate waters (CIW), formed during the autumn-winter

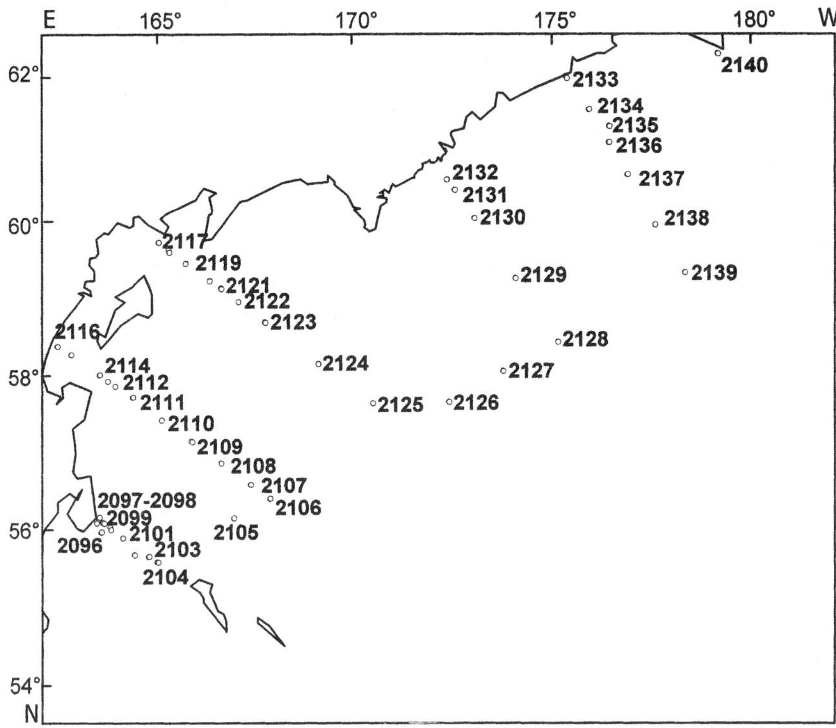

Figure 1.   Oceanographic stations occupied in the western Bering Sea in summer
1992 (R/V Akademik Alexander Nesmeyanov, cruise AN21, June 15-25).

convection, and warm intermediate waters (WIW), which were transferred
from the Pacific.

The vertical distribution of Alk/Cl (Fig. 2) revealed significant diversity which was especially apparent in upper layers. There was a tendency
for an increase of the alkalinity/chlorine coefficient with depth in the WIW.
Variations of Alk/Cl in the upper layers were apparently associated with
both the layer formation (significant river discharge and melting sea ice),
and with biological processes of carbonate-ion consumption.

In the seaward direction, the UQL showed a decrease in pH values,
whereas the saturation of CO_2 and specific inorganic carbon increased
(Figs. 3-5) because of the late timing of photosynthesis. The characteristic
curve of isolines for pH, CO_2 (%), and C_{tot}/Cl obtained at stations 2120-
2122, 2137 could have been caused by local anticyclonic eddies (Verkhunov
et al. 1995) (Fig. 6).

The spatial distribution of Alk/Cl in surface waters was characterized
by elevated values of Alk/Cl in the coastal zone northeastward off Cape

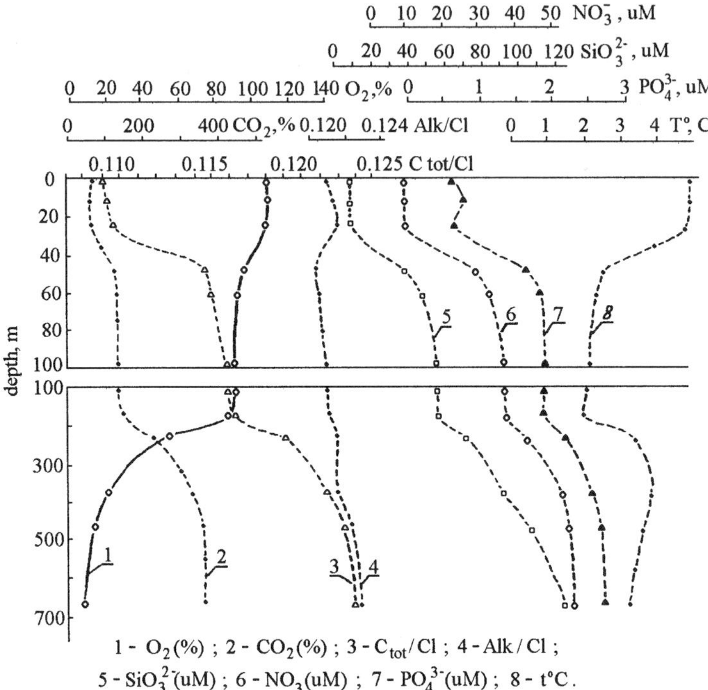

Figure 2.   Vertical profiles of water properties at station 2138.

Navarin (Figs. 7, 8) (stations 2133 and 2132) and may have been caused by the input of saline waters from Anadyr Bay. The coastal zone of southern Karagin Bay and up to Kamchatka Strait (stations 2112-2114, 2099) also showed an increase in the Alk/Cl ratio, likely the result of saline waters.

Despite these complex water dynamics, all stations in the surveyed basin reported vertical distribution of pH, CO$_2$ (%), and C$_{tot}$/Cl with similar regularities: highest gradients of these parameters occurred at the boundaries of the UQL/CIW and the CIW/WIW.

Three groups of stations could be defined by variability in pH, CO$_2$ (%), and C$_{tot}$/Cl in the UQL. Pelagic stations within the CO$_2$ isoline of 100% comprise the first group (Fig. 9). These stations were located in waters where photosynthesis had just started, surface waters were oversaturated with CO$_2$ and there was a CO$_2$ flux to the atmosphere. For the other two groups, surface water was undersaturated with CO$_2$ because of photosynthesis. By comparison with stations on the Kamchatka Bank, stations located to the northeast off the Shirshov Chain reported a greater uniformity of the upper layer and somewhat smaller gradients of hydrochemical parameters at the UQL-CIW boundary.

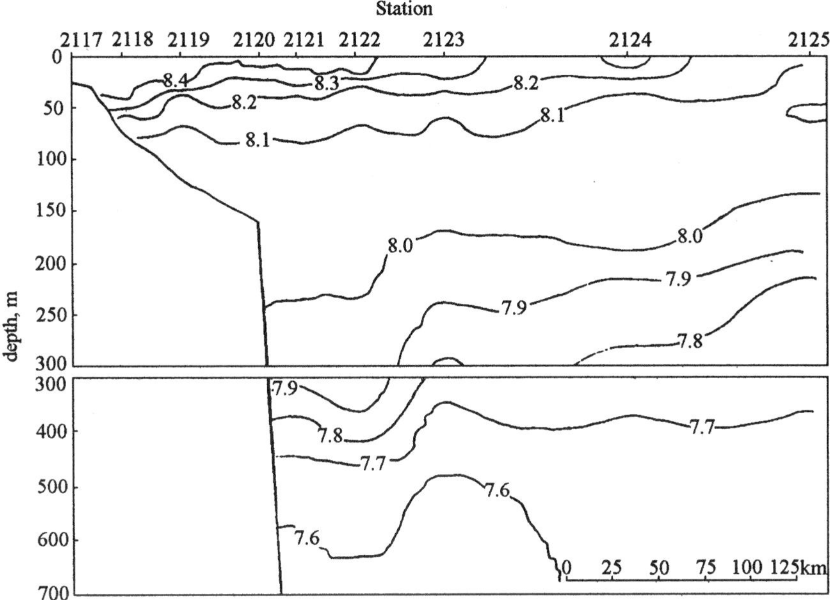

Figure 3.   Distribution of pH in situ for the section from station 2117 to station 2125.

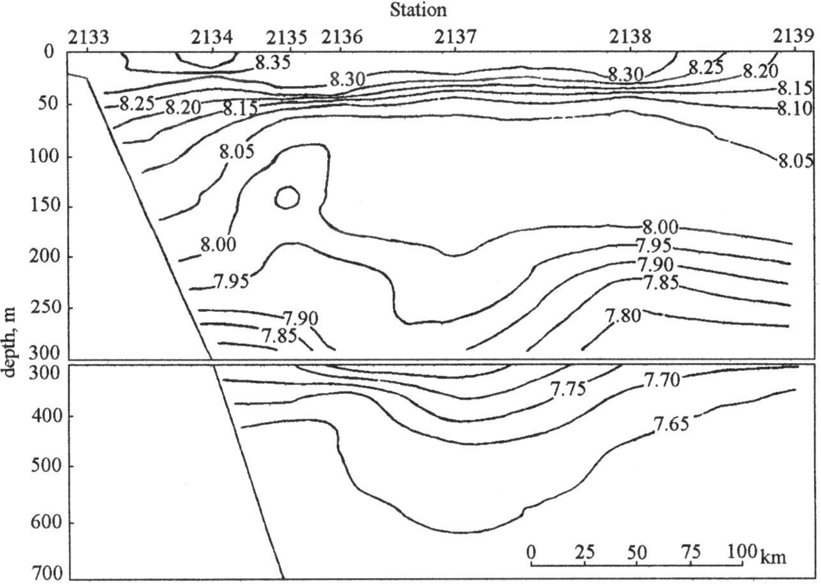

Figure 4.   Distribution of pH in situ for the section from station 2133 to station 2139.

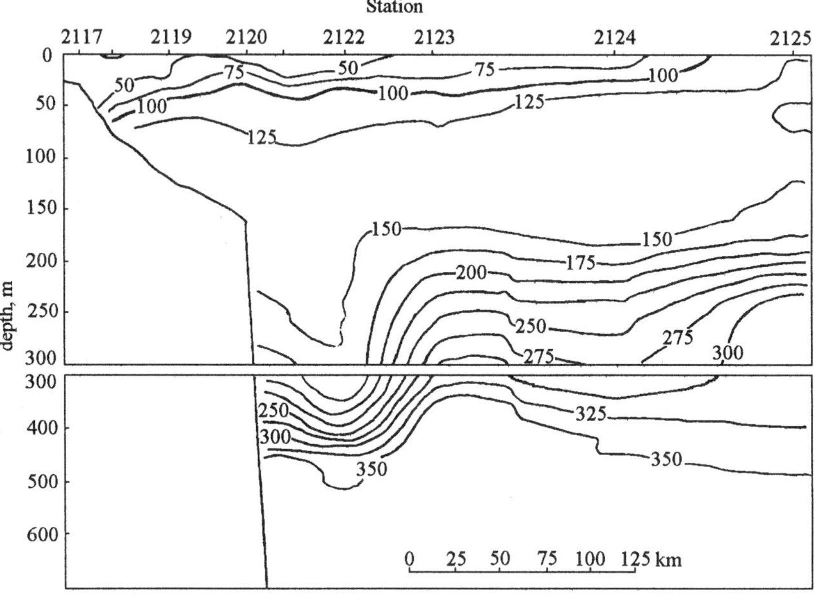

Figure 5.   Distribution of CO_2 (%) for the section from station 2117 to station 2125.

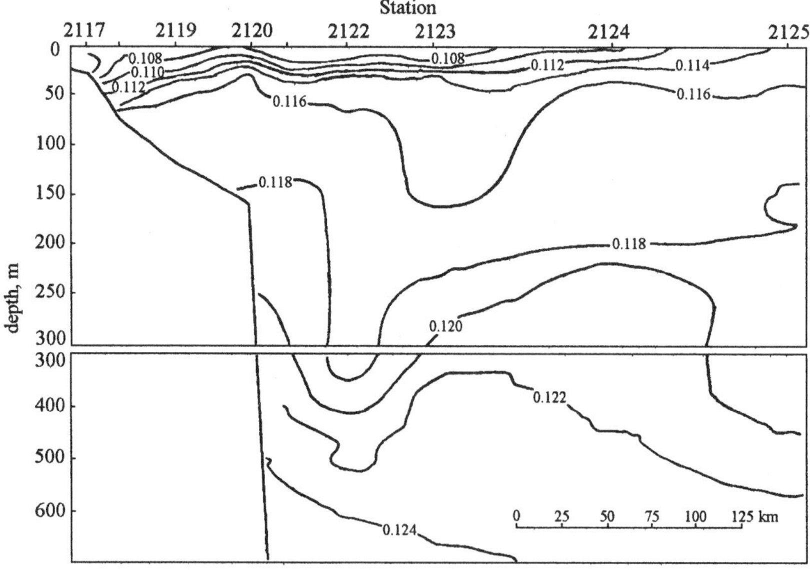

Figure 6.   Distribution of C_{tot}/Cl for the section from station 2117 to station 2125.

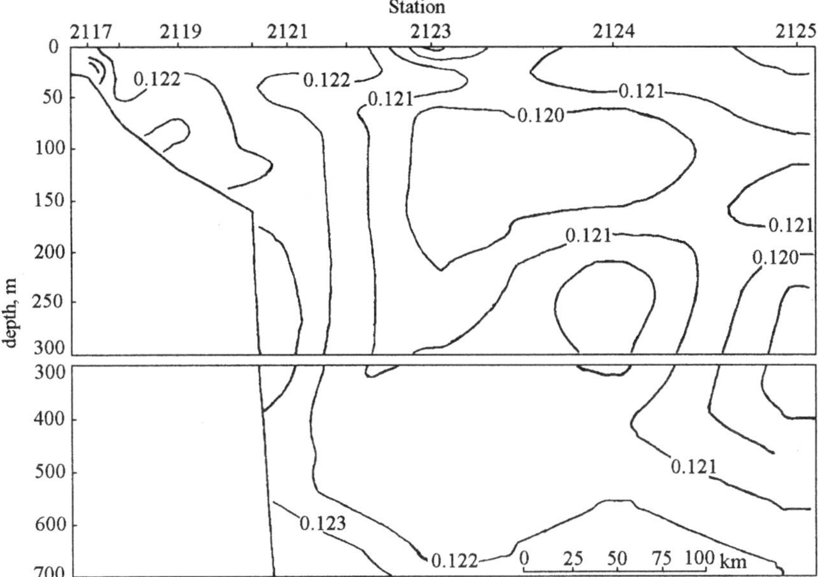

Figure 7.   *Distribution of Alk/Cl for the section from station 2117 to station 2125.*

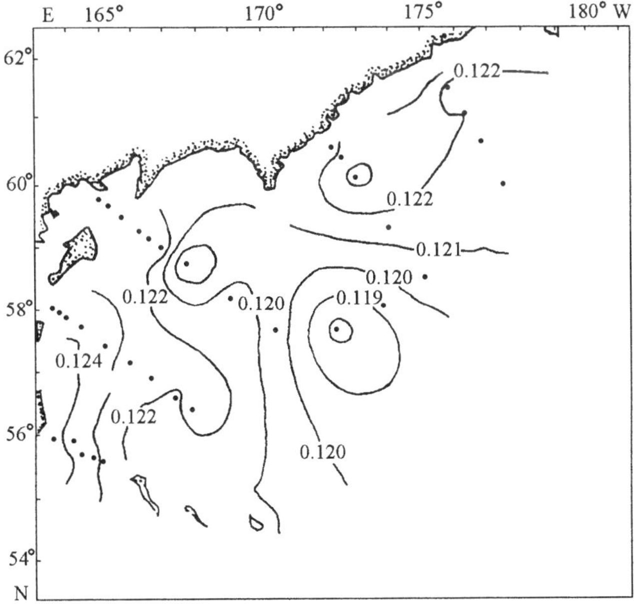

Figure 8.   *The distribution of specific alkalinity (Alk/Cl).*

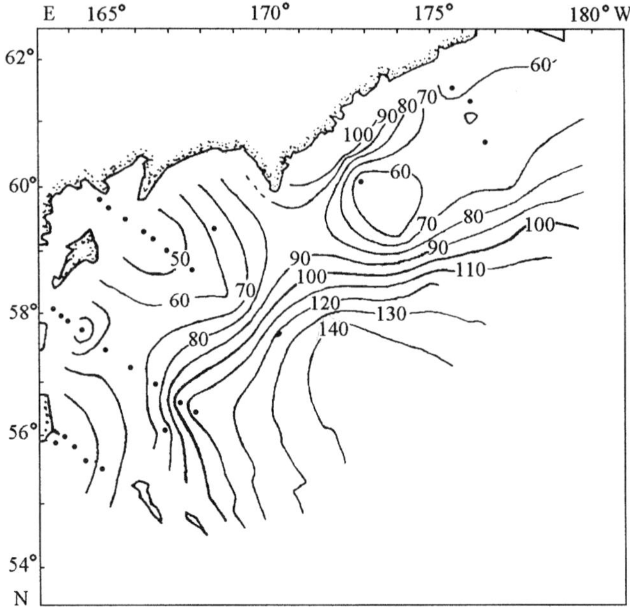

Figure 9.   The distribution of saturation degree for CO$_2$ (%).

To understand vertical distribution of the carbonate system components, the observed total inorganic carbon variations were compared with:
(1) variations of nutrients and oxygen deficit; (2) variations of C$_{tot}$ which
forecasted from the development of production/destruction processes;
and (3) variations of C$_{tot}$ related with changes of alkalinity and dissolution.
Except for a small number of stations influenced by coastal freshwater
discharge, the alkalinity variability in the surveyed area could be associated with the development/dissolution of CaCO$_3$ and the production/de-
struction of organic matter. Then, following Poisson and Chen (1987), the
C$_{tot}$ variations can be considered by applying function F:

$$F = C_{tot} - 0.5\left(\text{Alk} + 1.0625 \times 10^{-3}\,\text{NO}_{3^-}\right) \tag{1}$$

where C$_{tot}$ (mM) and Alk (mg-eq) are total inorganic carbon and total alkalinity adjusted to the mean chlorinity (18.27% for the UQL and 18.71% for
the CIW and WIW), expressed in mM and mg-eq, respectively; NO$_3$ equals
the concentration of NO$_3$ in µm. We use the symbol $\Delta C'$ to indicate variations of total inorganic carbon (except variations connected with the development of carbonates and chlorinity changes). It is then apparent that
$\Delta C' = \Delta F$.

Table 1. **The ratios of $\Delta O_2/\Delta C'$ and $\Delta C'/\Delta NO_3$ at the boundary between the CIW and the WIW, after correction on the formation/dissolution of calcium carbonate.**

Area	$\Delta O_2/\Delta C' \times 10^2$ (ml/μM)	$\Delta C'/\Delta NO_3$ (M/M)
Entire area	5.6 ± 0.6	6.3 ± 0.5
Kamchatka Bank	5.4 ± 0.5	6.7 ± 0.7
Aleutian Bank	5.9 ± 0.6	5.7 ± 0.8
Redfield ratio (Redfield et al. 1963)	2.92	6.63

For most stations a linear dependency was observed between $\Delta C'$, the oxygen deficit (ΔO_2), and variations of nitrate concentrations (ΔNO_3) at the boundary between the CIW and the WIW (Table 1). Results showed that the average $\Delta O_2/\Delta C'$ ratio significantly (1.9 times) exceeded the Redfield ratio (Redfield et al. 1963), while $\Delta C'/\Delta NO_3$ differed little from theoretical values.

It is thought that during the formation of the CIW, the CO_2 exchange between the ocean and atmosphere goes much slower than O_2 exchange. Consequently, variations of C_{tot}, resulting from oxidation of organic matter, were not influenced by gas exchange, and the ratio of $\Delta C'/\Delta NO_3$ was close to the theoretical value. A quicker oxygen exchange resulted in the CIW mean oxygen deficit of 0.62 ml/L versus 3.4 ml/L in the absence of gas exchange, consequently the $\Delta O_2/\Delta C'$ ratio significantly exceeded the Redfield ratio. Variability of total inorganic carbon at the CIW-WIW boundary was determined not only by the increase of its concentrations in the WIW due to oxidation of organic matter, but also by the intensity of CO_2 exchange between the ocean and atmosphere at the formation of the CIW.

To determine the major factors of variability of total inorganic carbon in the surface layer, we compared the variations of $\Delta C'$ (Q_c^{exp}) with variations of inorganic carbon, theoretically forecasted from production processes which continued since the initiation of photosynthesis (Q_c^{bio}). The values of Q_c^{exp} and Q_c^{bio} were calculated from the vertical distribution of specific inorganic carbon and silica by the following equations:

$$Q_c^{ex} = \int_0^{h_1} \left[F(h_1) - F \right] dh \tag{2}$$

$$Q_c^{bio} = \alpha \int_0^{h_1} \left[C_{Si}(h_1) - C_{Si} \right] dh \tag{3}$$

where F is the value in equation (1); C_{Si} is the dissolved silicon concentration; h_1 is the depth of the CIW upper boundary; and α is the atomic ratio

Table 2.   The variability of inorganic carbon in the surface layer from the end of the autumn-winter convection.

Area	pH in situ	CO_2 (%)	Q_c^{exp} (gC/m^2)	Q_c^{exp}/Q_c^{bio}	Q_c^{gas} (gC/m^2)
Entire area	8.35 ± 0.13	69 ± 25	50 ± 24	0.87 ± 0.50	7 (invasion)
Kamchatka Strait	8.54 ± 0.01	40 ± 1	82 ± 27	1.10 ± 0.30	– (evasion)
Kamchatka Bank	8.38 ± 0.08	61 ± 14	52 ± 22	0.85 ± 0.23	8 (invasion)
Aleutian Basin	8.35 ± 0.07	64 ± 11	54 ± 11	0.45 ± 0.09	30 (invasion)
Pelagic stations (2106, 2107, 2125, 2128, 2139)	8.12 ± 0.02	111 ± 9	22 ± 8	1.46 ± 0.74	10 (evasion)

$Q_c^{exp} = Q_c^{experimental}$; $Q_c^{bio} = Q_c^{biological}$

of carbon to silicon (4.61) in the natural populations of phytoplankton (Spravochnik 1991).

When values of Q_c^{exp} and Q_c^{bio} were calculated it was assumed that the levels of inorganic carbon and dissolved silicon in the photic layer before photosynthesis were equal to their levels in the CIW. We also assumed that calculations of Q_c^{bio} using the dissolved silicon data, were more reliable than values obtained on nitrogen and phosphorus, primarily because the dissolved silicon already consumed was removed more efficiently from the photosynthetic zone (Arzhanova 1982). Assuming that the difference between Q_c^{exp} and Q_c^{bio} was associated with CO_2 exchange across the air-sea surface, one could estimate the integral variation of C_{tot} in the UQL resulting from the processes of CO_2 invasion and evasion processes since the end of the autumn-winter convection (Q_c^{gas}) (Table 2). In Kamchatka Strait, the ratio of Q_c^{exp} and Q_c^{bio} exceeded unity in the surface layer, despite the low content of CO_2. These results could be incorrect estimations of theoretical variation of C_{tot}, which were made using the assumption that the initial content of nutrients in the UQL equaled that in the CIW. Possibly, melting sea ice and freshwater discharge, which contribute to the formation of the surface layer at these stations, introduce surplus nutrients. If so, Q_c^{bio} was underestimated. Actually, station 2099 contained 20 μM of SiO_3^{2-} in surface waters, while neighboring stations and underlying layers of this very station were undetectable. At pelagic stations where upwelling occurs, CO_2 evasion increases average variation of inorganic carbon by 46%. Variations of inorganic carbon in the UQL made up only 87% of the expected results of bioassimilation. Consequently, despite the observed summer oversaturation of surface waters with CO_2, the resulting effect was CO_2 consumption from the atmosphere totaling 7 gC/m^2 (average value for study area) for the entire period from the end of winter convection to the survey period. The influence of gas exchange on inorganic carbon variability was most prominent in the Aleutian Basin, likely because it was associated with more intensive wind mixing.

Conclusions

Analysis of hydrochemical data collected in the western Bering Sea in June 1992 indicates that maximum vertical variability of inorganic carbon occurred at the boundaries of the UQL-CIW and the CIW-WIW at all stations of the survey area. Variability of inorganic carbon at the boundary of the CIW-WIW was determined by the decomposition of organic matter in the WIW and the intensity of CO_2 exchange between the ocean and the atmosphere at the formation of the CIW during autumn-winter convection. We also found that variability of inorganic carbon in the surface layer was mainly determined by productivity and gas exchange with the atmosphere. Despite the initial evasion of CO_2 due to oversaturation of surface waters, the resulting process was CO_2 consumption from the atmosphere, which averaged 7 gC/m² in the waters surveyed from the end of convection in winter to the study period.

References

Arsenev, V.S. 1967. Currents and water masses of the Bering Sea. Nauka, Moscow. 133 pp. (In Russian.)

Arzhanova, N.V. 1982. Regeneration of nutrients at the bacterial decay of organic matter of dead plankton in the Atlantic. In: Uslovya sredi i bioproductivnost, Moscow, Lyogkaya i pishchevaya promyshleumost', pp. 7-15. (In Russian.)

Bordovskiy, O.K., and P.N. Makkaveev. 1991. The exchange of CO_2 with the atmosphere and the carbon balance in the Pacific. DAN SSSR 320(6):1470-1474. (In Russian.)

Ivanenkov, V.N. 1964. Hydrochemistry of the Bering Sea. Nauka, Moscow. 137 pp. (In Russian.)

Ivanenkov, V.N. 1985. The exchange of oxygen and carbon dioxide between the World Ocean and the atmosphere. In: O.K. Bordovskiy and V.N. Ivanenkov (eds.), Hydrochemical processes in the ocean. Moscow, IO AN SSSR. 121 pp. (In Russian.)

Lyman, J. 1956. Buffer mechanisms of seawater. Ph.D. dissertation, University of California, Los Angeles. 196 pp.

Mehrbach, C., C.H. Culberson, J.E. Hawley, and R.M. Pytkowicz. 1973. Measurements of the apparent dissociation constants of carbonic acid in seawater at atmospheric pressure. Limnology and Oceanography 18:897-907.

Millero, F.J. 1979. Chemical thermodynamics of the carbonate system in seawater. Geochimica Cosmochimica Acta 43:1651-1661.

Millero, F.J. 1983. Chemical oceanography, Vol. 8. Academic Press, New York, pp. 2-88.

Poisson, A., and C.-T.A. Chen. 1987. Why is there little anthropogenic CO_2 in the Antarctic bottom water? Deep-Sea Research 34:1255-1275.

Redfield, A.C., B.H. Ketchum, and F.A. Richards. 1963. The influence of organisms on the composition of seawater. In: M.N. Hill (ed.), The sea: Ideas and observations, Vol. 2. Interscience, New York, pp. 26-77.

Spravochnik. 1991. Reference book of hydrodynamics: Fisheries. Agropromizdat, Moscow. 224 pp. (In Russian.)

Verkhunov, A.V., A.V. Reed, Yu.Yu. Tkachenko, and V.V. Krukov 1995. Large-scale variability of the Bering Sea circulation. In: Complex studies of the Bering Sea ecosystem: Collected papers. VNIRO, Moscow, pp. 39-51. (In Russian.)

Weiss, R.F. 1974. Carbon dioxide in water and seawater: The solubility of nonideal gas. Marine Chemistry 2:203-215.

CHAPTER 15

Effect of Nutrients on Phytoplankton Size in the Bering Sea Basin

Akihiro Shiomoto
National Research Institute of Far Seas Fisheries, Shizuoka, Japan

Abstract

Surface nutrient [NO_2+NO_3, $Si(OH)_4$, PO_4] concentrations are generally higher in winter than in summer in the Bering Sea basin, whereas concentrations in both seasons are nearly equal at water depths greater than 200-300 m. Surface silicates are relatively low compared to other nutrient concentrations in summer, whereas a linear relationship is found between nitrite + nitrate and phosphate. Ammonium concentrations are 0.1-0.3 µM at the surface; a subsurface maximum of 1-3 µM is found in summer. Small-sized phytoplankton (<10 µm), especially less than 2 µm (picoplankton), dominate and contribute much to summer and winter production, but large-sized phytoplankton (>10 µm) sometimes contribute to biomass and production. The role of nutrients in controlling phytoplankton size structure in the Bering Sea basin is discussed. There is a possibility that ammonium inhibition of nitrate uptake by phytoplankton plays an important role in controlling phytoplankton size structure, at least in summer. Moreover, a steep pycnocline formed between 20 m and 40 m seems to be related to phytoplankton size structure in summer.

Introduction

The Bering Sea is one of the more eutrophic and high primary productive marginal seas (Taniguchi 1969, Taguchi 1972, McRoy and Goering 1974, Saino et al. 1979). Diatoms are known to be the dominant phytoplankton species in the Bering Sea basin (Motoda and Minoda 1974, Goering and Iverson 1981). One of the characteristics of a high contribution of diatoms in the phytoplankton community in surface waters of the Antarctic Ocean (e.g., Nelson et al. 1989, Treguer et al. 1990) is that molar ratios (0.3-0.65) of biogenic silica (BSi) to particulate organic carbon (POC) are several times higher than the typical cellular Si/C ratio for culture diatoms (mean ± 95% confidence limits; 0.13 ± 0.04) reported by Brzezinski (1985). The same result was observed in the Bering Sea basin in summer (BSi/POC = 0.3-0.6; Shiomoto and Ogura 1994). This implies that diatoms contribute to the high primary productivity there.

In the oceans, large-sized phytoplankton (i.e., >10 μm) generally are diatoms and dinoflagellates, and thus large-sized phytoplankton may be important primary producers in the Bering Sea basin. On the other hand, Odate (1996) showed a high contribution of small-sized phytoplankton, especially picoplankton (i.e., <2 μm) (Sieburth et al. 1978), in chlorophyll *a* (Chl *a*) concentrations in the Bering Sea basin during summer. The size structure of the phytoplankton community plays an important role in deciding and controlling members of the marine ecosystem (e.g., Parsons et al. 1984). Considering that little is known about the size structure of the phytoplankton community in the Bering Sea basin, more information is necessary to clarify the material and energy flow within the ecosystem.

The concentration and composition of nutrients, including ammonium, nitrate (nitrite + nitrate), silicate, and phosphate, have a significant influence on the size structure of the phytoplankton community (Parsons and Takahashi 1973, Malone 1980, Furuya and Marumo 1983, Maita and Odate 1988, Maita 1996). Large-sized phytoplankton dominate in eutrophic conditions and small-sized ones in oligotrophic conditions. Silicate/nitrate (Si/N) ratios of at least 1.7 (by atoms) are a necessary condition for the predominance of large-sized diatoms. Moreover, the frequency of addition of the limiting nutrient, as a sporadic or continuous supply (Turpin and Harrison 1980), the inhibition of nitrate uptake by ammonium (Wheeler and Kokkinakis 1990), and nitrate flux from deeper layers (Taguchi et al. 1992) affect phytoplankton size structure. A sporadic supply of nutrients, the lock on ammonium inhibition, and a large nitrate flux are advantageous to the predominance of large-sized phytoplankton, whereas a continuous supply of nutrients, ammonium inhibition, and a small nitrate flux are advantageous to small-sized phytoplankton.

In this chapter, additional data on size-fractionated phytoplankton biomass and production are shown, and the role of nutrients in controlling phytoplankton size structure in the Bering Sea basin is discussed. Moreover, the influence of other factors (e.g., iron, water column conditions, and light intensity) are briefly mentioned.

Data Set

Summertime data on size-fractionated (i.e., picoplankton [<2], medium-sized plankton [2-10], and large-sized plankton [>10 μm] fractions) Chl *a* concentration and phytoplankton production were measured at the surface during the R/V *Wakatake Maru* cruises in July 1992 and 1993, and wintertime data were measured during the R/V *Kaiyo Maru* cruise in January to March 1993. Chl *a* concentrations were measured from the surface to 100 m during the R/V *Wakatake Maru* cruises in July 1993 and 1994, and during the R/V *Kaiyo Maru* cruise in January to March 1993. The detailed methods for measuring size-fractionated Chl *a* concentration and phytoplankton production are described in Shiomoto et al. (1997). Daily primary production rates in summer were cited from the summary of Saino et al. (1979) and the rates in winter were cited from McRoy et al. (1972).

Nutrient concentrations were measured by using a Bran and Luebbe Traacs 800 during the R/V *Wakatake Maru* cruises and by using a Bran and Luebbe Auto Analyzer II during the R/V *Kaiyo Maru* cruise. Further data on the concentrations were cited from the reports during the R/V *Hakuho Maru* cruises in July 1975 and 1978 (Hattori 1977, 1979) and the R/V *Oshoro Maru* cruise in July 1993 (Hokkaido University 1994). The concentrations were measured manually during the *Hakuho Maru* cruise in 1975, and by using the Auto Analyzer II during other two cruises.

Distribution and Behavior of Nutrients in the Bering Sea Basin

During the summer, surface silicate concentrations range from 3 to 45 µM, nitrite + nitrate concentrations from 5 to 22 µM, and phosphate concentrations from 0.9 to 2.1 µM. Surface winter silicate concentrations range from 35 to 60 µM, nitrite + nitrate concentrations from 19 to 30 µM, and phosphate concentrations from 1.7 to 2.4 µM (Fig. 1). The surface nutrient concentrations are generally higher in winter than in summer.

A Si/N ratio of more than ~1.7 (by atoms) can be one of the nutrient conditions necessary for a predominance of large-sized diatoms (Maita 1996). Si/N ratios less than 1.7 are frequently observed in summer (Fig. 1). In contrast, a linear relationship is found between the nitrite + nitrate concentration and the phosphate concentration (Fig. 1). These results imply that, compared to other nutrients, more silicate is taken up preferentially by large-sized diatoms. As a result, it may be that silicate frequently limits the growth of large-sized diatoms in summer.

An intermediate cold layer is observed at around 50-200 m in summer and the depth of the surface-mixed layer that is defined by the pycnocline is found at around 30 m (Fig. 2). Nutrient concentrations were nearly uniform in the surface-mixed layer and increased with depth below the surface-mixed layer in summer. In winter, an intermediate cold layer is not observed and the surface-mixed layer reaches down to around 120 m. Nutrient concentrations are almost uniform within the surface-mixed layer. Nutrient concentrations in summer and winter are similarly deeper than 200-300 m: silicate concentrations are more than 50 µM, nitrite + nitrate concentrations are more than 20 M, and phosphate concentrations are more than 2 µM.

There are few ammonium data, compared with the other nutrients. Summertime ammonium concentrations are generally less than 1 µM in the surface layer and maximum values are observed in the subsurface layer (30-100 m), which corresponds to the intermediate cold layer (Fig. 3). The subsurface ammonium concentrations are about 1-3 µM. Wintertime

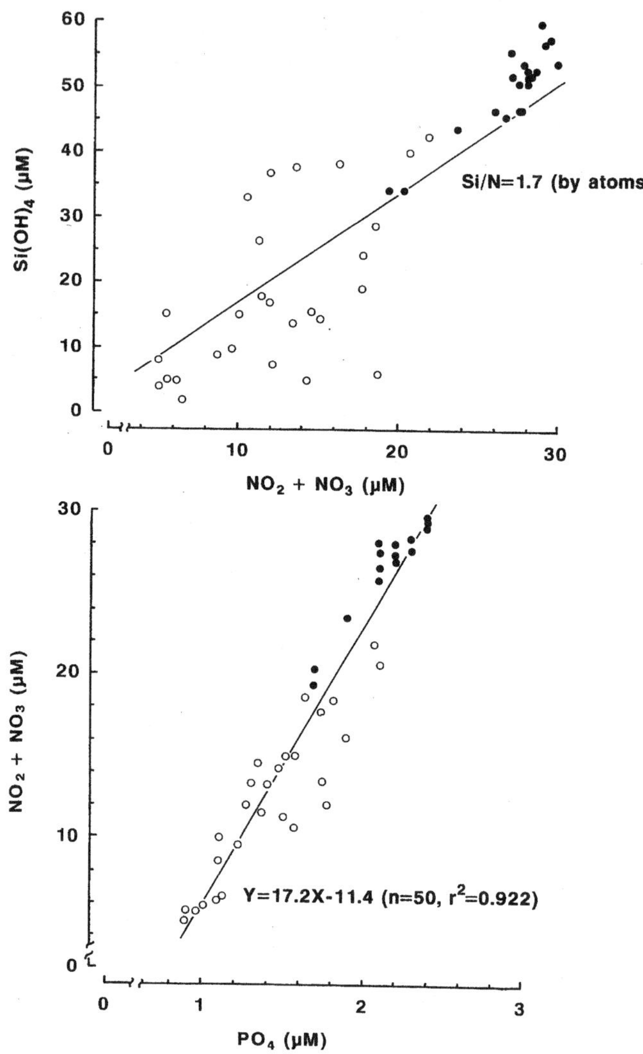

Figure 1.   *(Above) The relationship between silicate [Si(OH)₄] and nitrite + nitrate (NO₂ + NO₃) concentrations at the surface, and (below) the relationship between nitrite + nitrate and phosphate (PO₄) concentrations at the surface. Open and solid circles indicate summer and winter data, respectively. The summer data were obtained during the R/V Wakatake Maru cruises in July 1992 and 1993, and the T/V Oshoro Maru cruise in July 1993 (Hokkaido University 1994); winter data were obtained during the R/V Kaiyo Maru cruise in January-March 1993. The line in the upper panel indicates Si/N = 1.7 (by atoms). The line in the lower panel indicates the regression line obtained by the least-squares method.*

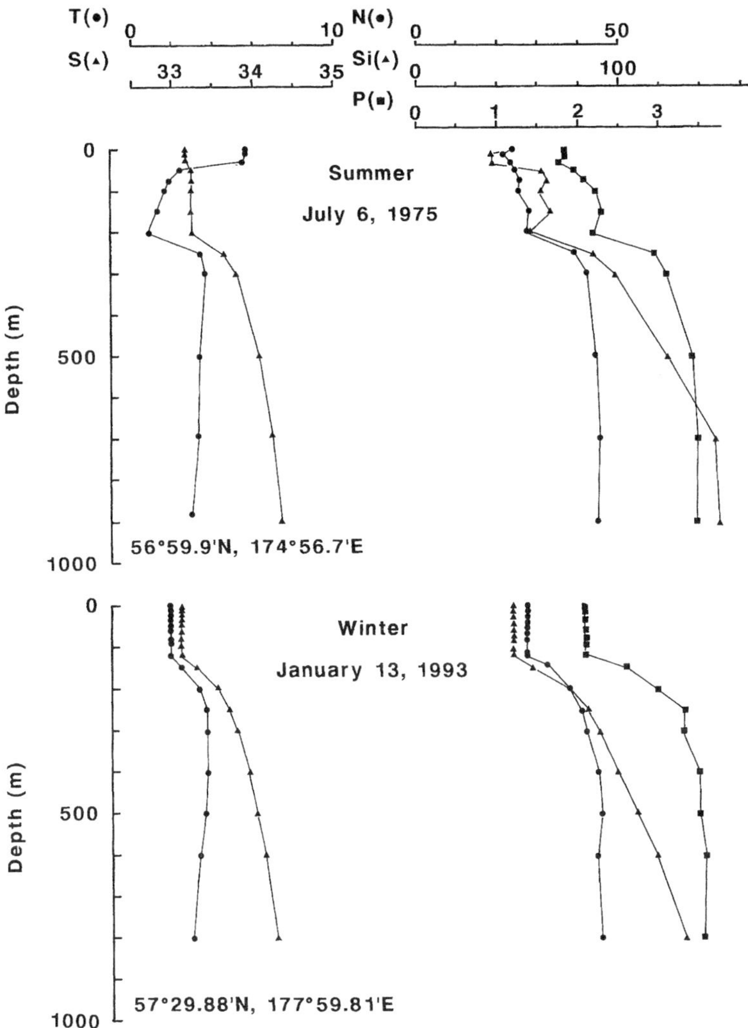

Figure 2. *Vertical profiles of temperature (T; °C), salinity (S), and concentrations (μM) of NO_2 + NO_3 (N), $Si(OH)_4$ (Si), and PO_4 (P) in the Bering Sea basin in summer and winter. The profiles in summer were obtained during the R/V Hakuho Maru cruise in July 1975 (Hattori 1977) and the profiles in winter were obtained during the R/V Kaiyo Maru cruise in January-March 1993.*

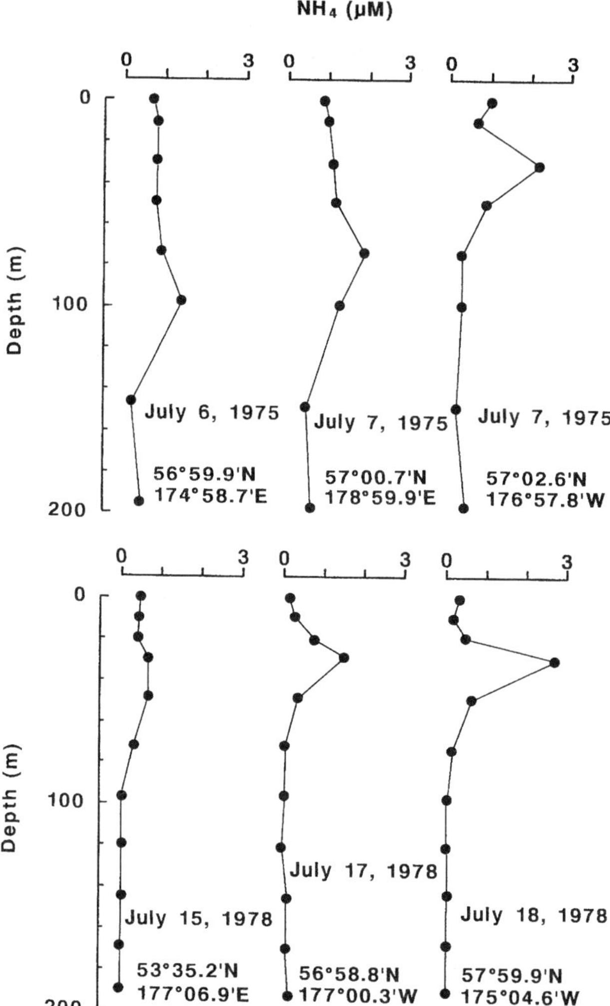

Figure 3.   Vertical profiles of ammonium concentrations (NH₄) in the Bering Sea basin in summer. The profiles above were obtained during the R/V Hakuho Maru cruise in July 1975 (Hattori 1977) and the profiles below were obtained during the R/V Hakuho Maru cruise in July 1978 (Hattori 1979).

ammonium concentrations were reported to be 0.3-3 µM at the surface on the Bering Sea shelf and slope (McRoy et al. 1972, Whitledge et al. 1986) and have not been reported in the basin.

Size-Fractionated Phytoplankton Biomass (Chl a Concentration) and Production in the Bering Sea Basin

Mean surface Chl *a* concentrations are significantly higher in summer than in winter (*U*-test, $P < 0.05$); the mean value (0.66 µg per L) in summer is twice as large as the value (0.33 µg per L) in winter (Fig. 4a). The same result is found for Chl *a* standing stock integrated above 100 m (mean stock is 47 mg per m^2 in summer and 27 mg per m^2 in winter) (Fig. 4b). The total phytoplankton production at the surface is significantly higher in summer than in winter (*U*-test, $P < 0.05$); the mean value (1.46 µgC per L per h) in summer is about three times as large as the value (0.57 µgC per L per h) in winter (Fig. 4c). Daily primary production is roughly one order of magnitude larger in summer (range = 34-630 mgC per m^2 per day; mean = 334 mgC per m^2 per day) than in winter (range = 7-71 mgC per m^2 per day; mean = 21 mgC per m^2 per day) (Fig. 4d).

The picoplankton fraction generally contributes substantially to the total biomass in summer (the fraction accounts for about 50% of the total) (Fig. 5a). The medium-sized and large-sized phytoplankton fractions account for about 25% of the total. In winter, the picoplankton fraction also generally contributes much to the total biomass (the fraction accounts for about 55% of the total) (Fig. 5a). The large-sized fraction accounts for about 30% of the total and the medium-sized fraction accounts for 15% of the total on average. The contribution to the total biomass of the large-sized fraction is sometimes nearly equal to or more than the contribution of the picoplankton fraction.

The picoplankton fraction generally contributes much to the total production in summer (the fraction accounts for about 45% of the total) (Fig. 5b). The medium-sized and large-sized fractions account for about 30% of the total. In winter, however, the three fractions contribute about equally to total production, accounting for 30% of the total on average (Fig. 5b). The same pattern as with the Chl *a* size structure was found in summer but not in winter. The contribution to production of the large-sized fraction is sometimes nearly equal to or more than the contribution of the picoplankton fraction.

Influence of Nutrients on Phytoplankton Size Structure in the Bering Sea Basin

Nutrient concentrations are generally abundant in the Bering Sea basin in summer and winter (Fig. 1). Nevertheless, it is evident that picoplankton are generally dominant in the phytoplankton community and contribute the most to primary production in the Bering Sea basin in these seasons

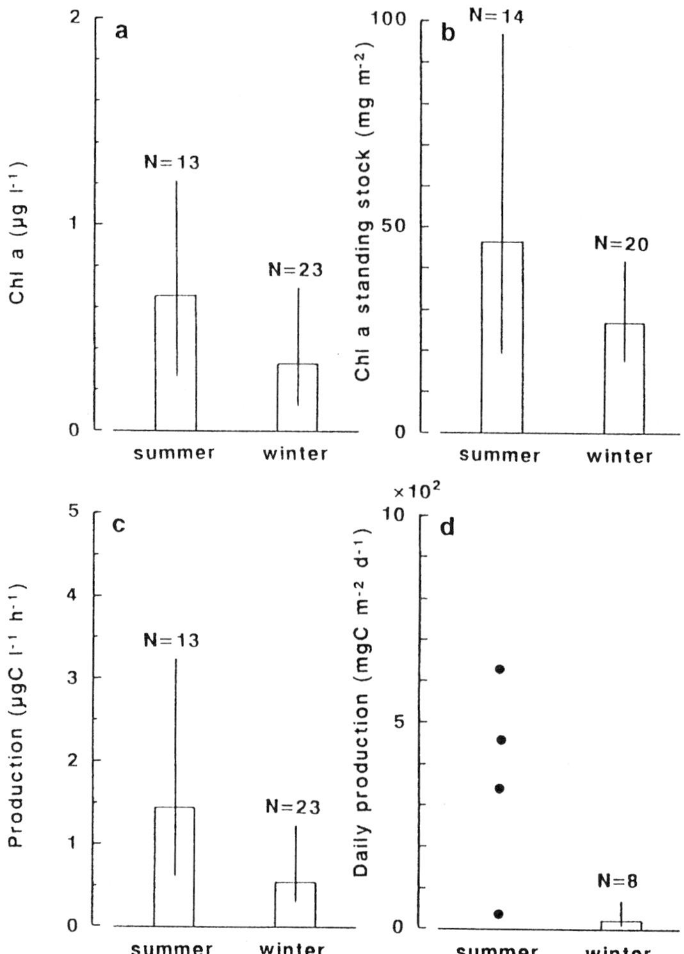

Figure 4.   Chlorophyll a and production measurements in the Bering Sea basin in summer and winter: (a) Mean surface chlorophyll a concentration; (b) chlorophyll a standing stock above 100 m; (c) surface production; and (d) daily production. Chlorophyll a standing stock was calculated to be integrated from the surface down to 100 m using the chlorophyll a concentrations obtained at 10-25 m intervals. The daily production rates in winter were obtained on the shelf and slope. Bars indicate the range between maximum and minimum values. N indicates the number of data. Summer data were obtained during the R/V Wakatake Maru cruises in July 1992 and 1993; winter data were obtained during the R/V Kaiyo Maru cruise in January-March 1993. Daily production data in the summer were cited from the summary of Saino et al. (1979), and the data in the winter from McRoy et al. (1972).

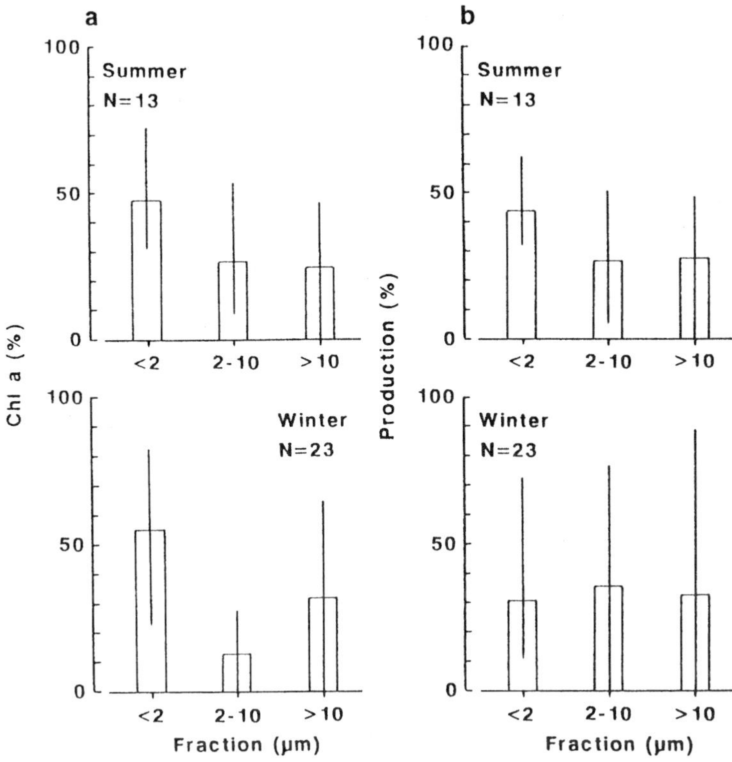

Figure 5. *Surface measurements in the Bering Sea in summer and winter: (a) Mean percentage contributions of the <2, 2-10, and >10 µm fractions to the total chlorophyll a concentration, and (b) total production. Bars indicate the range between maximum and minimum values. N indicates the number of data. Summer data were obtained during the R/V Wakatake Maru cruises in July 1992 and 1993; winter data were obtained during the R/V Kaiyo Maru cruise in January-March 1993.*

(Fig. 5). Silicate concentrations are sometimes relatively low compared to the needs of large-sized diatoms (Fig. 1a), and it was thus proposed that silicate may be frequently limiting for large-sized diatom growth in summer. Hence, the influence of silicate on large-sized phytoplankton biomass and production is examined. As a result, Chl *a* concentration and production of large-sized phytoplankton are not significantly related to silicate concentrations (Fig. 6), and furthermore, no significant relationship is found between the biomass and production of the fraction and the Si/N ratio (Fig. 7). These results show that silicate probably does not take part in controlling the change of large-sized phytoplankton biomass and production in summer. Diatoms are major members of the large-sized phytoplankton that require silicon for their growth in the Bering Sea. Silicate is thus unlikely to be a limiting factor in the growth of large-sized diatoms in the Bering Sea basin in summer. The nutrients $NO_2 + NO_3$, $Si(OH)_4$, and PO_4 are unlikely to be related to the general predominance of picoplankton in the Bering Sea basin.

During summer in the Bering Sea basin, it is assumed that nutrients are supplied continuously by diffusion to the euphotic zone from deeper layers, because the water column is well stratified in the surface layers (Fig. 8). A continuous supply of nutrients is advantageous to the predominance of small-sized phytoplankton (Turpin and Harrison 1980). Surface nutrients, however, are probably not limiting for large-sized phytoplankton growth even in summer. Small-sized phytoplankton have a smaller half-saturation constant for nutrient uptake than do large-sized ones (e.g., Malone 1980). It is believed that a half-saturation constant may determine the minimum nutrient concentration at which a species can grow (e.g., Parsons et al. 1984). Surface nutrients are also probably not limiting for the small-sized phytoplankton (<10 μm) growth in summer. The continuous supply of nutrients is unlikely related to the general predominance of picoplankton in the Bering Sea basin.

Nitrate and ammonium are important nitrogen sources for phytoplankton. It is well known that nitrate uptake may be inhibited at ammonium concentrations of 0.5-1.0 μM (McCarthy 1980). Wheeler and Kokkinakis (1990) further observed that concentrations of ammonium between 0.1-0.3 μM caused complete inhibition of nitrate uptake in the oceanic subarctic North Pacific. Ammonium concentrations exceeding 0.1-0.3 μM are observed frequently at the surface in the basin in summer (Fig. 3), and on the shelf and slope in winter (McRoy et al. 1972, Whitledge et al. 1986). It is therefore possible that ammonium inhibition of nitrate uptake is frequent and ammonium is generally used by phytoplankton. Small-sized phytoplankton, nano- and picoplankton size classes (<10 or 20 μm), prefer ammonium and use it more efficiently than do netplankton (>10 or 20 μm) (Rönner et al. 1983, Probyn and Painting 1985, Koike et al. 1986). As a result, the nutrient environment seems to be advantageous to small-sized phytoplankton in the Bering Sea basin in summer and on the shelf and slope in winter.

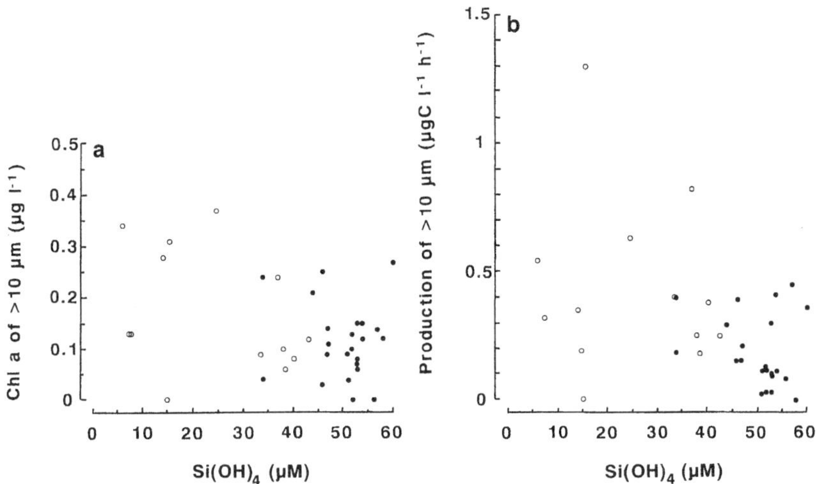

Figure 6. (a) The relationship between the chlorophyll a concentration of the >10 μm fraction and silicate concentrations at the surface, and (b) the relationship between the production of the >10 μm fraction and silicate concentrations at the surface, in the Bering Sea basin. Open and solid circles indicate summer and winter data, respectively. Summer data were obtained during the R/V Wakatake Maru cruises in July 1992 and 1993; winter data were obtained during the R/V Kaiyo Maru cruise in January-March 1993.

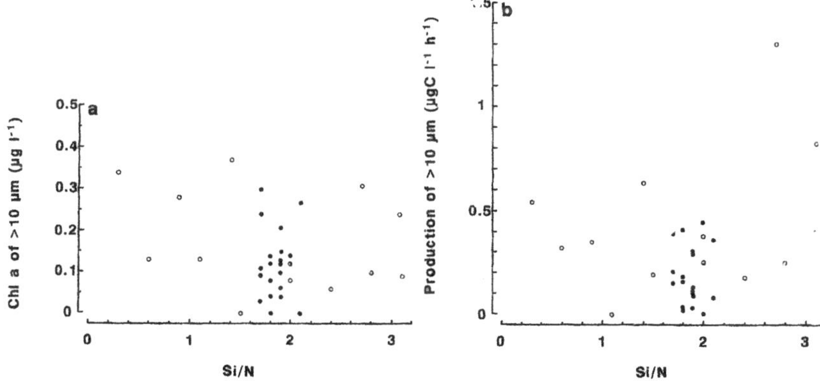

Figure 7. (a) The relationship between the chlorophyll a concentration of the >10 μm fraction and the Si/N ratio at the surface, and (b) the relationship between the production of the >10 μm fraction and Si/N ratio at the surface, in the Bering Sea basin. Open and solid circles indicate summer and winter data, respectively. Summer data were obtained during the R/V Wakatake Maru cruises in July 1992 and 1993; winter data were obtained during the R/V Kaiyo Maru cruise in January-March 1993.

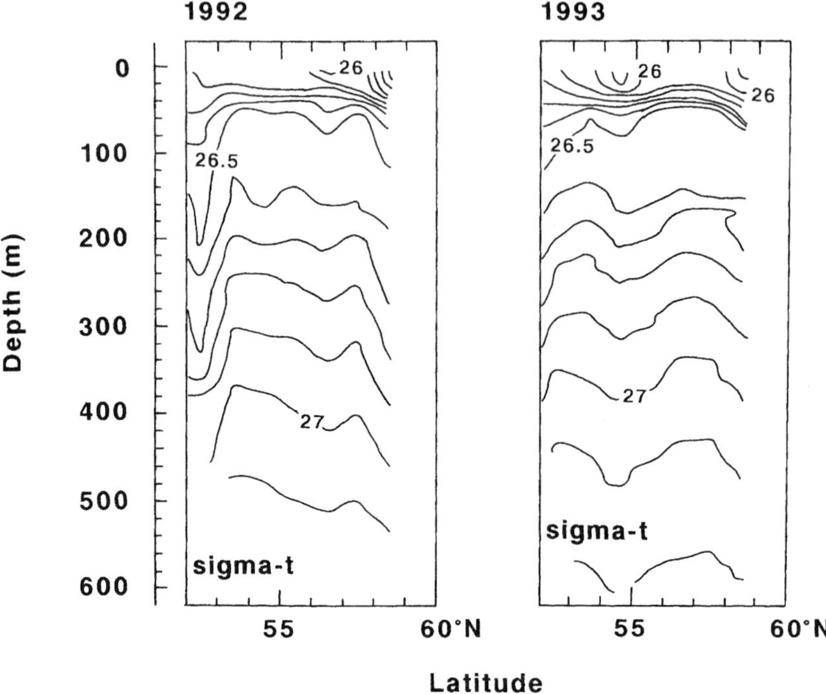

Figure 8.   Vertical sections of sigma-t along 179°30'W in July 1992 and 1993. The data were obtained during the R/V Wakatake Maru cruises.

In the Bering Sea basin in summer, the nitracline is mostly observed between 30 m and 50 m (Hattori 1979; Fig. 2). The depths of the euphotic zone (Shiomoto, unpublished results) and the surface-mixed layer (Fig. 8) are observed at around 40 m. The nitracline is located near the bottom of the euphotic zone and the surface-mixed layer. Therefore, a large flux of nitrate to the euphotic zone is expected in the basin even in summer. This large flux may affect the phytoplankton size structure in the surface layer (Taguchi et al. 1992). The upward flux of nitrate through the seasonal thermocline (30-35 m) was estimated to be 12 mmol per m^2 per day in the subarctic region in summer (Saino et al. 1983). The standing stock of nitrate in the surface layer shallower than 40 m (euphotic zone and surface-mixed layer) is calculated to be 560 mmol per m^2 if the mean nitrate concentration at the surface in summer is taken as 14 µM (Fig. 1). The daily vertical flux accounts for only 2% of the standing stock in the surface layer. From these considerations, it is difficult to think that the size structure of the phytoplankton community is seriously affected by the flux of nitrate from the deeper layer in summer, because of the effect of the high ambient concentrations of nutrient in the surface layer.

Influence of Other Factors on Phytoplankton Size Structure in the Bering Sea Basin

Miller et al. (1991) stated that iron limitation establishes a phytoplankton community dominated by small-sized phytoplankton in the subarctic North Pacific. Iron must be supplied primarily by the atmosphere in the ocean (Duce 1986). The eolian supply was estimated to be of the same order of magnitude throughout the Bering Sea (Donaghay et al. 1991). Iron, moreover, also may be supplied from islands. It was suggested, for example, that the total phytoplankton productivity index (production per unit of Chl *a*) increases near islands in the equatorial region because of the effect of iron supply from islands (Barber and Chavez 1991). If iron were a limiting factor for the phytoplankton community in the Bering Sea basin, the proportion of small-sized phytoplankton should increase with the distance from the Aleutian Islands, and total productivity should decrease. An increasing trend, however, is not found in the contribution of the <2 µm fraction in the Chl *a*, and neither an increase nor a decrease is found in the productivity index (Fig. 9). Iron may thus fail to be a limiting factor in the Bering Sea basin.

The location of the Bering Sea basin stations corresponds to the Ridge Domain where upwelling is observed (Favorite et al. 1976). Upwelling and seasonal thermocline and halocline (Fig. 2) probably form the great density gradient, that is, the steep pycnocline, between 20 m and 40 m in summer (Fig. 8). The steep pycnocline slows the sinking rates of large-sized phytoplankton. Consequently, large-sized phytoplankton possibly reside in the surface layer for a relatively long time, and plankton have a chance to flourish accordingly. I therefore speculate that the physical conditions of the water column are advantageous to large-sized phytoplankton survival in the surface layer in the Bering Sea basin, at least in summer. The result that the contribution of the large-sized phytoplankton (>10 µm) fraction to the total biomass is sometimes nearly equal to or more than that of the picoplankton (<2 µm) fraction (Fig. 5a) supports my speculation.

It may safely be said that the light intensity is regarded to be about the same in the Bering Sea basin. Nevertheless, large-sized phytoplankton sometimes contribute to the biomass and production (Fig. 5). Light intensity is thus unlikely to be related to the size structure of phytoplankton in the Bering Sea basin.

Summary

Picoplankton are usually dominant in the phytoplankton community in the Bering Sea basin in summer and winter, and these plankton are the

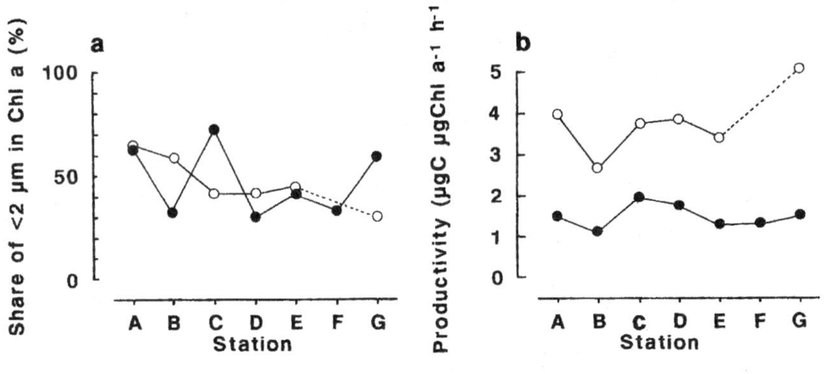

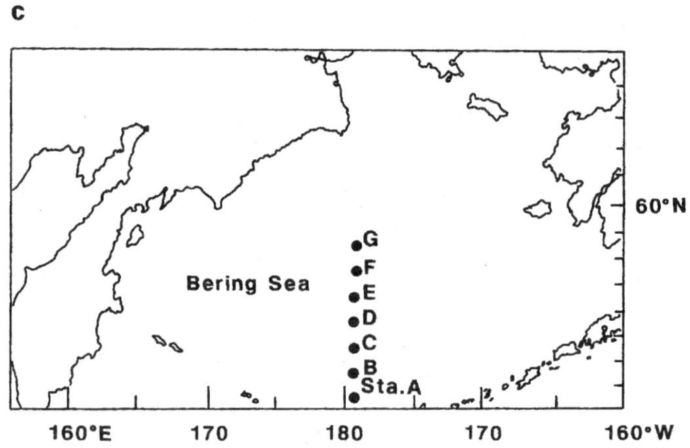

Figure 9. (a) Surface variations in the percentage of the <2 μm fraction in the chlorophyll a concentration, (b) phytoplankton productivity (production per unit of chlorophyll a: μgC μg per Chl a per h) of the total, and (c) the location of stations in the Bering Sea basin in summer. The data were obtained during the R/V Wakatake Maru cruises in July 1992 (open circles) and 1993 (solid circles). Spearman rank correlation coefficient (r_s) between the share of the <2 μm fraction in the chlorophyll a concentration and the location of the station (distance from the Aleutian Islands) was calculated as –0.83 in 1992 (n = 6) and –0.83 in 1993 (n = 7). The value in 1992 was not significant at P < 0.05 (two-tailed test) and the value in 1993 was significant at P < 0.05 (two-tailed test). r_s between the phytoplankton productivity of the total and the station was 0.23 in 1992 (n = 6) and 0.49 in 1993 (n = 7). These values were not significant at P < 0.05 (two-tailed test).

most important primary producers. Sometimes large-sized phytoplankton dominate and contribute substantially to primary production. Nitrite + nitrate, silicate, and phosphate are unlikely to be limiting factors for the phytoplankton community nor do they play an important role in the predominance of picoplankton. It is highly possible that ammonium inhibition of the nitrate uptake plays a part in the predominance of picoplankton, at least in summer. There is an urgent need to verify experimentally the role of the inhibition in controlling the phytoplankton size structure in the Bering Sea basin. Moreover, iron and light may not be related to the predominance of picoplankton in the phytoplankton community. The steep pycnocline formed between 20 m and 40 m seems to be related to the occasional predominance of large-sized phytoplankton in the Bering Sea basin in summer.

Acknowledgments

I thank the captains and crews of R/V *Wakatake Maru* and *Kaiyo Maru*, as well as Drs. Y. Ishida, K. Nagasawa, and K. Tadokoro of the National Research Institute of Far Seas Fisheries, for their kind help in collecting samples. I am grateful to Professor Y. Maita of Hokkaido University for his critical reading of and comments on my manuscript. I also thank Drs. T.E. Whitledge and L.A. Codispoti for comments that improved the manuscript.

References

Barber, R.T., and F.P. Chavez. 1991. Regulation of primary productivity rate in the equatorial Pacific. Limnology and Oceanography 36:1803-1815.

Brzezinski, M.A. 1985. The Si/N ratio of marine diatoms: Inter-specific variability and the effect of some environmental variables. Journal of Phycology 21:345-357.

Donaghay, P.L., P.S. Liss, R.A. Duce, D.R. Kester, A.K. Hanson, T. Villareal, N.W. Tindale, and D.J. Gifford. 1991. The role of episodic atmospheric nutrient inputs in the chemical and biological dynamics of oceanic ecosystems. Oceanography 4:62-70.

Duce, R.A. 1986. The impact of atmospheric nitrogen, phosphorus, and iron species on marine biological productivity. In: P. Buat-Menard (ed.), The role of air-sea exchange in geochemical cycling. D. Reidel Publishing Company, Dordrecht, pp. 497-529.

Favorite, F., A.J. Dodimead, and K. Nasu. 1976. Oceanography of the subarctic Pacific region, 1960-1971. Bulletin of International North Pacific Fisheries Commission 33:1-187.

Furuya, K., and R. Marumo. 1983. Size distribution of phytoplankton in the western Pacific Ocean and adjacent waters in summer. Bulletin of Plankton Society of Japan 30:21-32.

Goering, J.J., and R.L. Iverson. 1981. Phytoplankton distribution in the southeastern Bering Sea shelf. In: D.W. Hood and J.A. Calder (eds.), The eastern Bering Sea shelf: Oceanography and resources, Vol. 2. Published by the Office of Marine Pollution Assessment, National Oceanic and Atmospheric Administration and Bureau of Land Management. Distributed by the University of Washington Press, Seattle, WA 98105, pp. 933-946.

Hattori, A. (ed.) 1977. Preliminary report of the *Hakuho Maru* cruise KH-75-4. Ocean Research Institute, University of Tokyo, Tokyo. 87 pp.

Hattori, A. (ed.) 1979. Preliminary report of the Hakuho Maru cruise KH-78-3. Ocean Research Institute, University of Tokyo, Tokyo. 87 pp.

Hokkaido University. 1994. Data record of oceanographic observations and exploratory fishing No. 37. Hakodata, Hokkaido, Japan. 374 pp.

Koike, I., O. Holm-Hansen, and D.C. Biggs. 1986. Inorganic nitrogen metabolism by Antarctic phytoplankton with special reference to ammonium cycling. Marine Ecology Progress Series 30:105-116.

Maita, Y. 1996. Characteristics on the distribution and composition of nutrients in the Okhotsk Sea. In: Proceedings of the international workshop on the Okhotsk Sea and Arctic; the physics and biogeochemistry implied to the global cycles. Influence of sea ice on climate and marine ecosystems. JAMSTEC, Science and Technology Agency of Japan, pp. 140-148.

Maita, Y., and T. Odate. 1988. Seasonal changes in size-fractionated primary production and nutrient concentrations in the temperate neritic water of Funka Bay, Japan. Journal of Oceanographical Society of Japan 44:268-279.

Malone, T.C. 1980. Algal size. In: I. Morris (ed.), The physiological ecology of phytoplankton. Blackwell Scientific Publications, London, pp. 433-464.

McCarthy, J.J. 1980. Nitrogen. In: I. Morris (ed.), The physiological ecology of phytoplankton. Blackwell Scientific Publications, London, pp. 191-233.

McRoy, C.P., and J.J. Goering. 1974. The influence of ice on the primary productivity of the Bering Sea. In: D.W. Hood and E.J. Kelley (eds.), Oceanography of the Bering Sea with emphasis on renewable resources. Occasional Publication No. 2, Institute of Marine Science, University of Alaska, Fairbanks, pp. 403-421.

McRoy, C.P., J.J. Goering, and W.E. Shiels. 1972. Studies of primary production in the eastern Bering Sea. In: A.Y. Takenouti et al. (eds.), Biological oceanography of the northern North Pacific Ocean. Idemitsu Shoten, Tokyo, pp. 199-216.

Miller, C.B., B.W. Frost, P.A. Wheeler, M.R. Landry, N. Welschmeyer, and T.M. Powell. 1991. Ecological dynamics in the subarctic Pacific, a possibly iron-limited ecosystem. Limnology and Oceanography 36:1600-1615.

Motoda, S., and T. Minoda. 1974. Plankton of the Bering Sea. In: D.W. Hood and E.J. Kelley (eds.), Oceanography of the Bering Sea with emphasis on renewable resources. Occasional Publication No. 2, Institute of Marine Science, University of Alaska, Fairbanks, pp. 207-241.

Nelson, D.M., W.O. Smith Jr., R.D. Muench, L.I. Gordon, C.W. Sullivan, and D.W. Husby. 1989. Particulate matter and nutrient distributions in the ice-edge zone of the Weddell Sea: Relationship to hydrography during late summer. Deep-Sea Research 36:191-209.

Odate, T. 1996. Abundance and size composition of the summer phytoplankton communities in the western North Pacific Ocean, the Bering Sea, and the Gulf of Alaska. Journal of Oceanography 52:335-351.

Parsons, T.R., and M. Takahashi. 1973. Environmental control of phytoplankton cell size. Limnology and Oceanography 18:511-515.

Parsons, T.R., M. Takahashi, and B. Hargrave. 1984. Biological oceanographic processes, 3rd edition. Pergamon Press, Oxford. 330 pp.

Probyn, T.A., and S.J. Painting. 1985. Nitrogen uptake by size-fractionated phytoplankton in Antarctic waters. Limnology and Oceanography 30:1327-1332.

Rönner, U., F. Sorensson, and O. Holm-Hansen. 1983. Nitrogen assimilation by phytoplankton in the Sotia Sea. Polar Biology 2:137-147.

Saino, T., K. Miyata, and A. Hattori. 1979. Primary productivity in the Bering and Chukchi seas and in the northern North Pacific in 1978 summer. Bulletin of Plankton Society of Japan 26:96-103.

Saino, T., H. Otobe, E. Wada, and A. Hattori. 1983. Subsurface ammonium maximum in the northern North Pacific and the Bering Sea in summer. Deep-Sea Research 30:1157-1171.

Shiomoto, A., and M. Ogura. 1994. Distribution of particulate matter with special reference to biogenic silica of surface waters in the Bering Sea Gyre in summer 1991. Proceedings of National Institute of Polar Research Symposium on Polar Biology 7:10-16.

Shiomoto, A., K. Tadokoro, K. Monaka, and M. Nanba. 1997. Productivity of picoplankton compared with that of larger phytoplankton in the subarctic region. Journal of Plankton Research 19:907-916.

Sieburth, J.M., V. Smetacek, and J. Lenz. 1978. Pelagic ecosystem structure: Heterotrophic compartments of the plankton and their relationship to plankton size fractions. Limnology and Oceanography 23:1256-1263.

Taguchi, S. 1972. Mathematical analysis of primary production in the Bering Sea in summer. In: A.Y. Takenouti et al. (eds.), Biological oceanography of the northern North Pacific Ocean. Idemitsu Shoten, Tokyo, pp. 253-262.

Taguchi, S., H. Saito, H. Kasai, T. Kono, and Y. Kawasaki. 1992. Hydrography and spatial variability in the size distribution of phytoplankton along the Kuril Islands in the western subarctic Pacific Ocean. Fisheries Oceanography 1:227-237.

Taniguchi, A. 1969. Regional variations of surface primary production in the Bering Sea in summer and vertical stability of water affecting the production. Bulletin of Faculty of Fisheries, Hokkaido University 20:169-179.

Tréguer, P., D.M. Nelson, S. Gueneley, C. Zeyons, J. Morvan, and A. Buma. 1990. The distribution of biogenic and lithogenic silica and the composition of particulate organic matter in the Scotia and the Drake Passage during autumn 1987. Deep-Sea Research 37:833-851.

Turpin, D.H., and P.J. Harrison. 1980. Cell size manipulation in natural marine planktonic diatoms communities. Canadian Journal of Fisheries and Aquatic Science 37:1193-1195.

Wheeler, P.A., and S.A. Kokkinakis. 1990. Ammonium recycling limits nitrate use in the oceanic subarctic Pacific. Limnology and Oceanography 35:1267-1278.

Whitledge, T.E., W.S. Reeburgh, and J.J. Walsh. 1986. Seasonal inorganic nitrogen distributions and dynamics in the southeastern Bering Sea. Processes and Resources of the Bering Sea Shelf (PROBES). Continental Shelf Research 5:109-132.

CHAPTER 16

Seasonal Variation in the Process of Marine Organism Production Based on Downward Fluxes of Organic Substances in the Bering Sea

Yoshiaki Maita and Mitsuru Yanada
Faculty of Fisheries, Hokkaido University, Hakodate, Japan

Kozo Takahashi
Faculty of Science, Kyushu University, Fukuoka, Japan

Abstract

A time-series sediment trap experiment was conducted at 3,200 m in the Bering Sea Basin (53°30'N, 177°00'W; 3,788 m) during 1991 and 1992 to determine seasonal variation in the processes of biological production in the surface water of the Bering Sea. The seasonal pattern of downward fluxes of organic matter in the Bering Sea was characterized by three high peaks in flux (May, July-August, and October), each with different sinking properties. The mass flux showed maximum value in May, and the sinking particles were of small size with low organic content, composed mostly of diatoms. The bulk of sinking particles in July-August was small diatoms, though there were contributions of crustacean zooplankton remains. In October, sinking particles showed a relatively high contribution of foraminifera. In winter the particles were characterized by both foraminifera and crustacean zooplankton. We propose that seasonal variations in the flux and composition of sinking particles in the Bering Sea were directly influenced by the proportion of mixing among materials originating from smaller diatoms, larger foraminifera, and crustacean zooplankton. Judging from the seasonal variation of mass flux and the nature of the particles obtained from the sediment trap experiment, the biological production in the surface water of the Bering Sea was characterized by high primary production (about 250 g C per m^2 per yr) maintained by smaller diatoms and the low grazing pressure of larger crustacean zooplankton throughout a year.

Introduction

The Bering Sea has relatively high primary production in the world open seas (Koblenz-Mishke et al. 1970). However, information on the scale and variations in production, and the assemblage of marine organisms occurring in the oceanic areas of the Bering Sea are limited because most research has been conducted in summer.

Recently, the processes of biological production, with reference to the relations between various parameters of primary production, have been investigated in the northern North Pacific and the adjacent seas by many investigators (e.g., Taniguchi 1972). The results on the particulate organic matter or suspended organic matter in the subarctic sea have offered significant information regarding biological processes and other mechanisms, such as production, decomposition, distribution, and transport in the ocean (Tanoue and Handa 1980; Iseki 1981; Tsunogai 1982; Takahashi 1986, 1989; Honjo 1990). Recent evidence from sediment trap studies suggests that: (1) the flux of organic substances on sinking particles directly correlates to surface productivity (Deuser and Ross 1980; Deuser et al. 1981, 1983; Karl et al. 1991); and (2) the minerals in sinking particles are generally composed of biogenic carbonate and silicate. This process reflects biological events in the food webs such as crustacean material and fecal matter derived from marine zooplankton, and terrestrial sources including the clay component and other fine particles derived from surrounding land or marine sediment (Bishop et al. 1977; Honjo 1978, 1980; Knauer et al. 1979; Urrere and Knauer 1981).

Long-term experiments in productive areas provide information on biological events taking place in overlying waters through the sinking of particles driven from the surface to deep layers. In this sense, a sediment trap experiment may be the most valid approach to monitor the seasonal and/or annual variation of the primary production, and the assemblage of the higher trophic levels. Since 1990, we studied particle flux using a time-series sediment trap in the subarctic Pacific and the Bering Sea. The aim of this report is to provide our observations on the biological processes in primary and secondary trophic levels from a chemical point of view, based on our results in 1991 and 1992.

Material and Methods

A time-series sediment trap array (PARFLUX Mark 7G-13) was successively deployed at the depth of 3,200 m in the Bering Sea Basin (53°30′N, 177°00′W, water depth 3,788 m) for two years, both 1991 and 1992 (Fig. 1). The sample bottles were filled with in situ seawater (2,000 m depth) containing a 5% glutaraldehyde solution which was buffered to pH 7.6 with sodium borate. In order to target high resolution during highly production, we set the sampling intervals as follows: 20 days during late March through early October and 56 days from early October through late March based

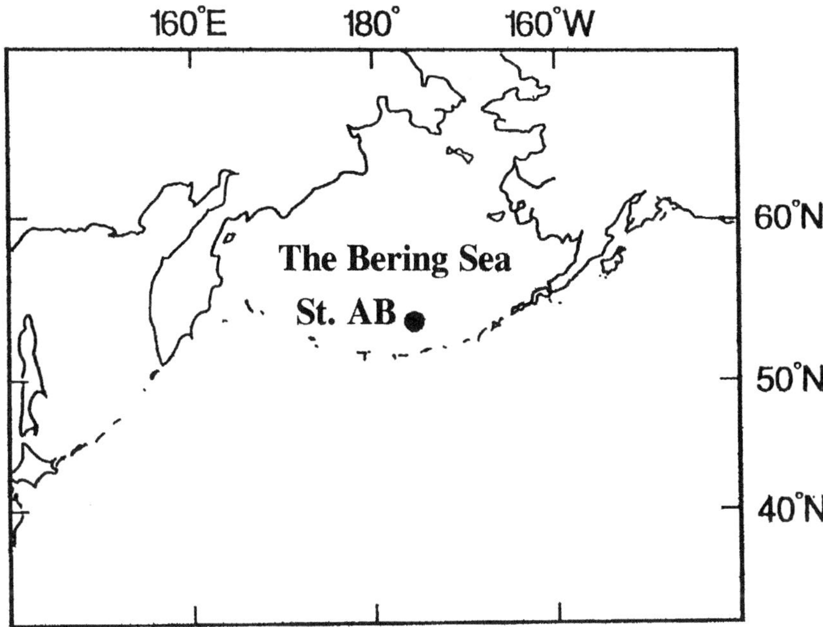

Figure 1. *Location of the sediment trap station at Station Aleutian Basin (AB) in the Bering Sea.*

on flux data from the Gulf of Alaska (Honjo et al. 1995). The trap samples were sieved with 1 mm and 63 µm mesh screen to remove some large, swimming zooplankton and to fractionate the source material immediately after recovering the trap array. Aliquots of the dried samples less than 63 µm and the 63 µm-1 mm size fractions were analyzed for determination of the content of carbon and nitrogen using a CHN analyzer. The other parts of the samples were analyzed for determination of the content and composition of amino acids and hexosamine by a ophthalaldehyde (OPA) method using High Performance Liquid Chromatography (Model 503, Shimazu Co.).

Results and Discussion

Seasonal Variation in the Flux of Sinking Particles

In 1991, total mass flux and the fluxes of organic and inorganic particles showed the highest peak in summer (July), and a secondary peak in autumn, whereas minimum values were obtained during winter and spring (Fig. 2). Temporal variations in total carbon and nitrogen flux were consistent with

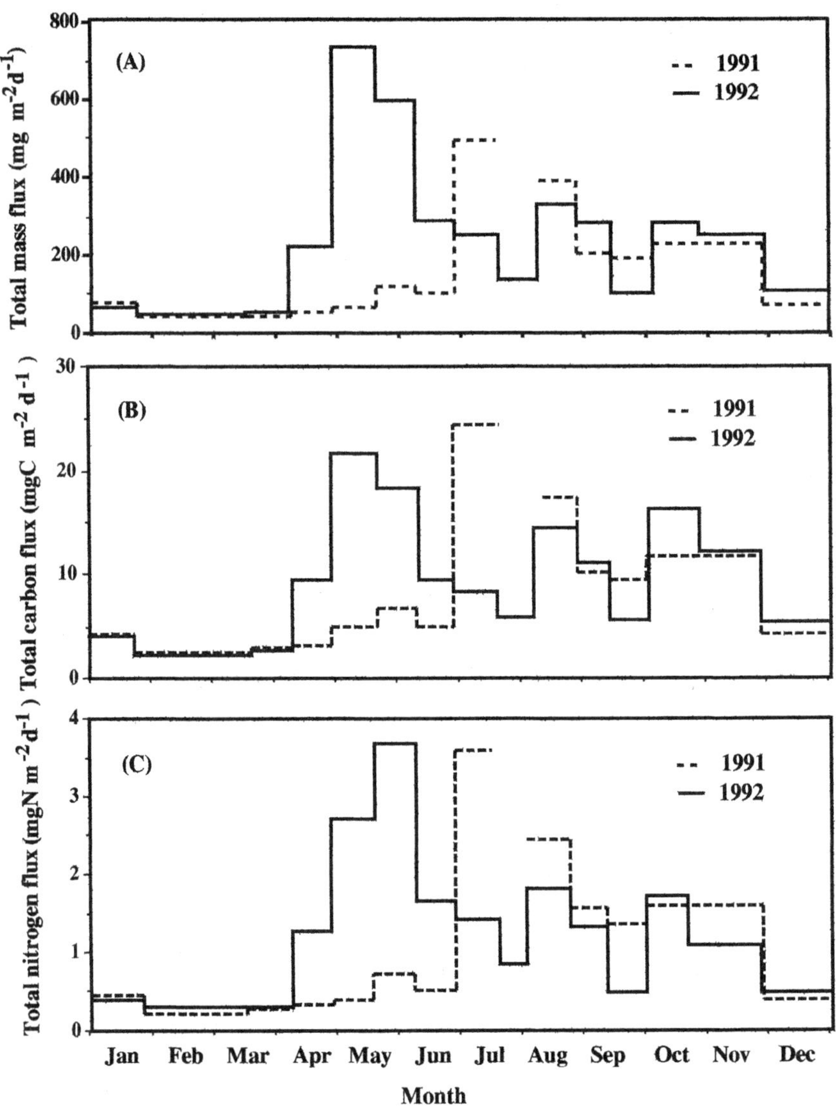

Figure 2.   Seasonal variations in (A) total mass flux, (B) total carbon flux, and (C) total nitrogen flux of sinking particles at the 3,200 m depth in the Bering Sea (53°30'N, 177°00'W).

total mass flux. In 1992, the maximum peak of fluxes occurred in May and secondary peaks occurred in both late summer and autumn; low values were obtained during winter and early spring. We suggest that annual changes in the flux of sinking particles in the Bering Sea basin are characterized with one or two peaks, i.e., May and July-August. The secondary peak found in October in both years has not been observed in the north-central area of the North Pacific, the Sargasso Sea (Deuser et al. 1981, Ittekkot et al. 1984b), and the Panama Basin (Ittekkot et al. 1984a).

Seasonal Variation in the Chemical Constituents of Sinking Particles

In general, the sinking particles caught by the sediment trap can be roughly classified into the source materials produced in the overlying water such as phytoplankton, zooplankton, and inorganic materials derived from terrestrial origin.

The contents of total carbon and nitrogen in sinking particles can be used as an indicator for the amount of mixing between organic and inorganic materials. The flux and content of amino acids, the ratios of amino acid carbon to total carbon, and amino acid nitrogen to total nitrogen, have been used as the chemical indicators to speculate on the decomposition process and/or the source of sinking organic matter (Lee and Cronin 1982, 1984; Wefer et al. 1982; Ittekkot et al. 1984a, 1984b; Wakeham et al. 1984; Müller et al. 1986; Liebezeit and Bodungen 1987; Cowie and Hedges 1992). Most recently, hexosamine, which is the main organic constituent in the crustacean zooplankton shell, has been used in the analysis of lower trophic levels (Ittekkot et al. 1984a, 1984b; Maita 1985; Müller et al. 1986; Liebezeit and Bodungen 1987; Haake et al. 1993). Therefore, the flux and content of hexosamine was adopted in our study as an indicator for estimating the relative contribution between phytogenic or zoogenic origins.

The temporal variation in the flux of total amino acids (AA) was consistent with those in the fluxes of total carbon and total nitrogen with strong positive correlation coefficients ($r = 0.97$ and 0.90 for total carbon and total nitrogen, respectively; Fig. 3). In contrast, the temporal variation in the flux of hexosamine (HA) was significantly correlated with the fluxes of total carbon and total nitrogen with weak positive correlation coefficients ($r = 0.65$ and 0.68 for total carbon and total nitrogen, respectively). These results reflect seasonal differences in the relative contribution of zooplankton in precipitating particles. The chemical properties of sinking particles in some typical periods when total mass flux forms the undulations are shown in Table 1. In February-March of both 1991 and 1992, the fluxes of total carbon, total nitrogen, protein amino acids, and hexosamine were obviously low compared to the total mass flux. This means that the flux of materials derived from terrestrial origin is the lowest in this period throughout the year. However, the high content of hexosamine and low

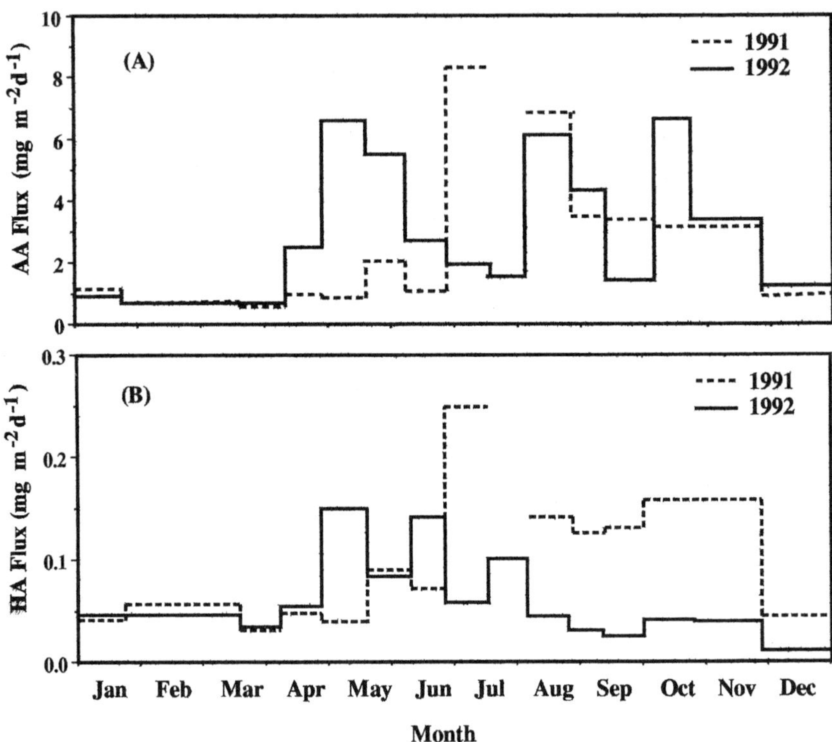

Figure 3.   *Seasonal variations in (A) total amino acid flux and (B) hexosamine flux of sinking particles at the depth of 3,200 m in the Bering Sea (53°30′N, 177°00′W).*

amino acids/hexosamine (AA/HA) ratios suggests that the contribution of zooplankton in sinking materials is relatively high compared to other periods. In July 1991, the parameters representing the flux of terrestrial materials were highest, implying the greatest supply of materials from phytoplankton and zooplankton. However, the hexosamine content was lower and AA/HA ratio was higher than those in February-March. This result implies that the bulk of sinking particles in this period may derive from phytoplankton origin although there was also a large supply of crustacean zooplankton. In October-November 1991 when the mass flux showed a secondary peak, the contents of total carbon, nitrogen, total protein amino acids, and hexosamine were approximately equivalent to those in July. This means that the characteristics of sinking particles in this period approximates those in July.

In May 1992, when total mass flux was highest, the parameters representing the fluxes and contents of marine origin were lower in compari-

Table 1. **The chemical properties of precipitating particles obtained from time-series sediment trap at the depth of 3,200 m in the Bering Sea Basin.**

Year	Month	Mass flux (mg m^{-2}d^{-1})	TC-flux (mg C m^{-2}d^{-1})	TN-flux (mg N m^{-2}d^{-1})	AA-flux (mg m^{-2}d^{-1})	HA-flux (mg m^{-2}d^{-1})
1991	Feb.-Mar.	41	2	0.2	0.7	0.06
	July	493	24	3.6	8.4	0.25
	Oct.-Nov.	221	11	1.5	3.1	0.16
1992	Feb.-Mar.	45	2	0.3	0.7	0.05
	May	663	20	3.2	6.1	0.12
	Aug.	313	14	1.8	6.1	0.05
	Oct.	286	17	1.7	6.5	0.04

Year	Month	C-con. (mg C g^{-1})	N-con. (mg N g^{-1})	AA-con. (mg g^{-1})	HA-con. (mg g^{-1})	C/N (A.R.)	AA-C/TC (%)	AA-N/TN (%)	AA/HA
1991	Feb.-Mar.	60	5.7	18	1.4	12.2	13	44	13
	July	49	7.3	18	0.5	7.9	17	38	36
	Oct.-Nov.	50	6.6	14	0.7	8.9	12	29	20
1992	Feb.-Mar.	50	6.2	16	1.0	9.5	14	36	16
	May	30	4.9	9	0.2	7.6	13	26	45
	Aug.	45	5.8	19	0.1	9.2	19	46	19
	Oct.	58	6.0	23	0.1	11.3	17	52	23

TC = total carbon; TN = total nitrogen; AA = amino acid; HA = hexosamine; con. = content; A.R. = atomic ratio.

son with those in other seasons. These results suggest that the contribution of marine organisms in sinking particles is much less than that in the other season. The high AA/HA ratio also suggests that the contribution of zooplankton material may be low. In August 1992, sinking particles had a higher contribution of marine organism materials, including a higher contribution of zooplankton compared to that in May. However, judging from the levels of hexosamine, the contribution of zooplankton was lower compared to those in May 1992 and July 1991. The contribution of zooplankton, however, was lower in October 1992 than in October 1991.

We estimated the fluxes and the contents of the chemical parameters in the <63 μm size fraction and the 63 μm-1 mm size fraction for the sediment trap samples in 1992. In the <63 μm size fraction the proportion of the mass flux to the total mass flux accounted for 70-89% (average: 80±6%), 56-83% (average: 70±8%) for carbon flux and 56-88% (average: 73±9%) for nitrogen flux. The proportions tended to become relatively high when the fluxes of sinking particles became high. These results suggest that the bulk of sinking particles was composed of the <63 μm size fraction particles and that the increase of total flux of sinking particles was influenced by the increase of the flux in <63 μm size fraction. Comparison of the parameters representing the content of biogenic origin of the <63 μm size fraction and the 63 μm-1 mm size fraction suggests that the contents of all parameters in the 63 μm-1 mm size fraction are higher than those in the <63 μm size fraction. It is likely that temporal variations in the flux and the characteristics of sinking particles are directly influenced by the mixing proportion between the 63 μm and 1 mm size fractions. Microscopic observation showed that the predominant organisms were the siliceous diatoms for the <63 μm size fraction and calcareous foraminifera for the 63 μm-1 mm size fraction.

Consequently, the sinking particles in May of 1992, when total mass flux showed maximum value, were characterized by some smaller particles with lower organic content, primarily diatoms. The bulk of the sinking particles in July-August was principally smaller diatoms although there was some contribution of crustacean zooplankton origin. On the other hand, the characteristics of sinking particles in October and February-March included a relatively large proportion of foraminifera, and relatively small proportions of both foraminifera and crustacean zooplankton, respectively. It is possible then that seasonal variations in the flux and the characteristic of sinking particles in the Bering Sea were directly influenced by mixing among smaller diatoms, larger foraminifera, and crustacean zooplankton materials

Characteristics of Marine Organism Production in the Bering Sea

Variations in the flux and composition of sinking particles in the Bering Sea Basin reflect the processes of phytoplankton production and successive zooplankton grazing pressure in surface water. Therefore, it may be possible to predict the change of processes of biological production that is caused by global climate change from a time-series sediment trap experiment. We estimated the annual average fluxes and contents of the parameters representing the biogenic source in the Bering Sea in both 1991 and 1992 (Table 2).

Suess (1980) has proposed that the downward organic carbon flux in the open ocean could be expressed as a function with primary production and depth in situ. From Suess's experimental equation and our total carbon flux, the annual estimated primary production in the Bering Sea was 230 g C per m^2 (average 630 mg C per m^2 per day) in 1991 and 250 g C per m^2 (average 680 mg C per m^2 per day) in 1992. The values estimated from carbon fluxes are fairly high compared to that in the Bering Sea described by Koblentz-Mishke et al. (1970). Namely, primary production in the Bering Sea may reveal the highest values in the world's open oceans as described by Suess (1980). Judging from sediment trap studies, the high primary production in the Bering Sea may be maintained by the phytoplank-

Table 2. **Annual fluxes and contents of the chemical parameters at a station in the Bering Sea Basin in 1991 and 1992**

Parameters		1991		1992	
		Average	Range	Average	Range
Mass flux	(mg m^{-2}d^{-1})	161	41-490	222	45-730
Total carbon flux	(mg C m^{-2}d^{-1})	8.2	2.5-24.3	8.9	2.3-21.5
Total nitrogen flux	(mg N m^{-2}d^{-1})	1.1	0.2-3.6	1.1	0.3-3.7
Amino acid flux	(mg m^{-2}d^{-1})	2.6	0.6-8.4	2.8	0.7-6.6
Hexosamine flux	(mg m^{-2}d^{-1})	0.099	0.03-0.25	0.056	0.01-0.15
Total carbon content	(mg C g^{-1})	56	45-75	45	29-59
Total nitrogen content	(mg N g^{-1})	6.3	5.5-7.3	5.2	3.7-6.2
C/N ratio	(atom base)	10.5	7.9-14.7	10.4	5.8-14.7
Amino acid content	(mg g^{-1})	16	13-20	13	8-23
Hexosamine content	(mg g^{-1})	0.77	0.37-1.35	0.40	0.09-1.00
Amino acid-C/total C	(%)	13	7-17	13	10-19
Amino acid-N/total N	(%)	36	30-44	36	20-52
Amino acid/hexosamine	(by weight)	21	13-49	33	14-153

ton assemblage which is largely composed of smaller diatoms with two or three peak productions. The values of total carbon, total nitrogen, total amino acids, C/N ratio, AA-C/TC ratio, and AA-N/TN ratio in 1991 are nearly the same as those in 1992 or slightly higher, except the parameters presenting the contribution of zooplankton origin materials (i.e., hexosamine content and AA/HA ratio). These differences of the hexosamine parameters between both years may depend on the participation of zooplankton in surface layers. The hexosamine parameters in the Bering Sea were one order of magnitude lower than those in the northeastern Pacific Ocean (Haake et al. 1993), whereas total amino acid parameters were similar to those in the northeastern Pacific Ocean. In the northeastern Pacific Ocean, the precipitating marine origin of largely copepod materials accounted for about 90% of zooplankton production (Miller et al. 1984), and the high flux of hexosamine was found in summer when the standing stock of copepod reach maximum (Haake et al. 1993). The low values of hexosamine parameters in the Bering Sea suggest that the grazing pressure of crustacean zooplankton taking place in the overlying water column may be lower than that in the northeastern Pacific Ocean (Frost et al. 1983). Consequently, this low grazing pressure of crustacean zooplankton may be responsible for the high phytoplankton production in the Bering Sea.

Conclusions

Judging from the seasonal variation in the flux and the characteristics of sinking particles obtained from the sediment trap experiment deployed successively in the Bering Sea in both 1991 and 1992, the biological production in the surface water in the Bering Sea Basin was characterized by high primary production maintained by smaller diatoms with two to three peak and the low production of larger crustacean zooplankton throughout a year.

Acknowledgments

We thank Captains K. Masuda and G. Anma, officers, and crew of T/V *Oshoro Maru*, Hokkaido University, for their assistance in mooring the sediment trap array. We are also very grateful to Dr. Hideo Miyake, Faculty of Fisheries, Hokkaido University, and Rindy Ostermann, Woods Hole Oceanographic Institution, for their support of our sediment trap experiment. This study was supported by the Fund of JSPS and NFS (YM and KT), cooperative Studies for Particle Flux in the Central Subarctic Pacific between the United States and Japan. The study was also supported by project No. 801-3524-07680562 of the Ministry of Education, Science and Culture of Japan to KT.

References

Bishop, L.K., J.M. Edmond, D.R. Kettens, M.P. Bacon, and W.B. Silker. 1977. The chemistry, biology, and vertical flux of particulate matter from the upper 400 m of the equatorial Atlantic Ocean. Deep-Sea Research 241:511-548.

Cowie, G.L., and J.I. Hedges. 1992. Sources and reactivities of amino acids in a coastal marine environment. Limnology and Oceanography 37:703-724.

Deuser, W.G., and E.H. Ross. 1980. Seasonal change in the flux of organic carbon to the deep Sargasso Sea. Nature 283:364-365.

Deuser, W.G., E.H. Ross, and R.F. Anderson. 1981. Seasonality in the supply of sediment to the deep Sargasso Sea and implications for the rapid transfer of matter to the deep ocean. Deep-Sea Research 28:495-505.

Deuser, W.G., P.G. Brewer, T.D. Jickells, and R.D. Commeau. 1983. Biological control of abiogenic particles from the surface ocean. Science 219:388-391.

Frost, B.W., M.R. Landry, and R.P. Hassett. 1983. Feeding behavior of large calanoid copepods *Neocalanus cristatus* and *N. plumchrus* from the subarctic Pacific Ocean. Deep-Sea Research 30:1-13.

Haake, B., V. Ittekkot, S. Honjo, and S. Manganini. 1993. Amino acids, hexosamine and carbohydrate fluxes to the deep subarctic Pacific (Station P). Deep-Sea Research 40:547-560.

Honjo, S. 1978. Sedimentation of materials in the Sargasso Sea at a 5,367 m deep station. Journal of Marine Research 36:469-492.

Honjo, S. 1980. Material fluxes and mode of sedimentation in the mesopelagic and bathypelagic zone. Journal of Marine Research 38:53-97.

Honjo, S. 1990. Particle fluxes and modern sedimentation in the polar oceans. In: W.O. Smith Jr. (ed.), Polar oceanography, vol. 2. Academic Press, New York, pp. 322-353.

Honjo, S., J. Dymond, R. Collier, and S.J. Manganini. 1995. Export production of particles to the interior of the equatorial Pacific Ocean during the 1992 EqPac experiment. Deep-Sea Research 42:831-870.

Iseki, K. 1981. Vertical transport of particulate organic matter in the deep Bering Sea and Gulf of Alaska. Journal of Oceanographic Society of Japan 37:101-110.

Ittekkot, V., E.T. Degens, and S. Honjo. 1984a. Seasonality in the fluxes of sugars, amino acids, and amino sugars to the deep oceans: Panama Basin. Deep-Sea Research 31:1071-1083.

Ittekkot, V., W.D. Deuser, and E.T. Degens. 1984b. Seasonality in the fluxes of sugars, amino acids, and amino sugars to the deep oceans: Sargasso Sea. Deep-Sea Research 31:1057-1069.

Karl, D.M., B.D. Tilbrook, and G. Tien. 1991. Seasonal coupling of organic matter production and particle flux in the western Bransfield Strait, Antarctic. Deep-Sea Research 38:1097-1123.

Knauer, G.A., J.H. Martin, and K.W. Bruland. 1979. Fluxes of particulate carbon, nitrogen, and phosphorus in the upper water column of the northeast Pacific. Deep-Sea Research 26:97-108.

Koblenz-Mishke, O.J., V.V. Volkovinsky, and J.G. Kobanova. 1970. Plankton primary production of the world ocean. In: Scientific exploration of the South Pacific. National Academy of Science, Washington, DC, pp. 183-193.

Lee, C., and C. Cronin. 1982. The vertical flux of particulate organic nitrogen in the sea: Decomposition of amino acids in the Peru upwelling area and the equatorial Atlantic. Journal of Marine Research 40:227-251.

Lee, C., and C. Cronin. 1984. Particulate amino acids in the sea, effects of primary productivity and biological decomposition. Journal of Marine Research 42:1075-1097.

Liebezeit, G., and B.V. Bodungen. 1987. Biogenic fluxes in the Bransfield Strait, planktonic versus macroalgal sources. Marine Ecology Progress Series 36:23-32.

Maita, Y. 1985. Decomposition processes of organic matter in seawater, with special reference to protein and amino acids. In: Nippon Biseibutsu Seitai Gakkai (Society of Japan Microorganism Ecology) (eds.), Ecology of Microorganisms 13:95-110. (In Japanese.)

Miller, C.B., B.W. Frost, H.P. Batchelder, M.J. Clemons, and R.E. Conway. 1984. Life histories of larger grazing copepods in a subarctic ocean gyre: *Neocalanus plumchrus, Neocalanus cristatus,* and *Eucalanus bungii* in the northeast Pacific. Progress in Oceanography 13:201-243.

Müller, P.J., E. Suess, and C.A. Ungerer. 1986. Amino acids and amino sugars of surface particulate and sediment trap material from waters of the Scotia Sea. Deep-Sea Research 33:819-838.

Suess, E. 1980. Particulate organic carbon flux in the oceans: Surface productivity and oxygen utilization. Nature 288:260-263.

Takahashi, K. 1986. Seasonal fluxes of pelagic diatoms in the subarctic Pacific, 1982-1983. Deep-Sea Research 33:1225-1251.

Takahashi, K. 1989. Silicoflagellates as productivity indicators: Evidence from long temporal and spatial flux variability responding to hydrography in the northeastern Pacific. Global Biogeochemical Cycles 3:43-61.

Taniguchi, A. 1972. Geographical variation of primary production in the western Pacific Ocean and adjacent seas with reference to the inter-relations between various parameters of primary production. Memoirs of the Faculty of Fisheries Hokkaido University 19:1-33.

Tanoue, E., and N. Handa. 1980. Distribution of particulate organic carbon and nitrogen in the Bering Sea and northern North Pacific Ocean. Journal of Oceanographic Society of Japan 35:47-62.

Tsunogai, S., M. Uematsu, S. Noriki, N. Tanaka, and M. Yamada. 1982. Sediment trap experiment in the northern North Pacific: Undulation settling particles. Geochemical Journal 16:129-147.

Urrere, M.A., and G.A. Knauer. 1981. Zooplankton fecal pellet fluxes and vertical transport of particulate organic material in the pelagic environment. Journal of Plankton Research 31:369-387.

Wakeham, S.G., C. Lee, J.W. Farrington, and R.B. Gagosian. 1984. Biogeochemistry of particulate organic matter in the oceans: Results from sediment trap experiments. Deep-Sea Research 31:509-528.

Wefer, G., E. Suess, W. Balzer, G. Liebezeit, P.J. Müller, C.A. Ungerer, and W. Zenk. 1982. Fluxes of biogenic components from sediment trap deployment in circumpolar waters of the Drake Passage. Nature 299:145-147.

CHAPTER 17

Trends in the Distribution of Organic Nitrogen, Ammonium, and Urea in the Bering Sea

Victor V. Sapozhnikov, Victor L. Zubarevich, and Natalya V. Mordasova
Russian Federal Research Institute of Fisheries and Oceanography (VNIRO), Moscow, Russia

Abstract

Results of studies in the western Bering Sea during 1990-1993 are presented for different forms of organic nitrogen, ammonium and urea, and nitrates, including those for which little is known. The principal trends of urea distribution in summer are discussed. Dissolved organic nitrogen made up 65-95% of the total dissolved nitrogen (TDN) at the shelf and shelf break, while it was 15-30% in open surface waters free of a phytoplankton bloom, and declined to 5-15% with depth. In summer, the shelf and shelf break were characterized by a high content of urea and ammonium (3.0-4.0 µM) while nitrates were below 1.0 µM in the surface layer. Primary production relied on recycling of labile forms of nitrogen.

Introduction

Phytoplankton is an active consumer of both inorganic and organic forms of nitrogen. Some researchers (Glibert et al. 1982) hold that ammonium nitrogen is more favorable, compared to other forms. The same is true with urea, which is second after ammonium for the majority of diatoms (Harvey and Caperon 1967).

Recent studies of urea have allowed us to improve our understanding of its role in production and destruction processes and have revealed general trends in the spatial distribution of urea (Sapozhnikov and Zubarevich 1989, Sapozhnikov and Propp 1990, Kavakina and Sapozhnikov 1993). The importance of urea in the nitrogen cycle was shown by Cooper (1937), who estimated hydrolysis rates and dissociation constants of urea in seawater. This allowed him to conclude that urea rapidly decomposes to ammonium and cyanate-ion, facilitating urea nitrogen assimilation by phytoplankton. Studies on urea were advanced when Newell et al. (1967)

suggested the use of diacetilmonooxime (DAM) to increase the sensitivity of urea measurements in blood. Using this technique, determination of low concentrations of urea in seawater are now possible.

Recent studies of the transformation and utilization of nitrogen compounds in seawater have shown a growing interest in poorly studied forms of nitrogen, i.e. dissolved organic nitrogen, ammonium, and urea. There are scarce data on these forms of nitrogen in the western Bering Sea primarily due to methodological problems. Better data are available for nitrate and ammonium nitrogen and their role in production processes in the eastern part of the basin (Whitledge et al. 1986, 1988).

During research in the Bering Sea on board the R/Vs *Mlechniy Put'*, *Professor Soldatov,* and *Akademik Alexander Nesmeyanov* in 1990-1993, we studied inorganic and organic forms of nitrogen over various seasons. The data on nitrates, nitrites, and ammonium were obtained in winter, spring, summer, and autumn; urea data covered only summer; and total dissolved organic nitrogen data were obtained in summer and autumn.

Materials and Methods

Ammonium nitrogen was determined with hypochlorite and phenol by the Sagi-Solorzano method (Sapozhnikov 1987). Dissolved organic nitrogen was calculated as the difference between total and inorganic nitrogen. Total dissolved nitrogen (TDN) was measured after the sample was treated with Valderrama reagents in titanium tubes under 130°C. The nitrite analysis was made on an autoanalyzer RFA-300 ALPKEM with sulfanilamide and N-1 naphthilethylendiamine dihydrochloride. Nitrates were determined on an AA-IIC Technicon autoanalyzer after nitrite reduction with a cadmium column. Urea measurements were made with DAM (Spravochnik 1991) on an RFA-300 ALPKEM autoanalyzer.

Results and Discussion

In April-May 1990, nitrogen stocks in pelagic waters of the western Bering Sea were dominated by inorganic forms; annual nitrate concentrations were maximal (24-26 µM) in the surface layer and were characteristic of the winter period. On the shelf, where the spring bloom started at the ice edge, nitrate concentrations decreased to 14-16 µM, and at some stations to 4 µM.

The level of ammonium nitrogen, as the product of metabolism and the primary product of the mineralization of organic matter, was low. Values did not exceed 0.5-0.6 µM in the maximum ammonium layer and increased to 1 µM on the shelf and on the continental slope. The vertical distribution of ammonium was diverse, caused by the daily migration of its major producers, including zooplankton, squid, and other animals (Fig. 1). Maximum values were observed at 200-1,500 m during the daytime, and at 0-200 m at night. Similar values of ammonium nitrogen were found in the eastern Bering Sea in May 1979 (Dagg et al. 1982).

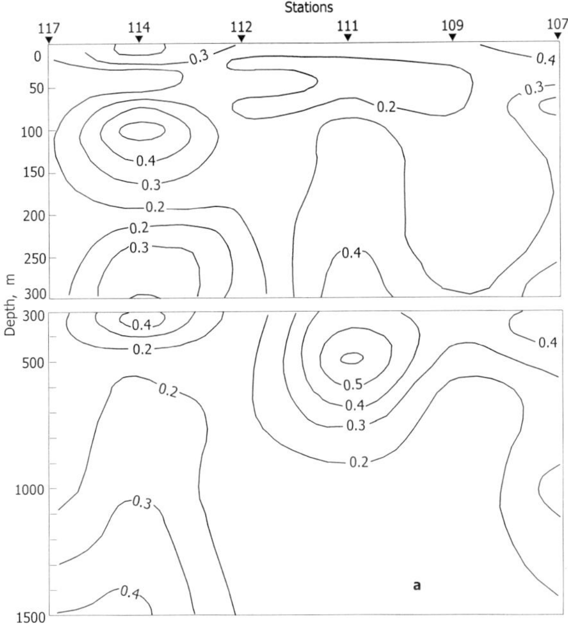

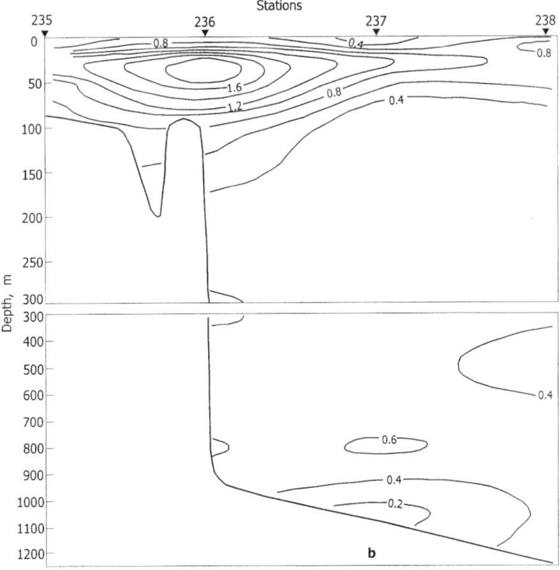

Figure 1. *The distribution of ammonium nitrogen (µM): (a) at the section of Stations 117-107 (12-14 April 1990); (b) at the section of Stations 235-238 (02-03 June 1990). (Station depths in m.)*

A sharp decrease of nutrients (nitrates were depleted to nil) was accompanied by a marked abundance of ammonium (1-2 µM) during surface layer warming and by the increase in photosynthesis by phytoplankton in late May-early June 1990. These changes occurred in the zone of phytoplankton accumulation above the thermocline at 10-25 m. In calm weather, ammonium nitrogen concentrations rose to the maximum value of 2.8 µM in the bottom layer of shallow waters (Fig. 2b), while the content of nitrates totaled 0.2-0.5 µM. That is, the ammonification processes prevailed while nitrification was hampered by low temperatures.

In June 1992, ammonium nitrogen was observed in the entire water column (0-1,000 m), indicating intensive production and remineralization processes. Ammonium nitrogen concentrations generally varied in the surface layer from 0.5 to 2.5 µM and attained maximum values of about 6.0 µM in Kamchatka Strait (Fig. 2b).

The high concentrations of urea (up to 7-11 µM) and ammonium (up to 6-12 µM) in the entire water column were observed in shallow shelf waters. This was associated with the breakdown of sedimentary organic matter (OM) in the benthos and with intensive water mixing.

In summer, shelf nitrates were almost completely utilized by phytoplankton (Fig. 2b). Maximum surface values (20-24 µM) were observed in pelagic areas where the upper quasi-uniform layer had not yet developed and the intensity of phytoplankton photosynthesis was low (oxygen saturation totaled 102-104%). Seasonal minimal concentrations of dissolved organic nitrogen (<8 µM) and urea (2.0-2.4 µM) were also observed (Fig. 2c,d). Nitrate stocks were small (0.1-0.3 µM) along the entire survey area.

The complex dynamic conditions at the continental edge interrupted the continuous distribution of maximum concentrations of urea seaward. The zones of anticyclonic eddies were characterized by a higher content of urea and dissolved organic nitrogen (Fig. 2).

The highest values of ammonium (5-6 µM) were observed in the upper 0-200 m. The maximum occurred at the upper boundary of the seasonal thermocline and sank from 10 to 50 m seaward (Fig. 3). This was due to the accumulation and mineralization of OM sinking from the euphotic zone. Maximum stocks of TDN (up to 16-20 µM) occurred in the upper 50 m and decreased to 4-6 µM with depth (Fig. 3b).

In areas of intense production (i.e., waters of maximum accumulations of phytoplankton, particularly on the shelf and at the shelf break) the amount of dissolved organic nitrogen (DON) made up 65-95% of the total, and reached 14-20 µM, while nitrates did not exceed 3-4 µM.

In pelagic waters, where the phytoplankton bloom was not observed, the contribution of DON decreased to 15-30% in the surface layer: as a rule, the content of DON and nitrates did not exceed 10 and 16-24 µM, respectively. Below the photic layer, where breakdown processes prevailed over production, the contribution of inorganic nitrogen increased with depth, while organic forms decreased to 5-15% of the total amount. The concentrations of DON totaled about 4 µM.

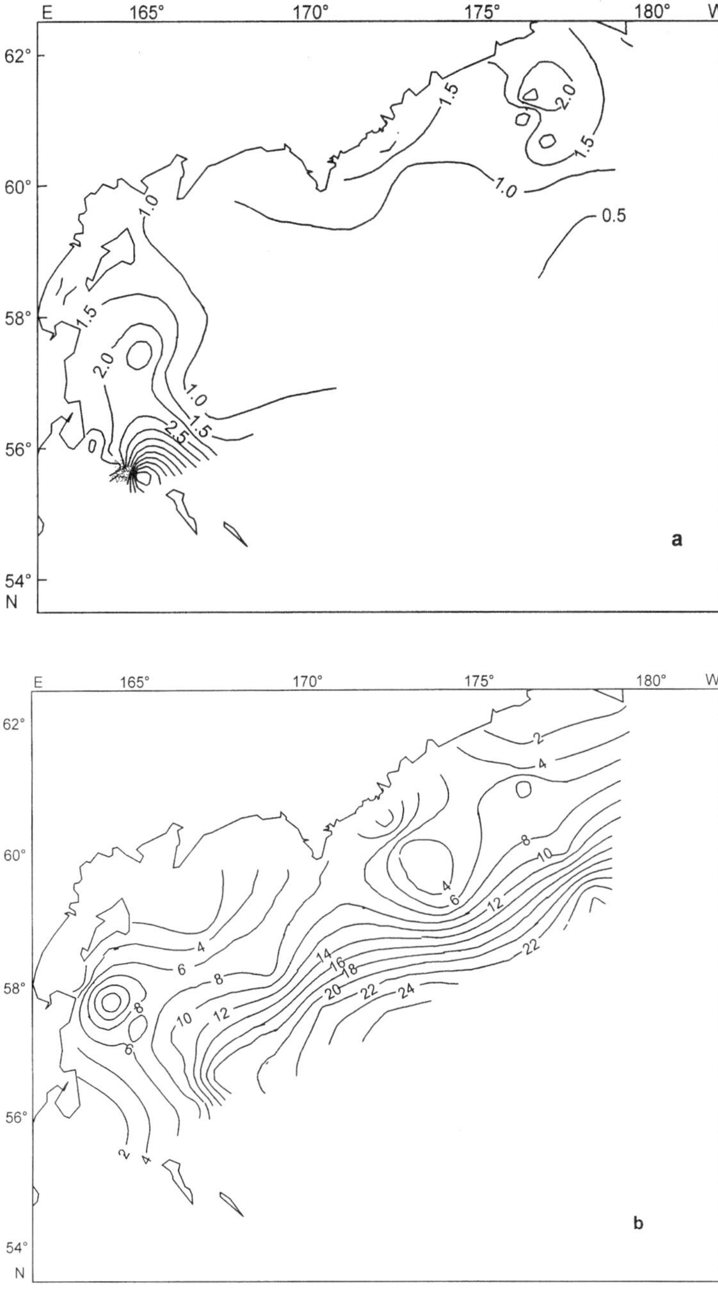

Figure 2. *The distribution of nitrogen compounds (μM) in the surface layer of the western Bering Sea (June 1992): (a) ammonium nitrogen; (b) nitrates; (c) organic nitrogen; (d) urea.*

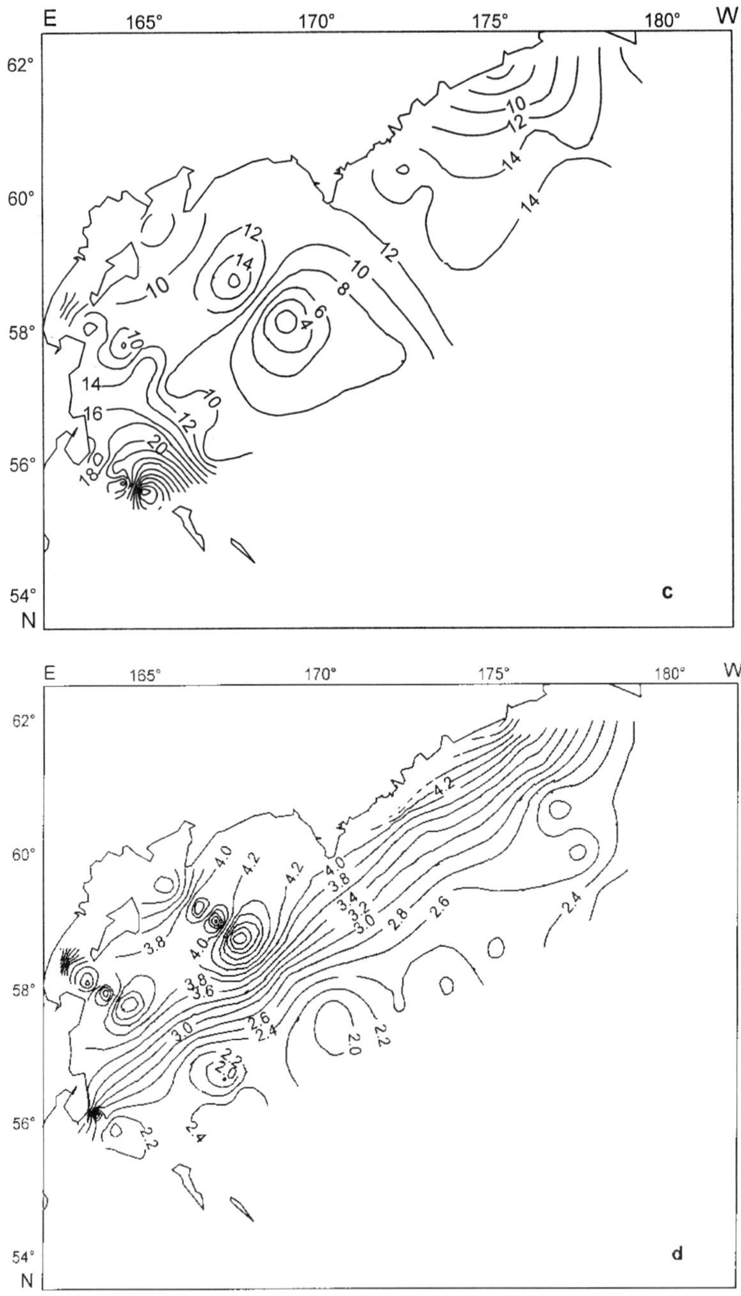

Figure 2.   (Continued.) The distribution of nitrogen compounds (μM) in the surface layer of the western Bering Sea (June 1992): (a) ammonium nitrogen; (b) nitrates; (c) organic nitrogen; (d) urea.

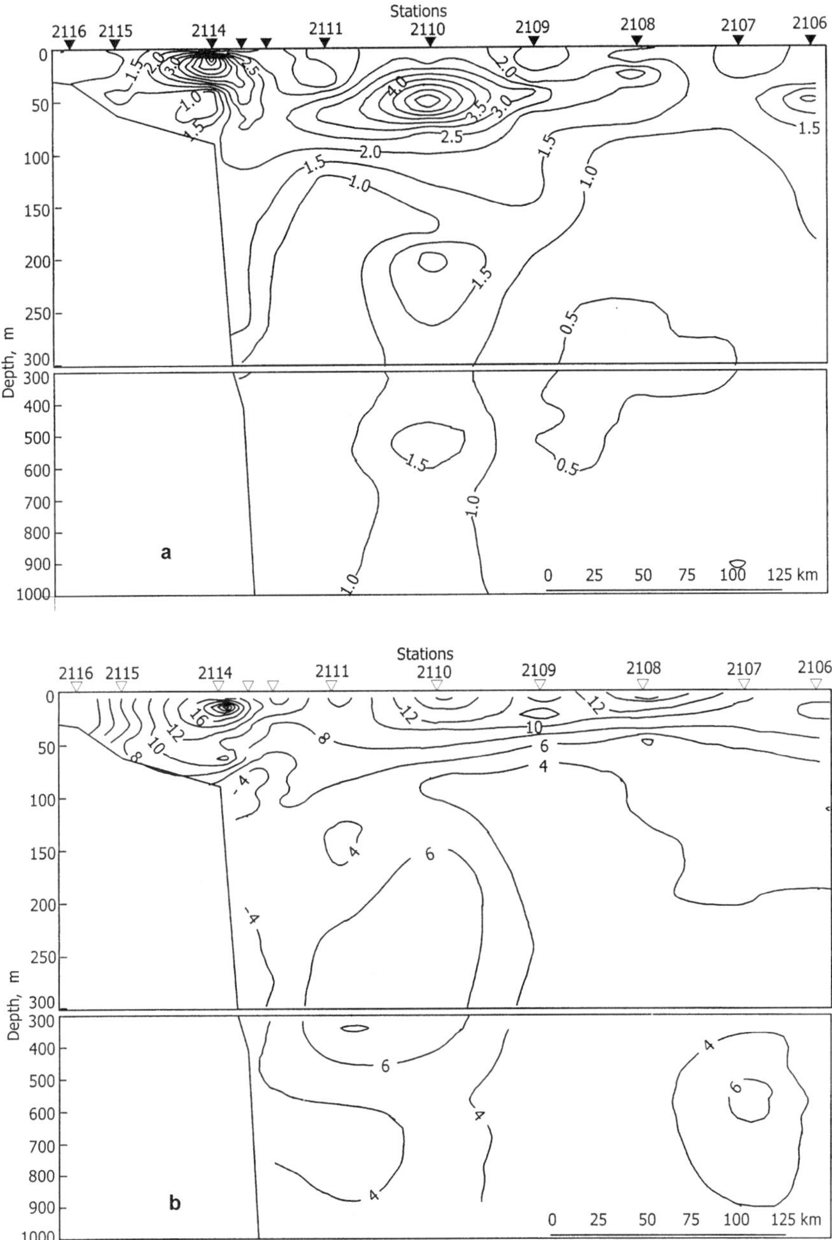

Figure 3. The distribution of nitrogen compounds (μM) at the section of Stations 2116-2106 (June 1992): (a) ammonium nitrogen; (b) organic nitrogen; (c) urea.

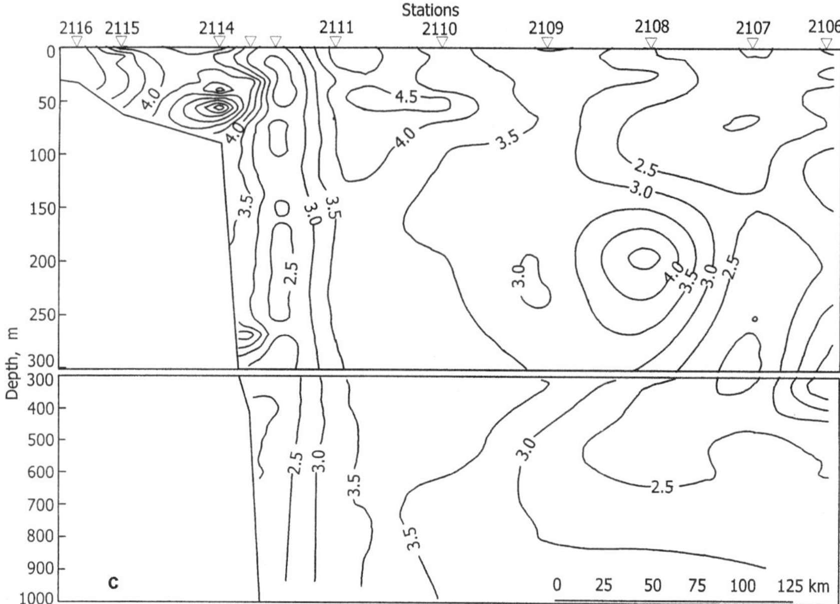

Figure 3.   *(Continued.) The distribution of nitrogen compounds (μM) at the section of Stations 2116-2106 (June 1992): (c) urea.*

The vertical distribution of urea presents a more complex pattern with a number of maxima at various depths (Fig. 3c): some maxima were found in the upper layer (above the pycnocline), and some at 200-300 m and deeper. Maximum urea values (about 8 μM) and nitrates (more than 50 μM) were observed in the bottom layer of the shelf edge (Station 2114). These large amounts of nitrates indicate advection processes in the area. The isoclines of urea concentrations in surface layers agreed with the distribution of chlorophyll and primary production (PP). The biomass of phyto-, zoo-, and bacterioplankton also increased toward the coast. We conclude that the increase of urea concentrations was associated with an increase in production and decomposition processes.

In summer, concentrations of nitrates in the surface layer did not exceed 1.0 μM (often nearly nil). The only bound nitrogen available to phytoplankton was ammonia and urea. These observations suggest that the main fraction was derived from remineralization, indicating a complete shift of the surface ecosystem to recycling.

In deep waters the total amount of ammonium and urea (3.0-3.5 μM) comprised 6-10% of the content of nitrates (40-43 μM) and 40-50% of the total amount of DON (5-10 μM).

In summer, surface waters are a habitat of abundant accumulations of infusoria (12,670 cells per liter, biomass of 199 mg/m^3), bacteria (about 3,106 cells per ml), and nanoheterotrophs (Moiseev 1995, Sorokin 1995, Sorokin et al. 1995). This community of microplankton processes OM and excretes ammonium and urea in large quantities.

In summer, high urea content (more than 2.5 µM) in both surface and deep layers corresponded well with concentrations of ammonium from the increase of productivity observed earlier. Similar high concentrations of urea were observed in Antarctic waters and in upwelling zones of Peru and California (up to 2.5 µM) (McCarthy 1970, Remsen 1971, DeManche et al. 1973).

Despite the continuous intensive development of phytoplankton in autumn (October-November 1990), the distribution of ammonium and DON in surface layers nearly followed the pattern of surface geostrophic circulation. Large cyclonic eddies off the shelf in Olyutorsk Bay and southwestward off Cape Navarin were revealed by maximum concentrations of nutrient salts. These eddies transport waters with relatively low concentrations of ammonia (1.0-1.2 µM) to the surface which are considerably lower than in those coastal waters, or anticyclonic formations, which occur in adjacent deep waters of these basins (Fig. 4).

In the core of the anticyclonic eddy in the northern Aleutian basin, surface waters are depleted of inorganic nitrogen but are high in ammonium and DON (2.0-2.4 and 12-16 µM, respectively). These waters also show a high percentage of oxygen saturation (up to 116-120%) indicating intense photosynthesis in surface layers.

In autumn, high concentrations of ammonium (0.8-1.2 µM) were observed in deep layers. This is more than double the winter-spring period (0.4-0.6 µM). During autumn vertical mixing, phytoplankton development couples with the input of phyto- and microzooplankton, and various organic substance into the subsurface layer. From the shallow zone, where vertical mixing usually includes bottom waters, waters with elevated content of POM and high concentrations of ammonium are transported to the stationary convergence zone on the continental slope and there sink to the cold intermediate waters.

We also note extremely high concentrations of ammonium nitrogen (3.0-6.0 µM) in the surface layer of shallow waters in Karagin Bay and southern Olyutorsk Bay. Maximum values of ammonium were recorded in autumn in the surface layer and could be explained by vertical mixing, onshore currents, offshore winds, and compensatory currents. This observation differs from the vertical distribution of ammonium in summer when maximum values occur in bottom waters

When nitrates were completely consumed in the euphotic layer on the shelf of the western Bering Sea, phytoplankton satisfied its nitrogen demands with urea and ammonium, amounting to 6-11 µM. This provided PP values of about 4-6 gC/m^2 daily, but this production relied on recycling and could not provide a long-term transfer to higher trophic levels.

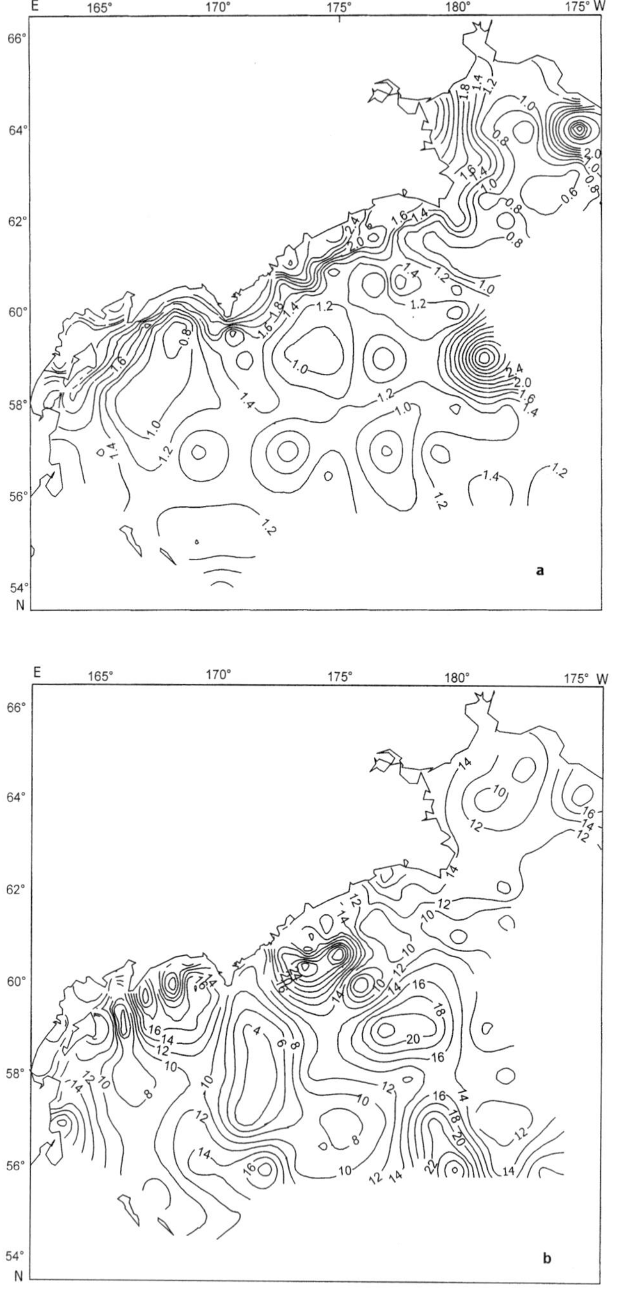

Figure 4.   *The distribution of nitrogen compounds (µM) in the surface layer of the western Bering Sea (October-November 1990): (a) ammonium nitrogen; (b) organic nitrogen.*

Studies in the western Bering Sea revealed the following annual pattern: In shelf waters, maximum quantities of ammonium nitrogen in the photic layer (up to 8-10 µM) were observed in July. When nitrates were almost absent, concentrations of DON were about 10 µM, and nitrites were about 0.1 µM. In pelagic waters, the content of nitrate nitrogen increased to 12-24 µM, while ammonium and DON slightly decreased. Maximum nitrate values in surface waters (20-25 µM) were typical of the winter-spring period when all production utilized inorganic nitrogen, particularly nitrates and nitrites. A similar seasonal distribution of ammonium nitrogen characterized the eastern Bering Sea (Whitledge et al. 1986).

Conclusions

Maximum quantities of ammonium nitrogen were observed in summer when it was present in significant amounts in the entire water column (0-1,000 m). However, maximum concentrations occurred in the 0-200 m layer, and values of up to 6.5 µM were observed at the boundary of the seasonal thermocline, suggesting intense production and destruction processes. Maximum amounts of urea were usually found in the pycnocline, or directly underneath, due to the accumulation of heterotrophic biota in the pycnocline. In summer, organic nitrogen made up 65-95% of the total, and attained values of 14-20 µM. Nitrates did not exceed 3-4 µM in areas of maximum phytoplankton accumulations, particularly on the shelf and at the shelf break. In pelagic areas, where there was no intensive phytoplanktonic development, the contribution of DON in the surface layer decreased to 15-30%. The contribution of inorganic nitrogen increased with depth, while the organic forms declined to 5-15% of the total amount. During summer, primary productivity on the shelf and on the continental slope developed almost entirely from recycled nitrogen, especially when surface concentrations of urea and ammonium were 3.0-4.0 µM and nitrate concentrations did not exceed 1.0 µM. Also during summer (and autumn), significant concentrations of urea, ammonium, and DON were found in the Bering Sea which provided for the development of large quantities of nitrates in the deep Bering Sea. Large concentrations of urea in both surface and deep waters in summer were closely related to high biological productivity of Bering Sea waters. Similar concentrations of urea were observed in Antarctic waters and in the upwelling zones of Peru and California.

References

Cooper, L.N.H. 1937. The nitrogen cycle in the sea. Journal of the Marine Biological Association, U.K. 22:183-204.

Dagg, M.J., J. Vidal, T.E. Whitledge, R.L. Iverson, and J.J. Goering. 1982. The feeding, respiration, and excretion of zooplankton in the Bering Sea during a spring bloom. Deep-Sea Research 29:45-63.

DeManche, J.M., H. Curl Jr., and D.D. Coughenower. 1973. An automated analysis for urea in seawater. Limnology and Oceanography 18:686-689.

Glibert, P.M., D.C. Biggs, and J.J. McCarthy. 1982. Utilization of ammonium and nitrate during austral summer in the Scotia Sea. Deep-Sea Research 29:837-850.

Harvey, W.A., and J. Caperon. 1967. The rate of utilization of urea, ammonium, and nitrate by natural population of marine phytoplankton in an eutrophic environment. Pacific Science 30:329-340.

Kavakina, S.V., and V.V. Sapozhnikov. 1993. The major regularities of the urea distribution in the Bering Sea and the urea role in the nitrogen cycle. Oceanology 33:871-877. (In Russian.)

McCarthy, J.J. 1970. A urease method for urea in seawater. Limnology and Oceanography 15:309-315.

Moiseev, E.V. 1995. The particularities of the distribution of nanoheterotrophic organisms in the western Bering Sea in June, 1992. Kompleksnie issledovanya ekosistemi Beringova morya, Sbornik trudov, VNIRO, Moscow. (In Russian.)

Newell, B.S., B. Morgan, and J. Cundy. 1967. The determination of urea in seawater. Journal of Marine Research 25:201-202.

Remsen, C.C. 1971. The distribution of urea in coastal and oceanic water. Limnology and Oceanography 16:732-740.

Sapozhnikov, V.V. 1987. The determination of ammonium nitrogen in seawater (ammonia analysis by Sedgi-Solorzano): Methods of hydrochemical studies of the ocean. Nauka, Moscow 1987:179-185. (In Russian.)

Sapozhnikov, V.V., and L.N. Propp. 1990. The major regularities of the distribution of urea in the antarctic waters. Oceanology 37:144-147. (In Russian.)

Sapozhnikov, V.V., and V.L. Zubarevich. 1989. The distribution of urea in the Black Sea. Tezisi doklada na Mezhdunarodnom biogeokhim. kongress, Moscow. (In Russian.)

Sorokin, P.Y. 1995. Planktonic infusorias in the Bering Sea and the northern Pacific. Complex research on the Bering Sea, collected studies, VNIRO. Moscow, pp. 287-293. (In Russian.)

Sorokin, Y.I., T.L. Mamaeva, and P.Y. Sorokin. 1995. On the characteristics of bacterioplankton in the Bering Sea and the adjacent northern Pacific. Complex research on the Bering Sea, collected studies, VNIRO. Moscow, pp. 280-286. (In Russian.)

Spravochnik. 1991. Reference book of hydrodynamics: Fisheries. Agropromizdat, Moscow. 224 pp. (In Russian.)

Whitledge, T.E., W.S. Reeburgh, and J.J. Walsh. 1986. Seasonal inorganic nitrogen distribution and dynamics in the southeastern Bering Sea. Continental Shelf Research 5:109-132.

Whitledge, T.E., R.R. Bidigare, S.I. Zeeman, R.N. Sambrotto, P.F. Roscigno, P.R. Jensen, J.M. Brooks, C. Trees, and D.M. Veidt. 1988. Biological measurements and related chemical features in Soviet and United States regions of the Bering Sea. Continental Shelf Research 8:1299-1319.

CHAPTER 18

Paleoceanographic Changes and Present Environment of the Bering Sea

Kozo Takahashi
Faculty of Science, Kyushu University, Fukuoka, Japan

Abstract

The process of efficient biological pumping with high biological productivity was determined for the Bering Sea by employing a long-term time-series sediment trap deployed during 1990-1995. The present Bering Sea is an effective atmospheric CO_2 sink with significant drawdown of the CO_2. The high efficiency attributes to markedly high opal content of 69% in total mass flux, which is followed by 13% calcium carbonate content. The nodal location of the Bering Sea, in terms of the Pacific-Arctic-Atlantic gateway connection, makes this marginal sea significantly important for water circulation, balances of heat and salt, and various chemical properties, many of which affect global climate and mass balance. The present situations of the "opal" Pacific Ocean and the "carbonate" Atlantic Ocean were different during the glacial low stands primarily due to closure of the Bering Strait gateway caused by sea level drop. With a longer time scale than the Milankovitch cycles, Beringia (or the Bering land bridge) subsided a number of times due to tectonic movement, allowing intrusion of Atlantic signals. Based on mollusk faunal distribution in northern Japan, the Bering gateway must have been open initially at earlier than 5.1 Ma in the Late Miocene, followed by a number of openings and closures due to tectonics during the Pliocene. Detailed history of such openings and closures must be investigated with scientific drilling in the Bering Sea. The formation of the North Pacific Intermediate Water is established by evidence including stable isotopes, microfossils, and detritus sediments. The Meiji Drift is composed of sediments derived from the past Bering Sea, forming contourite deposits on the flank of the northern Emperor Seamount region just south of Kamchatka Strait.

Introduction

Subpolar regions, including marginal seas, play significant roles in the global carbon cycle and hence are important factors of global climate change (e.g., Tans et al. 1990, Wong et al. 1995). Surface waters in these regions, which have the potential to absorb atmospheric CO_2, contribute to this change due to high biological productivity. There are three principal belts of high biological productivity in the world oceans, including the subarctic belt (both Pacific and Atlantic oceans), the equatorial upwelling belt (Pacific, Atlantic, and Indian oceans), and the circumpolar subantarctic belt (Berger et al. 1987). Moreover, the high biological productivity in the upper ocean involves either emission or absorption of atmospheric CO_2. It is generally concluded that the equatorial belt is the largest natural source of atmospheric CO_2 (Tans et al. 1990, Murray 1995). The remaining two subpolar belts are generally regarded as CO_2 sinks. Based on measured carbon and opal particle fluxes using sediment traps, Wong et al. (1995) showed that the western subarctic Pacific is also a CO_2 sink with a fairly effective biological pump. Analogous information from the northern marginal seas of the Pacific region, such as the Bering Sea, has been meager (Takahashi et al. 1997). However, available evidence suggests that the Bering Sea plays a large role in the global material balance and, in turn, climate change (Takahashi et al. 1997; Takahashi et al., in press).

In this paper I review the current knowledge and discuss the importance of the Bering Sea as a northern marginal sea of the North Pacific, in terms of global mass balance and Atlantic-Pacific connections in the geologic past. Present-day high productivity in the Bering Sea based on biogenic particle fluxes will be presented and compared with those in other marginal seas and pelagic regions. The processes of water mass exchange between the marginal sea and the Pacific Ocean and/or the Arctic Ocean are important for understanding material and heat balances as well as climate change. Studies of paleoceanographic changes recorded in the sea provide pertinent information concerning the evolution of Northern Hemisphere glaciation in association with the Milankovitch orbital cycles, and other high frequency cycles such as Dansgaard-Oeschger cycles. The past climatic-paleoceanographic changes and the need for further studies in these regions will also be discussed

Geomorphology

The Bering Sea has a surface area of 2.29×10^6 km^2, a volume of 3.75×10^6 km^3, and is the third largest marginal sea in the world, surpassed only by the Mediterranean and the South China seas (Hood 1983). Three major rivers empty into the Bering Sea: the Kuskokwim and Yukon, draining central Alaska, and the Anadyr draining western Siberia (Fig. 1). The Yukon is the longest, supplying the largest discharge. Its discharge from the land is 4×10^4 m^3/s, peaking in August, which roughly equals the amount of the

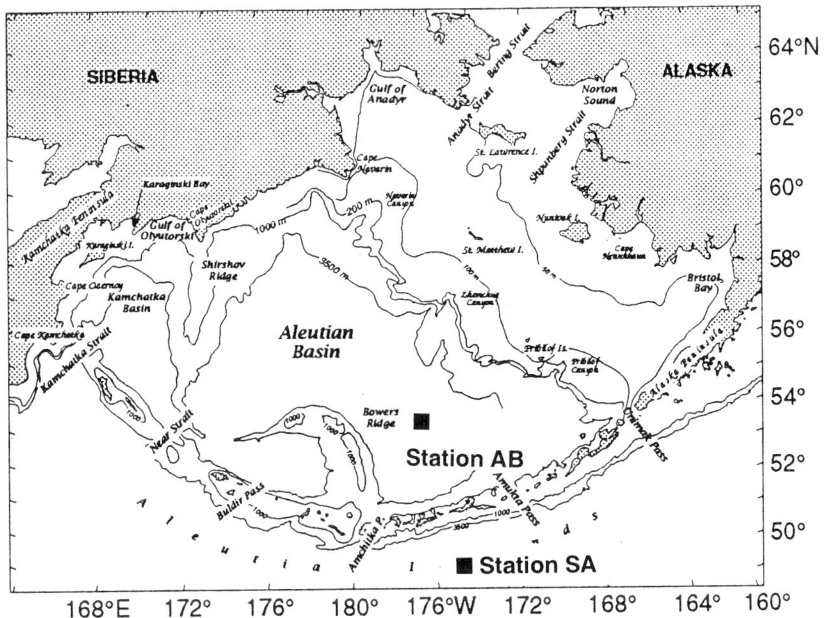

Figure 1. *Major topographic features of the Bering Sea and Aleutian Islands. Contours of 100, 200, 1,000, and 3,500 m are shown. (Basic map from U.S. GLOBEC 1996.)*

Mississippi River. The Yukon's annual mean flow is 5×10^3 m³/s, which is about two-thirds the annual flow of the Columbia River (Hood 1983).

Approximately one-half of the Bering Sea is covered by a shallow (0-200 m) neritic area (Fig. 1). The majority of the continental shelf spans the eastern side of the sea off Alaska, from Bristol Bay in the south to the Bering Strait in the north. The northern continental shelf is seasonally covered by ice, while ice is rarely present over the deep southwest areas (Japan Meteorological Agency 1990-1994). The continental slope occupies only 13% of the total Bering Sea area and has a slope of generally 4-5 degrees (Hood 1983).

Other than the shelf regions, two significantly high topographic features provide better calcium carbonate preservation than the basins (Supko 1973). One is the Shirshov Ridge which extends south from Kamchatka along 170°E, separating the Aleutian Basin into eastern and western parts. The second is Bowers Ridge (sometimes referred to the North Rat Island Ridge/Bank) and extends 300 km north from the Aleutian Island Arc (Fig. 1). The Aleutian Basin is a vast plain lying at a depth of 3,800-3,900 m with sloping hollows to depths of as much as 4,151 m (Hood 1983).

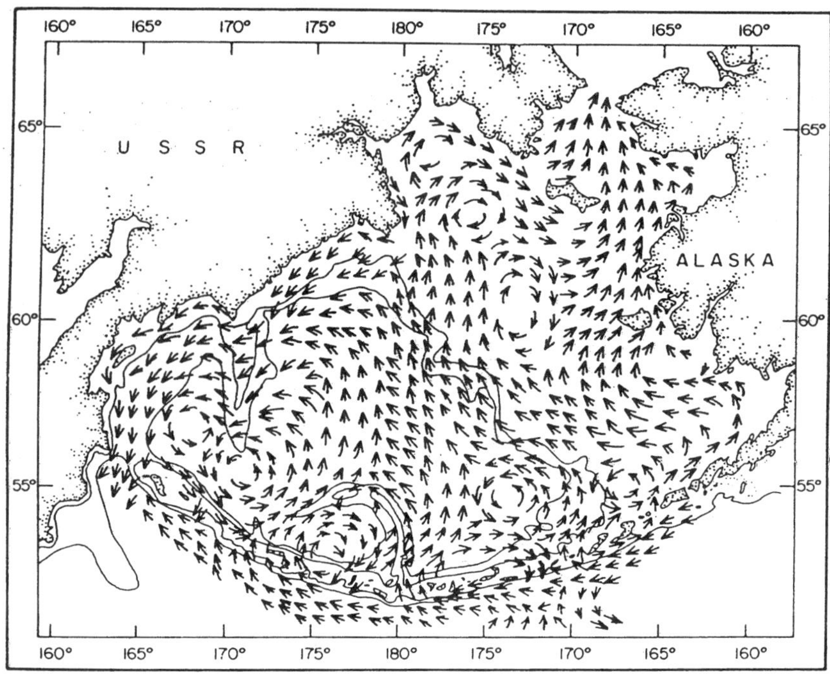

Figure 2.   Map showing surface currents in the Bering Sea. (From Arsenev 1967, cited in Hood 1983.)

Physical Oceanography

The Alaskan Stream, which is an extension of the Alaskan Current flowing westward along the Aleutian Islands in the subarctic Pacific, mainly enters the Bering Sea through the Amchitka Pass and the pass west of Attu Island in the eastern Aleutian Islands (Favorite et al. 1976, Arsenev 1967 cited in Hood 1983). A part of the Subarctic Current also joins the northward flow coming from the Alaskan Stream, resulting in a combined volume transport of 11 Sv (Ohtani 1965, 1973). Much of the Pacific water masses entering the Bering Sea leaves through passes in the Aleutian Islands (Fig. 2). The most significant one is through the Kamchatka Strait whose present depth is 4,420 m. If the glacial North Pacific Intermediate Water mass was formed in the Bering Sea, the Kamchatka Strait was the major passage through which it flowed out, followed by a secondary one at the Commander-Near Strait at 2,000 m present-day depth. A part of the water in the Bering Strait flows out to the Chukchi Sea in the Arctic, whose detail will be discussed in a later section since it has a paramount importance to global mass balance and climate. Inside of the Bering Sea a large scale counterclockwise surface water circulation is recognized along the continental slope and the Aleutian Islands in the Aleutian and Kamchatka basins (Fig. 2; Ohtani et al. 1972, Arsenev 1976 cited in Hood 1983). However, in the Bowers Basin the direction of the surface water flow is clockwise. Furthermore, on the Bering Sea Shelf there are at least three clockwise surface water circulation patterns recognized (Fig. 2).

According to Ohtani et al. (1972), the Bering Sea is a source region for Western Subarctic Pacific water which plays a major role in the circulation of the western subarctic Pacific. The Western Subarctic Pacific water is characterized by marked stratification with cold upper layers in winter and a remarkable dichothermal layer at around 100 m in summer. He further stated that vertical structure of temperature and salinity in the Aleutian Basin varies widely, from significant stratification to homogenous gradient in the upper 500 m. During winter, the patterns of vertical profiles become near vertical with much less stratification due to strong mixing which is caused by severe wind stress. Although there are varieties of summer vertical structures, they are characterized by the presence of a dichothermal layer. One example of the variation is that double dichothermal layers are found in the northeastern corner of the Aleutian Basin along the continental shelf (Ohtani et al. 1972).

Significance of Present High Biological Productivity

The Bering Sea is one of the most biologically productive areas in the world (Berger et al. 1987), as evidenced by large quantities of fish caught annually in the region (Ohtani and Azumaya 1995). A measure of this productivity was made with a time-series sediment trap, deployed at a fixed station. Particle fluxes were continuously measured beginning in August 1990 through the present (1998) in the Bering Sea (Fig. 1; Station AB: 53.5°N, 177°W; water depth: 3,788 m; trap depth: 3,198 m). The 5-yr record for 1990-1995 shows a mean daily total mass flux or export production of 177 mg/m^2 per day (see Takahashi et al. 1997; Takahashi et al., in press; Maita et al., chapter 16, this volume) which is one of the highest in the world. This value contrasts with a pelagic station just outside the Bering Sea, Station SA (49°N, 174°W; water depth: 5,406 m; trap depth: 4,812 m), in the central subarctic Pacific. At Station SA a 5-yr mean total mass flux of 91 mg/m^2 per day was measured (Takahashi et al. 1997; Takahashi et al., in press). Thus, Bering Sea data reveal that this sea produces approximately twice as much total mass flux as found in the pelagic Pacific station. The same trend is also seen in diatom flux (Fig. 3; Takahashi et al. 1996, 1997).

Furthermore, based on species list and percentage contribution of each taxon of siliceous and calcareous shell-bearing plankton fluxes, it

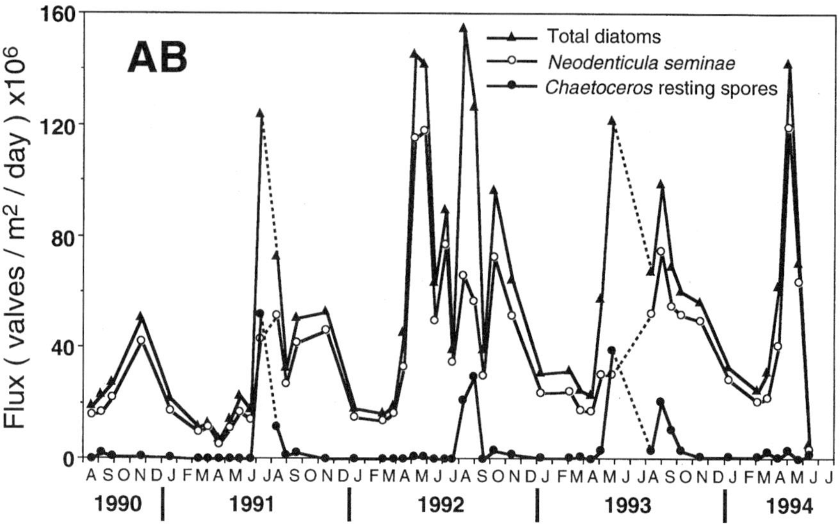

Figure 3.   Four-year-long fluxes of total diatoms, Neodenticula seminae, and Chaetoceros resting spores measured with a sediment trap at Station AB in the
Aleutian Basin of the Bering Sea, August 1990-July 1994. (From Takahashi et al. 1996.)

appears that biological ecosystems operating in the two regions are slightly
different in detail, but major components and structures are similar to
each other. The number of constituting diatom species in flux at Stations
AB and SA are 53 and 29 taxa, respectively, which appear substantially
different. However, many minor diatom species only occurring at Station
AB do not contribute significantly to total diatom fluxes. Therefore, the
minor species are numerically unimportant, although they have values as
tracers of neritic, benthic, and/or hemipelagic environmental conditions.
Chaetoceros radicans and other *Chaetoceros* resting spores are exceptional in that they can be included as major species category only at Station
AB, but not at Station SA. *Chaetoceros radicans* resting spores on the average contributed 4.0% at Station AB (1990-1994 record: Takahashi et al., in
press). In contrast they contributed only negligible fluxes during fall 1991
and fall-winter 1993-1994 at Station SA (Takahashi et al. 1996). *Chaetoceros*
resting spores have more hemipelagic and neritic characteristics rather
than pelagic characteristics reflecting the differences at the two stations.

Moreover, *Neodenticula seminae,* a dominant pennate diatom taxon,
contributed similarly 81.4% and 70.2% of the total diatom valve fluxes at
Stations AB and SA, respectively (1990-1994 record: Takahashi et al. 1996;
Takahashi et al., in press). Furthermore, flux percentage contribution of
Coccolithus pelagicus, a dominant coccolithophore species in the Bering

Sea, to total coccolithophore assemblage, is similar at the two stations (Station AB: 56.2%; and Station SA: 39.7%), followed by *Emiliania huxleyi* making up the remaining percentage at these sites; the minor species *Gephyrocapsa oceanica* and *Calcidiscus leptoporus* were negligible (<0.01%: 1990-1995 record: Takahashi et al., in press). Thus, it is fair to conclude that although the Bering Sea has more hemipelagic to neritic characteristics and is twice as productive relative to the pelagic counterpart, flux percentages of the major diatoms and coccolithophores are similar, suggesting qualitatively similar ecosystems are operating in the two regions.

Sediment trap data from the northern Aleutian Basin (58°N, 179°E, north of Station AB) had a total mass flux of 144 mg/m^2 per day (Honjo et al. 1995) during 1991-1992, 34% less than the total mass flux measured at Station AB (193 mg/m^2 per day during the same 1991-1992). This difference indicates a geographic variability within the basin that is smaller than between the basin and the North Pacific. In addition, Honjo and his colleagues deployed sediment traps at Shoyo Station in the northern Okhotsk Sea and measured a mean flux of 129 mg/m^2 per day during the 1990-1991 period (Honjo et al. 1995), illustrating a similarly high flux in the northern marginal sea of the Pacific.

These data from traps clearly indicate higher total mass fluxes in marginal seas with respect to the world's oceans (Honjo 1990, Honjo et al. 1995). One should recognize that the high fluxes are associated with high opal contribution (Fig. 4) mainly due to diatoms (Takahashi 1991, Takahashi et al. 1997: Fig. 3). Five-year means for the percent weight contribution of the total mass fluxes measured at Station AB are as follows: biogenic opal (opal hereafter): 69%; calcium carbonate: 13%; and organic matter: 10% (Fig. 4). Such a high opal contribution is unprecedented in flux measurements. For a comparison, the 5-year mean of opal values at Station SA in the central subarctic Pacific is 53%, which is substantially lower than that of the Bering Sea, but still significantly high for the world standard. Furthermore, opal flux contributions in the Equatorial Pacific ranging from 9°N to 12°S represent 12-35% of total mass flux, which ranges from 14 to 132 mg/m^2 per day (Honjo et al. 1995).

The particle flux constituents from the Bering Sea, which are made of high opal and relatively low calcium carbonate contents, are ideal for an effective biological pump. The major reason for this is that the production of siliceous plankton with opal shells involves only cytoplasmic organic carbon which draws CO_2 into the upper ocean, without incorporating carbon in their shells. This organic carbon production does not cause CO_2 emission. Production of shells by the calcareous plankton, on the other hand, involves both CO_2 incorporation by organic matter formation and CO_2 emission by calcium carbonate shell formation. Ratios depend on various factors such as taxon, season, and physiological conditions. A ratio of organic carbon to inorganic carbon (C_{org}/C_{inorg}) is an important indicator when determining whether a CO_2 sink or source situation exists (Berger and Kier 1984). The C_{org}/C_{inorg} ratios measured at Station AB were almost

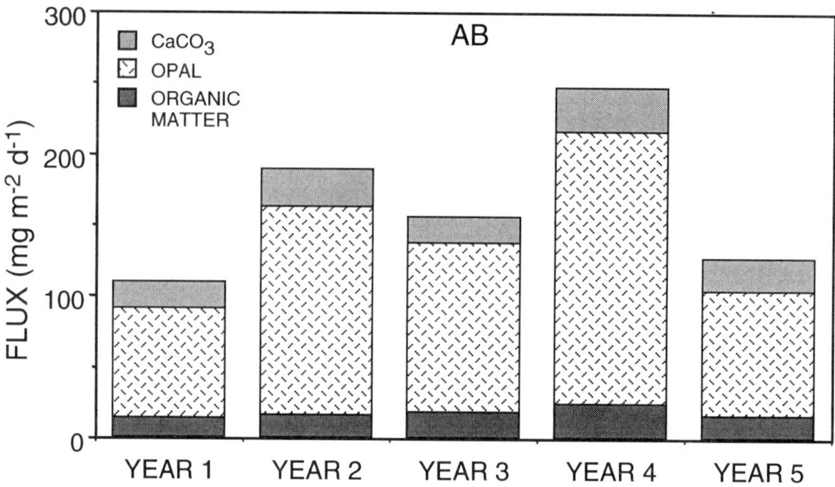

Figure 4.   Five-year record of annual mean fluxes of organic matter, opal, and CaCO$_3$ at Station AB in the Bering Sea. (From Takahashi et al. 1997.)

always greater than one throughout the 5 years of the measurements, indicating that the Bering Sea is an efficient atmospheric CO_2 sink (Takahashi et al. 1997; Takahashi et al., in press). Furthermore, although values are slightly less than those at Station AB, Station SA in the central subarctic Pacific also represents a CO_2 sink. Therefore, both the Bering Sea and the central subarctic Pacific play a significant role as CO_2 sinks in the carbon cycle.

It is of interest to delineate how the Bering Sea may have behaved during past glacial and interglacial cycles in terms of the carbon cycle and climate change. Much has to be considered: siliceous and calcareous microfossils, carbon isotopes, organic compounds and other relevant indicators that are preserved in the sediments of these regions.

Present Gateway to the Arctic Ocean and Its Significance

As the largest semi-enclosed marginal sea of the Pacific rim, the Bering Sea's indisputable importance has been recognized in various oceanographic processes (Tsunogai et al. 1979, Lisitzin 1972, Ohtani et al. 1972, Edmond et al. 1979, Craig et al. 1981, Sancetta 1981, Sambrotto et al. 1984). The northerly outflow of Bering Strait surface water is important since it flows one way out to the Chukchi Sea in the Arctic Ocean, although the amount is less than water exchange through the Aleutian passes. The amount of water passing into the Chukchi Sea is estimated to be 0.8 Sv (Coachman and Aagaard 1981). This "Pacific" origin water eventually flows out to the Atlantic via the Arctic Ocean. The Bering Strait also provides one of the highest biologically productive areas in the world approaching 324 gC/m^2 per yr over a wide area (2.12 × 10^4 km^2: Sambrotto et al. 1984). Much of the biological production of organic matter and associated nutrients flowing into the Arctic Ocean are due to today's northerly current direction.

One-way flow into the Arctic Ocean, however, may not have been always true in the past, and this high nutrient, biologically productive flow may have had a profound effect in the nature of the greater carbonate production in the Atlantic and opal production in the Pacific: the carbonate ocean versus silica ocean hypothesis (Honjo 1990). During glacial periods the Bering Strait, which is about 50 m deep today, was aerially exposed due to sea level drop closing the Bering-Arctic gateway. What was the impact on global water circulation then? The Bering water mass circulation and river discharge during the glacial periods was exclusively southward. The glacial Yukon River discharge, for example, had no alternative but to eventually come out of the Bering Sea to the North Pacific, affecting the salinity of Pacific waters.

A unidirectional flow of Bering Strait water which reaches the Atlantic should affect not only heat balance, but also salt balance and hence the formation of deep water masses in the Atlantic. It is known that during glacial intervals, the Atlantic Ocean became more Pacific-like, and that the Pacific Ocean became more Atlantic-like with regard to circulation as well as carbonate and siliceous microfossil preservation. This situation is known as the basin-to-basin fractionation model by Berger (1970). Thus, the siliceous and carbonaceous sedimentary record provides one of the clues to changes in the past water circulation. Glacial intervals are characterized by better preservation of calcareous plankton in the Pacific, which can be considered as a shift to more of an Atlantic type (Berger 1970). Such a shift made the two great oceans far more even in water mass chemistry than they are today, as evidenced in the microfossil records. Since the global circulation of water masses significantly affects climate, we need to investigate paleoceanographic changes recorded in Bering Sea and perhaps Arctic Ocean sediments in order to solve how this type of excursion occurred and how quickly it shifted from one state to the other.

Beringia and the Bering Gateway in the Past: Atlantic Connection through the Arctic Ocean

Prior to presenting the Quaternary glacial events, let us briefly discuss major tectonic events which occurred with a longer time scale than the Milankovitch cycles. This includes the Bering land bridge called Beringia as well as the intrusion of the Bering marine gateway. Detailed paleogeography of the Tertiary Period, especially the Paleogene, has not been well understood (Worrall 1991). However, the history of the Neogene Bering gateway is slightly better understood. The North Atlantic-Arctic Ocean-North Pacific connection can be argued for on the basis of paleo-biogeography of mollusks. Uozumi et al. (1986) found *Tridonta alaskensis* and *Tridonta borealis* in the Atsuga Formation in central Hokkaido, Japan. These bivalve species are known to have intruded from the North Atlantic. The age of the Atsuga Formation has been determined to be older than 5.1 Ma, employing a fission truck method (Uozumi et al. 1986). Sagayama et al. (1992) examined diatoms and placed the Atsuga Formation in the lower part of the *Neodenticula kamtschatica* partial-range zone, which ranges from 6.3 to 3.1 Ma (Barron 1985). Thus, the Atlantic-Pacific connection dates back to the Late Miocene. Furthermore, Ogasawara and Gladenkov (1995), analogous to the above, showed that the marine gateway connection through the Bering region initially occurred around 4.2-3.0, and subsequently at 2.5 and 2.2 Ma, based on biogeographic occurrences of mollusks originating from the northern Atlantic or northern Pacific regions.

Bering Sea Deepwater Exchange with the Pacific Ocean: Past and Present

There is one piece of geological evidence suggesting that bottom water of the North Pacific Ocean was generated in the Bering Sea in the past. Mammerickx (1985) discussed the possibility of bottom thermohaline circulation as a cause of the Meiji sediment tongue (Ewing et al. 1968) or the Meiji Drift (Scholl and Stevenson 1997), whose features are similar to the North Atlantic drifts in their general shape, length, and thickness (Scholl et al. 1977). In this scenario, the sediments in the Meiji Drift were supplied from the Bering Sea through Kamchatka Strait. Kamchatka Strait represents the only deep strait (4,420 m) where deepwater masses can exit from the Kamchatka Basin (i.e., Commander Basin) to the North Pacific. Other deep straits include the Commander-Near Strait (2,000 m), and the rest of the straits in the Aleutian Islands which are less than 1,155 m deep. The fact that the Meiji Drift thickens toward the Kamchatka Strait implies that the Bering Sea source of the past bottom water flow (D. Scholl, pers. comm., 1998). According to Scholl et al. (1997), from approximately 30 to at least 5 Ma, the Oligocene to the early Pliocene, the sediments from the Bering Sea were carried into the Pacific and distributed along the Meiji Drift by bottom water impinging against the northern flank of the submerged Emperor Seamounts. In addition, evidence for recent contact of the bottom waters with the atmosphere in the past 40 years is provided by Warner and Roden (1995). They found anthropogenic chlorofluorocarbons in bottom waters of the Aleutian Basin, an indication of recent ventilation of the deep Bering Sea.

Paleoceanography of North Pacific Marginal Seas

Sancetta (1981) derived four diatom assemblages based on factor analysis on diatoms from the surface sediments of the Bering Sea (Fig. 5). She demonstrated that diatom assemblages show close correlation to the distribution of major water masses. The most important is the Bering Basin assemblage, which occupies the entire Aleutian Basin and extends to the Commander Basin (Fig. 5a). This assemblage is also found on the eastern continental slope, but it decreases rapidly with decreasing depth and is absent above 200 m. This assemblage is dominated by *Neodenticula seminae* with minor contributions from *Rhizosolenia hebetata* forma *hiemalis* and *Thalassiosira trifulta*. The distribution of this assemblage matches closely with the water mass typical of the Alaskan Stream (9°C at the surface) and its derivative (6°C) with admixture of waters coming from tidal mixing in the Aleutian passes and the Bering Shelf (Sancetta 1981).

Moreover, she showed three additional assemblages, Bering Shelf, productivity, and sea-ice assemblages (Fig. 5b-d), each of which has significantly characteristic distribution with water masses. The Bering Shelf assemblage is associated with benthic littoral genera such as *Melosira/Paralia* and *Delphineis*. The productivity assemblage is dominated by *Chaetoceros* resting spores, which are indicative of high productivity with large surface water temperature fluctuations from spring to fall. The sea-ice assemblage is found on the northern Bering Shelf and the Chukchi Sea and is principally composed of *Thalassiosira nordenskioldii, Nitzschia grunowii,* and *Nitzschia cylindrus* (Sancetta 1981).

Gorbarenko (1996) recently published paleoceanographic changes in the Bering Sea and adjacent regions during the Late-Glacial and Holocene based on stable isotopic and sediment measurements on a number of cores raised in the regions. Based on core K-119 (50°25′N, 167°44′E, 2,440 m) from the Detroit Seamount in the western subarctic Pacific, he concluded that with better ventilation during the glacial period, the formation of the intermediate water occurred in the Bering Sea and spread into the western subarctic Pacific to a depth of 2,440 m. He also discussed that ice-rafted debris observed in the Bering Sea and western subarctic Pacific were transported by sea ice, but not icebergs, and left a strong sediment signature in the Kamchatka Strait and immediately adjacent parts of the western subarctic Pacific. One of the major conclusions which Gorbarenko (1996) presented was that the North Pacific low salinity surface water signal did not propagate from the North Atlantic; instead, the low saline surface waters of both oceans independently and simultaneously spread quickly in the northern regions. Moreover, calcium carbonate peaks observed at the terminations 1A and 1B in core 2594 raised from the Shirshov Ridge in the Bering Sea (56°56′N, 169°53′E; Fig. 6) are synchronous in the western subarctic Pacific and its marginal seas as well as in the North Atlantic and

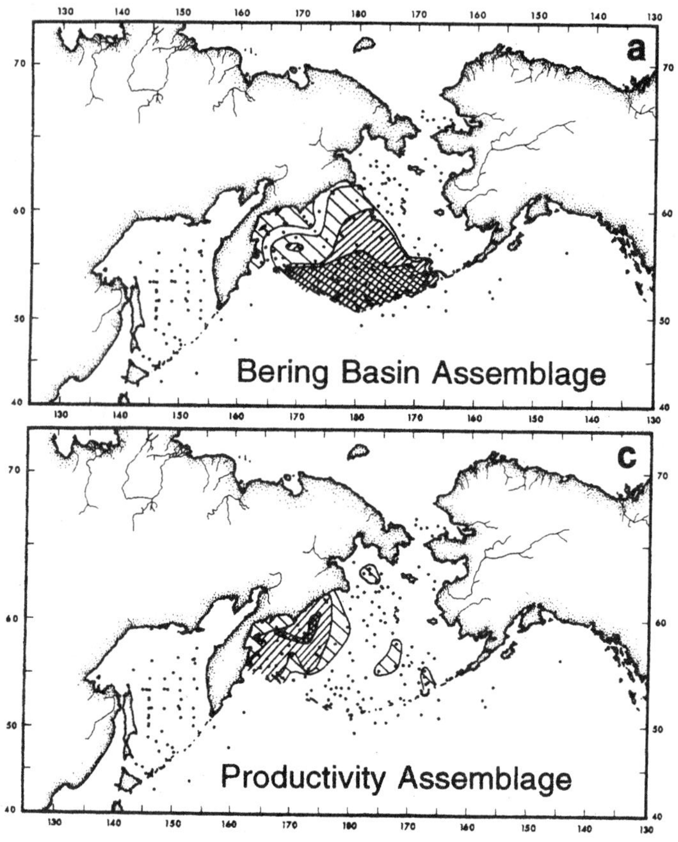

Figure 5. Distribution of four diatom assemblages derived by a factor analysis in the Bering Sea: Bering Basin assemblage, Bering Shelf assemblage, productivity assemblage, and sea-ice assemblage. Contours at factor loadings of 0.900 (cross hatch), 0.600 (fine hatch), and 0.300 (coarse hatch) (from Sancetta 1981).

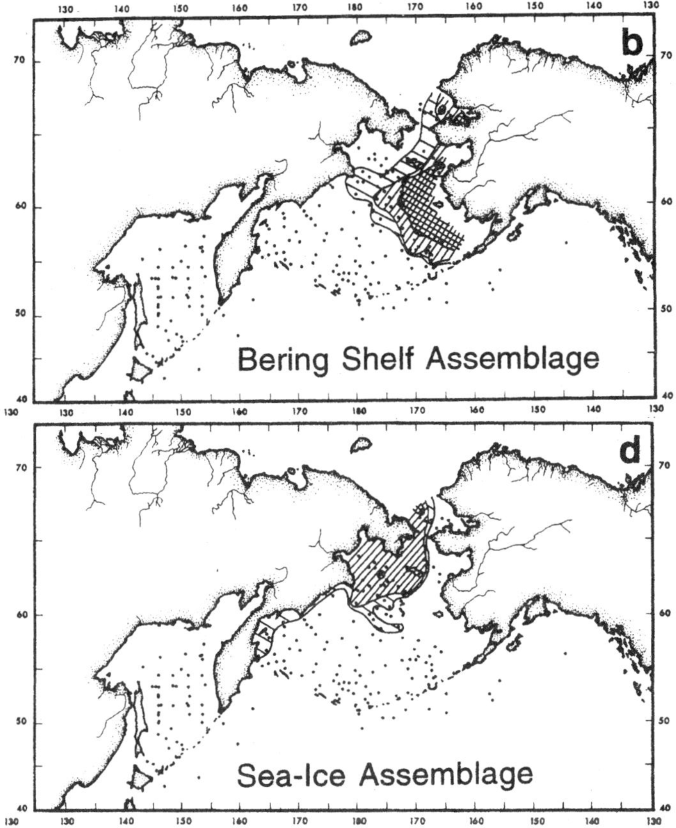

Figure 5. (Continued.)

have ages of 12,500 and 9,300 yr B.P., respectively. Paleoproductivity in the Bering Sea was lower during the glacial period than today, based on organic carbon contents. This is demonstrated by low glacial diatom contents (Fig. 6). Such a decreased glacial productivity is due to partial ice cover which prevented insolation during the glacial period. The diatom contents increased after termination 1A and decreased once, but increased again after termination 1B which continued into the present (Fig. 6).

Further, the northern marginal seas of the North Pacific (Bering, Okhotsk, and Japan seas) experienced major climate changes during the late Pleistocene. Because of the semi-closed nature of the marginal seas, strong signatures due to environmental changes of millennial time scale, which have a higher frequency than the Milankovitch cycles, can be faithfully recorded in the sediments. For instance, a sequence of laminated and non-laminated sediments in the Japan Sea testifies that exchange of

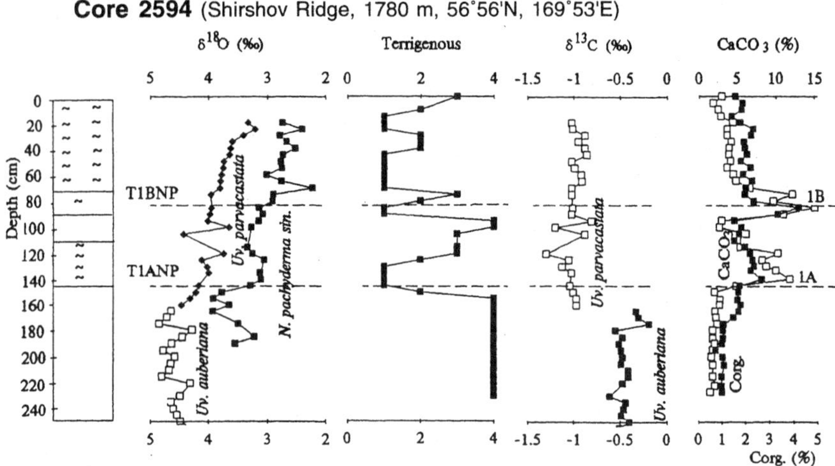

*Figure 6. Paleoceanographic measurements on core 2594 raised from the Shir-
shov Ridge (56°56'N, 169°53'E, 1,780 m), Bering Sea. Illustrated are layers of diatomaceous sediments, δ¹⁸O and δ¹³C of planktonic foraminifera
Neogloboquadrina pachyderma sinistral and benthic foraminifera Uvi-
gerina parvacastata, percent terrigenous matter, percent CaCO₃, and
percent Cₒᵣ₉ (from Gorbarenko 1996).*

Japan Sea water with that of open pelagic Pacific water occurred and was
periodically restricted during the Pleistocene (Tada et al. 1992). The laminated sequence can be correlated with Dansgaard-Oeschger cycles which
were identified in the Greenland ice core record (Dansgaard et al. 1993)
and represents global events rather than regional phenomena (Broecker
1994, Tada et al. 1992). According to Broecker (1994), there are seven
locations in the world (outside the North Atlantic) which can be correlated
on the basis of Dansgaard-Oeschger cycles, Heinrich events, or the Younger Dryas. The atmospheric temperature recorded in Greenland ice cores,
as well as lake sediment records, can be linked with high resolution marine sedimentary records in the marginal seas. Moreover, Behl and Kennett (1996) recently found that a cyclic sedimentary sequence in the Santa
Barbara Basin, analogous to the Japan Sea laminated and non-laminated
sequence, was formed under the influence of a northern source of oxygenated Pacific Intermediate Water. A part of the North Pacific Intermediate Water is thought to be presently formed in the Bering Sea (Tally 1991).
Riser (1997), however, suggests the likelihood of NPIW formation in the
Okhotsk Sea, instead.

Paleoenvironmental work generated very useful predictions for future climate. Scholl and Stevenson (1997) argue that the condition that
once existed in the Aleutian Islands may return. The paleoenvironmental

conditions in the Aleutian Arc and adjacent marine realm include the growth of redwood with open skies and increased surface water evaporation during the Oligocene to the early Pliocene, 30 to at least 5 Ma.

They explain their scenario as follows. The present transport volume of relatively warm Alaskan Stream water, entering the Bering Sea through Aleutian Island passes such as the Near Pass, is sufficient to keep Bering Sea Basin surface water from winter freeze-over. The salinity of Bering Sea surface water is low because there is less evaporation than precipitation chiefly due to foggy and overcast summer weather conditions. The density of such low saline surface water does not sufficiently increase to initiate vertical thermohaline circulation in winter today. Sea ice at present only extends to the southern margin of the Bering Shelf. The Aleutian Arc is presently covered with a carpet of tundra.

Conditions can be changed, however, and the winter freeze-over of Bering Sea surface water can occur as follows. If and when Bering Sea deepwater channels, such as Kamchatka Pass, become shallow or narrow by tectonic movement, which will occur within 2 million years, the rate of water exchange between the Bering Sea and the Pacific will decrease. This will allow winter freeze-over, which subsequently forces the Aleutian Low to move toward the Gulf of Alaska. Such an atmospheric shift would further weaken the Bering/Pacific water exchange rate as well as weaken the Bering gyre. This will allow warmer Pacific water to drift into the Bering Sea, causing the return of the coniferous forest, increased surface water evaporation, and warmer summer days to the Aleutian Arc and adjacent region (Scholl and Stevenson 1997).

Future Work

In future studies of the Bering gateway region, a scientific drilling program should be initiated so that the history of water mass interchange between the Pacific and the Arctic/Atlantic can be thoroughly studied. Studies of calcium carbonate and biogenic opal microfossils will also shed light on the evolution of the Pacific (opal) and Atlantic (carbonate) type oceans. Production of the carbonaceous and siliceous plankton involve organic carbon production; hence their process is closely coupled with the global CO_2 system. Thus, with this type of work on the past material cycles, a better understanding of the paleoclimate will be achieved.

High resolution paleoceanographic investigations, including the North Pacific Intermediate Water formation during the entire Pleistocene glacial and interglacial intervals, are needed to decipher details of climatic change, with emphasis on understanding land-ocean links. On a longer time scale, large climatic changes such as the detailed evolution of Northern Hemisphere glaciation can also be studied due to high sedimentation rates in the Bering Sea. Pertinent parameters that should be investigated include ice algae, planktonic diatoms of shallow Bering Shelf origin, and coccolithophores such as *Coccolithus pelagicus.*

Conclusions

The Bering Sea, the largest semi-closed marginal sea of the Pacific Ocean, is located in the subpolar region and it plays a significant role in the global carbon cycle. This mainly stems from the highly efficient biological pump operating in the region, which results in the production of greater amounts of opal than calcium carbonate. Measured biogenic particle flux data indicate the Bering Sea has twice the annual production of adjacent Pacific waters. Therefore, the biological pump manifested in the marginal sea is efficient, not only in the quality of high opal content, but also in quantity in terms of high fluxes. It is concluded here that the Bering Sea acts as a CO_2 sink.

The nodal location of the Bering Sea between the Pacific Ocean and the Arctic Ocean mediates transport of Pacific water into the Arctic Ocean. The contrast between the opal ocean of the Pacific and the carbonate ocean of the Atlantic today is attributed to the one-way water flow of the Bering Shelf water through the Bering Strait to the Arctic and eventually to the Atlantic. During past glacial low stands the modern deep sill of the Bering Strait (50 m) was aerially exposed and the gateway closed. This had a major impact on water mass exchange between the Atlantic and Pacific. A large amount of runoff, such as from the Yukon River, had to flow out to the Pacific during the glacial low stand, whereas a part of it eventually reaches the Chukchi Sea today. Such a shift in circulation must have caused significant changes in intermediate and deepwater formation in the past.

With limited geological data, biological productivity of the Bering Sea was lower during the glacial period than today, judging from the lower organic carbon contents. Such a decreased glacial productivity is due to partial sea-ice cover. Sea ice did not fully cover the glacial Bering Sea since the formation of intermediate or deep waters such as North Pacific Intermediate Water by atmospheric cooling of the surface water was possible. A somewhat similar trend of the lower glacial productivity was seen in the Okhotsk Sea, but detailed comparisons with the Bering Sea have to be made in future work. The timing of terminations 1A and 1B in the Bering Sea and adjacent regions are synchronous with those in the Atlantic, suggesting that the deglaciation process occurred independently in the Pacific and Atlantic sides.

The Pacific-Atlantic connections existed several times on a longer geological time scale than the Milankovitch cycles. Based on the discovery of the Atlantic mollusk faunas in northern Japan, the initial connection due to a tectonic event was earlier than 5.1 Ma, the Late Miocene. Since then the Pacific-Atlantic connections recurred several times. After the series of the tectonic events, Beringia and the Bering gateway have been subjected to shorter time scale events such as the Milankovitch cycles of 100 ka frequency.

Water exchange rates between the Bering Sea and the subarctic Pacific significantly influence the weather conditions of the region. The Aleutian Arc is presently covered with a carpet of tundra. During the Oligocene to the early Pliocene, the Aleutian Arc was covered with redwood forests with clear summer skies. This may occur again when and if the Bering-Pacific water exchange decreases and the Aleutian Low moves toward the Gulf of Alaska. The geomorphology of the Kamchatka Strait has a key to this possible change since it allows substantial water exchange between the Bering Sea and the Pacific today.

Acknowledgments

The basic thoughts of this review chapter sprouted from our ongoing collaborative joint field program employing sediment traps, which has been carried out together with Prof. Yoshiaki Maita, Dr. Mitsuru Yanada, Captain Gen Anma and crew of T/S *Oshoro Maru,* and Hokkaido University. I also thank a number of my graduate students who aided in various phases of the joint program—in particular, Naoki Fujitani, Kenichi Hisamichi, and Toshihiro Tominaga, who are responsible for some of the results cited in this chapter. Another dimension of this chapter was developed from discussions on future ODP drilling in the marginal seas and I thank my colleagues: Prof. Itaru Koizumi, Dr. Sergei Gorbarenko, and Dr. Ralph Tiedemann. I thank Dr. David Scholl, USGS and Stanford University, who provided useful information on the paleoceanography and the future environment of the Bering Sea region, and Dr. Toshio Takagi, the Power Reactor and Nuclear Fuel Development Corporation, Japan, for his kind supply of the useful information on mollusk paleobiogeographic distribution, concerning the Atlantic-Pacific connections. This chapter benefited by critical reviews on an early draft provided by Drs. Tom Loughlin, Terry Whitledge, and a peer reviewer. E. Sarah Takahashi offered editorial assistance. This work was partially funded by the Ministry of Education, Science, Sports and Culture (Grant-in-Aid for Scientific Research B project number 10480128).

References

Barron, J.A. 1985. Miocene to Holocene planktonic diatoms. In: H.M. Bolli, J.B. Saunders, and K. Perch-Nielsen (eds.), Plankton stratigraphy, Vol. 2. Cambridge University Press, Cambridge, U.K., pp. 763-809.

Behl, R.J., and J.P. Kennett. 1996. Brief interstadial events in the Santa Barbara basin, N.E. Pacific, during the past 60 kyr. Nature 379:243-246.

Berger, W.H. 1970. Biogenous deep-sea sediments: Fractionation by deep-sea circulation. Bulletin of Geological Society of America 81:1385-1402.

Berger, W.H., and R. S. Kier. 1984. Glacial and Holocene changes in atmospheric CO_2 and the deep-sea record. In: J.E. Hansen and T. Takahashi (eds.), Climate processes and climate sensitivity. Geophysical Monograph Series 29:337-351.

Berger, W.H., K. Fischer, C. Lai, and G. Wu. 1987. Ocean productivity and organic carbon flux, Part I: Overview and maps of primary production and export production. Scripps Institution of Oceanography SIO Ref. 87-30. 67 pp.

Broecker, W.S. 1994. Massive iceberg discharges as triggers for global climate change. Nature 372:421-424.

Coachman, L.K., and K. Aagaard. 1981. Reevaluation of water transports in the vicinity of Bering Strait. In: D.W. Hood and J.A. Calder (eds.), The eastern Bering Sea shelf: Oceanography and resources, Vol. 2. Published by the Office of Marine Pollution Assessment, NOAA and BLM. Distributed by the University of Washington Press, Seattle, WA 98105, pp. 95-110.

Craig, H., W.S. Broecker, and D.W. Spencer. 1981. Sections and profiles. GEOSECS Pacific expedition, Vol. 4. IDOE, National Science Foundation. 251 pp.

Dansgaard, W., H.B. Clausen, N. Gundestrup, S.J. Johnsen, and C. Rygner. 1993. Dating and climatic interpretation of two deep Greenland ice cores. In: C.C. Langway Jr., H. Oeschger, and W. Dansgaard (eds.), Greenland ice core: Geophysics, chemistry, and the environment. Geophysical Monograph 33:71-76.

Edmond, J.M., S.S. Jacobs, A.L. Gordon, A.W. Mantyla, and R.F. Weiss. 1979. Water column anomalies in dissolved silica over opaline pelagic sediments and the origin of the deep silica maximum. Journal of Geophysical Research 84:7809-7826.

Ewing, J., M. Ewing, T. Aitken, and W.J. Ludwig. 1968. North Pacific sediment layers measured by seismic profiling. American Geophysical Union Monograph 12:147-472.

Favorite, F., A.J. Dodimead, and K. Nasu. 1976. Oceanography of the subarctic Pacific region, 1960-71. Bulletin of the International North Pacific Fisheries Commission 33. 187 pp.

Gorbarenko, S.A. 1996. Stable isotope and lithological evidence of late-glacial and Holocene oceanography of the Northwest Pacific and its marginal seas. Quaternary Research 46:230-250.

Honjo, S. 1990. Particle fluxes and modern sedimentation in the polar oceans. In: W.O. Smith (ed.), Polar oceanography, Part B: Chemistry, biology, and geology. Academic Press, New York, pp. 687-739.

Honjo, S., J. Dymond, R. Collier, and S.J. Manganini. 1995. Export production of particles to interior of the equatorial Pacific Ocean during the 1992 EqPac Experiment. Deep-Sea Research, Part II 42:831-870.

Hood, D.W. 1983. The Bering Sea. In: B.H. Ketchum (ed.), Estuaries and enclosed seas. Elsevier Scientific Publication Co., pp. 337-373.

Japan Meteorological Agency. 1990. Sea ice charts in the Okhotsk Sea and the north of Japan Sea. In: The results of sea ice observations. No. 8. Japan Meteorological Agency, pp. 27-41.

Japan Meteorological Agency. 1991. Sea ice charts in the Sea of Okhotsk and the north of Japan Sea. In: The results of sea ice observations. No. 9. Japan Meteorological Agency, pp.25-39.

Japan Meteorological Agency. 1992. Sea ice charts in the Sea of Okhotsk and the north of Japan Sea. In: The results of sea ice observations. No. 10. Japan Meteorological Agency, pp. 32-45.

Japan Meteorological Agency. 1993. Sea ice conditions in the Sea of Okhotsk and adjacent seas in November 1992 through May 1993. In: The results of sea ice observations. No. 11. Japan Meteorological Agency, pp. 29-45.

Japan Meteorological Agency. 1994. Sea ice conditions in the Sea of Okhotsk and adjacent seas in November 1993 through May 1994. In: The results of sea ice observations. No. 12. Japan Meteorological Agency, pp. 28-43.

Lisitzin, A.P. 1972. Sedimentation in the World Ocean. Society of Economic Paleontologists and Mineralogists, Special Publication No. 17. 218 pp.

Mammerickx, J. 1985. A deep-sea thermohaline flow path in the Northwest Pacific. Marine Geology 65:1-19.

Murray, J.W. 1995. Tropical studies in oceanography: A U.S. JGOFS process study in the equatorial Pacific. Deep-Sea Research, Part II 42:275-903.

Ogasawara, K., and Y.B. Gladenkov. 1995. Review and comments on the late Neogene climatic fluctuation and intermittence of the Bering Land Bridge. In: Oji seminar on Neogene evolution of Pacific Ocean gateways, International Geological Correlation Programme (IGCP) Project-355, Abstracts, p. 26.

Ohtani, K. 1965. On the Alaskan Stream in summer. Bulletin of Faculty of Fisheries, Hokkaido University 15:260-273. (In Japanese.)

Ohtani, K. 1973. Oceanographic structure in the Bering Sea. Memoirs of Faculty of Fisheries, Hokkaido University 21:65-106.

Ohtani, K., and T. Azumaya. 1995. Influence of interannual changes in ocean conditions on the abundance of walleye pollock (*Theragra chalcogramma*) in the eastern Bering Sea. In: R.J. Beamish (ed.), Climate change and northern fish populations. Canadian Special Publication of Fisheries and Aquatic Sciences 121:87-95.

Ohtani, K., Y. Akiba, and A.Y. Takenouti. 1972. Formation of western subarctic water in the Bering Sea. In: A.Y. Takenouti (ed.), Biological oceanography of the northern North Pacific Ocean. Idemitsu Shoten Publishing, Tokyo, pp. 32-44.

Riser, S.C. 1997. Ventilation and formation of North Pacific Intermediate Water. In: S. Tsunogai (ed.), Biogeochemical processes in the North Pacific: Proceedings of the International Marine Science Symposium, Mutsu, Japan, 1996. Japan Marine Science Foundation, pp. 12-20.

Sagayama, T., K. Hoyanagi, and S. Miyasaka. 1992. Diatom biostratigraphy and the stage of Neogene coarse-grained deposits in the Hidaka coastal land, central Hokkaido, Japan. Journal of Geological Society of Japan 98:309-321. (In Japanese with English abstract.)

Sambrotto, R.N., J.J. Goering, and C.P. McRoy. 1984. Large yearly production of phytoplankton in the western Bering Strait. Science 225:1147-1150.

Sancetta, C. 1981. Oceanographic and ecological significance of diatoms in surface sediments of the Bering and Okhotsk seas. Deep-Sea Research 28A:789-817.

Scholl, D.W., and A.J. Stevenson. 1997. Geologic controls on the flow of water into and out of the Bering Sea: Thoughts about the past, present, and future climate effects on Beringia. In: J. Brigham-Grette and S. Elias (eds.), Program and abstracts, Beringian Paleoenvironmental Workshop, Florissant, Colorado, 20-23 September 1997. INSTARR, University of Colorado, Boulder.

Scholl, D.W., J.R. Hein, M.S. Marlow, and E.C. Buffington. 1977. Meiji sediment tongue: North Pacific evidence for limited movement between the Pacific and North American plates. Bulletin of Geological Society of America 88:1567-1576.

Supko, P.R. (ed.). 1973. Initial reports of the Deep Sea Drilling Project, Volume XIX, Kodiak, Alaska to Yokohama, Japan Jul-September 1971. U.S. Government Printing Office, Washington, DC. 913 pp.

Tada, R., I. Koizumi, A. Gramp, and A. Rahman. 1992. Correlation of dark and light layers, and the origin of their cyclicity in the Quaternary sediments from the Japan Sea. In: K.A. Pisciotto, J.C. Ingle Jr., M.T. von Breyman, and J. Barron (eds.), Proceedings of the Offshore Drilling Project, scientific results, 127/128, Pt. 1. Ocean Drilling Program, College Station, TX, pp. 577-601.

Takahashi, K. 1991. Mineral flux and biogeochemical cycles of marine planktonic protozoa. In: P.C. Reid, C.M. Turley, and P.H. Burkill (eds.), Protozoa and their role in marine processes. Springer-Verlag, Berlin, pp. 347-360.

Takahashi, K., N. Fujitani, M. Yanada, and Y. Maita. 1997. Five year long particle fluxes in the central subarctic Pacific and the Bering Sea. In: S. Tsunogai (ed.), Biogeochemical processes in the North Pacific, Proceedings of the International Marine Science Symposium, Mutsu, Japan, 1996. Japan Marine Science Foundation, pp. 277-289.

Takahashi, K., N. Fujitani, M. Yanada, and Y. Maita. In press. Long term biogenic particle fluxes in the Bering Sea and the central subarctic Pacific Ocean. Deep-Sea Research.

Takahashi, K., K. Hisamichi, M. Yanada, and Y. Maita. 1996. Seasonal change of phytoplankton productivity: Results from sediment traps. Kaiyo Monthly, Special Volume 10:109-115. (In Japanese with English figure and table captions.)

Tally, L.D. 1991. An Okhotsk Sea water anomaly: Implication for ventilation in the North Pacific. Deep-Sea Research 38(Supplement 1):S171-S190.

Tans, P.P, I.Y. Fung, and T. Takahashi. 1990. Observational constraints on the global atmospheric CO_2 budget. Science 247:1431-1438.

Tsunogai, S., M. Kusakabe, H. Iizumi, I. Koike, and A. Hattori. 1979. Hydrographic features of the deep-water of the Bering Sea: The sea of silica. Deep-Sea Research 26:641-659.

Uozumi, S., M. Akamatsu, and T. Takagi. 1986. Takikawa-Honbetsu and Tatsunokuchi faunas (*Fortipecten takahashii*-bearing Pliocene faunas). Palaeontological Society of Japan, Special Publication 29:211-226.

U.S. GLOBEC. 1996. Global Ocean Ecosystems Dynamics: Report on climate change and carrying capacity of the North Pacific ecosystem. U.S. GLOBEC Report No. 15. 95 pp.

Warner, M.J., and G.L. Roden. 1995. Chlorofluorocarbon evidence for the recent ventilation of the deep Bering Sea. Nature 373:409-412.

Worrall, D.M. 1991. Tectonic history of the Bering Sea and the evolution of Tertiary strike-slip basins of the Bering Shelf. Geological Society of America Special Paper 257. 165 pp.

Wong, C.S., F.A. Whitney, I. Tsoy, and A. Bychkov. 1995. The opal pump and subarctic carbon removal. In: S. Tsunogai (ed.), Global fluxes of carbon and its related substances in the coastal sea-ocean-atmosphere system, Proceedings of the 1994 Sapporo IGBP Symposium. M. & J. International, Yokohama, Japan, pp. 339-344.

CHAPTER **19**

Summary of Biology and Ecosystem Dynamics in the Bering Sea

Thomas R. Loughlin
Alaska Fisheries Science Center, Seattle, Washington

Irina N. Sukhanova
Shirshov Institute of Oceanology, Russian Academy of Sciences, Moscow, Russia

Elizabeth H. Sinclair and Richard C. Ferrero
Alaska Fisheries Science Center, Seattle, Washington

Introduction

The combination of a broad continental shelf, extensive winter ice coverage, and convergence of nutrient-rich current systems position the Bering Sea as one of the most productive ecosystems in the world. The Bering Sea is seasonal or year-round home to some of the largest marine mammal, bird, fish, and invertebrate populations among the world's oceans and supports some of the world's largest commercial harvests of seafood, including groundfish, salmon, and crabs. For centuries, Bering Sea resources have provided subsistence, food, clothing, and cultural substrate to the native inhabitants of coastal communities, and remain essential to their livelihood today.

Archaeological records demonstrate that Bering Sea resources have been subject to periodic localized depletion through subsistence harvests (Simenstad et al. 1978). However, since at least the mid-1970s, system-wide shifts in the biomass and composition of the marine community have occurred due to the synergistic effects of environmental fluctuation and disproportionate commercial fishing pressures (NRC 1996). Reductions of biomass at lower trophic levels have precipitated depletions of top-level predators such as seabirds and marine mammals, to the point where some species may be driven toward extinction, and nowhere are the effects more apparent than to local and global communities reliant on the sustainability of Bering Sea resources.

Much of the earliest research in the Bering Sea was conducted to address questions related to international fisheries and resource exploitation. Since the mid-1980s, the scope of research has broadened to address the interrelationship between the unique physical, climatological, and biological characteristics of the Bering Sea; an ecosystem-based approach driven largely by the need to define those mechanisms forcing the rise and decline of key species (e.g., chapters 30-37). Much of the research described in the chapters that follow is the result of such directed multidisciplinary efforts. A background for the chapters to follow is provided below.

Summary of Biological Resources

Phytoplankton

Phytoplankton synthesize carbon through photosynthesis using dissolved nutrients in surface-sunlit waters and provide most of the organic matter required by the rest of the marine food web. The amount and rate of carbon production are defined by the species composition of the dominant phytoplankton groups which, in turn, define the role of phytoplankton in pelagic and benthic food webs. Pelagic herbivores (zooplankton) consume varying proportions of phytoplankton biomass directly; the rest sinks to the bottom as detritus where it is eaten or accumulated in bottom sediments. The efficiency of phytoplankton consumption by pelagic herbivorous organisms is directly linked to algal cells sizes, abundance of the respective size groups (Parsons and LeBrasseur 1970), and the types of grazers.

Presently there are 266 species, 5 forms, and 3 varieties of marine phytoplankton in the Bering Sea phytoplankton community. The community can be categorized within eight classes (e.g., Table 1 in Chapter 22; Semina 1953; Flint et al. 1994, 1996; Ventsel 1994). Among the Dinophyceae, 40 species belong to non-photosynthesizing phytoplankton (Lessard and Swift 1985). These are all *Gyrodinium* (6 species), *Podolampas* (1 species) and *Protoperidinium* (33 species) species. The extensive species list demonstrates the diversity of the Bering Sea phytoplankton community. The list includes ice flora species mentioned by Saito and Taniguchi (1978), Alexander and Cooney (1979), and Goering and Iverson (1981). The list does not cover freshwater and estuarine-water species.

The high level of taxonomic diversity of Bering Sea phytoplankton is due to the substantial extension from the Arctic down to about 50°N, a vast shelf area with a unique community, and the penetration of tropical species from the North Pacific carried by the Alaska Coastal Current and Alaskan Stream into the northern and central (up to 60°N) areas of the Bering Sea. Of the 179 Bering Sea species with described distributions, 61% range into boreal areas, 33% are cosmopolitan, and 6% are tropical (Semina 1974, 1981; Hasle 1976; Konovalova et al. 1989; Medlin and Priddle 1992). Thirty-three percent of those species with boreal ranges are arctic-boreal, 15% have tropical-boreal ranges, 10% have arcto-boreal-tropical ranges, and 3% have bipolar ranges. The only endemic Pacific boreal species is *Neodenticula seminae.* Bacillariophyceae and Dinophyceae, classes that include the highest numbers of species, exhibit different ratios between species with different biogeographic patterns. Seventy-one percent of the Bacillariophyceae are widespread in the boreal area, and 23% are cosmopolitan in distribution while the Dynophyceae are nearly equally boreal (46%) and cosmopolitan (49%). Tropical species were observed in the deep basin north of the Aleutian straits, but were not found north of 60°N.

Goering and Iverson (1981) describe four major phytoplankton communities: (1) a temperate-neritic community ranging from the Aleutian Islands to inner Bristol Bay, (2) a boreal-oceanic community extending from the deep central basin onto the eastern continental shelf, (3) an arctic community near the coasts of Siberia and Kamchatka, and (4) a seasonal ice community. Distinguishing the temperate-neritic community of Bristol Bay and the Aleutian Islands from the arctic community near the coasts of Siberia and Alaska, in light of present data, seems questionable. In the opinion of Russian phytoplankton biologists (Sukanova et al., chapter 22, this volume) these communities should be considered as a single shelf community. The differences described by Goering and Iverson (1981) depend only on the seasonal successional condition of phytoplankton. The arctic community is composed of spring bloom species which develop at low temperatures and high concentration of nutrients. The temperate-neritic community includes neritic species that develop on the shelf in summer (from the middle of June to late August), prefer relatively high (7-10°C) temperature, and are adapted to moderate or even low concentration of nutrients.

Zooplankton

Approximately 300 species of zooplankton occur in the Bering Sea, of which copepods, coelenterates, and amphipods are the most abundant taxa (Coyle et al. 1996). Cooney (1981) summarizes the fauna into 341 species of both holoplankton and microplankton. Four general groupings of zooplankton based on calanoid copepods have been described for the Bering Sea: oceanic and outer shelf, mixed oceanic-shelf, middle shelf and coastal, and near shore (Cooney 1981). Zooplankton biomass demonstrates strong seasonal signals, and distribution varies with a much larger biomass over the outer domain and decreasing biomass over the middle and coastal domains of the eastern Bering Sea (NRC 1996). Estimates of annual zooplankton production in the southeastern Bering Sea are as high as 64 g C per m^2 per yr for the shelf edge down to a low of about 4 g C per m^2 per yr for the coastal domain (Cooney 1981, Vidal and Smith 1986). The greatest biomass of zooplankton occurs in the "green belt" from the southeastern Bering Sea into the Chukchi Sea (Cooney 1981, Springer et al. 1996), and zooplankton biomass is also high along the coastline of the Kamchatka Peninsula (NRC 1996).

The general patterns of zooplankton distribution were summarized by Coyle et al. (1996) based on Russian literature for both sides of the Bering Sea. During winter, biomass averages reach about 100 mg/m^3. Over the shelf about 14 species reach a density near 10 specimens per m^3 while over the deep basin densities average 1,200 specimens per m^3. During spring, zooplankton development varies substantially in different regions based on climate, depth, and hydrography. The greatest biomass of phytoplankton and zooplankton occurs off the eastern Bering Sea shelf (1,000-2,000 mg/m^3), and in the Gulf of Anadyr, Norton Sound, and Bering Strait where biomass can reach 3,000-3,200 mg/m^3. The pattern in summer is similar to spring, but with somewhat lower values (near 400 mg/m^3). Biomass over the upper 100 m in some areas of the deep basin remain high in fall (300-400 mg/m^3) relative to substantially lower values over the shelf.

Neocalanus cristatus, N. plumchrus, N. flemingerii, Eucalanus bungii, and *Metridia pacifica* are the predominant herbivorous zooplankton of the oceanic Bering Sea. They graze on large diatoms typical of the spring bloom flora, and in turn are important prey for fishes, birds, and mammals (NRC 1996). Inshore of the outer shelf (middle and coastal domains) they are replaced by *Calanus marshallae, Thysanoëssa raschii,* and various species of small copepods (e.g., *Pseudocalanus* sp.).

Cephalopods and Finfish

The Bering Sea includes over 450 species of fish and invertebrates (NRC 1996), yet our knowledge of seasonal and temporal distribution (latitudinal and vertical), habitat, abundance, and life history, is limited to approximately 25 species that are commercially important. The lack of baseline distribution and abundance data on cephalopods such as gonatid and chiroteuthid squid and forage fishes such as capelin *(Mallotus villosus),* eulachon *(Thaleichthys pacificus),* deep sea smelts *(Bathylagidae),* myctophiids, Pacific sand lance *(Ammodytes hexapterus),* and Atka mackerel *(Pleurogrammus monopterygius)* limit current interpretation of apparent changes in the Bering Sea ecosystem, as well as the future direction of resource management.

Forage fishes and juvenile cephalopods have much in common in terms of their role in the Bering Sea ecosystem. They form locally abundant schools, typically in the top 50 m of the water column, and at an average length of 12 cm form the greatest portion of the prey base for marine mammals, seabirds, and larger, commercially important cephalopods (e.g., *Berryteuthis magister*) and teleost fishes (e.g., walleye pollock [*Theragra chalcogramma*], salmonids, Pacific cod [*Gadus macrocephalus*], and Pacific halibut [*Hippoglossus stenolepis*]). Forage fishes spawn in both freshwater and saltwater, pelagic, and nearshore littoral areas. Cephalopods spawn in intertidal waters, over the continental shelf and the shelf edge, pelagically, and in benthic zones over the deep ocean. It is in large part due to their variable dispersion as eggs, larvae, juveniles, and adults that makes forage fishes and cephalopods critical prey to many faunal groups throughout the oceanic habitat (Alaska Sea Grant 1993).

An exception to the lack of attention to noncommercial species has been in the western Bering Sea where Russian researchers have collected extensive data in efforts to characterize midwater fish and invertebrate communities. Some of the syntheses resulting from those efforts are presented in the chapters that follow. Chapters 23-26 present background data and the current status of research efforts on mesopelagic fish, forage fish, groundfish, and walleye pollock; each underscoring the critical need for baseline data.

Seabirds

Thirty-eight species of seabirds inhabit the Bering Sea and adjacent waters (Table 1). The group includes principally the Procellariiformes (albatrosses, shearwaters, and petrels), Pelecaniformes (cormorants), and two families of the Charadriiformes: Laridae (gulls) and Alcidae (auks, such as puffins, murres, auklets, and murrelets). Trends are reasonably well known for species that nest on cliffs or flat ground (fulmars, cormorants, some gulls, kittiwakes, and murres) and for storm petrels and tufted puffins *(Fratercula cirrhata).* Data are limited for most other species; however, available data suggest that the overall abundance of kittiwakes and murres has declined in recent years (Table 2). Murre populations appear to be declining in a majority of the surveyed rookeries over the past 5-15 years in the northern Bering Sea but are increasing in the Chukchi Sea; kittiwake abundance trends there are stable or increasing. However, murre and kittiwake populations on the large Pribilof Islands rookeries appear to be declining, while kittiwake populations in the western Aleutian Islands have increased (Alaska Sea Grant 1993, NMFS 1998).

Seabirds (like marine mammals) are characterized by low reproductive rates, low annual mortality, and long life span. Population trends can result from changes in either productivity or survival, but most trends that have been adequately studied were attributed to changes in productivity (NMFS 1998). Annual kittiwake and murre productivity is variable and is negatively correlated with sea surface temperature and pollock biomass at several sites, but data are insufficient to link the observed seabird population declines to long-term climatic cycles or climatic changes affecting primary or secondary productivity (Alaska Sea Grant 1993). The natural factor most often associated with low breeding success is scarcity of food (NMFS 1998). A discussion of species composition, distribution, and abundance trends for seabirds in the Bering Sea follows in chapters 28 and 29.

Marine Mammals

The Bering Sea and Gulf of Alaska support one of the richest assemblages of marine mammals in the world. Twenty-five species are present in the

Table 1. Estimated populations and principal diets of seabirds that breed in the Bering Sea and Aleutian Islands.

Species	Population[a,b]		Diet[c,d]
	BSAI	GOA	
Northern fulmar *(Fulmarus glacialis)*	1,500,000	600,000	Q,M,F,Z,I
Fork-tailed storm petrel *(Oceanodroma furcata)*	4,500,000	1,200,000	Z,Q,C
Leach's storm petrel *(Oceanodroma leucorrhoa)*	4,500,000	1,500,000	Z,Q
Double-crested cormorant *(Phalacrocorax auritis)*[e]	9,000	8,000	F,I
Pelagic cormorant *(Phalacrocorax pelagicus)*	80,000	70,000	S,C,P,H,F,I
Red-faced cormorant *(Phalacrocorax urile)*	90,000	40,000	C,S,H,F,I
Brandt's cormorant *(Phalacrocorax penicillatus)*	0	100	?
Pomarine jaeger *(Stercorarius pomarinus)*	Common	Common	C,S
Parasitic jaeger *(Stercorarius parasiticus)*	Common	Common	C,S
Long-tailed jaeger) *(Stercorarius longicaudus*	Common	Common	C,S
Bonaparte's gull *(Larus philadelphia)*	Rare	Common	?
Mew gull *(Larus canus)*[e]	700	40,000	C,S,I,D
Herring gull *(Larus argentatus)*[e]	50	300	C,S,H,F,I,D
Glaucous-winged gull *(Larus glaucescens)*	150,000	300,000	C,S,H,F,I,D
Glaucous gull *(Larus hyperboreus)*[e]	30,000	2,000	C,S,H,I,D
Black-legged kittiwake *(Rissa tridactyla)*	800,000	1,000,000	C,S,P,F,M,Z
Red-legged kittiwake *(Rissa brevirostris)*	150,000	0	M,C,S,Z,P,F
Sabine's gull *(Xema sabini)*	Common	Common	?
Arctic tern *(Sterna paradisaea)*[e]	7,000	20,000	C,S,Z,F
Aleutian tern *(Sterna aleutica)*	9,000	25,000	C,S,Z,F

Source: USDC 1998.

Table 1.   (Continued.)

Species	Population[a,b]		Diet[c,d]
	BSAI	GOA	
Common murre (*Uria aalge*)	3,000,000	2,000,000	C,S,H,P,F,Z
Thick-billed murre (*Uria lomvia*)	5,000,000	200,000	C,S,P,Q, Z,M,F,I
Pigeon guillemot (*Cepphus columba*)	100,000	100,000	S,C,F,H,I
Marbled murrelet (*Brachyramphus marmoratus*)	Uncommon	Common	C,S,P,F,Z,I
Kittlitz's murrelet (*Brachyramphus brevirostris*)	Uncommon	Uncommon	S,C,H,Z,I,P,F
Ancient murrelet (*Synthliboramphus antiquus*)	200,000	600,000	Z,F,C,S,P,I
Cassin's auklet (*Ptychoramphus aleuticus*)	250,000	750,000	Z,Q,I,S,F
Least auklet (*Aethia pusilla*)	9,000,000	50	Z
Parakeet auklet (*Cyclorrhynchus psittacula*)	800,000	150,000	F,I,S,P,Z
Whiskered auklet (*Aethia pygmaea*)	30,000	0	Z
Crested auklet (*Aethia cristatella*)	3,000,000	50,000	Z,I
Rhinoceros auklet (*Cerorhinca monocerata*)	50	200,000	C,S,H,A,F
Tufted puffin) (*Fratercula cirrhata*	2,500,000	1,500,000	C,S,P,F,Q,Z,I
Horned puffin (*Fratercula corniculata*)	500,000	1,500,000	C,S,P,F,Q,Z,I
Total		36,000,000	12,000,000

[a] Source of population data for colonial seabirds that breed in coastal colonies: modified from USFWS 1998. Estimates are minima, especially for storm petrels, auklets, and puffins.

[b] Numerical estimates are not available for species that do not breed in coastal colonies. Approximate numbers: abundant $\geq 10^6$; common = 10^5-10^6; uncommon = 10^3-10^5; rare $\leq 10^3$.

[c] Abbreviations of diet components: M = Myctophid; P = walleye pollock; C = capelin; S = sand lance; H = herring; A = Pacific saury; F = other fish; Q = squid; Z = zooplankton; I = other invertebrates; D = detritus; ? = no information for Alaska. Diet components are listed in approximate order of importance. However, diets depend on availability and usually are dominated by one or a few items (see text).

[d] Sources of diet data: see species accounts in text.

[e] Species breeds both coastally and inland; population estimate is only for coastal colonies.

[f] The principal source for the information in this table was Dr. Vivian Mendenhall, U.S. Department of Interior, U.S. Fish and Wildlife Service, Anchorage, Alaska.

Table 2. Recent population trends of Alaskan seabirds: kittiwakes, murres. Trends are shown for the last 5 years for species monitored over 4 years or more.

	Northern fulmar	Storm petrels	Pelagic cormorant	Red-faced cormorant	Glaucous-winged gull	Black-legged kittiwake	Red-legged kittiwake	Common murre	Thick-billed murre
Chukchi Sea			?			0		+	+
N. Bering Sea	0		?			0		0	–
Central and SE Bering Sea	0		?	0	?	0	0	+	0
Bristol Bay			0	–	?	0		0	
W. Aleutian Islands		0	–	?	–	+	0	?	+
C. Aleutian Islands	?	0	?	?	?	0		?	?
E. Aleutian Islands		0	–	–	0	?		0	0

Stable = 0; increase = +; decline = –; trend unknown = ?; species not present = blank.

Source: USDC 1998.

Bering Sea from the orders Pinnipedia (sea lions, walrus, and seals), Carnivora (sea otter), and Cetacea (whales, dolphins, and porpoises) (Lowry et al. 1982). Since none of the chapters that follow provide an overview of marine mammal species, abundance trends, and distribution in the Bering Sea, we provide summary data in tabular form here. Some species are resident throughout the year, while others migrate into or out of the area seasonally (Table 3). They occur in diverse habitats, including island and coastal haul-outs, deep oceanic waters, and over the continental shelf (Table 4). The majority of species are found over the continental shelf and in coastal waters. Ice-associated species are found in the Bering Sea mostly during the fall and winter as seasonal ice moves in and out of the area (NRC 1996).

Reliable estimates of abundance and population trends for most species are known only in general terms and winter distributions are poorly understood. The abundance of most whales and dolphins is low, primarily because the Bering Sea is the northern extent of their range. However, species such as the northern right whale *(Balaena glacialis)* are rare (Table 3). Bowhead *(Balaena mysticetus)* and gray whale *(Eschrictius robustus)* numbers overall are relatively low (Table 3) but are increasing (gray whales were removed from the U.S. Endangered Species list in 1994). Beluga whale *(Delphinapterus leucas)* abundance is probably high and stable except the Cook Inlet population where subsistence harvests are likely reducing population size. Dall's porpoise *(Phocoenoides dalli)* is likely the most abundant cetacean (>10,000 individuals) occurring in the Bering Sea/Aleutian Islands area.

Pinniped populations are generally high with many species exceeding 100,000 animals (NRC 1996). Exceptions include the endangered western population of Steller sea lions *(Eumetopias jubatus)* which has declined by over 90% since the 1960s in the Bering Sea and adjacent waters. The northern fur seal *(Callorhinus ursinus)* population has been stable since about 1984 after declines of 50% resulted in its being declared as "depleted" under the U.S. Marine Mammal Protection Act. But even at the reduced level the population in 1992 was estimated at almost one million animals (Loughlin et al. 1994). Recent surveys of harbor seals *(Phoca vitulina)* in Aleutian Islands and Bering Sea waters suggest significant declines in this species also.

The diet of marine mammals includes a variety of animal forms (Table 5; Fig. 1; Loughlin and Miller 1995) but encompasses the full spectrum of the food chain, including copepods and euphausiids (e.g., baleen whales), small to medium sized shallow and midwater fishes and cephalopods (e.g., otariids and phocids), and bottom dwelling invertebrates (e.g., gray whale and walrus). For some species diet varies by season and location and often by age and sex. For instance, as female northern fur seals migrate south out of the Bering Sea in October/November their prey changes. In the Bering Sea they feed principally on walleye pollock, capelin, Atka mackerel, and squid; in the Gulf of Alaska they feed on Pacific sand lance, Pacific

Table 3.   Abundance and trend of marine mammals in the Bering Sea.

Common name	Scientific name	Abundance/trend[a]	Occurrence
Pinnipeds			
Steller sea lion	*Eumetopias jubatus*	Moderate/declining	All seasons
Northern fur seal	*Callorhinus ursinus*	High/stable	Summer/fall (winter/spring)
Bearded seal	*Erignathus barbatus*	High/unknown	Spring/winter
Harbor seal	*Phoca vitulina*	Moderate/declining	All seasons
Spotted seal	*P. largha*	High/unknown	Spring/winter
Ribbon seal	*P. fasciata*	Moderate/unknown	Spring/winter
Ringed seal	*P. hispida*	High/unknown	Spring/winter
Walrus	*Odobenus rosmarus*	High/stable	All seasons, but principally spring/winter
Baleen whales			
Gray whale	*Eschrictius robustus*	Moderate/increasing	Spring/summer/fall
Fin whale	*Balaenoptera physalus*	Low/unknown	Spring/summer/fall
Minke whale	*B. acutorostrata*	Low/unknown	All seasons
Blue whale	*B. musculus*	Low/unknown	Summer
Sei whale	*B. borealis*	Low/unknown	Spring/summer/fall
Humpback whale	*Megaptera novaeangliae*	Low/increasing?	Spring/summer/fall
Northern right whale	*Balaena glacialis*	Low/unknown	Summer/fall
Bowhead whale	*Balaena mysticetus*	Low/increasing	Spring/winter
Toothed whales and porpoises			
Sperm whale	*Physeter macrocephalus*	Moderate/unknown	Summer
Cuvier's beaked whale	*Ziphius cavirostris*	Low/unknown	Unknown
Baird's beaked whale	*Berardius bairdi*	Low/unknown	Unknown
Stejneger's beaked whale	*Mesoplodon stejnegeri*	Low/unknown	All seasons
Beluga whale	*Delphinapterus leucas*	Moderate/stable	All seasons
Killer whale	*Orcinus orca*	Low/unknown	All seasons
Dall's porpoise	*Phocoenoides dalli*	Moderate/unknown	All seasons
Harbor porpoise	*Phocoena phocoena*	Low/unknown	All seasons
Other			
Sea otter	*Enhydra lutris*	Moderate/stable	All seasons

[a] Low = <10,000; moderate = 10,000-100,000; high = >100,000.

Source: Swartzman and Hofman 1991, NRC 1996.

Table 4.   General oceanographic/habitat associations of Bering Sea marine mammals.

Oceanic/ deep water	Continental slope/ slope break	Continental shelf/ coastal waters
Sei whale	Fin whale	Gray whale
Fin whale	Minke whale	Humpback whale
Minke whale	Blue whale	Minke whale
Blue whale	Humpback whale	Bowhead whale
Humpback whale	Dall's porpoise	Beluga
Sperm whale	Stejneger's beaked whale	Killer whale
Cuvier's beaked whale	Killer whale	Harbor porpoise
Baird's beaked whale	Steller sea lion	Pacific walrus
Dall's porpoise	Northern fur seal	Steller sea lion
Killer whale	Ribbon seal	Northern fur seal
Northern fur seal		Harbor seal
		Spotted seal
		Ringed seal
		Ribbon seal
		Bearded seal
		Sea otter

Modified from NRC 1996.

herring *(Clupea harengus),* and capelin. Historically, midwater forage fishes and squid were a principal component of those marine mammal (and seabird) species presently or recently declining in population numbers (Steller sea lion, northern fur seal, harbor seal) but currently they consume primarily walleye pollock in the Bering Sea (Antonelis and Sinclair 1996).

Human Activities

The Periods

The history of interactions between humans and resources of the Bering Sea began soon after man's arrival into the area and can be separated into four distinct "periods." (Loughlin and Jones 1984; Table 6). The first was the "subsistence period" and dates to about 28,000 years ago when, as indicated by archeological and geological evidence, primitive hunting groups arrived by diffusion or migration in the Bering Sea region and crossed the emerging Bering Sea land bridge into Alaska (Laughlin 1967, Müller-Beck 1967). The entire coastal area of the Bering Sea was occupied by Mongoloid peoples and their migration into North America was across either the coastal or interior portions of the land bridge, or both; none

Table 5. Species, habitat, diet, and food consumption of ice-associated phocid seals and sea otters in the Bering Sea.

Common name	Scientific name	Habitat	Diet	Estimated food consumed[a]
Spotted seal	*Phoca largha*	Pack ice	Crustaceans, cephalopods, fish	89.1×10^3
Bearded seal	*Erignathus barbatus*	Pack ice, continental shelf	Crustaceans, molluscs, fish	2.65×10^3
Ringed seal	*Phoca hispida*	Seasonal and permanent ice	Crustaceans, fish	256.7×10^3
Ribbon seal	*Phoca fasciata*	Pack ice/shore-fast ice, open ocean	Crustaceans, cephalopods, fish	70.7×10^3
Sea otter	*Enhydra lutris*	Shallow, coastal ice-free waters	Benthic invertebrates, octopus, fish	157×10^3

[a] Values are estimates derived by Perez and McAlister (1993).

Table 6. Four periods of interaction between humans and the Bering Sea ecosystem including their duration and type of interaction.

Period	Duration	Type of interaction
Subsistence	ca. 28,000 years ago to present	Subsistence hunting by Natives
Northern fur seal	1786 to 1984	Commercial harvest of northern fur seals
Whaling	1845 to ca. 1914	Commercial harvest of whales and walruses
Commercial fishing	ca. 1952 to present	Commercial harvest of fish and shellfish

Source: Loughlin and Jones 1984.

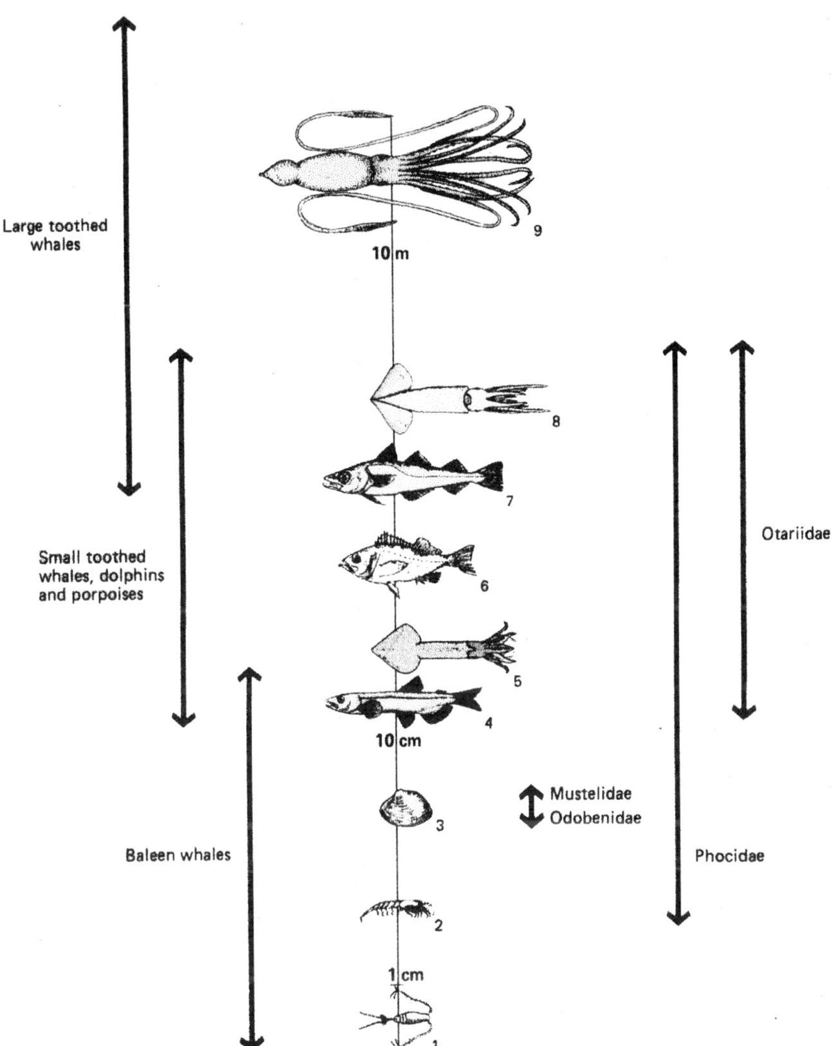

Figure 1. Diagram depicting the generalized size ranges of prey for marine mammals in the Bering Sea. Prey size increases on a logarithmic scale. (1) Calanoid copepods; (2) euphausiids; (3) benthic invertebrates; (4) shallow and midwater fishes; (5) market squid; (6) bottom fishes; (7) cods (principally juveniles); (8) bottom and midwater squids; and (9) giant squids (and smaller marine mammals). (From Loughlin and Miller 1995.)

seemed to have entered by way of the Commander and Aleutian islands (Laughlin 1967). The subsistence period persists today, evidenced by the hunting and use of fish, seabirds, and marine mammals by Russian and Alaskan coastal and island Natives. The number of animals taken has generally been considered to be low and minimal impact to local prey resources is assumed, except in a few cases where the take of some marine mammal species may have affected their abundance (Simenstad et al. 1978). Currently, the Bering Sea coast is home to approximately 65,000 Aleut, Koryak, Chukchi, Inupiat, and Yupik peoples (NRC 1996).

Loughlin and Jones (1984) termed the next period the "fur seal period," characterized as the period when northern fur seals were discovered by the crew of the Russian ship *St. George,* under command of Gerasim Pribilof on the islands that now bear his name. Fur seals were commercially harvested on these islands from their discovery in 1786 to 1984; presently there is a native subsistence harvest of a few thousand immature animals each year. The harvest of northern fur seals (and sea otters and some walrus) was the sole commercial activity between humans and marine resources in the Bering Sea until 1845 when Yankee whaling ships moved from whaling grounds near Kodiak Island and south of the Aleutian Islands into the Bering Sea to hunt bowhead whales and right whales (Hunt 1975). The discovery of large numbers of harvestable whales touched off the third period termed the "whaling period" (Loughlin and Jones 1984) which lasted until about 1914 when whaling in the Bering Sea became uneconomical for most companies.

The fourth, and perhaps most significant in terms of its impact on the Bering Sea ecosystem, is the "commercial fishing period." It began in about 1952 when commercial trawl fisheries from Japan and elsewhere started to exploit groundfish in the southeastern Bering Sea soon after the end of World War II (Bakkala et al. 1979, Alverson 1992). The large-scale commercial fisheries for groundfish which developed in the early 1950s presently exceed 2 million t annually and generate ex-vessel revenues of over $600 million in the U.S. alone. The explosive growth involved other fisheries as well, including Japanese and U.S. salmon fisheries, Japanese, Soviet, and U.S. herring fisheries, and fisheries for crab, shrimp, and various line fisheries (Alverson 1992).

The historical composition of the pelagic fish community (and thus fisheries) of the Bering Sea (principally the western section) has been classified into three periods (Naumenko et al. 1990 cited in NRC 1996). These include the initial period (1950s and early 1960s), the transitional period from the mid-1960s to 1973, and the recent period from 1973 to the present. The initial period was dominated by herring, but later in the period it declined as smelt and capelin biomass increased; pollock biomass was low at that time. There was no single dominant species in the transitional period; then during the present period, walleye pollock began to dominate and comprised more than 85% of the biomass of commercially important, pelagic species (NRC 1996). The dynamics of western Bering Sea pelagic communities are strongly connected with the intertwined effects of both human-induced and climatic/oceanographic conditions (Alverson 1992, NRC 1996).

Fisheries Interactions

Interactions between commercial fisheries and top-level predators inhabiting Bering Sea waters vary widely, given the predators' diverse life history and spatial distribution patterns. In general, the impacts are likely to be apparent among those predators with the greatest dependence either on prey species that are harvested commercially, or on prey that consume those commercially important species at some life stage. Indirect interactions between top-level predators (e.g., marine mammals and seabirds) and commercial fisheries, such as competition for similar prey resources, may result in local prey depletions and disturbance by fishing activities. A potential mechanism by which predators may be disadvantaged through competition with commercial fisheries for food resources is caused by localized depletion of prey. Whereas the overall abundance of prey may not be affected by fishing activity, reduction in local abundance or dispersion of schools could prove more energetically costly to foraging predators. Additional impacts have been suggested, including (1) alteration of the actual and relative abundance of fish stocks in the ecosystem contributing to the dominance of less desirable forage species, and (2) alteration of the age structure of fish stocks targeted by a fishery, resulting in a shift in biomass from older to younger age classes. For example, northern fur seals forage in shallow to midwater depths (0-250 m) of both nearshore and pelagic regions throughout their migratory range (Loughlin et al. 1987, Goebel et al. 1991). Female and young male fur seals generally consume juvenile and small-sized (5-20 cm) schooling fishes and squids. The species of prey consumed varies with oceanographic subregions along their migration routes (Perez and Bigg 1986) and around breeding locations (Sinclair et al. 1994, Antonelis et al. 1997). In the eastern Bering Sea, primary prey species include the fish families Gadidae (walleye pollock, Pacific cod), Bathylagidae (deep sea smelts), Myctophidae (lanternfish), and squids of the family Gonatidae. Historically, eulachon, capelin, herring, and Atka mackerel were principal prey species of northern fur seals in the Bering Sea; however, those species are currently rare to absent in the fur seal diet (Sinclair et al. 1994, 1996). Declines of some otariid and seabird populations in the Bering Sea and North Pacific Ocean are currently attributed to this reduction in diet diversity indirectly related to anthropogenic factors such as commercial fishing and historical whaling operations (Sinclair et al. 1994, 1996; Merrick and Calkins 1996; Merrick et al. 1997). However, the extent to which any particular reduction in fisheries removals translates to changes in pinniped or seabird natality or mortality cannot be determined with currently available information. Marine mammals serve as indicators of ecosystem health. But their slow maturation and low productivity means that reductions in populations due to environmental or anthropogenic perturbation may not be observed for years after the fact. The relative impact of fisheries perturbations compared to broad, regional events such as climate shifts are uncertain, but given the potential importance of localized prey availability, they warrant close consideration. The timing and location of fisheries relative to foraging patterns of predators may be a more relevant management concern than total removals.

Other Human Activities

Other human activities in the Bering Sea that presently that have had no (appreciable) influence on the biological resources of the Bering Sea include oil and gas development, national defense, livestock and fox farming, tourism, and transportation. Oil and gas development on the Bering Sea outer continental shelf has not been on the scale of that in the U.S. Beaufort Sea and is summarized in Chapter 36. Regarding military actions, the Aleutian Islands were the focus of large U.S. and Japanese military activities during WW II and were occupied in part by Japan. Amchitka Island in the central Aleutian Islands was the site of underground nuclear experiments related to the "Cannikin" event detonated on November 6, 1971, and U.S. military bases currently exist on numerous Aleutian Islands. There have been numerous efforts to farm cattle, reindeer, and fox on many Aleutian Islands over the past 50+ years. However, none have met with economic success (e.g., Bailey 1993). The U.S. Fish and Wildlife has begun removing introduced foxes from these islands to eliminate deleterious impacts to endemic seabird colonies.

Important Unresolved Questions

Research questions critical to understanding ecosystem response to variability in climate and physical forcing mechanisms, nutrient flux, shifts in food web dynamics, and direct anthropogenic influences such as habitat degradation, contaminants, and fishery removals are outlined below. These questions represent a synthesis of recent research recommendations by a team of management and research agencies, academic institutions, and independent groups calling for a commitment to long-term coordinated research and management efforts in the Bering Sea. They are a continuation of numerous efforts to coordinate and focus international research in the Bering Sea. For example, the questions posed by one such group, the PICES Bering Sea Working Group (Appendix I), are key components to this thrust. The questions below are summarized from a recent initiative in the United States to develop a unifying Bering Sea Ecosystem Research Plan (Draft Bering Sea Ecosystem Research Plan, September 1998, unpublished report, available from Alaska Fisheries Science Center, 7600 Sand Point Way, NE, Seattle, WA, USA, 98115).

1. How does climate variability influence circulation, stability of currents, formation of fronts and eddies, seasonal production and extent of sea ice, sources and amounts of nutrients transported to the eastern

Bering Sea shelf, and the resulting primary and secondary production regimes? How do each of these factors in turn influence biophysical processes throughout the food web?

2.   How are decadal and annual atmospheric and oceanographic changes associated with periods of strong and weak recruitment of fish, invertebrates, birds, and mammals? What are the implications of these changes for multispecies and ecosystem-based management including commercial fisheries? How do different stocks of the same species in the Bering Sea respond to physical and anthropogenic perturbations?

3.   What are the effects of removals (fishing) and additions (fishery discard) of tons of carbon from the Bering Sea on long-term system productivity? What influence do predator-prey interactions have on ecosystem structure and dynamics at various time and space scales, including different developmental stages of predator and prey? What are population demographics, associations, and trends of noncommercial fishes, invertebrates, and apex predators? Do climatic changes or fishing removals contribute to declines in marine mammal and seabird populations through localized depletion of prey or changes in prey species diversity? Are observed declines related to factors affecting populations on their breeding grounds in the Bering Sea, or to factors during the non-breeding season outside the Bering Sea? What are the interrelationships between productivity in pelagic and benthic realms of the ecosystem?

4.   What are the habitat characteristics of various life stages of fish and invertebrates? What are the nearshore habitat requirements of otherwise pelagic marine mammals and seabirds, both in terms of their role as buffer zones and as sources of prey availability for young of the year, lactating or nesting adults? How does fishing impact physical habitat attributes including benthic biota?

The complexity of these questions brings to mind an analogy suggested by Gregory Bateson (1979) when contemplating a pot of water placed over a hot stove. On the broad scale, we know that when sufficient heat is applied the water will boil, just as we know that severe perturbations to an ecosystem will cause major changes (e.g., El Niño). But on a finer scale, the movement and interactions of water molecules and the probability that certain bubbles will touch or affect one another is impossible to predict. And what happens under conditions that are not severe, but merely alterations away from "normal" conditions (moderate heat on the burner)? Marine ecosystems are at least as difficult to understand as it is to predict the stochastic movement of water molecules. The more we learn about the complexities of natural systems, the more likely we are to frame our ecosystem recommendations in certainties and our regional and local recommendations in probabilities (Pastor 1995). Care is required before pronouncing that we understand cause-and-effect relationships, or that anthropogenic or natural forces are likely to result in predicted effects. Equal care is required when there is an apparent absence of any effect from human or environmental perturbation. In this case, we must be cautious in our assumptions that because we see no effects, that indeed there have been none.

These and other issues need to be resolved for optimal use and conservation of the Bering Sea ecosystem. As outlined above, only recently have we begun to explore the full complexity of the Bering Sea. In an interesting review of the general use of oceans by humans, Ray (1970) characterized this thrust as the "Marine Revolution," an event that follows the Agricultural and Industrial revolutions as a significant change in man's relationship with his environment. The Marine Revolution consists of five major aspects: fisheries, minerals and management, military interests, science and technology, and conservation and recreation. Many of the activities and conflicts arising in the Bering Sea discussed above are included in these five aspects. In his paper Ray makes a strong argument for consideration of ecosystem ecology and environmental concerns as man's technological ability optimizes his exploitation of marine ecosystems. Ray also argues for close international cooperation in science, economics, and technology to conserve our natural marine resources. We concur.

References

Alaska Sea Grant. 1993. Is it food? Addressing marine mammal and seabird declines: Workshop Summary. University of Alaska Sea Grant, AK-SG-93-01, Fairbanks. 59 pp.

Alexander, V., and R.T. Cooney. 1979. Ice edge ecosystem study: Primary productivity, nutrient cycling and organic matter transfer. Final Report to BLM/NOAA, Outer Continental Shelf contract number 03-5-022-56. 182 pp.

Alverson, D. 1992. A review of commercial fisheries and the Steller sea lion *(Eumetopias jubatus):* The conflict arena. Reviews in Fisheries Science 6:203-256.

Antonelis, G.A., and E.H. Sinclair. 1996. Comparison of pinniped feeding habits in the North Pacific Ocean. In: International symposium on the role of forage fishes in marine ecosystems: Abstracts. University of Alaska Sea Grant, Fairbanks, p. 21.

Antonelis, G.A., E.H. Sinclair, R.R. Ream, and B.W. Robson. 1997. Inter-island variation in the diet of female northern fur seals *(Callorhinus ursinus)* in the Bering Sea. Journal of Zoology 242:435-451.

Bailey, E.P. 1993. Introduction of foxes to Alaskan islands: History, effects on avifauna, and eradication. U.S. Fish and Wildlife Service, Resource Publication 193. 53 pp.

Bakkala, R., W. Hirshberger, and K. King. 1979. The groundfish resources of the eastern Bering Sea and the Aleutian Islands region. Marine Fisheries Review 41(11):1-24.

Bateson, G. 1979. Mind and nature, a necessary unity. Bantam Books, New York. 255 pp.

Cooney, R.T. 1981. Bering Sea zooplankton and micronekton communities with emphasis on annual production. In: D.W. Hood and J.A. Calder (eds.), The eastern Bering Sea Shelf: Oceanography and resources, vol. II. University of Washington Press, Seattle, pp. 947-974.

Coyle, K.O., V.G. Chavtur, and A.I. Pinchuk. 1996. Zooplankton of the Bering Sea: A review of Russian-language literature. In: O.A. Mathisen and K.O. Coyle (eds.), Ecology of the Bering Sea. University of Alaska Sea Grant, AK-SG-96-01, Fairbanks, pp. 97-133.

Flint, M.V., A.V. Drits, M.B. Emelianov, A.I. Koplov, I. Merculieff, P.V. Rybnikov, I.N. Sukhanova, and T.E. Whitledge. 1996. Investigations of the Pribilof marine ecosystem, third year: Significance of oceanographic and biological processes at the shelf break, outer shelf domain, and middle shelf front for biological productivity of the Pribilof region. P.P. Shirshov Institute of Oceanology, Moscow. 504 pp.

Flint, M.V., A.V. Drits, M.B. Emelianov, A.I. Koplov, I. Merculieff, S.G. Pojarkov, P.V. Rybnikov, I.N. Sukhanova, and T.E. Whitledge. 1994. Investigations of the Pribilof marine ecosystem, second year: Study of the marine ecosystem in the vicinity of the Pribilof Islands for general ecological evaluation and elucidation of the most valuable ecological zones. P.P. Shirshov Institute of Oceanology, Moscow. 504 pp.

Goebel, M.E., J.L. Bengtson, R.L. DeLong, R.L. Gentry, and T.R. Loughlin. 1991. Diving patterns and foraging location of northern fur seals. Fishery Bulletin, U.S. 89:171-179.

Goering, J.J., and R.L. Iverson. 1981. Phytoplankton distribution on the southeastern Bering Sea Shelf. In: D.W. Hood and J.A. Calder (eds.), The eastern Bering Sea Shelf: Oceanography and resources, vol. II. University of Washington Press, Seattle, pp. 933-946.

Hasle, G.R. 1976. The biogeography of some marine planktonic diatoms. Deep-Sea Research 23:319-338.

Hunt, W.R. 1975. Arctic passage. Charles Scribners Sons, New York. 395 pp.

Konovalova, G.V., T.Y. Orlova, and L.A. Pautova. 1989. Atlas of the phytoplankton of the Sea of Japan. Nauka Publishing, Leningrad. 160 pp. (In Russian.)

Laughlin, W.S. 1967. Human migration and permanent occupation in the Bering Sea area. In: D. Hopkins (ed.), The Bering land bridge. Stanford University Press, Palo Alto, CA, pp. 409-450.

Lessard, E.J., and E. Swift. 1985. Species-specific grazing rates of heterotrophic dinoflagellates in oceanic waters, measured with a duel-label radio-isotope technique. Marine Biology 87(3):289-296.

Loughlin, T.R., and L.L. Jones. 1984. Review of existing data base and of research and management programs for marine mammals in the Bering Sea. In: Proceedings of the workshop on biological interactions among marine mammals and commercial fisheries in the southeastern Bering Sea. University of Alaska Sea Grant, AK-SG-84-01, Fairbanks, pp. 79-99.

Loughlin, T.R., and R.V. Miller. 1995. Marine Mammals, Chapter 17. In: F.G. Johnson and R.R. Stickney (eds.), Fisheries: Harvesting life from water (2nd edn.). Kendall/Hunt Publishing Co., Dubuque, IA, pp. 296-333.

Loughlin, T.R., J.L. Bengtson, and R.L. Merrick. 1987. Characteristics of feeding trips of female northern fur seals. Canadian Journal of Zoology 65:2079-2084.

Loughlin, T.R., G.A. Antonelis, J.D. Baker, A.E. York, C.W. Fowler, R.L. DeLong, and H.W. Braham. 1994. The status of the northern fur seal population on the U.S. west coast in 1992. In: E. Sinclair (ed.), Northern fur seal investigations for 1992. NOAA Technical Memo NMFS-AFSC-45, pp. 9-28.

Lowry, L.F., K.J. Frost, D.G. Calkins, G.L. Swartzman, and S. Hills. 1982. Feeding habits, food requirements, and status of Bering Sea marine mammals. North Pacific Fishery Management Council, 605 W. 4th Ave., Suite 306, Anchorage, AK 99501. Documents No. 19 and 19A. 574 pp.

Medlin, L.K., and J. Priddle (eds.). 1992. Polar marine diatoms. British Antarctic Society, Natural Environment Research Council, London. 199 pp.

Merrick, R.L., and D.G. Calkins. 1996. Importance of juvenile walleye pollock in the diet of Gulf of Alaska Steller sea lions. NOAA Technical Report NMFS 126:153-166.

Merrick, R.L., M.K. Chumbley, and G. Vernon Byrd. 1997. Diet diversity of Steller sea lions *(Eumetopias jubatus)* and their population decline in Alaska: A potential relationship. Canadian Journal of Fisheries and Aquatic Science 54:1342-1348.

Müller-Beck, H. 1967. On migrations of hunters across the Bering land bridge in the upper Pleistocene. In: D. Hopkins (ed.), The Bering land bridge. Stanford University Press, Palo Alto, CA, pp. 373-408.

National Research Council. 1996. The Bering Sea ecosystem. National Academy Press, Washington, DC. 307 pp.

Parsons, T.R., and R.J. LeBrasseur. 1970. The availability of food to different trophic levels in the marine food chain. In: J.H. Steele (ed.), Marine food chains. University of California Press, Berkeley, CA, pp. 325-343.

Pastor, J. 1995. Ecosystem management, ecological risk, and public policy. BioScience 45:286-288.

Perez, M.A., and M.A. Bigg. 1986. Diet of northern fur seals, *Callorhinus ursinus,* off western North America. Fishery Bulletin, U.S. 84:957-971.

Perez, M.A., and W.B. McAlister. 1993. Estimates of food consumption by marine mammals in the eastern Bering Sea. NOAA Technical Memo NMFS-AFSC-14. 36 pp.

Ray, C. 1970. Ecology, law, and the "Marine Revolution." Biological Conservation 3(1):7-17.

Saito, K., and A. Taniguchi. 1978. Phytoplankton communities in the Bering Sea and adjacent seas. II. Spring and summer communities in seasonally ice-covered areas. Astarte 11:27-35.

Semina, H.I. 1953. Phytoplankton in the western Bering Sea. Ph.D. dissertation, P.P. Shirshov Institute of Oceanology, Academy of Sciences USSR, Moscow. 189 pp. (In Russian.)

Semina, H.I. 1974. Pacific phytoplankton. Nauka Publishing, Moscow. 237 pp. (In Russian.)

Semina, H.I. 1981. Specific composition of phytoplankton in the western part of the Bering Sea and adjacent part of the Pacific Ocean: Diatoms. In: Ecology of marine phytoplankton. Academy of Sciences, USSR, P.P. Shirshov Institute of Oceanology, Moscow, pp. 6-31.

Sinclair, E., T.R. Loughlin, and W. Pearcy. 1994. Prey selection by northern fur seals *(Callorhinus ursinus)* in the eastern Bering Sea. Fishery Bulletin, U.S. 92:144-156.

Sinclair, E.H., G.A. Antonelis, B.R. Robson, R. Ream, and T.R. Loughlin. 1996. Northern fur seal, *Callorhinus ursinus,* predation on juvenile walleye pollock, *Theragra chalcogramma.* NOAA Technical Report NMFS 126:167-178

Simenstad, C.A., J.A. Estes, and K.W. Kenyon. 1978. Aleuts, sea otters, and alternate stable-state communities. Science 200:403-411.

Springer, A.M., C.P. McRoy, and M.V. Flint. 1996. The Bering Sea Green Belt: Shelf-edge processes and ecosystem production. Fisheries Oceanography 5(3/4):205-223.

Swartzman, G.L., and R.J. Hofman. 1991. Uncertainties and research needs regarding the Bering Sea and Antarctic marine ecosystems. U.S. National Technical Information Service PB91-201731. 106 pp.

United States Fish and Wildlife Service (USFWS). 1998. Beringian seabird colony catalog: Computer database and colony status record archives. U.S. Fish and Wildlife Service, Migratory Bird Management, Anchorage, AK.

United States Department of Commerce (USDC). 1998. Groundfish total allowable catch specifications and prohibited species catch limits under the authority of the fishery management plans for the groundfish fishery of the Bering Sea and Aleutian Islands area and groundfish of the Gulf of Alaska. Draft supplemental environmental impact statement. 687 pp. + 6 appendices.

Venttsel, M.V. 1994. Plankton phytocoenosis of the oceanic and shelf areas of the Bering Sea. Ph.D. dissertation, VNIRO, Moscow. 106 pp.

Vidal, J., and S.L. Smith. 1986. Biomass, growth, and development of populations of herbivorous zooplankton in the southeastern Bering Sea during spring. Deep-Sea Research 33:523-556.

CHAPTER 20

Modeling and Management of the Bering Sea Ecosystem

Robert C. Francis and Kerim Aydin
University of Washington, Seattle, Washington

Richard L. Merrick
Northeast Fisheries Science Center, Woods Hole, Massachusetts

Stephen Bollens
San Francisco State University, San Francisco, California

All models are wrong. However, some are useful. —Anonymous

Abstract

This chapter addresses the issue of modeling and marine ecosystems with a particular focus on the Bering Sea. It is divided into three sections. In the first, we attempt to lay a conceptual foundation for ways to think about ecosystems. Next, we describe three kinds of models (trophic, process, and conceptual) and how they have been used to gain various levels of insight about the Bering Sea ecosystem. Finally, we discuss the concept of ecosystem management as applied to the Bering Sea and how models of various kinds might help facilitate the process. Our major point is that policies and management which apply static, fixed rules lead to systems which gradually lose resilience—systems that gradually break down in the face of disturbances that previously could have been absorbed. Consideration of these characteristics in management requires modeling. To best assist managers, models should be viewed as predictors (not prescriptions) of the outcome of various management outcomes, and the outcomes must then be monitored, evaluated, and revised as information becomes available.

Narrowing the focus to Bering Sea fishery management from an ecosystem perspective, we emphasize that managers should: (1) strive to retain critical types and ranges of natural variation in the Bering Sea ecosystem; (2) set allowable catches taking into account interactions between a target species and its competitors, predators, and prey; (3) set allowable catches using a balanced approach to the levels of harvest on various target species; and (4) undertake restoration efforts using active adaptive strategies.

Introduction and Overview

This chapter addresses the issue of modeling and marine ecosystems with a particular focus on the Bering Sea. It is divided into three sections. In the first we attempt to lay a conceptual foundation for ways to think about ecosystems. Next we describe three kinds of models (trophic, process, and conceptual) and how they have been used to gain various levels of insight about the Bering Sea ecosystem. Finally we discuss the concept of ecosystem management as applied to the Bering Sea and how models of various kinds might help facilitate the process.

One key issue which arises in this paper and which is beautifully discussed by Kingsland (1985) is the fundamental dichotomy that many mathematical models create between historical and ahistorical thinking. Most ecosystem science is intended to address historical issues which are culminations of unique sequences of events. To a great extent these kinds of processes can only be described to the extent that present processes can be inferred from ordered sequences of past processes. However, the very act of imposing a model (often mathematical) on nature often involves the rejection of history in favor of some unifying concept, generally providing a quasi-equilibrium view of nature—a view that has been common whenever analytic methods taken from the physical sciences have been applied to the study of the dynamics of biological populations. And because of their predictive nature, most models are forward looking rather than backward looking. It is this dilemma which is at the center of our chapter.

Marine Ecosystems—A Conceptual Framework

Marine ecosystems are complex structures in which significant (re-)ordering occurs in both top-down and bottom-up directions. The key aspects are the roles of the basic rates governing processes and the distribution of infrequent dramatic events associated with physical forcing due to climate change. In addition, most of the marine fisheries ecosystems with which we are presently concerned have very substantial pieces missing; i.e., major components with which the current fish community evolved and interacted are no longer there in any significant numbers (Apollonio 1994). In the Bering Sea and North Pacific Ocean, some populations, such as large whales, have been harvested to the point of extinction. Others, such as marine fish, have both been reduced in absolute abundance and had their population structure altered by the significant truncation of their age structure. This combination of factors has clearly had significant effects on the Bering Sea ecosystem.

This section describes four major conceptual challenges to scientists and resource managers interested in evaluating ecological pattern and order and its possible consequences on management. These are scale, history, symmetry, and succession.

Scale

Ricklefs (1990) defines scale as the characteristic distance or time associated with variation in natural systems. He goes on to make three important points about why the concept of scale is so important to developing an understanding of ecosystem structure and dynamics. These points are: (1) every process and pattern has a temporal and spatial extent; (2) patterns and processes that occur on different scales of time and space are linked in such a way that the rules governing these relationships may help to define ecosystem boundaries; and (3) scales change as signals propagate through ecosystems, particularly as they move from physical to biological components of the ecosystem.

Levin (1992) in his seminal paper on pattern and scale in ecosystems says that "Scale is ..., I will argue, the fundamental conceptual problem in ecology, if not in all of science." He too makes three points which, we think, are of interest to our discussions of marine ecosystems and their management. These are: (1) understanding patterns in terms of the processes that produce them is the essence of science, and is the key to the development of principles for management; (2) the identification of pattern is an entree into the identification of scales. Once patterns are detected and described, we can seek to discover the determinants of pattern, and the mechanisms that generate and maintain those patterns; and (3) no single mechanism explains pattern on all scales.

Perhaps one of the real difficulties we seem to have with ecosystem properties has to do with our struggle to deal with scale. Clearly many processes that affect ecosystem structure and dynamics occur on different time and space scales. Levin (1992) and Carpenter (1990) provide some clues of scientific directions that may help us with problems associated with scale. Levin (1990, 1992) suggests that quantitative modeling is a useful tool for developing an understanding of how information is transferred across scales. He says that "the essence of modeling is, in fact, to facilitate the acquisition of this understanding, by abstracting and incorporating just enough detail to produce observed patterns." Carpenter (1990) proposes that because of the nature of ecosystem dynamics, in many cases manifesting themselves in rather abrupt "sledgehammer blows" in the words of Schindler (1987), the classical domain of replicable experimental science is not available to the ecosystem analyst. He goes on to recommend a number of relatively new statistical approaches which show promise for the analysis of large scale ecosystem properties (e.g., intervention analysis, a time series method designed to detect abrupt discontinuous shifts in time series, and empirical Bayesian analysis which allows one to reach quantitative conclusions from the combined results of different studies). Levin (1992) adds to the list some powerful new methods of spatial statistics which provide the capacity to describe how patterns change across scales. They both point out that ecosystem scientists (and managers) must look to modern developments in quantitative modeling and statistics if they want to seriously come to grips with fundamental ecosystem properties associated with scale.

Perhaps the major crosscutting challenge to scientists interested in evaluating ecological patterns is the integration of the multiple spatial and temporal scales involved (Levin 1992). There is no single "correct" scale at which to view ecosystems. Processes that occur at one scale have impacts observable at other scales. Patchiness on virtually all scales is a characteristic of ecosystems that complicates their complete description and poses an implicit challenge to ecologists to explain these patterns (Denman and Powell 1984). Our snapshot views of temporally dynamic ecological systems represent one of the most serious limitations to proper understanding of causes of ecosystem dynamics.

Central to conventional ecosystem theory is the concept that structure (the way process distributes itself in space) results from differences in process rates. Thus, according to this theory, ecosystems are structured according to the relationships between processes occurring at different time and space scales. Although organisms as taxonomic entities are not the essential "stuff" of ecosystems, their populations, as variable conduits, clearly define the fundamental nature of ecosystem dynamics. It is the populations of species that participate in creating or modifying energy fluxes that, as a result, occur at different rates in space and time. What we are talking about here is ecology in four dimensions. An ecosystem cannot be defined or conceived of without clear reference to time. In the words of Allen and Hoekstra (1992), "... complex behavior will arise when very different rates are pressed together in the formulation of a population equation ... complexity arises from the interaction of differently scaled processes ...."

History

The evolutionary history and the more immediate past in ecological time are important contributors to the patterns observed in nature. Yet the historical memories present in any ecosystem are difficult or impossible to extract in the short time horizons of both management institutions and scientific study plans. Because of the significance of processes on multiple spatial and temporal scales to the generation of ecological pattern, and especially the unknown contributions of historical events, scientists studying ecosystems are always faced with incomplete and imperfect information. As a result, ecosystem processes reside, in what Magnuson (1990) called "the invisible present," hidden from view or understanding because they occur slowly or because effects lag years behind causes.

Development of the science necessary to support ecosystem management probably will require a new synthesis of previously separate approaches to the study of ecology, a melding of relevant aspects of population ecology, community ecology, and traditional ecosystem ecology. The goal is to be able to predict, understand, and interpret changes in populations of component species of an ecosystem, especially those that have value to human society. Such understanding must be based in large part on knowledge of the interaction of those important populations with other species in the ecosystem. Such questions are the essence of community ecology, and the experimental work done on trophic cascades by Paine (1980) in marine intertidal systems, the research done to evaluate the indirect effects of removal of sea otters from subtidal communities (Estes and Palmisano 1974, Simenstad et al. 1978), and the experimental manipulations by Carpenter and colleagues (e.g., Carpenter et al. 1985) in lakes represent the best examples of how community ecology must be involved to develop appreciation of processes controlling dynamics of individual species.

Parsons (1992) presented examples from the Atlantic, Pacific, and Antarctic oceans where the removals of marine predators by fisheries have had various cascading short- and long-term effects on ecosystems (e.g., the effects of the harvest of Antarctic blue whales *(Balaenoptera musculus).* These changes affect abundance of other krill-eating and krill-dependent species, and the inverse relationship between the abundance of planktivorous right whales *(Eubalaena glacialis)* and the abundance of planktivorous Pacific sand lance *(Ammodytes hexapterus)* in the Gulf of Maine. These reports tend to substantiate the existence and importance of top-down effects in structuring marine ecosystems. In addition, they demonstrate that these effects can occur at a number of different time scales.

Symmetry and Succession

Ecologists and resource scientists have long debated the means by which trophic interactions affect the distribution and abundance of organisms in an ecosystem. The debate centers on whether primary control is by resources (bottom-up) or predators (top-down). A special feature of the journal *Ecology* (Vol. 73[3] 1992) gives an excellent review of this controversy. The editors came to the conclusion that (1) top-down and bottom-up forces act on populations and communities simultaneously, and (2) the issue is no longer about what occurs, but rather what controls the strength and relative importance of the various forces under varying conditions and what drives feedback and interactions among multiple trophic levels. Power (1992) provides an excellent review of factors that affect the relative strength of top-down and bottom-up forces in food webs as well as discussing methodological problems that color our perceptions of the importance of these forces. She also reviews specific examples of such things as dynamic feedbacks between adjacent and nonadjacent trophic levels, and concludes that due to these complexities and nonlinear relationships, there appear to be real difficulties in applying food web theories to the real world.

As illustrated in Francis et al. (1998), in the large marine ecosystems of the northeastern Pacific, the primary effect of physical forcing is on lower trophic levels working its way up through the trophic ladder. There are also top-down effects as changes in distribution and abundance of higher levels affect predators, competitors, and prey at other levels. Changes in abundance at higher trophic levels may result from fishing as well as from climate variation, and can cause significant top-down responses due to changes in predation.

A number of recent works on ecosystem order have grappled with the relationships between symmetry (top-down, bottom-up forcing) and succession. One view is of ecosystems as hierarchies involving asymmetric interactions (top-down) where system components with large amounts of stored capital (biomass) and operating at slow rates maintain control over components with small amounts of stored capital operating at fast (turnover) rates (Allen and Starr 1982, O'Neill et al. 1986). In this view, succession is controlled by two functions—exploitation in which rapid colonization of recently disturbed areas occurs (r-phase, Fig. 1) and conservation in which a slow accumulation and storage of energy occurs (K-phase, Fig. 1). Lower-level components (those operating at short time scales) respond significantly to disturbances at higher levels. However, higher levels "see" only the averaged or integrated responses of lower-level components. This does not mean that everything in an ecosystem occurs in a top-down fashion. O'Neill et al. (1986) pointed out that in hierarchical organizations, lower-level behaviors are essential to the functioning and persistence of higher-level structure, although, because of their slower rates, they are unable to affect the behavior of the higher level.

Apollonio (1994) brought hierarchy theory into the realm of marine fisheries. He distinguished between long- and short-lived species in terms of their ecosystem-organizing capacities as follows: "[Long-lived] species, because of their biological attributes, have a role in structuring or shaping their systems ... That is to say, their presence induces predictability into the system. And, by inference, their reduction or removal reduces internal control mechanisms and increases variability and unpredictability." He points out that the same can be said for migratory species which transfer energy among regions and, consequently, tend to stabilize marine communities.

The point that Apollonio (1994) was making is that, according to hierarchy theory, ecosystems are structured in part by the natural frequencies (both spatially and temporally) of their component parts. These species serve as space- and/or time-binders. "Structuring" tends to occur as a function of these natural frequencies. The implication is that ecosystems that are dominated at the top by long-lived highly migratory species tend to be more "structured" and predictable than those that are not.

A more recent view of ecosystem symmetry and succession (Holling Infinity Loop) is expressed by Holling (1995). He identifies two additional successional functions (Fig. 1): release (Ω–phase) in which a tightly bound accumulation of stored capital (biomass, nutrients) is suddenly released in response to perturbation, and reorganization (α-phase) in which re-

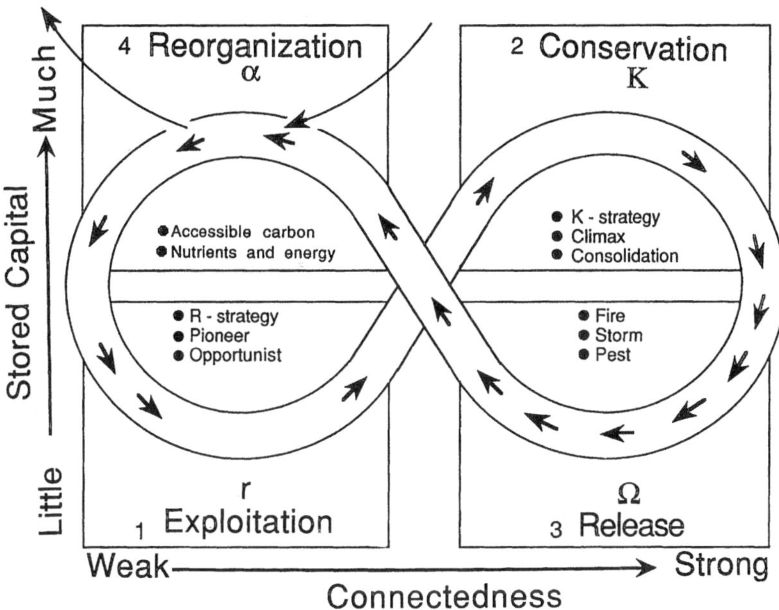

Figure 1.   The four ecosystem functions and the flow of events between them. The arrows show the speed of that flow in the ecosystem cycle, where arrows close to each other indicate a rapidly changing situation and arrows far from each other indicate a slowly changing situation. The cycle reflects changes in two attributes; that is: (1) the Y axis—the amount of accumulated capital (nutrients, carbon) stored in variables that are the dominant keystone variables at the moment—and (2) the X axis—the degree of connectedness among variables. The exit from the cycle indicated at the left of the figure suggests the stage where a flip is most likely into a less or more productive and organized system (i.e., devolution or evolution as revolution!).

leased nutrients are rechanneled to become available for the next phase of exploitation. In these two phases, which tend to occur fairly rapidly, components with small amounts of stored capital operating at fast rates can exert control (bottom-up) over components operating at slow rates with large amounts of stored capital. Release tends to occur in systems which have become overconnected, rigid, and brittle, and can occur in response to relatively small forcing events operating on fast processes. Reorganization occurs in systems that are underconnected and weakly organized. It is at this stage where unexpected jumps to completely new states of organization can occur. In the words of Holling (1995), "… variables can become reshuffled, perhaps to establish different relationships, perhaps to be open to foreign and entirely novel entrants." He goes on to point out that "the least understood and probably the most difficult and critical portions to understand are the discontinuities resulting from the endogenous dynamics of human systems." Using a conceptual model, Merrick (1995) and the National Research Council (1996) provide an example (described in the following section) of this kind of succession occurring in the Bering Sea over the past four or five decades.

The Bering Sea Ecosystem and Modeling

Introduction

In this section, we describe three kinds of models (trophic, process, and conceptual) and how they have been used to gain various levels of insight into the Bering Sea ecosystem. In particular, we attempt to show how these various types of models help to address some of the key conceptual issues discussed above. In particular, trophic models, while they reveal important "unseen" aspects of marine ecosystems, tend to be ahistorical and incapable of dealing with succession. In this context, historical representation or explanation does not rest solely on harmonious unifying concepts which repeat themselves over and over—rather it also takes into account an unpredictable sequence of antecedent states where the final result is dependent, or contingent, on everything that came before (Francis and Hare 1994). Process models, very focused abstractions, reveal aspects of important links between the physical environment and the natural frequencies of select life history stages of important players. In addition they help to recreate historical scenarios of lower trophic level processes. Conceptual models, essentially back-of-the-envelope calculations, flow freely with history but lack the quantitative rigor that many assessment scientists and resource managers associate with convincing science. In the Bering Sea example reported here, they are used to determine if historical interpretations of ecosystem dynamics "add up" from an energy balance point of view.

Trophic Models

Trophic, or food web, models are meant to examine fishery ecosystem issues from the point of mass-balance trophic dynamics. In essence these models are driven by the nature and magnitude of species trophic interactions within the ecosystem. There are two fundamental approaches to this kind of modeling. In the bottom-up approach, calculations are made of the primary production necessary to sustain various ecosystem configurations, or, conversely, the kind of ecosystem structure, particularly at the top where fisheries tend to operate, sustainable by known levels of primary production. The most notorious of these is ECOPATH (Christensen and Pauly 1995, Pauly and Christensen 1995) which has been used in various incarnations to review the status and potential of world fisheries (see

Pauly and Christensen [1995] for estimates of the amount of global aquatic primary production which is required to sustain world fisheries). The top-down approach starts with apex predators, including fishing by man, and determines what lower trophic structure and abundances are necessary to sustain them. In this approach, termed "back down the food chain" by Laevastu and Favorite (1988), the food chain is descended until zooplankton production is reached and estimated. Over the past 20 years, a major application of this approach has been made in the Bering Sea by Taivo Laevastu and colleagues at the NMFS Alaska Fisheries Science Center (Laevastu and Larkins 1981, Laevastu and Favorite 1988) in their PROBUB and DYNUMES simulation models. Some of the results are summarized below.

In the late 1970s and early 1980s, the major problem for the North Pacific Fishery Management Council was to assess the status of individual stocks of groundfish which were the target of harvest and attempt to set annual quotas which, in some way, adhered to the concepts of sustainable yield theory. The scientific riddles which were solved in order to facilitate this annual decision process involved classical single species stock assessments which were thoroughly reviewed within the council "family." At the same time, ecosystem science was being developed in a very sophisticated manner and focused on these very marine ecosystems of the northeastern Pacific as a major task within the National Marine Fisheries Service, Northwest and Alaska Fisheries Center (e.g., Laevastu and Larkins 1981, Laevastu and Favorite 1977). Most of this research involved spatially resolved biomass-based ecosystem modeling (DYNUMES) of large regions of the northeastern Pacific and Bering Sea. A number of interesting conclusions were reached:

1. Changes in predator biomass could significantly alter ecosystem structure.

2. Forage fish (e.g., capelin, *Mallotus villosus*; Pacific herring, *Clupea harengus*) play an important role in structuring these large marine ecosystems. The ecosystem models of Laevastu required the presence of "considerable quantities of pelagic fish ... dictated mainly by the food composition of predators ... The consumption of these species requires that they be relatively widely distributed over the area" (Laevastu and Favorite 1977).

3. Forage fish abundance may fluctuate inversely to that of walleye pollock *(Theragra chalcogramma)* and "have periods ranging from about 10 to greater than 20 years" (Laevastu and Larkins 1981).

4. "Both mammals and birds are apex predators, their effects on the ecosystem is [sic] similar, but not identical to the effect of fishing by man" (Laevastu and Larkins 1981).

5. It was not possible to balance (equilibrium) ecosystem computations driven by known mammal biomasses and consumption rates.

6. The effects of fishing on the production of finfish populations might "work in the favor of marine mammals who feed predominantly on smaller fish than taken by the fishery" (Laevastu and Larkins 1981).

Laevastu and colleagues point out that "there are, however, two specific subjects in which our knowledge on nearly all species and areas is deficient; the abundance and migrations of prefishery juveniles and the biomasses of noncommercial dispersed species" (Laevastu and Larkins 1981). In essence, they were able to show that the large marine ecosystems of the northeastern Pacific are extremely dynamic (they had trouble making equilibrium ecosystem models balance properly), are significantly influenced from the top by marine mammals, appear to be structured both above (top down) and below (bottom up) by the dynamics in space and time of forage fish, and respond to fishing in (intuitively) strange ways.

Process Models

Process models—operationally defined here as ecosystem models that explicitly include links between specific biological processes (e.g., feeding dynamics, birth and mortality rates, primary and secondary production) and/or physical processes (e.g., heat and momentum transfer, mixed layer dynamics)—allow investigators to examine the underlying mechanisms, or processes, that generate the patterns we observe in nature. This allows for a much greater insight into ecosystem function, thereby providing managers (at least in theory) with a more detailed view of the effects of different anthropogenic perturbations (e.g., fishing) relative to the natural variability of the system. Process models thus increase the predictability of consequences of different management scenarios, leading ultimately to more effective management. In the subsection that follows we provide two examples of process modeling studies in the Bering Sea—the PROBES and Bering Sea FOCI programs—that have improved our basic understanding of ecosystem function and offered the promise of improved management techniques.

The Processes and Resources of the Bering Sea Shelf project (PROBES) was a ten-year study of the mechanisms of biological productivity on the southeast Bering Sea shelf (Hood, chapter 32, this volume). The project used the life history of walleye pollock as a biological tracer for physical and biological ecosystem processes across the region.

The first essential work of PROBES was an extensive oceanographic study of the shelf region (Coachman 1986). Rather than revealing the previously supposed large cyclonic circulation pattern, the study showed three highly-structured water masses divided by oceanographic fronts. These fronts separated the shelf into outer, middle, and coastal domains, with little water mixing between the regions (see National Research Council 1996).

Within the domains, PROBES distinguished two distinct zooplankton communities. The first, living on the middle and inner shelves, consists mainly of small *Pseudocalanus* spp. and *Acartia* spp.—year-round residents reproducing only after spring blooms. These smaller zooplankton are inefficient grazers of large phytoplankton due to the lag between phytoplankton growth and zooplankton production. The second, more oceanic, zooplankton community lives in deep waters off the shelf during winter, and moves onto the outer shelf during spring, prevented by the middle front from advancing into shallower waters. These larger species (especially *Calanus* spp. and *Eucalanus* spp.) are efficient grazers, and their overwintering stages eliminate the time lag between phytoplankton and zooplankton development (Smith and Vidal 1986).

The discovery of two distinct zooplankton communities led to replacing the hypothesized model of a single food web with a model which divided the shelf into two systems with physically separated food webs. In the inner- and middle-shelf food webs, most primary production sinks out of the surface waters and enters the benthic community leading to high benthic production. In the outer-shelf food web, most phytoplankton are consumed by zooplankton and thus are consumed by pollock creating a region of high pelagic production (McRoy et al. 1986). Changing the model to reflect these separated systems led to the development of hypotheses which better tied modeled variations to physical and biological observations.

The new process models of circulation, primary, and secondary production were then used to examine the effect of interannual climate variation on the success of larval pollock. Comparing modeled pollock dispersal in warm and cold years with zooplankton abundances, Walsh and McRoy (1986) found that during warm years (e.g., 1978), high zooplankton abundance in both the middle and outer shelves led to a strong larval pollock year-class. However, during cold years (e.g., 1976), only the outer shelf would contain enough zooplankton for pollock survival. The simulated difference between pollock year class strengths between the coldest and warmest observed years (1976 and 1978, respectively) was approximately sevenfold.

Process modeling studies undertaken as part of the Bering Sea Fisheries Oceanography Cooperative Investigations (BS FOCI) program extended the work of the PROBES program in several important regards. BS FOCI is an interdisciplinary study of the possibly different aggregates of pollock in the central basin and the continental shelf of the eastern Bering Sea. In particular, BS FOCI is an attempt to understand how the different oceanographic domains affect productivity of larval pollock populations, especially the degree to which larval production is coupled/decoupled with the spring phytoplankton bloom.

Bollens (unpublished data) adapted and extended the nutrient-phytoplankton-zooplankton (NPZ) model of Frost (1987) to include stage-structured dynamics of copepod populations and growth of larval pollock as a function of abundance of appropriately sized prey. The NPZ model, in its most basic form, is a model of the population dynamics of the lowest trophic levels, phytoplankton and herbivorous zooplankton, as affected by nutrient and light levels (Polovina et al. 1995). Protozoan microplankton and dynamically modeled mixed layer depths were also added to the model, in the latter case using the Price et al. (1986) mixed-layer model (PWP). Input data were taken from the Bering Sea (e.g., PROBES; Sambrotto et al. 1986, Vidal and Smith 1986) whenever possible, but were at other times taken from Ocean Station Papa in the North Pacific. The model was applied to two distinct oceanographic domains in the Bering Sea—the mid-continental shelf and deep open basin—each with its own physical and lower-trophic level biological dynamics.

Model results indicated that three processes were particularly important to the growth of early-feeding pollock larvae:

1. The species composition and dynamics of meso- and macrozooplankton. Larvae dependent on *Calanus* as prey experienced much better growth conditions later in the spring, whereas those dependent on *Neocalanus* were likely to grow more slowly generally, although somewhat better earlier in the year when young *Neocalanus* are first recruited to the surface layer.

2. The role of protozoans. Larval pollock in this model were highly dependent on these organisms for sustaining high growth rates, especially during times of year when young copepods are not abundant (i.e., early spring on the shelf and late spring-summer in the basin).

3. Short-term mixed-layer depth variability. This process had significant impacts on all trophic levels (phytoplankton, protozoans, copepods, and larval fish). Most striking were the rapid and intense shoaling events during winter and spring that last from one to several days and which led to earlier net phytoplankton and protozoan growth. This in turn led to higher larval pollock growth rates (by a factor of two on the shelf and a factor of three in the basin) early in the production season (March-April), due almost entirely to increased protozoan biomass at this time of year.

Although these "process models" (i.e., PROBES) were not used to make explicit predictions of ecosystem response to management practices, they nevertheless served to dramatically improve our understanding of ecosystem response to biological and physical processes and "perturbations." These modeling studies were also important in identifying several data gaps and processes and phenomena that need to be addressed in future field, modeling, and laboratory studies. Moreover, certain management implications arise from these studies. First, the existence of two distinct physical and biological systems in the southeast Bering Sea, bridged by common species, shows that a management plan based on species rather than region may not accurately reflect the dynamics of managed populations. Second, the connection to the benthos suggests that production and/or predator-prey linkages between the pelagic (e.g., pollock) and benthic (e.g., king crab) fisheries may lead to interrelated variation in both.

Finally, a detailed process model has a greater potential for examining the range of natural, climate-driven variation in the system, which can then be used to examine the relative effect of anthropogenic perturbations such as fishing.

Conceptual Models

One of the important issues for us raised by complex but temporally static dynamic models has to do with what Kingsland (1985) refers to as the conflict between historical and ahistorical thinking. She points out that "the very act of imposing mathematics (or any model) on nature often involve a rejection of history (contingency) in favor of a harmonious, unifying concept" (i.e., the equilibrium view of nature). Unfortunately, mathematical ecology was intended to address historical issues. And therein lies a major dilemma. In this context, a simple "back-of-the-envelope" model, what we call a conceptual model, is created which takes into account an unpredictable sequence of antecedent states where the final result is dependent, or contingent, on everything that came before (Francis and Hare 1994). In the Bering Sea example reported here, a conceptual model is used to determine if historical interpretations of ecosystem dynamics "add up" from an energy balance point of view. The historical issue which we modeled is described as follows.

Large population declines (>50%) since the early 1970s of some eastern Bering Sea (EBS) and Aleutian Island apex predator populations (e.g., Steller sea lions, *Eumetopias jubatus*; murres, *Uria* spp.; kittiwakes, *Rissa* spp.) may be an indication that major changes have occurred in the structure of the Bering Sea ecosystem. One possible cause of the decline in predator populations has been a decrease in abundance of preferred prey (e.g., capelin, juvenile pollock).

At least two factors may have contributed to either the increase in adult groundfish or decrease in forage fish; these are: (1) a trophic cascade (Carpenter et al. 1985) resulting from the depletion of large whales, fur seals, and some fin fish during 1955-1975; and (2) an oceanic regime shift in 1976-1977 which led to dramatic increases in a number of piscivorous fish populations (total adult groundfish biomass is estimated to have increased from 13 million t to 22 million t from 1978-1984) (Wespestad 1993). Prior to exploitation, a suite of *K*-selected (long-lived, low production) species (large whales, fur seals, sea lions, harbor seals, ice seals, seabirds, some fish) may have exerted considerable control over the Bering Sea ecosystem. Consumption of zooplankton and predation on groundfish by these species may have kept juvenile groundfish abundance at low levels, which restricted the possible stock size of adult groundfish. This limited the impacts of groundfish on forage fish populations.

Heavy commercial harvest of a number of these *K*-selected populations (baleen whales; northern fur seals, *Callorhinus ursinus*; Pacific Ocean perch, *Sebastes alutus*; and Pacific herring) during 1955-75 may have led to increased zooplankton availability and reduced predation on juvenile groundfish (particularly juvenile pollock) thus triggering a sequence of events which, in combination with oceanic changes (1976-77 regime shift) and current fishing practices (low exploitation rates), could have produced the sharp increases in adult groundfish biomass observed in the 1980s. The resulting increased consumption of forage fish may have reduced the alternate prey available to other predators and, coupled with highly variable juvenile pollock abundance, may have led indirectly to the recent declines in some EBS and northeastern Pacific marine mammal and seabird populations. Thus, the large increases in groundfish biomass that occurred in the EBS after the early 1960s, while good for commercial groundfish fisheries, may have been to the detriment of other members of the ecosystem.

Conceptual models presented by Merrick (1995) and the National Research Council (1996) reconstruct a skeleton of this recent history of trophic dynamics in the EBS, demonstrating that this 50-year scenario is at least possible from an energetics point of view, and provides a possible explanation for the decline in pinniped and seabird prey abundance. Several hypotheses are key to these conceptual models. The first is that the recent levels of adult groundfish are unusually high and may be sufficiently large to deplete mammalian and avian prey (i.e., small schooling fish). The second is that the high groundfish biomass was initially stimulated by the release of zooplankton and reduced predation resulting from reductions in whale, fur seal, and some fish populations during the 1950s, 1960s, and early 1970s. Finally, groundfish productivity may have been further stimulated by changes in climate and ocean conditions (e.g., the 1976-77 regime shift; Francis and Hare 1994, Miller et al. 1994, Trenberth and Hurrell 1995).

A conceptual model such as this supports the theory that the reduction in *K*-selected species (e.g., marine mammals and *Sebastes* spp.) abundance coupled with an increase in abundance of *r*-selected (short-lived, high production) species (e.g., pollock) shifts ecosystem stability toward greater variability in population sizes (Dickie 1973, Margalef 1975, Estes 1979, Ray 1981, Allen 1985, O'Neill et al. 1986). This variability makes fishery stocks progressively less amenable to conventional management and progressively more vulnerable to the uncontrollable and largely unpredictable forces of the environment (Apollonio 1994). This is essentially what may have occurred in the EBS since the 1950s.

This view of the Bering Sea also illustrates how symmetry and succession may operate within such a system. Historically, much of the capital of the Bering Sea may have been bound up in or conserved by the *K*-selected species. Heavy exploitation of these species released this capital, and may have resulted in a reorganization of the system, such that it is presently dominated by *r*-selected species (e.g., pollock) in an exploitation phase (Holling 1995; Fig. 1). Whether the system will return to its historical *K*-phase or has jumped to a new state of organization cannot be predicted.

Linkages

Clearly the hypothesis put forth by Merrick (1995) and the National Research Council (1996), sometimes referred to as the "Cascade Hypothesis," can neither be proved nor disproved. However, based on conceptual modeling on the "back of an envelope," the numbers are in the realm of possibility. In particular, the top-down release of zooplankton prey, as well as the decline in juvenile pollock predation, both in response to heavy exploitation, could have allowed increased recruitment of pollock in the 1960s. Recently, a group of resource scientists met at the National Center for Ecological Analysis and Synthesis (NCEAS) in Santa Barbara to consider the issue of assessing the impact of the fishing of apex predators on marine ecosystem dynamics. They suggested that the application of models such as ECOPATH (Christensen and Pauly 1995, Pauly and Christensen 1995) and its dynamic offspring ECOSIM (Walters et al. 1997) to the North Pacific and Bering Sea could provide insight into some fundamental ecosystem questions.

ECOPATH provides a framework for summarizing information over trophic levels, emphasizing natural rates of growth and consumption of marine populations. This allows small-scale studies or models (such as fish bioenergetics models or diet composition data) to be viewed in a common currency, in the context of the ecosystem as a whole. While the equilibrium assumptions of the model do not allow for the determination of "true" biomasses or rates in a changing system, the modeler may use the bookkeeping of ECOPATH to examine the basic trophic processes in the system; for example, to learn whether a predator or fishery is potentially consuming more forage than is hypothesized to be available, and thus if its existence is putting pressure on other ecosystem components.

ECOSIM, as an extension to ECOPATH, provides a dynamic extension to test hypotheses such as the "Cascade Hypothesis." While its large number of difficult-to-measure parameters allow experimenters to overspecify many situations, the process of testing the model can provide an indication of the robustness of any particular hypothesis. The quick examination of many components of the system essentially extends the concept of maximum sustainable yield (MSY) to multiple interacting species, suggesting a management framework for creating ideal species assemblages.

A number of conservation organizations have claimed that "the Gulf of Alaska/Bering Sea ecosystem is seriously threatened by industrial fishing."

(Letter to *Seattle Times* from Fred Munson, Greenpeace). ECOPATH and ECO-SIM formalize data and common-sense ideas of ecosystem function into an idea of what might happen or what may have happened in response to human exploitation. ECOPATH, ECOSIM, and their variants provide ecosystem modelers with a useful set of tools. They rely heavily on "back of the envelope" style calculations and stable, simplified trophic webs, and thus cannot model sudden ecosystem shifts to new (possibly undesirable) states. Yet this generality and overstability may be their greatest recommendation.

The potential of these models may lie in developing metrics (e.g., the amount of primary production removed by fishing) relating measurable ecosystem quantities to levels of risk incurred by exploitation. At this time and despite great effort, metrics of ecosystem stability and resilience, which are both measurable and meaningful, have been elusive. As a tool for developing management "rules of thumb" for avoiding ecological catastrophes, the value of comparing simple, stable pre- and post-manipulation equilibrium or functional models with historic records of change in exploited ecosystems cannot be dismissed.

The Bering Sea and Ecosystem Management

Introduction

The wealth of information on the Bering Sea ecosystem available from modeling has been little used in management of Bering Sea resources. This may be due to the difficulty of incorporating this information into the traditional approach to marine resource management (i.e., a single species, monitoring-based exploitation strategy; Royce 1989). Walters and Collie (1988) argue that the best way to manage such fisheries is resource monitoring-based, and that environmental data would not measurably improve our abilities to exploit fish stocks. Despite the current depressed status of North Pacific shellfish and marine mammal stocks, Aron et al. (1993) have argued that the recent history of North Pacific fisheries management has been a success story.

Others have been less sanguine. Greenpeace (1996) claims that the continuing declines of many seabird and marine mammal populations of the Pacific and Bering Sea is an indicator of "an over-exploited ecosystem which can no longer support historic levels of many wildlife species. The failure of the NPFMC (North Pacific Fishery Management Council) to recognize warning signs of overfishing stems from the failure of single-species management models to recognize the ecosystem processes required to sustain those fisheries." In a more general vein, Ludwig et al. (1993) have described how the bleak record of fishery collapses worldwide shows that sustainable development is not possible under the present management regime.

Clearly, fishery exploitation will continue. If a better approach to sustainable development is possible, then the challenge to management agencies is to design management regimes that allow fisheries to be developed while maintaining (or rebuilding) healthy ecosystems. Holling (1978), Walters (1986, 1992), Holling and Meffe (1996), Steele (1996), and numerous others (see Ecological Applications, 1993, 3[4]) suggest that it is possible to sustain development of resources if management is approached differently. A key element of this new approach is a redefinition of sustainability. Rather than attempt to force the ecosystem into a state that supports a certain level of exploitation over a long period, this approach proposes that fish exploitation should be consciously and actively adapted to accommodate an ecosystem's inherent variability. Holling and Meffe (1996) propose a golden rule for natural resources management: "Natural resource management should strive to retain critical types and ranges of natural variation in ecosystems. That is, management should facilitate existing processes and variabilities rather than changing and controlling them."

Shifting the focus away from the sustainability of a single resource to that of an ecosystem increases the demand for knowledge of the ecosystem, its resilience (Holling and Meffe 1996), and its natural frequencies (structure and dynamics in both time and space) as it impacts or is impacted by development. Modeling can help ecosystem managers by assisting them to make better decisions and provide a better (albeit provisional) understanding of how ecosystems work by structuring knowledge, exploring potential outcomes of management actions, and evaluating the relative merits of alternative actions. Here we wish to briefly address how modeling could benefit the ecosystem management process and perhaps improve management of the Bering Sea.

An Approach to Ecosystem Management in the Bering Sea

Presently, ecosystem management of marine resources remains largely at the conceptual level, although key elements of an approach have been outlined (Grumbine 1994, Ecological Society of America 1996). It is tempting to consider a single, comprehensive process for managing the Bering Sea ecosystem, as is being attempted in the Antarctic Ocean (Nicol and de la Mare 1993). However, it may be presumptuous to believe that such an approach is immediately applicable to ecosystems and fisheries as complex as the Bering Sea. In considering the application of ecosystem concepts to the management of Bering Sea fishery resources, two major issues arise: actual levels of resource exploitation and restoration of depleted ecosystem components.

With regard to levels of resource exploitation, two problems arise. First, a single species approach to setting allowable catches largely ignores interactions between a target species and its competitors, predators, and prey. Consequently, unforeseen impacts (e.g., reduction of prey) may have occurred within the Bering Sea because of the singular focus of state-of-the-art management on populations which are the principal targets for extraction. Second, the setting of allowable catches based partially on economic and bycatch considerations has resulted in uneven exploitation of target species. Those species with relatively high value (e.g., pollock) are harvested at their approximate biologically acceptable upper limits, while species, like some flatfish, which have low value or high bycatch of prohibited species are harvested at levels much below acceptable biological limits. This differential removal may have actually enhanced the trophic cascade described above, allowing less favored species to maintain high biomasses, while reducing biomasses of more favored species (in the face of predation by both humans and traditional marine predators).

The second ecosystem issue faced by Bering Sea managers is restoration of reduced or declining marine mammal, seabird, and fish populations (National Research Council 1996). Rebuilding plans are in place for one (i.e., Pacific Ocean perch), are implied by regulation for two (i.e., Steller sea lions and northern fur seals), and are proposed for others (e.g., king crab). Typically, these plans have involved the reduction of directed fishing effort, bycatch of the species, or fishery pressure in areas inhabited by the species. At best, this approach could be considered as a passive adaptive strategy since historical data is used to produce a single best approach (Walters 1986). The current approach is flawed for at least two reasons. First, little monitoring and evaluation of the results of the management actions has occurred. Second, and perhaps more seriously, little attention has been given to competing hypotheses (e.g., the inability of crab to recover could be because of environmental conditions and not simply fishery bycatch). None of the management actions have involved an active adaptive strategy. One plausible approach to implementation of an active adaptive strategy could include: (1) use of data and hypotheses to set a range of alternative actions; (2) a policy choice is then made reflecting a balance between alternatives; and (3) experimental manipulations are used to gather additional data on the efficacy of the actions (Walters 1986). Under the current approach, resource managers may never know with any confidence why the implemented actions did or did not work.

In summary, ecosystem management in the Bering Sea may be viewed as management philosophy focused on ecosystem sustainability with two operational elements—resource exploitation and restoration ecology. The current management approach probably achieves neither element well from an ecosystem perspective, so how can ecosystem management and modeling improve the situation?

Modeling the Bering Sea under an Ecosystem Management Approach

Resource modeling should have two basic goals: to assist managers in decision making and to improve the understanding of the system. To best assist managers, models should be viewed as predictors (not prescriptions) of the outcome of various management outcomes, and the outcomes must then be monitored, evaluated, and revised as information becomes available. This is a fundamental part of adaptive management. Models which support adaptive management should have: (1) realism—be closely connected to biological theory or hypotheses; (2) precision—be able to explain real world phenomena; (3) testable predictions about consequences of management actions; and (4) a formal mechanism to modify both our understanding, and selection of future management alternatives (Levins 1966, Conroy 1993).

Models should also improve our understanding of ecosystems at all levels of organization. Knowledge and understanding of ecosystem function and organization is incomplete and provisional. Recent research on ecosystem dynamics has shown the need to recognize that:

1. Connectedness, diversity, and complexity are natural in an ecosystem, and it appears that they strengthen an ecosystem against disturbance.

2. Change is inherent in ecosystems because they are dynamic in both space and time. Ecological change is not always continuous and gradual; rather it can be sudden and episodic (Connell and Sousa 1993, Holling and Meffe 1996).

3. Multiple temporal and spatial scales operate, and hierarchies of scales are important. Critical processes function at radically different rates and at spatial scales covering several orders of magnitude, and these scales tend to cluster around several dominant frequencies (O'Neill et al. 1986, Steele 1991, Levin 1992, Holling and Meffe 1996).

4. Ecosystems do not have single equilibria, governed by stable functions. Rather, multiple equilibria, destabilizing forces far from equilibria, and absence of equilibria define functionally different stable states, and movement between states maintains an overall structure and diversity (Holling and Meffe 1996).

Ecosystem management is the process of incorporating these concepts into the development of management policy. Policies and management which apply static, fixed rules lead to systems which gradually lose resilience—systems that gradually break down in the face of disturbances that previously could have been absorbed (Holling and Meffe 1996). Consideration of these characteristics in management requires modeling. This modeling can range from trophic models which identify the connections between commercially important target species and associated conspecifics to process models which explore specific ecological connections to conceptual models which attempt to identify the historical dynamics of the ecosystem which may have contributed to present conditions.

Summary

How does all this translate into an operational approach to ecosystem management in the Bering Sea? Perhaps a brief review of some of the important points that have already been made will reveal some clues.

Most ecosystem science is intended to address historical issues which are culminations of unique sequences of events. Yet the historical memories present in any ecosystem are difficult or impossible to extract in the short time horizons of both management institutions and scientific study plans. In order to facilitate ecosystem management we need to both study and manage over longer time horizons.

Ecosystems are structured according to the relationships between processes occurring at different time and space scales and tend to occur as a function of natural frequencies. One important implication is that ecosystems that are dominated at the top by long-lived and/or highly migratory species tend to be more structured and predictable than those that are not.

Existing models of the Bering Sea ecosystem have revealed a number of important points:

1. Trophic models have shown that the large marine ecosystems of the northeastern Pacific are extremely dynamic (they had trouble making equilibrium ecosystem models balance properly), are significantly influenced from the top by marine mammals, appear to be structured both above (top down) and below (bottom up) by the dynamics in space and time of forage fish (including prefishery juveniles of commercially important species), and respond to fishing in (intuitively) strange ways.

2. Process models have shown that a management plan based on species rather than region may not accurately reflect the dynamics of managed populations, and that there may be important connections between pelagic (e.g., pollock) and benthic (e.g., king crab) fisheries in the Bering Sea.

3. Conceptual models, such as those presented by Merrick (1995) and the National Research Council (1996), reconstruct a plausible skeleton of the recent history of trophic dynamics in the EBS, demonstrating that the "Cascade Hypothesis" of the past 50 years is at least possible

from an energetics point of view, and provide a possible explanation for the recent declines in pinniped and seabird prey abundance.

The focus of Bering Sea fishery management from an ecosystem perspective should:

1. Strive to retain critical types and ranges of natural variation in ecosystems. That is, management should facilitate existing processes and variabilities rather than changing and controlling them.

2. Set allowable catches taking into account interactions between a target species and their competitors, predators, and prey.

3. Set allowable catches using a balanced approach to the levels of harvest on various target species.

4. Use an active adaptive strategy to restore depleted stocks. This would require the results of management actions to be monitored and evaluated as well as paying close attention to competing hypotheses.

In summary, we should always keep in mind that policies and management which apply static, fixed rules lead to systems that gradually break down in the face of disturbances that previously could have been absorbed (i.e., the systems gradually lose resilience). Consideration of these characteristics in management requires modeling. To best assist managers, models should be viewed as predictors (not prescriptions) of the outcome of various management outcomes, and the outcomes must then be monitored, evaluated, and revised as information becomes available. Models should be used as an integral part of adaptive management.

References

Allen, P.M. 1985. Ecology, thermodynamics, and self-organization: Toward a new understanding of complexity. In: R.E. Ulanowicz and T. Platt (eds.), Ecosystem theory for biological oceanography. Canadian Bulletin of Fisheries and Aquatic Sciences 213:2-26.

Allen, T.F.H., and T.W. Hoekstra. 1992. Toward a unified ecology. Columbia University Press, New York. 384 pp.

Allen, T.F.H., and T.B. Starr. 1982. Hierarchy: Perspectives for ecological complexity. University of Chicago Press, Chicago. 310 pp.

Apollonio, S. 1994. The uses of ecosystem characteristics in fisheries management. Reviews of Fisheries Science 2:157-180.

Aron, W., D. Fluharty, D. McCaughran, and J.F. Roos. 1993. Fisheries management. Science 261:813-814.

Carpenter, S.R. 1990. Large-scale perturbations: Opportunities for innovation. Ecology 7:2038-2043.

Carpenter, S.R., J.F. Kitchell, and J.R. Hodgson. 1985. Cascading trophic interactions and lake productivity. BioScience 35:634-639.

Christensen, V., and D. Pauly. 1995. Fish production, catches and the carrying capacity of the world oceans. ICLARM Quarterly 18:34-40.

Coachman, L. 1986. Circulation, water masses, and fluxes on the southeastern Bering Sea shelf. Continental Shelf Research 5:23-108.

Connell, J.H., and W.P. Sousa. 1993. On the evidence needed to judge ecological stability or persistence. American Naturalist 111:1119-1144.

Conroy, M.J. 1993. The use of models in natural resource management: Prediction, not prescription. Transactions of the 58th North American Wildlife and Natural Resources Conference 58:509-519.

Denman, K.L., and T.W. Powell. 1984. Effects of physical processes on planktonic ecosystems in the coastal ocean. Oceanography and Marine Biology: An Annual Review 22:125-168.

Dickie, L.M. 1973. Management of fisheries: Ecological subsystems. Transactions of the American Fisheries Society 102:470-480.

Ecological Society of America. 1996. The scientific basis for ecosystem management.

Estes, J.A. 1979. Exploitation of marine mammals: *r*-selection of *K*-strategists? Journal of the Fisheries Research Board of Canada 36:1009-1017.

Estes, J.A., and J.E. Palmisano. 1974. Sea otters: Their role in structuring nearshore communities. Science 185:1058-1060.

Francis, R.C., and S.R. Hare. 1994. Decadal-scale regime shifts in the large marine ecosystems of the Northeast Pacific: A case of historical science. Fisheries Oceanography 3:279-291.

Francis, R.C., S.R. Hare, A.B. Hollowed, and W.S. Wooster. 1998. Effects of interdecadal climate variability on the oceanic ecosystems of the NE Pacific. Fisheries Oceanography 7(1).

Frost, B.W. 1987. Grazing control of phytoplankton stock in the open subarctic Pacific Ocean: A model assessing the role of mesozooplankton, particularly the large calanoid copepods, *Neocalanus* spp. Marine Ecology Progress Series 39:49-68.

Greenpeace. 1996. Sinking fast: How factory trawlers are destroying U.S. fisheries and marine ecosystems. Greenpeace, Washington, DC. 10 pp.

Grumbine, R.E. 1994. What is ecosystem management? Conservation Biology 8:27-38.

Holling, C.S. 1978. Adaptive environmental assessment and management. J. Wiley and Sons, New York.

Holling, C.S. 1995. What barriers? What bridges? In: L.H. Gunderson, C.S. Holling, and S.S. Light (eds.), Barriers and bridges to the renewal of ecosystems and institutions. Columbia University Press, New York, pp. 3-36.

Holling, C.S., and G.K. Meffe. 1996. Command and control and the pathology of natural resource management. Conservation Biology 10:328-337.

Kingsland, S.E. 1985. Modeling nature: Episodes in the history of population ecology. University of Chicago Press, Chicago. 306 pp.

Laevastu, T., and F. Favorite. 1977. Preliminary report on dynamical numerical marine ecosystem model (DYNNME II) for the eastern Bering Sea. NOAA report.

Laevastu, T., and F. Favorite. 1988. Fishing and stock fluctuations. Fishing News Books, Ltd., Surrey, England. 239 pp.

Laevastu, T., and H.A. Larkins. 1981. Marine fisheries ecosystems: Its quantitative evaluation and management. Fishing News Books, Surrey, England. 162 pp.

Levin, S.A. 1990. Physical and biological scales and the modeling of predator-prey interactions in large marine ecosystems. In: K. Sherman, L.M. Alexander, and B.D. Gold (eds.), Large marine ecosystems: Patterns, processes and yields. American Association for the Advancement of Science, Washington, DC, pp. 179-187.

Levin, S.A. 1992. Pattern and scale in ecology. Ecology 73:1943-1967.

Levins, R. 1966. The strategy of model building in population biology. American Scientist 54:421-431.

Livingston, P.A. 1991. Groundfish food habits and predation on commercially important prey species in the eastern Bering Sea from 1984 to 1986. NOAA Technical Memo NMFS F/NWC-207. 240 pp.

Ludwig, D., R. Hilborn, and C. Walters. 1993. Uncertainty, resource exploitation, and conservation: Lessons learned from history. Science 260:17-36.

Magnuson, J.J. 1990. Long-term ecological research and the invisible present. BioScience 40:495-501.

Margalef, R. 1975. Diversity, stability, and maturity in natural ecosystems. In: W.H. van Dobhen and R.H. Lowe-McConnell (eds.), Unifying concepts in ecology. W. Junk. B. V. Publishers, The Hague, pp. 151-160.

McRoy, C., D. Hood, L. Coachman, J. Walsh, and J. Goering. 1986. Processes and resources of the Bering Sea shelf (PROBES): The development and accomplishments of the project. Continental Shelf Research 5:5-21.

Merrick, R.L. 1995. The relationship of the foraging ecology of Steller sea lions *(Eumetopias jubatus)* to their population decline in Alaska. Ph.D. Thesis, University of Washington, Seattle. 175 pp.

Miller, A.J., D.R. Cayan, T.P. Barnett, N.E. Graham, and J.M. Oberhuber. 1994. The 1976-77 climate shift of the Pacific Ocean. Oceanography 7:21-26.

National Research Council. 1996. The Bering Sea ecosystem. National Academy Press. Washington, DC. 307 pp.

Nicol, S., and W. de la Mare. 1993. Ecosystem management and the Antarctic krill. American Scientist 81:36-47.

O'Neill, R.V., D.L. DeAngelis, J.B. Waide, and T.F.H. Allen. 1986. A hierarchical concept of ecosystems. Princeton University Press, Princeton, NJ. 312 pp.

Paine, R.T. 1980. Food webs: Linkage, interaction strength and community infrastructure. Journal of Animal Ecology 49:667-685.

Parsons, T.R. 1992. The removal of marine predators by fisheries and the impact of trophic structure. Marine Pollution Bulletin 25:51-53.

Pauly, D., and V. Christensen. 1995. Primary production required to sustain global fisheries. Nature 374:255-257.

Polovina, J.J., G.T. Mitchum, and G.T. Evans. 1995. Decadal and basin-scale variation in mixed layer depth and the impact on biological production in the Central and North Pacific, 1960-88. Deep-Sea Research 42:1701-1716.

Power, M.E. 1992. Top-down and bottom-up forces in food webs: Do plants have primacy? Ecology 73:733-746.

Price, J.F., R.A. Weller, and R. Pinkel. 1986. Diurnal cycling: Observations and models of the upper ocean response to diurnal heating, cooling, and wind mixing. Journal of Geophysical Research 91:8411-8427.

Ray, G.C. 1981. The role of large organisms. In: A.R. Longhurst (ed.), Analysis of marine ecosystems. Academic Press, New York, pp. 397-413.

Ricklefs, R.E. 1990. Scaling pattern and process in marine ecosystems. In: K. Sherman, L.M. Alexander, and B.D. Gold (eds.), Large marine ecosystems: Patterns, processes and yields. American Association for the Advancement of Science, Washington, DC, pp. 169-178.

Royce, W.F. 1989. A history of Maine fishery management. Review of Aquatic Sciences 1:27-44.

Sambrotto, R.N., H.J. Niebauer, J.J. Goering, and R.L. Iverson. 1986. Relationships among vertical mixing, nitrate uptake, and phytoplankton growth during the spring bloom in the southeast Bering Sea middle shelf. Continental Shelf Research 5:161-198.

Schindler, D.W. 1987. Detecting ecosystem response to anthropogenic stress. Canadian Journal of Fisheries and Aquatic Science 44(Suppl.):6-25.

Simenstad, C.A., J.A. Estes, and K.W. Kenyon. 1978. Aleuts, sea otters and alternate stable state communities. Science 200:403-411.

Smith, S.L., and J. Vidal. 1986. Variations in the distribution, abundance and development of copepods in the southeastern Bering Sea in 1980 and 1981. Continental Shelf Research 5:215-239.

Steele, J.H. 1991. Marine ecosystem dynamics: Comparison of scales. Ecological Research 6:175-183.

Steele, J.H. 1996. Regime shifts in fisheries management. Fisheries Research 25:19-23.

Trenberth, K.E., and J.W. Hurrell. 1995. Decadal coupling atmospheric-ocean variations in the North Pacific Ocean. In: R.J. Beamish (ed.), Climate change and northern fish populations. Canadian Special Publication of Fisheries and Aquatic Sciences 121:15-24.

Vidal, J., and S.L. Smith. 1986. Biomass, growth, and development of populations of herbivorous zooplankton in the southeastern Bering Sea during spring. Deep-Sea Research 33:523-556.

Walsh, J., and P. McRoy. 1986. Ecosystem analysis in the southeastern Bering Sea. Continental Shelf Research 5:259-288.

Walters, C.J. 1986. Adaptive management of renewable resources. McMillan Publishing Company, New York. 374 pp.

Walters, C.J. 1992. Perspectives on adaptive policy design in fisheries management. In: S.K. Jian and L.W. Botsford (eds.), Applied population biology. Kluwer Academic Publication, Netherlands, pp. 249-262.

Walters, C.J., and J.S. Collie. 1988. Is research on environmental effects on recruitment worthwhile? Canadian Journal of Fisheries and Aquatic Sciences 45:1848-1854.

Walters, C., V. Christensen, and D. Pauly. 1997. Structuring dynamic models of exploited ecosystems from trophic mass-balance assessments. Review of Fisheries Biology and Fisheries 7:1-34.

Wespestad, V.G. 1993. Walleye pollock. In: Stock assessment and fishery evaluation report for the groundfish resource of the Bering Sea/Aleutian Island regions as projected for 1994. Bering Sea and Aleutian Island Plan Team, North Pacific Fishery Management Council, P.O. Box 103136, Anchorage, AK 99510, pp. 1-26.

CHAPTER 21

Seasonal Sea Ice Variability and the Bering Sea Ecosystem

Tina Wyllie-Echeverria
Joint Institute for the Study of Atmosphere and Ocean, University of Washington, Seattle, Washington

Kiyotaka Ohtani
Laboratory of Physical Oceanography, Hokkaido University, Hokkaido, Japan

Abstract

Two methods of indexing seasonal sea ice cover in the Bering Sea are evaluated and the consequences of sea ice variability to the ecosystem are discussed. The residence time index (RTI) reflects the duration of ice each year in a gridded area while the seasonal sea ice index (SSII) is the southernmost position of sea ice along longitude 169°W. Both indices were derived from, and confirm the use of, satellite data for understanding biotic as well as abiotic variables. Both indices show interannual and multi-annual variability between heavy and light ice cover and a subsequent shift from heavy ice cover pre-1977 to lighter ice cover. The SSII can be used to forecast summer conditions from the previous winter's ice extent. Two summertime conditions prevail: cool, when ice extends beyond 57°30′N and warm, when ice remains northward. Populations of fish, birds, and mammals on the shelf respond differently to interannual warm or cool conditions, and when conditions are sustained for several years. Distribution of walleye pollock *(Theragra chalcogramma)* and reproductive success and food habits of thick-billed murres *(Uria lomvia)* and northern fur seals *(Callorhinus ursinus)* change on a multi-annual scale, while distribution of Arctic cod *(Boreogadus saida)* and predation pressure on age-1 pollock varies interannually with warm or cool conditions. The relationship between atmospheric, oceanic, and biotic factors explored in this chapter show that the Bering Sea system can vary on multi-annual time scales as well as interannually.

Current address for T. Wyllie-Echeverria is Department of Geography, Brigham Young University, Provo, UT.

Introduction

A dominant feature over much of the Bering Sea shelf is the annual appearance of seasonal sea ice. Its presence simulates arctic conditions far beyond the ice-capped Arctic Ocean, in that water temperatures are near freezing (–1.7°C), wave and wind effects are modulated, sunlight is reduced, and species associated with cold water are present. However, once the ice melts, conditions are subarctic. Concomitant with seasonal environmental change are changes in species abundance and assemblages. Migrations of subarctic seabirds, water birds, fish, and marine mammals follow the retreating ice northward. Summer on the shelf is not unlike that in high latitude terrestrial systems where birds and mammals seasonally migrate to feed and reproduce. This chapter discusses the consequences of seasonal sea ice on the ecosystem. There is variability not only in the seasonal presence of a species, but also in abundance and distribution on annual and decadal scales. Effects of warm or cool conditions on the ecosystem range from the timing and amount of primary productivity (Schumacher and Alexander, chapter 6, this volume) to the distribution of species. We will focus on the effects of sea ice on higher trophic levels, particularly the consequences of changing environmental conditions on species distribution and interaction.

Physical State

Seasonal sea ice is generated in the shallow Chukchi Sea and in polynyas of the Bering Sea. Its formation is dependent upon a combination of air and water temperatures, and wind direction over the shelf. Interannual variability in ice extent stems from changes in position of two pressure systems, the Siberian High and the Aleutian Low (Niebauer et al., chapter 2, this volume). Winds generated by these pressure systems affect the shelf through surface mixing, bringing relatively warm air (from the North Pacific) or cold air (from the Arctic) to the shelf, and by moving the boundary of seasonal sea ice (Overland and Pease 1982, Niebauer and Day 1989). The predominant situation occurs with northeast winds, which push sea ice southward until water temperatures are high enough to melt it.

Seasonal sea ice acts as a "conveyor belt" which moves low-salinity sea ice from the location of freezing southward to the location of melting (Pease 1980). This results in an influx of low-salinity water into the middle and outer shelf areas. A combination of low bottom relief and slow advection results in a residence time of about one year for water on the middle shelf (Schumacher et al. 1979). The middle shelf is deep enough to stratify the water column into two layers as surface waters are warmed and diluted in spring. This stratification, in combination with low flow rates, traps the residual water formed in winter below the pycnocline. Evidence of the influence of winter conditions is seen in the temperature (<2.0°C) and salinity (32 psu) of the subsurface water. This cold pool (Maeda 1977) is a persistent feature in the middle domain of the Bering Sea shelf (Takenouti and Ohtani 1974; Ohtani 1969, 1973).

Dynamics of the cold pool have been discussed since the first multi-annual scientific voyages documented bottom temperatures of the Bering Sea shelf (Barnes and Thompson 1938). The cold pool varies in size annually at its southeastern edge where the shelf is wide and rises gently toward Bristol Bay. It extends from the bottom to the thermocline, varying in thickness from 30 to 70 m. The gently sloping bottom topography between 50 and 100 m approximates its northern and southern boundaries. On average, the cold pool extends from the Gulf of Anadyr (Khen 1988) to 163°W (Ingraham 1981, Reed 1995), and may extend all the way into Bristol Bay. Its significance lies in its size and persistence, and the associated species assemblages.

The hydrographic effects of seasonal sea ice in turn affect the Bering Sea ecosystem. The species complex changes seasonally, as species associated with sea ice retreat northward in spring while subarctic species move into waters newly freed of ice. Interannually, the species complex varies between an arctic/subarctic mix when species such as Arctic cod *(Boreogadus saida)* are present (Wyllie-Echeverria 1995b), to a predominately subarctic complex when water temperatures suit the physiological preference of species such as walleye pollock *(Theragra chalcogramma).*

Scales of Variability

Two independent investigations into the relationships between atmosphere, ocean, and sea-ice conditions and the potential consequences to the ecosystem are presented in this chapter. Ohtani (1996) measured sea ice variability using a gridded region ($1°\times2.5°$) of the Bering Sea shelf with a smaller portion as the research area (Fig. 1a). The sea ice concentration in each grid was evaluated from historical ice charts for the western Arctic (User Service Office, National Snow and Ice Data Center, University of Colorado, 1972-1995). The residence time index (RTI) of sea ice was represented as a total frequency of resident time that exceeds the concentration of 4th class in the research area (RTIarea) or the monitoring grid (RTIgrid) during the latter half of the ice season (February to April) (Fig. 2a). The minimum summer temperature in the bottom layer (T_b) inside the monitoring grid provided an index of the cold pool (Fig. 2b). Air temperature (T_a) at St. Paul Island (January to March mean) provided an index of winter atmospheric conditions (Figs. 2c and 3a) (Ohtani and Azumaya 1995, Azumaya and Ohtani 1995). Wyllie-Echeverria (1998) used the position of the leading ice edge at 30% concentration along 169°W to track the annual extent of ice. This seasonal sea ice index (SSII) reflects the areal distribution of ice in comparison to the data of Chapman and Walsh (1993) ($r = 0.91$) and has the additional benefit of simplicity (i.e., a single data point represents the annual extent of sea ice). The SSII and/or wind direction over the shelf during April can be used to forecast the summertime conditions of the

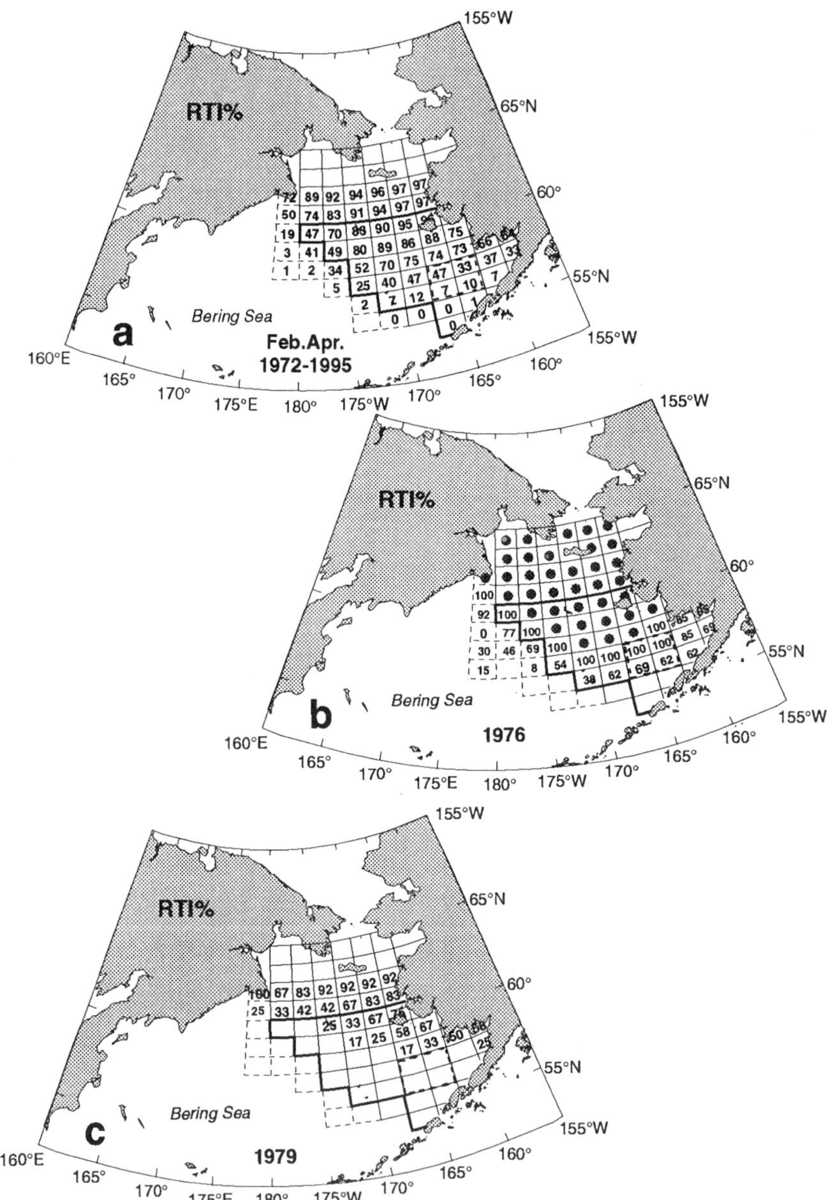

Figure 1.   *(a) Averaged RTI of each grid (1° latitude × 2.5° longitude) from 1972 to 1995. Research area is enclosed by solid lines and monitoring grid is a broken line inside research area. RTI in each grid in the (b) coolest year of 1976 and (c) in the warmest year of 1979. Black points in grids north of 100% in (b) indicate all 100% RTI.*

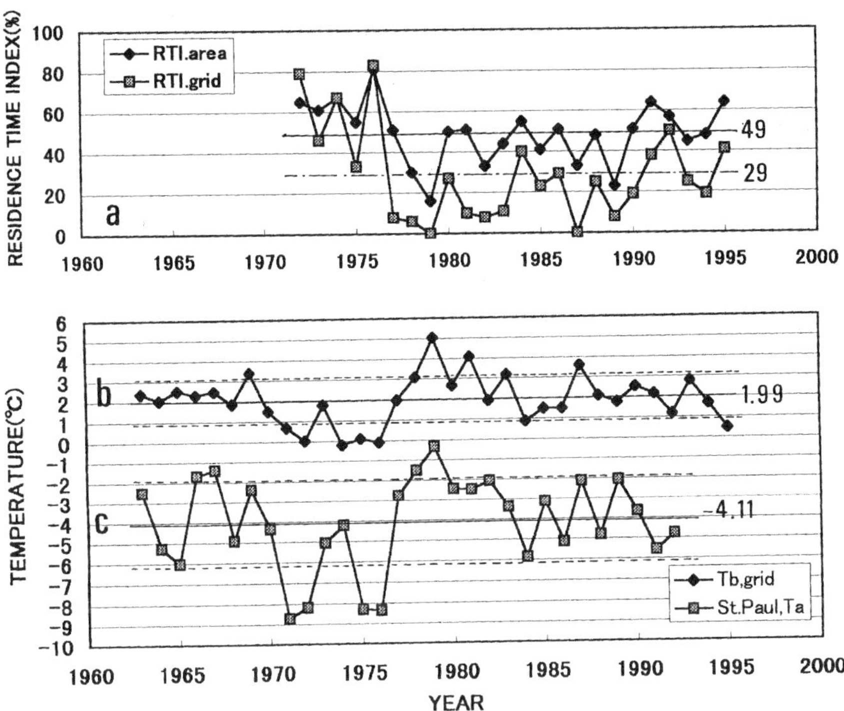

Figure 2. Time series of (a) residence time index (RTI) for overall area and monitoring grid in percent during a period from maximum ice extent (February) to the beginning of primary production (April) on the eastern Bering Sea shelf (from Ohtani 1996), (b) bottom temperature inside monitoring grid (from Azumaya and Ohtani 1995), and (c) averaged air temperature at St. Paul Island in the middle of the ice season (February-March) (from Azumaya and Ohtani 1995). Broken lines in (b) and (c) indicate standard deviation.

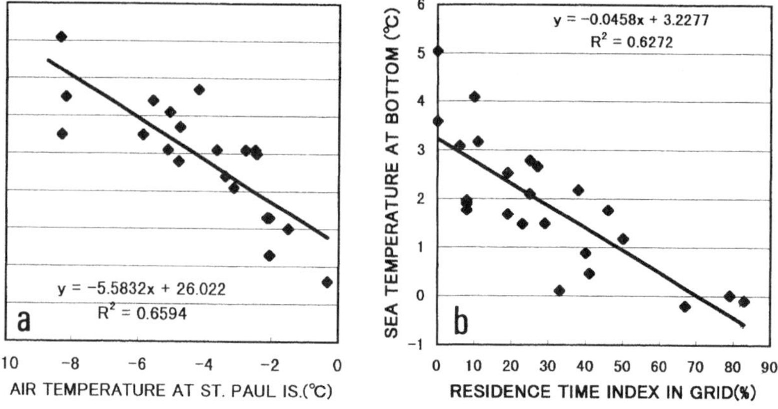

Figure 3.   Relationship between air temperature at St. Paul Island and residence time index in the research area (a) and between the residence time index and bottom layer water temperature in the monitoring grid (b).

eastern extent of the cold pool (2°C isotherm) and the average bottom temperature of the southeastern shelf (Table 1). Significant correlations (*F*-test, $P < 0.01$) were found between the SSII and the direction of wind to the shelf in April ($r = 0.78$), the eastern boundary of the cold pool ($r = 0.68$), and the average bottom temperature ($r = 0.76$). The data were separated into three groups which correspond to cool, warm, or intermediate conditions of the Bering Sea shelf system. When a cool condition exists, continental or arctic air sweeps over the shelf from the northeast, seasonal sea ice extends below 57°30′N, and the cold pool is strengthened, extending eastward of meridian 163°W (Fig. 4). A warm condition occurs when maritime air sweeps over the shelf from the North Pacific, ice remains north of 57°30′N, and the cold pool is weakened, extending only to a position somewhere between St. Matthew Island and 163°W (Fig. 5). Occasionally, continental or arctic air sweeps the shelf in April, and sea ice conditions and cold pool extent are intermediate between the two conditions (Table 1).

The RTI from 1972-1995 shows the southern limit of drift ice along 56°N in the coolest year (Fig. 1b) and 58°N in the warmest year (Fig. 1c), a difference exceeding 200 km. High interannual variability of ice distribution is implied when the RTI is less than 50%, as along the shelf break and east of the Pribilof Islands. Ice over the southeastern part of the cold pool, which extends east of the Pribilof Islands, is one of the most variable (Fig. 1a). The relationships between RTIarea and T_a ($r = 0.81$), and between RTIgrid and T_b ($r = 0.83$), are significantly correlated (*F*-test, $P < 0.01$; Fig. 3), as are RTIarea and RTIgrid ($r = 0.78$), T_a and T_b ($r = 0.75$), and RTIarea and T_b ($r = 0.69$). The connections between RTI, sea, and atmospheric con-

Table 1. **The environmental condition of the Bering Sea shelf can be determined from the seasonal sea ice index (SSII).**

Condition	Wind direction	SSII (latitude)	Cold pool (longitude)	Bottom temperature
Cool	Continental	57°30′-56°	163-158°W	1.2-3.0°C
Intermediate	Continental	57°-57°30′	166-163°W	3.0-3.8°C
Warm	Maritime	60°-57°30′	170-166°W	3.8-4.6°C

A cool summer, with an extensive cold pool and lower average bottom temperatures, follows a winter of extensive ice cover and winds of continental origin sweeping the shelf in April. A warm summer, with a reduced cold pool and high average bottom temperatures, follows a winter of reduced ice cover and winds of maritime origin sweeping the shelf in April. An intermediate summer is defined by conditions bordering the two principal conditions.

ditions suggest that the influence of sea conditions on distribution of young fishes in winter, primary productivity, and other biological conditions could be monitored by satellite information of sea ice.

In addition to the interannual variability evident in Figs. 2a and 6, a multi-annual state of 4-6 years duration with either cool or warm conditions can also be discerned. Shifts in the climate of the Bering Sea shelf occurred in 1977 (from cool conditions), and in 1984 (from warm conditions). Similar shifts in sea surface temperature were shown in the North Pacific from predominately cool (1972-1977), to warm (1978-1983), and mixed (1984-1990) conditions (Wooster and Hollowed 1995).

Biotic System

Variation in the abundance and distribution of fish species is affected by environmental conditions (Maeda 1977, Bakkala and Alton 1986, Azumaya and Ohtani 1995, Wyllie-Echeverria 1995a, Wyllie-Echeverria et al. 1997). On the Bering Sea shelf, yellowfin sole *(Limanda aspera)* had large year classes in warm years (1955, 1958, 1963, and 1967) and poor year classes in cool years (1956, 1960, 1962), as manifested by northern winds, less saline water, lower sea temperatures, and increased ice (Maeda 1977). The linear relationship between year-class strength and the position of the 2°C isotherm (cold pool boundary) was significant for the years 1958-1967 ($r = 0.63$). Bottom temperatures and salinity decreased during 1955-1958, steadily increased between 1961 and 1967, and decreased again between 1968 and 1976. The largest year classes for pollock occurred in 1972, 1978, 1982, and 1984, years that mark the transition between multi-annual states (except for 1984) (Hollowed and Wooster 1995). Interannual variation in the cold pool affects the abundance of age-2 pollock as well (Azumaya and Ohtani 1995). They found that the number of age-2 pollock was related to a combination of numbers of age-1 pollock and the winter

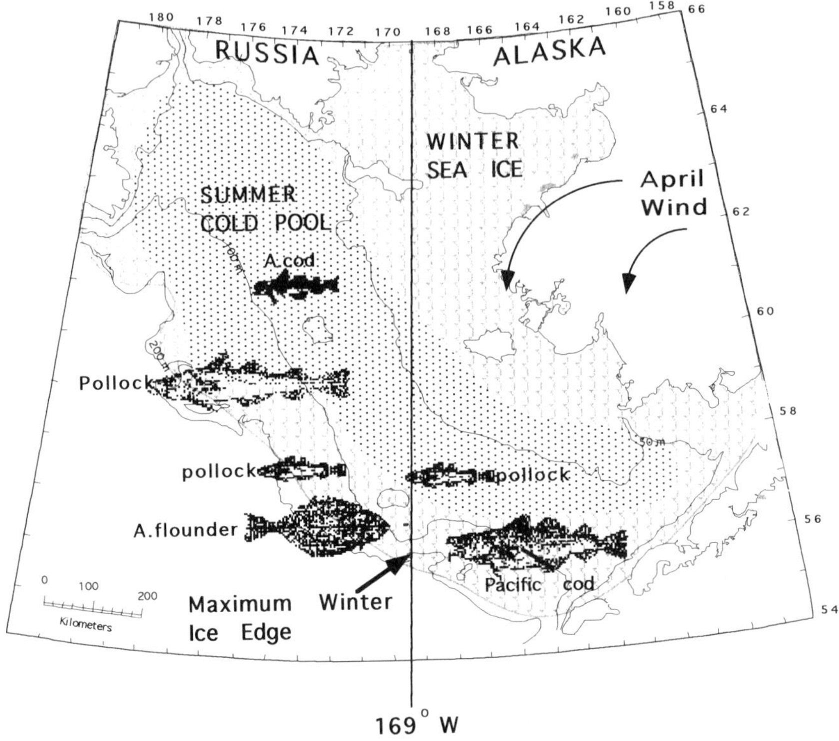

*Figure 4.   Cool conditions on the Bering Sea shelf follow winters when seasonal sea
ice extends beyond 57°30′ and the direction of wind to the shelf during
April is of continental origin. In summer, the cold pool (<2.0°C) extends
eastward, Arctic cod (A.cod) occurs on the shelf and adult Pollock, age-1
pollock, arrowtooth flounder (A.flounder), and Pacific cod are distributed
in the outer shelf. Age-1 pollock also occurs in the middle shelf.*

conditions faced by age-1 pollock. Cold bottom water restricts age-1 pollock to the outer shelf, increasing their density and hence their vulnerabil-
ity to predators. Azumaya and Ohtani (1995) suggest that since pollock,
ages 0 and 1, are a forage fish, predation pressure during the first two
years is instrumental in shaping the size of the year class recruited to the
fishery.

Shifts in the pattern of distribution of pollock on the Bering Sea shelf
occur on multi-annual scales (Wyllie-Echeverria 1995b, 1996). During cool
years, age-1 pollock were primarily distributed in the middle shelf, but
were also concentrated on the outer shelf, shifting to the middle and inner
shelves during the warm and mixed states. Older pollock varied on the
same scale, but with a different distribution pattern than age-1. Older

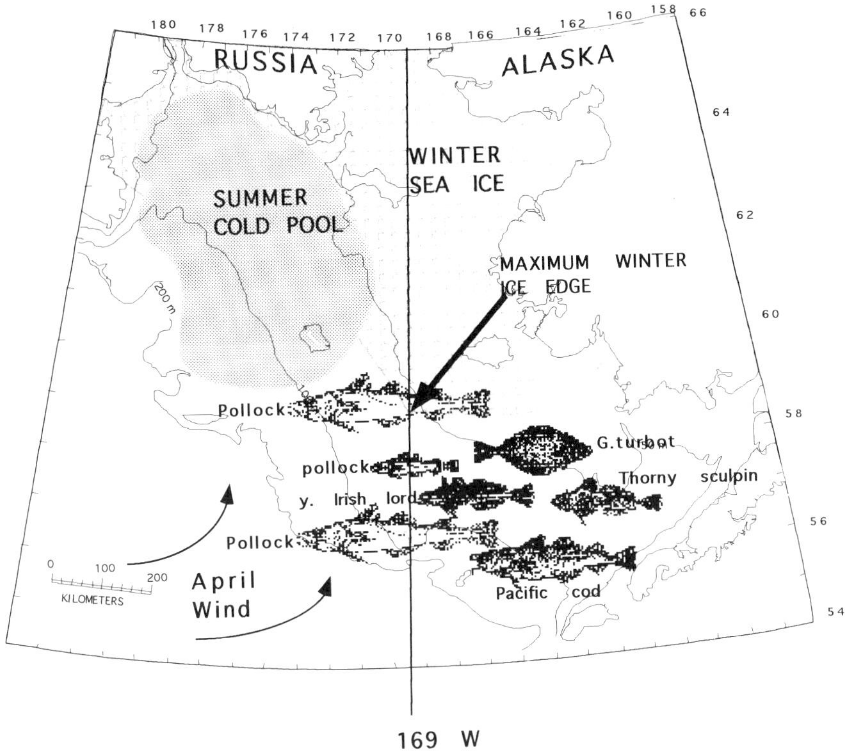

Figure 5. Warm conditions on the Bering Sea shelf follow winters when seasonal sea ice remains above 57°30' and the direction of wind to the shelf during April is of maritime origin In summer, the cold pool (<2.0°C) is reduced, adult pollock, age-1 pollock, yellow Irish lord (Y. Irish lord), thorny sculpin, and Greenland turbot (G.turbot) overlap in distribution in the middle shelf, with adult pollock and Pacific cod concentrated in the outer shelf.

pollock were primarily present on the outer shelf during the cool state (1972-1977), expanding into the middle shelf during the warm state (1978-1982), and the inner shelf in the mixed state after 1983 (Wyllie-Echeverria 1995b). As found by Ohtani and Azumaya (1995), age-1 pollock and cannibalistic older pollock co-occur on the outer shelf during cool years.

Predator-Prey Interactions

Just as the abundance and distribution of individual fish species may change on annual or multi-annual scales, predator-prey interactions will also be related to environmental conditions and will vary on similar scales. The distribution of each predator, as well as their prey, needs to be evalu-

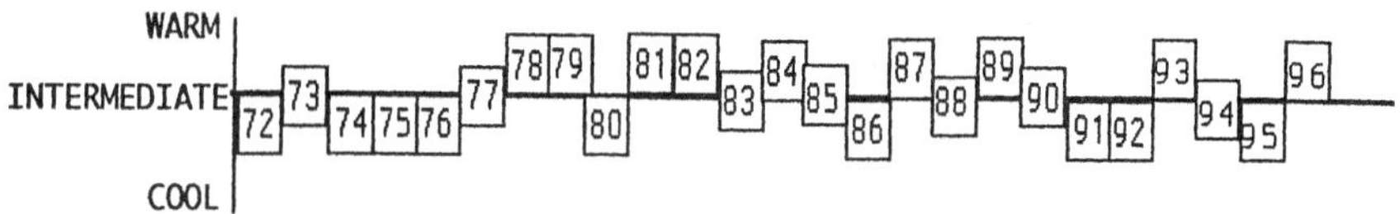

Figure 6. Timeline of annual conditions on the Bering Sea shelf, revealing warm and cool states. Intermediate conditions are midway between warm and cool states.

ated when considering their interactions. In this section, we will consider the variability exhibited by pollock due to its pivotal position in the Bering Sea ecosystem. Young pollock are the principal prey of many marine mammals, seabirds, and fish on the shelf (Kihara and Shimada 1988a, Livingston 1991, Livingston et al. 1993, Sinclair et al. 1994, Springer 1992) and shifts in their distribution impact the food web.

Shifts in the distribution of older pollock will change the encounter rate for cannibalism on age-0 and age-1 pollock, and also change the area of the commercial fishery on pollock. Predation by seabirds and marine mammals, which are limited in their foraging areas because of nesting and pupping requirements, is another consideration (Springer 1992, Sinclair et al. 1994). An understanding of the scales of variability of species distribution and environmental conditions, and the response of both predator and prey species, is needed when investigating ecosystem interactions.

Pollock, arrowtooth flounder *(Atheresthes stomias),* and Pacific cod *(Gadus macrocephalus)* are the principal fish predators of age-1 pollock (Livingston et al. 1993, Shuntov and Dulepova 1996). Area overlap between predatory Pacific cod and their pollock prey increases under cool conditions (Kihara and Shimada 1988b). Predation for the years 1984-1992 amounted to 100-15,000 million fish per year by pollock, 100-1,400 million fish per year by arrowtooth flounder, and 60-1,600 million fish per year by Pacific cod (Livingston 1991, Livingston et al. 1993; P.A. Livingston, National Marine Fisheries Service, Alaska Fisheries Science Center, Seattle WA, 98115, pers. comm., April 1995). We divided these data into three categories to evaluate relative predation pressure from each predator on age-1 pollock, and demonstrated that predation by cannibalism is much higher than that by either of the other major predators (Table 2). Predation pressure during this time period is related to the interannual warm or cool conditions rather than multi-annual states.

Predation pressure on age-1 pollock by fish is linked to water temperature and to the distribution of prey pollock (Kihara and Shimada 1988a, Livingston 1991, Livingston et al. 1993). The occurrence of juvenile pollock within cold water masses has led to the hypothesis that juvenile pollock are less restricted by cold water than age-2 and older pollock, which

Table 2. **Relative (low, medium, high) level of predation by age-2 and older walleye pollock, arrowtooth flounder, and Pacific cod on age-1 walleye pollock for 1984-1992.**

Year	State	Cannibalism	Arrowtooth flounder	Pacific cod
1984	C	No data	High	Low
1985	I	High	Med	Med
1986	C	Med	High	High
1987	W	Low	Low	Med
1988	C	High	Med	Low
1989	W	Low	Low	Low
1990	I	High	High	Med
1991	C	Med	Med	Med
1992	C	High	Low	Low

Cold (C), warm (W) and intermediate (I) environmental conditions.

would reduce species overlap and cannibalism (Francis and Bailey 1983). Age-1 pollock not only change their distribution on the horizontal plane by locating in different areas of the shelf, but also in the vertical plane by moving between water masses. This tolerance may be crucial where a cold pool of water varies in size each year producing a less desirable habitat for either predators or prey. For instance, during cool conditions, predation pressure on age-1 pollock was intense by their major piscine predators (age-2 and older pollock, arrowtooth flounder, and Pacific cod) (Table 2; Kihara and Shimada 1988a). On the other hand, when the cold pool was reduced during warm conditions, predation on age-1 pollock increased with overlapping distributions of Greenland turbot *(Reinhardtius hippoglossoides),* yellow Irish lord *(Hemilepidotus jordani),* and thorny sculpin *(Icelus spiniger)* (Kihara and Shimada 1988a). Predation pressure was highly variable from Pacific cod and a clear pattern was not seen for the period 1984-1992 (Table 2). Conditions that increase predation from one suite of predators may reduce predation from another.

The reproductive success and prey species of Steller sea lions *(Eumetopias jubatus)* (National Marine Fisheries Service 1992), black-legged kittiwakes *(Larus tridactyla),* and thick-billed murres *(Uria lomvia)* (Decker et al. 1995, Hunt et al. 1996) is not related to interannual shifts in the distribution of age-1 pollock but rather to multi-annual shifts. Pollock (Perez and Bigg 1986, National Marine Fisheries Service 1992, Sinclair et al. 1994), capelin *(Mallotus villosus),* and Pacific herring *(Clupea harengus)* are important prey items for lactating northern fur seals *(Callorhinus ursinus)* near the Pribilof Islands. Between 1958 and 1974 (Kajimura 1984), pollock

Table 3. Reproductive success and food habits of thick-billed murres and northern fur seals in the Bering Sea vary on multi-annual scale.

Years	Mode	Age-2+ pollock predominant domain	Northern fur seal		Thick-billed murres	
			Reproduction	Food	Reproduction	Food
1972-1977	Cold	Outer	Good	Capelin Pollock	Good	Capelin
1978-1984	Warm	Middle	Poor	Pollock Sand lance	Poor	Pollock
1985-1991	Mixed	Middle and Inner	Poor	Pollock Sand lance	Poor	Pollock

and sand lance *(Ammodytes hexapterus)*, but not herring or capelin, were eaten by seals collected in 1981, 1982, and 1985 (Sinclair et al. 1994). The distribution and abundance of principal prey species could affect the feeding success and nutritional condition of these seals during lactation and thus the success of pups when they begin feeding. Although the data do not exist for every year between 1972 and 1991, a shift appears to have occurred in the diet of seals sometime between 1974 and 1981, during a transition between a cool state and a warm state (Table 3).

Most of the information on the distribution of fish on the Bering Sea shelf comes from demersal trawl surveys. While this is adequate for species associated with the bottom, benthic trawls do not provide us with an estimate for species that are wholly, or in part, pelagic. Age-1 pollock and Arctic cod inhabit both the pelagic and demersal environments (Rose and Walters 1990), and that segment of the population which is pelagic must be taken into account when evaluating food availability. In addition, concentrations of Arctic cod also occur northwest of the Pribilof Islands in some years (e.g., 1975, 1976, 1980, 1984, 1986, and 1988) (Wyllie-Echeverria 1995b), and may be utilized by northern fur seals when they occur within their feeding range.

Reproductive success of thick-billed murres on the Pribilof Islands between 1974 and 1991 (Decker et al. 1995) was not correlated with inter-annual measures of sea ice condition, age-1 or age-0 pollock abundance, or sea surface temperature (Hunt et al. 1996). Decker et al. (1995) propose that there was an ecosystem-wide change on the Bering Sea shelf with the regime shift of 1976-77, and that a shift back has not yet occurred. Cool, warm, and mixed states (Table 3), defined using P_s, are in agreement with their conclusions. A return to pre-1978 conditions has not occurred and consequently, reproductive and feeding behavior in thick-billed murres and northern fur seals have not returned to their pre-1978 levels. Possibly the cooler conditions that affect the distribution of age-1 pollock are also changing the distribution of pelagic capelin and herring populations, decreasing their availability to thick-billed murres and northern fur seals. Pascal and Adkison (1994) designed a model to test the effects of sea lion harvest, short-term environmental fluctuations, and transient variations of age-class structure and concluded that Steller sea lion declines were not related to short-term environmental fluctuations, but to significant long-term environmental degradation, perhaps on the scale of regime shifts.

Summary

A defining characteristic of winter on the Bering Sea shelf is the amount of seasonal sea ice. The degree of sea ice cover is an indicator of environmental conditions that will persist from winter through summer. Sea ice affects the summer shelf hydrography by the influx of low-salinity cold water. When sea ice extends south of 57°30′N, this low-salinity cold water appears on the shelf the next summer. Thus, an environment suitable for arctic species appears on this subarctic shelf in those years with extensive ice cover during the winter. Reproductive success and spatial distribution of subarctic species will vary between warm and cool conditions. In addition to the interannual effects, multi-annual warm and cool states persist on the shelf and influence different aspects of the ecosystem than interannual fluctuations. Sorting out the scale of variability for each portion of the ecosystem is fundamental in understanding the system and predicting the types of responses the ecosystem will have to varying climatic scenarios.

When cool conditions exist, the shelf is a marginal habitat for subarctic species, which concentrate in the outer shelf and slope areas (Fig. 4). Arctic species, such as Arctic cod, may extend their summertime range southward onto the shelf. Predator-prey interactions change, as was the case for age-1 pollock. When warm conditions exist, age-1 pollock are concentrated in the middle shelf, and age-2 and older pollock extend their area of concentration to include the middle shelf (Fig. 5). Predation pressures from subarctic species increase and occur over a wider area on the shelf. Predation by older pollock, Pacific cod, and arrowtooth flounder was greater during cool and intermediate years. Insights into multiple levels of the state of the demersal habitat and species interactions can be derived from a single variable expressing the extent of sea ice cover along meridian 169°W.

The correspondence of regime-scale variability in sea ice cover, sea surface temperatures, and species distributions spanning the North Pacific system is suggestive of a single area-wide cause for the physical state of the ocean. The physical states defined for the Bering Sea, on the basis of the amount of seasonal sea ice controlled by the position and strength of the Aleutian low pressure system, have a timing and duration similar to the regimes defined by sea surface temperatures in the northeastern Pacific (Wooster and Hollowed 1995). The behavior and interactions of species, while having an interannual component, also respond on the scale of the observed regimes. Year-class strength of pollock in the Bering Sea is higher during warmer regimes (Hollowed and Wooster 1995). Also, during warm regimes, age-2 and older pollock are distributed more widely over the outer, middle, and inner areas of the Bering Sea shelf. This increase in available habitat and the warmer conditions appear to be conducive to larval survival in this subarctic species. The links between the physical and biological systems enable us to predict events in the biological system from the physical system.

Insights into the relationship among atmospheric, oceanic, and biotic conditions on the Bering Sea shelf will aid in the understanding of this system and provide baseline data useful in projecting future conditions through modeling. Cold or warm conditions dominate the shelf each summer, and bottom temperatures, extent of the cold pool, and distribution of species that vary on an annual basis can be determined from the previous winter ice conditions. The predictive value of sea ice may extend to other harvestable stocks, with the result that the location of species whose distribution varies annually could be forecast.

Acknowledgments

This publication is funded by the Joint Institute for the Study of the Atmosphere and Ocean (JISAO) under NOAA Cooperative Agreement No. NA67RJ0155, Contribution No. 422. The views expressed herein are those of the authors and do not necessarily reflect the views of NOAA of any of its subagencies.

Research was supported in part by the Pacific Marine Environmental Laboratory (PMEL), sponsored by NOAA's Coastal Ocean Program for the Fisheries-Oceanography Coordinated Investigations (FOCI), through JISAO, University of Washington, Seattle, WA, and the Alaska Outer Continental Shelf Region of the Minerals Management Service, U.S. Department of the Interior, Anchorage, AK, contract number 14-35-0001-3-559. This manuscript has contribution numbers 422 for JISAO, B288 for FOCI, and 1788 for PMEL.

References

Azumaya, T., and K. Ohtani. 1995. Effect of winter meteorological conditions on the formation of the cold bottom water in the eastern Bering Sea shelf. Journal of Oceanography 51:665-680.

Bakkala, R.G., and M.S. Alton. 1986. Evaluation of demersal trawl survey data for assessing the condition of eastern Bering Sea pollock. In: Symposium on biology, stock assessment, and management of pollock, Pacific cod and hake in the North Pacific region. International North Pacific Fishery Commission Bulletin 45:90-120.

Barnes, C.A., and T.G. Thompson. 1938. Physical and chemical investigations in the Bering Sea and portions of the North Pacific Ocean. University of Washington Publications in Oceanography 3:35-79.

Chapman, W.L., and J.E. Walsh. 1993. Recent variations of sea ice and air temperature in high latitudes. Bulletin American Meteorological Society 74(1):33-47.

Decker, M.B., G.L. Hunt Jr., and G.V. Byrd Jr. 1995. The relationships among sea-surface temperature, the abundance of juvenile walleye pollock *(Theragra chalcogramma),* and the reproductive performance and diets of seabirds at the Pribilof Islands, southeastern Bering Sea. In: R.J. Beamish (ed.), Climate change and northern fish populations. Canadian Special Publication of Fisheries and Aquatic Science 121:425-437.

Francis, R.C., and K.M. Bailey. 1983. Factors affecting recruitment of selected gadoids in the northeast Pacific and east Bering Sea. In: W.S. Wooster (ed.), From year to year: Interannual variability of the environment and fisheries of the Gulf of Alaska and the eastern Bering Sea. Washington Sea Grant, University of Washington Press, Seattle, pp. 35-60.

Hollowed, A.B., and W.S. Wooster. 1995. Decadal-scale variations in the eastern subarctic Pacific: II. Response of northeast Pacific fish stocks. In: R.J. Beamish (ed.), Climate change and northern fish populations. Canadian Special Publication of Fisheries and Aquatic Science 121:373-385.

Hunt, G.L., Jr., A.S. Kitaysky, M.B. Decker, D.E. Dragoo, and A.M. Springer. 1996. Changes in the distribution and size of juvenile walleye pollock *(Theragra chalcogramma)* as indicated by seabird diets at the Pribilof Islands and by bottom trawl surveys in the eastern Bering Sea, 1975-1993. In: R.D. Brodeur, P.A. Livingston, T.R. Loughlin, and A.B. Hollowed (eds.), Ecology of juvenile walleye pollock *(Theragra chalcogramma).* NOAA Technical Report NMFS-126, pp. 125-139.

Ingraham, W.J., Jr. 1981. Temperature and salinity observations at surface and near bottom over the eastern Bering Sea shelf, averaged by $1° \times \frac{1}{2}°$ squares. Northwest Alaska Fishery Center, Processed Report 81-09:51. (Available from Alaska Fisheries Science Center, 7600 Sand Point Way NE, Seattle, WA 98115.)

Kajimura, H. 1984. Opportunistic feeding of the northern fur seal *(Callorhinus ursinus)* in the northeastern Pacific Ocean and eastern Bering Sea. NOAA Technical Report NMFS SSRF-779. 49 pp.

Khen, G.V. 1988. Oceanographic conditions and Bering Sea biological productivity. In: Proceedings of the International Symposium on the Biology and Management of Walleye Pollock. University of Alaska Sea Grant, AK-SG-89-01, Fairbanks, pp. 79-94.

Kihara, K., and A.M. Shimada. 1988a. Prey-predator interaction of walleye pollock, *Theragra chalcogramma,* and water temperature. Bulletin of the Japanese Society of Scientific Fisheries 54:1131-1135.

Kihara, K., and A.M. Shimada. 1988b. Prey-predator interactions of Pacific cod, *Gadus macrocephalus,* and water temperature. Bulletin of the Japanese Society of Scientific Fisheries 54:2085-2088.

Livingston, P.A. (ed.). 1991. Groundfish food habits and predation on commercially important prey species in the eastern Bering Sea from 1984-1986. NOAA Technical Memo NMFS F/NWC-207. 240 pp. (Available from NMFS, Scientific Publications Office, 7600 Sandpoint Way N.E., Seattle, WA, 98115-0070.)

Livingston, P.A., A. Ward, G.M. Lang, and M.-S. Yang. 1993. Groundfish food habits and predation on commercially important prey species in the eastern Bering Sea from 1987-1989. NOAA Technical Memo NMFS-AFSC-11. 192 pp. (Available from NMFS, Scientific Publications Office, 7600 Sandpoint Way N.E., Seattle, WA, 98115-0070.)

Maeda, T. 1977. Relationship between annual fluctuation of oceanographic conditions and abundance of year classes of the yellowfin sole in the eastern Bering Sea. In: Fisheries biological production in the subarctic Pacific region. Hokkaido University, Hakodate, Japan. Special Volume, pp. 259-268.

National Marine Fisheries Service. 1992. Recovery plan for the Steller sea lion *(Eumetopias jubatus)*. Prepared by the Steller Sea Lion Recovery Team of the National Marine Fisheries Service, Silver Spring, MD. 92 pp.

Niebauer, H.J., and R.H. Day. 1989. Causes of interannual variability in the sea ice cover of the eastern Bering Sea. GeoJournal 18:45-59.

Ohtani, K. 1969. On the oceanographic structure and the ice formation on the continental shelf in the eastern Bering Sea. Bulletin of the Faculty of Fisheries, Hokkaido University 20:94-117. (In Japanese with English abstract.)

Ohtani, K. 1973. Oceanographic structure in the Bering Sea. Memoirs of Faculty of Fisheries Hokkaido 21:65-106.

Ohtani, K. 1996. Variations in sea ice distribution and recruitment of walleye pollock in the eastern Bering Sea. In: T. Sugimoto (ed.), Application to modeling on response of marine resources to global changes in the environment and to management for fish stock. Ocean Research Institute, University of Tokyo. Report for grant-in-aid for scientific research. Ministry of Education, Science and Culture, Japan 06304018, pp. 38-49. (In Japanese.)

Ohtani, K., and T. Azumaya. 1995. Influence of interannual changes in ocean conditions on the abundance of walleye pollock *(Theragra chalcogramma)* in the eastern Bering Sea. In: R.J. Beamish (ed.), Climate change and northern fish populations. Canadian Special Publication of Fisheries and Aquatic Science 121, pp. 87-95.

Overland, J., and C.H. Pease. 1982. Cyclonic climatology of the Bering Sea and its relation to sea ice extent. Monthly Weather Review 110:5-13.

Pascal, M., and M. Adkison. 1994. The decline of the Steller sea lion in the northeast Pacific: Demography, harvest or environment? Ecological Applications 4:393-403.

Pease, C.H. 1980. Eastern Bering Sea ice processes. Monthly Weather Review 108(12):2015-2023.

Perez, M.A., and M.A. Bigg. 1986. Diet of northern fur seals, *Callorhinus ursinus*, off western North America. Fishery Bulletin, U.S. 84:957-971.

Reed, R.K. 1995. Water properties over the Bering Sea shelf: Climatology and variations. NOAA Technical Report ERL 452-PMEL 42. 15 pp.

Rose, C.S., and G.E. Walters. 1990. Trawl width variation during bottom trawl surveys: causes and consequences. In: L. Low (ed.), Proceedings of the symposium on application of stock assessment techniques to gadids. International North Pacific Fisheries Commission, Bulletin 50:57-67.

Schumacher, J.D., T.H. Kinder, D.J. Pashinski, and R.L. Charnell. 1979. A structural front over the continental shelf of the eastern Bering Sea. Journal Physical Oceanography 9:79-87.

Shuntov, V.P., and E.P. Dulepova. 1996. Walleye pollock in the western Pacific Ocean and Bering Sea ecosystems. In: R.D. Brodeur, P.A. Livingston, T.R. Loughlin, and A.B. Hollowed (eds.), Ecology of juvenile walleye pollock *(Theragra chalcogramma)*. NOAA Technical Report NMFS-126, p. 196.

Sinclair, E., T. Loughlin, and W. Pearcy. 1994. Prey selection by northern fur seals *(Callorhinus ursinus)* in the eastern Bering Sea. Fishery Bulletin, U.S. 92:144-156.

Springer, A.M. 1992. A review: Walleye pollock in the North Pacific—how much difference do they really make? Fisheries Oceanography 1:80-96.

Takenouti, A.Y., and K. Ohtani. 1974. Currents and water masses in the Bering Sea: A review of Japanese work. In: D.W. Hood and E.J. Kelley (eds.), Oceanography of the Bering Sea with emphasis on renewable resources. Occasional Publication No. 2, Institute of Marine Science, University of Alaska, Fairbanks, pp. 39-57.

Wooster, W.S., and A.B. Hollowed. 1995. Decadal-scale variations in the eastern subarctic Pacific. Part I. Winter ocean conditions. In: R.J. Beamish (ed.), Climate change and northern fish populations. Canadian Special Publication of Fisheries and Aquatic Sciences 121:81-85.

Wyllie-Echeverria, T. 1995a. Sea-ice conditions and the distribution of walleye pollock *(Theragra chalcogramma)* on the Bering and Chukchi Sea shelf. In: R.J. Beamish (ed.), Climate change and northern fish populations, Canadian Special Publication of Fisheries and Aquatic Sciences 1217:131-136.

Wyllie-Echeverria, T. 1995b. Seasonal sea ice, the cold pool, and gadid distribution on the Bering Sea shelf. Ph.D. thesis, University of Alaska Fairbanks. 318 pp.

Wyllie-Echeverria, T. 1996. The relationship between the distribution of one-year-old walleye pollock, *Theragra chalcogramma,* and sea ice characteristics. In: R.D. Brodeur, P.A. Livingston, T.R. Loughlin, and A.B. Hollowed (eds.), Ecology of juvenile walleye pollock *(Theragra chalcogramma)*. NOAA Technical Report NMFS-126, pp. 47-56.

Wyllie-Echeverria, T., S. Wyllie-Echeverria, and W.B. Barber. 1997. Water masses and transport of age-0 Arctic cod and age-1 Bering flounder into the northeastern Chukchi Sea. In: J. Reynolds (ed.), Fish ecology in Arctic North America. American Fisheries Society Symposium 19, Bethesda, MD, pp. 60-67.

Wyllie-Echeverria, T. 1998. Remotely sensed seasonal sea ice and oceanology of the Bering Sea. In: Satellite remote sensing of the Pacific Ocean. Earth, Ocean and Space Pty. Ltd., Sydney, Australia.

Spatial Distribution and Temporal Variability of Phytoplankton in the Bering Sea

Irina N. Sukhanova and Halina J. Semina
Shirshov Institute of Oceanology, Russian Academy of Sciences, Moscow, Russia

Mikhail V. Venttsel
Russian Federal Research Institute of Fisheries and Oceanography (VNIRO), Moscow, Russia

Abstract

In this chapter we describe variation in the distribution of phytoplankton species associations, total phytoplankton numbers, biomass, and chlorophyll *a* concentration in oceanic and shelf phytoplankton communities. In the oceanic domain, numbers of picophytoplankton dominate during all periods of seasonal succession, followed by nanophytoplankton. The latter are dominated by *Phaeocystis pouchetii, Chromulina* sp., and diatoms in spring and flagellates and cryptomonads in summer. Large "spots" of phytoplankton composed mostly of diatoms occur in the oceanic domain resulting from phytoplankton transport by mesoscale anticyclonic eddies from the shelf slope area. High biomass values up to several mg/L are typical for these "spots."

The shelf phytoplankton community occupies more than 50% of the sea. Characteristic of the community is the predomination of diatoms in both numbers and biomass during spring and by biomass during all phases of seasonal succession. The time and intensity of the spring phytoplankton bloom over the shelf, seasonal cycles of abundant species, and sequence and duration of community succession are determined by differences in hydrophysical and hydrochemical conditions, currents, daylight duration, and ice conditions.

In shelf areas not covered by ice, the beginning of the bloom is determined by an increase in solar radiation. The first stage of spring bloom ("prebloom") is associated with a small increase in diatoms, *Phaeocystis pouchetii,* and flagellates. The spring bloom itself is determined by centric colonial diatoms. Spring bloom over the southeastern shelf not covered by ice may begin both simultaneously or within 3 to 4 weeks; annual variability in the beginning of the spring bloom is common. The duration of the bloom over the outer shelf is shorter than over the middle shelf. In areas covered by ice in winter, phytoplankton seasonal growth starts with an increase in ice-associated flora abundance. The spring bloom there starts near the ice edge and occurs in the upper 10-15 m. Further development of the bloom is going on in ice-free water and extends as deep as 30 m. Spatial distribution and temporal variability of phytoplankton in the Gulf of Anadyr and other locations are discussed in detail.

Introduction

The important role phytoplankton plays in marine ecosystems is due to its ability to synthesize primary production. Production rate and the amount of newly synthesized organic matter largely depend on the species composition of the dominating phytoplankton groups. Phytoplankton biomass is partially consumed by pelagic herbivores, and the remainder sinks to the bottom where it is converted into benthic food or accumulated in bottom sediments in the form of detritus. Phytoplankton involvement in pelagic or bottom food webs is defined by the species composition of the dominating phytoplankton groups, while the efficiency of phytoplankton consumption by herbivorous organisms is directly linked to algae cell sizes and the abundance of the respective size groups (Parsons and LeBrasseur 1970).

In this chapter we review the spatial and temporal variability of phytoplankton in the Bering Sea in regard to different hydrophysical structures. Our review is based on data obtained in the past two decades, primarily thanks to large-scale research efforts under OCSEAP, PROBES, ISHTAR, and BERPAC programs, four Russian VNIRO-TINRO-Academy expeditions in 1990-1992, and the Pribilof Marine Ecosystem Research Program 1992-1994.

Methods

We made an attempt to take advantage of all data on phytoplankton published during the past two decades. We also include information on phytoplankton of the middle and the outer shelf in the region of the Pribilof Islands obtained in 1992-1994 by the senior author (in Flint et al. 1994, 1996) as well as data on the deep basin, the western coastal waters, and the Gulf of Anadyr obtained by the junior authors.

A step forward in Russian studies of phytoplankton since the early 1980s was a new method of soft reverse filtration to concentrate phytoplankton samples (Sukhanova 1983). The method enabled working with unpreserved material, or samples preserved in low concentrations of preservatives which allow counting algae groups that cannot usually withstand fixation. Samples for counting micro- and nanophytoplankton (diameter >16 μm, volume >2,000 μm^3; diameter 3-16 μm, volume 10-2,000 μm^3, respectively) were filtered through 1 μm nuclear filters. Picophytoplankton (diameter <2-2.5 μm, volume <7-8 μm^3) cells were counted with a luminescent microscope over nuclear filters (0.17 μm) prepared using the Caron method (Caron 1983). The method allowed us to identify phytoplankton cells with a size range >1 μm. Luminescent microscopy also enabled us to identify and select heterotrophic phytoplankton.

We also found it possible to use data obtained with the help of a sedimentation method (UNESCO 1978) during the 5th, 8th, and 10th cruises of the R/V *Vityaz* (1950-1952) because the data on phytoplankton in coastal and open areas of the western Bering Sea were scarce. Using the data obtained using sedimentation was justified to compare taxonomic composition and distribution patterns of microphytoplankton and diatomic nanophytoplankton.

Phytoplankton Associations

Three well distinguished associations (communities) may be determined in the Bering Sea phytoplankton: (1) boreal oceanic community, which is mainly composed of panthalassic boreal species; (2) seasonal ice phytoplankton community where pennate diatoms predominate; and (3) shelf community composed mostly of neritic species.

Distinguishing groups of abundant species within associations was one of the most convenient forms of data representation. It was also useful for studying spatial distribution patterns, seasonal succession, and quantitative features of phytoplankton, as well as for monitoring purposes. These groups included species in which total algae numbers and biomass were the most significant. We deemed species abundant in cases where 3-5 samples exhibited numbers over 50,000 cells/L or biomass over 50 μg/L. According to these criteria, 59 species were distinguished as abundant species (Table 1). The table shows that diatom algae make up the basis of the three community associations. In the dominating group they were responsible for 85% of all species. This value was over 1.5 times as high as diatom percentage in the Bering Sea flora. In the ice community, diatoms make up 100%, while in the shelf and oceanic communities they are responsible for ~80%.

Spatial Distribution and Temporal Variability of Phytoplankton

Spatial distribution of phytoplankton is defined by numerous factors, some of which are stationary (e.g., depth, bottom profile, large-scale circulation), some act seasonally (e.g., ice cover, temperature, insolation), and some act only over a short period (e.g., wind, mesoscale eddies). Depth is a factor of the greatest importance because it breaks the Bering Sea into shelf and deepwater areas. Each area is populated by a specific community.

Table 1.　Dominant species of Bering Sea phytoplankton.

Species	Ecological status	Biogeographic status	Ice	Shelf	Oceanic
Achnanthes spp.		A-B	+		
Fragilaria islandica		A-B	+		
F. striatula		A-B	+		
Nitzschia cylindrus		BIP	+		
N. grunowii		A-B	+		
Navicula pelagica		A-B	+		
Amphiprora paludosa var. *hyperborea*	n	A-B		+	
Bacterosira bathomphala	n	A-B			+
Chaetoceros atlanticus	p	C			+
C. compressus	n	C		+	+
C. concavicornis	p	A-B		+	+
C. constrictus	n	C		+	
C. convolutus	p	BIP		+	+
C. debilis	n	C		+	
C. furcellatus	n	A-B		+	
C. gracilis	n	A-B		+	
C. laciniosus	n	C		+	
C. socialis	n	C		+	
C. subsecundus	n	C		+	
C. wighamii	n	A-B-T		+	
Corethron criophilum	p	C			+
Coscinodiscus oculus-iridis		A-B		+	+
Cylindrotheca closterium		C		+	
Detonula confervacea	n	A-B-T		+	
Dinophysis norvegica		A-B			
Eucampia zoodiacus		C		+	+
Eutreptia lanowii	n	B ?		+	
Fragilariopsis oceanica	n	A-B		+	
F. cylindriformis	n	A-B		+	
Gymnodinium wulfii	p	C			+
Halosphaera viridis	p	C			+
Leptocylindrus danicus	n	C		+	
L. minimus	n	A-B-T		+	
Neodenticula seminae	p	B			+

Table 1. (Continued.)

Species	Ecological status	Biogeographic status	Type of community		
			Ice	Shelf	Oceanic
Nitzschia longissima	p	C			+
Phaeocystis pouchetii		A-B-T		+	+
Porosira glacialis	n	BIP		+	
Pseudonitzschia delicatissima	n	C		+	
P. seriata	p	A-B			+
Rhizosolenia alata	p	C		+	+
R. fragilissima	n	C		+	
R. hebetata f. *hiemalis*	p	C		+	+
R. hebetata f. *semispina*	p	C		+	+
R. styliformis	p	C		+	+
Thalassionema nitzschioides	n	A-B-T		+	
Thalassiosira anguste-lineata	n	A-B-T		+	
T. antarctica	n	BIP		+	
T. conferta	n	A-B		+	
T. gravida	n	BIP		+	
T. hyalina	n	A-B		+	
T. nordenskioldii	n	A-B		+	
T. sp. #1	n			+	
Tropidoneis antarctica var. *polyplasta*	p	BIP			+
Flagellatae				+	+
Chromulina sp. #1				+	+
Chromulina sp. #2				+	+
Synechococcus sp.				+	+
Gyrodinium lachrima	n	A-B		+	
Protoperidinium pellucidum	n	C		+	

n = neritic species; p = panthalassic species; A-B = arctoboreal type; A-B-T = arcto-boreal-tropical type; B = boreal type; BIP = bipolar type; C = cosmopolitan type.

Abundant species of the "ice" community according to Saito and Taniguchi (1978) and Alexander and Cooney (1979). Abundant species of the shelf and oceanic communities according to original data and to: Allen (1927, 1929); Aikawa (1932); Phifer (1934); Semina (1953, 1981, 1997); Kawarada (1957); Kawarada and Ohwada (1957); Karohji (1958, 1959); Ohwada and Kon (1963); Motoda and Minoda (1974); Taniguchi et al. (1976); Goering and Iverson (1978); Saito and Taniguchi (1978); Alexander and Cooney (1979); Iverson et al. (1979); Flint et al. (1994, 1996); Venttsel (1994).

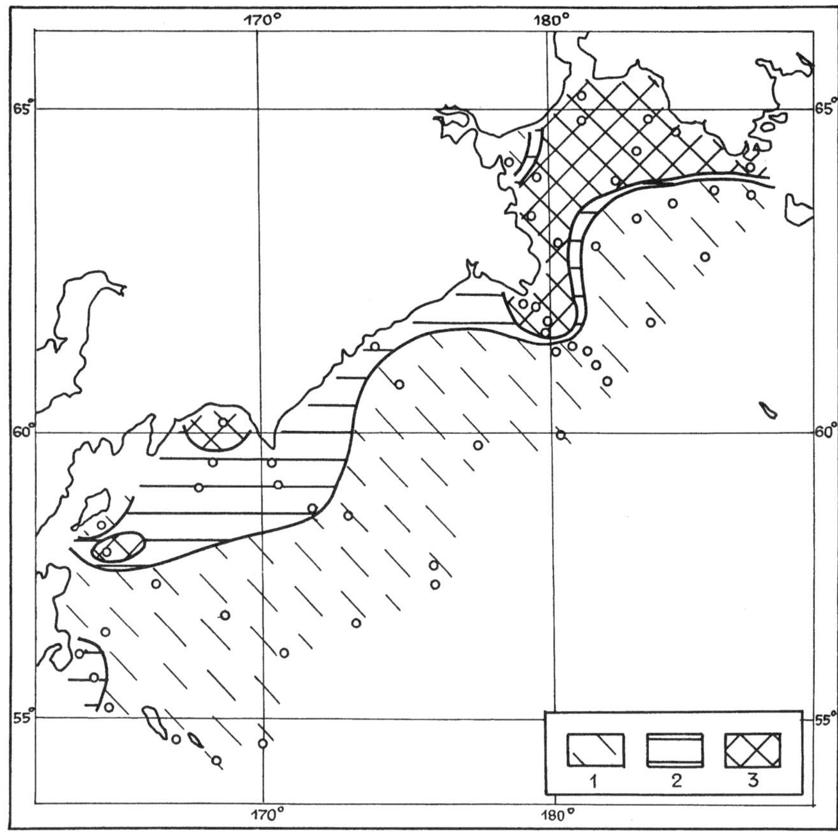

Figure 1.   Phytoplankton numbers (cells/L) in the 0-100 m layer in the western Bering Sea taken on the R/V Vityaz during September 1950. (1 = <10^4; 2 = 10^4-10^5; 3 = >10^5).

The border between the communities lies in the zone over the 200 m isobath. The shelf community differs from the oceanic one by wider species diversity, higher phytoplankton biomass, and higher primary production values. Comparative description of the shelf and the oceanic communities of the Bering Sea is given by Kiselev (1937), Semina (1955), Taniguchi (1969), Venttsel (1991, 1994), Flint et al. (1994, 1996), Mordasova et al. (1995), and Sorokin (1995).

Deep Basin

The data obtained in 1950-1952 during three cruises of the R/V *Vityaz* suggest that in spring (late May through mid-June) and in fall (late August

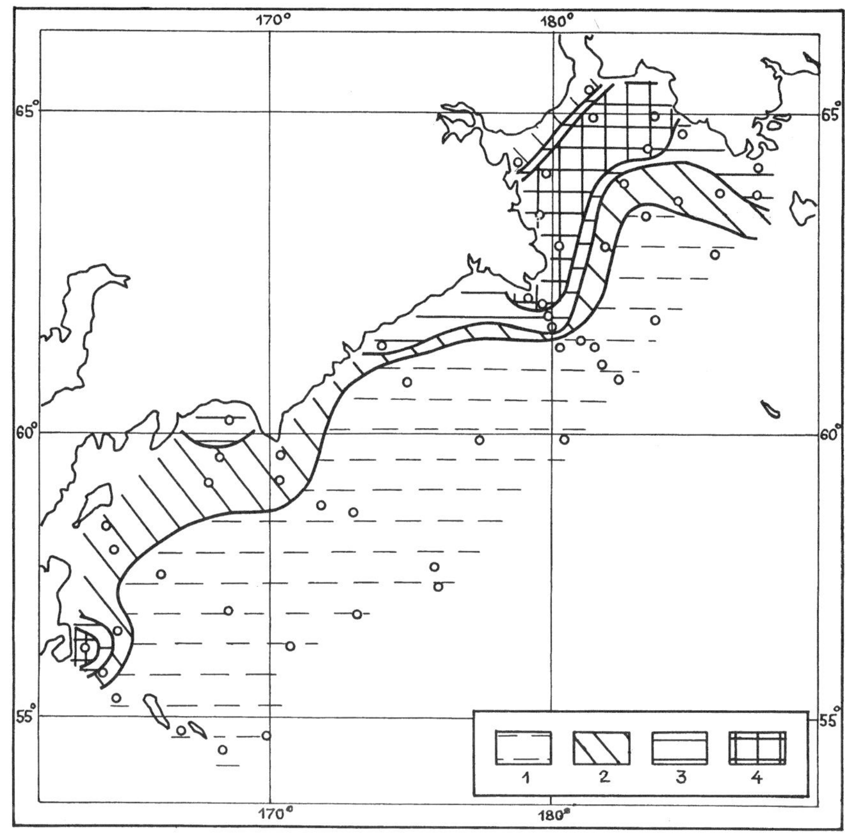

*Figure 2.   Phytoplankton biomass (mg/m³) in 0-100 m layer in the western Bering
Sea taken by the R/V Vityaz during September 1950. (1 = <50; 2 = 50-100;
3 = 100 – 1,000; 4 = >1,000).*

through September and October) the number of microphytoplankton and
diatomic nanophytoplankton in open areas was less than a few thousand
cells/L and that biomass was <100 µg/L. At the same time on the northern
shelf, in the Gulf of Anadyr and in Olyutorsk Bay, the number and biomass
were higher by one or two orders of magnitude (Figs. 1, 2). It should be
emphasized that differences were most substantial in spring and early
fall. In October, differences in phytoplankton cell numbers and biomass
were insignificant between the shelf and the open sea areas. Noteworthy,
in the western deepwater areas in fall numbers were nearly the same as in
spring, but they were accompanied by much higher biomass: 10 µg/L in
the spring and 100 µg/L in the fall. This difference was due to seasonal

Table 2.   Mean biomass values in different regions of the Bering Sea.

Areas	April-May	June	July	August
Shelf	7,000	2,500	510	1,200
Shelf break	3,100	170	120	100
Deep basin	160	80	70	180

Units = (µg/L).

changes in the taxonomic composition of phytoplankton. In June the following species were found: neritic species—*Thalassiosira nordenskioldii, Chaetoceros furcellatus, C. debilis,* and *C. socialis;* and panthalassic species—*C. atlanticus, C. decipiens,* and *Corethron criophilum.* In September the following species were found: neritic species—*Pseudonitzschia seriata* was the most common; panthalassic species—*Thalassiotrix longissima, Coscinodiscus osculus-iridis, C. marginatus, Corethron criophilum, Chaetoceros concavicornis,* and *Asteromphalus robustus.* In September, high individual cell volume resulted in higher biomass, compared to that in spring, at the same low numbers of cells.

Data collected during 1981, 1984, 1990, and 1992 enabled us to compare micro- and nanophytoplankton biomass on the northern shelf to that in the oceanic section and to that over the shelf slope in the western part of the sea in various seasons (Table 2). In April and early May phytoplankton biomass values on the shelf and in deep water differed by an order of magnitude, while in August the difference was 3.5 times.

The same data significantly improved our knowledge of seasonal changes in the taxonomic structure of the oceanic association in microphytoplankton and diatom nanophytoplankton. In early spring (April and early May) diatoms in most areas of the western Bering Sea were in low abundance and included: *Chaetoceros atlanticus, C. decipiens, C. debilis, C. socialis,* and *Fragilariopsis cylindriformis.* The situation was very similar to that observed in early June 1952. The locations of high biomass were due to the presence of spring bloom species from the shelf (i.e., *Thalassiosira nordenskioldii, T. antarctica, Chaetoceros furcellatus, C. debilis, C. socialis,* and *Bacterosira bathomphala).* In spring 1990, the average biomass and abundance in the open ocean were an order of magnitude above those observed in 1952. Where shelf species were in high abundance the biomass was over 3 g/m^3 (Fig. 3). In early summer (late June) the species with small cell sizes dominated oceanic phytoplankton (i.e., *Fragilariopsis cylindriformis, Neodenticula seminae,* and *Chaetoceros compressus).* Background numbers of micro- and nanophytoplankton were ~1.8 × 10^6 cells/L and the biomass was less than 210 µg/L. During the same period (the second half of June), phytoplankton numbers within the surface

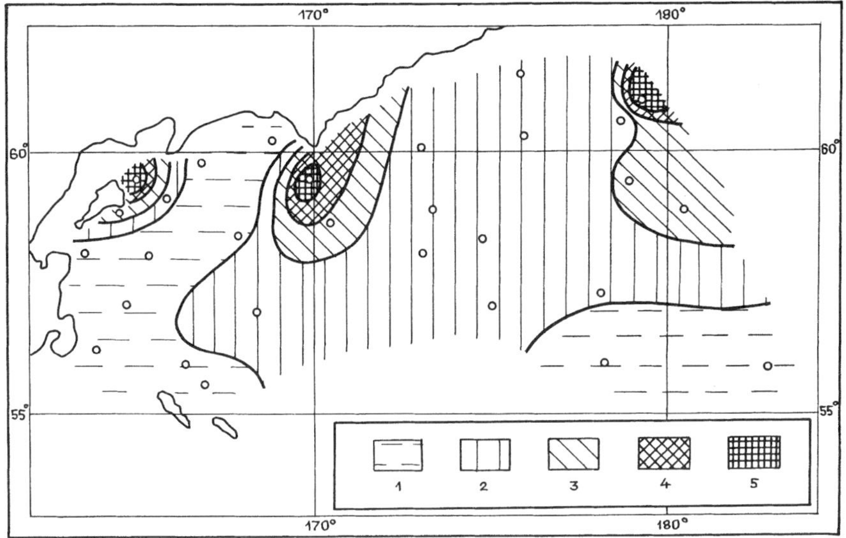

Figure 3.   Phytoplankton biomass (μg/m³) in 0-100 m layer in the northwestern Bering Sea. May 1990, R/V Mlechni Put. (1 = <100; 2 = 100-300; 3 = 300-1,000; 4 = 1,000-3,000; 5 = >3,000).

layer in the open sea varied from 17×10^3 to 35×10^3 cells/L (Taniguchi 1969). The most abundant species were *Chaetoceros debilis* and *Neodenticula seminae*. Dominating phytoplankton species in summer (the second half of July and early August) were *Cylindrotheca closterium* and *Rhizosolenia hebetata* f. *semispina*. In the southeastern part of the Aleutian Basin, *R. hebetata* f. *semispina* reached the bloom level ($\sim 350 \times 10^3$ cells/L).

The data on chlorophyll *a* concentrations also suggest substantial differences in phytoplankton concentrations between shelf and oceanic regions, as well as seasonal changes. Chlorophyll *a* concentrations over the shelf and in deepwater areas were: 6.4 μg/L and 0.4 μg/L in spring, respectively, and 1.9 μg/L and 0.9 μg/L in summer, respectively (Mordasova et al. 1995).

Estimates of phytoplankton numbers in the open ocean and on the shelf have changed recently based on new measurements of all size ranges of algae (Flint et al. 1994, 1996; Venttsel 1994; Venttsel et al. 1995). These new methods of sample collection and processing enabled more detailed evaluation of phytoplankton composition in boreal oceanic and neritic communities. The most important feature of the oceanic community seems to be wide distribution of picophytoplankton and small flagellates. They form a kind of permanent background, dominated by numbers of cells through all phases of seasonal succession (Table 3). In April 1990 and

Table 3.   Size composition of phytoplankton on the western Bering Sea shelf, in the Gulf of Anadyr, and in the deep basin.

Zones	Micro	Nano	Pico
August 1988			
Western shelf	– / –	– / –	– / –
Gulf of Anadyr	5 / 340	275 / 60	19 / <1
Deep basin	20 / 182	565 / 122	487 / <1
April-May 1990			
Western shelf	– / –	– / –	– / –
Gulf of Anadyr	60 / 890	750 / 1,055	45 / <1
Deep basin	5 / 92	520 / 35	540 / <1
End of June 1992			
Western shelf	960 / 1,685	375 / 47	165 / <1
Gulf of Anadyr	430 / 3,200	1,000 / 970	130 / <1
Deep basin	75 / 120	1,688 / 88	2,118 / 2

Number of cells $\times$ 10^3 per liter/biomass, µg per liter.

June 1992, picophytoplankton in open ocean accounted for >50% of total phytoplankton numbers and for <1.0% of total biomass (Table 3).

Nanoplankton algae made up a numerous group. At most stations in early spring of 1990, nanoplankton was represented by *Phaeocystis pouchetii* and *Chromulina* sp. #2; in some areas the group was composed of *Chaetoceros debilis, C. socialis, Fragilariopsis cylindriformis,* and *F. oceanica.* In summer 1992 the main components of nanophytoplankton were flagellates and cryptomonads.

Analysis of the taxonomic composition of open ocean phytoplankton across the entire size range revealed a mosaic distribution of abundant species. Figure 4 shows the distribution of abundant phytoplankton species in spring, which dramatically differs from the scheme given by Motoda and Minoda (1974). In most parts of the Kamchatka and Aleutian basins and near the slope of the western shelf, a high level of vegetation was exhibited by *Phaeocystis pouchetii* and *Chromulina* sp. #2. On the shelf near Karagin Island over Shirshov Ridge and the slope of the northern shelf, species of early spring bloom dominated (i.e., *Thalassiosira nordenskioldii, T. hyalina, T. antarctica, Chaetoceros debilis, C. socialis, Fragilariopsis oceanica, F. cylindriformis, Navicula septentrionalis, Bacterosira bathomphala,* and *Nitzschia frigida*). In the central part of the Aleutian Basin and in the deepwater area near the slope of the eastern shelf the most abundant species were *Chaetoceros debilis* and *C. compressus* (sec-

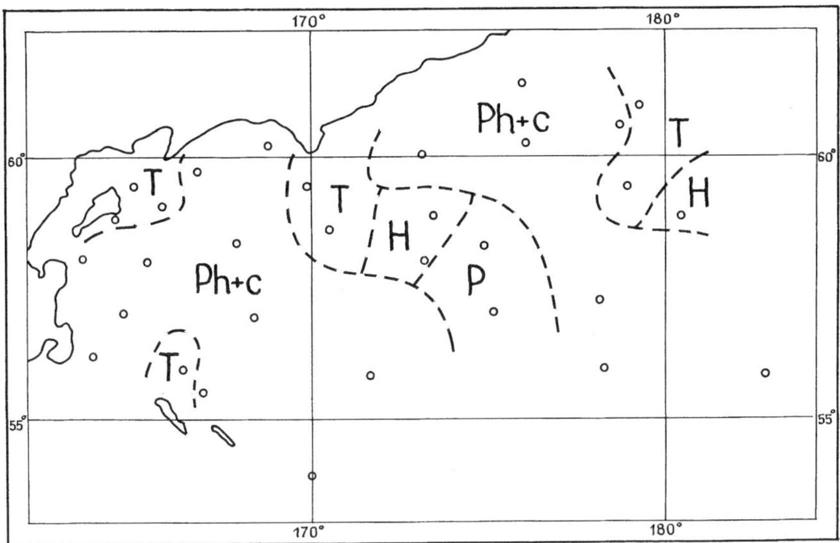

Figure 4.   *Areas of predomination of the following genera in the northwestern Ber-*
ing Sea. T = Thalassiosira, Ph+C = Phaeocystis and Chromulina, H = Chaeto-
ceros (section Hyalochaeta), P = Chaetoceros (section Phaeoceros).

tion Hyalochaeta). *Chaetoceros atlanticus* (section Phaeoceros) dominated
in a few locations in the Aleutian Basin. Neritic species were observed far
away from the shelf edge due to formation of mesoscale eddies near the
outer border of the continental slope (Sapozhnikov 1995, Verkhunov 1995).
This may contribute to the transport of shelf phytoplankton into deepwater areas (Kotenev 1995). In some cases the neritic phytoplankton community may be found almost as far as the eastern border of the Kamchatka
Basin. A similar system of eddies forms over the slope of the eastern shelf.
But a priori one can suppose a more intensive invasion of neritic species
into deepwater areas in the western part of the sea. This stems from differences in shelf geomorphology, which heavily affects the structure and
dynamics of shelf waters and the Bering Slope Current. The outer domain
of the wide eastern shelf is a zone where the boreal oceanic and the shelf
communities mix together. Numbers of the shelf species here are dramatically lower than in the middle and inner domains. Therefore, the outer
domain may be considered as a wide barrier zone. On the narrow and
deep western shelf there are no analogs of the middle and outer domains
and the middle and the shelf break fronts merge. These conditions are
favorable for spreading neritic species across the entire shelf and their
transport by anticyclone eddies along the slope and into the deep basin.

Shelf Areas

The Bering Sea is among the unique marine basins in which the shelf covers over 50% of the area. This feature largely defines its high productivity. Different parts of the Bering Sea shelf have similar taxonomic composition of phytoplankton and the same groups of species dominating in certain seasons. The sequence of phases of the spring bloom, the phytoplankton seasonal succession cycle, and seasonal cycles of the abundant species are all affected by: (1) different hydrophysical and hydrochemical conditions in certain shelf areas; (2) the extent to which the primary currents move them; (3) length of daylight; and (4) conditions and the time of ice cover (Konishi and Saito 1974; McRoy and Goering 1974; Coachman and Charnell 1979; Niebauer et al. 1981; Coachman 1986; Sambrotto et al. 1986; Walsh and McRoy 1986; Hansell et al. 1989, 1993; Flint et al. 1996).

Phytoplankton growth period on the southeastern shelf, and in Bristol, Olyutorsk, and Karagin bays lasts 180-200 days. On the northeastern shelf and on the shelf along the Koryak coast it lasts 150-170 days, on the northern shelf it lasts 120-140 days, and in the northern part of the Chirikov Basin it lasts 100-110 days.

The beginning of phytoplankton growth in ice-free sections is linked to an increase of insolation in spring. Sambrotto et al. (1986) referred to this stage of the phytoplankton seasonal succession as prebloom. At that stage they observed a slight increase in the growth rate in a few *Thalassiosira* species. Alexander and Cooney (1979) also reported on an increase of microflagellate numbers in early spring. According to Goering and Iverson (1981) prebloom is due to a minor advantage in numbers of dinoflagellates and *Phaeocystis pouchetii* along with numerous spring-bloom species present in small numbers. During that period, chlorophyll *a* concentration was not higher than 1 µg/L.

Necessary conditions for the spring bloom of phytoplankton to start are: high concentration of nutrients, sufficient illumination, sufficient length of photic period, and stabilization within the winter mixed-layer with formation of the seasonal pycnocline at the depth above the compensation point. The spring bloom of centric diatoms makes the principal contribution to annual primary production. According to various authors, from 45 to 70% of the phytoplankton annual production is produced during two spring months (Ivanenkov 1964, McRoy et al. 1972, Niebauer et al. 1981, Whitledge et al. 1986, Hansell et al. 1993). It is assumed that all primary production synthesized during the spring bloom is "new," i.e., maintained with nutrients, available in the photic layer at the time the spring phytoplankton bloom begins (Whitledge et al. 1986, Hansell et al. 1993, Arzhanova et al. 1995).

Colonial diatoms start blooming on the southeastern shelf in early or mid-April at the mid-shelf and inner-shelf fronts and spreads from these loci across the coastal, mid-shelf, and outer-shelf domains (Goering and Iverson 1981). Data on chlorophyll *a* distribution, collected in 1981 at six points on the main PROBES line, suggest simultaneous blooms in the three eastern Bering Sea domains (Whitledge et al. 1986). Four years of observations (1978-1981) on changes in chlorophyll *a* concentrations at standard locations on the middle shelf and outer shelf showed that there was a 3-4 week gap between phytoplankton bloom time in the first 2 years and a simultaneous bloom in the domains in the next 2 years. Concentration of chlorophyll *a* on the middle shelf never dropped below 16 µg/L, while on the outer shelf it was never above 16 µg/L. On the outer shelf the bloom was shorter than that on the middle shelf (Sambrotto et al. 1986, Walsh and McRoy 1986).

Interannual differences in bloom time, duration, and intensity within each domain were also well pronounced. In 1979 on the middle shelf the bloom began on 15 April; in 1981 it began on 2 May. The bloom is considered to begin when chlorophyll *a* concentration reaches ~4 µg/L. According to our data this level of chlorophyll *a* concentration corresponds to the number of spring shelf micro- and nanophytoplankton at ~1 × 10⁶ cells/L and a biomass of ~2 g/m³. The bloom lasted from 25 to 45 days. The interannual difference in chlorophyll *a* concentrations were ~20 µg/L (from 16 to 36 µg/L). In 1978 and 1979 on the outer shelf the bloom began on 9-11 May and lasted for 2.5 weeks and 5 days, respectively. In 1980 and 1981 the bloom began three weeks earlier, ~20 April, and lasted for ~6 weeks. The highest concentration of chlorophyll *a* varied from 8 to 16 µg/L from year to year.

Sambrotto et al. (1986) suggested the possibility of a two-stage development of the spring bloom on the middle shelf. The "second wave" of the bloom comes after intensive storm mixing and is a response to enrichment of the surface layer with nutrients. The average uptake of $^{15}NO_3^-$ made up 700 mg-at/m² over 38 days (from 25 April through 2 June) in 1979-1981. On average, the initial bloom leading to the exhaustion of NO_3^- from the mixed layer accounted for ~320 mg-at of NO_3^- uptake. Nitrate uptake before stabilization accounts for 30-60 mg-at NO_3^-/m² and post-bloom diffusive flux can account for an additional 50-70 mg-at NO_3^-/m². The enhancement of new production by wind mixing, therefore, is responsible for the remaining 260 mg-at/m² of NO_3^- uptake, or approximately 37% of the yearly spring bloom average. Interannual variations of the contribution of postbloom wind mixing enhancement to NO_3^- uptake ranged from 10% in the 1981 season to 50% in 1979. The storm events, which occur over a short period of time, are an important factor influencing the annual primary production cycle on the middle shelf.

In areas covered with ice in winter, vegetation begins when cryophilic flora starts developing. At that time the thickness and transparency of ice cover enables penetration of light, sufficient for the growth of an algal community living in the lower portions of sea ice (Niebauer et al. 1981). Pennate diatoms dominate the cryophilic flora, particularly *Fragilaria islandica, F. striatula, Nitzschia cylindrus, N. grunowii, Navicula pelagica?* (Homer and Alexander 1972, Saito and Taniguchi 1978). Vegetation of cryophilic flora results in high concentrations of chlorophyll *a* (6.8 µg/L; McRoy and Goering 1974) and high primary production (5.55 µg C/m³ per hr, ibid.) within a very thin 5 cm layer at the lower ice surface. Calculated values for the entire water column are low (0.34 mg chlorophyll *a*/m² and 0.28 mg C/m² per hr).

The spring bloom of colonial diatoms (mostly centric) in these areas begins near the ice edge, often spreading as far as 50-100 km within the tail of the retreating ice (Niebauer et al. 1981). The most favorable situation for phytoplankton growth can be found in water with remains of melting ice. Seasonal increases of insolation and the remains of melting ice, which decrease surface salinity and reduce wind mixing, facilitate stabilization across the water column and phytoplankton blooming within the upper 10-15 m. Some cases of such bloom exhibited high rates of carbon fixation at 600 mg C/m² per hr.

The next phase of the spring bloom occurs in water that is completely ice free. The upper mixed layer thickens up to 20-30 m. This is a continued near-edge bloom and is defined by development of the same spring complex of diatoms. The intensity of the phase is largely defined by the intensity of the bloom near the ice edge, or, in other words, by available nutrients.

The dominating species of the early spring bloom on the shelf, which synthesizes new primary production, includes: *Thalassiosira nordenskioldii, T. hyalina, T. antarctica, T. gravida, T. anguste-lineata, Chaetoceros furcellatus, C. socialis, C. debilis, Fragilariopsis oceanica, F. cylindriformis, Bacterosira bathomphala,* and *Porosira glacialis.* At the shelf break front and in the adjacent area these are largely *Phaeocystis pouchetii.* The species are adapted to low temperature in a range from ~ –1° to 4°C. Population growth can be seen on the background of excessive amounts of nutrients and silicon. Mordasova et al. (1995) found high concentrations in spring of chlorophyll *a* in *Thalassiosira* sp. cells (75-85% from the total amount of chlorophyll *a, b,* and *c*), which is a sign of high physiological activity. According to our calculations, the assimilation values for the late spring bloom were 5-7 mg C/mg Chl *a* per hr, which also stems from intensive photosynthesis. Some spring assemblage species, such as *Chaetoceros debilis, C. socialis, Phaeocystis pouchetii,* and *Thalassiosira nordenskioldii,* are also abundant in conditions of higher temperatures. We observed high numbers of *C. debilis* (~0.5 × 10⁶ cells/L) and *Thalassiosira nordenskioldii* (>0.1 × 10⁶ cells/L) in early August in close vicinity to St. Paul Island at 6.0°C (Flint et al. 1996). This followed the "injection" of deep basin water into the coastal zone of the island and the related increase of phosphates, nitrates, and silicon concentrations in the euphotic layer. The early-spring bloom of *Phaeocystis pouchetii* on the middle and outer shelf fronts lasted over a relatively long period through late May (Iverson et al. 1979). In mid-June 1993 and 1994 at 5.0-6.4°C, we observed a bloom of *Phaeocystis pouchetii* in the outer domain, at the shelf break front, and in the coastal zone of St. Paul Island (Flint et al. 1994, 1996).

Well pronounced interannual or spatial variations in population growth rate of early-spring bloom species, as well as variations in time when the populations reach their highest abundance, can be seen for early-spring diatoms. Dominant species in Olyutorsk Bay, the Gulf of Anadyr, and on the western shelf in early spring were *Thalassiosira nordenskioldii, Fragilariopsis oceanica, Bacterosira bathomphala, Chaetoceros debilis, C. furcellatus,* and *C socialis* (Semina 1953). On the eastern and the northern shelves and in Chirikov Basin the most numerous were *Thalassiosira* species (Saito and Taniguchi 1978). In spring 1981 in the northeastern part of the Gulf of Anadyr the most abundant were *C. furcellatus* and *F. oceanica,* and in 1990 these were *C. socialis* and *C. furcellatus.* On the western shelf and near Karagin Island the dominant species were *T. nordenskioldii, T. antarctica, C. socialis, F. oceanica,* and *B. bathomphala.* Goering and Iverson (1981) report that the most numerous species on the middle shelf in the eastern Bering Sea were *Chaetoceros debilis,* and those on the outer shelf were *Phaeocystis pouchetii.* According to Flint et al. (1994, 1996) during the spring bloom in 1993 a few species of large colonial *Thalassiosira* species and *Chaetoceros debilis* dominated over the middle shelf near the Pribilof Islands, while in 1994 *Chaetoceros furcellatus, C. socialis, Fragilariopsis oceanica,* and *Thalassiosira* were most abundant. One can assume that the initial stage of the early spring bloom, as well as the bloom near the ice edge, was often attributed to the rapid increase in abundance of *Thalassiosira* species; *Chaetoceros* species are subdominant then.

During the "second wave" of the spring bloom in the ice-free areas and during the bloom in the open water after the ice is gone *Chaetoceros* dominate (Taniguchi et al. 1976; Saito and Taniguchi 1978; Alexander and Cooney 1979; Flint et al. 1994, 1996). These differences in abundance of certain species are partially because studies often rely upon a formal calendar time rather than upon a particular phase of the spring succession of phytoplankton. The observations conducted in the northeastern part of the Gulf of Anadyr in the first half of June 1952, 1981, and 1990 show how different the condition of phytoplankton can be on the same calendar dates. In 1952 most of the Gulf of Anadyr was covered with ice. The surface water temperature near the ice edge, where the ice separates into smaller floes, varied from $-1.3°$ to $1.4°$C. *Thalassiosira nordenskioldii* dominated, while *Chaetoceros socialis* and *C. furcellatus* were subdominant. Phytoplankton numbers reached millions per liter; the average biomass was 1.8 g/m^3 (maximum value was as high as 6 g/m^3). In 1981 the gulf was free from ice. Surface water temperature varied from $2.3°$ to $3.9°$C. Dominating in the phytoplankton were spring neritic *C. furcellatus* and *Fragilariopsis oceanica* as well as panthalassic *Coscinodiscus oculus-iridis.* Average phytoplankton numbers were 3.5×10^6 cells/L, while the average biomass was 0.8 g/m^3. The warmest year was 1990 when surface temperature varied from $4.0°$ to $5.2°$C. The most abundant phytoplankton were two spring neritic species, *Chaetoceros socialis* and *C. furcellatus,* as well as panthalassic late-spring and summer species, *Corethron criophilum* and *Rhizosolenia alata.* Average phytoplankton numbers were $\sim0.8 \times 10^6$ cells/L, and biomass was ~2.0 g/m^3. Composition of the phytoplankton indicated that in

June 1952 the survey covered the earliest stage of the spring bloom. Most likely, in 1981 the survey period occurred during the end of the spring bloom. Phytoplankton composition in 1990 matched the transitional period from spring to summer.

During the transitional spring-summer period different parts of the shelf may exhibit assemblages of phytoplankton species typical for different succession phases, or a mixed spring-summer species composition. For instance, from 10 through 15 June 1994 phytoplankton communities were at three different phases of seasonal succession (Figs. 5, 6). We found this over the eastern Bering Sea shelf, southwest and northeast (toward the continent) of St. Paul Island, as well as south of St. George Island. Northeast of St. Paul Island late-spring succession phase was observed on the middle shelf. *Chaetoceros furcellatus* dominated in numbers of cells; about one half of the population was in the spore formation stage. *Thalassiosira* and *Coscinodiscus* made up 45% of the biomass. Water temperature within the upper 10 m varied from 2.1° to 4.2°C, and in the layer below it varied from 0.1° to 1.7°C. To the southwest of St. Paul Island the most numerous species were small-size *Thalassiosira (T. conferta, T. angulata, and T. hispida* with cell diameter <15 μm,) and *Eutreptia lanowii*. About 50% of the biomass was *Rhizosolenia alata, R. fragilissima,* and *R. hebetata* f. *semispina*—typical species for late spring and summer. Water temperature within the upper 10 m varied from 4.5° to 5.0°C; under the thermocline the temperature was <2.5°C. The middle shelf near St. George Island featured a mixed assemblage of species which dominated in the coastal zone of the island (three *Rhizosolenia* sp.) as well as that of outer shelf domain species (*Phaeocystis pouchetii* and small colonial Pennatae). Water temperature within the upper 20 m was 5.3°C. Deeper down to the bottom temperature varied from 3.0° to 4.0°C. In the coastal zone of St. Paul Island the spring assemblage of *Chaetoceros, Thalassiosira,* and *Coscinodiscus* dominated at the temperature range of 1.5-2.5°C. In the coastal zone of St. George Island *Rhizosolenia* sp. dominated in late spring and summer.

A similar situation was observed in the second half of June 1992 in the Gulf of Anadyr across a 96 × 96 km area (Fig. 7). In the southern part of the area (station numbers 2141-2146) early spring bloom species dominated (i.e., *T. nordenskioldii, C. furcellatus, C. debilis, C. socialis,* and *Bacterosira bathomphala*). In this southern region the average number of phytoplankton was 3.1×10^6 cells/L and the average biomass was 11 mg/L. Water temperature at the surface was 3.2°C. The concentration of chlorophyll *a* in that part of the gulf varied from 8 to 35 μg/L (Mordasova et al. 1995). According to Sorokin (1995) primary production within the upper layer reached a record level of 1.0-2.2 mg C/m^3 per day. Its integral values across the water column within the bloom area were 3.1-6.7 g C/m^2 per day. Peridinae dominated in the northeastern part of the area, particularly *Goniaulax* sp. The dominant group also included *Leptocylindrus danicus* and *Meringosphaera mediterranea*—summer assemblage species. A typical feature was high numbers of heterotrophic Peridinae, namely *Gyrodinium*

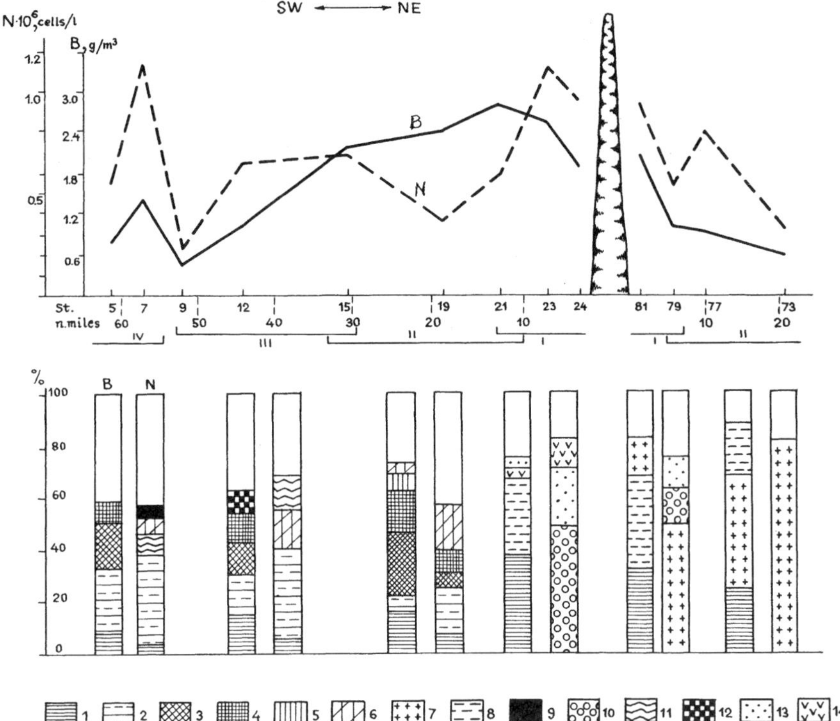

Figure 5. Distribution and proportion of phytoplankton numbers (N) and biomass (B) of abundant species in areas southwest and northwest off St. Paul Island. I = coastal zone; II = middle domain; III = outer domain; IV = deep basin. 1 = Thalassiosira spp. (diameter > 15 μm); 2 = Thalassiosira spp. (diameter < 15 μm); 3 = Rhizosolenia alata; 4 = R. fragilissima; 5 = R. hebetata f. semispina; 6 = Eutreptia lanowii; 7 = Chaetoceros furcellatus; 8 = Coscinodiscus spp.; 9 = Phaeocystis pouchetii; 10 = Chaetoceros socialis; 11 = Pennatae sp. #l; 12 = Corethron criophilum; 13 = Fragilariopsis oceanica; 14 = Chaetoceros debilis.

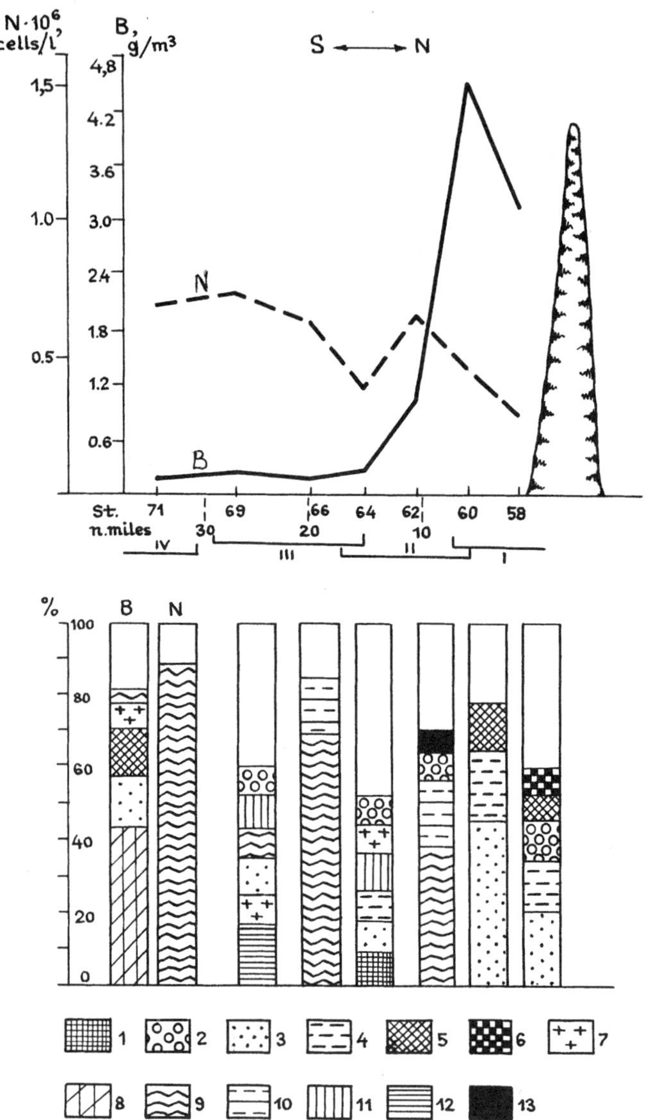

Figure 6. Distribution and proportion of phytoplankton numbers (N) and biomass (B) of abundant species in areas south off St. George Island. I = coastal zone; II = middle domain; III = outer domain; IV = deep basin. 1 = Thalassiosira spp. (diameter > 15 µm); 2 = Thalassiosira spp. (diameter < 15 µm); 3 = Rhizosolenia alata; 4 = R. fragilissima; 5 = R. hebetata f. semispina; 6 = Eutreptia lanowii; 7 = Neodenticula seminae; 8 = Coscinodiscus spp.; 9 = Phaeocystis pouchetii; 10 = Pennatae sp. #2; 11 = Rhizosolenia styliformis; 12 = Dinoflagellatae; 13 = Pennatae sp. #1.

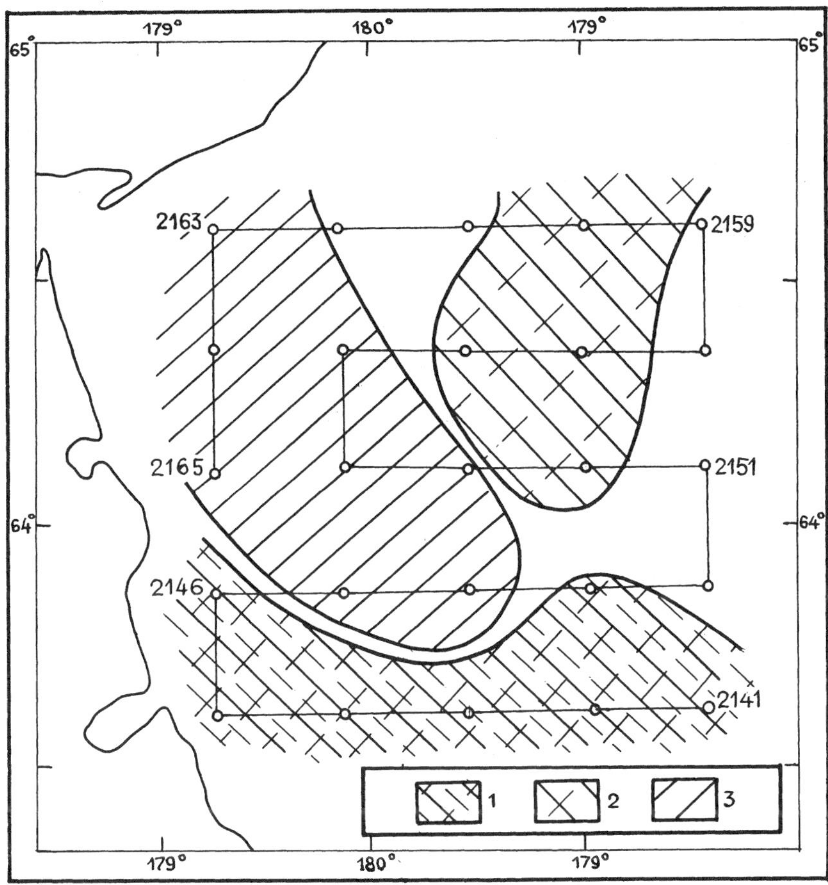

Figure 7.   Regions with domination of different phytoplankton complexes in the
Gulf of Anadyr. 1 = southern region; 2 = northern region; 3 = western
region.

lachrima and *Protoperidinium pellucidum*. Total phytoplankton numbers (8.2×10^5 cells/L) and biomass (1.1 mg/L) values were significantly lower than in the southern part of the gulf. This part of the area was within the water mass of the Anadyr cold spot (Verkhunov 1995) or within the water mass of the central part of the Gulf of Anadyr (Coachman and Shigaev 1988). The western part featured low salinity and relatively high temperature (7.6°C). Phytoplankton biomass was the lowest (i.e., 0.4 mg/L) and numbers were 8.4×10^3 cells/L. Summer species of the diatoms and Peridinae dominated the phytoplankton. A few *Melosira* spp. were observed, which are common for waters less saline than typical Bering Sea water. The area was influenced by Anadyr coastal river water.

Summer phytoplankton generally differed from those in spring by lower microphytoplankton numbers and, consequently, by lower biomass (Figs. 8, 9). The summer shelf assemblage of the dominating species included: *Rhizosolenia alata*, *R. fragilissima*, *R. hebetata* f. *semispina*, *R. hebetata* f. *hiemalis*, *R. styloformis*, *Eutreptia lanowii*, Pennatae sp. #1, Pennatae sp. #2, *Leptocylindrus danicus*, *L. minimus*, *Chaetoceros compressus*, *C. concavicornis*, *C. subsecundus*, *C. laciniosus*, *C. gracilis*, *C. wighamii*, *Detonula confervaceae*, *Pseudonitzschia delicatissima*, *Thalassiosira* sp. #1, a few Peridinae species, two of which are heterotrophic— *Peridinium pellucidum* and *Gyrodinium lachrima*, and phytoflagellates. Succession of summer species occurs as mineral forms of nitrogen and phosphorus begin to decline in the euphotic layer. At that time ammonium and urea were found in the mixed layer in relatively high concentrations. That means nearly all primary production was synthesized through recycling (Mordasova et al. 1995; Sapozhnikov et al. 1995a,b; Whitledge et al. 1986). At the same time, in summer (especially in the first half) the middle eastern shelf and the northern shelf occasionally feature high abundance of certain species typical for summer or, in some cases, species typical during the spring phase of succession. In most situations this is a response to "injections" of nutrient-rich subthermocline waters into the euphotic layer.

In a relatively cold year versus a warm one, differences in phytoplankton taxonomic composition, numbers, and biomass temporal changes were well pronounced. From 14 May through 25 July 1993 (a relatively warm year) three peaks of high phytoplankton numbers and biomass were observed: (1) the "second wave" of spring bloom (last 10 days of May), (2) a period of high abundance of *Phaeocystis pouchetii* (the second 10 days of June), and, (3) a period of high numbers and biomass of summer-autumn species, largely that of diatoms (*Chaetoceros wighamii*, *Pseudonitzschia delicatissima*, and *Thalassiosira* sp. #1; mid-July) (Figs. 9, 10). In the second half of July a rapid increase of picophytoplankton numbers was observed (Figs. 9, 11). High numbers and biomass of summer-autumn species exist in the background during occasions of extremely low concentrations of nutrients and four-fold increase of NH_4 concentration. In 1994, water temperature near St. Paul Island was 2°C below average. Observations be-

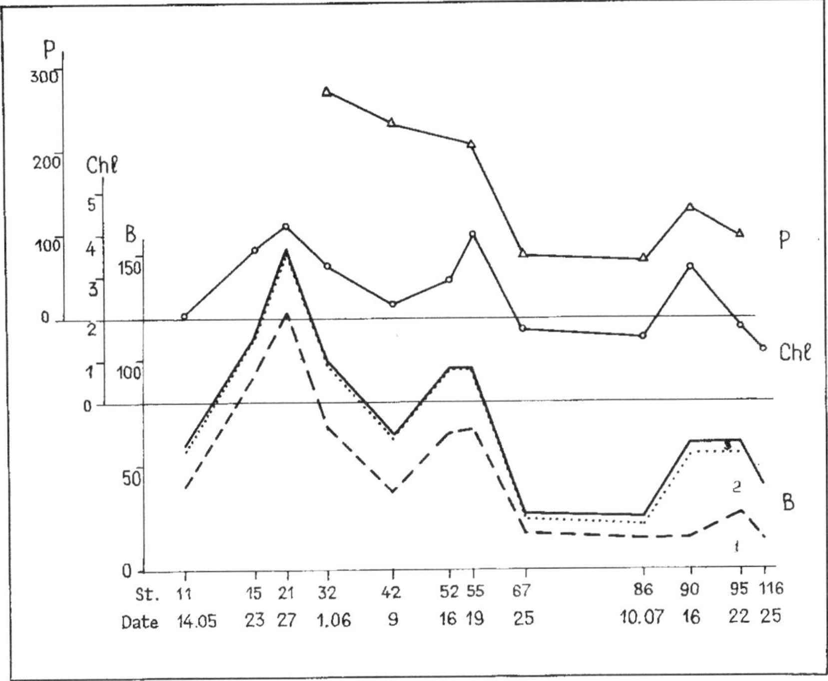

Figure 8. Temporal changes in phytoplankton biomass (B, µg C/L; solid line = total phytoplankton biomass; 1 = microphytoplankton; 2 = nanophytoplankton; 3 = picophytoplankton), chlorophyll a (Chl, µg/L), and primary production (P, mg C/m³ per day) in the St. Paul Island coastal zone.

ginning on 10 June coincided with the end of the spring bloom. At that time high numbers and biomass of phytoplankton were observed even though nitrates and silicon were nearly depleted from the euphotic layer. The peak in numbers and biomass observed in late July was due to a subsequent bloom of *Phaeocystis pouchetii* and of the spring species—*Chaetoceros debilis* (0.4 × 10⁶ cells/L), *Thalassiosira nordenskioldii* (0.15 × 10⁶ cells/L), and *T. hyalina* (55 × 10³ cells/L). The bloom of spring species occurred when most *P. pouchetii* died and sank, followed by the "injection" of deep water into the euphotic layer, a tenfold growth in silicon concentration, and a doubling in nitrate concentration. Such short phytoplankton blooms in summer are significant additions to the "new" production synthesized on the shelf.

In late summer and autumn some summer species remain in high numbers, namely: both *Leptocylindrus* species, *Chaetoceros compressus, C. wighamii, Pseudonitzschia delicatissima, Thalassiosira* sp. #1, abundant

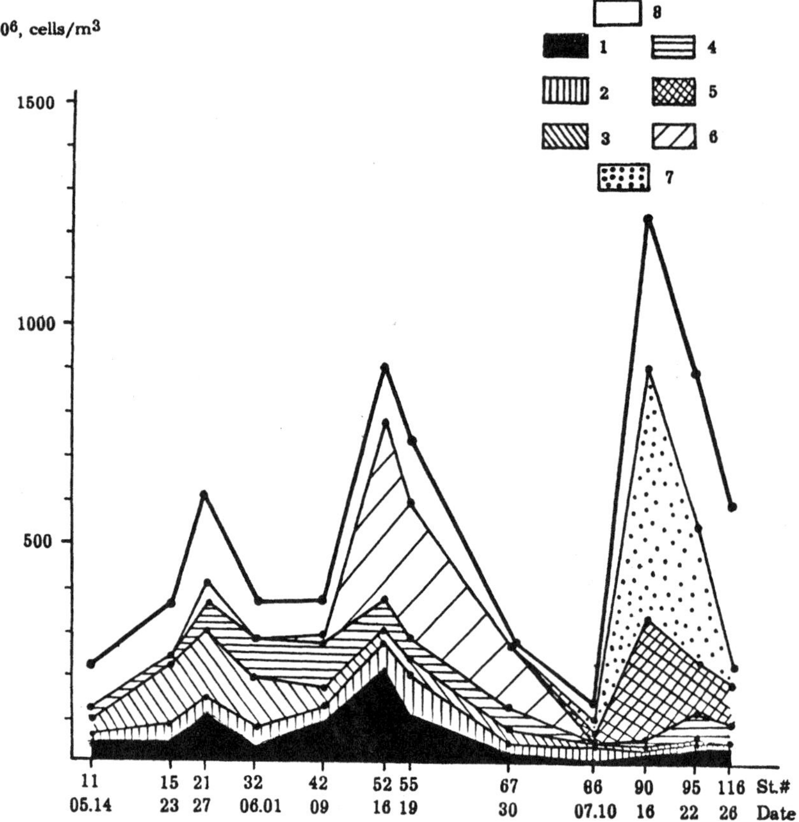

Figure 9.   *Temporal changes in phytoplankton taxonomic composition near St. Paul Island in May-July 1993. 1 = Chaetoceros debilis; 2 = Rhizosolenia alata; 3 = Thalassiosira spp. (diameter < 15 μm); 4 = Thalassiosira spp. (diameter > 15 μm); 5 = Chaetoceros wighamii; 6 = Phaeocystis pouchetii; 7 = Pseudonitzschia delicatissima; 8 = others.*

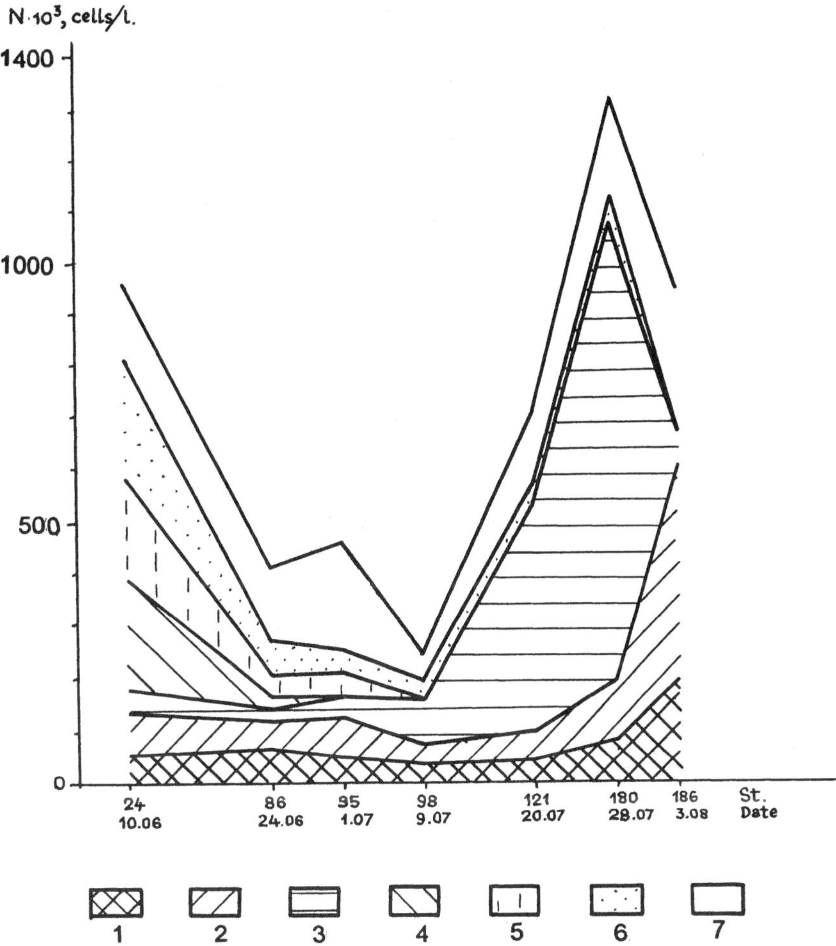

Figure 10. Temporal changes in phytoplankton taxonomic composition near St. Paul Island in July-August 1994. 1 = Thalassiosira hyalina + T. nordenskioldii; 2 = Chaetoceros debilis; 3 = Phaeocystis pouchetii; 4 = Chaetoceros socialis; 5 = C. socialis (spores); 6 = Fragilariopsis oceanica; 7 = others.

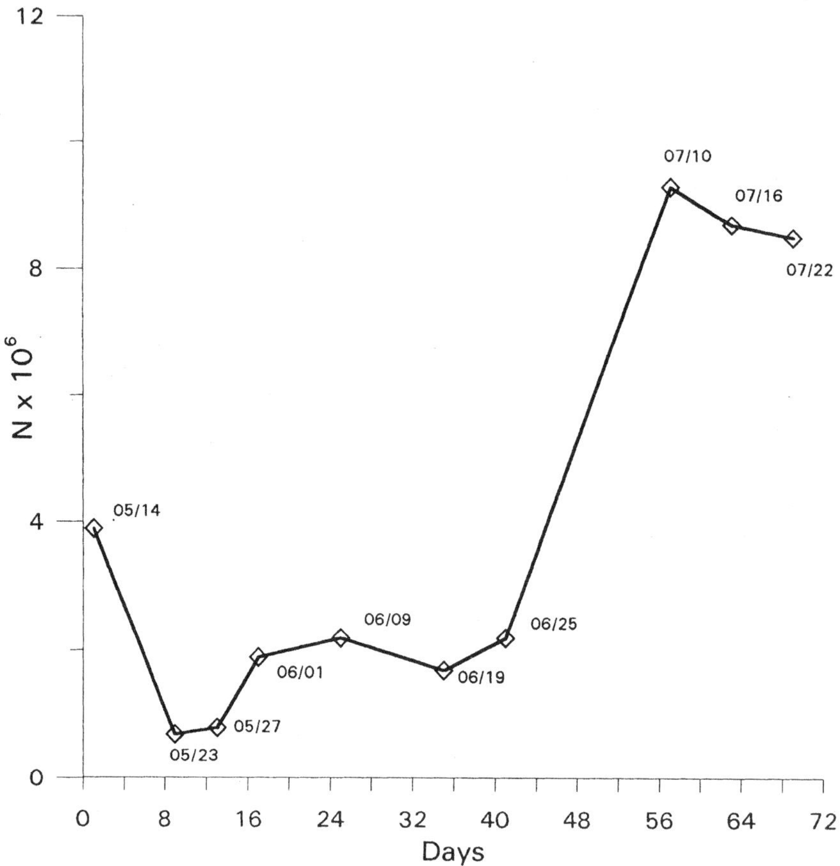

Figure 11. Temporal changes in picophytoplankton numbers (N × 10⁶ cells/L) in the St. Paul Island coastal zone.

development of *Thalassionema nitzschioides, Nitzschia longissima, Gymnodinium wulfii, Eucampia zoodiacus,* and *Amphiprora paludosa* var. *hyperborea.* High abundance of *Chaetoceros debilis* and *Phaeocystis pouchetii* was often observed.

In the Gulf of Anadyr during August 1988 the following species dominated by number: *Phaeocystis pouchetii, Prymnesiales* sp., *Synechococcus* sp., *Chroomonas* sp., *Fragilariopsis oceanica, Chaetoceros compressus,* and *C. socialis* (Venttsel 1991). Dominating in biomass were *Gymnodinium wulfii* and *Leptocylindrus danicus.* At some stations in the central part of the gulf *T. nordenskioldii* was predominant. The total number of phytoplankton ranged from 110 × 10³ to 550 × 10³ cells/L and biomass varied in wider limits from 20 to 2,000 µg/L. The highest biomass (2.0 mg/L) was found in the central part of the gulf and in Anadyr Strait and the lowest biomass (20.0 µg/L) was observed in the southern part of the gulf. Summer species of diatoms dominated in the Chirikov Basin then. Numbers at the surface ranged from 50×10^3 to 1.75×10^6 cells/L; those at 45 m deep ranged from 85×10^3 to 1.0×10^6 cells/L. The biomass varied from 270 to 3,150 µg/L at the surface and from 125 to 940 µg/L at 45 m.

Data on phytoplankton in the Bering Sea during fall are very limited. Phytoplankton numbers and biomass have been evaluated for fall using data obtained in 1950-1951 during the 5th and the 8th cruises of the R/V *Vityaz*. The bloom was observed in the western and southern parts of the Gulf of Anadyr, over the western shelf, and in Olyutorsk Bay, and was principally *Thalassionema nitzschioides* in September. The number of species in the Gulf of Anadyr was highest and reached 1×10^6 cells/L. The highest biomass within 0-25 m was 5 mg/L (Figs. 1, 2). At the north of the Gulf of Anadyr and in Anadyr Strait two *Chaetoceros* species dominated, *C. compressus* ($<2 \times 10^5$ cells/L) and C. *debilis* ($<4 \times 10^4$ cells/L), and total phytoplankton numbers were ~0.5×10^6 cells/L. Phytoplankton in October exhibited extremely low numbers and biomass. On the northern and the western shelves in the Gulf of Anadyr, the composition, numbers, and biomass of phytoplankton were the same as in the open basin. Neritic *Thalassionema nitzschioides, Chaetoceros constrictus, C. debilis,* and *Leptocylindrus danicus* numbers did not exceed a few thousand per liter near Chukotskii, Navarin, and Olyutorsk capes.

The mechanism supporting high density and production of phytoplankton (Sorokin 1995) in the Gulf of Anadyr and Chirikov Basin relied upon nutrients provided by advection with the Anadyr Current, a northern branch of the Bering Slope Current (Coachman and Shigaev 1988, Coachman 1993). When the spring bloom is over this source maintains high rates of phytoplankton growth in certain parts of these areas during almost all the vegetation period. Even during the summer succession phase, which had the lowest numbers and biomass over the whole succession period, phytoplankton abundance in the central part of the Gulf was still high.

The Edge of the Continental Shelf

One of the most productive areas in the Bering Sea is the continental shelf edge area (Iverson et al. 1979, Hansell et al. 1989). Since the 1960s Russian and U.S. researchers have undertaken occasional surveys in the area. The area exhibited high concentrations of chlorophyll *a* (Iverson et al. 1979, Hansell et al. 1989), high primary production (Azova 1964, Starodubtsev 1970, Goering and Iverson 1981, Hansell et al. 1993), as well as high concentrations of zooplankton (Iverson et al. 1979; Cooney 1981; Flint et al. 1994, 1996; Springer et al. 1996). All researchers observed abundant stocks of marine fishes, birds, and mammals in the shelf break area as well. Perhaps the least studied issue is the composition of phytoplankton and the number and biomass of the dominating species at different stages of the seasonal succession in various parts of the shelf break area.

It is not known how large (up to 100 km) and small (up to 30 km) eddies and meanders of the Bering Slope Current affect the distribution and abundance of phytoplankton.

In late May 1978 under the PROBES program a 170 km section of the shelf break front was surveyed (Iverson et al. 1979). A 15 km belt of high chlorophyll *a* concentrations was found over the shelf break front at depths of 180-200 m. *Phaeocystis pouchetii* dominated within the surface layer in the northwest of the surveyed area. At 15 m deep a diatom *Biddulphia* sp. and a Peridinae *Prorocentrum* sp. dominated. Chlorophyll *a* concentration was as high as 6-8 µg/L. At the opposite, southeast side of the surveyed area the dominating species included only diatoms, particularly *Thalassiosira aestivales*. Chlorophyll *a* concentration did not exceed 2 µg/L.

In the first 10 days of May 1972 the bloom of spring diatoms was found on the shelf break front (Taniguchi et al. 1976). Numbers at some depths reached 2.4×10^6 cells/L and the concentration of chlorophyll *a* was 3.2 µg/L. *T. hyalina, T. nordenskioldii,* and *Fragilaria islandica* dominated. In 6 weeks in the same area the numbers decreased to 0.03 times the previous numbers, while the chlorophyll *a* dropped to 0.25 times the previous concentration. The dominating species was *Thalassiosira decipiens.*

In the first 10 days of June 1993 over 85% of total phytoplankton numbers were made up by *Phaeocystis pouchetii* at the shelf break front and in the adjacent areas of the deep basin (Flint et al. 1994). The highest numbers of this species (2×10^5 cells/L) were observed at a deepwater station near the shelf break. In early June 1994 maximum phytoplankton numbers were also observed in the deep basin beyond the shelf break front (Flint et al. 1996). Small-size *Thalassiosira* species dominated by numbers and biomass. *Phaeocystis pouchetii* made up ~15% from the total numbers of phytoplankton. Of the biomass, 25% were made up of *Rhizosolenia alata* and *R. fragilissima.* At the shelf break front the lowest numbers and biomass of phytoplankton were observed (Fig. 5).

In July 1993 a different taxonomic composition of phytoplankton was found with maximum numbers and biomass located in the deep-basin area close to the shelf break front. *Pseudonitzschia delicatissima* and *Cylindotheca closterium* were dominant in cell numbers, while *Corethron criophilum* dominated in biomass with up to 320 µg/L. In the shelf break area to the south of St. George Island in 1994, *P. pouchetii* dominated, making up 80-90% of the total numbers. This population was probably at the end of the bloom based on low fluorescence of cells, destruction of the colonies, and numerous mucous chucks with dead cells (Flint et al. 1996).

It is noteworthy that in the dying colonies of *P. pouchetii* clusters of heterotrophic choanoflagellates, largely *Salpincoeca natans,* were found. This hints the role that *P. pouchetii* may play in the pelagic trophic webs.

References

Aikawa, H. 1932. On the summer plankton in the waters of the west Aleutian Islands in 1928. Bulletin of the Faculty of Fisheries, Hokkaido University 6:191-200.

Allen, W.E. 1927. Surface catches of marine diatoms and dinoflagellates made by U.S.S. *Pioneer* in Alaskan waters in 1923. Bulletin of the Scripps Institution of Oceanography Tech. Ser. 1(4):39-48.

Allen, W.E. 1929. Surface catches of marine diatoms and dinoflagellates made by U.S.S. *Pioneer* in Alaskan waters in 1924. Bulletin of the Scripps Institution of Oceanography Tech. Ser. 2:139-153.

Alexander, V., and T. Cooney. 1979. A quantitative study of the phytoplankton from the eastern Bering Sea. In: Environmental assessment of the Alaska continental shelf. NOAA/OCSEAP Annual Report.

Arzhanova, N.V., V.L. Zubarevich, and V.V. Sapozhnikov. 1995 . Seasonal variability of nutrient stocks in the euphotic zone and assessment of primary production in the Bering Sea. In: B.N. Kotenev and V.V. Sapozhnikov (eds.), Complex studies of the Bering Sea ecosystem. VNIRO, Moscow, pp. 162-179. (In Russian.)

Azova, N.V. 1964. Primary productivity of the Pribilof-Bristol area of the Bering Sea. In: P.A. Moiseev (ed.), Soviet fisheries investigation in the northeastern Pacific, Part III. Pishchevaya Promyshlennost Publishing, Moscow, VNIRO Proceedings 53 and TINRO Proceedings 52:149-154. (In Russian.)

Caron, D. 1983. Technique for enumeration of heterotrophic and phototrophic nanoplankton, using epifluorescence microscopy, and comparison with other procedures. Applied Environmental Microbiology 46:491-498.

Coachman, L.K. 1986. Circulation, water masses, and fluxes on the southeastern Bering Sea Shelf. Continental Shelf Research 5:23-108.

Coachman, L.K. 1993. On the flow field in the Chirikov Basin. Continental Shelf Research 13:481-509.

Coachman, L.K., and R.L. Charnell. 1979. On lateral water mass interaction—a case study: Bristol Bay, Alaska. Journal of Physical Oceanography 9:278-297.

Coachman, L.K., and V. Shigaev. 1988. Northern Bering-Chukchi Sea ecosystem: The physical basis. In: Results of the third joint U.S.-U.S.S.R. Bering and Chukchi seas expedition (BERPAC), pp. 17-27.

Cooney, R.T. 1981. Bering Sea zooplankton and micronekton communities with emphasis on annual production. In: D.W. Hood and J.A. Calder (eds.), The eastern Bering Sea shelf: Oceanography and resources, Vol. 2. Published by the Office of Marine Pollution Assessment, NOAA and BLM. Distributed by the University of Washington Press, Seattle, WA, pp. 947-974.

Flint, M.V., A.V. Drits, M.V. Emelianov, A.J. Kopylov, I. Merculieff, P.V. Rybnikov, I.N. Sukhanova, and T.E. Whitledge. 1996. Investigations of the Pribilof marine ecosystem, third year: Significance of oceanographic and biological processes at the shelf break, outer shelf domain, and middle shelf front for biological productivity of the Pribilof region. P.P. Shirshov Institute of Oceanology, Moscow. 407 pp.

Flint, M.V., A.V. Drits, M.V. Emelianov, A.J. Kopylov, I. Merculieff, S.G. Pojarkov, P.V. Rybnikov, I.N. Sukhanova, and T.E. Whitledge. 1994. Investigations of the Pribilof marine ecosystem, second year: Study of the marine ecosystem in the vicinity of the Pribilof Islands for general ecological evaluation and elucidation of the most valuable ecological zones. P.P. Shirshov Institute of Oceanology, Moscow. 504 pp.

Goering, J.J., and R.L. Iverson. 1978. Primary production and phytoplankton composition of the southeast Bering Sea. In: C.P. McRoy (ed.), Process and Resources of the Bering Sea Shelf (PROBES), Annual Report, pp. 203-240.

Goering, J.J., and R.L. Iverson. 1981. Phytoplankton distribution on the southeastern Bering Sea shelf. In: D.W. Hood and J.A. Calder (eds.), The eastern Bering Sea shelf: Oceanography and resources, Vol. 2. Published by the Office of Marine Pollution Assessment, NOAA and BLM. Distributed by the University of Washington Press, Seattle, WA, pp. 933-946.

Hansell, D.A., T.E. Whitledge, and J.J. Goering. 1993. Patterns of nitrate utilization and production over the Bering-Chukchi shelf. Continental Shelf Research 13:601-629.

Hansell, D.A., J.J. Goering, J.J. Walsh, C.P. McRoy, L.K. Coachman, and T.E. Whitledge. 1989. Summer phytoplankton production and transport along the shelf break in the Bering Sea. Continental Shelf Research 9:1085-1104.

Horner, R., and V. Alexander. 1972. Algal populations in Arctic sea ice: An investigation of heterotrophy. Limnology and Oceanography 17:454-458.

Ivanenkov, V.N. 1964. Hydrochemistry of the Bering Sea. Nauka Publishing, Moscow. 137 pp. (In Russian.)

Iverson, R.L., T.E. Whitledge, and J.J. Goering. 1979. Chlorophyll and nitrate fine structure in the southeastern Bering Sea shelf break front. Nature, London 281:664-666.

Karohji, K. 1958. Diatom standing crops and the major constituents of the populations as observed by net sampling. 4: Report from the *Oshoro Maru* on oceanographic and biological investigations in the Bering Sea and northern North Pacific in the summer of 1955. Bulletin of the Faculty of Fisheries, Hokkaido University 9:243-252.

Karohji, K. 1959. Diatom associations as observed by underway sampling. 6: Report from the *Oshoro Maru* on oceanographic and biological investigations in the Bering Sea and northern North Pacific in the summer of 1955. Bulletin of the Faculty of Fisheries, Hokkaido University 9:259-267.

Kawarada, Y. 1957. A contribution of microplankton observations to the hydrography of the northern North Pacific and adjacent seas. 2: Plankton diatoms in the Bering Sea in summer of 1955. Journal of the Oceanographic Society of Japan 13:151-155.

Kawarada, Y., and M. Ohwada. 1957. A contribution of microplankton observations to the hydrography of the northern North Pacific and adjacent seas. 1: Observations in the western North Pacific and Aleutian waters during the period from April to July 1954. Oceanogr. Mag. 14:149-158.

Kiselev, I.A. 1937. Composition and distribution of the phytoplankton in the northern Bering Sea and in southern Chukchi Sea. Investigations of the Seas of the U.S.S.R. 25:217-245. (In Russian.)

Konishi, R., and M. Saito. 1974. The relationship between ice and weather conditions in the Bering Sea. In: D.W. Hood and E.J. Kelley (eds.), Oceanography of the Bering Sea with emphasis on renewable resources. Occasional Publication No. 2, Institute of Marine Science, University of Alaska, Fairbanks, pp. 425-451.

Kotenev, B.N. 1995. Water dynamics as the major factor of the long-term variability of marine bioproductivity and reproduction of fishery resources in the Bering Sea. In: B.N. Kotenev and V.V. Sapozhnikov (eds.), Complex studies of the Bering Sea ecosystem. VNIRO, Moscow, p. 739. (In Russian.)

McRoy, C.P., and J.J. Goering. 1974. The influence of ice on the primary productivity of the Bering Sea. In: D.W. Hood and E.J. Kelley (eds.), Oceanography of the Bering Sea with emphasis on renewable resources. Occasional Publication No. 2, Institute of Marine Science, University of Alaska, Fairbanks, pp. 403-421.

McRoy, C.P., J.J. Goering, and W. Shiels. 1972. Studies in primary productivity in the eastern Bering Sea. In: A.Y. Takenouti (ed.), Biological oceanography of the northern North Pacific Ocean. Motida Commemorative Volume. Idemitsu Shoten, Tokyo, pp. 199-216.

Mordasova, N.V., M.P. Metreveli, and M.V. Venttsel. 1995. Phytoplankton enzymes in the western Bering Sea. In: B.N. Kotenev and V.V. Sapozhnikov (eds.), Complex studies of the Bering Sea ecosystem. VNIRO, Moscow, pp. 256-264. (In Russian.)

Motoda, S., and T. Minoda. 1974. Plankton of the Bering Sea. In: D.W. Hood and E.J. Kelley (eds.), Oceanography of the Bering Sea with emphasis on renewable resources. Occasional Publication No. 2, Institute of Marine Science, University of Alaska, Fairbanks, pp. 207-410.

Niebauer, H.J., V. Alexander, and R.T. Cooney. 1981. Primary production at the eastern Bering Sea ice edge: The physical and biological regimes. In: D.W. Hood and J.A. Calder (eds.), The eastern Bering Sea shelf oceanography and resources, Vol. 2. Published by the Office of Marine Pollution Assessment, NOAA and BLM. Distributed by the University of Washington Press, Seattle, WA, pp. 763-772.

Ohwada, M., and H. Kon. 1963. A microplankton survey as a contribution to the hydrography of the North Pacific and adjacent seas. 2: Distribution of the microplankton and their relation to the character of water masses in the Bering Sea and northern North Pacific Ocean in the summer of 1960. Oceanogr. Mag. 14:87-99.

Parsons, T.R., and R.J. LeBrasseur. 1970. The availability of food to different trophic levels in the marine food chain. In: J.H. Steele (ed.), Marine food chains, pp. 325-343.

Phifer, L.D. 1934. The occurrence and distribution of planktonic diatoms in the Bering Sea and Bering Strait, July 26-August 24, 1934. Rep. Oceanogr. Cruise U.S. Coast Guard Cutter *Chelan,* 1934, Part II(a), pp. 1-44.

Saito, K., and A. Taniguchi. 1978. Phytoplankton communities in the Bering Sea and adjacent seas, II. Spring and summer communities in seasonally ice-covered areas. Astarte 11:27-35.

Sambrotto, R.N., H.J. Niebauer, J.J. Goering, and R.L. Iverson. 1986. Relationships among vertical mixing, nitrate uptake, and phytoplankton growth during the spring bloom in the southeast Bering Sea middle shelf. Continental Shelf Research 5:161-199.

Sapozhnikov, V.V. 1995. Mesoscale anticyclonic eddies at the shelf break and their impact on the formation of hydrochemical structure of the Bering Sea. In: B.N. Kotenev and V.V. Sapozhnikov (eds.), Complex studies of the Bering Sea ecosystem. VNIRO, Moscow, pp. 149-155. (In Russian.)

Sapozhnikov, V.V., V.L. Zubarevich, and N.V. Mordasova. 1995a. General trends of the distribution of organic and ammonium forms of nitrogen and urea in the Bering Sea. In: B.N. Kotenev and V.V. Sapozhnikov (eds.), Complex studies of the Bering Sea ecosystem. VNIRO, Moscow, pp. 134-149. (In Russian.)

Sapozhnikov, V.V., N.V. Arzhanova, A.K. Gruzevich, V.L. Zubarevich, M.A. Karpushin, N.V. Mordasova, and D.O. Tolmachev. 1995b. Hydrochemistry of the west Bering Sea. In: B.N. Kotenev and V.V. Sapozhnikov (eds.), Complex studies of the Bering Sea ecosystem. VNIRO, Moscow, pp. 96-134. (In Russian.)

Semina, H.J. 1953. Phytoplankton in the western Bering Sea. Ph.D. dissertation, P.P. Shirshov Institute of Oceanography, Academy of Sciences USSR, Moscow. 189 pp. (In Russian.)

Semina, H.J. 1955. About two zonal groups of phytoplankton (an example of the Bering Sea). Proceedings of the Academy of Sciences, U.S.S.R. 101:363-366. (In Russian.)

Semina, H.J. 1981. Specific composition of phytoplankton in the western part of the Bering Sea and adjacent part of the Pacific Ocean: Diatoms. In: Ecology of marine phytoplankton. Acad. of Sci. USSR, P.P. Shirshov Inst. of Ocean., pp. 6-32. (In Russian.)

Semina, H.J. 1997. An outline of the geographical distribution of oceanic phytoplankton. Advances in Marine Biology 32:527-563.

Sorokin, Y.J. 1995. Primary production in the Bering Sea. In: B.N. Kotenev and V.V. Sapozhnikov (eds.), Complex studies of the Bering Sea ecosystem. VNIRO, Moscow, pp. 264-276. (In Russian.)

Springer, A.M., C.P. McRoy, and M.V. Flint. 1996. The Bering Sea Green Belt: Shelf age processes and ecosystem production. Fisheries Oceanography 5(3/4):205-223.

Starodubtsev, E.G. 1970. Seasonal changes of primary production in the southeastern Bering Sea. In: Soviet fisheries investigations in the north-eastern part of the Pacific Ocean. Part V. VNIRO Proceedings Vol. 70, TINRO Proceedings Vol. 72, Pishevaya Promyshlennost Publishing, Moscow, pp. 93-97. (In Russian.)

Sukhanova, I.N. 1983. Concentrating of phytoplankton in a sample. In: Contemporary methods of quantitative estimation of marine plankton distribution. Nauka Publishing, Moscow, pp. 97-104. (In Russian.)

Taniguchi, A. 1969. Regional variations of surface primary production in the Bering Sea in summer and the vertical stability of water affecting the production. Bulletin of the Faculty of Fisheries, Hokkaido University 20:169-179.

Taniguchi, A., K. Saito, A. Koyama, and M. Fukuchi. 1976. Phytoplankton communities in the Bering Sea and adjacent seas. I: Communities in early warming season in southern areas. Journal of the Oceanographic Society of Japan 32:99-106.

UNESCO. 1978. Phytoplankton manual. Monograph on oceanographic methods. 272 pp.

Venttsel, M.V. 1991. Phytoplankton of the open and shelf areas of the Bering Sea. Oceanology 31:252-258. (In Russian.)

Venttsel, M.V. 1994. Plankton phytocenosises of the oceanic and shelf areas of the Bering Sea. Ph.D. thesis. VNIRO, Moscow. 106 pp. (In Russian.)

Venttsel, M.V., A.S. Michaelyan, and E.N. Kokurkina. 1995. Phytocenosis biomass and variability in the Bering Sea. In: B.N. Kotenev and V.V. Sapozhnikov (eds.), Complex studies of the Bering Sea ecosystem. VNIRO, Moscow, pp. 305-311. (In Russian.)

Verkhunov, A.V. 1995. The role of hydrological and hydrochemical processes in the formation of bioproductivity of the Bering Sea shelf. In: B.N. Kotenev and V.V. Sapozhnikov (eds.), Complex studies of the Bering Sea ecosystem. VNIRO Publishing, Moscow, pp. 39-52.

Whitledge, T.E., W.S. Reeburgh, and J.J. Walsh. 1986. Seasonal inorganic nitrogen distributions and dynamics in the southeastern Bering Sea. Continental Shelf Research 5:109-132.

Walsh, J.J., and C.P. McRoy. 1986. Ecosystem analysis in the southeastern Bering Sea. Continental Shelf Research 5:259-288.

CHAPTER **23**

Distribution and Ecology of Mesopelagic Fishes and Cephalopods

Elizabeth H. Sinclair
Alaska Fisheries Science Center, Seattle, Washington

Andrey A. Balanov
Institute of Marine Biology, Russian Academy of Sciences of the Far East, Vladivostok, Russia

Tsunemi Kubodera
National Science Museum Tokyo, Tokyo, Japan

Vladimir I. Radchenko and Yury A. Fedorets
Pacific Research Institute of Fisheries and Oceanography (TINRO), Vladivostok, Russia

Abstract

The distribution and ecology of mesopelagic fauna are presented based on collections made by Russian, Japanese, and U.S. researchers in the western and eastern Bering Sea between 1979 and 1991. The findings from these three independent data sets represent the most extensive description of mesopelagic fish and cephalopod distribution across the Bering Sea to date, and further demonstrate the importance of mesopelagic nekton in the ecology of the Bering Sea.

Based on this and previous studies, 61 species of mesopelagic fishes in 57 genera and 38 families are known to inhabit the Bering Sea. The Myctophidae (lanternfishes) and Bathylagidae (deepsea smelts) are consistently the most highly represented mesopelagic fish families in trawl surveys and studies on predator diets across the Bering Sea. Bering Sea mesopelagic cephalopods include 18 species in 10 genera and seven families. The squid families Gonatidae and Cranchiidae appear in greatest numbers in trawl surveys and predator diets across the Bering Sea. Patterns of distribution and abundance of the most common mesopelagic nekton concur with records from the northern North Pacific Ocean.

Introduction

Until recently there have been very limited data available on the biomass, distribution, and life histories of mesopelagic fauna (animals inhabiting depths of 200-1,000 m by day, but often migrating into the upper 200 m at night) in the Bering Sea. Historically, mesopelagic nekton are not well studied because they are patchy in distribution and some species of cephalopods actively evade small nets. Thus, with some notable exceptions (Fedorets 1986, Bizikov and Arkhipkin 1996), mesopelagic nekton lack direct commercial value. As a result, basic information is still lacking on the taxonomic structure of the mesopelagic zone and species-specific distributional data.

In order to address these questions, Russian researchers initiated a series of surveys on the mesopelagic fauna of the western Bering Sea in the late 1980s. The volume of data collected in those surveys is substantial, and includes 17 species of mesopelagic fishes and two species of cephalopods never before recorded in the Bering Sea. Some aspects of this extensive database have been reported previously (Radchenko 1992; Balanov 1994, 1995), but much is reported or summarized here for the first time.

Japanese scientists have conducted research on squid systematics, abundance, and distribution in subarctic waters for many years (Kubodera and Jefferts 1984, Okutani et al. 1988). However, samples were collected with conventional micronekton trawls that targeted larval stages of squid and focused primarily on the western and northern North Pacific. A description of juvenile squid collected in 1988 from the eastern Bering Sea, first reported in Japanese (Kubodera 1993) is translated into English here and presented along with new data obtained in the same area in 1989.

U.S. researchers have opportunistically collected mesopelagic fishes from the eastern and central Bering Sea during groundfish studies. As bycatch, the relative abundance values of mesopelagic fishes in these collections are very low, but indicate distribution trends. The distribution of Myctophidae (lanternfishes) and Bathylagidae (deepsea smelts) collected from this region since 1979 are here reported for the first time.

The lack of commercial interest in mesopelagic organisms contrasts sharply with their ecological importance. From larval through adult stages, mesopelagic fishes and squids are integral to the trophic balance of the world oceans. Mesopelagic fishes are among the primary consumers of zooplankton (Tyler and Pearcy 1975, Balanov 1995) and squids (Balanov and Gorbatenko 1995). Mesopelagic squids are primary consumers of zooplankton, larval and juvenile fishes such as walleye pollock *(Theragra chalcogramma)* (Naito et al. 1977, Kuznetsova and Fedorets 1987), and mesopelagic fishes (Naito et al. 1977, Sinclair 1991). As strong vertical migrators, the nekton of the mesopelagic zone play a substantial role in the transport and redistribution of organic matter from the rich surface to the oligotrophic waters of the ocean depths (Parin 1968, Vinogradov 1968, Hopkins and Baird 1977, Willis and Pearcy 1982, Robison 1984).

Mesopelagic nekton form over 90% of the seasonal prey base of some marine mammals (Betesheva and Akimushkin 1955, Okutani and Nemoto 1964, Wada 1971, Pitcher 1981, Stroud et al. 1981, Fiscus 1982, Kajimura and Loughlin 1988, Antonelis et al. 1997) and are among the primary prey of seabirds (Ogi and Tsujita 1977, Ogi et al. 1980, Ogi and Hamanaka 1982, Sanger 1988, Decker et al. 1996, Hunt et al. 1996) and commercially important fishes (Ito 1964, LeBrasseur 1966, Takeuchi 1972, Yang and Livingston 1988, Pearcy 1992, Lang and Livingston 1996). Our limited understanding of the structure of the mesopelagic community and its importance as a prey base contributes to our inability to define many of the key factors involved in recent changes in the Bering Sea ecosystem, including the dramatic population declines of some marine mammals and birds.

Methods

The distribution and abundance data presented here were derived from a directed study of mesopelagic fishes and cephalopods conducted in the western Bering Sea in 1989 and 1990 (Data set I), mesopelagic cephalopods in the eastern Bering Sea in 1988 and 1989 (Data set II), and opportunistic collections of mesopelagic fishes in the eastern Bering Sea between 1979 and 1991 (Data set III). Collection methods and analyses of each data set are described separately below. Nomenclature follows Nelson (1994) for fishes and Nesis (1987) for cephalopods.

Data set I

Mesopelagic fishes and cephalopods were collected by researchers of the Pacific Research Institute of Fisheries and Oceanography, Laboratory of Applied Biocoenology, during three western Bering Sea cruises in 1989 (May-July) and 1990 (April-June and October-November). Sampling methodology was designed to target midwater nekton. A total of 490 trawls were conducted with a non-closing pelagic rope trawl (vertical mouth opening 50-60 m; horizontal mouth opening 70 m) with a 10-15 m codend liner of 10 mm mesh. All measures of mesh size are for stretched mesh. At each station (Fig. 1) a 1-hr oblique tow was conducted within the upper mesopelagic (200-500 m) and lower mesopelagic (500-1,000 m) zones. Average trawling speed was approximately 3 kts. Vessel speed decreased and net retrieval speed increased to minimize bycatch between layers.

In 490 trawls, 70,900 fishes were identified to species and weighed (kg). Biomass (B) was determined by catch per unit volume:

$$B = \frac{Vqt}{Kv}$$

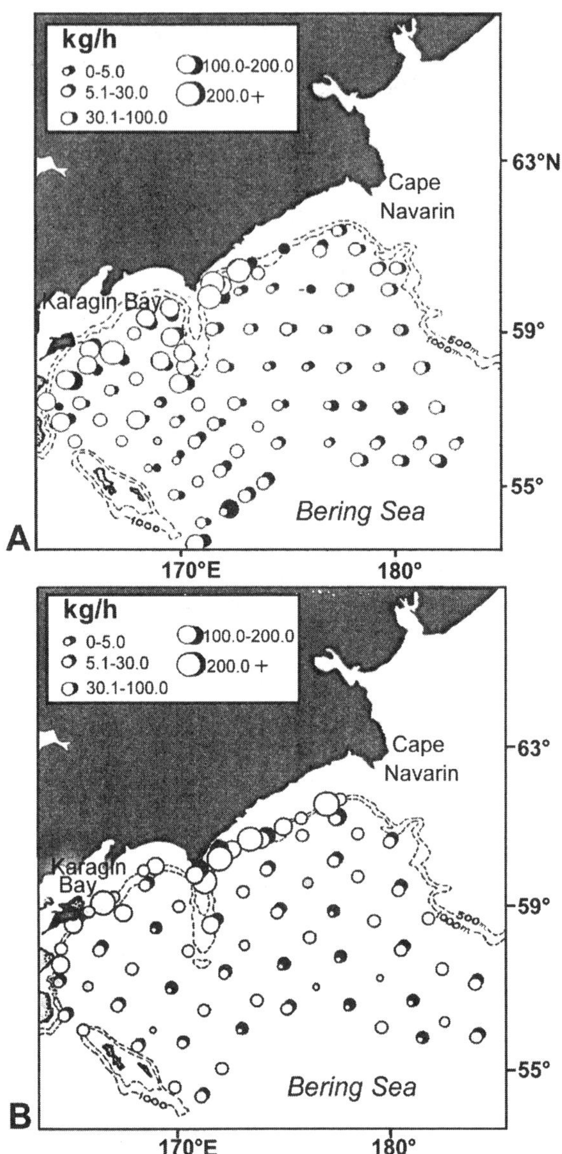

Figure 1. *Study area and distribution of catches of mesopelagic fishes in the Bering Sea in June 1989 (A); and April-May 1990 (B). Open symbols represent the upper mesopelagic zone. Filled symbols represent the lower mesopelagic zone.*

where *V* is the total volume of the sampled layer; *v*, the volume of water filtered per trawling; *t*, trawling time (1 h); *q*, average catch; and *K*, net efficiency coefficient. Net efficiency coefficients were calculated as 0.1 for mesopelagic fishes and nektonic squids, and 0.3 for planktonic squids (Balanov and Ilyinsky 1992, Radchenko 1992).

A Sorenson-Czekanowski Similarity Index (Andreev 1980) was used to calculate the percent similarity between mesopelagic fish species composition in the western Bering Sea and other oceanic regions of the northern North Pacific. This index is a variation of the Czekanowski Mean Character Difference formula (Nei 1987):

$$Cz = \frac{1}{m} \sum_{i=1}^{m} |xi - yi|$$

where *x* and *y* denote score numbers (1 for presence, 0 for absence of a species) at the *i*th oceanic location and *m* denotes the total number of locations compared.

Data set II

Cephalopods were collected in the eastern Bering Sea in August-October 1988 and 1989. Collections were made by researchers of the National Research Institute of Far Seas Fisheries during Japanese and U.S. joint research on juvenile pollock. A large midwater trawl with two otter boards and a 43 × 34 m mouth opening was used to sample 41 stations off the eastern continental shelf in 1988 (Fig. 2). The mesh size ranged 210-90 mm from the mouth and the codend was lined with 3 mm mesh. The net was towed several hours after sunset for 30 min at 4 kts and depths of 25-70 m. In 1989, trawls were conducted at 38 stations and the research area expanded to include areas over the eastern continental shelf (Fig. 2). The mouth size of the trawl net used in 1989 (20 m × 30 m) was slightly smaller than that used in 1988, but codend mesh size and towing operations were the same.

Data set III

Mesopelagic fishes were collected opportunistically in the eastern Bering Sea by researchers of the U.S. National Marine Fisheries Service during surveys for pollock. Groundfish surveys were conducted along the eastern Bering Sea continental slope, between late May and late August, in 1979, 1981, 1982, 1988, and 1991. Sampling grids were consistent between years with some variation in sampling stations.

An eastern otter trawl (type 83-112) has been used in groundfish surveys on the continental slope since 1982 (Karp and Walters 1994). It is a non-closing bottom trawl with 102 mm-89 mm mesh in the body and a 32 mm codend liner. The doors measure 2.7 × 1.8 m in height with a dandyline length of 54.9 m. While fishing, the trawl measured 12-20 m width and 2-2.3 m vertical height. The headrope length is 25.3 m and

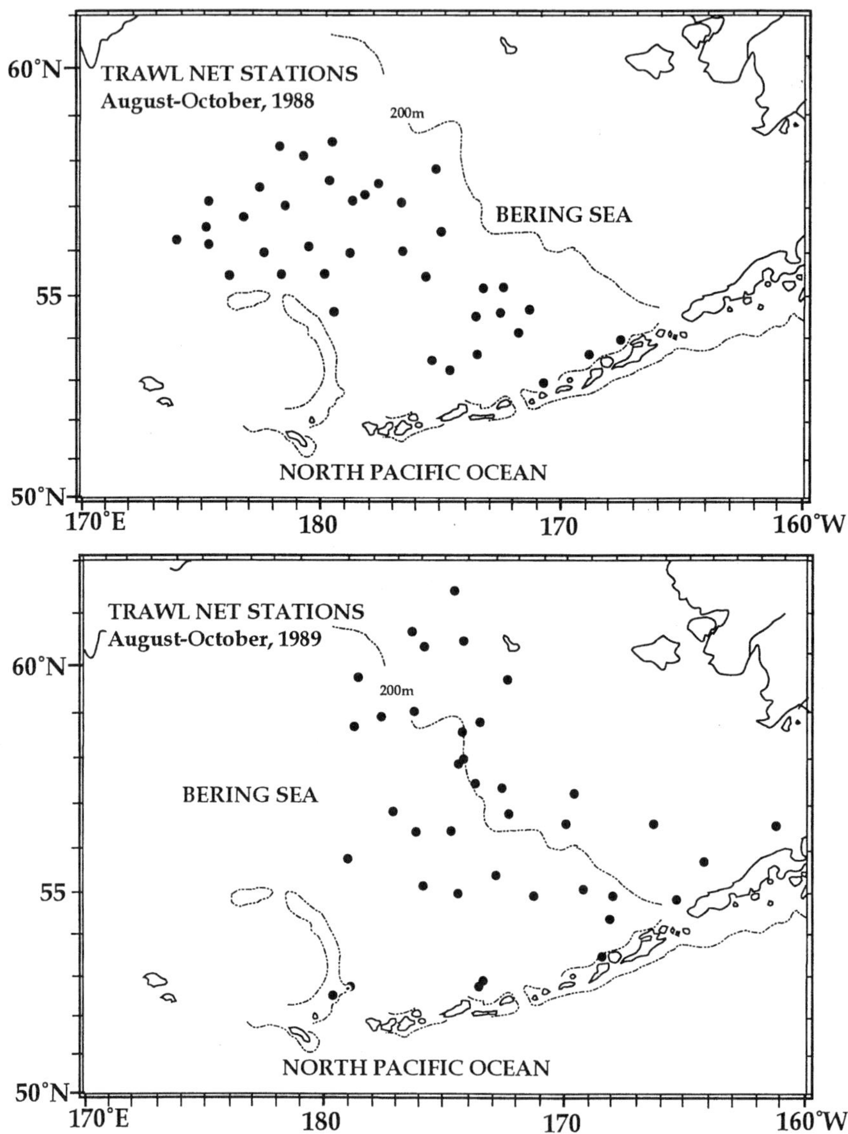

Figure 2.   Midwater trawl stations for eastern Bering Sea cephalopod studies in 1988 and 1989.

footrope length 34.1 m. The trawl was fished on the bottom for an average of 30 minutes at approximately 3 kts. Prior to 1982 a similar but smaller net (400 mesh eastern trawl) of the same mesh size was used.

These surveys were designed primarily to document distribution and biomass levels of groundfishes, particularly pollock. Consequently, documentation of the numbers and identity of mesopelagic fishes varied between cruises. Therefore, we have limited our analyses to the relative distribution and biomass of the two predominant families of mesopelagic fishes identified. Mean catch per unit effort ($\overline{CPUE}$) by species was standardized as follows (Smith and Bakkala 1982):

$$\overline{CPUE}_{ik} = \frac{\sum_{j=1}^{n_i} CPUE_{ijk}}{n_i}$$

where $\overline{CPUE}_{ik}$ is the mean catch per area trawled (kg/ha) for species k for the jth station in the ith strata, and n_i is the number of successfully trawled stations.

Results and Discussion

Fish Species Composition, Distribution, and Abundance

Based on this and other studies (Pearcy et al. 1979b, Willis 1984, Willis et al. 1988, Balanov 1992, Balanov and Ilyinsky 1992, Balanov 1995), the fishes resident in the western Bering Sea mesopelagic zone total 61 species in 55 genera and 33 families (Table 1). Of these, 17 species, 15 genera, and 6 families were recorded for the first time from the intensive collections of the 1980s (Balanov 1992). In this study, approximately 40% of the species recorded in the western Bering Sea occurred in as few as 1-3% of trawls (Table 1) and many families are represented by a single species. Seven species from just three families (Myctophidae, Bathylagidae, and Chauliodontidae) composed over 91% of the fish biomass. These were *Stenobrachius leucopsarus, S. nannochir, Bathylagus schmidti, B. ochotensis, B. milleri, B. pacificus,* and *Chauliodus macouni.* The greatest number of species, and the most common by frequency and biomass occurred among the Myctophidae (eight species) and Bathylagidae (four species). *Stenobrachius leucopsarus* occurred in 100% of the trawls and composed approximately 64% of the biomass of all fishes caught in the western Bering Sea. The relatively low representation of so many species is characteristic of a rich ecosystem. As part of the permanent complex of the mesopelagic fauna, species occurring frequently in low numbers are potentially important as predators and prey. It emphasizes the requirement that large numbers of trawls must be conducted in order to adequately characterize the community.

Table 1. **Fish species composition and percent frequency of occurrence in the upper (200-500 m) and lower (500-1,000 m) mesopelagic zones of the western Bering Sea from 1989 and 1990.**

| | Percent frequency | |
Taxon	200-500 m	500-1,000 m
Petromyzontidae		
Lampetra tridentata	22.0	10.8
Lethenteron japonica	0.7	1.2
Squalidae		
Somniosus pacificus	4.0	4.8
Nemichthyidae		
Avocettina infans	2.0	6.0
Microstomatidae		
Nansenia candida	2.0	6.0
Bathylagidae		
Bathylagus pacificus	33.3	100.0
Bathylagus milleri	42.0	97.6
Bathylagus ochotensis	69.0	97.6
Bathylagus schmidti	99.3	–
Opisthoproctidae		
Dolichopteryx sp.	99.3	98.8
Macropinna microstoma	4.7	60.2
Platytroctidae		
Holtbyrnia innesi	3.3	15.7
Maulisia argipalla	–	0.5
Maulisia acuticeps	–	0.5
Sagamichthys abei	7.3	0.6
Gonostomatidae		
Cyclothone atraria	2.0	75.9
Gonostoma gracile	16.0	8.4
Sternoptychidae		
Argyropelecus sladeni	0.7	–
Chauliodontidae		
Chauliodus macouni	98.0	95.0
Melanostomiidae		
Tactostoma macropus	2.7	–
Pachystomias microdon	0.7	0.5

Table 1.   (Continued.)

| | Percent frequency | |
Taxon	200-500 m	500-1,000 m
Malacosteidae		
Aristostomias scintillans	*	–
Malacosteus niger	0.7	–
Scopelarchidae		
Benthalbella dentata	34.0	80.7
Notosudidae		
Scopelosaurus adleri	27.3	67.5
Scopelosaurus harryi	25.3	27.7
Paralepididae		
Magnisudis atlantica	–	0.5
Arctozenus rissoi	13.3	4.8
Lestidiops ringens	2.0	–
Anotopteridae		
Anotopterus pharao	0.7	–
Alepisauridae		
Alepisaurus ferox	0.7	–
Neoscopelidae		
Scopelengys tristis	0.7	1.2
Myctophidae		
Diaphus theta	38.0	19.3
Lampanyctus jordani	19.0	54.2
Lampanyctus regalis	22.0	84.3
Lampanyctus ritteri	–	*
Protomyctophum thompsoni	20.3	15.7
Stenobrachius leucopsarus	100.0	100.0
Stenobrachius nannochir	60.7	100.0
Tarletonbeania crenularis	*	–
Moridae		
Halargyreus johnsonii	–	19.3
Laemonema longipes	0.7	0.5
Gadidae		
Theragra chalcogramma	85.3	73.5
Macrouridae		
Albatrossia pectoralis	24.0	77.1
Coryphaenoides cinereus	0.7	70.0

* Unpublished records from V.V. Fedorov, Pacific Research Institute of Fisheries and Oceanography (TINRO), Vladivostok, Russia.

Table 1.   (Continued.)

Taxon	Percent frequency	
	200-500 m	500-1,000 m
Bythitidae		
Thalassobathia pelagica	–	0.5
Ceratiidae		
Ceratias holboelli	–	0.5
Oneirodidae		
Oneirodes bulbosus	2.0	36.1
Oneirodes thompsoni	0.5	42.0
Bertella idiomorpha	–	0.5
Melamphaidae		
Melamphaes lugubris	19.3	45.8
Poromitra crassiceps	2.0	53.0
Cetomimidae		
Gyrinomimus sp.	–	*
Psychrolutidae		
Malacocottus sp.	4.0	4.8
Cyclopteridae		
Aptocyclus ventricosus	75.3	74.7
Liparidae		
Nectoliparis pelagicus	–	0.5
Paraliparis sp.	–	1.2
Caristiidae		
Caristius sp.	0.7	–
Chiasmodontidae		
Kali indica	*	–
Icosteidae		
Icosteus aenigmaticus	0.7	2.4
Pleuronectidae		
Reinhardtius hippoglossoides	0.7	–
Number of species	48	53

* Unpublished records from V.V. Fedorov, Pacific Research Institute of Fisheries and Oceanography (TINRO), Vladivostok, Russia.

Table 2. **Sorensen-Czekanowski Similarity Indices (Andreev 1980) for mesopelagic nekton occuring in frequencies greater than 10% from trawl surveys in regions of the Bering Sea and North Pacific Ocean (N=51).**

	1	2	3	4
2	0.71			
3	0.59	0.67		
4	0.55	0.65	0.84	
5	0.52	0.62	0.84	0.87

1 = Eastern Transition Zone; 2 = Western Transition Zone; 3 = Alaska Gyre; 4 = Western Subarctic Gyre; 5 = Bering Sea (see Willis et al. 1988 for boundary descriptions).

With the exception of *Thalassobathia pelagica,* all new records for the western Bering Sea (Balanov 1992) have been identified in the North Pacific as well (Willis 1984, Willis et al. 1988). This supports past speculations of a lack of endemic mesopelagic fishes in the Bering Sea (Pearcy et al. 1979b). It remains to be seen whether the species found in the Bering Sea are members of a reproductively viable population or expatriates from the northern North Pacific. The species composition of the western Bering Sea most closely resembles regions of the western Subarctic Gyre as defined by Willis (1984) and Willis et al. (1988). When upper and lower mesopelagic zones were considered together, percent similarity values between the western Bering Sea and the western Subarctic Gyre were 87% while values ranged between 52% and 84% with other zoogeographic regions of the North Pacific (Table 2). Differences in species composition between the western Bering Sea and more southerly zoogeographic regions of the North Pacific, may be related to reproductive and spatial distribution patterns of some species. The Bering Sea is characterized by a subarctic water mass structure (Arseniev 1967, Ohtani 1973) in which seasonal temperature and salinity changes within the pelagic region occur primarily within the epipelagic zone (0-200 m) (Favorite et al. 1976, Pearcy et al. 1979b). Eggs and larvae of many midwater fishes develop in the epipelagic and upper mesopelagic zones of more southerly areas of the North Pacific, presumably dying off if they enter or are advected into the colder Bering Sea. By contrast, juvenile and adult fishes of the same species residing principally in the lower mesopelagic zone are exposed to more stable environmental conditions throughout the North Pacific into the

Bering Sea, and are consequently more common to both regions. Hence, an affinity for the cold and stable temperatures of the lower mesopelagic zone, and relatively long life span (6-10 years) among many midwater fish species, probably contribute to the similarity in their distribution patterns throughout the North Pacific and Bering Sea.

Overall, the frequency of occurrence and number of fish species was greater in the lower mesopelagic zone than the upper mesopelagic. Only eight species occurred in the upper mesopelagic at a frequency of 50% or higher, compared with 18 species in the lower mesopelagic (Table 1). By family, the Myctophidae made up 89% and Bathylagidae 8% of wet biomass of fishes in the upper mesopelagic zone. In the lower mesopelagic zone, the Myctophidae made up 63% and Bathylagidae 24% of total fish biomass. *Stenobrachius leucopsarus* dominated the wet biomass of fishes collected in both zones, but was most prevalent in the upper mesopelagic (86%). In contrast, the remaining six species that composed most of the biomass were more prevalent in the lower mesopelagic. *Stenobrachius leucopsarus* is more common in the upper mesopelagic than the lower mesopelagic zone of the North Pacific as well (Frost and McCrone 1979, Pearcy et al. 1979b, Parin and Fedorov 1981). It migrates vertically into the epipelagic area at night with variability in the extent of migration depending upon body size (Willis and Pearcy 1980). *Bathylagus schmidti* was the only bathylagid more frequently found in the upper mesopelagic (Table 1). Biomass values for *B. schmidti* were slightly higher in the lower mesopelagic zone relative to the upper mesopelagic, again probably because larger fish (adults) tend to be found deeper in the water column (Willis and Pearcy 1980). This illustrates a potential bias in our data which does not account for fish length. Also, the hour of collection was averaged over day and night. Thus, sampling the upper mesopelagic at night may result in low biomass values for species that are diel migrants into the epipelagic. Potential biases from these two factors may be minimized by the large sample sizes considered here.

Earlier studies have concluded that mesopelagic fishes are primarily zooplanktivorous and feed on the most seasonally abundant species (Gjosaeter 1973, Tyler and Pearcy 1975, Hopkins and Baird 1977, Pearcy et al. 1979a, Cailliet and Ebeling 1990). The dietary analyses of fishes collected during these studies (Balanov et al. 1994, Balanov 1995) indicate that like other studies, copepods and euphausiids were the main prey items of the Myctophidae, while the Bathylagidae fed mainly on coelenterates, appendicularians, ctenophores, pteropods, and heteropods (Adams 1979, Pearcy et al. 1979a, Balanov et al. 1994, Balanov 1995). For *Stenobrachius leucopsarus,* the consumption of copepods decreased and euphausiids increased between summer and winter, possibly related to the seasonal vertical migration patterns of calanoid copepods in the North Pacific (Vinogradov 1968, Marlowe and Miller 1975).

Across all regions in the Bering Sea, the greatest concentration of myctophid and bathylagid fishes occurred near the continental slope, underwater elevations, and canyons (Figs. 1 and 3). In the upper mesopelagic zone of the western Bering Sea, maximum catches (500 kg/h) were along the continental slope between Cape Olyutorsk and Cape Navarin, but decreased to 30-50 kg/hr in open water regions. Catches were also relatively large in areas adjoining Shirshov Ridge (Fig. 1). This may be caused by higher productivity near land margins and ridges, as well as a concentration of fishes by prevailing currents.

A similar pattern was apparent in eastern Bering Sea collections. Estimated densities of mesopelagic fishes were low in the eastern Bering Sea slope survey data due to the use of large trawls designed for groundfish assessment. Nonetheless, persistent areas of mesopelagic fish concentrations were apparent between years. The southern extreme of the survey consistently had the highest concentrations of fishes in all five sample years, illustrated here by data collected in 1988 and 1991 (Fig. 3). This portion of the sample region skirts the Bering Sea Canyon, which, in addition to its great depths, is influenced by three strong current systems (Aleutian North Slope Current, Bering Slope Current, and the Alaskan Stream). The Bering Slope Current is a consistent feature in the area. It is very high in upwelled nutrients and rife with eddies helping to create what is considered one of the most physically dynamic regions in the eastern Bering Sea (Stabeno et al., chapter 1, this volume).

Cephalopod Species Composition, Distribution, and Abundance

Eighteen species of mesopelagic squids have been identified in the Bering Sea, 11 of which belong to the family Gonatidae (Table 3). The Gonatidae are also predominant members of the cephalopod community in the northern North Pacific based on larval and adult surveys (Bublitz 1980, Kubodera and Jefferts 1984, Nesis 1987, Okutani et al. 1988). The most frequently occurring squids (>50% frequency of occurrence) in the western Bering Sea were members of the families Cranchiidae *(Galiteuthis phyllura* and *Belonella borealis)* and Gonatidae *(Gonatopsis borealis, G. octopedatus,* and *Gonatus middendorffi). Japetella diaphana* and *Histioteuthis dofleini* were identified for the first time in the Bering Sea from the 1990 survey data (Radchenko 1992).

Similarly to midwater fishes, squids occurred overall more frequently in the lower than in the upper mesopelagic zone (Table 3). Also, the greatest concentrations were found near the outer continental shelf and slope. Over one-quarter of the mesopelagic squid biomass in the western Bering Sea was found in these boundary regions. Biomass values were dominated by *Berryteuthis magister* (42%) and *Gonatopsis borealis* (18%) in the upper layer and *Belonella borealis* (27%) in the lower layer of the western Bering

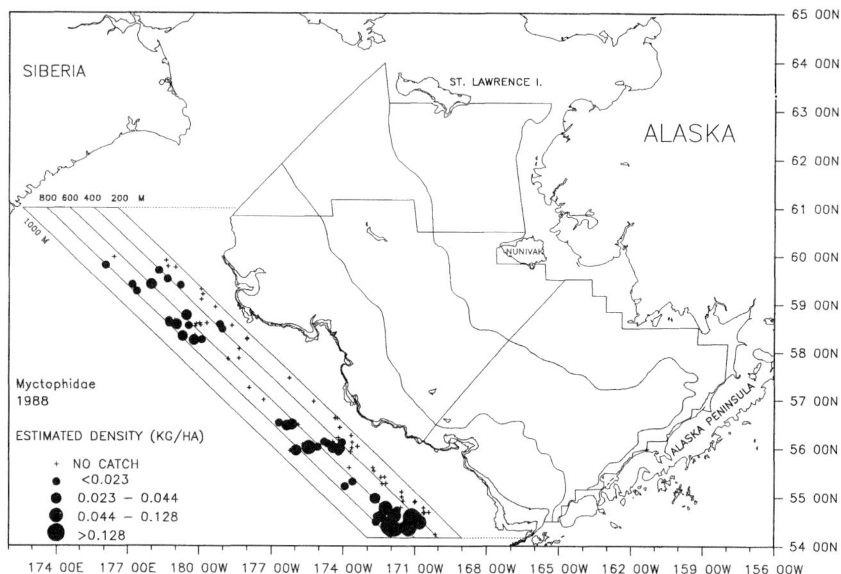

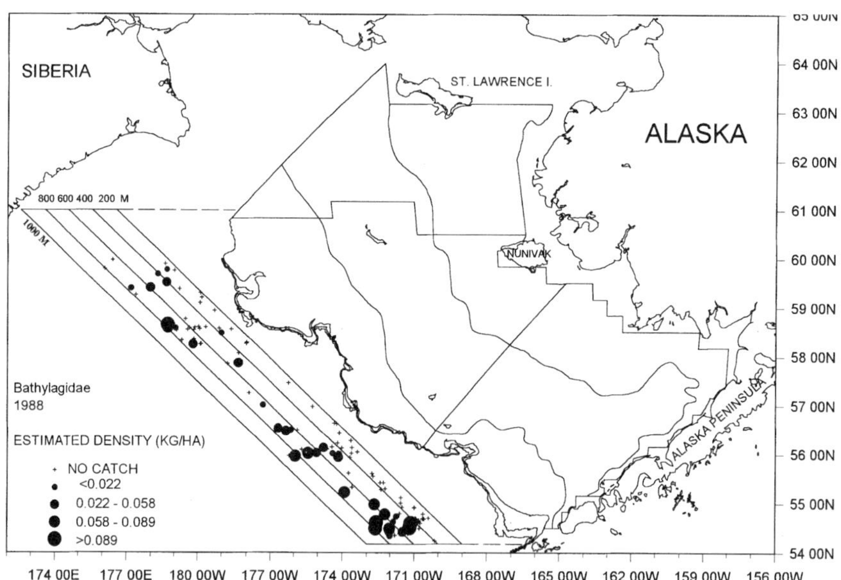

Figure 3.   Catch-per-unit-effort of the Myctophidae (lanternfishes) and Bathylagidae (deepsea smelts) in the eastern Bering Sea continental slope groundfish surveys. Sample years 1988 and 1991 are shown as representatives of all five sampling years (1979, 1981, 1982, 1988, and 1991).

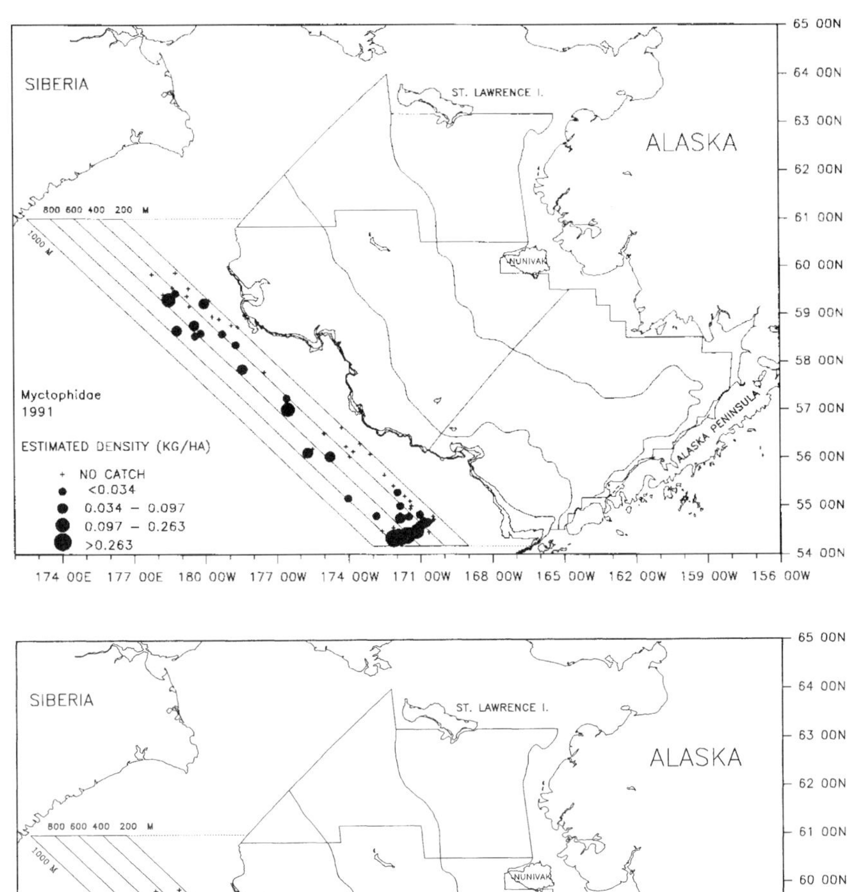

Figure 3. (Continued.)

Table 3. **Cephalopod species composition and percent frequency of occurrence in the upper (200-500 m) and lower (500-1,000 m) mesopelagic zones of the western Bering Sea from samples collected in 1989 and 1990.**

	Percent frequency	
Taxon	200-500 m	500-1,000 m
Gonatidae		
Berryteuthis magister	28.0	13.6
Gonatus (Eogonatus) tinro	0.7	1.2
Gonatus onyx	16.0	26.1
Gonatus madokai	5.3	17.0
Gonatus pyros	22.7	12.5
Gonatus berryi	15.0	15.9
Gonatus middendorffi	53.0	18.2
Gonatopsis okutanii	3.3	12.5
Gonatopsis octopedatus	24.0	70.5
Gonatopsis borealis	84.0	81.8
Gonatopsis makko	3.3	22.7
Cranchiidae		
Galiteuthis phyllura	64.7	61.4
Belonella borealis	50.0	93.2
Onychoteuthidae		
Moroteuthis robusta	1.3	1.1
Bolitaenidae		
Japetella diaphana	8.0	23.7
Chiroteuthidae		
Chiroteuthis calyx	1.3	5.7
Histioteuthidae		
Histioteuthis dofleini	–	1.1
Architeuthidae		
Architeuthis spp.	–	*
Number of species	17	18

* Unpublished occurrence records contributed by K.N. Nesis, Shirshov Institute of Oceanology, Russian Academy of Sciences, Moscow, Russia.

Table 4. Cephalopod species composition from samples collected in the eastern Bering Sea in 1988 (n = 41) and 1989 (n = 38).

Species	1988			1989		
	Number (N)	% Frequency of Occurrence (%FOO)	N/FOO	Number (N)	% Frequency of Occurrence (%FOO)	N/FOO
Gonatidae						
Gonatopsis borealis	297	41.5	17.5	1350	57.9	61.36
Gonatopsis type A	29	4.9	14.5	2	5.3	1.00
Gonatus berryi	7	7.3	2.3	0	0.0	0.00
Gonatus onyx	57	14.6	9.5	90	39.5	6.00
Gonatus pyros	95	9.8	23.8	1	2.6	1.00
Gonatus madokai	105	39.0	6.6	48	34.2	3.69
Gonatus middendorffi	114	53.7	5.2	429	57.9	19.50
Gonatus type A	12	7.3	4.0	0	0.0	0.00
Gonatus ursabrunae	0	0.0	0.0	1	2.6	1.00
Gonatus spp.	500	0.0	0.0	261	0.0	0.00
Gonatus (Eogonatus) tinro	7	26.8	0.6	205	31.6	17.08
Berryteuthis magister	5	2.4	5.0	92	15.8	15.33
Berryteuthis anonychus	64	4.9	32.0	699	47.4	38.83
Cranchiidae						
Galiteuthis phyllura	3	4.9	1.5	0	0.0	0.00
Taonius sp.	0	0.0	0.0	1	2.6	1.00
Octopodidae						
Octopus (Paroctopus) dofleini	2	4.9	1.0	2	5.3	1.00
Total	1297	87.8	36.0	3181	84.2	99.41

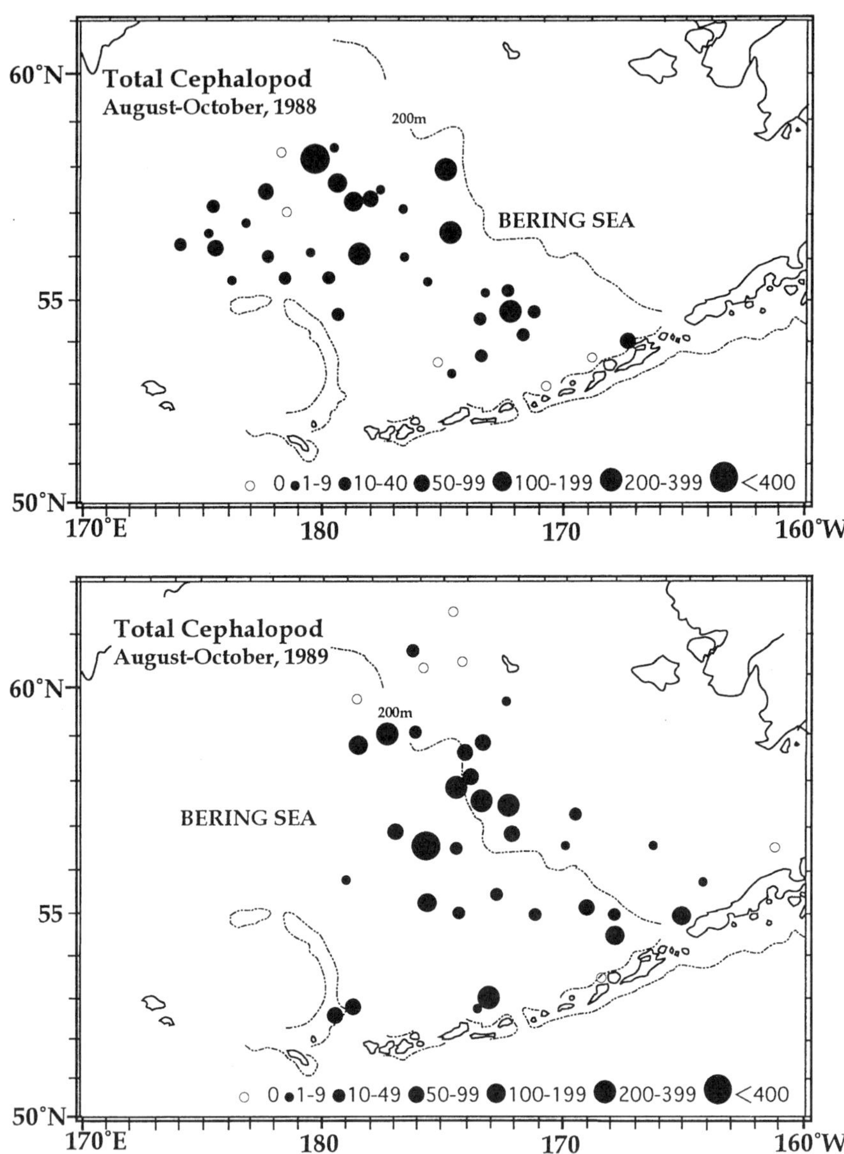

Figure 4.   Distribution and relative abundance (individuals per tow) of cephalopods collected during eastern Bering Sea cephalopod studies in 1988 and 1989.

Sea mesopelagic zone. As with analyses of the fish data, the cephalopod data may be biased because they were averaged for the time of day. Some species of cephalopods also move deeper into the water column as they mature, potentially inflating relative measures of importance based on biomass values at depth.

In the eastern Bering Sea, the Gonatidae and Cranchiidae were again the two most common families. The Gonatidae dominated squid catches by numbers and frequency of occurrence (Table 4). *Gonatopsis borealis* was the most numerous cephalopod in both 1988 and 1989; however, *Gonatus middendorffi* and *G. madokai* occurred in frequencies similar to *Gonatopsis borealis.* The frequency of squids caught was similar between years, but the number of individuals caught per haul in 1989 was nearly triple that in 1988 (Table 4). Also, the catches of some species (i.e., *Gonatus pyros* and *Berryteuthis* spp.) in the eastern Bering Sea were markedly variable between years, probably related to sampling positions which included the continental shelf and shelf edge in 1989 (Table 4, Fig. 4).

As an index of patchiness for each species, the number of squid caught at each station of occurrence was calculated (Table 4) demonstrating that although the frequency of occurrence of some squid species may be low, the number of individuals caught when the species is encountered may be quite high. Similar patterns of patchiness are apparent in the diet composition of apex predators in the Bering Sea. For example, when juvenile squids and juvenile or small fishes that occur infrequently in northern fur seal *(Callorhinus ursinus)* diet do occur in gastrointestinal and scat remains, they are often found in great numbers (Sinclair et al. 1994, Antonelis et al. 1997). Very similar patterns in the distribution and abundance of mesopelagic fishes and cephalopods are indicated by midwater trawls run concurrently with captures of northern fur seals for gastrointestinal analysis (Sinclair et al. 1994). This returns to the premise that infrequently occurring nekton, as well as those that occur frequently in low densities, are integral as permanent members of the ecosystem. It is only with directed, large scale sampling efforts of the type presented here that we will continue to characterize the fauna of the mesopelagic zone and more fully define their critical role as both predator and prey.

Acknowledgments

The authors extend special thanks to G.E. Walters and P. Goddard for compiling the eastern Bering Sea database records; J. Little and K. Zecca for graphical support; W.G. Pearcy, W.A. Walker, P. Browne, T.W. Pietsch, M. Busby, R. Brodeur, and G.J. Duker for reviews; shipboard personnel of the vessels *Gissar, Professor Soldatov, Mlechniy Put, Novokotovsk,* and *Miller Freeman;* and to the many unnamed researchers and technicians without whom this publication would not have been possible.

References

Adams, A.E. 1979. The food habits, age and growth of three midwater fishes *(Stenobrachius leucopsarus, S. nannochir* and *Leuroglossus schmidti)* from the southeastern Bering Sea. M.S. thesis, University of Alaska, Fairbanks. 302 pp.

Andreev, V.L. 1980. Classification reconstructions in topology and systematics. Nauka, Moscow. 142 pp. (In Russian.)

Antonelis, G.A., E.S. Sinclair, R.R. Ream, and B.W. Robson. 1997. Inter-island variation in the diet of female northern fur seals *(Callorhinus ursinus)* in the Bering Sea. Journal of Zoology, London. 242:435-451.

Arsenev, V.S. 1967. Currents and water masses of the Bering Sea. Nauka, Moscow. 135 pp.

Balanov, A.A. 1992. New records of deep-sea fishes from the pelagic zone of the Bering Sea. Journal of Ichthyology (Moscow) 32(4):151-154. (In Russian.)

Balanov, A.A. 1994. Feeding of dominant mesopelagic fishes in the Bering Sea. Journal of Ichthyology (Moscow) 34(6):73-82.

Balanov, A.A. 1995. Composition and structure of mesopelagic nekton community of the Bering Sea. Ph.D. thesis, Institute of Marine Biology, Russian Academy of Sciences of the Far East, Vladivostok, Russia. 24 pp.

Balanov, A.A., and K.M. Gorbatenko. 1995. On the feeding of the abundant species of mesopelagic fishes in the Bering Sea in autumn. In: B.N. Kotenev and V.V. Sapozhnikov (eds.), Complex studies of the Bering Sea ecosystem. VNIRO, Moscow, pp. 344-349.

Balanov, A.A., and E.N. Ilyinsky. 1992. Species composition and biomass of mesopelagic fishes in the Sea of Okhotsk and the Bering Sea. Journal of Ichthyology 32(1):56-63. (Transl. from Russian to English by Scripta Technica, Inc.)

Balanov, A.A., K.M. Gorbatenko, and T.A. Gorelova. 1994. Daily dynamics of feeding of mesopelagic fishes of the Bering Sea. Journal of Ichthyology (Moscow) 34(4):534-541.

Betesheva, Y.I., and I.I. Akimushkin. 1955. The feeding habits of the sperm whale *(Physeter catodon* L.) in the region of the Kuril chain. Trudy Instituta Okeanologii 18:86-94. (In Russian.)

Bizikov, B.A., and A.I. Arkhipkin. 1996. Distribution, stock structure and fishery outlooks of the Commander squid. Rybnoe Khozyaistvo 1:42-45. (In Russian.)

Bublitz, C. 1980. Systematics of the cephalopod family Gonatidae from the southeastern Bering Sea. M.S. thesis, University of Alaska, Fairbanks. 171 pp.

Cailliet, G.M., and A.W. Ebeling. 1990. The vertical distribution and feeding habits of two common midwater fishes *(Leuroglossus stilbius* and *Stenobrachius leucopsarus)* off Santa Barbara. California Cooperative Oceanic and Fisheries Investigative Reports 31:106-123.

Decker, M.B., G.L. Hunt Jr., and G.V. Byrd Jr. 1996. The relationship between sea surface temperature, the abundance of juvenile walleye pollock *(Theragra chalcogramma),* and the reproductive performance and diets of seabirds at the Pribilof Islands, southeastern Bering Sea. In: R.J. Beamish (ed.), Climate change and northern fish populations. Canadian Special Publication of Fisheries and Aquatic Sciences 121, pp. 425-437.

Favorite, F., A.J. Dodimead, and K. Nasu. 1976. Oceanography of the subarctic Pacific region, 1960-1971. International North Pacific Fisheries Commission Bulletin No. 33. 187 pp.

Fedorets, Y.A. 1986. Biology and stock value of squid *Berryteuthis magister* (Gonatidae) near Commander Islands. In: Resources and fishery perspectives of squids of the world ocean. VNIRO, Moscow, pp. 57-66. (In Russian.)

Fiscus, C.H. 1982. Predation by marine mammals on squids of the eastern North Pacific Ocean and the Bering Sea. Marine Fisheries Review 44:1-10.

Frost, B.W., and L.E. McCrone. 1979. Vertical distribution, diel vertical migration, and abundance of some mesopelagic fishes in the eastern subarctic Pacific Ocean in summer. Fishery Bulletin, U.S. 76(4):751-770.

Gjosaeter, J. 1973. The food of the myctophid fish, *Bentosema glaciale* (Reinhardt), from western Norway. Sarsia 52:53-58.

Hopkins, T.H., and R.C. Baird. 1977. Aspects of the feeding ecology of midwater fishes. In: N.R. Andersen and B.J. Zahuranec (eds.), Oceanic sound scattering prediction. Plenum Press, New York, pp. 325-360.

Hunt, G.L., Jr., M.B. Decker, and A.S. Kitaysky. 1996. Fluctuations in the Bering Sea ecosystem as reflected in the reproductive ecology and diets of kittiwakes on the Pribilof Islands, 1975 to 1990. In: S. Greenstreet and M. Tasker (eds.), Aquatic predators and their prey. Blackwell, London, pp. 142-153.

Ito, J. 1964. Food and feeding habit of Pacific salmon (genus *Oncorhynchus*) in their oceanic life. Bulletin of the Hokkaido Regional Fisheries Research Lab 29:85-97.

Kajimura, H., and T.R. Loughlin. 1988. Marine mammals in the oceanic food web of the eastern subarctic Pacific. In: T. Nemoto and W.G. Pearcy (eds.), The biology of the subarctic Pacific. Bulletin of the Ocean Research Institute, University of Tokyo, No. 26 (Part II):87-223.

Karp, W.A., and G.E. Walters. 1994. Survey assessment of semi-pelagic Gadoids: The example of walleye pollock, *Theragra chalcogramma,* in the eastern Bering Sea. Marine Fisheries Review 56(1):8-22.

Kubodera, T. 1993. Bering Sea cephalopods collected by mid-water trawl (1988): A short report. Japan Sea Fisheries Research Laboratory No. 28:115-124. (In Japanese.)

Kubodera, T., and K. Jefferts. 1984. Distribution and abundance of the early life stages of squid, primarily Gonatidae (Cephalopoda, Oegopsida), in the northern North Pacific. Bulletin of Natural Science Museum Series A (Zool.) (Tokyo) 10(4).

Kuznetsova, N.A., and Y.A. Fedorets. 1987. On feeding of Commander squid *Berryteuthis magister.* Biol. Morya 1:71-73. (In Russian.)

Lang, G.M., and P.A. Livingston. 1996. Food habits of key groundfish species in the eastern Bering Sea slope region. NOAA Technical Memo NMFS-AFSC-67. 111 pp.

LeBrasseur, R.J. 1966. Stomach contents of salmon and steelhead trout in the northeastern Pacific Ocean. Journal of Fisheries Research Board (Canada) 23:85-100.

Marlowe, C.J., and C.B. Miller. 1975. Patterns of vertical distribution and migrations of zooplankton at Ocean Station "P." Limnology and Oceanography 20:824-844.

Naito, M., K. Murakami, and T. Kobayashi. 1977. Growth and food habit of oceanic squids *(Ommastrephes bartrami, Onychoteuthis borealijaponicus, Berryteuthis magister,* and *Gonatopsis borealis)* in the western subarctic Pacific region. In: Research Institute of North Pacific Fisheries, Faculty of Fisheries, Hokkaido University, Special Volume, Hakodate, Japan, pp. 339-351.

Nei, M. 1987. Molecular evolutionary genetics. Columbia University Press, New York.

Nelson, J.S. 1994. Fishes of the world, 3rd edn. John Wiley. 600 pp.

Nesis, K.N. 1987. Cephalopods of the world. T.F.H. Publications, New Jersey. 351 pp.

Ogi, H., and T. Hamanaka. 1982. The feeding ecology of *Uria lomvia* in the northwestern Bering Sea region. Journal of Yamashina Institute for Ornithology 14:270-280.

Ogi, H., and T. Tsujita. 1977. Food and feeding habits of common murre and thick-billed murre in the Okhotsk Sea in summer, 1972 and 1973. In: Research Institute of North Pacific Fisheries, Faculty of Fisheries, Hokkaido University, Special Volume 52, Hakodate, Japan, pp. 459-517.

Ogi, H., T. Kubodera, and K. Nakamura. 1980. The pelagic feeding ecology of the short tailed shearwater, *Puffinus tenuirostris,* in the subarctic Pacific region. Journal of Yamashina Institute for Ornithology 12:157-182.

Ohtani, K. 1973. Oceanographic structure in the Bering Sea. Memoirs of the Faculty of Fisheries, Hokkaido University 21(1):65-106.

Okutani, T., and T. Nemoto. 1964. Squids as the food of sperm whales in the Bering Sea and Alaskan Gulf. Scientific Report of the Whales Research Institute, Tokyo 18:111-122.

Okutani, T., T. Kubodera, and K. Jefferts. 1988. Diversity, distribution, and ecology of gonatid squids in the subarctic Pacific: A review. Bulletin of the Ocean Research Institute, University of Tokyo (26):159-192.

Parin, N.V. 1968. Fishes of the epipelagic zone of the ocean. Nauka, Moscow. 186 pp. (In Russian.)

Parin, N.V., and V.V. Fedorov. 1981. Comparison of the midwater fish faunas of the western and eastern North Pacific. In: N.G. Vingradova (ed.), Biology of the Pacific Ocean depths. Far Eastern Science Center, Academia Nauk, Vladivostok, pp. 72-78. (In Russian.)

Pearcy, W.G. 1992. Ocean ecology of North Pacific salmonids. University of Washington Sea Grant, Seattle, WA. 179 pp.

Pearcy, W.G., H.V. Lorz, and W. Peterson. 1979a. Comparison of the feeding habits of migratory and non-migratory *Stenobrachius leucopsarus* (Myctophidae). Marine Biology 51:1-8.

Pearcy, W.G., T. Nemoto, and M. Okiyama. 1979b. Mesopelagic fishes of the Bering Sea and adjacent northern North Pacific Ocean. Journal of the Oceanographical Society (Japan) 35(3,4):127-135.

Pitcher, K.W. 1981. Prey of the Steller sea lion, *Eumetopias jubatus,* in the Gulf of Alaska. Fishery Bulletin, U.S. 79:467-471.

Radchenko, V.I. 1992. The role of squids in the pelagic ecosystem of the Bering Sea. Oceanology 32(6):1093-1101. (In Russian.)

Robison, B.H. 1984. Herbivory by the myctophid fish *Ceratoscopelus warmingii.* Marine Biology 84:119-123.

Sanger, G.A. 1988. Review of the distribution and feeding ecology of seabirds in the oceanic subarctic North Pacific Ocean. Bulletin of the Ocean Research Institute, University of Tokyo, No. 26 (Part II):161-186.

Sinclair, E.H. 1991. Review of the biology and distribution of the neon flying squid *(Ommastrephes bartrami)* in the North Pacific Ocean. In: J.A. Wetherall (ed.), Biology, oceanography, and fisheries of the North Pacific transition zone and subarctic frontal zone. NOAA Technical Report, NMFS 105, pp. 57-68.

Sinclair, E.H., T.R. Loughlin, and W.G. Pearcy. 1994. Prey selection by northern fur seals *(Callorhinus ursinus)* in the eastern Bering Sea. Fishery Bulletin, U.S. 92:144-156.

Smith, G.B., and R.G. Bakkala. 1982. Demersal fish resources of the eastern Bering Sea: Spring 1976. NOAA Technical Report, NMFS-SSRF-754. 129 pp.

Stroud, R.K., C.J. Fiscus, and H. Kajimura. 1981. Food of the Pacific white sided dolphin, *Lagenorhynchus obliquidens,* Dall's porpoise, *Phocoenoides dalli,* and northern fur seals, *Callorhinus ursinus,* off California and Washington. Fishery Bulletin, U.S. 78:951-959.

Takeuchi, I. 1972. Food animals collected from the stomachs of three salmonid fishes *(Oncorhynchus)* and their distribution in the natural environments in the northern North Pacific. Bulletin Hokkaido Regional Fisheries Research Laboratory 38:1-119.

Tyler, H.R., and W.G. Pearcy. 1975. The feeding habits of three species of lanternfishes (family Myctophidae) off Oregon, USA. Marine Biology 32:7-11.

Vinogradov, M.E. 1968. Vertical distribution of the oceanic zooplankton. Instituta Okeanologii Akademii Nauk SSSR, Moscow. 339 pp. (Transl. from Russian by A. Mercado and J. Salkind, Isr. Progm. Sci. Transl., Jerusalem, 1970.)

Wada, K. 1971. Food and feeding habit of northern fur seals along the coast of Sanriku. Bulletin Tokai Regional Fisheries Research Lab 64:1-37.

Willis, J.M. 1984. Mesopelagic fish faunal regions of the northeast Pacific. Biological Oceanography 3(2):167-185.

Willis, J.M., and W.G. Pearcy. 1980. Spatial and temporal variations in the population size structure of three lanternfishes (Myctophidae) off Oregon, USA. Marine Biology 57:181-191.

Willis, J.M., and W.G. Pearcy. 1982. Vertical distribution and migration of fishes of the lower mesopelagic zone of Oregon. Marine Biology 70(1):87-98.

Willis, J.M., W.G. Pearcy, and N.V. Parin. 1988. Zoogeography of midwater fishes in the subarctic Pacific. Bulletin Ocean Research Institute, University of Tokyo 26(Part II):79-142.

Yang, M-S., and P.A. Livingston. 1988. Food habits and daily ration of Greenland halibut, *Reinhardtius hippoglossoides,* in the eastern Bering Sea. Fishery Bulletin, U.S. 86(4):675-690.

CHAPTER 24

Forage Fishes in the Bering Sea: Distribution, Species Associations, and Biomass Trends

Richard D. Brodeur, Matthew T. Wilson, and Gary E. Walters
Alaska Fisheries Science Center, Seattle, Washington

Igor V. Melnikov
Pacific Research Institute of Fisheries and Oceanography (TINRO), Vladivostok, Russia

Abstract

Relatively little is known about distribution patterns and species associations of forage fishes in the Bering Sea despite their importance as major prey of many higher trophic level organisms, such as seabirds, marine mammals, and predatory fishes. In this study, we examined survey data on some dominant pelagic forage fishes (Pacific herring *Clupea pallasi*, capelin *Mallotus villosus*, and eulachon *Thaleichthyes pacificus*) and the juvenile stages of major commercial groundfish species (walleye pollock *Theragra chalcogramma* and Pacific cod *Gadus macrocephalus*). We analyzed two main data sets: (1) a 1987 Russian survey that covered most of the Bering Sea, and (2) National Marine Fisheries Service (NMFS) summer surveys (1982-95) in the eastern Bering Sea which sampled the same grid of stations each year. In the Russian survey, age-0 pollock had the highest biomass and were the most widely distributed forage fish, although jellyfish and age-2+ pollock dominated the biomass overall. Several geographically distinct assemblages were recognized in both the eastern and western Bering Sea. Age-0 pollock were associated with warmer bottom temperatures and capelin with colder bottom temperatures, compared with other species. Distributions of all species from the NMFS surveys during a cold year (1986) were more widespread and overlap among species was greater than during a warm year (1987). Herring showed the most dramatic fluctuations in their biomass index over 14 years of NMFS trawl surveys and was the dominant forage fish caught in most years, although when their biomass index was low, they were exceeded by age-1 pollock, eulachon, and capelin.

Introduction

The Bering Sea shelf is highly productive and contains some of the largest populations of fishes, crabs, marine mammals, and seabirds in the world. Although more than 300 species of fish are known to inhabit this shelf, the 20 most abundant species account for more than 98% of the total abundance of survey catches (Bakkala 1993). Our knowledge of the small non-commercial pelagic fishes and juvenile stages of commercially important demersal fishes is meager, in spite of the fact that these species are an important food base for many of the higher trophic levels, including fishes, birds, and mammals (Minerals Management Service 1987, Springer 1992, Alaska Sea Grant 1997). The limited long-term dietary information on these top-level predators (Hunt et al. 1996, Sinclair et al. 1996) and survey data (Naumenko et al. 1990, Fritz et al. 1993, Naumenko 1996) that do exist suggest that the abundance and distribution of many of these forage fishes have undergone dramatic changes through time, which may have important implications for the apex predators which depend upon these resources (Springer 1992, Alaska Sea Grant 1997).

One of the important data gaps recognized by the National Research Council study of the Bering Sea ecosystem was that our understanding of forage species is inadequate and that there is almost no monitoring of the forage species upon which top predators rely (National Research Council 1996). Reliable estimates of how species biomass and distribution patterns change through time and how they are affected by environmental conditions are needed for effective management and as inputs to ecosystem modeling studies.

Despite the importance of juvenile walleye pollock *(Theragra chalcogramma)* and other forage fishes in the Bering Sea ecosystem (Springer 1992, Livingston 1993, Brodeur et al. 1996), relatively little is known about the large-scale distribution patterns of these fish in the Bering Sea. This situation contrasts sharply with that in the Gulf of Alaska where the large-scale distribution patterns of juvenile pollock and other small fishes are known (Walters et al. 1985, Hinckley et al. 1991, Bailey and Spring 1992, Brodeur et al. 1995, Brodeur and Wilson 1996, Wilson et al. 1996). Although these fishes are often captured incidentally in standard bottom trawl surveys such that an index of the relative size of the year class can be obtained (Walters 1989, Wyllie-Echeverria 1996, Hunt et al. 1996), directed surveys are often required to get more precise estimates of their abundance (Koehler et al. 1986, Wilson et al. 1996).

This paper reports results from late-summer Russian surveys conducted in 1987 of the entire Bering Sea and analyzes distribution and species associations of the dominant forage fishes. We also present data on incidental catches of forage fishes in the eastern Bering Sea from multi-year (1982-95) surveys carried out by U.S. fisheries scientists and analyze interannual variation in the distribution patterns, species associations, and trends in biomass in the eastern Bering Sea, focusing on two adjacent but highly contrasting years. Our objective was to determine how the geographic and interannual patterns of biomass distribution relate to variations in depth, temperature, and composition of co-occurring species. It is beyond the scope of this paper to present detailed information on the life history and ecology of the species examined, and the reader is referred to existing literature on forage fishes (e.g., papers in Minerals Management Service 1987; Alaska Sea Grant 1997).

Methods

Juvenile Fish Surveys

We examined data from two surveys designed to determine the distribution of age-0 pollock in the Bering Sea. The first cruise took place aboard the Russian Pacific Research Fisheries Centre (PFRC) research vessel (R/V) *Darwin* during August and September 1987. Scientists from the Alaska Fisheries Science Center (AFSC) participated on this cruise and trawl collections were processed using standard AFSC sampling methodology. The main fishing gear was a pelagic rope trawl (77.4/212) with small mesh (10 mm) inserts extending from the codend up 15 m into the trawl. Horizontal opening (HO) of the trawl was measured as 40 m and vertical opening (VO) was 20-25 m. Approximately 1.3×10^7 m^3 of water was filtered in a typical 1 hour tow. A few stations were sampled with a 108/528 pelagic trawl (HO = 65 m, VO = 48-55 m) and their catches were standardized relative to the main trawling gear based on differences in mouth area. Sampling was conducted on a predetermined grid of 149 stations over much of the eastern Bering Sea shelf and Aleutian Basin (Fig. 1) in a step-oblique fashion. A second cruise was conducted during August and September of 1987 aboard the R/V *Gnevny*. This survey consisted of 183 trawls carried out in the western Bering Sea shelf and basin using two different trawl types. The main gear used was the 108/528 trawl described previously, but at about one-third of the stations, a 118/620 trawl (HO = 75-80 m, VO = 60 m) was used.

The fish catch was identified to species and the invertebrate catch was classified only to major taxonomic categories (e.g., Cnidaria, Cephalopoda, Natantia). Pollock made up the majority of the fish biomass caught and was therefore subdivided by age group. Lengths were measured on a subsample of all pollock collected and individuals were assigned as age-0, age-1, and age-2+ based upon modes in the length-frequency distribution. The biomass caught from both gear types was standardized relative to the smaller trawl used on the R/V *Darwin* and converted to kg per hour trawled. Although the extensive coverage of the 1987 Russian surveys required two months to complete, no mortality or growth corrections were applied prior to making areal comparisons of fish density and size because of a lack of available information on these parameters for most species included here.

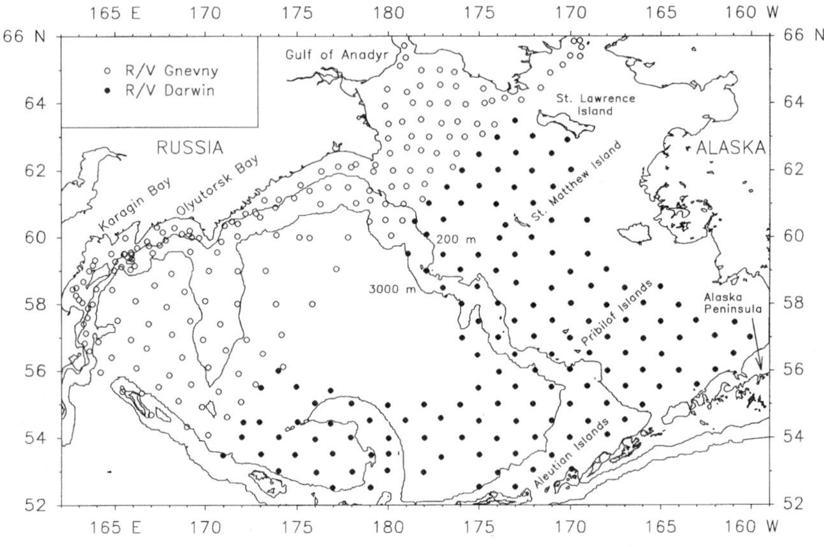

Figure 1.   Location of stations sampled during the R/V Gnevny (open circles) and R/V Darwin (closed circles) cruises during August and September 1987. Also shown are the 200 m and 3,000 m isobaths.

To associate the distributions of forage fishes to the biotic community, we used a polythetic clustering technique (Clifford and Stephenson 1975) to detect existing natural associations among species and among stations. Species clusters were used to provide some indication as to how similar the distributions of the species were, whereas the station clusters were useful in determining the spatial extent of "natural communities." Because rare taxa may adversely afftect the analyses, only those that occurred in >10% of the tows in each survey were included. Biomass estimates (kg/h) of the dominant species at each station was $\ln(x+1)$ transformed to reduce the sensitivity to large catches. From these log-transformed species-station matrices, dissimilarity indices were calculated using the QSK measure (Faith et al. 1987) and clusters were formed using a group-average fusion strategy (Clifford and Stephenson 1975). For each cruise, a station and species dendrogram was produced. Station groupings were made for various numbers of groups and these groups were plotted to find levels of dissimilarity that gave geographically cohesive groupings. Two-way tables (species × station groups) were constructed to show how species were distributed within and among the station groups. Indices of constancy (Boesch 1977) were calculated for each taxon to see how distinct each grouping was. Constancy is defined as the ratio of the occurrences in a group to the total possible number of occurrences and ranges from 0 (taxon not found at any stations) to 1 (found at all stations).

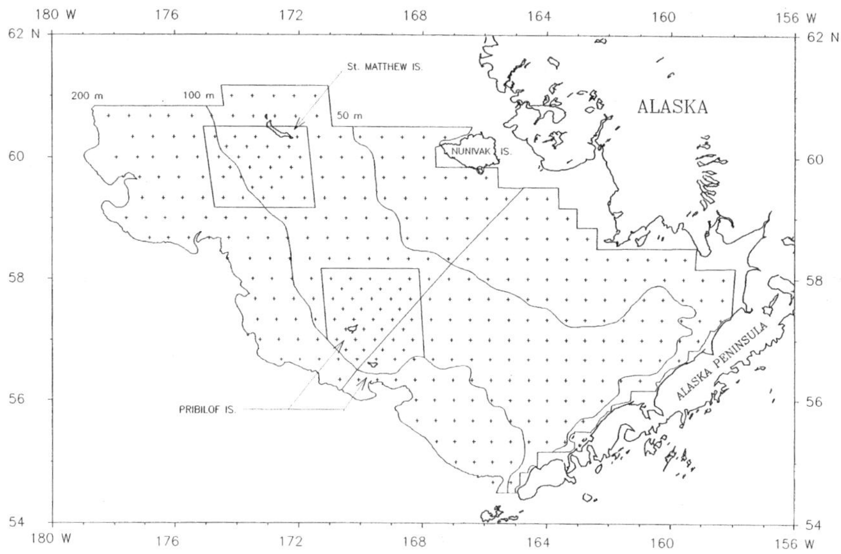

Figure 2. Station locations for the standard Bering Sea bottom trawl survey grid occupied each summer since 1982 and the six areas used to stratify the sampling area by depth and location (northwest or southeast). The smaller boxes denote areas of increased sampling density.

Finally, we examined bottom temperature and depth data for the R/V *Darwin* survey to determine how these variables relate to the observed fish distribution patterns. We statistically compared the cumulative distributions of select taxa and these variables following the methods described in Perry and Smith (1994) and Smith (1996). We compared the actual cumulative distribution of each taxon along with 1,000 randomizations of the data to the habitat available during each cruise.

Retrospective Analysis of Eastern Bering Sea Forage Fishes

We examined the catch of five species of forage fishes (age-1 pollock, age-1 Pacific cod *Gadus macrocephalus*, eulachon *Thaleichthyes pacificus*, Pacific herring *Clupea pallasi*, and capelin *Mallotus villosus*) collected during the annual AFSC summer surveys in the eastern Bering Sea (Bakkala 1993). Virtually the same systematic sampling grid has been occupied since 1982 and consists of approximately 355 stations spaced 37 km apart for a total survey area of 465,000 km² (Fig. 2). The same gear (83/112 eastern bottom trawl; 2.3 m VO; 32 mm codend liner), towing method (0.5 hour tows at 3 knots), and processing methods were used throughout the sampling period (Bakkala 1993). The surveys were generally conducted from June 1

Table 1. **Dominant taxa collected during R/V *Darwin* survey in the eastern Bering Sea in 1987 used in cluster analysis. Shown are the frequency of occurrence in 149 tows and the mean biomass collected per hour towed.**

Scientific name	Common name	Freq. occ.	Mean kg/h
Cnidaria	Jellyfish	139	44.2
Theragra chalcogramma (age 2+)	Walleye pollock	120	166.7
Theragra chalcogramma (age 0)	Walleye pollock	100	41.8
Oncorhynchus keta	Chum salmon	85	12.3
Zaprora silenus	Prowfish	84	0.3
Cephalopoda	Unidentified squid	67	1.5
Gadus macrocephalus	Pacific cod	48	2.1
Pleurogrammus monopterygius	Atka mackerel	45	4.2
Clupea pallasi	Pacific herring	40	1.6
Mallotus villosus	Capelin	35	0.6
Stenobrachius leucopsarus	Northern lampfish	32	0.9
Theragra chalcogramma (age 1)	Walleye pollock	28	25.2
Pleuronectes asper	Yellowfin sole	25	1.5
Leuroglossus schmidti	Northern smoothtongue	23	0.2
Blepsias bilobus	Crested sculpin	22	<0.1
Podothecus acipenserinus	Sturgeon poacher	20	0.1
Natantia	Unidentified shrimp	20	<0.1
Oncorhynchus tshawytscha	Chinook salmon	16	0.4

to August 5 on NOAA research vessels and chartered fishing vessels. Although the trawl survey was designed to assess the abundance and distribution of bottom fishes, some pelagic species were captured mainly during the deployment and retrieval of the net. Thus, the abundances of the pelagic schooling species (capelin, eulachon, and herring) can only be considered a relative index of the total population abundance.

We mapped the biomass distribution patterns (kg/ha) for each of the major forage fishes by year to look at long-term changes in spatial distribution. We were interested in examining differences in distributions for cold and warm years within a particular species. We chose to compare and contrast 1986 (cold year) and 1987 (warm year) which were quite different oceanographically. An added benefit to using the year 1987 is that we could then compare the AFSC distributions for that year with those of the R/V *Darwin*. We subsequently compared the catches of all five species for both years with the bottom depth and temperature at each station as described in the previous section to understand how these variables related to the observed distributions.

Table 2. **Dominant taxa collected during R/V *Gnevny* survey in the western Bering Sea in 1987 used in cluster analysis. Shown are the frequency of occurrence in 183 tows and the mean biomass caught per hour towed.**

Scientific name	Common name	Freq. occ.	Mean kg/h
Cnidaria	Jellyfish	150	3.7
Theragra chalcogramma (age 2+)	Walleye pollock	144	232.9
Oncorhynchus keta	Chum salmon	122	3.0
Theragra chalcogramma (age 0)	Walleye pollock	109	0.8
Cephalopoda	Unidentified squid	87	0.7
Oncorhynchus nerka	Sockeye salmon	68	0.4
Oncorhynchus tshawytscha	Chinook salmon	66	0.3
Mallotus villosus	Capelin	48	2.8
Clupea pallasi	Pacific herring	42	12.8
Aptocyclus ventricosus	Smooth lumpsucker	39	0.2
Theragra chalcogramma (age 1)	Walleye pollock	35	7.5
Lumpenus maculatus	Daubed shanny	31	<0.1
Stenobrachius leucopsarus	Northern lampfish	28	0.5
Eumicrotremus sp.	Unidentified lumpsucker	27	<0.1
Ammodytes hexapterus	Pacific sand lance	26	<0.1
Leuroglossus schmidti	Northern smoothtongue	25	0.4
Lampetra tridentata	Pacific lamprey	21	<0.1
Zaprora silenus	Prowfish	21	<0.1
Natantia	Unidentified shrimp	20	0.2
Reinhardtius hippoglossoides	Greenland halibut	19	<0.1
Blepsias bilobus	Crested sculpin	19	<0.1
Hippoglossoides elassodon	Flathead sole	18	<0.1
Hemilepidotus papilio	Butterfly sculpin	18	<0.1

Results

Forage Fish Distributions

Out of 62 distinct taxa/age groups identified during the R/V *Darwin* survey in the eastern Bering Sea, 18 occurred in >10% of the hauls and were included in the cluster analysis (Table 1). Of these, the three age groups of pollock and scyphozoan jellyfish comprised 91.4% of the total mean biomass. Because of their relatively high individual weight, age-2+ pollock comprised nearly 55% of the total biomass of these dominant species.

For the R/V *Gnevny* survey, 64 taxa/age groups were found and 23 were included in the clustering (Table 2). There was substantial overlap in the dominant groups in the two surveys in that 14 of the 18 commonly occurring groups in the R/V *Darwin* survey were also dominant in the R/V *Gnevny* survey. Age-2+ pollock dominated the catch in terms of weight (85.9% of total) and was second only to jellyfish in number of occurrences. Herring, the next most important species in weight, constituted only 4.7% of the total biomass.

The distribution of age-0 pollock biomass was skewed toward the eastern Bering Sea shelf region with highest concentrations found around the Pribilof Islands and along the Alaska Peninsula (Fig. 3a). In the western Bering Sea, high biomass of age-0 pollock was found only in Karagin Bay. Age-1 pollock were found mainly on the outer eastern Bering Sea shelf, almost entirely north of the Pribilof Islands, with high biomasses found in only a few isolated areas of the Gulf of Anadyr and Bay of Olyutorsk in the western Bering Sea (Fig. 3b). Herring biomass showed a similar distribution to age-1 pollock but extended farther south to the Aleutian Islands in the east and Karagin Bay in the west (Fig. 3c). Capelin were distributed along the same latitudinal range as herring but tended to be found in the inner shelf regions on both sides (Fig. 3d).

Species Associations

Three major species groupings were identified using the dissimilarity index at the 0.65 level in the eastern Bering Sea (Fig. 4a). Five taxa were associated with no others at this level (Table 3). Station Group 1 consisted of mainly open ocean pelagic and mesopelagic nekton (Species Group 3) but also showed high constancy and biomass relationships with age-2+ pollock, jellyfish, and prowfish *(Zaprora silenus)* (Table 3). Station Group 2 occurred entirely on the shelf and was dominated by all age-0 pollock and jellyfish. Station Group 3 was dominated in terms of constancy and biomass by age-2+ pollock and jellyfish (Table 3). All three age groups of pollock and Pacific cod were important to the Station Group 4, which was confined to a relatively narrow part of the shelf north of the Pribilof Islands (Table 3, Fig. 4a). Station Groups 5 and 6 were confined to the northeast corner of the study area and were associated mainly with capelin, chinook salmon *(Oncorhynchus tshawytscha),* and age-0 pollock. Station Group 7 was found in the westernmost part of the study area and comprised mainly Atka mackerel *(Pleurogrammus monopterygius)* and jellyfish (Table 3, Fig. 4a).

For the western Bering Sea, five species groups and four nonaligned taxa groups were identified using the dissimilarity index at the 0.75 level along with six station groups (Fig. 4b). Station Groups 2 and 4 contained only one station each (Table 4) and were not considered in our analysis. Station Group 1 occurred mainly on the shelf and showed high constancy and biomass estimates for the pelagic species groups (5 and 8) which included all the Pacific salmon species and pollock age-0 and 2+ classes (Table 4). In contrast, the relatively minor Station Group 3 was dominated by herring and age-1 pollock. Age-1 pollock and Species Group 5 (especial-

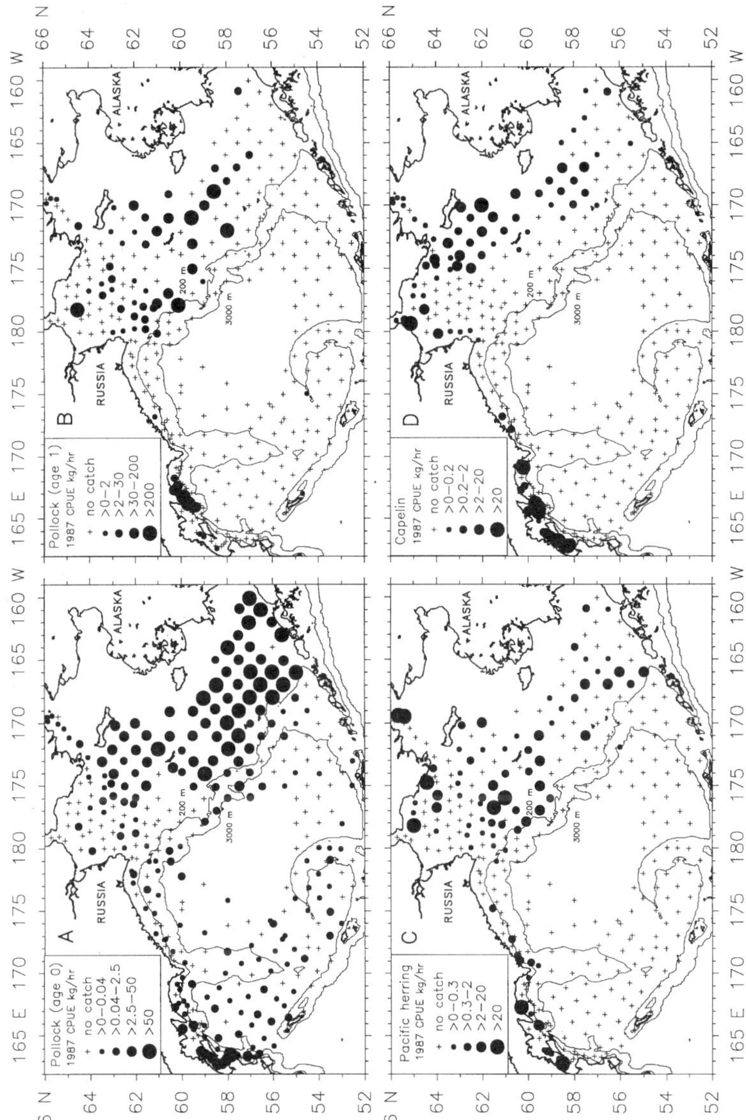

Figure 3.   Distribution of age-0 pollock (a), age-1 pollock (b), herring (c), and capelin (d) from combined R/V Darwin and R/V Gnevny surveys conducted in 1987.

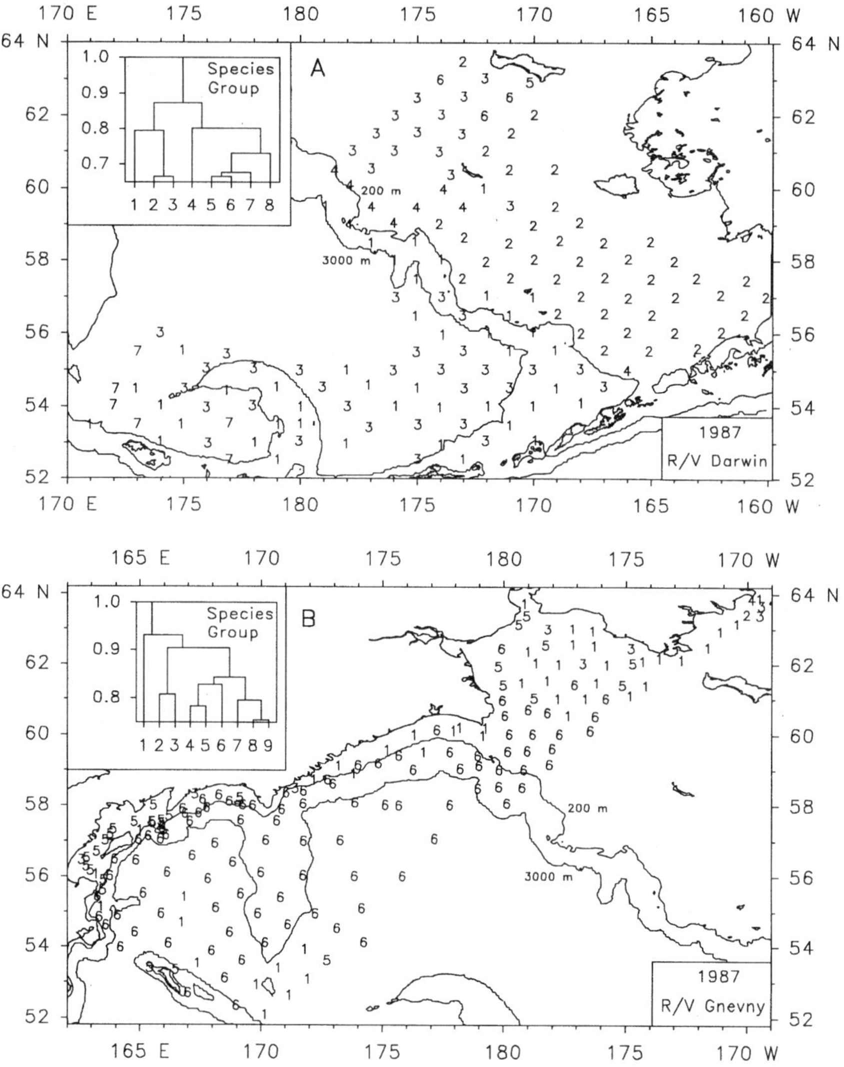

Figure 4. (a) Distribution of station groups resulting from numerical classification of biomass of the dominant species during the R/V Darwin cruise. Also shown is the species dendrogram up to the 0.65 dissimilarity level (inset). The species groups are given in Table 3. (b) Distribution of station groups resulting from numerical classification of biomass of the dominant species during the R/V Gnevny cruise. Also shown is the species dendrogram up to the 0.75 dissimilarity level (inset). The species groups are given in Table 4.

Table 3. Constancy (C) and mean biomass (B) of species in station groups for R/V *Darwin* 1987 data. Biomass is expressed as kg per hour towed.

Species group	Station group Number of hauls Taxon	1 44 C	 B	2 43 C	 B	3 44 C	 B	4 8 C	 B	5 & 6 4 C	 B	7 6 C	 B
1	*Pleurogrammus monopterygius*	0.4	0.6	0.2	0.1	0.3	0.9	0.3	0.1	0.0	0.0	1.0	93.0
2	*Zaprora silenus*	0.9	0.5	0.4	0.3	0.5	0.1	0.3	0.2	0.0	0.0	0.7	0.2
3	*Oncorhynchus keta*	0.9	21.7	0.4	2.6	0.5	16.0	1.0	4.3	0.0	0.0	0.3	4.9
	Leuroglossus schmidti	0.5	0.6	0.0	0.0	0.1	<0.1	0.0	0.0	0.0	0.0	0.0	0.0
	Cephalopoda	0.9	4.6	0.2	<0.1	0.4	0.3	0.4	<0.1	0.0	0.0	0.2	0.1
	Stenobrachius leucopsarus	0.7	2.9	0.0	0.0	0.0	0.0	0.0	0.0	0.0	0.0	0.0	0.0
4	*Oncorhynchus tshawytscha*	0.1	0.1	0.3	1.0	0.0	0.0	0.0	0.0	0.3	3.8	0.0	0.0
5	*Clupea pallasi*	0.0	0.0	0.4	0.5	0.2	3.9	1.0	5.5	0.8	0.3	0.0	0.0
	Theragra chalcogramma 2+	0.6	42.8	0.8	56.3	1.0	111.7	1.0	1,941.1	0.8	21.9	0.3	0.2
	Theragra chalcogramma 1	0.0	0.0	0.3	61.0	0.2	9.6	0.5	87.9	0.5	0.7	0.0	0.0
6	*Pleuronectes asper*	0.0	0.0	0.5	5.1	0.1	0.1	0.1	0.1	0.2	0.0	0.0	0.0
	Theragra chalcogramma 0	0.6	0.3	1.0	74.4	0.5	2.7	0.5	358.7	1.0	7.6	0.5	0.1
	Cnidaria	1.0	32.8	1.0	93.0	1.0	21.1	0.2	1.7	0.5	0.3	1.0	32.7
	Podothecus acipenserinus	0.0	0.0	0.4	0.3	0.0	0.0	0.0	0.0	0.0	0.0	0.0	0.0
	Gadus macrocephalus	0.1	<0.1	0.7	3.0	0.1	0.6	0.6	18.4	0.8	0.1	0.0	0.0
	Blepsias bilobus	0.0	0.0	0.5	0.1	0.0	0.0	0.0	0.0	0.0	0.0	0.0	0.0
7	*Mallotus villosus*	0.0	0.0	0.5	1.5	0.1	0.1	0.0	0.0	1.0	6.2	0.0	0.0
8	Natantia	0.1	<0.1	0.2	<0.1	0.1	0.1	0.1	<0.1	0.0	0.0	0.0	0.0

Table 4. Constancy (C) and mean biomass (B) of species in station groups for R/V *Gnevny* 1987 data. Biomass is expressed as kg per hour towed.

Species group	Station group Number of hauls Taxon	1 46 C	 B	2 1 C	 B	3 8 C	 B	4 1 C	 B	5 29 C	 B	6 98 C	 B
1	*Ammodytes hexapterus*	0.3	0.1	1.0	<0.1	0.3	0.0	1.0	<0.1	0.2	0.1	0.0	0.0
2	*Clupea pallasi*	0.2	0.2	0.0	0.0	1.0	285.5	0.0	0.0	0.2	1.8	0.2	0.1
	Lumpenus maculatus	0.2	<0.1	0.0	0.0	0.1	0.5	1.0	<0.1	0.3	0.0	0.1	<0.1
3	*Hippoglossoides elassodon*	0.1	<0.1	0.0	0.0	0.1	<0.1	1.0	0.1	0.2	<0.1	0.1	<0.1
	Theragra chalcogramma 1	0.1	0.2	0.0	0.0	0.6	13.1	1.0	0.1	0.2	24.7	0.2	5.5
	Reinhardtius hippoglossoides	0.2	<0.1	0.0	0.0	0.3	0.0	0.0	0.0	0.2	<0.1	0.1	<0.1
	Hemilepidotus papilio	0.2	<0.1	0.0	0.0	0.1	<0.1	1.0	0.1	0.1	<0.1	0.1	<0.1
4	Natantia	0.0	0.0	0.0	0.0	0.3	<0.1	1.0	0.3	0.2	0.7	0.1	0.2
5	*Mallotus villosus*	0.3	0.2	1.0	<0.1	0.6	0.9	1.0	<0.1	0.8	17.3	0.0	0.0
	Theragra chalcogramma 0	0.5	0.2	1.0	<0.1	0.5	0.1	1.0	<0.1	0.7	4.7	0.6	0.1
	Cnidaria	0.7	1.7	0.0	0.0	0.6	0.3	0.0	0.0	0.8	12.4	0.9	2.3
	Blepsias bilobus	0.0	0.0	0.0	0.0	0.1	<0.1	0.0	0.0	0.5	<0.1	0.0	0.0
	Oncorhynchus tshawytscha	0.3	0.2	1.0	1.3	0.1	0.1	0.0	0.0	0.6	0.8	0.3	0.2
6	*Eumicrotremus* sp.	0.2	<0.1	1.0	0.1	0.3	<0.1	0.0	0.0	0.3	<0.1	0.1	<0.1
7	*Zaprora silenus*	0.0	0.0	0.0	0.0	0.0	0.0	0.0	0.0	0.1	0.1	0.2	<0.1
8	*Aptocyclus ventricosus*	0.1	<0.1	1.0	0.2	0.0	0.0	0.0	0.0	0.0	0.0	0.3	0.4
	Theragra chalcogramma 2+	0.6	1.7	0.0	0.0	0.3	0.4	0.0	0.0	0.6	6.9	1.0	432.0
	Cephalopoda	0.3	0.4	0.0	0.0	0.0	0.0	0.0	0.0	0.2	0.1	0.7	1.1
	Lampetra tridentata	0.1	0.0	0.0	0.0	0.0	0.0	0.0	0.0	0.0	0.0	0.2	<0.1
	Oncorhynchus keta	1.0	8.1	0.0	0.0	0.6	2.8	0.0	0.0	0.5	0.4	0.6	1.4
	Oncorhynchus nerka	0.4	0.7	0.0	0.0	0.1	<0.1	0.0	0.0	0.4	0.1	0.4	0.5
9	*Stenobrachius leucopsarus*	0.1	0.2	0.0	0.0	0.0	0.0	0.0	0.0	0.0	0.0	0.3	0.9
	Leuroglossus schmidti	0.1	0.1	0.0	0.0	0.0	0.0	0.0	0.0	0.0	0.0	0.2	0.7

ly capelin and jellyfish) were important to the inshore Station Group 5 (Table 4, Fig. 4b). Finally, Station Group 6 comprised mainly age-2+ pollock and the mesopelagic Species Group 9 (Table 4).

Retrospective Analysis of Eastern Bering Sea Forage Fishes

The distribution of bottom temperatures showed dramatic differences between 1986 and 1987 (Figs. 5a and 5b). During 1986, a cold pool of <2°C bottom water penetrated almost to the Alaska Peninsula (56°N), whereas in 1987, it extended only a short distance south of 60°N. Age-1 pollock showed a much broader distribution during the cold year of 1986 compared with 1987 when they were relatively sparse and confined mainly to the middle shelf region (Figs. 5c and 5d). Age-1 Pacific cod exhibited a less dramatic change in their distribution than pollock but were found more on the middle shelf during 1986 compared with the more inner shelf distribution of 1987 (Figs. 5e and 5f). Herring were substantially more widely distributed in 1986 and were found well offshore compared with 1987 (Figs. 5g and 5h). Capelin were found mainly in the northeast part of the study area but extended farther south and west in 1986 than in 1987 (Figs. 5i and 5j). Finally, eulachon were restricted almost entirely to the southern part of the shelf (south of 57°N) but their distribution was also more widespread in 1986 than in 1987 (Figs. 5k and 5l).

As can be seen in these figures, there was some overlap in the distribution ranges of these five forage fish species. We compiled the number of species occurrences at each station for the same two years (Fig. 6) and observed that the overlap was higher in a colder year (1986; Fig. 6a), when the species distributions are expanded, than in a warmer year (1987; Fig. 6b).

For all 14 years combined, age-1 pollock were the most widespread in their distribution and most likely to be found when only one species was present (i.e., single species found at 34% of the sites, of which 70% of these were age-1 pollock; Fig. 7) due to their wider distribution than the other species (Fig. 6). Eulachon were found only in a restricted area in the southern part of the Bering Sea and had a distribution different from the other species. Thus, eulachon have a low probability of being found with the remaining species. The mean probability for the 1982-95 series that none of the species was present was around 14% while the mean availability of all five being found at the same station was 1.6%.

Substantial interannual variability was evident in the biomass index for the five forage species in the AFSC bottom trawl collection series (Fig. 8a). Herring showed the most variability and also made up the majority of biomass during most years. Age-1 pollock were generally of secondary importance and showed relatively little change over the period compared to herring; however, their biomass estimates exceeded that of herring during years of low herring biomass (e.g., 1982, 1986, 1990). Capelin was relatively important only during one year (1993), whereas the biomass index of age-1 Pacific cod and eulachon represented a small fraction of the

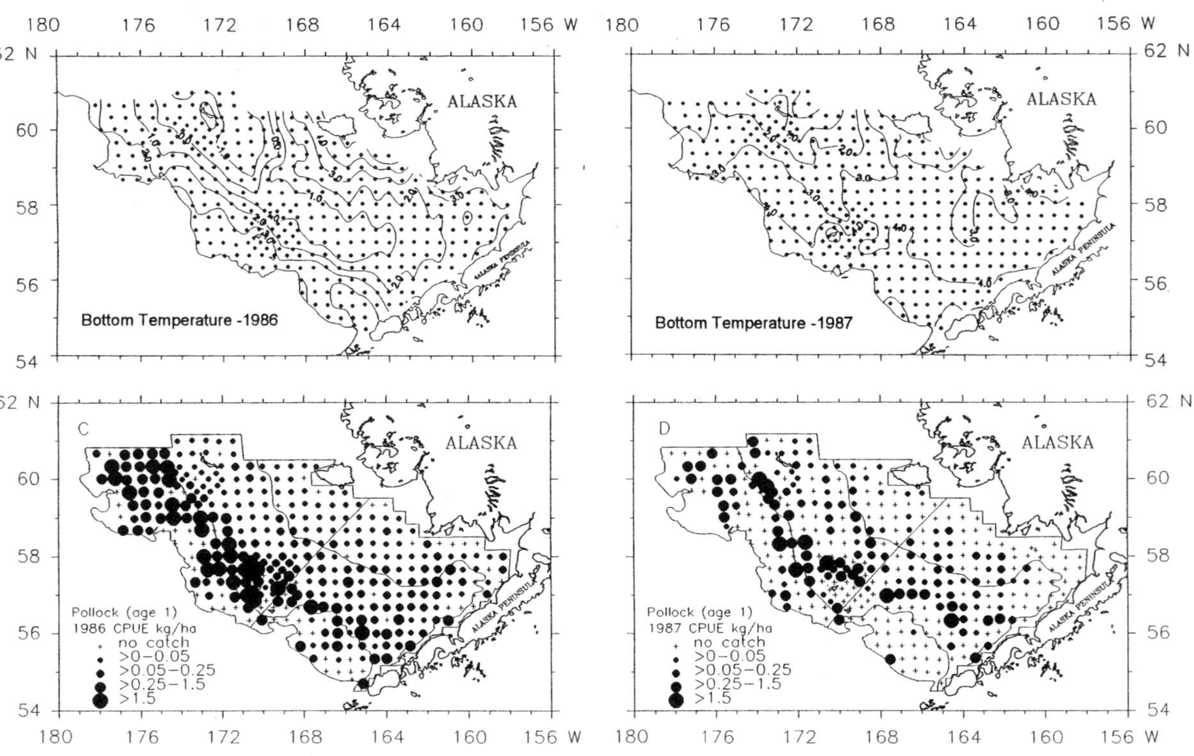

Figure 5.   Distribution of bottom temperature and standardized catches of the dominant forage fishes from a cold year (1986) and a warm year (1987) collected in AFSC summer surveys of the Bering Sea.

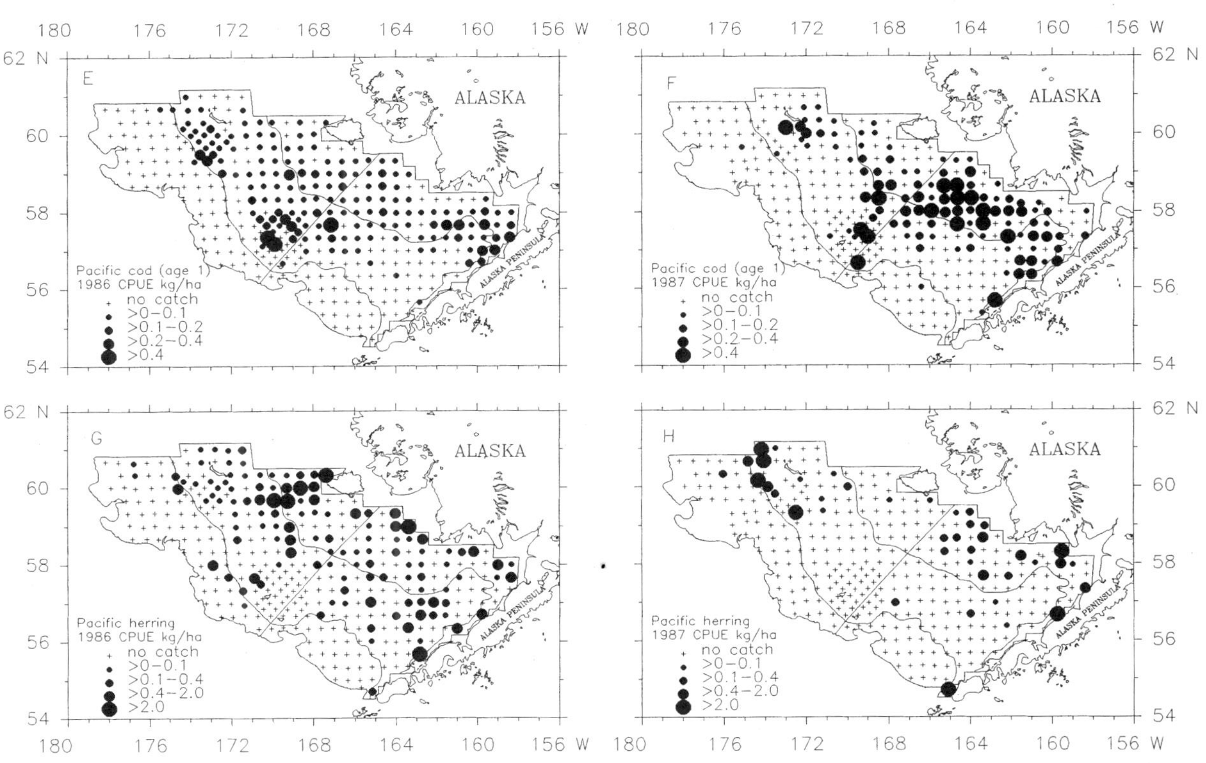

Figure 5.   (Continued.)

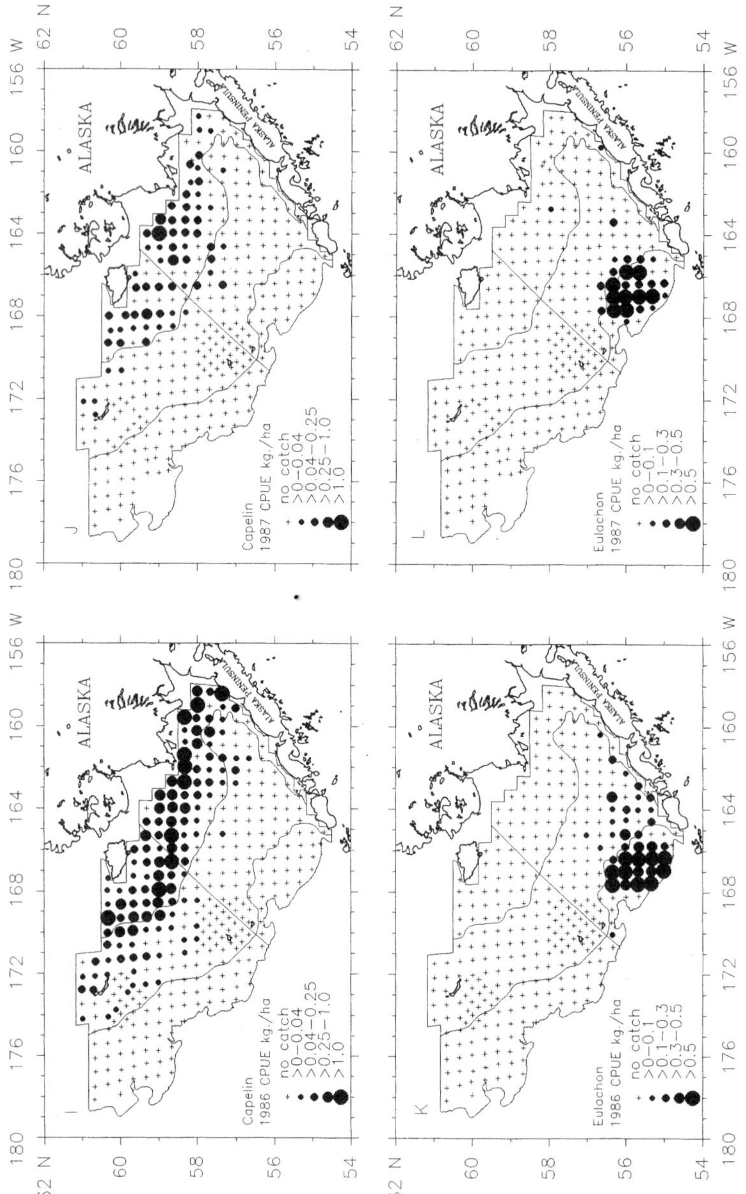

Figure 5.   *(Continued.) Distribution of bottom temperature and standardized catches of the dominant forage fishes from a cold year (1986) and a warm year (1987) collected in AFSC summer surveys of the Bering Sea.*

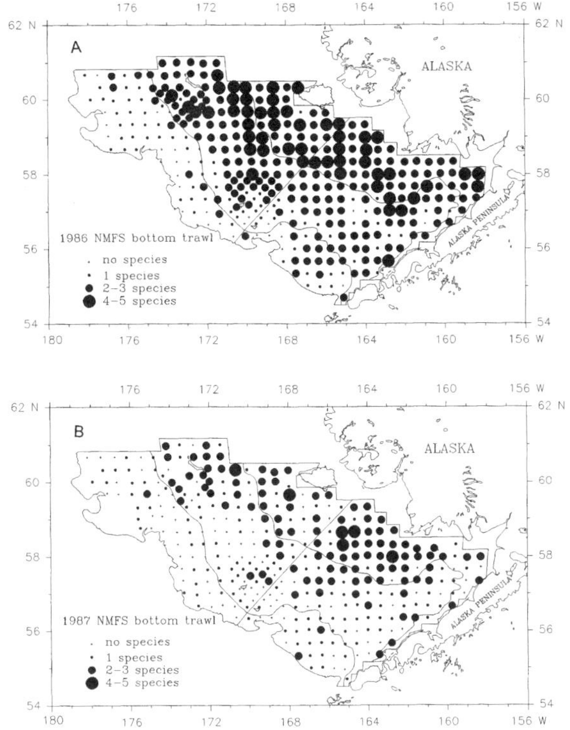

Figure 6. Number of forage species occurring at each sampling site from the (a) 1986 and (b) 1987 AFSC bottom trawl surveys.

total biomass index of these five species during all years (Fig. 8a). There was no indication of any synchrony in the peak biomass index of these forage fishes, nor was there any apparent relation of the peaks to either mean surface or bottom temperature measured during each survey (Fig. 8).

Forage Fish Distribution Relative to Depth and Temperature

We compared the cumulative distributions of habitat variables sampled during the survey with those of the dominant species collected during the 1987 R/V *Darwin* survey (Fig. 9). Because of the large number of deep-water stations which do not appear to be good habitat for most of these shelf species (Fig. 3), we included only the 75 stations that were <200 m in depth for this analysis. Greater than 60% of the capelin were collected at depths <50 m and this species showed a significant response to depth

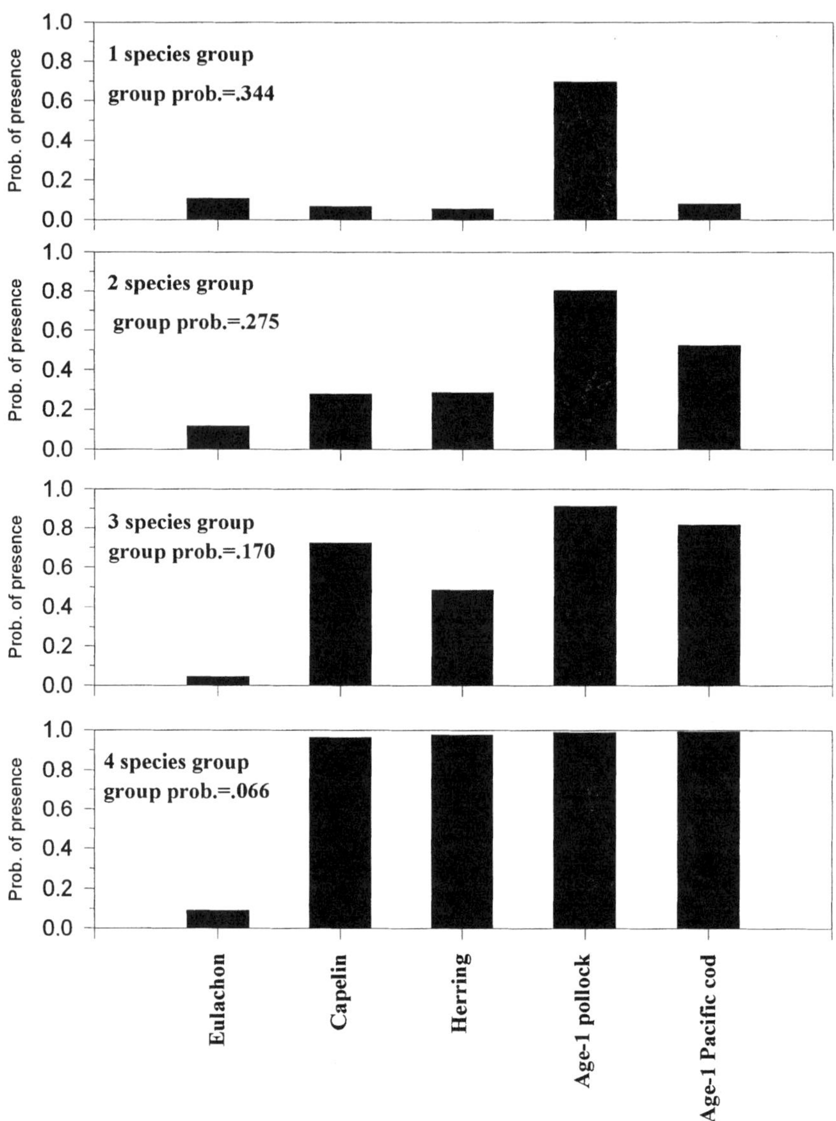

Figure 7. Probability of a species being present at a site as a function of the number of total species present based on averages for the 1982-95 AFSC bottom trawl surveys. Group size probabilities are shown for each grouping.

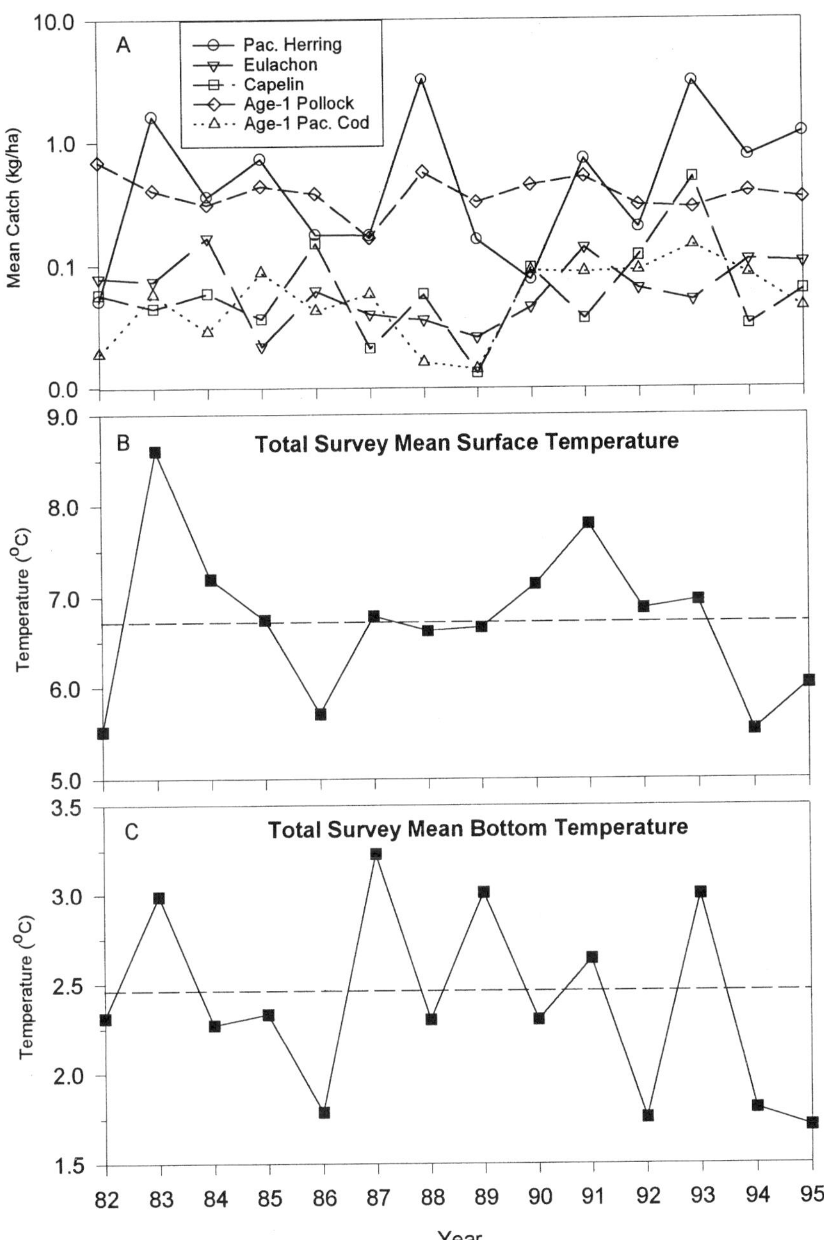

Figure 8. Log biomass of the dominant forage fishes collected in AFSC bottom trawl surveys from 1982 to 1995 and annual averages of surface and bottom temperatures recorded throughout the survey area for the same period.

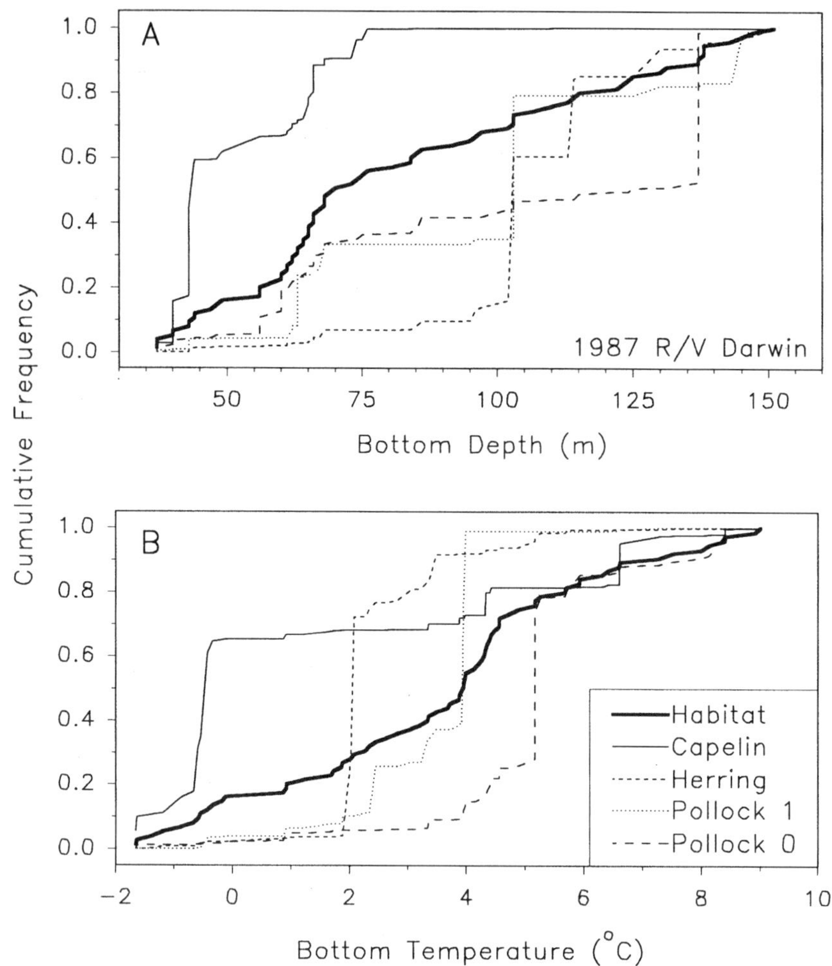

Figure 9.   Cumulative distribution functions for the observed (habitat) and the four species examined from the R/V Darwin surveys for (a) depth and (b) temperature. See Table 5 for statistical comparison of the distributions.

**Table 5. Results of Kolmorogorov-Smirnov test
for statistical differences in catch-
weighted cumulative frequency of each
species and the cumulative frequency
of habitat (temperature or bottom depth)
for R/V *Darwin* trawl survey data for
1987.**

Species	Depth	Temperature
Age-0 pollock	0.368	0.009
Age-1 pollock	0.562	0.228
Pacific herring	0.056	0.174
Capelin	0.006	0.002

The *P*-values given are the proportion of the 1,000 test-statistic values that were greater than or equal to the observed test statistic.

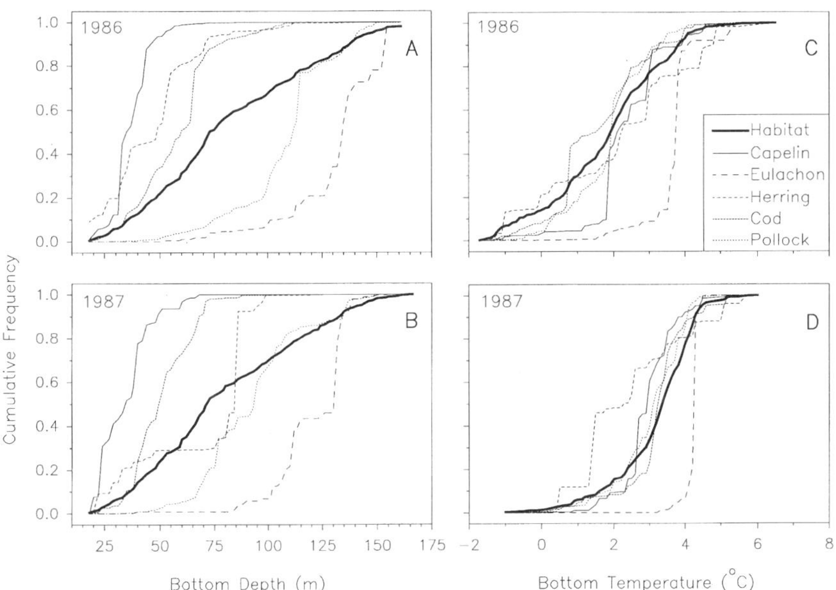

*Figure 10. Cumulative distribution functions for the observed (habitat) and the five
species examined from the AFSC bottom trawl surveys for both years by
depth (a and b) and temperature (c and d). See Table 6 for statistical
comparison of the distributions.*

Table 6.   Results of Kolmorogorov-Smirnov test for statistical differences in catch-weighted cumulative frequency of each species and the cumulative frequency of habitat (temperature or bottom depth) for AFSC bottom trawl survey data for 1986 and 1987.

Species	1986		1987	
	Depth	Temperature	Depth	Temperature
Age-1 pollock	0.001	0.872	0.002	0.372
Age-1 Pacific cod	0.001	0.593	0.001	0.079
Pacific herring	0.001	0.444	0.468	0.089
Capelin	0.001	0.110	0.001	0.123
Eulachon	0.001	0.001	0.001	0.001

The P-values given are the proportion of the 1,000 test-statistic values that were greater than or equal to the observed test statistic.

(Fig. 9a; Table 5). Capelin were significantly ($P = 0.002$) associated with colder temperatures in the northern part of the study area, whereas age-0 pollock were associated ($P = 0.009$) with warmer temperatures than those sampled overall (Fig. 9b; Table 5). None of the remaining species were significantly associated with either bottom temperature or depth.

Based upon the 1986 and 1987 AFSC bottom trawl surveys, most species were found over a relatively narrow depth range compared to what was sampled in the survey (Figs. 10a and 10b). Pollock and eulachon tended to be found deeper than the average depth of the survey during both years, whereas Pacific cod and capelin were found significantly shallower than the depth distribution sampled (Table 6). Herring were found shallower in 1986 but not significantly different from their depth distribution in 1987 (Fig. 10, Table 6). The bottom temperature occupation curves of most species were not significantly different from those found in the environment with the exception of eulachon which was found only in the warmer temperatures at the southern part of the sampling area (Fig. 10c; Table 6). Herring was the only species to show interannual differences in distribution with respect to temperature, with 1987 showing a distribution in cooler waters compared with 1986 (Fig. 10d).

Discussion

This study presents the first examination of the large-scale distribution patterns of several prominent forage fish species throughout the Bering Sea. We also present new data on the association of these forage fishes with other taxa that may compete with these forage fishes and, in some cases, prey upon them (e.g., age-2+ pollock). Although other forage species were collected during these surveys (e.g., Pacific sand lance *Ammodytes hexapterus*, rainbow smelt *Osmerus mordax*, Atka mackerel, and Arctic cod *Boreogadus saida*), these other species were either not adequately sampled by our gear or did not occur over a broad enough range of the sampling area to be included here. Pacific salmon (*Oncorhynchus* spp.) were commonly caught and occurred in high biomass in the Russian surveys, but their distribution was more oceanic and they are generally not utilized as forage fish except in rare circumstances. Also, the data for the AFSC bottom trawl surveys included only age-1 fish for pollock and cod since age-0 fish were not captured due to the relatively large mesh used in the trawls, although the importance of age-0 juveniles in the diets of seabirds and predatory fishes of the Bering Sea is clearly recognized (Livingston 1993).

Many of these forage fishes form layered aggregations and structured schools which facilitates capture by predators but may introduce some bias in estimates of population size in trawl surveys. Most also undergo seasonal horizontal and diel vertical migrations which further complicate quantitative sampling. Some of the interspecific differences in distribution may be related to spawning habits in that some species (e.g., eulachon, herring, capelin) spawn in freshwater, intertidal or nearshore areas, whereas others (pollock, Pacific cod) spawn in relatively deep water off the continental shelf (Wespestad 1987, Fritz et al. 1993). Interannual differences in abundance of the species that undergo spawning migrations may result merely from temperature differences altering the time of spawning relative to the survey. Because of the substantial number of stations and wide geographic area covered in these surveys, the distributions portrayed here are not synoptic and may actually mask smaller-scale migrations occurring during the surveys.

Differences in observed distributions between two adjacent but thermally contrasting years in the AFSC bottom trawl data suggests that temperature is important in affecting the horizontal extent and overlap in distribution of these forage fishes. The consistency among species is somewhat surprising because some species (e.g., herring and capelin) are typically more pelagic than others (e.g., Pacific cod) but all were responsive to the variability in bottom temperature. However, we were unable to determine temperature effects on the vertical distribution or absolute biomass of these species due to the lack of depth-specific sampling and incomplete coverage of their distribution area. For example, what may appear to be a decline in fish populations during the warm year of 1987 may actually be a shift in the distribution to the region north of that sampled during the annual surveys (e.g., near St. Lawrence Island and in the Gulf of Anadyr), where high biomass of age-1 pollock, herring, and capelin was found in the 1987 Russian surveys (Fig. 4).

The distributions of these species are important by themselves, but what makes them most critical for their role as forage is their proximity to marine mammal and seabird rearing or feeding sites. Although capable of foraging at sea for extended periods, the practical foraging radius of many land-based predators may be quite limited (Schneider and Hunt 1984). Access to one or more species of forage fishes within a reasonable distance of a nest or rookery may be imperative if the offspring are to survive to maturity. The presence of two or more species at the same location not only increases the biomass available for a generalist feeder, but also improves the probability of finding suitable prey for the specialist for whom some food items may be preferred over other more abundant ones. This may be due to higher fat content, protein content, or total caloric value in some of the species (Payne et al. 1997). Many of the dominant mammal and seabird predators on these species have undergone dramatic shifts in their abundance (National Research Council 1996; Hunt and Byrd, chapter 28, this volume), with several species currently listed as endangered. Reproductive success in seabirds nesting on the Pribilof Islands has been shown to be strongly linked to the availability of appropriate forage fishes (Decker et al. 1995). Merrick et al. (1997) have found that population declines in Steller sea lions *(Eumetopias jubatus)* are less severe at rookeries where a higher diversity of prey is available than those where only one or two species are available.

The abundance of these major forage fishes fluctuated over the period studied and no long-term trends were evident. It should be noted, however, that our data were collected entirely after the major regime shift which occurred around 1975-76 (National Research Council 1996). Although some trawl surveys were conducted in the 1970s in the eastern Bering Sea, the data are not directly comparable to what we present here due to differences in gear, survey design, and areal coverage. Naumenko (1996) analyzed 36 years (1958-1993) of Russian pelagic trawl data from the western Bering Sea and found four distinct periods based upon ichthyofaunal communities which were related to environmental conditions. Beginning in 1975, herring, capelin, and other smelts showed substantial declines in abundance and pollock dominated the catches. This pattern held through the late 1980s, but in the 1990s there has been a resurgence in the pelagic species, particularly herring and capelin, in the western Bering Sea and Okhotsk Sea (Naumenko 1996, Shuntov et al. 1996). Although systematic survey data are less complete for the eastern Bering Sea, these trends in forage fish abundances are reflected in fisheries bycatch (Fritz et al. 1993) and seabird diet (Hunt et al. 1996) data.

Our analysis of the relationship between these forage fishes and two easily measured habitat variables is an admittedly simple approach to a complex ecological situation. For example, bottom temperature and depth are often correlated and a more powerful approach may be to examine both variables at the same time using a bivariate distribution (Perry and Smith 1994). However, our finding that most of these species do not appear to actively select a particular temperature range was surprising, considering that the distributions of these forage fishes appeared to vary so much between warm and cold years. Although we presented distributions only from two adjacent years which showed the greatest change in bottom temperature in the 14-year time series (Fig. 8) and also a substantial difference in sea ice and cold pool extent (Wyllie-Echeverria 1996), other years not presented showed similar patterns in warm and cold years. Most taxa were associated with, or excluded from, a particular bottom depth range, particularly for the U.S. bottom trawl survey data, where there is some segregation of the dominant forage fish by depth strata. The mechanism underlying why depth, or some factor related to it, is important to these fishes needs to be more fully explored.

Acknowledgments

This study could not have been undertaken without the diligent and conscientious efforts of the many U.S. and Russian scientists who collected the data during the many cruises included in our analysis. We thank Kathy Mier for her assistance in the clustering analysis. Steve Syrjala assisted with the statistical comparison between the fish biomass and environmental data. We thank Dr. Vladimir Radchenko for assistance in data collection and preparation. Art Kendall, Alan Springer, Lowell Fritz, and an anonymous referee provided helpful comments on previous drafts of the manuscript. Funding for this study was provided by the Coastal Ocean Program, Bering Sea Fisheries Oceanography Coordinated Investigations (BS FOCI). This is FOCI Contribution No. B294.

References

Alaska Sea Grant. 1997. Forage fishes in marine ecosystems. Proceedings of the International Symposium on the Role of Forage Fishes in Marine Ecosystems. University of Alaska Sea Grant, AK-SG-97-01, Fairbanks. 816 pp.

Bailey, K.M., and S.M. Spring. 1992. Comparison of larval, age-0 juvenile and age-2 recruit abundance indices of walleye pollock, *Theragra chalcogramma*, in the western Gulf of Alaska. ICES Journal of Marine Science 49:297-304.

Bakkala, R.G. 1993. Structure and historical changes in the groundfish complex of the eastern Bering Sea. NOAA Technical Report NMFS 114. 91 pp.

Boesch, D.F. 1977. Application of numerical classification in ecological investigations of water pollution. Environmental Protection Agency Ecological Research Series, EPA-600/3-77-033. 115 pp.

Brodeur, R.D., M.S. Busby, and M.T. Wilson. 1995. Summer distribution of late-larval and early juvenile walleye pollock *(Theragra chalcogramma)* and associated species in the western Gulf of Alaska. Fisheries Bulletin, U.S. 93:603-617.

Brodeur, R.D., and M.T. Wilson. 1996. A review of the distribution, ecology and population dynamics of age-0 walleye pollock in the Gulf of Alaska. Fisheries Oceanography 5:148-166.

Brodeur, R.D., P.A. Livingston, T.R. Loughlin, and A.B. Hollowed (eds.). 1996. Ecology of juvenile walleye pollock, *Theragra chalcogramma*. NOAA Technical Report NMFS 126. 227 pp.

Clifford, H.J., and W. Stephenson. 1975. An introduction to numerical classification. Academic Press, Inc., New York. 229 pp.

Decker, M.B., G.L. Hunt Jr., and G.V. Byrd. 1995. The relationships among sea-surface temperature, the abundance of juvenile walleye pollock (*Theragra chalcogramma*), and the reproductive performance and diet of seabirds at the Pribilof Islands, southeastern Bering Sea. In: R.J. Beamish (ed.), Climate change and northern fish populations. Canadian Special Publication of Fisheries and Aquatic Sciences 121, pp. 425-437.

Faith, D.P., P.R. Minchin, and L. Belbin. 1987. Compositional dissimilarity as a robust measure of ecological distance. Vegetatio 69: 57-68.

Fritz, L.W., V.G. Wespestad, and J.S. Collie. 1993. Distribution and abundance trends of forage fishes in the Bering Sea and Gulf of Alaska. In: Is it food? Addressing marine mammal and seabird declines: Workshop summary. University of Alaska Sea Grant, AK-SG-93-01, Fairbanks, pp. 30-44.

Hinckley, S., K.M. Bailey, S.J. Picquelle, J.D. Schumacher, and P.J. Stabeno. 1991. Transport, distribution, and abundance of larval and juvenile walleye pollock *(Theragra chalcogramma)* in the western Gulf of Alaska. Canadian Journal of Fisheries and Aquatic Sciences 48:91-98.

Hunt, G.L., Jr., M.B. Decker, and A. Kitaysky. 1996. Fluctuations in the Bering Sea ecosystem as reflected in the reproductive ecology and diets of kittiwakes on the Pribilof Islands, 1975 to 1991. In: S.P.R. Greenstreet and M.L. Tasker (eds.), Aquatic predators and their prey. Fishing News Books, Oxford, pp. 142-153.

Koehler, P.A., P.C.F. Hurley, P. Perley, and J.D. Neilson. 1986. Juvenile fish surveys on the Scotian shelf: Implications for year-class size assessment. Journal du Conseil International pour l'Exploration de la Mer 43:59-76.

Livingston, P.A. 1993. Importance of predation by groundfish, marine mammals and birds on walleye pollock *Theragra chalcogramma* and Pacific herring *Clupea pallasi* in the eastern Bering Sea. Marine Ecology Progress Series 102:205-215.

Merrick, R.L., M.K. Chumbley, and G.V. Byrd. 1997. Diet diversity of Steller sea lions (*Eumetopias jubatus*) and their population decline in Alaska: A potential relationship. Canadian Journal of Fisheries and Aquatic Sciences 54:1342-1348.

Minerals Management Service. 1987. Forage fishes of the southeastern Bering Sea. OCS MMS 87-0017. 90 pp.

National Research Council. 1996. The Bering Sea Ecosystem: Report of the Committee on the Bering Sea Ecosystem, National Research Council. National Academy Press, Washington, DC. 324 pp.

Naumenko, N.I. 1996. Long-term fluctuations in the ichthyofauna of the western Bering Sea. In: O.A. Mathisen and K.O. Coyle (eds.), Ecology of the Bering Sea: A review of Russian literature. University of Alaska Sea Grant, AK-SG-96-01, Fairbanks, pp. 143-158.

Naumenko, N.I., P.A. Balykin, E.A. Naumenko, and E.R. Shaginyan. 1990. Year-to-year variation of stocks and community structure of western Bering Sea pelagic fishes. In: International Symposium on Bering Sea Fisheries, Khavarovsk, U.S.S.R, pp. 223-224.

Payne, S.A., B.A. Johnson, and R.S. Otto. 1997. Proximate analysis of some northeastern Pacific forage fish species. In: Forage fishes in marine ecosystems. University of Alaska Sea Grant, AK-SG-97-01, Fairbanks, pp. 721-724.

Perry, R.I., and S.J. Smith. 1994. Identifying habitat associations of marine fishes using survey data and application to the northwest Atlantic. Canadian Journal of Fisheries and Aquatic Sciences 51:589-602.

Schneider, D., and G.L. Hunt Jr. 1984. A comparison of seabird diets and foraging distribution around the Pribilof Islands, Alaska. In: D.N. Nettleship, G.A. Sanger, and P.F. Springer (eds.), Marine birds: Their feeding ecology and commercial fisheries relationships. Canadian Wildlife Service Special Publication, Ottawa, pp. 86-95. (Available from Canadian Wildlife Service, 400 Laurier Ave. W., Ottawa, ON, K1A 0H3, Canada.)

Shuntov, V.P., E.P. Dulepova, V.I. Radchenko, V.V. Lapko. 1996. New data about communities of plankton and nekton of the far eastern seas in connection with climate-oceanological reorganization. Fisheries Oceanography 5:38-44.

Sinclair, E.H., G.A. Antonelis, B.W. Robson, R.R. Ream, and T.R. Loughlin. 1996. Northern fur seal, *Callorhinus ursinus*, predation on juvenile walleye pollock, *Theragra chalcogramma*. In: R.D. Brodeur, P.A. Livingston, T.R. Loughlin, and A.B. Hollowed (eds.), Ecology of juvenile walleye pollock, *Theragra chalcogramma*. NOAA Technical Report NMFS 126, pp. 167-178.

Smith, S.J. 1996. Analysis of data from bottom trawl surveys. In: H. Lassen (ed.), Assessment of groundfish stocks based on bottom trawl survey results. NAFO Scientific Council Studies, Northwest Atlantic Fisheries Organization 28:25-53.

Springer, A.M. 1992. A review: Walleye pollock in the North Pacific—how much difference do they really make? Fisheries Oceanography 1:80-96.

Walters, G.E. 1989. The value of pre-recruit abundance estimates from resource assessment surveys in forecasting year class strength of commercially important species in the eastern Bering Sea. In: S. Sundby (ed.), Year class variations as determined from pre-recruit investigations. Institute of Marine Research, Bergen, Norway, pp. 7-38.

Walters, G.E., G.B. Smith, P.A. Raymore Jr., and W. Hirschberger. 1985. Studies on the distribution and abundance of juvenile groundfish in the northwest Gulf of Alaska, 1980-82: Part II, Biological characteristics in the extended region. NOAA Technical Memo NMFS F/NWC-77. 95 pp. (Available from National Marine Fisheries Service, Scientific Publications Office, 7600 Sand Point Way N.E., Seattle, WA, 98115-0070.)

Wespestad, V.G. 1987. Population dynamics of Pacific herring (*Clupea pallasi*), capelin (*Mallotus villosus*), and other coastal pelagic fishes in the eastern Bering Sea. In: Forage fishes of the southeastern Bering Sea. Minerals Management Service OCS MMS 87-0017, pp. 55-60.

Wilson, M.T., R.D. Brodeur, and S. Hinckley. 1996. Distribution and abundance of age-0 walleye pollock, *Theragra chalcogramma*, in the western Gulf of Alaska during September 1990. In: R.D. Brodeur, P.A. Livingston, T.R. Loughlin, and A.B. Hollowed (eds.), Ecology of juvenile walleye pollock, *Theragra chalcogramma*. NOAA Technical Report NMFS 126, pp. 11-24.

Wyllie-Echeverria, T.W. 1996. A relationship between the distribution of one year old pollock and sea ice characteristics. In: R.D. Brodeur, P.A. Livingston, T.R. Loughlin, and A.B. Hollowed (eds.), Ecology of juvenile walleye pollock. NOAA Technical Report NMFS 126, pp. 47-56.

CHAPTER 25

Ecology of Groundfishes in the Eastern Bering Sea, with Emphasis on Food Habits

Kei-ichi Mito
Seikai National Fisheries Research Institute, Ishigaki, Japan

Akira Nishimura and Takashi Yanagimoto
National Research Institute of Far Seas Fisheries, Shimizu, Japan

Abstract

The food habits of groundfishes in the Bering Sea are presented based on data collected by the National Research Institute of Far Seas Fisheries of Japan. The ecology and biology of important species caught in fisheries also are described. Juvenile walleye pollock *(Theragra chalcogramma)*, the most abundant species, preyed primarily on zooplankton such as euphausiids and copepods. Zooplankton was also the primary prey of Pacific ocean perch *(Sebastes alutus)* and Pacific herring *(Clupea pallasi)*. Adult pollock preyed mainly on small pollock in the eastern Bering Sea, but on euphausiids and copepods in the Aleutian Basin. Pacific cod *(Gadus macrocephalus)* also exhibited an ontogenetic diet shift from shrimp and Tanner crabs to pollock with increasing cod size. Large-mouthed piscivorous flatfishes such as arrowtooth flounder *(Atheresthes stomias)*, Greenland turbot *(Reinhardtius hippoglossoides)*, and Pacific halibut *(Hippoglossus stenolepis)* preyed primarily on pollock, while smaller-mouthed flatfishes such as flathead sole *(Hippoglossoides elassodon)*, rock sole *(Lepidopsetta bilineata)*, and yellowfin sole *(Limanda aspera)* preyed on benthic invertebrates. In addition, many other species such as sculpins and skates preyed on juvenile pollock. Juvenile pollock occupy a key position in the Bering Sea ecosystem by transmitting energy from zooplankton to large-sized fishes.

Introduction

The parts of the continental shelf and the basin considered in this chapter are located in the eastern and the middle-western parts of the Bering Sea, respectively. Japanese fishing boats caught groundfishes in waters of the eastern continental shelf of the Bering Sea from the latter half of 1950s. The current catch of groundfishes in this area is by U.S. fishing vessels. Total catch of groundfishes on the eastern Bering Sea shelf and the Aleutian Basin is large, 2-3 million metric tons per year. Over 300 species of fish among 36 families have been recorded in the Bering Sea. However, the number of target species for fisheries and the number of species having large biomass is few.

Ecological and biological studies on groundfishes in the Bering Sea in recent times have been conducted principally by scientists from Japan, Russia, and the United States. Surveys continue to be conducted by the National Research Institute of Far Seas Fisheries (NRIFSF) of Japan and the Alaska Fisheries Science Center (AFSC) of the United States every year. Surveys of groundfish resources and their results up to 1986 are summarized by Bakkala (1993).

In this paper, results of studies on food habits of groundfishes conducted by NRIFSF are summarized. Also, characteristics of their ecology and biology are described for commercially important species. Food habit studies are important for analyzing interspecific relationships. Clarification of these relationships enhances our understanding of food web structure and the ecosystem. These results become the basic information for future multispecies population dynamics models.

Materials and Methods

Data on Food Habits

Contents of fish stomachs were sampled from groundfish resource surveys conducted by NRIFSF in the Bering Sea, including data analyzed by Mito (1977). Both data sets were included in the ecosystem model development work which the Fisheries Agency of Japan conducted in 1985-1988. These results of food habits are reported by Mito (1990a,b,c) and Mito et al. (1994a,b, 1996a,b,c).

Data on Ecology and Fisheries Biology

To clarify ecological and biological characteristics of important species, we relied on results from Ikeda (1980) and Bakkala (1993). Furthermore, we referred to the stock assessment documents published in recent years by the AFSC.

Analytical Procedures and Calculation of Stomach Contents

The length and weight of fish were measured, and the stomach was removed and preserved in formalin. The wet weight of each prey in the stomach was sorted and recorded. The stomach contents by year, season, area, depth zone, and the size of fish were then totaled. Seasons were defined as winter (January-March), spring (April-June), summer (July-September), and autumn (October-December). The Bering Sea was divided into five areas: the southern (1) and northern (2) areas of eastern shelf, the Aleutian Islands region (3), and the southern (4) and northern (5) areas of the Aleutian Basin. Depending on maximum body size of the predator, samples were divided into length intervals of 5, 10, and 20 cm. Prey categories were summarized according to prey abundance. That is, walleye pollock *(Theragra chalcogramma)* prey was divided into age groups, the most abundant fish was identified to species, other fish were categorized to family or order, and prey other than fish were classified to order or class except Tanner crabs. The age of pollock was determined by length; age-0 = 0-10 cm, age-1 = 10-20 cm, age-2 = 20-30 cm, age-3 = 30-40 cm, age-4 and 5 = 40-50 cm, and age-6 and more = 50 cm or greater. The total number of samples used in the analysis was from 47 species and 64,652 individuals.

Estimate of Daily Ration

The equation of Jones (1974) (evacuation rate of food in the stomach: ERS) and the equation of Winberg (1956) (energy budget: MEB) were used to estimate the daily ration.

$$DR = aBW^{0.44}SW^{0.46} \text{ and } DR = 1.25[(BG+GG)+3SM]CF$$

Here, DR = daily ration (g), a = coefficient of evacuation rate in the stomach, BW = predator body weight (g), SW = stomach content weight (g), BG = body growth, GG = gonad growth, SM = standard metabolism, and CF = ratio of prey caloric value to predator value.

Food Relationship and Analytical Procedures

Stomachs were collected from fish caught by Japanese trawlers (Mito 1995) during October-November 1972, July-September 1974, and February-June 1975. There were 175 sampling stations distributed over the eastern Bering Sea shelf edge, ranging from 130-330 m in depth. The species and length composition of the catch was estimated, consisting of 52 species and 45,000 length measurements. Stomach contents were analyzed from about 14,000 individuals. Thus, food relationships reported here were determined between individuals distributed in the same area and time.

To compare between species, samples were divided into 21 groups by area, depth, and season. Stomach content composition by species and length was calculated for each group and then the similarity index of stomach content composition was calculated. A food niche was determined and food web drawn for each group.

Results

Ecology and Biology of Commercially Important Species

The ecological and biological characteristic of 13 species, which includes two codfish species and seven flatfish species, are shown in Table 1. We found that most species spawn during the winter and spring, with two notable exceptions. Yellowfin sole *(Limanda aspera)* and Atka mackerel *(Pleurogrammus monopterygius)* spawn during the summer and Greenland turbot *(Reinhardtius hoppoglossoides)* in autumn. Most species spawn over the continental shelf except for Greenland turbot, Pacific halibut *(Hippoglossus stenolepis)*, Pacific ocean perch (POP) *(Sebastes alutus)*, and sablefish *(Anoplopoma fimbria)*, which spawn over the slope. Not surprising, their distribution is similar to their spawning areas. Pollock is distributed throughout the Bering Sea, but mostly on the shelf. Most other species occur primarily on the shelf except for arrowtooth flounder *(Atheresthes stomias)*, Greenland turbot, POP, and sablefish, which occur mostly on the slope.

Of the species that we examined, most lay pelagic eggs. However, POP exhibit viviparity, and Pacific herring *(Clupea pallasi)*, Pacific cod *(Gadus macrocephalus)*, rock sole *(Lepidopsetta bilineata)*, and Atka mackerel spawn demersal eggs. Note Table 1 for the size and approximate number of eggs spawned for most species.

Groundfish Food Habits

Codfishes, Gadidae

Walleye pollock *(Theragra chalcogramma)*. Pollock food habits data for the years 1970-1985 are summarized in Mito (1990a, b). Here we also included data from samples collected after 1986 (Table 2).

The daily ration decreases as fish mature. According to our calculations of ERS, the daily ration was high in spring and summer and low in winter and autumn on the eastern Bering Sea shelf. It was especially low in winter although the change by season was also similar in the Aleutian Basin. In fish 50 cm and greater, the rate was higher in the eastern Bering Sea shelf than in the Aleutian Basin. This relationship corresponds to the growth rate of large-sized fish which is higher in the eastern Bering Sea shelf than in the Aleutian Basin.

Stomach content composition by area, season and length in 1970-1985 are shown in Fig. 1. In the eastern Bering Sea shelf, there was a tendency for fish, such as pollock, to be more important as prey in winter, then copepods became important in spring and summer, and euphausiids in autumn. As body size increases, primary prey items changed from copepods, through euphausiids, to fishes such as pollock. When the southern area was compared with the northern area in the eastern Bering Sea shelf, there was a tendency for euphausiids to be more important in the southern area and copepods and walleye pollock more important in the northern area. In the Aleutian Islands region, euphausiids were preyed on most of the summer, and euphausiids and fishes were consumed in autumn. In the Aleutian Basin, euphausiids, squids and fishes were preyed on most in winter, copepods were consumed in spring, and copepods and euphausiids were consumed in summer, and euphausiids only in autumn. Walleye pollock cannibalism was not apparent and thus they were not observed as prey in the Aleutian Islands region and the Aleutian Basin.

The results of analysis after 1986 by NRIFSF are shown in Table 2. Stomach content composition of adult fish was similar to the result before 1985. Age-0 fish chiefly preyed on copepods in each area, and the large-sized individuals among age-0 fish preyed on euphausiids on the eastern Bering Sea shelf.

Nishiyama et al. (1986) analyzed stomach content composition of pollock larvae on the eastern Bering Sea shelf and found that they preyed on copepod nauplii as their primary first prey, and gradually preyed on larger-sized prey such as *Acartia* spp. and *Pseudocalanus* spp. According to Dwyer et al. (1987), prey changed as pollock grew. Euphausiids were more important in spring and pollock were important in autumn on the eastern Bering Sea shelf. Moreover, walleye pollock preyed on fishes and euphausiids mainly in winter, Appendicularia were consumed in spring, copepods and euphausiids in summer, and euphausiids in autumn in the Aleutian Basin. According to Kachina and Savicheva (1987), small pollock preyed on copepods and euphausiids, and large-sized pollock preyed on pollock and shrimp in the western Bering Sea.

Pacific cod *(Gadus macrocephalus).* The data we used here were summarized by Mito (1990c). We estimated the daily ration to be about 1.3-1.5% at age-4 and 0.9-1.3% at age-7. Stomach content composition by area, season, and length are shown in Fig. 2. When comparing contents between areas, we observed that polychaetes, shrimps, and Tanner crabs were relatively important in the southern area in spring, and pollock were important in the northern area in spring. When the data were totaled for all seasons, small cod preyed on shrimps, Tanner crabs, gammarus, and polychaetes, and the large-sized cod preyed on pollock. Seasonally, there was a tendency for polychaetes and shrimps to be important in summer, Tanner crabs in autumn, and pollock in winter. Pacific cod 30-40 cm long preyed mostly on age-1 pollock, cod 40-75 cm long preyed on age-2 walleye pollock, and cod 75 cm or greater preyed on age-2-5 walleye pollock.

Livingston et al. (1986) analyzed samples collected in 1982-1984 and found that small cod ($\leq$ 55 cm long) preyed on crustaceans, and large-sized cod preyed on fishes. Small cod preyed on small crustaceans in winter and autumn, and preyed on Tanner crabs in spring and summer.

Grenadiers, Macrouridae

Giant grenadier *(Albatrossia pectoralis).* These fish inhabit the continental slope in water deeper than 730 m in summer, and are the most abundant species at this depth in the eastern Bering Sea. Small individuals preyed primarily on squids, medium-sized individuals preyed primarily on squids and fishes, and large-sized individuals preyed on fishes.

Popeye grenadier *(Coryphaenoides cinereus).* We have very few data on this species and report only that the main food items were shrimps and fishes.

Flatfishes, Pleuronectidae

The food habits of flatfishes, except for rex sole, are summarized by Mito et al. (1994a,b).

Arrowtooth flounder *(Atheresthes stomias).* We estimated the daily ration by MEB and ERS and found that they were high in autumn and low in winter. Stomach content composition by year, season, and length are shown in Fig. 3. The main prey items were pollock, and to a lesser extent shrimps. Studies before 1980 showed small flounder (≤ 20 cm) preyed primarily on shrimps, medium flounder (20-40 cm) preyed on age-0 and age-1 pollock, and large-sized flounder (≥ 40 cm) preyed on age-2 pollock. Studies after 1980 showed medium-sized individuals preyed primarily on shrimps and fishes, and large-sized individuals preyed on fishes such as pollock. Seasonal changes were not apparent, but shrimp was preyed upon more in summer, age-0 pollock in autumn, and age-1 pollock in winter.

Greenland turbot *(Reinhardtius hippoglossoides).* The daily rations estimated by MEB and ERS were low in summer, and high in spring and autumn. Stomach content composition by year, season, and length are shown in Fig. 4. Main prey items were pollock, but squids and euphausiids were also present. Studies before 1980 showed small turbot (≤ 30 cm) chiefly preyed on age-0 and age-1 pollock and euphausiids, medium turbot (30-50 cm) preyed on age-1 and age-2 pollock, and larger turbot (≥ 50 cm) preyed on age-2 and age-3 pollock and herring. After 1980, large turbot preyed primarily on pollock, squids, and octopus. Seasonal change was not clear, but squids and euphausiids were consumed in spring and summer, age-1 pollock in winter, and age-2 pollock in autumn. There was no apparent tendency in the similarity of prey consumed by season, between depth zones, between areas, and between lengths; differences by length were not large.

Pacific halibut *(Hippoglossus stenolepis).* The daily rations estimated by MEB and ERS were low in winter and high in summer with small seasonal changes. Stomach content composition by season and length are shown in Table 2. Small individuals preyed mostly on crustaceans, such as hermit crabs. Large-sized halibut preyed mainly on pollock.

Flathead sole *(Hippoglossoides elassodon).* Daily rations were high in autumn and low in winter. Figure 5 shows stomach content composition by season and length from data collected in 1970-1975. Flathead sole 20-40 cm long preyed on ophiuroids mainly in winter; other prey items, such as Tanner crabs, were consumed by sole 20-30 cm long and pollock by sole 30-40 cm long. Individuals 10-20 cm long preyed on more euphausiids and individuals 40 cm long or greater preyed on ophiuroids, even though each length group preyed on pollock in spring. In summer, small sole (10-20 cm) preyed on shrimp, sole 20-30 cm long preyed on shrimp and smelt, 30-40 cm long preyed on shrimp and ophiuroids, and the large-sized individuals ($\geq$ 40 cm) preyed on ophiuroids. Sole 20 cm or more in length preyed primarily on ophiuroids in autumn. We conclude that stomach content composition between areas differs, although there a few differences between lengths.

Data collected after 1985 are shown in Table 2. The rate of ophiuroid consumption was lower than that of similar stomach content composition data collected in 1970-1975.

Rock sole *(Lepidopsetta bilineata).* Estimated daily rations were high in summer and low in winter. Figure 6 shows stomach content composition data by season and length for data collected in 1971-1976. Each rock sole length group preyed on polychaetes in winter, but rock sole 20-30 cm long preyed on more gammarus and those 30-40 cm long preyed on ophiuroids and other prey items. Each length group preyed mostly on polychaetes in spring, and large rock sole 40 cm long or greater preyed on more Pacific sand lance *(Ammodytes hexapterus)* as an alternate food. Many ophiuroids were consumed in autumn although numerous polychaetes were consumed in summer. We conclude that the difference in stomach content composition was large between depth zones although the difference between areas was small.

Yellowfin sole *(Limanda aspera).* Estimated daily rations were high in spring and summer and low in winter. Stomach content composition data by year, season, and length are shown in Fig. 7. Medium-sized yellowfin sole (20-30 cm) preyed primarily on shrimp and ophiuroids, and large individuals ($\geq$ 30 cm) preyed on polychaetes in winter before 1980. In spring, 10-30 cm long yellowfin sole preyed mainly on polychaetes and gammarus, and the large-sized ones ($\geq$ 30 cm) preyed on echiuroid worms and polychaetes. Small sole (10-20 cm) preyed primarily on polychaetes and gammarus, medium sole (20-30 cm) preyed on bivalves and polychaetes, and larger ones ($\geq$ 30 cm) preyed on bivalves and echiuroid worms in summer. In every length group, sea cucumbers were the main prey item in spring and summer, and gammarus were primarily consumed in autumn after 1980. In summer, large sole ($\geq$ 30 cm) preyed primarily on pollock. Comparison of stomach content composition data was large between years although the difference between areas was small. The difference by size of fish was not as large.

Alaska plaice *(Pleuronectes quadrituberculatus).* We have very few data on this species and report only that the main food items were polychaetes.

Rex sole *(Glyptocephalus zachirus).* We have very few data on this species and report only that the main food items were polychaetes and gammarus.

Summary of flatfish food habits: The young of large-mouthed and three species of large-sized flatfishes preyed mainly on crustaceans and the adults preyed on fish. Smaller-mouthed and five species of smaller-sized flatfishes preyed primarily on benthic invertebrates. The young of Greenland turbot (a large-mouthed species) preyed on pelagic animals, such as euphausiids. Flathead sole, which have a relatively large mouth compared to other smaller-mouthed species, preyed more on fishes than other species.

Wakabayashi (1986) analyzed samples from flounders collected in summer of 1970 and spring of 1971. He found no difference in the prey items consumed by yellowfin sole and Alaska plaice; however, large-sized rock sole preyed on more Pacific sand lance and polychaetes. Small Pacific halibut preyed mostly on shrimp.

Livingston et al. (1986) analyzed yellowfin sole stomachs sampled in 1982-1985. They found that yellowfin sole preyed mostly on bivalves in spring, and polychaetes, echiuroid worms, euphausiids, and shrimps in summer. However, they found very little predation on gammarid amphipods. In flathead sole, small individuals preyed on small crustaceans such as mysids and gammarus; the large-sized individuals preyed on ophiuroids (in summer they preyed on shrimp). Small Greenland turbot preyed on squids primarily and a relatively small amount of pollock in spring and autumn. Arrowtooth flounder preyed more on fish other than pollock in spring.

Rockfishes, Scorpaenidae

The food habits of rockfishes are summarized by Mito et al. (1996a).

Pacific ocean perch (POP) *(Sebastes alutus).* The percentage of stomach content weight compared to the body weight decreased with increasing body size. Our data suggest that feeding activity of small POP was high in spring and low in winter. Stomach content composition by year, season, and length are shown in Fig. 8. POP 20-40 cm long preyed primarily on euphausiids in autumn of 1972. Other food consumed consisted of mysids which were eaten by POP 20-30 cm long, and squids and mysids eaten by POP 30-40 cm long. In summer 1974, small POP preyed mainly on copepods while large POP preyed on euphausiids. POP 30 cm or less in length preyed mostly on euphausiids and those 30 cm or greater preyed on shrimps in winter of 1975. In spring 1975, small POP preyed mostly on copepods and euphausiids and the large-sized individuals preyed on euphausiids. During the summer in 1985, 20-40 cm long POP preyed primarily on euphausiids in the northern area of the upper slope. On the upper continental slope, POP 20-30 cm long preyed primarily on fishes and those 30 cm or more in length preyed on lanternfish and bristlemouth. On the outer shelf of the Aleutian Islands region in summer of 1986, POP 10-30 cm long preyed on euphausiids. When we summarize these data we conclude that small POP preyed mostly on copepods and euphausiids, medium-sized POP preyed on euphausiids, and large-sized POP preyed on fish, such as lanternfish.

Northern rockfish *(Sebastes polyspinis).* Small individuals preyed primarily on euphausiids and copepods, medium-sized individual preyed on squids and euphausiids, and large ones preyed on pollock.

Dusky rockfish *(Sebastes ciliatus).* Few data are available, but small fish preyed primarily on gammarus and the large ones preyed on age-0 pollock.

Rougheye rockfish *(Sebastes aleutianus).* The primary food items were fishes.

Blackgill rockfish *(Sebastes melanostomus).* The primary food items were pollock and Tanner crabs.

Shortraker rockfish *(Sebastes borealis).* The primary food items were fishes and shrimps.

The food habits of rockfishes can be summarized as follows. POP preyed on zooplankton, northern rockfish preyed on zooplankton and benthic invertebrates, and dusky rockfish preyed on benthic invertebrates. The larger-sized individuals of these three species preyed mostly on fishes.

Sculpins, Cottidae

The stomach contents of sculpins collected in the eastern Bering Sea were summarized by Mito et al. (1996b).

Thorny sculpin *(Icelus spiniger).* The primary food items were shrimps.

Spectacled sculpin *(Triglops scepticus).* The primary food items were euphausiids, gammarus, Hyperiidae, and squids.

Butterfly sculpin *(Melletes papilio).* The primary food items were Tanner crabs, gammarus, age-2 pollock and ostracods.

Yellow Irish lord *(Hemilepidotus jordani).* Small individuals preyed mostly on Tanner crabs and hermit crabs, and the ratio of Tanner crabs consumed increased with increasing fish length. Large-sized individuals preyed on pollock.

Darkfin sculpin *(Malacocottus zonurus)*. Small darkfin sculpins preyed primarily on polychaetes and gammarus; the larger ones consumed these prey plus numerous Tanner crabs and sea cucumbers.

Plain sculpin *(Myoxocephalus jaok)*. The primary food item was pollock and second Tanner crabs.

Great sculpin *(Myoxocephalus polyacanthocephalus)*. Small individuals (30-40 cm) preyed mostly on Tanner crabs, and the large-sized fish (40-70 cm) consumed pollock.

Warty sculpin *(Myoxocephalus verrucosus)*. The primary food items were age-2 pollock and Tanner crabs.

Spinyhead sculpin *(Dasycottus setiger)*. Small individuals (10-20 cm) preyed on shrimp, medium-sized fish (20-25 cm) preyed on shrimp and Tanner crabs, and large-sized individuals preyed on age-1 and age-2 pollock, shrimp, and Tanner crabs.

Bigmouth sculpin *(Hemitropterus bolini)*. Fish 20 cm or greater in length preyed primarily on pollock, small 20-30 cm long preyed on age-1 pollock, and large-sized individuals preyed on age-2 and age-3 pollock.

Thus, small species of sculpins preyed mostly on benthic invertebrates, except the spectacled sculpin which preyed on zooplankton and benthic invertebrates. Large-sized species preyed primarily on fish.

Eelpouts and pricklebacks, Zoarcidae and Stichaeidae

The stomach contents of eelpouts and pricklebacks collected in the eastern Bering Sea were summarized by Mito et al. (1996c).

***Bothrocara* spp.** The primary food items were mysids and blacksmelts.

Wattled eelpout *(Lycodes palearis)*. Small eelpouts preyed mostly on gammarus, Tanner crabs, and shrimps, and large-sized individuals preyed on eelpouts and shrimps.

Shortfin eelpout *(Lycodes brevipes)*. The primary food items were ophiuroids, polychaetes, and gammarus.

***Lycodes* spp.** The primary food items were fish and shrimp.

Longsnout prickleback *(Lumpenella longirostris)*. The primary food items were polychaetes.

Poachers, Agonidae

The stomach contents of poachers were summarized by Mito et al. (1996c).

Sturgeon poacher *(Podothecus acipenserinus)*. The primary food items were fish and gammarus.

Sawback poacher *(Sarritor frenatus)*. The primary food items were gammarus.

***Asterotheca* spp.** The primary food items were mysids and shrimp.

Snailfishes, Liparidae

The stomach contents of snailfishes collected in the eastern Bering Sea were summarized by Mito et al. (1996c).

Polka-dot snailfish *(Liparis cyclostigma).* Small individuals (20-40 cm) preyed chiefly on Tanner crabs and age-1 pollock, and large-sized individuals (40-60 cm) preyed on age-2 pollock and poachers.

Salmon snailfish *(Careproctus rastrinus).* The primary food items were gammarus.

Blacktail snailfish *(Careproctus melanurus).* Small individual (30-50 cm) preyed mostly on gammarus and shrimp, and large-sized individuals (50-60 cm) preyed on gammarus and fish.

Skates, Rajidae

Alaska skate *(Bathyraja parmifera).* Small skates (20-40 cm) preyed on gammarus, skates 40-50 cm long preyed on Tanner crabs, shrimps, and age-1 pollock, and those 50-120 cm long preyed on age-2 and age-3 pollock.

Bering skate *(Bathyraja interrupta).* The main prey items changed from gammarus, shrimp, and Tanner crabs, to age-2 pollock for skates 20-80 cm long and as length increases.

Other Families

Pacific herring *(Clupea pallasi).* The primary food items were euphausiids.

Sablefish *(Anoplopoma fimbria).* The primary food item was pollock in every length range, and the proportion of age-2 pollock increased in the larger individuals.

Atka mackerel *(Pleurogrammus monopterygius).* The primary food items were euphausiids.

Searcher *(Bathymaster signatus).* Small fish 20-25 cm long preyed on polychaetes, those 25-35 cm long preyed on shrimps and fish, and those 35-40 cm long preyed on age-1 pollock.

Prowfish *(Zaprora silenus).* The primary food items were medusas.

Red squid *(Berryteuthis magister).* The primary food items were fishes.

Food Relationships Among Groundfishes

We simplified the results on the food relationship of groundfishes in the eastern Bering Sea presented by Mito (1995), and summarize those results here.

Primary prey items

We grouped predators according to their primary prey and present them in Table 3. For example, the *Thysanoessa inermis* predator group consisted of small pollock, northern rockfish (medium), Pacific herring, and Atka mackerel (small). Many species such as the large-sized pollock, Pacific cod, Greenland turbot, and arrowtooth flounder were grouped as predators of pollock.

Food niche

Determination of a food niche indicates the position of each species within its trophic community, and is useful for understanding food relationships between species. The food niche was designed to show the position of each species within its community in two dimensions (Fig. 9). The vertical axis indicates the index of trophic level (ITL) and the horizontal axis is the index of vertical distribution (IVD). As for ITL, 0 is the producer, 1 is the first consumer, 2 is adjusted to the second consumer, etc. The ITL of a predator preyed on animals across two or more trophic levels had the value below the decimal point. IVD of 0 shows all prey consumed by a predator are pelagic and 1 shows that all prey consumed by the predator are demersal. The same species is connected in a line. For small pollock, the ITL and IVD were about 2.6 and 0.0 respectively, and for large pollock, the values were about 3.3 and 0.1 respectively. In flathead sole, the yellowfin sole and rock sole, which are benthic predators, the ITL and IVD were 2.5-3.2 and 0.7-1.0 respectively. In Pacific cod, Greenland turbot and arrowtooth flounder, which are predators of pollock, the values were 3.3-4.0 and 0.0-0.3 respectively. These analyses suggest, for example, that flathead sole preys on more pelagic prey and is at a higher trophic level than yellowfin sole and rock sole.

Structure of food web

The food web in the northern area during winter is shown in Fig. 10. The figure shows that age-3 pollock was the most abundant species in this area.

A basic food chain by season and area is shown in Fig. 11. The basic food chain was constructed from copepods or euphausiids, age-2 or -3 pollock to age-6 and more pollock. Copepods were more important in spring and in the northern area, age-1 pollock appeared in north area, and age-0 pollock were in the shallow southern area in autumn. We suggest that these structural differences were caused by the difference in species composition which defined the community. Moreover, the basic food chain was a grazing food chain which originates from phytoplankton.

[Discussion follows tables and figures.]

Table 1. Ecological and biological characteristics of commercially important species of groundfishes in the Bering Sea.

	Walleye pollock *(Theragra chalcogramma)*	Pacific cod *(Gadus macrocephalus)*
Spawning periods	March-August on the eastern Bering Sea shelf and January-March in the southeastern Aleutian Basin.	January-March on the eastern and January-May o the western Bering Sea shelf.
Spawning ground	Eastern Bering Sea shelf edge and southeastern Aleutian Basin.	Continental shelf edge.
Characteristics of egg	Pelagic	Demersal
Diameter of egg	About 1.7 mm	About 1.4 mm
Fecundity	100,000 eggs at length 35 cm, 1,400,000 eggs at length 75 cm	1.4-6.4 million eggs
Development	Eggs hatch in 28 days at 3°C. Length of larvae at hatching is about 5 mm, and will absorb the yolk in about 20 days. Larvae become 2 mm juveniles about 50 days after hatching, and the number of fin rays is completed.	–
Growth	Females grow faster than males and the fish on the eastern Bering Sea shelf grow faster than in the Aleutian Basin. Length at age-10 is 60 cm for males and 63 cm for females on the eastern Bering Sea shelf, and 50 cm for males and 52 cm for females in the Aleutian Basin.	30 cm at age-2, 54 cm at age-5, and 84 cm at age-9
Age at 50% maturity (female)	3-4 yr	6 yr
Average size at maturity (female)	36 cm	63 cm
Maximum age	16 yr on the eastern Bering Sea shelf and 25 yr in the Aleutian Basin.	14 yr
Natural mortality coefficient	0.3 on the eastern Bering Sea shelf and 0.2 in the Aleutian Basin.	0.22-0.45
Distribution and migration	Continental shelf and midwater layer of the basin (entire Bering Sea). Eggs and larvae are carried from the spawning ground by the current. In the eastern continental shelf, eggs and larvae extend northwest from Unimak Island. In the Aleutian Basin, the eggs and larvae drift from the southeast area of the basin to the eastern continental shelf. Age-0 walleye pollock is distributed in the Aleutian Basin though the density is low. Age-1 to age-4 fish are not distributed in the basin. On the eastern Bering Sea shelf, walleye pollock inhabit the upper slope in winter, spawn on the outer shelf in spring, then extend widely to the continental shelf and feed actively. In the Aleutian Basin, walleye pollock spawn in the southeast area, then move to the northwest in summer, and turn to the southeast in autumn.	Bering Sea shelf including the Aleutian Islands region, mainly at 50-200 m depth. After spawning at the continental shelf edge, they move to the shallower area in spring for feeding
Stock	Stock identification is not well clarified. Four stocks are identified for management: the eastern Bering Sea shelf, the western Bering Sea shelf, the Aleutian Basin, and the Aleutian Islands region.	–
Biomass	8.94-12.2 million tons by bottom trawl and acoustic/midwater trawl surveys in the eastern Bering Sea in 1978-1988. Range of 400,000-820,000 tons by bottom trawl surveys in the Aleutian Islands region in 1980-1986. About 2.5 million tons by acoustic/midwater trawl surveys in the Aleutian Basin in 1988.	710,000-1,180,000 tons in the eastern Bering Sea and 90,000-180,000 tons in the Aleutian Islands region in the 1980s.
Total allowable catch	1.3 million tons in the eastern Bering Sea after 1990.	Increased from 70,000 tons in 1980 to 280,000 tons in 1987.
Catch	1.2-1.3 million tons in the eastern Bering Sea after 1991.	Increased from 50,000 tons in 1980 to 200,000 tons in 1988.

Table 1. (Continued.) Ecological and biological characteristics of commercially important species of groundfishes in the Bering Sea.

	Arrowtooth flounder *(Atheresthes stomias)*	Greenland turbot *(Reinhardtius hippoglossoides)*	Pacific halibut *(Hippoglossus stenolepis)*
Spawning periods	December-February	October-December	November-March
Spawning ground	Not clarified	Continental slope	Continental slope
Characteristics of egg	Pelagic	Pelagic, distributed deep layer	Pelagic
Diameter of egg	About 3 mm	4.0-4.5 mm	3.0-3.5 mm
Fecundity	130,000-500,000	2,000-145,000	102,000 at length 75 cm and 2,801,000 at length 135 cm.
Development	–	–	Eggs hatch in about 20 days and the length is about 7 mm. After six months of drifting, the larvae become juvenile at length 3.5 cm.
Growth	About 8 cm at age-1, 34 cm at age-5, and 53 cm at age-10.	About 11 cm at age-1, 38 cm at age-5, 59 cm at age-10 and 74 cm at age-15.	Female grows faster than male. About 7 cm at age-1, 49 cm at age-5, 83 cm at age-10, and 106 cm at age-15.
Age at 50% maturity (female) or maturity age	The maturity age is 6-7 years old in male and 9-10 years old in female.	7-8 yr	12 yr
Average size at maturity (female)	–	66 cm	120 cm
Maximum age	About 15 yr	19 yr	42 yr
Natural mortality coefficient	0.20	0.18	0.20
Distribution and migration	Outer shelf to upper slope of the eastern Bering Sea, 100-500 m depth mainly including the Aleutian Islands region. Young fish distribute on the continental shelf in summer, adult fish in the continental slope.	Continental slope, 400-1,000 m depth including the Aleutian Islands region. Adults are distributed on the continental slope and young fish are distributed on the continental shelf in summer.	Continental shelf and upper slope, chiefly 50-250 m depth including the Aleutian Islands region. To 100-500 m depth between Unimak Island and Pribilof Islands in winter, moving to 50-150 m depth.
Stock	–	–	The stock is thought to be the same in the eastern Bering Sea and in the Gulf of Alaska to California. Another stock is distributed in the northern and northwestern Bering Sea.
Biomass	Increased from 70,000 tons in 1979 to 340,000 tons in 1988 in the eastern Bering Sea. Increased from 40,000 tons in 1980 to 130,000 tons in 1986 in the Aleutian Islands region.	Decreased from 350,000 tons in 1979 to 60,000 tons in 1988 in the eastern Bering Sea, 50,000-80,000 tons in the Aleutian Islands region in 1980-1986.	About 130,000 tons in the eastern Bering Sea including the Aleutian Islands region
Total allowable catch	5,500-20,000 tons in 1986-1990	Decreased from 33,000 to 7,000 tons in 1986-1990.	–
Catch	3,200-18,400 tons in 1980s.	Decreased from 57,000 to 7,000 tons in 1980s.	Increased from 257 tons in 1980 to 1,786 tons in 1986.

Table 1. (Continued.) Ecological and biological characteristics of commercially important species of groundfishes in the Bering Sea.

	Flathead sole (*Hippoglossoides elassodon*)	Rock sole (*Lepidopsetta bilineata*)	Yellowfin sole (*Limanda aspera*)
Spawning periods	February-March	March-June	June-September
Spawning ground	Mid-shelf	Mid-shelf	Inner shelf
Characteristics of egg	Pelagic	Attaching demersal	Pelagic
Diameter of egg	2.75-3.75 mm	0.92 mm	0.7-0.8 mm
Fecundity	52,000 eggs at length 22-26 cm and 160,000 eggs at length 40-43 cm.	152,000 eggs at length 22-24 cm and 404,000 eggs at length 40-43 cm.	1,300,000 eggs at length 25-30 cm and 3,300,000 eggs at length 40-45 cm.
Development	–	–	–
Growth	About 7 cm at age-1, about 21 cm at age-2, and about 33 cm at age-10.	Female grows faster than male, about 11 cm at age-2, about 23 cm at age-5, and about 37 cm at age-10.	Little difference in growth rate between sexes. About 6 cm at age-1, about 18 cm at age-5, and about 28 cm at age-10.
Age at 50% maturity (female) or maturity age	Maturity age is about 6 yr	Maturity age is about 5 yr	9 yr
Average size at maturity (female)	–	–	26 cm
Maximum age	20 yr	19 yr	22 yr
Natural mortality coefficient	0.20-0.23	0.20-0.23	0.12
Distribution and migration	Continental shelf including the Aleutian Islands region, chiefly 50-200 m depth. Distributed from the outer shelf to the upper slope in winter, move to 20-180 m depth in spring.	Chiefly distributed on the continental shelf at 10-100 m depth including the Aleutian Islands region; move to deeper area in winter and shallower area in summer.	Chiefly on the continental shelf at 10-100 m depth except the Aleutian Islands region. Adult fish form dense schools and inhabit the bottom layer of 100-300 m depth in winter. Wintering grounds are north of Unimak Island.
Stock	–	–	It is guessed that the stock is divided into two: southeast and northwest of St. George Island on the eastern Bering Sea shelf.
Biomass	Increased from 120,000 tons in 1980 to 650,000 tons in 1990 on the eastern Bering Sea shelf, though the biomass in the Aleutian Islands region is about 10,000 tons.	Increased from 280,000 tons in 1980 to 1,410,000 tons in 1990 in the eastern Bering Sea shelf; the biomass in the Aleutian Islands region is about 30,000 tons.	About 2-3 million tons in the eastern Bering Sea shelf in the 1980s.
Total allowable catch	–	–	117,000-254,000 tons in the eastern Bering Sea.
Catch	About 5,000 tons in the 1980s	Increased from 7,600 tons in 1980 to 63,000 in 1988, and decreased afterward.	Increased from 87,000 tons in 1980 to 227,000 tons in 1985, and about 200,000 tons until 1988, afterward decreased.

Table 1. (Continued.) Ecological and biological characteristics of commercially important species of groundfishes in the Bering Sea.

	Alaska plaice (*Pleuronectes quadrituberculatus*)	Pacific ocean perch (*Sebastes alutus*)	Pacific herring (*Clupea pallasi*)
Spawning periods	April-June	Egg viviparity, larvae spawned from March to June.	May-July
Spawning ground	Mid-shelf	Continental slope	Shore reef coast, shallower lagoon, creek, and bay. Eggs spawned in marine algae.
Characteristics of egg	Pelagic	–	Attaching demersal
Diameter of egg	1.95 mm	–	–
Fecundity	56,000 eggs at length 28-30 cm and 313,000 eggs at length 48-50 cm.	Number of larvae is 29,000-103,000 at length 32-44 cm.	27,000-78,000
Growth	About 21 cm at age-5, and about 35 cm at age-10.	About 7 cm at age-1, about 23 cm at age-5, about 31 cm at age-10, and about 41 cm at age-20.	About 9 cm at age-1 and about 25 cm at age-5.
Age at 50% maturity (female) or maturity age	6-7 yr	8 yr	Average maturity age is about 3 yr
Average size at maturity (female)	31 cm	28 cm	–
Maximum age	17 yr	90 yr	12 yr
Natural mortality coefficient	0.20-0.23	0.05	–
Distribution and migration	Chiefly distributed in the continental shelf at 40-110 m depth except the Aleutian Islands region; move to deeper area in winter and shallower area in summer.	Chiefly distributed on the upper slope at 200-300 m depth in the eastern Bering Sea including the Aleutian Islands region. Main dense schools in the eastern Bering Sea continental slope are distributed in the vicinity of Pribilof Islands.	Pelagic but caught by trawl. They are distributed on the continental shelf, and extend to the Aleutian Basin in summer. Inhabit the area northwest of the Pribilof Islands in winter, and move to coastal spawning ground in spring.
Stock	–	Divided into the eastern continental slope stock and the Aleutian Islands region stock.	Different stocks in the eastern and western Bering Sea. There are spawning grounds in three places (Bristol Bay, Yukon River mouth, and Norton Sound) in the eastern Bering Sea and it is possible there is another stock.
Biomass	350,000-940,000 tons on the eastern Bering Sea shelf in the 1980s.	Increased from about 10,000 tons in 1980 to about 40,000 tons in 1988 in the eastern Bering Sea shelf. Increased from 110,000 tons to 220,000 tons in the Aleutian Islands region.	350,000-940,000 tons on the eastern Bering Sea shelf in the 1980s.
Catch	Increased from 6,900 tons in 1980 to 61,600 tons in 1988.	The maximum catch was 109,100 tons in the Aleutian Islands region in 1965, and 47,000 tons in the eastern Bering Sea in 1961. Decreased and reached lowest number in the mid-1980s.	Increased from 6,900 tons in 1980 to 61,600 tons in 1988.

Table 1. (Continued.) Ecological and biological characteristics of commercially important species of groundfishes in the Bering Sea.

	Sablefish *(Anoplopoma fimbria)*	Atka mackerel *(Pleurogrammus monopterygius)*
Spawning periods	January-April	Summer
Spawning ground	Continental slope	–
Characteristics of egg	Separating pelagic	Attaching demersal
Diameter of egg	About 2.1 mm	About 2.5 mm
Fecundity	100,000 eggs at length 53 cm and 1,280,000 eggs at length 98 cm.	100,000 eggs at length 53 cm and 1,280,000 eggs at length 98 cm.
Growth	About 28 cm at age-1, 55 cm at age-5, 71 cm at age-10, and 86 cm at age-20.	About 26 cm at age-2, about 36 cm at age-5, and about 40 cm at age-10.
Age at 50% maturity (female) or maturity age	5 yr	Average maturity age is 3 yr
Average size at maturity (female)	58 cm	–
Maximum age	55 yr	12 yr
Natural mortality coefficient	0.10	0.30
Distribution and migration	Mainly distributed at 400-1,000 m depth on the continental slope which includes the Aleutian Islands region. Young fish are distributed shallower than 100 m depth; adult fish are deeper than 200 m depth.	Chiefly distributed in the Aleutian Islands region. Young fish are pelagic and adults are distributed in the bottom layer of the continental shelf.
Stock	–	–
Biomass	29,000-77,000 tons in the eastern Bering sea and 30,000-80,000 tons in the Aleutian Islands region in the 1980s.	200,000-550,000 tons in the Aleutian Islands region.
Catch	1,250-4,200 tons in the eastern Bering Sea and 270-3,800 tons in the Aleutian Islands region in the 1980s.	4,000-5,000 tons in the eastern Bering Sea and 11,000-38,000 tons in the Aleutian Islands region in the 1980s.

Table 2. Summary of food habit data on Bering Sea groundfishes.

Species	No. individuals	Range of length	Year and season	Area and depth	DR or RSW (% body weight)	Size of predator	Main food items
Codfishes (Gadidae)							
Walleye pollock (*Theragra chalcogramma*)	35,864	2-80 cm	1970-1994	Entire Bering Sea			
			1970-1985	Entire Bering Sea	DR: by ERS, 3.1-3.2% at age-1; 1.2-1.4% at age-6 and above		Fig. 1
			1986 spring and summer	AIR		≤35 cm	Euphausiids
						≥35 cm	Fishes such as Pacific sand lance
			1989 summer	Surface of southern shelf in EBS		3-6 cm	Copepods
						6-8 cm	Euphausiids
				Surface of northern shelf in EBS		5-8 cm	Copepods
						8-9 cm	Copepods, euphausiids
				Midwater of AB		4-8 cm	Copepods
			1993 winter	Midwater of AB		40-55 cm	Lanternfish
			1994 summer	Midwater of AB		45-50 cm	Fishes such as lanternfish, copepods
						50-55 cm	Euphausiids, fishes
						55-60 cm	Fishes such as lanternfish
Pacific cod (*Gadus macrocephalus*)	5,051	15-100 cm	1970-1985	EBS	DR: by MEB and ERS, 1.3% and 1.5% at length 50-60 cm; 0.9% and 1.3% at ≥75 cm		Fig. 2
Grenadiers (Macrouridae)							
Giant grenadier (*Albatrossia pectoralis*)	384	10-50 cm (anus length)	1985 spring and summer	Outer shelf to lower slope in EBS		10-20 cm	Squids
						20-40 cm	Squids, fishes
						40-50 cm	Fishes
Popeye grenadier (*Coryphaenoides cinereus*)	31	5-15 cm (anus length)	1985 summer	Upper slope in EBS		5-10 cm	Shrimps
						10-15 cm	Fishes

Table 2. (Continued.) Summary of food habit data on Bering Sea groundfishes.

Species	No. individuals	Range of length	Year and season	Area and depth	DR or RSW (% body weight)	Size of predator	Main food items
Flatfishes (Pleuronectidae)							
Arrowtooth flounder *(Atheresthes stomias)*	2,116	10-75 cm	1972-1985	EBS	DR: by MEB and ERS, 2.0% and 0.8-1.5% at length 20-30 cm; 1.3% and 0.5-1.2% at ≥55 cm		Fig. 3
Greenland turbot *(Reinhardtius hippoglossoides)*	1,234	10-95 cm	1972-1985	EBS	DR: by MEB and ERS, 1.6% and 1.2-1.7% at length 40-50 cm; 1.2-1.3% and 0.8-1.1% at ≥75 cm		Fig. 4
Pacific halibut *(Hippoglossus stenolepis)*	1,098	15-210 cm	1970-1975	EBS	DR: by MEB and ERS, 1.2% and 1.1-1.3% at length 40-60 cm; 0.8% and 0.6-0.7% at 120-150 cm		
			Winter			30-40 cm	Hermit crabs, age-1 walleye pollock, Tanner crabs
						40-60 cm	Age-2 walleye pollock
			Spring			30-40 cm	Tanner crabs, hermit crabs
						40-60 cm	Age-2 walleye pollock
						60-150 cm	Age-3 walleye pollock
			Summer			15-30 cm	Pacific sand lance, shrimps
						30-40 cm	Walleye pollock, Pacific sand lance
						40-100 cm	Walleye pollock, cods
						100-120 cm	Age-2 walleye pollock
						120-150 cm	Age-4 and 5 walleye pollock
			Autumn			40-60 cm	Fishes

AB = Aleutian Basin, AIR = Aleutian Island Region, DR = Daily Ration, EBS = Eastern Bering Sea, ERS = Evacuation Rate of food in Stomach, MEB = Method of Energy Budget, RSW = Ratio of Stomach content Weight

Table 2. (Continued.) Summary of food habit data on Bering Sea groundfishes.

Species	No. individuals	Range of length	Year and season	Area and depth	DR or RSW (% body weight)	Size of predator	Main food items
Flathead sole (*Hippoglossoides elassodon*)	3,304	5-50 cm	1970-1986				
			1970-1975		DR: by MEB and ERS, 2.0% and 0.4-1.2% at length 20-30 cm; 1.4% and 0.5-0.9% at ≥40 cm		Fig. 5
			1985 summer	Northern upper slope in EBS		20-30 cm	Ophiuroids, shrimps, gammarus
						30-50 cm	Ophiuroids
			1986 spring	Outer shelf in AIR		20-30 cm	Euphausiids
						30-40 cm	Tanner crabs, pricklebacks
Rock sole (*Lepidopsetta bilineata*)	2,303	10-50 cm	1971-1986				
			1971-1976		DR: by MEB and ERS, 1.9% and 0.6-1.2% at length 20-30 cm; 1.6% and 0.4-1.0% at 30-40 cm		Fig. 6
			1986 summer	Upper slope in AIR		30-40 cm	Isopods
Yellowfin sole (*Limanda aspera*)	4,653	5-45 cm	1971-1983	EBS	DR: by MEB and ERS, 1.6% and 0.4-1.0% at length 20-30 cm; 1.6% and 0.2-0.9% at ≥30 cm		Fig. 7
Alaska plaice (*Pleuronectes quadrituberculatus*)	1,006	5-50 cm	1970-1975		DR: by MEB, 1.7% at length 20-30 cm; 1.4% at ≥40 cm		
			Winter			20-30 cm	Polychaetes
			Summer				Polychaetes
Rex sole (*Glyptocephalus zachirus*)	154	20-60 cm	1972-1985	Outer shelf to upper slope in EBS	RSW: 0.98% at length 20-30 cm, and 0.54% at 30-40 cm		
			1972 autumn	Southern upper slope in EBS		20-40 cm	Polychaetes, gammarus
			1974 summer	Southern outer shelf in EBS		30-40 cm	Gammarus, polychaetes
Rockfishes (Scorpaenidae)							
Pacific ocean perch (*Sebastes alutus*)	1,022	10-50 cm	1972-1986	EBS and AIR	RSW: 0.39% at length 10-20 cm, 0.20% at 20-30 cm, 0.11% at 30-40 cm.		Fig. 8

Table 2. (Continued.) Summary of food habit data on Bering Sea groundfishes.

Species	No. individuals	Range of length	Year and season	Area and depth	DR or RSW (% body weight)	Size of predator	Main food items
Northern rockfish (*Sebastes polyspinis*)	266	10-50 cm	1972-1986		RSW: 0.23-0.26%		
			1972 autumn	Southern outer shelf in EBS		20-30 cm	Euphausiids, shrimps
						30-40 cm	Squids, hermit crabs
			1974 summer			20-30 cm	Shrimps
						30-40 cm	Squids
			1975 winter			20-30 cm	Euphausiids, copepods
			1975 spring			20-30 cm	Euphausiids, copepods
			1986 summer	AIR		20-40 cm	Euphausiids
Dusky rockfish (*Sebastes ciliatus*)	32	20-50 cm	1972-1975	Outer shelf to upper slope in EBS	RSW: 0.13%	30-40 cm	Gammarus
						40-50 cm	Age-0 walleye pollock
Rougheye rockfish (*Sebastes aleutianus*)	38	10-50 cm	1972-1986	Outer shelf to upper slope in EBS	RSW: 0.05%	40-50 cm	Fishes
Blackgill rockfish (*Sebastes melanostomus*)	46	10-50 cm	1972-1975	Outer shelf to upper slope in EBS	RSW: 0.38%		
			1974 summer			30-40 cm	Age-2 walleye pollock, Tanner crabs
			1975 spring			30-40 cm	Age-1 walleye pollock
Shortraker rockfish (*Sebastes borealis*)	80	20-80 cm	1986 summer	Slope in AIR	RSW: 0.07%	20-60 cm	Fishes, shrimps
						60-80 cm	Lanternfish
Sculpins (Cottidae)							
Thorny sculpin (*Icelus spiniger*)	57	10-25 cm	1972-1974	Outer shelf to upper slope in EBS	RSW: 2.68%		
			1972 autumn			20-25 cm	Shrimps
			1974 summer			10-25 cm	Shrimps

AB = Aleutian Basin, AIR = Aleutian Island Region, DR = Daily Ration, EBS = Eastern Bering Sea, ERS = Evacuation Rate of food in Stomach, MEB = Method of Energy Budget, RSW = Ratio of Stomach content Weight

Table 2. (Continued.) Summary of food habit data on Bering Sea groundfishes.

Species	No. individuals	Range of length	Year and season	Area and depth	DR or RSW (% body weight)	Size of predator	Main food items
Spectacled sculpin (*Triglops scepticus*)	62	5-25 cm	1972-1975	Outer shelf to upper slope in EBS	RSW: 0.75%		
			1974 summer			10-15 cm	Euphausiids, gammarus, Hyperiidae
						15-20 cm	Hyperiidae, squids
			1975 spring			20-25 cm	Euphausiids
Butterfly sculpin (*Melletes papilio*)	64	20-40 cm	1975	Outer shelf to upper slope in EBS	RSW: 0.93%		
			1975 winter			20-25 cm	Tanner crabs
						25-30 cm	Gammarus
						30-40 cm	Age-2 walleye pollock
			1975 spring			20-25 cm	Ostracods
						25-30 cm	Tanner crabs
Yellow Irish lord (*Hemilepidotus jordani*)	362	15-50 cm	1972-1975	Outer shelf to upper slope in EBS	RSW: 0.99% at length 20-25 cm, 2.50% at 40-45 cm		
			1972 autumn			30-40 cm	Tanner crabs
			1974 summer	Southern outer shelf in EBS		25-30 cm	Hermit crabs, Tanner crabs, age-1 walleye pollock, polychaetes
						30-35 cm	Tanner crabs
						35-40 cm	Gammarus, Tanner crabs
						40-50 cm	Age-2 walleye pollock
			1975 winter			20-40 cm	Tanner crabs
			1975 spring			20-30 cm	Tanner crabs, hermit crabs
						30-40 cm	Tanner crabs
						40-50 cm	Age-2 walleye pollock

Table 2. (Continued.) Summary of food habit data on Bering Sea groundfishes.

Species	No. individuals	Range of length	Year and season	Area and depth	DR or RSW (% body weight)	Size of predator	Main food items
Darkfin sculpin (*Malacocottus zonurus*)	267	10-35 cm	1972-1985	Outer shelf to upper slope in EBS	RSW: 2.93% at length 10-15 cm, 1.82% at 25-30 cm		
			1972 autumn			15-20 cm	Polychaetes
						20-25 cm	Tanner crabs, polychaetes
						25-30 cm	Squids, Tanner crabs
			1974 summer			15-25 cm	Polychaetes, gammarus
						25-30 cm	Polychaetes, shrimps
			1975 winter			Small fish	Polychaetes
						Medium fish	Gammarus, Tanner crabs
						Large fish	Gammarus
			1975 spring			Small fish	Gammarus, polychaetes
						Large fish	Tanner crabs, polychaetes
			1985 summer			10-15 cm	Squids, shrimps
						15-20 cm	Polychaetes
						20-25 cm	Polychaetes, shrimps
						25-30 cm	Polychaetes, sculpins
Plain sculpin (*Myoxocephalus jaok*)	107	20-60 cm	1975	Outer shelf to upper slope in EBS	RSW: 0.67% at length 30-40 cm, 1.27% at 40-50 cm, 2.13% at 50-60 cm		
			1975 winter			30-40 cm	Age-1 and age-2 walleye pollock
						40-60 cm	Tanner crabs, Greenland turbot
			1975 spring				Walleye pollock, Tanner crabs

AB = Aleutian Basin, AIR = Aleutian Island Region, DR = Daily Ration, EBS = Eastern Bering Sea, ERS = Evacuation Rate of food in Stomach, MEB = Method of Energy Budget, RSW = Ratio of Stomach content Weight

Table 2. (Continued.) Summary of food habit data on Bering Sea groundfishes.

Species	No. individuals	Range of length	Year and season	Area and depth	DR or RSW (% body weight)	Size of predator	Main food items
Great sculpin (*Myoxocephalus polyacanthocephalus*)	173	20-80 cm	1972-1985	Outer shelf to upper slope in EBS	RSW: 3.37% at length 30-40 cm, 4.94% at 40-50 cm, 3.99% at 50-60 cm, 3.14% at 60-70 cm		
			1972 autumn			30-40 cm	Tanner crabs
						40-50 cm	Walleye pollock, age-2 walleye pollock, Tanner crabs
						50-60 cm	Age-4 and 5 walleye pollock
			1974 summer			Small fish	Tanner crabs
						Large fish	Age-2 walleye pollock
			1975 winter				Age-2 walleye pollock
			1975 spring			Small fish	Tanner crabs
						Large fish	Age-2 and age-3 walleye pollock
Warty sculpin (*Myoxocephalus verrucosus*)	19	20-60 cm	1975	Outer shelf in EBS	RSW: 5.21%		
			1975 winter			30-40 cm	Tanner crabs
			1975 spring			40-50 cm	Tanner crabs, age-2 walleye pollock
Spinyhead sculpin (*Dasycottus setiger*)	223	10-40 cm	1972-1985	Outer shelf to upper slope in EBS	RSW: 3.37% at length 10-15 cm, 2.21% at 25-30 cm		
			1972 autumn			15-20 cm	Shrimps
						20-30 cm	Shrimps, Tanner crabs
			1974 summer			Small fish	Shrimps
						Large fish	Shrimps, Tanner crabs, fishes
			1975 winter			Small fish	Tanner crabs, shrimps
						Large fish	Tanner crabs, fishes
			1975 spring				Age-1 and age-2 walleye pollock
			1985 summer			10-20 cm	Shrimps

Table 2. (Continued.) Summary of food habit data on Bering Sea groundfishes.

Species	No. individuals	Range of length	Year and season	Area and depth	DR or RSW (% body weight)	Size of predator	Main food items
Bigmouth sculpin (*Hemitropterus bolini*)	320	10-70 cm	1972-1985	Outer shelf to upper slope in EBS	RSW: 3.37% at length 20-30 cm, 4.20% at 30-40 cm, 4.85% at 40-50 cm, 2.99% at 50-60 cm, 1.93% at 60-70 cm		
			1972 autumn			30-40 cm	Age-2 walleye pollock
						40-50 cm	Age-3 walleye pollock
						50-60 cm	Pacific cod, age-4, 5 walleye pollock
			1974 summer			Small fish	Eelpouts, flathead sole
						Medium fish	Eelpouts
						Large fish	Age-2 walleye pollock
			1975 winter			Small fish	Age-1 walleye pollock
						Medium fish	Age-2 walleye pollock
						Large fish	Fishes
			1975 spring			Small fish	Age-1 walleye pollock
						Medium fish	Age-3 and age-2 walleye pollock
						Large fish	Greenland turbot, age-2 walleye pollock
			1985 summer			20-30 cm	Fishes
Eelpouts and pricklebacks (Zoarcidae and Stichaeidae)							
Bothrocara sp.	28	20-70 cm	1985 summer	Upper slope in EBS	RSW: 1.47%	20-30 cm	Mysids
						30-40 cm	Deepsea smelts

AB = Aleutian Basin, AIR = Aleutian Island Region, DR = Daily Ration, EBS = Eastern Bering Sea, ERS = Evacuation Rate of food in Stomach, MEB = Method of Energy Budget, RSW = Ratio of Stomach content Weight

Table 2. (Continued.) Summary of food habit data on Bering Sea groundfishes.

Species	No. individuals	Range of length	Year and season	Area and depth	DR or RSW (% body weight)	Size of predator	Main food items
Wattled eelpout (*Lycodes palearis*)	131	20-60 cm	1972-1975	Outer shelf to upper slope in EBS	RSW: 0.41% at length 30-40 cm, 0.37% at 40-50 cm, 0.30% at 50-60 cm		
			1972 autumn			30-40 cm	Tanner crabs
						40-50 cm	Shrimps, gammarus
			1974 summer			Small fish	Shrimps
						Large fish	Eelpouts, gammarus
			1975 spring				Gammarus
Shortfin eelpout (*Lycodes brevipes*)	50	10-35 cm	1974-1975	Outer shelf to upper slope in EBS	RSW: 0.18%		
			1974 summer			15-25 cm	Ophiuroids
						25-35 cm	Polychaetes, gammarus
Lycodes sp.	20	21-37 cm	1985 summer	Upper slope in EBS	RSW: 0.31%		Fishes, shrimps
Longsnout prickleback (*Lumpenella longirostris*)	13	25-33 cm	1985 summer	Upper slope in EBS	RSW: 0.25%		Polychaetes
Poachers (Agonidae)							
Sturgeon poacher (*Podothecus acipenserinus*)	54	20-35 cm	1975	Outer shelf to upper slope in EBS	RSW: 0.32%		
			1975 winter			20-25 cm	Fishes
						25-30 cm	Gammarus
			1975 spring			25-30 cm	Gammarus
Sawback poacher (*Sarritor frenatus*)	113	15-30 cm	1972-1975	Outer shelf to upper slope in EBS	RSW: 0.60% at length 20-25 cm, 0.63% at 25-30 cm		
			1972 autumn			15-25 cm	Gammarus
			1974 summer				Gammarus
			1975 winter			Small fish	Shrimps, gammarus
						Large fish	Gammarus
			1975 spring			20-25 cm	Gammarus
Asterotheca sp.	30	16-23 cm	1985, summer	Upper slope in EBS	RSW: 0.51%		Mysids, shrimps

Table 2. (Continued.) Summary of food habit data on Bering Sea groundfishes.

Species	No. individuals	Range of length	Year and season	Area and depth	DR or RSW (% body weight)	Size of predator	Main food items
Snailfishes (Liparidae)							
Polka-dot snailfish (*Liparis cyclostigma*)	70	20-70 cm	1972-1975	Outer shelf to upper slope in EBS	RSW: 1.96%		
			1972 autumn			50-60 cm	Poachers, shrimps
			1975 winter			Small fish	Age-1 walleye pollock
						Large fish	Fishes, age-1 walleye pollock
			1975 spring			Small fish	Poachers, Tanner crabs
						Large fish	Age-2 walleye pollock
Salmon snailfish (*Careproctus rastrinus*)	275	10-45 cm	1972-1985	Outer shelf to upper slope in EBS	RSW: 3.21% at length 15-20 cm, 2.52% at 30-35 cm		
			1972 autumn			15-30 cm	Gammarus
			1974 summer				Gammarus
			197 winter				Gammarus
			1975 spring				Gammarus
			1985 summer			25-30 cm	Gammarus
Blacktail snailfish (*Careproctus melanurus*)	67	20-70 cm	1972-1985	Outer shelf to upper slope in EBS	RSW: 0.28%		
			1974 summer				Gammarus
			1975 spring			Small fish	Gammarus
						Large fish	Fishes
			1985 summer			30-50 cm	Shrimps

AB = Aleutian Basin, AIR = Aleutian Island Region, DR = Daily Ration, EBS = Eastern Bering Sea, ERS = Evacuation Rate of food in Stomach, MEB = Method of Energy Budget, RSW = Ratio of Stomach content Weight

Table 2. (Continued.) Summary of food habit data on Bering Sea groundfishes.

Species	No. individuals	Range of length	Year and season	Area and depth	DR or RSW (% body weight)	Size of predator	Main food items
Skates (Rajidae)							
Alaska skate (*Bathyraja parmifera*)	330	20-120 cm	1972-1975	Outer shelf to upper slope in EBS	RSW: 1.50% at length 20-30 cm, 2.00% at 40-50 cm, 2.19% at 60-70 cm, 1.48% at 80-100 cm		
			1972 autumn			30-70 cm	Shrimps, Tanner crabs
						70-80 cm	Age-2 walleye pollock
						80-100 cm	Age-2 and age-3 walleye pollock
			1974 summer			Small fish	Gammarus
						Medium fish	Tanner crabs
						Large fish	Age-2 walleye pollock
			1975 winter			Small fish	Gammarus
						Medium fish	Tanner crabs, age-1 walleye pollock
						Large fish	Age-2 walleye pollock
			1975 spring			Small fish	Gammarus, shrimps
						Medium fish	Age-1 walleye pollock
						Large fish	Age-2 walleye pollock
Bering skate (*Bathyraja interrupta*)	155	10-90 cm	1972-1975	Outer shelf to upper slope in EBS	RSW: 1.36% at length 20-30 cm, 2.11% at 50-60 cm, 1.37% at 70-80 cm		
			1972 autumn			Small fish	Shrimps
						Medium fish	Tanner crabs, shrimps
						Large fish	Tanner crabs, age-2 walleye pollock
			1974 summer			Small fish	Gammarus
						Large fish	Tanner crabs, gammarus
			1975 spring			Large fish	Age-2 walleye pollock

Table 2. (Continued.) Summary of food habit data on Bering Sea groundfishes.

Species	No. individuals	Range of length	Year and season	Area and depth	DR or RSW (% body weight)	Size of predator	Main food items
Other Families							
Pacific herring (*Clupea pallasi*)	250	15-35 cm	1974-1986	EBS and AIR	RSW: 0.47% at length 20-25 cm, 0.98% at 25-30 cm, and 0.93% at 30-35 cm		
			1974 summer	EBS		20-25 cm	Euphausiids
			1975 winter	EBS		20-30 cm	Euphausiids
			1975 spring	EBS		20-30 cm	Euphausiids
			1986 summer	AIR		20-30 cm	Euphausiids
Sablefish (*Anoplopoma fimbria*)	491	20-80 cm	1972-1985	EBS	RSW: 2.15% at length 40-60 cm, 0.86% at 60-80 cm in autumn of 1972, and 2.11% at 40-60 cm, 1.10% at 60-80 cm in summer of 1974		
			1972 autumn			40-60 cm	Age-2 walleye pollock
			1974 summer			60-80 cm	Squids, walleye pollock
			1985 summer			40-60 cm	Grenadiers, fishes
						60-80 cm	Fishes
Atka mackerel (*Pleurogrammus monopterygius*)	301	20-50 cm	1972-1986	EBS and AIR	RSW: 0.63% at length 20-30 cm, 0.77% at 30-40 cm, 0.48% at 40-50 cm		
			1974 summer			20-40 cm	Age-0 walleye pollock
			1975 winter			20-40 cm	Euphausiids
			1975 spring			20-40 cm	Euphausiids
			1986 summer	AIR		20-40 cm	Euphausiids

AB = Aleutian Basin, AIR = Aleutian Island Region, DR = Daily Ration, EBS = Eastern Bering Sea, ERS = Evacuation Rate of food in Stomach, MEB = Method of Energy Budget, RSW = Ratio of Stomach content Weight

Table 2. (Continued.) Summary of food habit data on Bering Sea groundfishes.

Species	No. individuals	Range of length	Year and season	Area and depth	DR or RSW (% body weight)	Size of predator	Main food items
Searcher *(Bathymaster signatus)*	175	15-40 cm	1972-1985	Outer shelf to upper slope in EBS	RSW: 0.25% at length 20-25 cm, 0.36% at 25-30 cm, 1.57% at 35-40 cm		
			1972 autumn			Small fish	Shrimps
						Medium fish	Tanner crabs, shrimps
						Large fish	Tanner crabs, age-2 walleye pollock
			1974 summer			Small fish	Shrimps, sea cucumbers
						Large fish	Hermit crabs, Tanner crabs, fishes
			1975 spring			Small fish	Fishes
						Large fish	Shrimps
			1985 summer			30-35 cm	Polychaetes
Prowfish *(Zaprora silenus)*	56	20-90 cm	1972-1975	Outer shelf to upper slope in EBS	RSW: 0.49%		
			1974 summer			50-70 cm	Medusas
			1975 winter				Medusas
			1975 spring				Medusas
Red squid *(Berryteuthis magister)*	30	15-35 cm (mantle length)	1985 summer	Upper slope in EBS	RSW: 0.51%	20-25 cm	Fishes

AB = Aleutian Basin, AIR = Aleutian Island Region, DR = Daily Ration, EBS = Eastern Bering Sea, ERS = Evacuation Rate of food in Stomach, MEB = Method of Energy Budget, RSW = Ratio of Stomach content Weight

Table 3.   Groundfishes and main prey species in the eastern Bering Sea, 1972-1975.

Prey species	Predator	
Scyphostoma	Prowfish	
Lumbriconereis sp.	Rex sole	
Sternaspis sp.	Alaska plaice	
Sabellidae	Shortfin eelpout	
Naticidae	*Malacocottus zonurus* (large)	
Pareuchaeta elongata	Northern rockfish (small)	
Metridia pacifica	Pacific ocean perch (small)	
Anonyx sp.	*Malacocottus zonurus* (small)	*Careproctus rastrinus*
Stegocephalus sp.	Blacktail snailfish	
Ampelisca spp.	Rock sole (small)	Alaska skate (small)
Syrrhoe sp.	Sturgeon poacher	Sawback poacher
Thysanoessa longipes	Pacific ocean perch (large)	
Thysanoessa inermis	Walleye pollock (small) Pacific herring	Northern rockfish (medium) Atka mackerel (small)
Pandalus borealis	Spinyhead sculpin	Thorny sculpin
Crangon dalli	Yellowfin sole (small)	Bering skate (small)
Pagurus spp.	Searcher (small)	
Chionoecetes spp.	Rock sole (large) Blackgill rockfish	Yellow Irish lord (small) Bering skate (large)
Ophiura sp.	Flathead sole	Rock sole (medium)
Ammodytes hexapterus	Wattled eelpout	
Theragra chalcogramma age 0	Northern rockfish (large) Polka-dot snailfish	Atka mackerel (large)
Theragra chalcogramma age 1	Greenland turbot (small) Plain sculpin (small)	Arrowtooth flounder Searcher (large)
Theragra chalcogramma age 2	Walleye pollock (large) Greenland turbot (large) Bigmouth sculpin Sablefish	Pacific cod Plain sculpin (medium) Pacific halibut Great sculpin
Theragra chalcogramma age 3	Yellow Irish lord (large)	Plain sculpin (large)

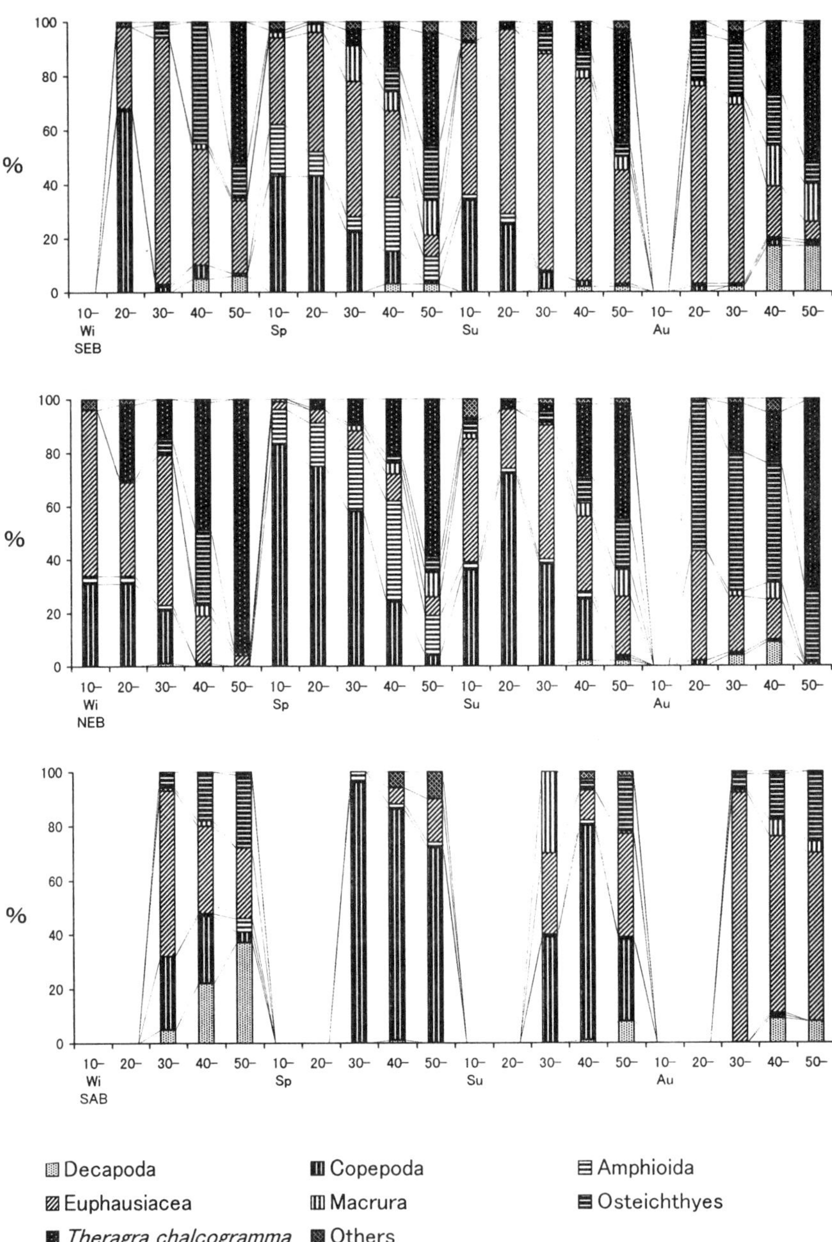

Figure 1.   Percentages by weight of main food items of walleye pollock by area, season, and size of predator in the Bering Sea.

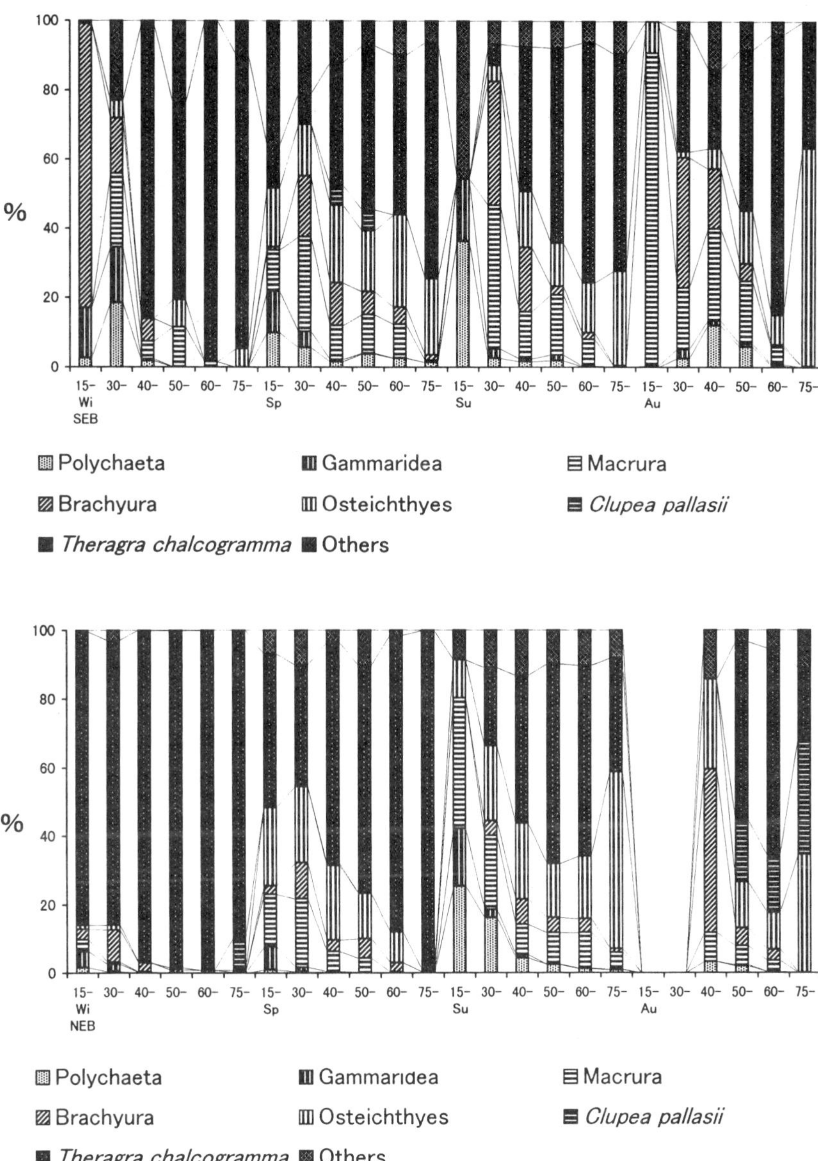

Figure 2. Percentages by weight of main food items of Pacific cod by area, season, and size of predator in the eastern Bering Sea.

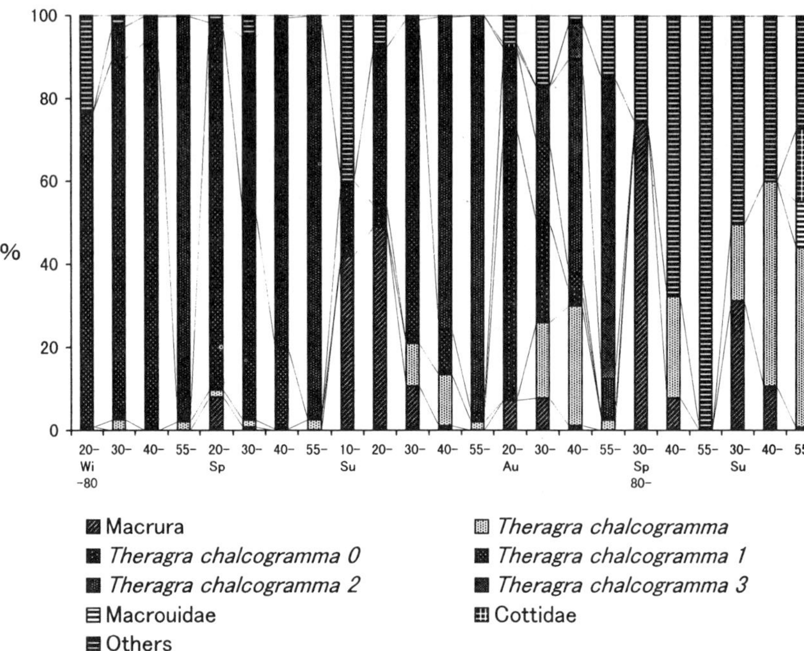

Figure 3.   *Percentages by weight of main food items of arrowtooth flounder by year, season, and size of predator in the eastern Bering Sea.*

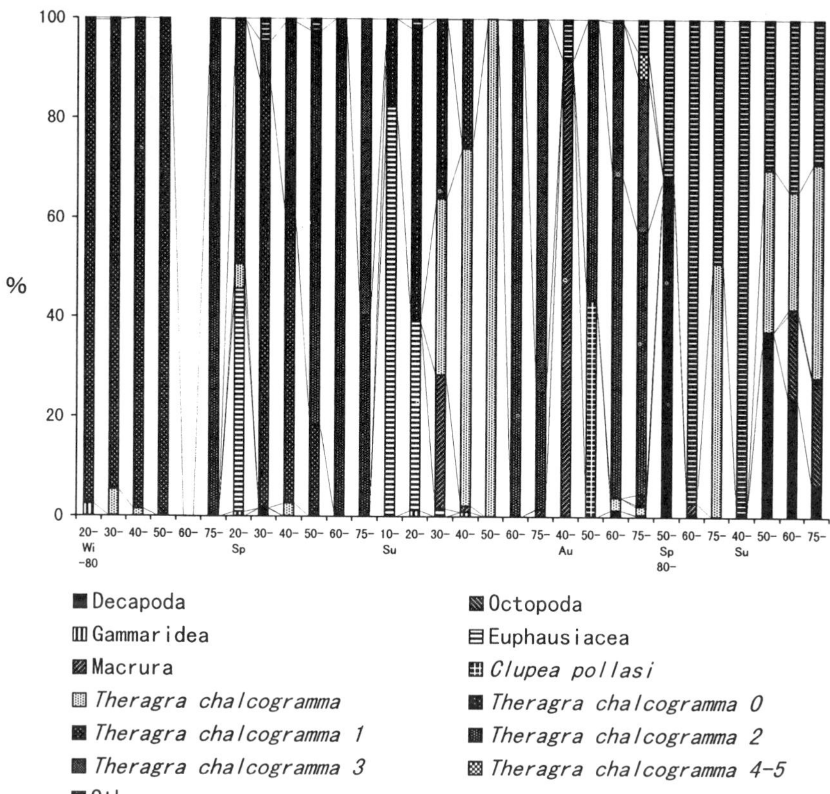

Figure 4. Percentages by weight of main food items of Greenland turbot by year, season, and size of predator in the eastern Bering Sea.

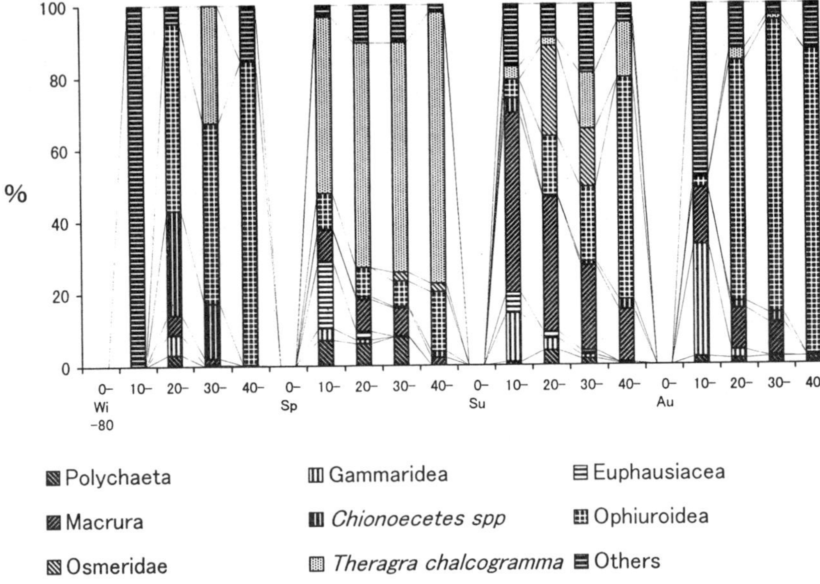

Figure 5. *Percentages by weight of main food items of flathead sole by season and size of predator in the eastern Bering Sea.*

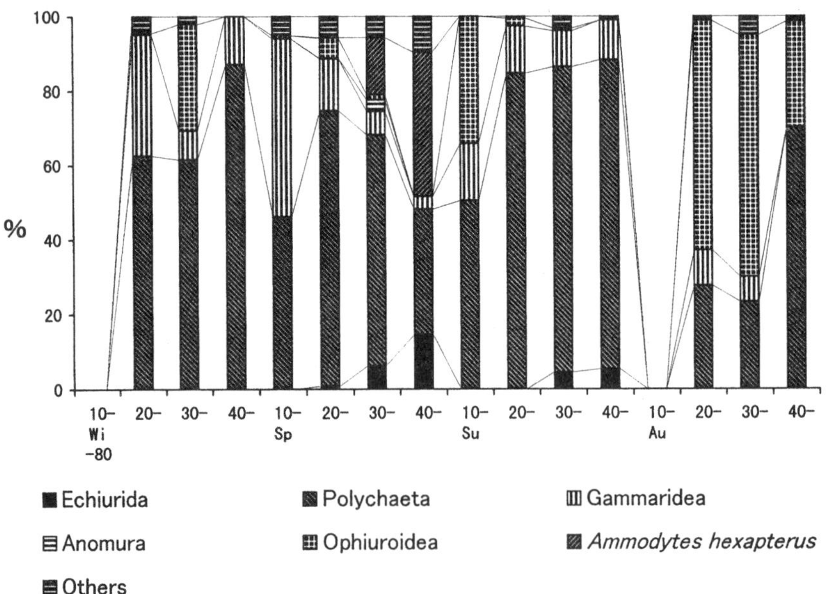

Figure 6. *Percentages by weight of main food items of rock sole by season and size of predator in the eastern Bering Sea.*

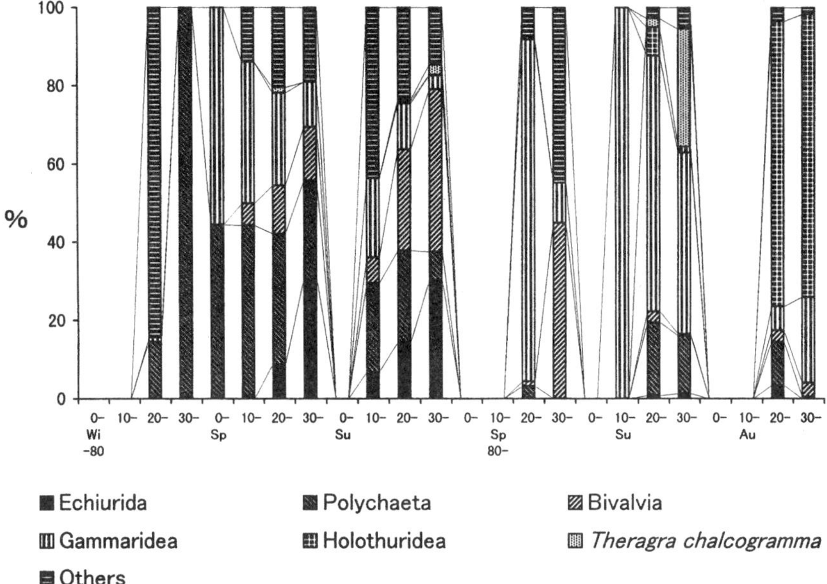

Figure 7. *Percentages by weight of main food items of yellowfin sole by year, season, and size of predator in the eastern Bering Sea.*

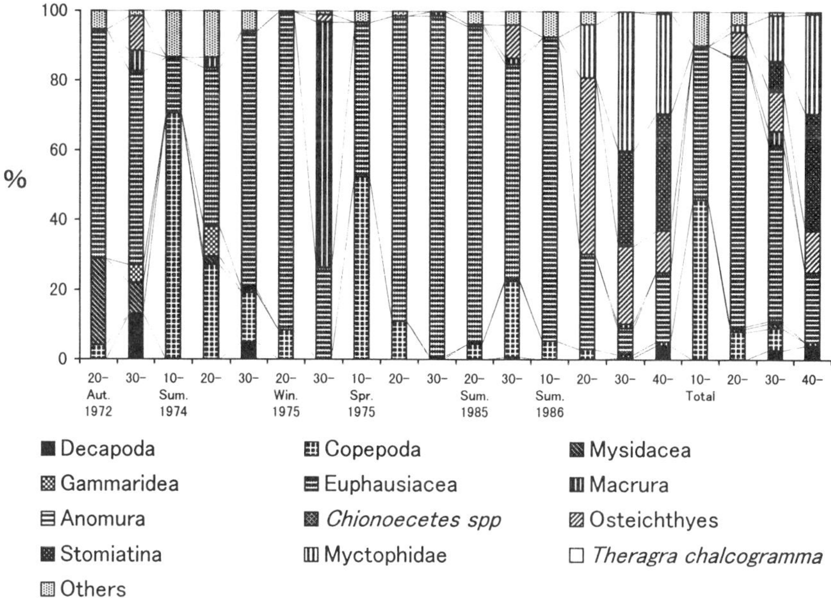

Figure 8. *Percentages by weight of main food items of Pacific ocean perch by year, season, and size of predator in the eastern Bering Sea.*

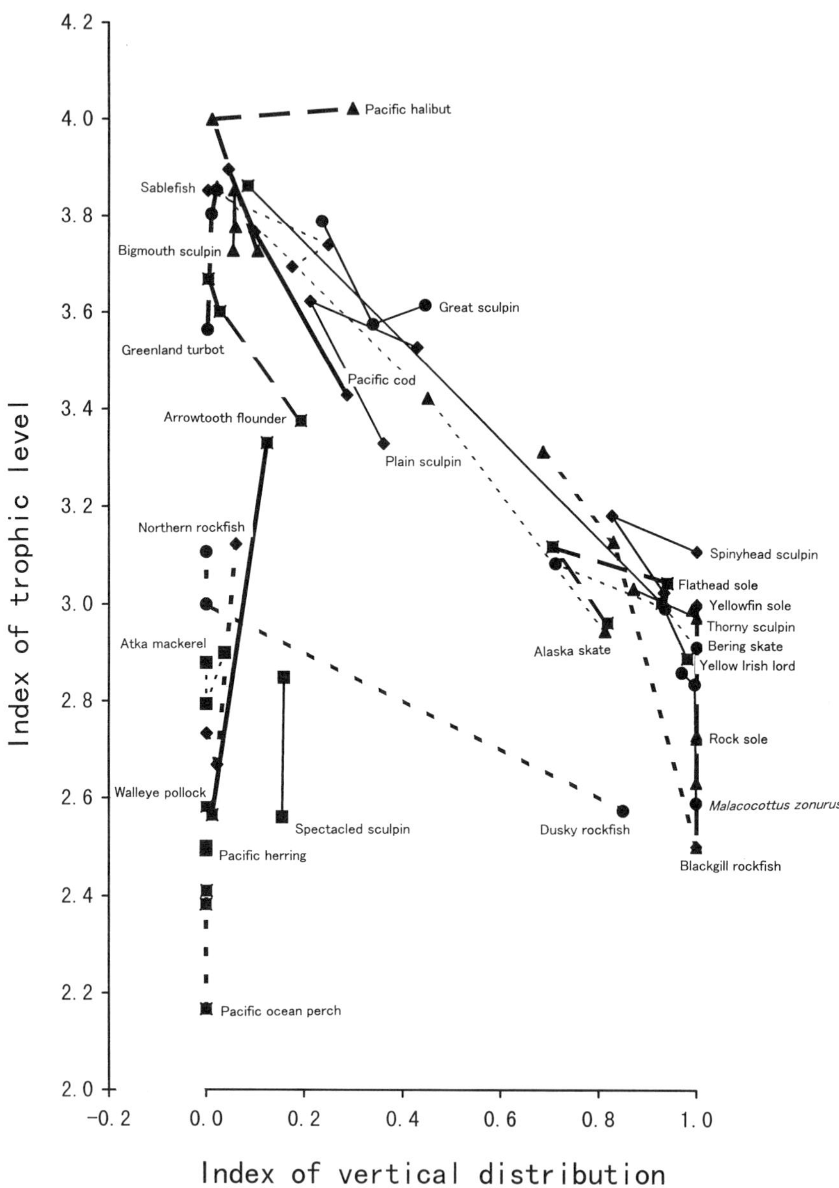

Figure 9.   *Food niche of species by body size in the groundfish community of the eastern Bering Sea in 1972-1975. See text for further explanation.*

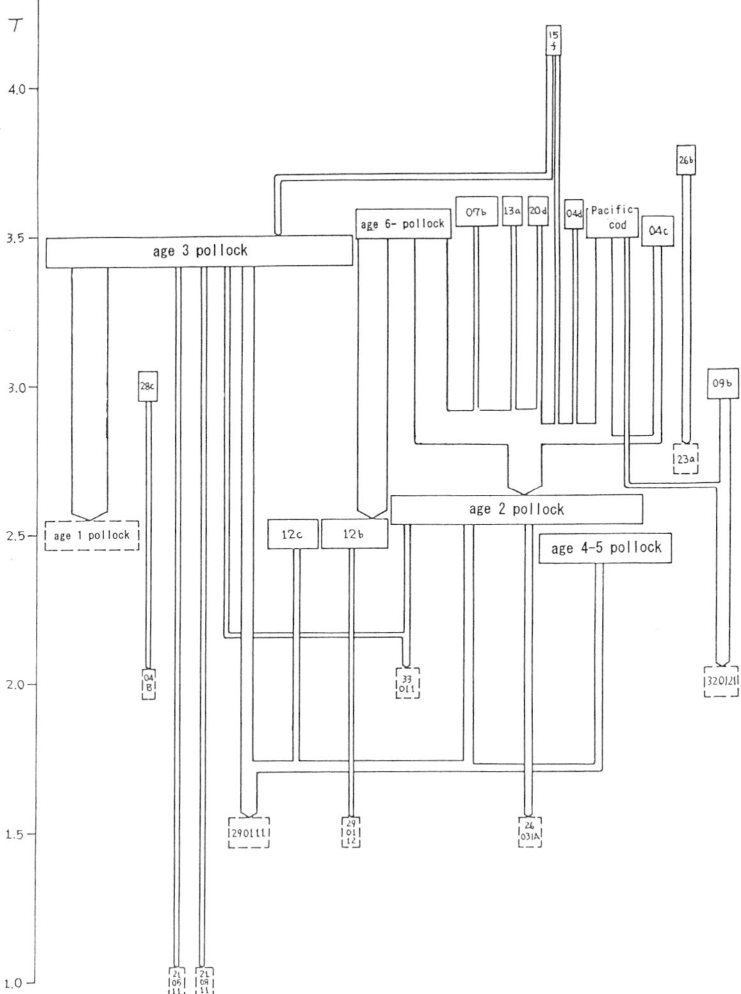

Figure 10. Example of the food web structure of the groundfish community in the northern area of the eastern Bering Sea during winter 1975. Vertical axis shows the index of trophic level. Though the horizontal axis shows the index of vertical distribution, the position depends on the order of the index. Boxes and arrows show animal and prey-predator relation respectively. Size and thickness correspond to a relative biomass and amount of predation. Numbers in boxes (solid line) show fish species code, and letters show length group (except for walleye pollock): 04 = Pacific cod, 07 = Greenland turbot, 09 = yellow Irish lord, 12= Pacific herring, 13 = bigmouth sculpin, 15 = Alaska skate, 26 = great sculpin. Letters later in the alphabet indicate a larger length group. Numbers in boxes (broken line) show food item code. 04B = Scyphostoma B, 210511 = Pareuchaeta elongata, 210811 = Candacia columbiae, 26031A = Ampelisca sp. A, 290111 = Thysanoessa longipes, 290112 = Thysanoessa inermis, 320121 = Chionoecetes opilio, 33011 = Sagitta sp.

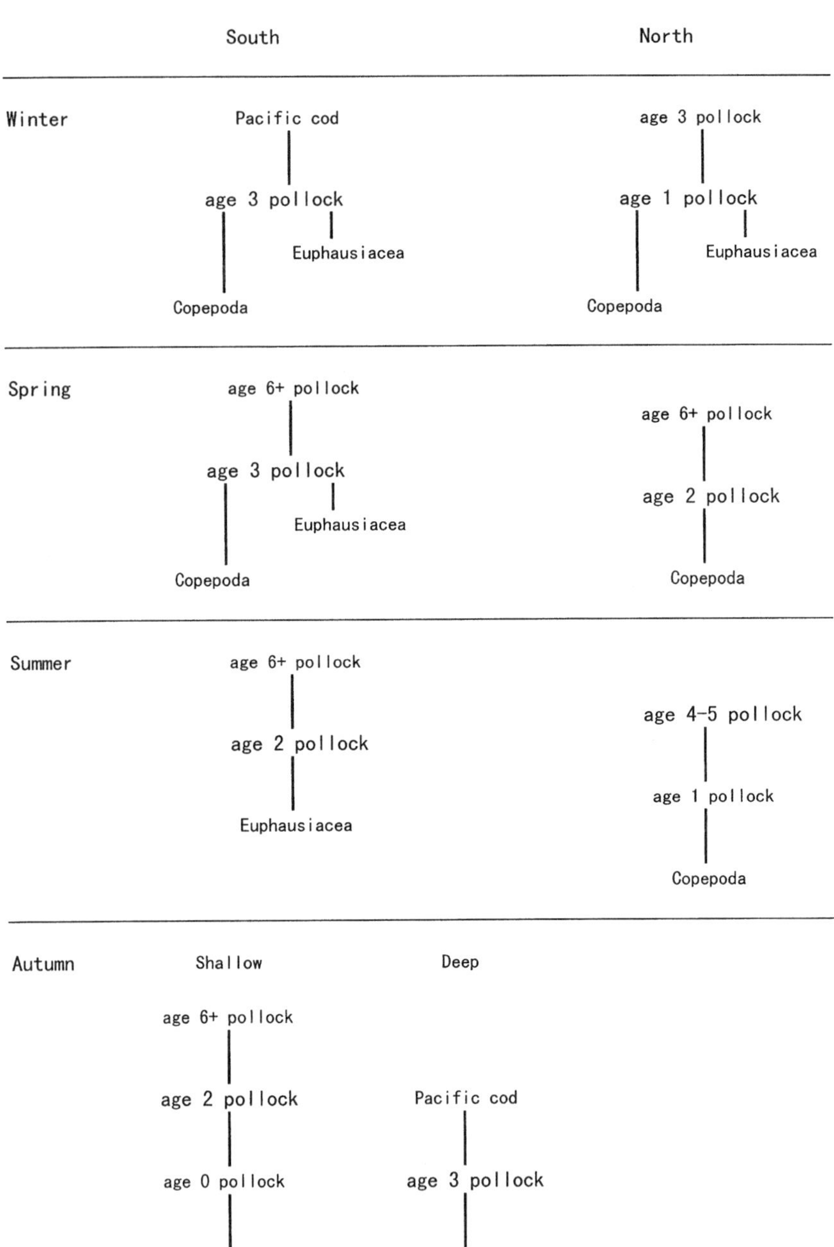

Figure 11. *Basic food chain of the groundfish community in the eastern Bering Sea in 1972-1975.*

Discussion and Conclusions

Food habit data collected in the past were limited by area and season, which made it difficult to adequately describe the food habits of groundfishes in the Bering Sea. It is also important to describe the biology and food habits for species that are not the target of fisheries.

When an ecosystem model is made, it is important that sufficient information is available on the structure of the food web. Laevastu and Larkins (1981) and Kishimoto et al. (1988) made an ecosystem model of the Bering Sea based on information on the food web structure of the groundfish community. There is an expectation for resource management to use these results of ecosystem models in the future. Further development of the model is necessary because stock dynamics of the primary species has not forecasted well by these models. To improve the models and to make them more practicable, it is necessary to fully understand food selection of the species included in the model.

We considered the structure of the food web in the eastern Bering Sea by season and area, and summarize it here. The food web structure in the southern shelf area is characterized by the key role of walleye pollock acting as both prey and predator in the community. The grazing food chain is dominant in this area while the detritus food chain is secondary. We suggest that the former will lead to higher diverse structures of this nearly single-species community. To us it seems more efficient to have a small detritus food chain and have the grazing food chain be dominant, especially since it is a more efficient way to transfer phytoplankton to upper trophic levels. The grazing food chain dominates the eastern Bering Sea outer shelf region and is associated with high groundfish production in this area.

The structure of the food web will be influenced by population density, species composition, and size composition of community members. A high density of copepods in the northern area in spring is predicted because there are many age-1 pollock in the northern area (Bakkala et al. 1992). However, specimens used for the analysis here were collected from pollock fishing areas closer to the outer shelf. Therefore our conclusions are restricted to this area.

As to problems in the future, we suggest that the estimation of a food selective coefficient for groundfishes is important. The coefficient is critical for ecosystem models, and extremely important for the analysis of ecosystem dynamics. For this, collection of data concerning the density of prey species in the environment is necessary.

Acknowledgments

We express our gratitude to the former staffs of the North Pacific Groundfish Section, NRIFSF, who collected the data on groundfishes, to the Japan NUS Co., Ltd., and the Fuyo Data Processing & Systems Development Ltd.

which calculated the stomach content data. We also express our gratitude to Dr. T.R. Loughlin, Dr. R.C. Francis, R. Paulson, and an anonymous reviewer who provided useful comments on an early draft of this manuscript.

References

Bakkala, R.G. 1993. Structure and historical changes in the groundfish complex of the eastern Bering Sea. NOAA Technical Report NMFS 114. 91 pp.

Bakkala, R.G., W.A. Karp, G.F. Walters, T. Sasaki, M.T. Wilson, T.M. Sample, A.M. Shimada, D. Adams, and C.E. Armistead. 1992. Distribution, abundance, and biological characteristics of groundfish in the eastern Bering Sea based on results of U.S.-Japan bottom trawl and hydroacoustic surveys during June-September, 1988. U.S. Department of Commerce, NOAA Technical Memorandum. NMFS F/NWC-213. 362 pp.

Dwyer, D.A., K.M. Bailey, and P.A. Livingston. 1987. Feeding habits and daily ration of walleye pollock *(Theragra chalcogramma)* in the eastern Bering sea, with special reference to cannibalism. Canadian Journal of Fisheries and Aquatic Science 44:1972-1984.

Ikeda, I. 1980. Hokuyo kaiiki sokouo shigen. In: T. Aoyama (ed.), Sokouo shigen. Kouseisha Kouseikaku, Tokyo, pp. 244-284. (In Japanese.)

Jones, R. 1974. The rate of elimination of food from the stomachs of haddock *(Melanogrammus aeglefinus),* cod *(Gadus morhua)* and whiting *(Merlangus merlangus).* Journal du Conseil International pour l'Exploration de la Mer 35:225-243.

Kachina, T.F., and E.A. Savicheva. 1987. Dynamics of food habit of pollock in the western Bering Sea. Population structure, dynamics and ecology of pollock. TINRO, Vladivostok.

Kishimoto, Y., Y. Iwata, K. Mito, and S. Tanaka. 1988. Development and preliminary calculation of ecosystem model for the eastern Bering Sea. In: Reports on the research and stock assessments of groundfish in the North Pacific in 1988. National Research Institute of Far Seas Fisheries, Shimizu, pp. 459-487. (In Japanese.)

Laevastu, T., and H.A. Larkins. 1981. Marine fisheries ecosystem: Its quantitative evaluation and management. Fishing News Books Ltd. 162 pp.

Livingston, P.A., D.A. Dwyer, D.L. Wencker, M.S. Yang, and G.M. Lang. 1986. Trophic interactions of key fish species in the eastern Bering Sea. Bulletin of International North Pacific Fisheries Commission 47:49-65.

Mito, K. 1977. Food relationships in the demersal fish community in the Bering Sea. In: I. Community structure and distributional patterns of fish species. Research Institute of North Pacific Fisheries, Hokkaido University, Special Volume, pp. 205-258. (In Japanese.)

Mito, K. 1990a. Differences in the weight and composition of stomach content of pollock between areas in the Bering Sea. In: Reports on the research and stock assessments of groundfish in the North Pacific in 1990. National Research Institute of Far Seas Fisheries, Shimizu, pp. 209-225. (In Japanese.)

Mito, K. 1990b. Estimate of daily ration of pollock in the Bering Sea. In: Reports on the research and stock assessments of groundfish in the North Pacific in 1990. National Research Institute of Far Seas Fisheries, Shimizu, pp. 227-245. (In Japanese.)

Mito, K. 1990c. Estimate of consumption of pollock by Pacific cod in the eastern Bering Sea. In: Reports on the research and stock assessments of groundfish in the North Pacific in 1990. National Research Institute of Far Seas Fisheries, Shimizu, pp. 247-277. (In Japanese.)

Mito, K. 1995. Food relationships among the groundfish community and factors affecting the structure in the continental shelf edge of the eastern Bering Sea in 1972-1975. In: Reports on the research and stock assessments of groundfish in the North Pacific in 1995. National Research Institute of Far Seas Fisheries, Shimizu, pp. 153-177. (In Japanese.)

Mito, K., A. Nishimura, and T. Yanagimoto. 1994a. Characteristics of stomach content composition of flounders and species grouping by the similarity index of prey in the eastern Bering Sea. In: Reports on the research and stock assessments of groundfish in the North Pacific in 1994. National Research Institute of Far Seas Fisheries, Shimizu, pp. 273-379. (In Japanese.)

Mito, K., A. Nishimura, and T. Yanagimoto. 1994b. Estimate of consumption of walleye pollock in the eastern Bering Sea—especially by flounders. In: Reports on the research and stock assessments of groundfish in the North Pacific in 1994. National Research Institute of Far Seas Fisheries, Shimizu, pp. 381-424. (In Japanese.)

Mito, K., A. Nishimura, and T. Yanagimoto. 1996a. Food habits of groundfish in the Bering Sea. 1. Rockfishes. In: Reports on the research and stock assessments of groundfish in the North Pacific in 1996. National Research Institute of Far Seas Fisheries, Shimizu, pp. 82-117. (In Japanese.)

Mito, K., A. Nishimura, and T. Yanagimoto. 1996b. Food habits of groundfish in the Bering Sea. 2. Sculpins. In: Reports on the research and stock assessments of groundfish in the North Pacific in 1996. National Research Institute of Far Seas Fisheries, Shimizu, pp. 118-203. (In Japanese.)

Mito, K., A. Nishimura, and T. Yanagimoto. 1996c. Food habits of groundfish in the Bering Sea. 3. Eelpouts, poachers and snailfishes. In: Reports on the research and stock assessments of groundfish in the North Pacific in 1996. National Research Institute of Far Seas Fisheries, Shimizu, pp. 204-257. (In Japanese.)

Nishiyama, T., K. Hirano, and T. Haryu. 1986. The early life history and feeding habits of walleye pollock, *Theragra chalcogramma* (Pallas), in the southeast Bering Sea. Bulletin of International North Pacific Fisheries Commission 45:177-227.

Wakabayashi, K. 1986. Interspecific feeding relationships on the continental shelf of the eastern Bering Sea, with special reference of yellowfin sole. Bulletin of International North Pacific Fisheries Commission 47:3-30.

Winberg, G.G. 1956. Rate of metabolism and food requirements of fish. Nauchn. Tr. Belorusskogo Gos. Univ. Imni V. I. Lenina, Minsk. 253 pp. (Fisheries Research Board of Canada Translation Series No. 194. 202 pp.)

CHAPTER 26

Population Ecology and Structural Dynamics of Walleye Pollock (*Theragra chalcogramma*)

Kevin M. Bailey
Alaska Fisheries Science Center, Seattle, Washington

Dennis M. Powers, Joseph M. Quattro, and Gary Villa
Hopkins Marine Station, Stanford University, Pacific Grove, California

Akira Nishimura
National Institute of Far Seas Fisheries, Orido, Shimizu, Japan

James J. Traynor and Gary Walters
Alaska Fisheries Science Center, Seattle, Washington

Abstract

In this paper, walleye pollock *(Theragra chalcogramma)* is characterized as a generalist species, occupying a broad niche and inhabiting a wide geographic range. Pollock's local abundance is usually high, dominating the fish biomass in many regional ecosystems. Thus, it appears to be an adaptable species capable of colonizing marginal habitats. The fields of macroecology, metapopulation dynamics, and genetic population structure are briefly reviewed and information on pollock is summarized within the framework of these concepts. Pollock show a pattern of apparent stock structure that has not always been indicated by genetic differences. Phenotypic differences between stocks, elemental composition of otoliths, and parasite studies indicate restricted mixing of adults. There are genetic differences between broad regions, but differences between adjacent stocks, especially within the eastern Bering Sea, are currently unresolved. The potential for gene flow mediated by larval drift is high between adjacent stocks. A generalized population structure for walleye pollock is proposed. The macro-population of walleye pollock is made of several major populations (such as the eastern Bering Sea and Sea of Okhotsk populations)

Current address for Joseph M. Quattro is Department of Biological Science, University of South Carolina, Columbia, SC 29208.

separated by large distances and geographic barriers with little gene flow between them, and numerous smaller populations with potential linkages among each other and the larger populations. Some populations may show local adaptations to their specific habitat, minimizing gene flow through reliance on larval retention features and natal homing. Application of high resolution genetic techniques, such as microsatellite polymorphisms, show promise for distinguishing adjacent pollock stocks. The architecture of stock structure, including metapopulation aspects of dispersal and colonization, are potentially important considerations in management of pollock fisheries.

Introduction

Walleye pollock *(Theragra chalcogramma)* is one of the world's largest commercial fisheries. Annual harvests have ranged from 4 to 7 million t in the North Pacific Ocean over the past decade. In U.S. waters catches are on the order of 1.5 million t with a value exceeding hundreds of millions of dollars. A large portion of the resource is bound for the export market, contributing to reduction of the U.S. trade deficit. Without question, this natural resource is of critical importance to the health of domestic fisheries. Furthermore, pollock is the dominant groundfish species in many of the regional ecosystems across the North Pacific, including the eastern Bering Sea, and has been implicated in the dynamics of higher trophic levels (see National Research Council 1996). As a key element in the North Pacific ecosystem, pollock is the target species for numerous research programs, past and present, in the eastern Bering Sea as well as in other seas.

The stock structure of pollock across the North Pacific, and especially in the Bering Sea, has been a topic of investigation for many years. Many studies have been based on pollock phenotypic characteristics, such as meristics and morphometrics, while other studies have been based on genotypic characters, such as allozyme frequencies. A broad characterization of pollock stock structure has come from these approaches. In general, allozyme studies have weakly distinguished eastern Pacific populations from western Pacific populations, but phenotypic characters suggest a more detailed stock structure. Only recently, using high resolution molecular techniques (Powers, unpubl. data reported here) has there been strong resolution between populations on either side of the Bering Sea and between Gulf of Alaska populations and eastern Bering Sea populations.

Fisheries resource management is based on the concept of renewable stocks. For pollock the regional stocks are managed separately; for example, the Gulf of Alaska population is managed separately from the Bering Sea, and the eastern and western Bering Sea and Aleutian Basin populations are managed separately. However, there may be isolated populations on even finer scales; within the Gulf of Alaska there may be separate

Table 1. Geographic distribution of walleye pollock stocks according to Wespestad (1996).

Stock	Stock characteristics
North American	
Southeast Alaska-Canada	Small stock, minor fisheries
Western-Central Gulf of Alaska	Variable stock, 50-200 thousand t catch
Eastern Bering Sea	Large stock, 1-2 million t catch
Asian	
Northwest Bering Sea	Mix of US and Russian fish, 0.5-1 million t catch
Western Bering Sea	Moderate stock, 0.5-1 million t catch
East Kamchatka	Small-medium stock, 100-300 thousand t catch
West Kamchatka	Large stock, near 1 million t catch
North Sea of Okhotsk	Moderate stock, 0.5-1 million t catch
Sakhalin	Small stock, 65 thousand t average catch
Kuril Islands	Small-moderate stock
Japan Sea	Heavily fished
Japan Pacific	Moderate catch to 0.5 million t

spawning populations in Shelikof Strait, Prince William Sound, and the Shumagin Islands region. Furthermore, the degree of intermingling of stocks, recolonization of depleted areas from healthy stocks, and other ecological questions about resource mixing within the greater population framework are almost completely unknown.

Although the overall catch of pollock on the U.S. side of the North Pacific is relatively stable, some populations have experienced declines and fisheries closures in recent years, including stocks in Puget Sound, Shelikof Strait, Donut Hole, and Bogoslof Island. More recently, there are reports that the western Pacific stocks are in a state of decline. The eastern Bering Sea shelf populations have been at healthy levels in the past, but there is concern about the sustainability of present harvest levels. Pollock harvests were especially high during the 1980s and 1990s resulting from relatively strong recruitment, high abundance levels, and unrestricted high seas fisheries. Wespestad (1996) lists 12 geographically distinct stock (although not necessarily genetically distinct) groupings and their catch trends (Table 1). Biomass and catch trends for the major stock groupings indicate declining levels since the mid-1980s in the major fishing

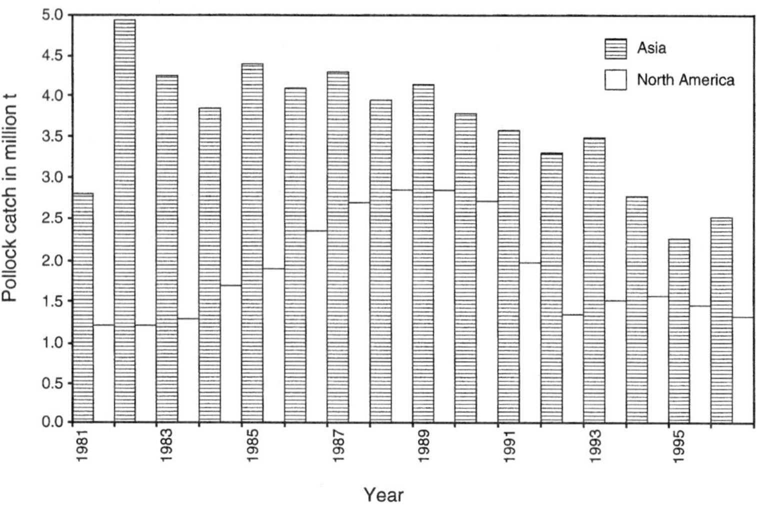

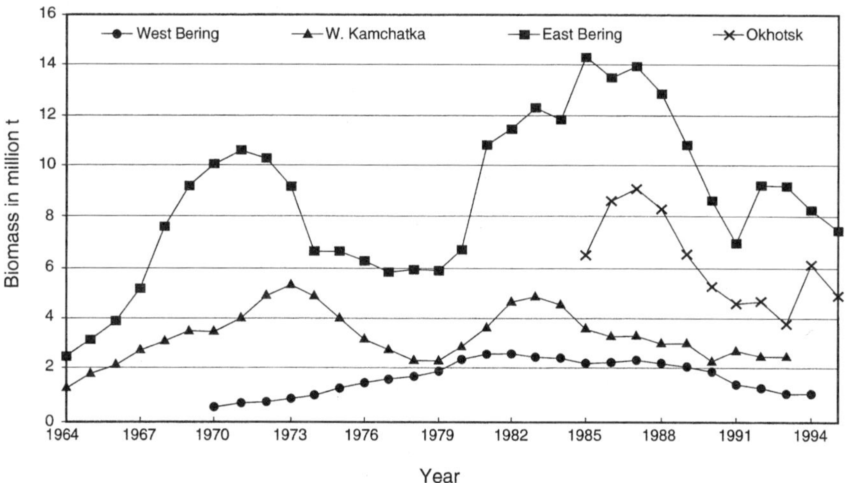

Figure 1.   Trends in major stock groupings of walleye pollock: (top) Abundance trends for major stocks, (bottom) Catch trends for Asian and North American stocks (from Wespestad 1996).

grounds (Fig. 1a,b). The greatest declines in biomass and catches for pollock seem to be away from the center of pollock's biomass. For example, in the Gulf of Alaska, catches peaked at 307 thousand t in 1984 and declined to 55 thousand t in 1996. At the extreme southern end of its range in the eastern North Pacific, pollock in south Puget Sound may almost be extinct (Palsson et al. 1996). Likewise for the southern end of its range in the western North Pacific, catches around northern Japan are reduced by 3-4 times from their maximal values in the 1970s (H. Yoshida, Hokkaido Central Fisheries Expt. Station, Hamanaka 238, Yoichi-cho, Hokkaido 046, Japan, pers. comm., February 1994).

There are many complex and unresolved issues involving the structure of the pollock population, including recent stock declines, discoveries of heretofore unknown stocks, potential consequences of fishing on selected components of the population, possible relationships between subunits, and potential fishing pressure on the same subpopulation at different times from different geographic and quota regions. What is needed, but currently lacking, is an understanding of the macroscale population ecology of pollock that incorporates the relationship among different subpopulations of the greater population, the rate of gene flow between these subpopulations, the relationship between juvenile and adult distributions as linked together by ontogenetic migrations, and the magnitude of colonization processes. A theoretical framework about the geographical structure and population ecology of pollock needs to include ecological concepts that can help us better understand the dynamics of this species and consider new approaches to management. One such concept is macroecology, the study of patterns of distribution of organisms in relation to their environment (Brown 1995). Many of the macroecological questions regarding pollock are unanswered, such as the relationship between range size and population abundance, and the effect of environmental factors on distribution of pollock. Another concept is that of metapopulation dynamics, the study of conditions by which the processes of population turnover and establishment of new populations are maintained in balance (Hanski and Gilpin 1991). Viewing pollock in the context of a metapopulation is a potential framework to study the linkage of stock structure, dispersal, and colonization events.

In this synthesis, we review the concepts of macroecology, metapopulations, and stock structure and point out areas where they intersect. We then summarize historical and current information on pollock, including stock structure and we propose a "strawman" conceptual model of the population macroecology of pollock.

Theoretical Framework: Structural Dynamics

The size of fish populations is determined by three major elements: (1) available habitat determines the potential size, (2) dynamics within local populations as influenced by the environment, competition, and predation determine the specific size of a subpopulation, and (3) movement and colonization determine the spread of a population, and modify local dynamics. In this model, distribution, dispersal and population structure are closely linked and interwoven characteristics of populations.

A population is a group of individuals of the same species living in a sufficiently restricted area such that any member can mate with any other; in large widespread populations it is recognized that individuals in the same locality are more likely to mate and share common ancestry. Local subpopulations may form intrabreeding units of geographically structured populations. Geographical structure in populations is the nonrandom pattern of spatial distribution that may be related to historical or current barriers between local populations, environmental patchiness, and/or environmental gradients. The term structural dynamics integrates change in population structure and aspects thereof, such as range, genetics, and dispersal patterns.

Distribution and Macroecology

Geographic range considerations are an important characteristic of populations. It is recognized that many species with broad niches become both widespread and locally abundant (Brown 1984), and that large ranges, abundance, and invasion ability are linked characteristics within a species (Lawton et al. 1994). Those species with extraordinary invasion abilities are generally those best adapted for marginal habitats (MacArthur and Wilson 1967). As described below, these concepts are especially relevant to pollock population biology. On the other hand, habitat specialization often limits colonization. For example specialized feeders such as benthic feeding flatfishes would appear to be tied to specific habitat types; thus they cannot move and colonize new habitats freely as adults or settled juveniles compared with species that have more generalized requirements. Furthermore, if the range of a species with specialized requirements is made of a mosaic of habitat patches, colonization by demersal stages may be impeded by regions of unfavorable habitat.

Some populations may expand their range as they become more abundant, although others do not show this trend but show increases in local density. As noted above, species with highly specialized niches may not expand their range readily, and indeed no range/abundance relationships have been found for some relatively specialized flatfish species such as rock sole *(Lepidopsetta bilineata),* Alaska plaice *(Pleuronectes quadrituberculatus)* (McConnaughey 1995), and American plaice *(Hippoglossoides platessoides)* (Swain and Morin 1996). Thus for these species, the range size/population size relationship due to density-dependent dispersal may not occur. More generalized species such as Atlantic cod *(Gadus morhua)* (Swain and Wade 1993) and arrowtooth flounder *(Atheresthes stomias)* (McConnaughey 1995) are observed to expand their range as abundance increases.

Physical impediments (i.e., temperature, salinity, and substrate availability) and biological factors (i.e., the presence of competitors and predators) may limit range expansions. There also may be historical structure in the environment, such as changes in the occupation of niches due to disease and environmental events. When competition or predation pressure is removed there can be an "ecological release" (MacArthur and Wilson 1967) resulting in an invasion event.

Metapopulations and Geographical Structure

Within the distributional range of animals, there may be considerable population structure. The realization that animal populations are composed of local populations has a rich history going back to Andrewartha and Birch (1954). In fisheries, knowledge of local populations goes back to the developments and ideas of Schmidt, Heincke, and Hjort in the early 1900s (Sinclair 1988). The concept of population structure was formalized as that of metapopulations by Levins (1970). In the metapopulation view, a local population is the spatial unit within which most interaction such as breeding occurs, and a metapopulation is an ensemble of local populations (Taylor 1991a) with some potential of interaction. Although many fish populations are not metapopulations in the strict sense, viewing them within a relaxed definition of metapopulations (i.e., populations with non-continuous distributions and some possibility of local extinction) (McCullough 1996) might be useful to applied fisheries science.

Harrison (1991) showed five potential types of metapopulations with examples of their degree of dependency and interaction (Fig. 2). These include: (A) the Levins metapopulation, where population sizes are nearly equal and all populations interact and habitats are patchy within the range area; (B) the core-satellite (Boorman and Levitt 1973) metapopulation, where habitats are patchy within the range, but one major population is the source of fringe populations; (C) the patchy population, where habitats are continuous throughout the specie's range, but local populations are aggregated; (D) a non-equilibrium metapopulation, which is like the Levins model except there is no movement between local populations, and (E) an intermediate model between B and C. According to Harrison (1991) the dominant metapopulation structure is that of a large local population in a large favorable habitat patch, surrounded by relatively unstable populations in smaller habitat patches.

In the metapopulation view, local subpopulations are connected by dispersal, which includes larval dispersal and colonizing movements of adults. Fish migrations, which are movements of individuals coordinated in space and time (Quinn and Brodeur 1991), can be ontogenetic, seasonal, or daily. These migrations are distinguished from dispersal or nomadic behavior, which is undirected and out of the natal or home range. Dispersal can occur due to environmental factors, such as El Niño events, and also due to density-dependent effects. As demonstrated for some fishes, fidelity to home range depends on habitat quality (Matthews 1990), which can depend on density. The advantages of dispersal are colonization of new habitats and avoidance of unfavorable local conditions (Quinn 1993). The

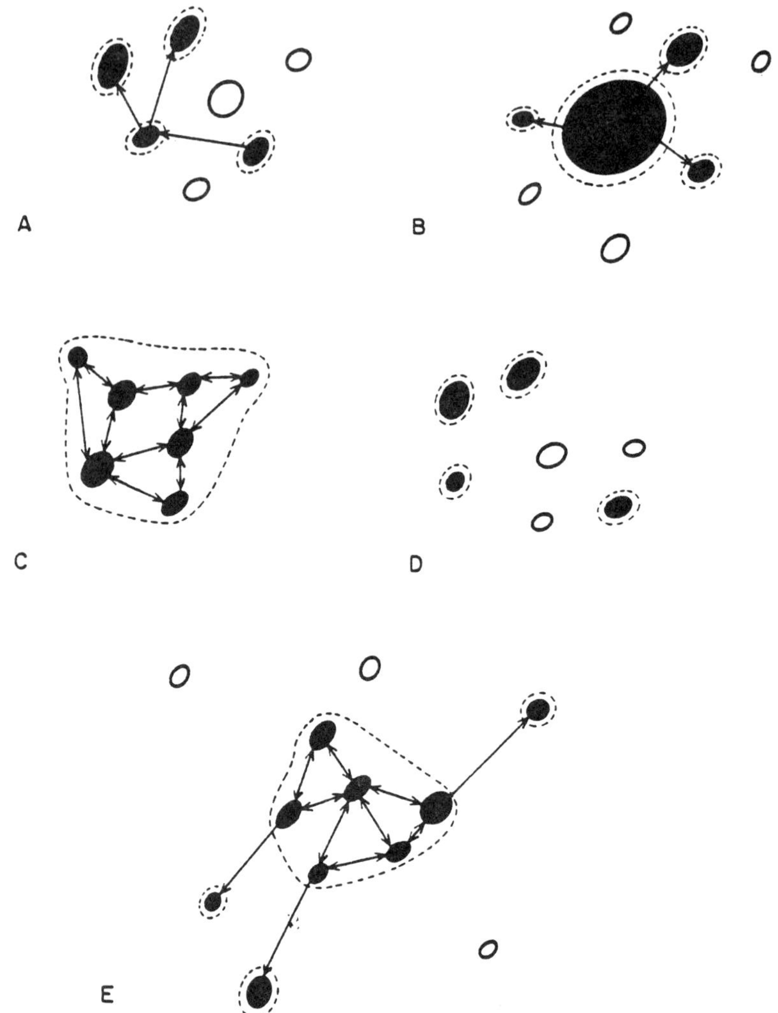

Figure 2.   Different kinds of metapopulations: (A) Levins metapopulation; (B) core-satellite metapopulation; (C) patchy population; (D) non-equilibrium metapopulation; and (E) an intermediate case that combines B and C. Closed circles represent habitat patches; filled = occupied, unfilled = vacant. Dashed lines indicate the boundaries of populations. Arrows indicate migration (colonization) (from Harrison 1991.)

effectiveness of movement between populations depends on the extent of the movement and the mortality of the dispersing individuals.

One of the major issues of the subpopulation concept is the fidelity to a home range or natal site (natal philopatry). Through natal homing, animals that stray away from their natal site may return for reproduction. In many fishes where juveniles and adults make seasonal and ontogenetic migrations related to feeding or reproduction, they may utilize one of three mechanisms, or a combination thereof, to return to their natal site and maintain philopatry. The three mechanisms are: (1) imprinting on environmental cues; (2) learning from other fishes (social tradition); and (3) genetically based homing. The nature of the homing mechanism, when it occurs, has some interesting implications for management issues, as discussed below.

Genetic Structure

How populations are organized as metapopulations and their genetic structure are closely linked, genetic structure being influenced by the amount of gene flow between local populations. There are widely differing opinions on the degree of structure in marine fish populations, ranging from a lot of subpopulation structure to almost none. At one end is Sinclair's (1988) member/vagrant hypothesis, whereby there are many discrete populations, genetically distinct through larval retention and mortality of vagrants. This concept was founded on observations of specific spawning locations that seem to favor retention of larvae. At the other end of the spectrum is Smith et al. (1990) whose dynamic population concept has stocks with no discrete genetic status as homogeneous isolates, but as dynamic heterogeneous components of the species; some genetic differences may occur from isolation by geographical distance and gene selection. This concept arises from observations of high genetic variability within an area and low genetic differentiation between areas.

Genetic divergence among subpopulations is promoted by mutation, genetic drift, and natural selection favoring adaptation to local environments and hindered by the homogenizing process of gene flow from movements of gametes, individuals, or populations (Slatkin 1987). Genetic drift is also minimized by the huge sizes of many marine fish populations. Phenotypic and genotypic characteristics have been used to distinguish fish subpopulations. The use of phenotypic characters is still widely practiced, but when used alone is generally received with some skepticism. Biochemical and molecular genetic techniques are powerful tools for fisheries; there have been many studies of protein polymorphisms and more recently molecular genetics, looking for differences between nearby geographical populations, with the goal of finding genetically isolated management units. Many studies of marine fishes fail to distinguish populations (Pawson and Jennings 1996) due to the state of technology as applied to population genetics, due to high levels of gene flow, either historical or ongoing, or due to poor resolution power due to small sample sizes (Taylor

1991b). However, there have also been some great successes; for example, distinguishing smelt populations on opposite sides of the St. Lawrence River (Bernatchez and Martin 1996), distinguishing mosquito fish populations within several hundred meters of shoreline (Kennedy et al. 1986) and distinguishing *Fundulus* populations along coastal transects (Gonzalez-Villasenor and Powers 1990).

In marine fishes, the importance of larval dispersal on gene flow and population structure has been shown by several studies (Waples and Rosenblatt 1987, Avise 1994). For example, Doherty et al. (1995) used isozymes to show that a reef species, the damselfish *(Acanthochromis polycanthus)*, whose offspring are reared in the parental territory and which lacks larval dispersal, demonstrates negligible gene flow over ca. 1,000 km. Doherty showed that other species with larval dispersal had genetically homogeneous subpopulations resulting from gene flow. However, high dispersal potential may not always translate into high gene flow and genetic homogeneity. There may be physical impediments to dispersal such as retention mechanisms, early life stages may not be passive drifters, and later stages may have natal homing behavior.

Characterization of Pollock

Distribution and Population Ecology

Walleye pollock is a species with a very broad niche. Although commonly associated with the outer shelf and slope regions of oceanic waters, as a species it is capable of utilizing a wide variety of habitats including nearshore eelgrass beds (J. Norris, School of Fisheries, University of Washington, Seattle, WA 98195, pers. comm.), large estuaries like Puget Sound, coastal embayments, and open ocean basins such as the Aleutian Basin of the Bering Sea. Although adults of the species are often described as semi-demersal, in some areas they are strictly pelagic (Bakkala 1993). Pollock commonly feed on a wide assortment of prey from pelagic copepods to epibenthic organisms as well as pelagic and demersal fishes.

Pollock has a broad range from Puget Sound to the northern Bering Sea and on across the North Pacific (Fig. 3). It is most abundant in the eastern Bering Sea and the Sea of Okhotsk. Besides an extensive range, the local abundance of pollock is usually high, and it often dominates regional groundfish communities. Based on its ecological plasticity, broad range and high levels of abundance, pollock can be characterized as a classical generalist species capable of invading and adapting to marginal habitats.

The distribution of pollock is closely linked to temperature. The vertical distribution of juvenile pollock, and horizontal distribution of schools of adults and juveniles are influenced by fronts, temperature, and depth (Bailey 1989, Swartzman et al. 1994). The interaction of temperature with behavior of pollock has been well-studied (Olla et al. 1996, Sograd and Olla 1996). The horizontal distribution of pollock is limited by cold tem-

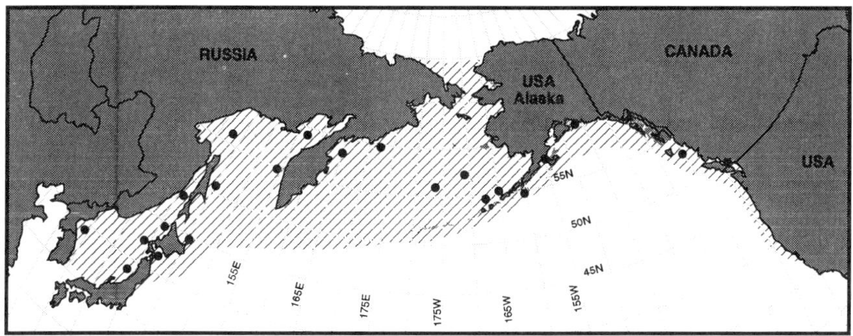

Figure 3.   The distribution range of walleye pollock shown with cross-hatching. Major spawning locations are shown with closed circles.

peratures, from 0-2°C, as shown by distribution of commercial catches, and catches of juveniles and adults in research surveys (Francis and Bailey 1983). Detailed studies have not been conducted at high temperatures, but the range of pollock appears to be limited by temperatures of 10-12°C.

We tracked the distribution pattern of several year classes in the eastern Bering Sea to examine ontogenetic dynamics in distribution of different year classes. The 1982 year class was found predominantly in the outer shelf region of the southeastern Bering Sea as larvae (Fig. 4A); as age-0 juveniles they had moved northward and inshore (Fig. 4B). As age-1 fish, they had distributed themselves farther northward and also a large portion of the population was found shoreward (Fig. 4C). As age-3 fish in summer, a portion of the 1982 year class returned to the southern outer shelf region, but a large number of fish remained in the northeastern outer shelf (Fig. 4D). The 1989 year class was not sampled as larvae, but as age-0 juveniles in summer it was broadly distributed across the middle shelf in the south and along the outer shelf region in the northwest (Fig. 5A). This pattern persisted for age-1, age-2, and age-3 fish (Figs. 5B-D). Portions of the year class distributed inshore tended to persist there through age-3. The 1993 year class was sampled as larvae along the outer shelf region of the south (Fig 6A). It was not sampled as age-0 juveniles, but as age-1's was found far northward in a band extending along the northern outer shelf (Fig. 6B,C). The distribution of age-3 fish appears similar to that of age-2, perhaps with a slight shift southward (Fig. 6D). Overall, these patterns indicate generally northward movements of age-0 and age-1 fish. However, there appears to be considerable interannual variability in distribution patterns; sometimes these two age classes move shoreward also. By age-3 it appears that pre-spawning fish shift their distribution southward again.

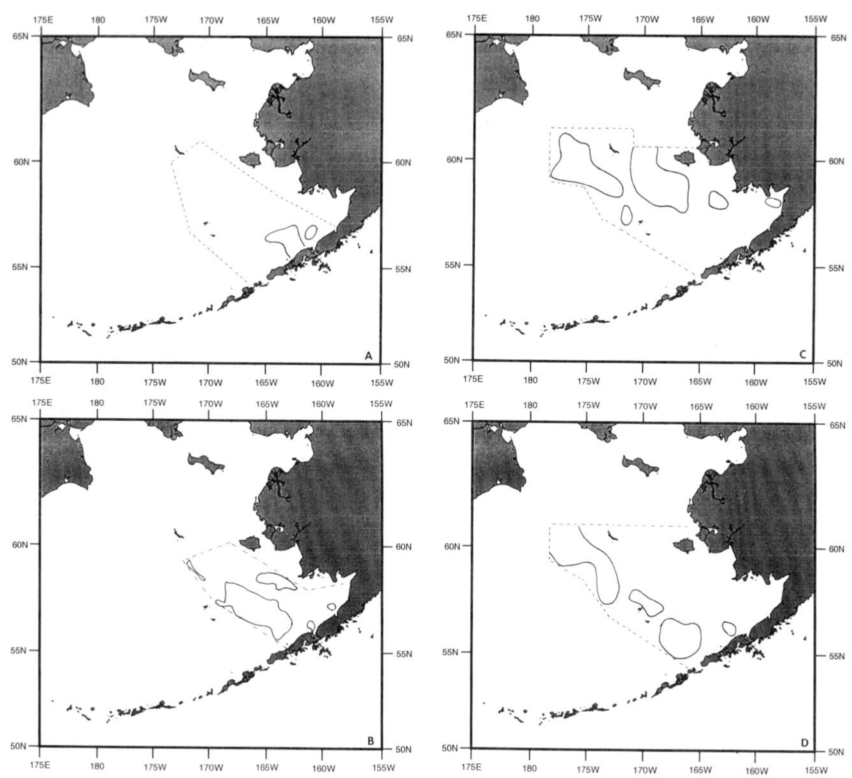

Figure 4.   *Relative distribution patterns of the 1982 year class of walleye pollock in the eastern Bering Sea: (A) larvae in June from ichthyoplankton surveys; (B) age-0 juveniles in autumn from midwater trawl surveys; (C) age-1 in summer from bottom trawl surveys; and (D) age-3 in summer from bottom trawl surveys. Contour shows area of highest abundance. Dashed line shows approximate region surveyed.*

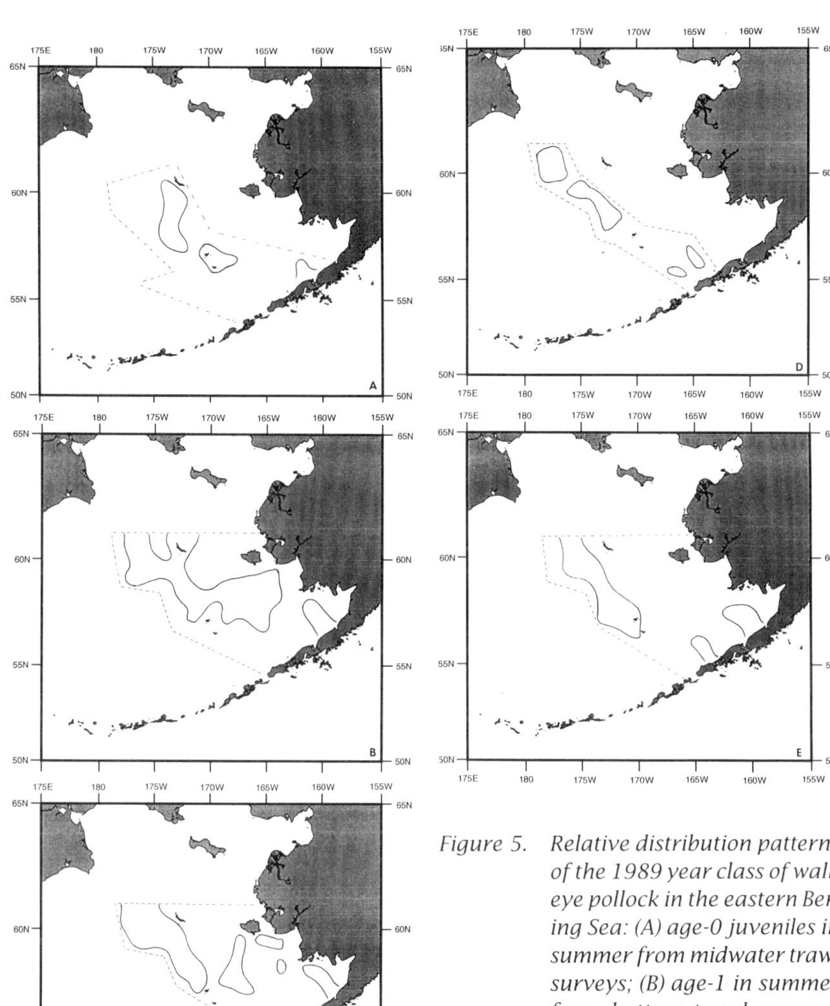

Figure 5. *Relative distribution patterns of the 1989 year class of walleye pollock in the eastern Bering Sea: (A) age-0 juveniles in summer from midwater trawl surveys; (B) age-1 in summer from bottom trawl surveys; (C) age-2 in summer from bottom trawl surveys; (D) age-2 in summer from midwater trawl surveys; and (E) age-3 in summer from bottom trawl surveys. Contour shows area of highest abundance. Dashed line shows approximate region surveyed.*

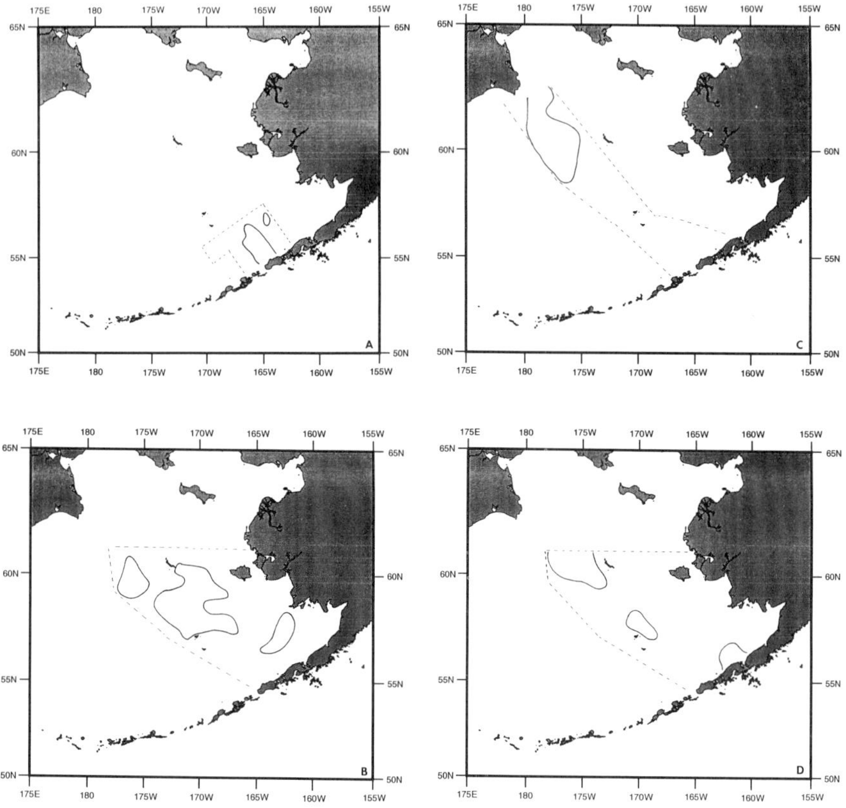

Figure 6.   Relative distribution patterns of the 1993 year class of walleye pollock in the eastern Bering Sea: (A) larvae in May from ichthyoplankton surveys; (B) age-1 in summer from bottom trawl surveys; (C) age-1 in summer from midwater trawl surveys; and (D) age-3 in summer from bottom trawl surveys (age estimated from length class). Contour shows area of highest abundance. Dashed line shows approximate region surveyed.

Characterization of pollock as a colonizing species, and knowing that physical and biological factors impede the invasion of colonizing species, lend support to historical suggestions that the rapid increase in pollock populations occurring in the late 1960s was due to an "ecological release" caused by removing competitive pressure of Pacific Ocean perch *(Sebastes alutus)* due to harvesting (Somerton 1978) and decreases in abundance of Pacific herring *(Clupea harengus).* Under this scenario, pollock were a classic *r*-selected species with rapid growth, early maturity, and high fecundity that were capable of rapidly taking over this niche opening; most likely, juvenile pollock were the stage capable of exploiting prey resources made available. The expanding population of juveniles could have made them the most available prey for pollock adults resulting in their cannibalistic nature in the eastern Bering Sea.

In the Gulf of Alaska, an increase of pollock in the mid-1980s coincided roughly with a regime shift occurring in 1977-78 (Hollowed and Wooster 1995) and possibly was associated with good conditions for juveniles of the 1976-78 year classes. Recently, however, pollock stocks in the Gulf of Alaska have declined markedly. Curiously, the age of maturity has increased (Megrey 1988) in spite of declining density. Furthermore, since the mid-1980s juvenile survival has been relatively poor; from 1980 to 1985, 5 of 6 year classes had relative age 0-2 mortality lower than the 1980-91 mean value. By contrast, from 1986 to 1991, 5 of 6 year classes had higher than the mean long term mortality (from data in Bailey et al. 1996). These data tend to indicate trends of increasing predation pressure on juveniles or eroding environmental conditions for juveniles and adults.

The relationship of distributional range and population abundance has not been formally examined for pollock because historical records over the whole range of pollock are not available. However, based on the historical declines of abundance in populations outside the major fishery area, while the central area of pollock biomass has remained relatively stable, pollock can be characterized as a species with a main central range and numerous fringe populations; the overall range of the population may not shrink with population decreases unless local fringe populations go extinct. Within local populations there may be a positive range and abundance association, similar to that found for Atlantic cod (Swain and Wade 1993). This is supported by Tsugi's (1989) assertion that during times of increasing commercial catch levels (and therefore abundance) pollock expand into adjacent waters.

Comparing the mean and variance of subpopulation abundances over time can characterize how populations are controlled in different areas of the species' range. In the case of pollock, a plot of mean subpopulation abundance versus variability shows a slope less than unity, indicating that peripheral, less abundant pollock stocks vary more than central, more abundant stocks (Fig. 7). These results could either indicate more density dependence in the central stocks, including density-dependent colonization

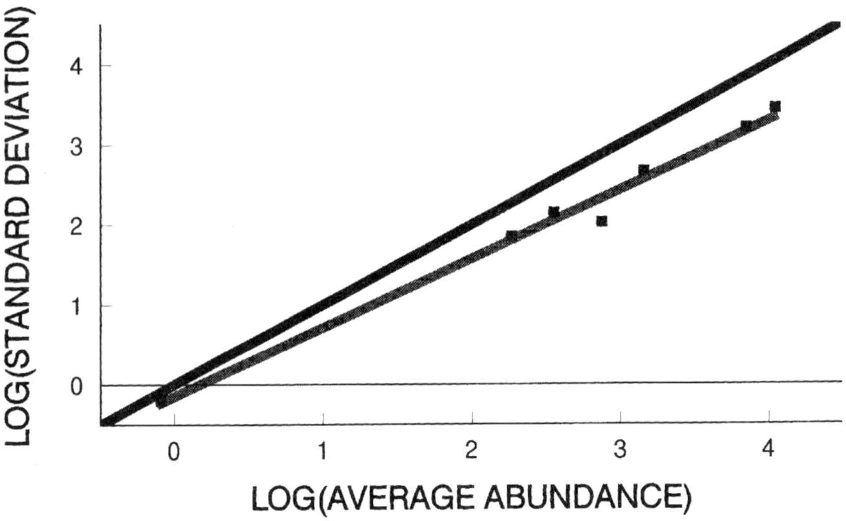

Figure 7.   *The standard deviation of pollock abundance against average abundance for seven stocks of walleye pollock, 1979-91, showing that peripheral, less abundant stocks vary more than central, more abundant stocks. Upper line has a slope of unity, bottom line is regression line of stock data (slope = 0.864).*

of peripheral populations, or that environmental conditions are more variable and have a greater effect on peripheral populations.

Year-Class Dynamics

For all of the major pollock groupings, stock fluctuations are strongly influenced by intermittent recruitment of strong year classes. For example, Fig. 8 shows the impact of a series of strong year classes on stock abundance in the Gulf of Alaska, as well as the drop in abundance related to subsequent recruitment of relatively poor year classes. In the eastern Bering Sea, the 1978 year class comprised 67% of the pollock population in 1981 and 53% of the population in 1982. Many regions share the same strong year classes; for example, 1978 was a strong year class in the Gulf of Alaska, Aleutian Basin, eastern Bering Sea, western Bering Sea, and Sea of Okhotsk. Likewise 1982, 1984, and 1989 were strong across the Bering Sea, although not in the Gulf of Alaska. Strong year classes in the Gulf of Alaska including 1976, 1977, 1979, and 1988 did not appear strong in the Bering Sea. Therefore there is not a consistent association of strong year classes among the Bering Sea and Gulf of Alaska populations, indicating a lack of density-dependent dispersal between these geographic regions. However, within the Bering Sea, there appears to be an association of strong

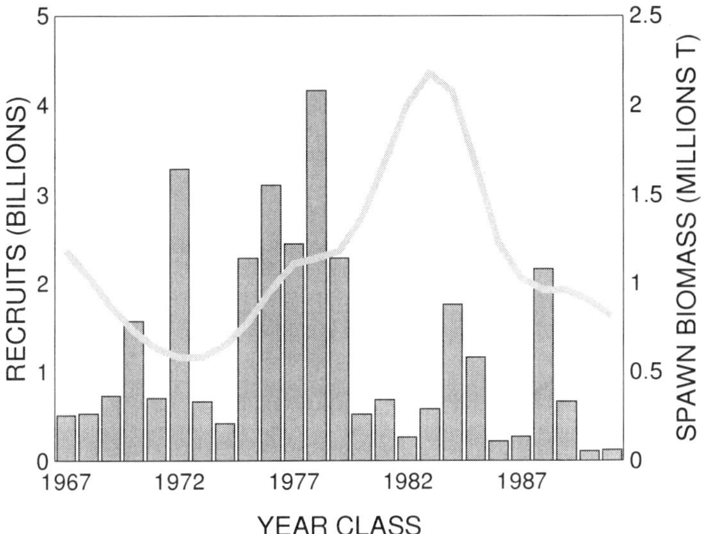

*Figure 8.   Year-class abundance compared to spawning biomass of walleye pollock
in the western Gulf of Alaska, showing the dramatic effects of year-class
strength on population biomass.*

year classes among the different regions. The occurrence of similar strong
year classes in different regions of the Bering Sea has been cited as evidence of panmixia within the Bering Basin (Dawson 1994). At the extreme
ends of the range of pollock (e.g., Puget Sound) the strong year classes
(1972-75; Palsson et al. 1996) were quite different from those in the centers of abundance.

Genetic and Metapopulation Structure

Pollock show repeated, predictable, and discrete spawning at specific sites
and times, which is a piece of evidence for natal philopatry. Some of these
persistent spawning sites are shown in Fig. 3. Among the best studied
spawning aggregations is that in Shelikof Strait which has been monitored
since 1981. The Shelikof Strait spawning aggregation is the largest spawning biomass in the Gulf of Alaska, and mostly spawns within a limited
area (40 × 80 km) during the first week in April (Kendall et al. 1996). The
geographic and temporal consistency of spawning for this and other spawning aggregations argues against the notion that the timing and position of
spawning is random, depending on suitable conditions. Mark-recapture
studies where pollock were tagged during the spawning season (April)
indicate dispersed feeding migrations but return homing to specific spawning locations (Tsuji 1989; Fig. 9). Thus, pollock show a relatively continuous

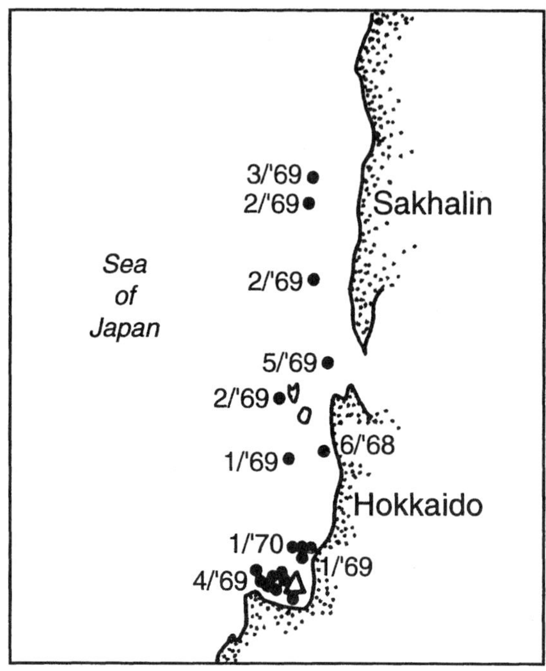

Figure 9.   *Recapture distribution for an experiment in which 666 pollock were tagged and released (triangle) in the northern Sea of Japan on 17 April 1968. Locations and dates of recapture are shown by closed circles (from Tsugi 1989).*

distribution during the feeding season, and a more patchy distribution during the spawning season. Over recent geologic times (i.e., the past 10,000 years), local pollock populations have presumably undergone local extinctions and recolonizations due to heavy glaciation of fjords and lower sea levels. As noted above, the local population in south Puget Sound has gone extinct for all practical purposes. Thus, some groups of pollock populations may be viewed as metapopulations under a relaxed definition of the term (McCullough 1996).

We have summarized studies on stock structure of pollock according the their phenotypic, acquired, or genetically based nature. Briefly, phenotypic characteristics of pollock both within small geographic regions and across much broader areas indicate considerable population structure (Table 2). For example, Koyachi and Hashimoto (1977) and Hashimoto and Koyachi (1977) used differences in vertebral counts and in growth rates to distinguish 11 groups of pollock across its range. Within a small area (i.e., around the islands of Japan) there are also distinct groups (Iwata and Hamai 1972).

Naturally acquired tags such as elemental composition of otoliths and parasite characteristics indicate restricted mixing among pollock juveniles and adults of different subpopulations (Table 3). For example, the chemical "fingerprint" of otoliths near the nucleus (deposited during early larval life) could be utilized to assign fish to their capture location as juveniles with 70-85% accuracy over broad regions of the eastern Bering Sea (Mulligan et al. 1989), indicating limited movement and mixing of fish from different geographic regions. Using parasite frequencies, adult pollock caught on the south side of Vancouver Island can be distinguished from those on the west side with about 75% accuracy (Arthur 1983). In the Sea of Okhotsk, several different populations of pollock were distinguished based on parasite frequencies (Avdeev and Avdeev 1989). By contrast, mark-recapture studies where pollock were tagged in summer indicate broad movement of individuals across areas of the Bering Sea (Fig. 10). However, critical studies of marking spawning pollock have not been conducted. Tagging results to date in the Bering Sea do not give information on spawning migrations or spawning site fidelity, but do indicate potential for dispersal during the summer feeding period. Nor have these studies provided information indicating whether individuals are moving with large schools as migrating populations, whether individuals are vagrants and mixing with local populations, or whether dispersal patterns in the Bering Sea are part of ordered seasonal migrations. Tagging studies around Japan do support dispersed feeding migrations and homing migrations to specific spawning areas (Tsuji 1989).

Studies using biochemical and molecular methods on pollock show mixed results for distinguishing stock structure of pollock, varying somewhat by technique (Table 4). Using allozymes, Grant and Utter (1980) detected minor differences between the Gulf of Alaska and Bering Sea, but no differences within either region. Using mitochondrial (mt) DNA sequence data from a 76 base pair (bp) spacer from the control region, Shields and Gust (1995) reported slight genetic differences between eastern and western Bering Sea pollock. Mulligan et al. (1992) employed mtDNA restriction fragment length polymorphism (RFLP) on fish collected during the spawning season and using pairwise comparisons of haplotype frequencies distinguished among stocks in the Gulf of Alaska, Donut Hole/Bogoslof Islands, and Aleutian Islands.

More recently, using sequences of mtDNA cytochrome *b* and ATPase-8, Powers (1996) and his colleagues (J. Quattro and D. Powers, unpubl. data, Hopkins Marine Station, Stanford University, Pacific Grove, CA 93950) identified haplotype differences between Asian and North American pollock. The unique DNA sequences allowed them to develop an enzyme method to rapidly identify individuals by RFLP. An analysis of molecular variance (AMOVA) on the combined data identified sharp distinctions between Asian

Table 2. Summary studies of pollock stock structure using phenotypic characteristics.

Author	Method	Area	Results
Ogata (1959)	Meristics: vertebrae counts	Sea of Japan and Pacific Ocean side of Japan	Sea of Japan has 3 different stocks. Sea of Japan differs from the Pacific Ocean side.
Iwata and Hamai (1972)	Meristics: vertebrae counts	Sea of Japan, Sea of Okhotsk, and Pacific Ocean near Hokkaido	8 "local forms": 2 groups in the Sea of Japan. 3 groups in the Okhotsk Sea. 3 groups in the Pacific Ocean.
Hashimoto and Koyachi (1969)	Morphometrics: body length and other morphological features	Northern Japan	3 groups discriminated.
Janusz (1994)	Meristics and morphometrics	Sea of Okhotsk	3 stocks distinguished.
Temnykh (1991)	Morphometrics	Sea of Okhotsk	Southern Kurils population distinguished from norhtern Sea of Okhotsk.
Temnykh (1994)	Morphometrics	Western Bering Sea and eastern Kamchatka	Western Bering and Eastern Kamchatka stocks distinguished.
Ishida (1954)	Morphometrics: otoliths	Northern Sea of Japan, Sea of Okhotsk, and northern Pacific Ocean coast of Japan	Otolith size is larger in Sea of Japan pollock than Okhotsk Sea. Otoliths are similar between Sea of Japan and Pacific Ocean pollock.
Shaw and McFarlane (1986)	Morphometrics: length-at-age	British Columbia: Dixon Entrance, Strait of Georgia	2 stocks discriminated: Strait of Georgia pollock are smaller. Little interaction between pollock north and south of Queen Charlotte Sound.
Thompson (1981)	Morphometrics: length-at-age	British Columbia: Dixon Entrance, Strait of Georgia, Queen Charlotte Sound	3 separate stocks: each area contains its own distinct stock. Little mixing occurred between them.
Saunders et al. (1989)	Morphometrics: life history	British Columbia	Separate stocks in Strait of Georgia, Hecate Strait/Dixon Entrance, Queen Charlotte Sound, and Western Vancouver Island.
Lynde et al. (1986)	Morphometrics: length-at-age	Eastern Bering Sea and Bering Sea basin	Northeastern shelf and slope and Aleutian Basin represent one stock distinct from other regions of the eastern Bering Sea.
Hinckley (1987)	Spawning time and location, morphometrics: length-at-age, fecundity	Aleutian Basin and eastern Bering Sea shelf and slope	3 spawning stocks in the eastern Bering Sea: basin, northeastern slope, and eastern shelf and slope.
Mulligan et al. (1989)	Spawning time and location	Eastern Bering Sea and Aleutian Basin	3 spawning areas separated in space and time: eastern Bering Sea southeastern shelf, eastern Bering Sea northwestern shelf, and Aleutian Basin.
Serobaba (1977)	Meristics and morphometrics	Northern, western, eastern, and southern Bering Sea	Different stocks occupy each region.
Dawson (1994)	Morphometrics	Bering Sea	3 stocks: eastern Bering Sea shelf, Aleutian Basin, and Aleutian Islands.
Janusz et al. (1989)	Meristics and morphometrics	Donut hole and eastern Bering sea shelf	2 stocks distinguished in Donut Hole and eastern Bering Sea.

Table 2. (Continued.) Summary studies of pollock stock structure using phenotypic characteristics.

Author	Method	Area	Results
Nitta and Sasaki (1990)	Morphometrics	Donut hole, eastern Bering Sea, near Japan	Characteristics distinguish 3 stocks, with about 90% classification accuracy.
Gong et al. (1990)	Meristics	Asian and Bering Sea	Asian stock and Bering Sea stocks distinguished but stocks within these regions not distinguished.
Wilimovsky et al. (1967)	Meristics: fin ray and vertebral counts; and morphometrics	Entire Pacific Ocean	Morphometrics: no strong evidence for discrete stocks. Meristics: no differences between Bering Sea and Puget Sound pollock. No differences between Gulf of Alaska and northern B.C. pollock.
Koyachi and Hasimoto (1977)	Meristics: fin ray, gill raker, and vertebrae counts	Entire Pacific Ocean	12 subpopulations, including the Bering Sea and Gulf of Alaska.

Table 3. Summary studies of pollock stock structure using acquired characteristics.

Author	Method	Area	Results
Nakano et al. (1991)	Otolith chemistry: adults, whole otolith homogenates	Eastern Bering Sea, western Bering Sea, Donut hole	Differences in 3 areas, little mixing.
Mulligan et al. (1989)	Otolith chemistry: juveniles, inner early life otolith increments	Eastern Bering Sea: southeastern shelf, northwestern shelf, and Aleutian Basin	Differences in 3 areas, some mixing.
Severin et al. (1995)	Otolith chemistry: juveniles, outer otolith increments	Eastern Bering Sea: Bristol Bay; and Gulf of Alaska	Distinguish 5 areas, some mixing.
Arthur (1983)	Parasites	British Columbia: Strait of Georgia, west side of Vancouver Island, Queen Charlotte Sound, and Dixon Entrance	3 stocks in this area: Strait of Georgia, Vancouver Island, and Queen Charlotte Sound/ Dixon Entrance.
Avdeev and Avdeev (1989)	Parasites	Sea of Okhotsk, Kommander Islands, and Kamachatka Peninsula	7 distinct groups within the northeastern Sea of Okhotsk. Distinct groups in Kommander Islands, eastern Kamchatka, and Shirshov Ridge.
Misc. authors, see Figs. 9, 10	Tagging studies	Western and eastern Bering Sea, Japan	Broad movements, homing migrations to spawning site.

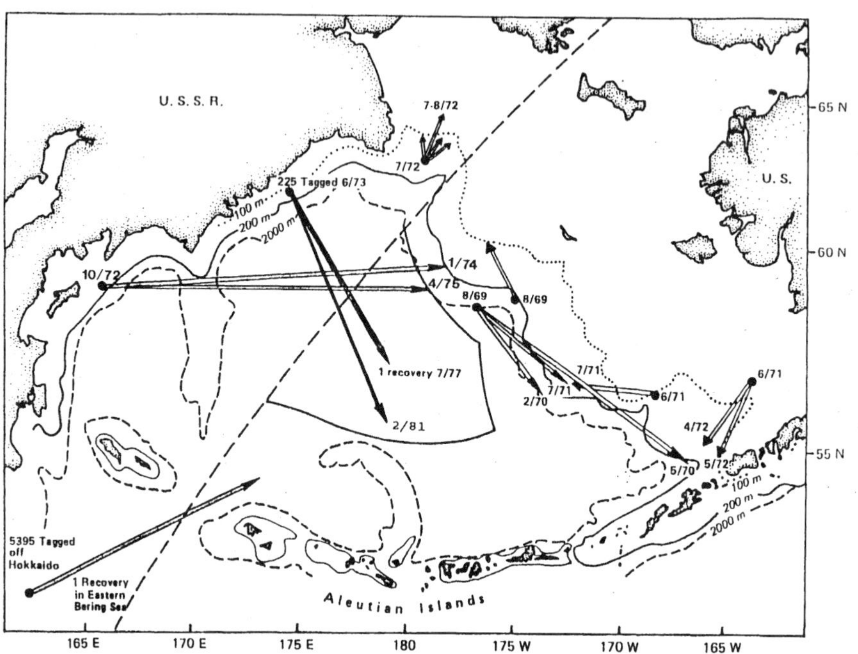

Figure 10. *Movement of walleye pollock tagged by Japanese scientists in the Bering Sea (from Dawson 1994). Note: Most tagging and recoveries occurred during the summer/autumn feeding season. Note cluster of April recoveries near Unimak Pass in the spawning season.*

and North American pollock and considerable substructure within the American and Asian clusters (Fig. 11).

Powers and Gary Villa (unpubl. data, Hopkins Marine Station, Stanford University, Pacific Grove, CA 93950) screened populations for DNA microsatellite loci that allowed the discrimination between populations on a finer scale than was possible by mtDNA analysis. Based on pairwise comparisons for the two microsatellite loci examined (Gmo-145 and Gmo-2; Brooker et al. 1994) significant differences between Asian and American stocks were confirmed and, furthermore, differences between Gulf of Alaska and Bering Sea stocks were resolved (Fig. 12). One microsatellite primer scored for a null allele in the western Bering Sea samples that was responsible for the major east-west differences, which warrants examination with more primers. Even though the statistical results are tentative, they are generally supportive of data from life history characteristics and phenotypic patterns that indicate considerable population heterogeneity in pollock. Avise (1994, 1995) cautions, however, that solid conclusions on

Table 4. Summary of stock structure studies on walleye pollock using biochemical genetics characteristics.

Author	Method	Area	Results
Mulligan et al. (1992)	mtDNA RFLP	Eastern Bering Sea basin, Aleutian Islands, and Gulf of Alaska	2 distinct stocks in Aleutian Islands and Donut Hole/Bogoslof; Gulf of Alaska and Donut Hole/Bogoslof have informative differences.
Shields and Gust (1995)	mtDNA sequencing	Across Bering Sea and Gulf of Alaska	Significant differences between eastern and western Bering Sea.
Grant and Utter (1980)	Allozyme	Southeastern Bering Sea and Gulf of Alaska	Minor genetic differences between the two areas. No differences within the areas.
Johnson (1977)	Allozyme	Eastern Bering Sea and Gulf of Alaska	No significant differences found.
Iwata (1973)	Allozyme	Northern Sea of Japan and North Pacific coast of Japan	No differences found.
Iwata (1975a,b)	Allozyme	Northern Sea of Japan and eastern Bering Sea	Highly significant differences found between the two areas.
Efremov et al. (1989)	Allozyme	Northern Sea of Okhotsk	Allozyme variability suggesting that aconitase could be genetic marker.
Powers (1996)	mtDNA RFLP and DNA microsatellite	Bering Sea and Gulf of Alaska	Eastern and western Bering Sea distinguished using mtDNA; Gulf of Alaska and eastern Bering Sea stocks have informative differences using microsatellite DNA.

population structure need to be based on multiple loci, setting up this future study as an important research priority.

There is a pattern of apparent stock structure in walleye pollock, that has not always been indicated by genetic differences. Phenotypic differences between stocks, elemental composition of otoliths and parasite studies indicate restricted mixing of adults. In addition, otolith elemental composition data indicate restricted mixing of juveniles between adjacent stocks (eastern Bering Sea shelf and basin). There are genetic differences between broad regions, but differences between adjacent stocks, especially within the eastern Bering Sea, is currently unresolved. The potential for gene flow mediated by larval drift is high between adjacent stocks. However, since there appears to be an unresolved degree of structuring within the Bering Sea, reduced gene flow due to larval retention mechanisms or strong natal homing and philopatry is also possible.

The standard metapopulation models of Harrison (1991; Fig. 2) do not fit the apparent population structure of pollock as known to date. We propose a generalized metapopulation structure that represents an intermediate model. The overall population of pollock is made of several large and major populations separated by large distances and geographic barriers (such as the eastern Bering Sea and Sea of Okhotsk populations) with little gene flow between them, and numerous smaller populations with potential linkages amongst each other and the larger populations (Fig.

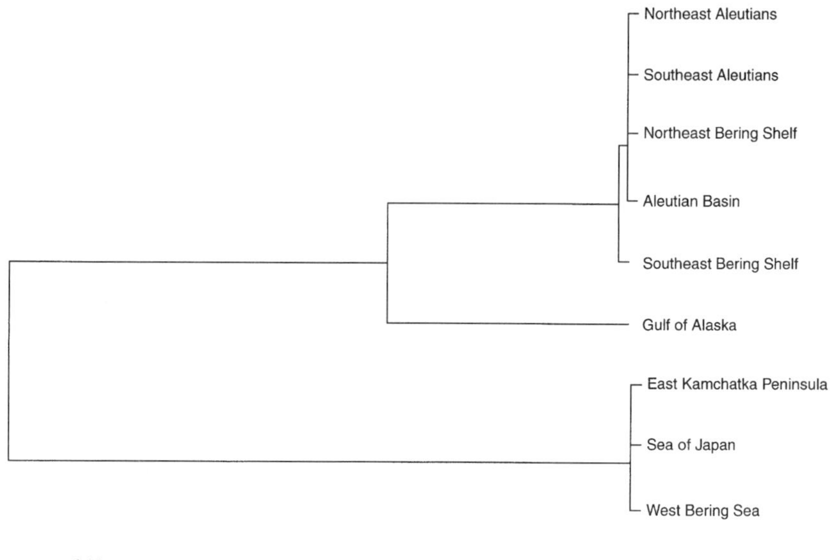

Figure 11. UPGMA dendrogram of depicting relationships of walleye pollock stocks from mtDNA RFLP data; pairwise distances were estimated with AMOVA (from Powers 1996). Significant differences are observed between American and Asian populations. Scale bars represent genetic distance.

13). Some populations may show local adaptations to their specific habitat, minimizing gene flow through reliance on larval retention features and natal homing.

Management Implications

When fisheries fluctuate, are local populations waxing and waning independently or are they connected in some way? This has been a central issue in fisheries since the time of Hjort (1914). Since that time, however, fisheries science has often been based on the concept of populations as closed and homogeneous systems. For example, prior to the collapse of the northern Atlantic cod *(Gadus morhua)* population in the Northwest Atlantic Ocean it was believed that there was no genetically based population structure due to extensive egg and larval drift, followed by opportunistic and nonphilopatric recruitment of juveniles to adult assemblages (deYoung and Rose 1993). Isozymes and mtDNA studies had little success in differentiating Northwest Atlantic cod stocks. Bentzen et al. (1996) used the relatively new technique of DNA microsatellite analysis which has, in this case at least, higher resolution and found that the northern cod pop-

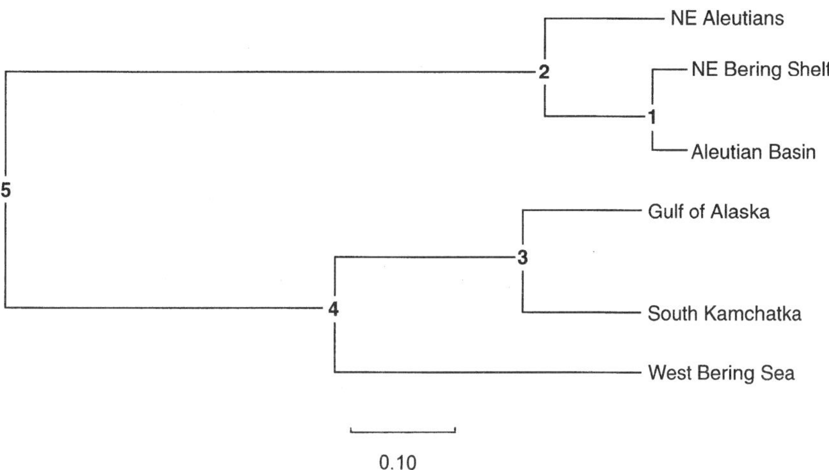

Figure 12. UPGMA clustering of genetic distances among walleye pollock stocks from allele frequencies determined with microsatellite primers Gmo-2 and Gmo-145. Pairwise ANOVA comparisons showed significant differences between the eastern and western Bering Sea and Gulf of Alaska (from Powers 1996). Scale bars represent genetic distance.

ulation does not comprise a single panmictic assemblage, but rather there are genetically distinguishable subunits, each of which is affiliated with a distinct spawning area. The development of technology that enables a higher level of resolution of stock structure will also need to be accompanied by new management strategies.

There are several issues in lumping genetically or geographically discrete subpopulations as single management units. The major issue is the depletion of unidentified local subpopulations and their possible extinction (Ovenden 1990). If genetically distinct subpopulations by definition have little movement among them, specific stock adaptations to local habitat conditions can impede recovery from extrinsic sources. The rate of recovery of geographically discrete populations will depend on the level of gene flow between subpopulations. Furthermore, Daan (1991) indicates that if Virtual Population Analysis is applied to a heterogeneous unit stock (a metapopulation), then fishing effort is underestimated and fishing mortality deviates from the recent true trend. This leads to the possibility of overfishing. Pawson and Jennings (1996) also indicate that lumping stocks leads to reductions in short-term yields.

In the case of pollock, whether broad-scale migrations can result in seasonal mixing of subpopulations is also a central issue. For example, if eastern or western Bering Sea fish migrate across or around the basin (as believed by Dawson 1994), they may be harvested on either side of the

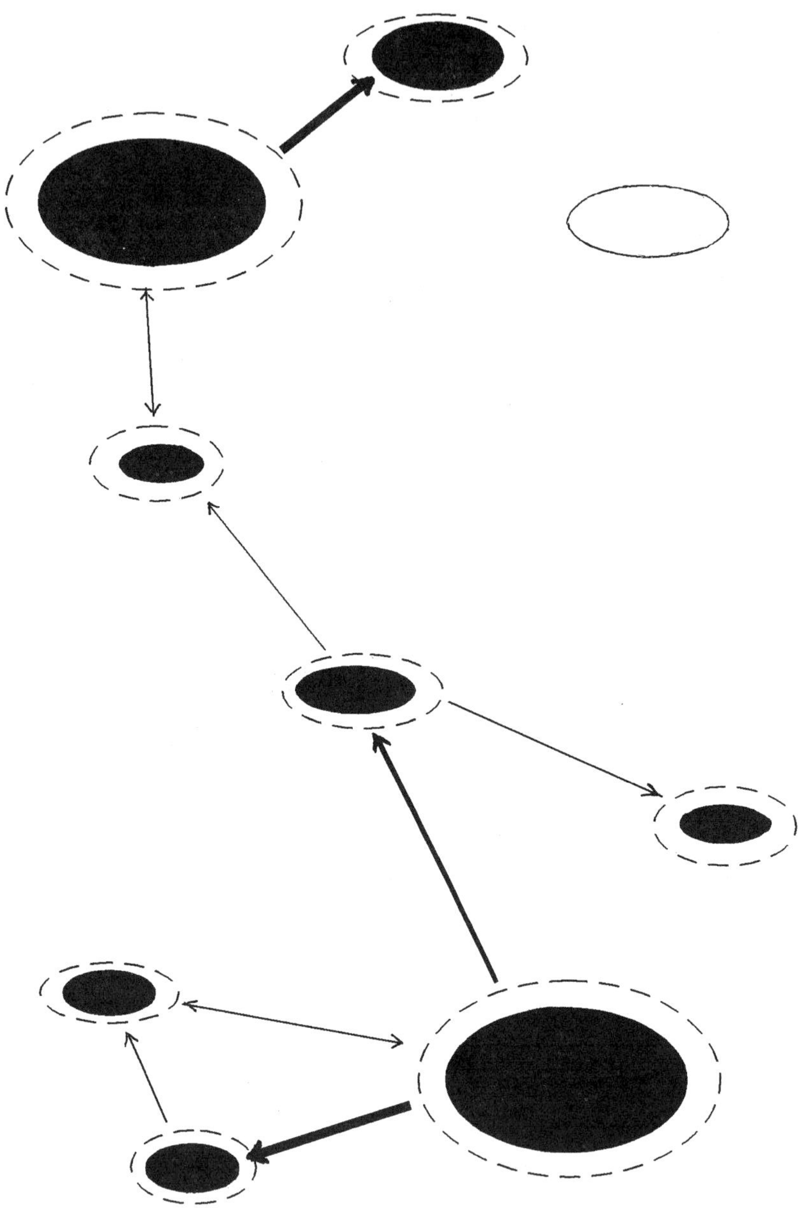

Figure 13. *Hypothetical nonspecific pictograph representing the population structure of walleye pollock. Closed circles represent habitat patches; filled = occupied, unfilled = vacant. Dashed lines indicate boundaries of populations. Arrows indicate relative gene flow through migration or larval transport. Probable main populations for pollock are Sea of Okhotsk and eastern Bering Sea shelf.*

Bering Sea basin each of which has its own harvest quota. Therefore, a migratory population would experience two independent sources of fishing mortality.

The mechanism of natal-site fidelity has broad implications to fisheries management and sustainable harvesting of these populations. For example, if fish learn their migration routes from older fish, currently thought to be important for cod (Rose 1993) and herring (Corten 1993), then removing older fish from the population can lead to increased straying of younger fish due to loss of social tradition. Furthermore, this mechanism implies that larvae dispersed from the range of adults can adopt new local populations. The mechanism of a "genetic memory" is that there are environmental clues which direct fish to return to an ancestral spawning location, and recognition of clues is inherent. If fish have this inherited sense, then colonization processes depend on natural selection and adaptation to local conditions. Vagrants from other populations are unlikely to spawn with local populations in the same region. Imprinting is a mechanism by which fish learn the clues of their spawning location by their own experience. Under this scenario, depleted populations can be restocked. In fact, marine fish species probably use a combination of all of these mechanisms for natal fidelity, and the response to any one stimulus is graded against others. It is not known which homing mechanisms are utilized by pollock during spawning migrations; however, based on the complex range of behaviors pollock are capable of exhibiting (Olla et al. 1996) it should not be surprising that they show considerable flexibility in their homing behavior.

Recently, new local populations of pollock have been discovered, and whether they can be maintained as self-sustaining populations or whether they are ephemeral stocks is a question that depends upon the mechanism of natal spawning and local adaptation for reproductive success. For example, in recent years a spawning population of 300,000 tons has been found around the Shumagin Islands in the western Gulf of Alaska (C. Wilson, Alaska Fisheries Science Center, 7600 Sand Point Way NE, Seattle, WA 98115, pers. comm., October 1996). This population is largely composed of the 1989 year class, which was strong in the Bering Sea, but weak in the rest of the Gulf of Alaska. It is unknown whether this population originated from density-dependent dispersal from the Bering Sea, or whether it represents an increase in a local subpopulation. Alternatively, vagrants from the Bering Sea could have joined a small local spawning population. In any case, for this population to sustain itself, there has to be local adaptation to the time and location of spawning to ensure reproductive success; or, larvae originating from the population can be dispersed vagrants, and maintenance of the population depends on a combination of recruitment from an upstream source and learning where to spawn from current inhabitants.

Given the high potential for gene flow in marine species, genetic approaches alone may not completely resolve stock structure issues. Low levels of diversity in marine fish populations have also been attributed to historical bottlenecks or recent founding of the population relative to the rate of mutation and genetic drift detectable in the particular genetic structure (mtDNA, microsatellite, etc.) analyzed. Tagging studies offer another way to study linkages between fish populations, providing information on stock separation and migration (Pawson and Jennings 1996). Carefully planned tagging experiments, or experiments utilizing naturally acquired tags are needed to establish relationships between fishes in different regions. Although not necessarily showing levels of gene flow between areas, tagging and/or transplant experiments offer a powerful approach to studying migration mechanisms. These studies of dispersal, in coordination with modern genetic studies, should contribute significantly to understanding links between population dynamics and structure of pollock populations.

Acknowledgments

We thank the following for critical comments on this manuscript or portions of it: P. Bentzen, W.S. Grant, R. Waples, M. Canino, C. Sibley, A. Kendall, and M. Canino.

References

Andrewartha, H.G., and L.C. Birch. 1954. The distribution and abundance of animals. University of Chicago Press, Chicago. 782 pp.

Arthur, J.R. 1983. A preliminary analysis of the discreteness of stocks of walleye pollock, *Theragra chalcogramma,* from the northeastern Pacific Ocean off Canada based on their parasites. Canadian Technical Report in Fisheries and Aquatic Sciences 1184. 15 pp.

Avdeev, V.V., and G.V. Avdeev. 1989. A study of walleye pollock population structure and migration routes using parasitological indicators. In: Proceedings of the International Symposium on the Biology and Management of Walleye Pollock. University of Alaska Sea Grant, AK-SG-89-01, Fairbanks, pp. 569-590.

Avise, J.C. 1994. Molecular markers, natural history and evolution. Chapman and Hall, NY. 511 pp.

Avise, J.C. 1995. Mitochondrial DNA polymorphism and a connection between genetics and demography of relevance to conservation. Conservation Biology 9:686-690.

Bailey, K.M. 1989. Interaction between the vertical distribution of juvenile walleye pollock, *Theragra chalcogramma,* in the eastern Bering Sea, and cannibalism. Marine Ecology Progress Series 53:205-213.

Bailey, K.M., R.D. Brodeur, and A.B. Hollowed. 1996. Cohort survival patterns of walleye pollock, *Theragra chalcogramma,* in Shelikof Strait, Alaska: A critical factor approach. Fisheries Oceanography 5:179-188.

Bakkala, R.G. 1993. Structure and historical changes in the groundfish complex of the eastern Bering Sea. NOAA Technical Report NMFS 114. 91 pp.

Bentzen, P., C.T. Taggart, D.E. Ruzzante, and D. Cook. 1996. Microsatellite polymorphism and the population structure of Atlantic cod *(Gadus morhua)* in the northwest Atlantic. Canadian Journal of Fisheries and Aquatic Sciences 53:2706-2721.

Bernatchez, L., and S. Martin. 1996. Mitochondrial DNA diversity in anadromous rainbow smelt, *Osmerus mordax* Mitchill: A genetic assessment of the member-vagrant hypothesis. Canadian Journal of Fisheries and Aquatic Sciences 53:424-433.

Boorman, S.A., and P.R. Levitt. 1973. Group selection on the boundary of a stable population. Theoretical Population Biology 4:85-128.

Brooker, A.L., D. Cook, P. Bentzen, J.M. Wright, and R.W. Doyle. 1994. Organization of microsatellites differs between mammals and cold-water teleost fishes. Canadian Journal of Fisheries and Aquatic Sciences 51:1959-1966.

Brown, J.H. 1984. On the relationship between abundance and distribution of species. American Naturalist 123:253-279.

Brown, J.H. 1995. Macroecology. University of Chicago Press, Chicago. 269 pp.

Corten, A. 1993. Learning processes in herring migrations. Conseil International pour l' Exploration de la Mer (ICES) C.M. 1993/H:18.

Daan, N. 1991. Bias in virtual population analysis when the unit stock assessed consists of sub-stocks. Conseil International pour l' Exploration de la Mer (ICES) C.M. 1991/D:17.

Dawson, P. 1994. The stock structure of Bering Sea walleye pollock *(Theragra chalcogramma)*. Master of Science thesis, University of Washington, Seattle. 220 pp.

deYoung, B., and G.A. Rose. 1993. On the recruitment and distribution of Atlantic cod *(Gadus morhua)* off Newfoundland. Canadian Journal of Fisheries and Aquatic Sciences 50:2729-2741.

Doherty, P.J., S. Planes, and P. Mather. 1995. Gene flow and larval duration in seven species of fish from the Great Barrier Reef. Ecology 76:2373-2391.

Efremov, V.V., E.V. Mikhajlova, and V.B. Molokanov. 1989. Aconitase, a new and promising marker to be used in genetic studies of *Theragra chalcogramma* populations. Genetika 25:2263-2265. (English abstract.)

Francis, R.C., and K.B. Bailey. 1983. Factors affecting recruitment of selected gadoids in the northeast Pacific and east Bering Sea. In: W. Wooster (ed.), From year to year. Washington Sea Grant, University of Washington, Seattle, pp. 35-60.

Gong, Y., Y.H. Hur, and S.S. Kim. 1990. Comparison of meristic characters of Alaska pollock, *Theragra chalcogramma,* from seven geographic areas of the North Pacific. In: Compilation of papers presented at the International Symposium on Bering Sea Fisheries, pp. 77-82. (Available from Alaska Fisheries Science Center, Seattle, WA 98115.)

Gonzalez-Villasenor, L.I., and D.A. Powers. 1990. Mitochondrial DNA restriction site polymorphisms in the teleost *Fundulus heteroclitus* support secondary intergradation. Evolution 44:27-37.

Grant, W.S., and F.M. Utter. 1980. Biochemical genetic variation in walleye pollock, *Theragra chalcogramma:* Population structure in the southeastern Bering Sea and Gulf of Alaska. Canadian Journal of Fisheries and Aquatic Sciences 37:1093-1100.

Hanski, I., and M. Gilpin. 1991. Metapopulation dynamics: Brief history and conceptual domain. In: M. Gilpin and I. Hanski (eds.), Metapopulation dynamics. Academic Press, Inc., San Diego, pp. 3-16.

Harrison, S. 1991. Local extinction in a metapopulation context: An empirical evaluation. In: M. Gilpin and I. Hanski (eds.), Metapopulation dynamics. Academic Press, Inc., San Diego, pp. 73-88.

Hashimoto, R., and S. Koyachi. 1969. Biology of the Alaska pollock, *Theragra chalcogramma* (Pallas), distributed on the fishery grounds of the Tohoku districts and the Pacific coast of Hokkaido, southward from the Erimo ground. I. The morphological differentiation of the three types and the comparison with the other fishery ground's groups. Bulletin of the Tohoku Regional Fisheries Research Laboratory 29:37-92. (In Japanese with English summary.)

Hashimoto, R., and S. Koyachi. 1977. Geographical variation of relative growth of walleye pollock, *Theragra chalcogramma* (Pallas). Bulletin of the Tohoku Regional Fisheries Research Laboratory 38:41-74. (In Japanese with English summary.)

Hinckley, S. 1987. The reproductive biology of walleye pollock, *Theragra chalcogramma,* in the Bering Sea, with reference to spawning stock structure. Fishery Bulletin, U.S. 85(3):481-498.

Hjort, J. 1914. Fluctuations in the great fisheries of northern Europe. Rapports et Proces-Verbaux des Reunions Conseil International pour l'Exploration de la Mer 20:1-228.

Hollowed, A.B., and W.S. Wooster. 1995. Decadal-scale variations in the eastern subarctic Pacific II: Responses of northeast Pacific fish stocks. Canadian Special Publication in Fisheries and Aquatic Sciences 121:373-385.

Ishida, T. 1954. On the age determination and morphometrical differences of the otolith of Alaska pollock in the Hokkaido coast. Bulletin of the Hokkaido Regional Fisheries Research Laboratory 11:36-67. (In Japanese with English summary.)

Iwata, M. 1973. Genetic polymorphism of tetrazolium oxidase in walleye pollock. Japanese Journal of Genetics 48:147-149.

Iwata, M. 1975a. Population identification of walleye pollock, *Theragra chalcogramma* (Pallas), population in the vicinity of Japan. Memoirs of the Faculty of Fisheries Hokkaido University 22:193-258.

Iwata, M. 1975b. Genetic identification of walleye pollock, *Theragra chalcogramma* (Pallas), populations on the basis of tetrazolium oxidase polymorphism. Comparative Biochemistry and Physiology 50B:197-201.

Iwata, M., and I. Hamai. 1972. Local forms of walleye pollock, *Theragra chalcogramma* (Pallas), classified by number of vertebrae. Bulletin of the Japanese Society of Scientific Fisheries 38(10):1129-1142.

Janusz, J. 1994. How many pollock *(Theragra chalcogramma)* stocks support the fishery in the Sea of Okhotsk? Bulletin of the Sea Fisheries Institute Gdynia 132:67-68. (English abstract.)

Janusz, J., T.B. Linkowski, and M. Kowalewska-Pahlke. 1989. Results of the population studies of walleye pollock, *Theragra chalcogramma,* in the Bering Sea. In: Proceedings of the International Scientific Symposium on Bering Sea Fisheries. NOAA Technical Memo, NMFS F/NWC-163, pp. 216-230. (Available from National Marine Fisheries Service, Scientific Publications Office, 7600 Sandpoint Way N.E., Seattle, WA 98115-0070.)

Johnson, A.G. 1977. A survey of biochemical variants found in groundfish stocks from the North Pacific and Bering Sea. Animal Blood Groups and Biochemical Genetics 8:13-19.

Kendall, A.W., J.D. Schumacher, and S. Kim. 1996. Walleye pollock recruitment in Shelikof Strait: Applied fisheries oceanography. Fisheries Oceanography 5:4-18.

Kennedy, P.K., M.L. Kennedy, E.G. Zimmerman, R.K. Chester, and M.H. Smith. 1986. Biochemical genetics of mosquito fish. V. Perturbation effects on genetic organization of populations. Copeia 1986:937-945.

Koyachi, S., and R. Hashimoto. 1977. Preliminary survey of variations of meristic characters of walleye pollock, *Theragra chalcogramma* (Pallas). Bulletin of the Tohoku Regional Fisheries Research Laboratory 38:7-40. (In Japanese with English summary.)

Lawton, J.H., S. Nee, A.J. Letcher, and P.H. Harvey. 1994. Animal distributions: Patterns and processes. In: P.J. Edwards, R.M. May, and N.R. Webb (eds.), Large-scale ecology and conservation. Blackwell Scientific Publications, Oxford, pp. 41-48.

Levins, R. 1970. Extinction. In: M. Gersternhaber (ed.), Some mathematical problems in biology. American Mathematical Society, Providence, R.I., pp. 77-107

Lynde, C.M., M.V.H. Lynde, and R.C. Francis. 1986. Regional and temporal differences in growth of walleye pollock *(Theragra chalcogramma)* in the eastern Bering Sea and Aleutian Basin with implications for management. Unpublished manuscript, 48 pp. (Available from Alaska Fisheries Science Center, 7600 Sand Point Way N.E., Seattle, WA 98115.)

MacArthur, R.H., and E.O. Wilson. 1967. The theory of island biogeography. Princeton University Press, Princeton. 203 pp.

Matthews, K.R. 1990. An experimental study of the habitat preferences and movement patterns of copper, quillback, and brown rockfishes (*Sebastes* spp.). Environmental Biology of Fishes 29:161-178.

McConnaughey, R.A. 1995. Changes in geographic dispersion of eastern Bering Sea flatfish associated with changes in population size. In: Proceedings of the International Symposium of North Pacific Flatfish. University of Alaska Sea Grant, AK-SG-95-04, Fairbanks, pp. 385-406.

McCullough, D.R. 1996. Introduction. In: D.R. McCullough (ed.), Metapopulations and wildlife conservation. Island Press, Washington, DC, pp. 1-10

Megrey, B.A. 1988. On the contribution of density-dependent factors to year class variations in Alaskan walleye pollock. In: Proceedings from workshop on year class variations as determined from pre-recruit investigations. Institute of Marine Research, Bergen, Norway, pp. 425-456.

Mulligan, T.J., K.M. Bailey, and S. Hinckley. 1989. The occurrence of larval and juvenile walleye pollock, *Theragra chalcogramma,* in the eastern Bering Sea with implications for stock structure. In: Proceedings of the International Symposium on the Biology and Management of Walleye Pollock. University of Alaska Sea Grant, AK-SG-89-01, Fairbanks, pp. 471-490.

Mulligan, T.J., R.W. Chapman, and B.L. Brown. 1992. Mitochondrial DNA analysis of walleye pollock, *Theragra chalcogramma,* from the eastern Bering Sea and Shelikof Strait, Gulf of Alaska. Canadian Journal of Fisheries and Aquatic Sciences 49:319-326.

Nakano, H., M. Takahashi, and K. Kikuchi. 1991. Geographical differences of walleye pollock caught from three Bering Sea areas from standpoint of metal and amino acid contents of otoliths. International Pacific Fisheries Commission Document No. 3668, 14 pp. (Available from Fisheries Agency of Japan, National Research Institute of Far Sea Fisheries, 7-1 Orido 5 chome, Shimizu, Shizuoka, Japan 424.)

National Research Council. 1996. The Bering Sea ecosystem. National Academy Press, Washington, DC. 307 pp.

Nitta, A., and T. Sasaki. 1990. Study on stock identification of walleye pollock based on morphometric data. In: Compilation of papers presented at the International Symposium on Bering Sea Fisheries, pp. 74-76. (Available from Alaska Fisheries Science Center, Seattle, WA 98115.)

Ogata, T. 1959. Population studies of the Alaska pollock in the Sea of Japan I. On the variation in the vertebral count. Annual Report of the Japan Sea Regional Fisheries Research Laboratory 5:119-125. (In Japanese with English summary.)

Olla, B.L., M.W. Davis, C.H. Ryer, and S.M. Sogard. 1996. Behavioral determinants of distribution and survival in early stages of walleye pollock, *Theragra chalcogramma:* A synthesis of experimental studies. Fisheries Oceanography 5:167-178.

Ovenden, J.R. 1990. Mitochondrial DNA and marine stock assessment: A review. Australian Journal of Marine and Freshwater Research 41:835-853.

Palsson, W.A., J.C. Hoeman, G.G. Bargmann, and D.E. Day. 1996. 1995 Status of Puget Sound bottomfish stocks. Washington Department of Fisheries, Olympia, WA. 98 pp.

Pawson, M.G., and S. Jennings. 1996. A critique of methods for stock identification in marine capture fisheries. Fisheries Research 25:203-217.

Powers, D.A. 1996. The use of molecular techniques to dissect the genetic architecture of pollock populations. Report to the Coastal Ocean Program. Hopkins Marine Station, Stanford University, Pacific Grove, CA 93950 (unpubl. report).

Quinn, T.P. 1993. A review of homing and straying of wild and hatchery-produced salmon. Fisheries Research 18:29-44.

Quinn, T.P., and R.D. Brodeur. 1991. Intra-specific variations in the movement patterns of marine animals. American Zoologist 31:231-241.

Rose, G.A. 1993. Cod spawning on a migration highway in the north-west Atlantic. Nature 366:458-461.

Saunders, M.W., G.A. McFarlane, and W. Shaw. 1989. Delineation of walleye pollock *(Theragra chalcogramma)* stocks off the Pacific coast of Canada. In: Proceedings of the International Symposium on the Biology and Management of Walleye Pollock. University of Alaska Sea Grant, AK-SG-89-01, Fairbanks, pp. 379-401.

Serobaba, I.I. 1977. Data on the population structure of the walleye pollock, *Theragra chalcogramma* (Pallas), from the Bering Sea. Voprosy Ikhtiologii 17:247-260. (In Russian.) (English translation 1978, Journal of Ichthyology 17:219-231.)

Severin, K.P., J. Carroll, and B.L. Norcross. 1995. Electron microprobe analysis of juvenile walleye pollock, *Theragra chalcogramma,* otoliths from Alaska: A pilot stock separation study. Environmental Biology of Fishes 43:269-283.

Shaw, W., and G.A. McFarlane. 1986. Biology, distribution and abundance of walleye pollock *(Theragra chalcogramma)* off the west coast of Canada. International North Pacific Fisheries Commission Bulletin 45:262-283.

Shields, G.F., and J.R. Gust. 1995. Lack of geographic structure in mitochondrial DNA sequences of Bering Sea walleye pollock, *Theragra chalcogramma.* Molecular Marine Biology and Biotechnology 4:69-82.

Sinclair, M. 1988. Marine populations: An essay on population regulation and speciation. Washington Sea Grant, University of Washington, Seattle. 252 pp.

Slatkin, M. 1987. Gene flow and the geographical structure of natural populations. Science 236:787-792.

Smith, P.J., A. Jamieson, and A.J. Birley. 1990. Electrophoretic studies and the stock concept in marine teleosts. Journal du Conseil International pour l' Exploration de la Mer 47:231-245.

Somerton, D. 1978. Competitive interaction of walleye pollock and Pacific Ocean perch in the northern Gulf of Alaska (Abstract). In: S.J. Lipovsky and C.A. Simenstad (eds.), Fish food habit studies. Washington Sea Grant, University of Washington, Seattle, p. 163.

Sogard, S.M., and B.L. Olla. 1996. Food deprivation affects vertical distribution and activity of a marine fish in a thermal gradient: Potential energy-conserving mechanisms. Marine Ecology Progress Series 133:43-55.

Swain, D.P., and R. Morin. 1996. Relationship between geographic distribution and abundance of American plaice *(Hippoglossoides platessoides)* in the southern Gulf of St. Lawrence. Canadian Journal of Fisheries and Aquatic Sciences 53:106-119.

Swain, D.P., and E.J. Wade. 1993. Density-dependent geographic distribution of Atlantic cod *(Gadus morhua)* in the southern Gulf of St. Lawrence. Canadian Journal of Fisheries and Aquatic Sciences 50:725-733.

Swartzman, G., W. Stuetzle, K. Kulman, and M. Powojowski. 1994. Relating the distribution of pollock schools in the Bering Sea to environmental factors. ICES Journal of Marine Science 51:481-492.

Taylor, A.D. 1991a. Studying metapopulation effects in predator-prey systems. In: M. Gilpin and I. Hanski (eds.), Metapopulation dynamics. Academic Press, Inc., San Diego, pp. 305-323.

Taylor, B. 1991b. Investigating species incidence over habitat fragments of different areas: A look at error estimation. In: M. Gilpin and I. Hanski (eds.), Metapopulation dynamics. Academic Press, Inc., San Diego, pp. 177-191.

Temnykh, O.S. 1991. The population structure of Alaska pollock in the Okhotsk Sea. Rybnoe Khozaistro 6:28-31. (English abstract.)

Temnykh, O.S. 1994. Morphological differentiation of Alaskan pollock, *Theragra chalcogramma,* in the west Bering Sea and Pacific waters of Kamchatka. Journal of Ichthyology 34:204-211.

Thompson, J.M. 1981. Preliminary report on the population biology and fishery of walleye pollock *(Theragra chalcogramma)* off the Pacific coast of Canada. Canadian Technical Report in Fisheries and Aquatic Sciences 1031. 157 pp.

Tsugi, S. 1989. Alaska pollock population, *Theragra chalcogramma,* of Japan and its adjacent waters, I: Japanese fisheries and population studies. Marine Behavior and Physiology 15:147-205.

Waples, R.S., and R.H. Rosenblatt. 1987. Patterns of larval drift in southern California marine shore fishes inferred from allozyme data. Fishery Bulletin, U.S. 85:1-11.

Wespestad, V.G. 1996. Prepared for Nor-fishing '96, Trondheim, Norway, 15-16 August 1996. See: http://www.refm.noaa.gov/norfish/norfish.htm.

Wilimovsky, N.J., A. Peden, and J. Peppar. 1967. Systematics of six demersal fishes of the North Pacific Ocean. Fisheries Research Board of Canada Technical Report 34. 95 pp.

CHAPTER **27**

Use of a Surface-Current Model and Satellite Telemetry to Assess Marine Mammal Movements in the Bering Sea

Thomas R. Loughlin and W. James Ingraham Jr.
Alaska Fisheries Science Center, Seattle, Washington

Norihisa Baba
National Research Institute of Far Seas Research, Shizuoka, Japan

Bruce W. Robson
Alaska Fisheries Science Center, Seattle, Washington

Abstract

We used satellite-linked time-depth recorders (SLTDRs) to monitor movements of eight free-ranging adult male northern fur seals *(Callorhinus ursinus)* in the Bering Sea and North Pacific Ocean during October through January 1993. We investigated whether the movements of adult male northern fur seals at sea during migration are independent of ocean surface currents by comparing fur seal movements obtained from SLTDRs to the movement of surface currents derived by the numerical model Ocean Surface Current Simulations (OSCURS). Results from this study suggest that movements of fur seals are independent of surface currents but that the seals likely cue on currents during migration and as an aid in finding prey. The eight male northern fur seals generally swam in the same direction as surface current trajectories calculated by the OSCURS model. Seals that left the Bering Sea toward the west, and all seals while in the Pacific or Gulf of Alaska, had movements similar to water surface current trajectories calculated by OSCURS. However, four seals that left the Bering Sea through passes in the eastern end of the Aleutian Islands chain swam opposite to the weak northerly water flow in the southeastern Bering Sea. Once in the North Pacific and Gulf of Alaska, their movements were similar to calculated surface water trajectories.

Introduction

There are 116 species of marine mammals in the world oceans of which 25 are known to inhabit the Bering Sea (Lowry et al. 1982, Kajimura and Loughlin 1988). Little is known about the ecology and life history of most of these mammals with the exception of a few species that received scientific study or that had been exploited by commercial whalers or sealers. Information on prey for most marine mammal species is based on small sample sizes and even less is known about their movement patterns and foraging behavior. Current information on marine mammals at sea is based primarily on opportunistic sightings; few directed studies of movements of marine mammals in the Bering Sea have occurred. One exception is the northern fur seal *(Callorhinus ursinus)*. Identification of its prey and feeding habits around the Pribilof Islands is well documented (Kajimura 1984, Sinclair et al. 1994) and information on foraging ecology (e.g., feeding locations and depths) has been studied (Loughlin et al. 1987, Goebel et al. 1991). However, identification of the oceanographic factors that influence movements of fur seals while at sea during migration remains largely undetermined. A major reason for this lack of information has been the difficulty of obtaining data on movements, foraging locations, and diving of free-ranging marine animals. Instruments which record diving depths over time (time-depth recorder or TDR) have existed since the 1970s and have allowed researchers to track some pinniped movements vertically in the water column during foraging trips (e.g., Gentry and Kooyman 1986). Subsequent studies coupled the TDR with a VHF radio transmitter and a ship or aircraft was used to track northern fur seals to obtain a partial picture of their pelagic movements (Loughlin et al. 1987, Antonelis et al. 1990, Goebel et al. 1991). Application of this technique has received limited use due in part to the high costs of aircraft and ship time, and the need to recapture the animal to recover the TDR data.

Recent developments in satellite telemetry allow tracking of marine animals using satellite-linked transmitters (e.g., Hill et al. 1987, Hills 1987, Stewart et al. 1989). Through the Service-Argos system on board NOAA Tiros-series satellites, it is possible to track and retrieve data from free-ranging animals using uplink communications between satellite transmitters attached to animals and receivers on board satellites. Locations at sea are determined from the Doppler shift of a series of signals received by the satellite (Fancy et al. 1988, Stewart et al. 1989).

By combining a satellite transmitter and TDR it is possible to determine locations and collect diving information while the animal is at sea. The TDR collects dive data which can be transmitted while the animal is at sea or saved for later transmission while the animal is on land. This is particularly important for animals like male northern fur seals which inhabit relatively inaccessible areas and return to their tagging site on an infrequent basis.

Table 1. **Deployment record of satellite-linked time-depth recorders (SLTDR) attached to male northern fur seals (estimated between 7 and 11 years of age) at St. Paul Island, Alaska, during October 1992.**

SLTDR	Date deployed	Final transmission	No. days	Date left Bering Sea
14100	10/22/92	01/06/93	76	[a]
14101	10/24/92	01/31/93	>99	11/14/92
14102	10/24/92	12/23/92	60	11/07/92
14103	10/26/92	01/31/93	>97	12/07/92
14104	10/27/92	12/13/92	47	12/07/92
14105	10/27/92	11/26/92	30	11/16/92
14106	10/27/92	01/31/93	>96	12/15/92
14107	10/28/92	01/21/93	85	11/30/92

[a] Seal 14100 did not leave the Bering Sea.

We used a satellite-linked time-depth recorder (SLTDR) developed for pinnipeds (Merrick et al. 1994) to provide information gained on free-ranging adult male northern fur seals in the Bering Sea and North Pacific Ocean. We describe their movements during October 1992 through January 1993. We report on the relationship between adult male northern fur seal movements at sea during migration and ocean surface currents (Fig. 1; Ohtani 1973). We compared fur seal movements derived from SLTDR locations to the movement of daily surface currents at the fur seal's location derived by the numerical model Ocean Surface Current Simulations (OS-CURS) (Ingraham and Miyahara 1988).

Methods

SLTDR Protocol

We deployed Type 3 (½ watt) SLTDRs from Wildlife Computers, Redmond, WA, USA, during October 1992 on eight adult male northern fur seals at St. Paul Island, Alaska (Table 1). All eight were estimated to be between 7 and 11 years of age, based on physical size, coloration, and form of the head and neck. Each showed indications of having fasted for a long period (breeding period) and were likely holders of territories before being instrumented. The male fur seals were immobilized with Telazol injected remotely with a 5 cc dart fired from a pneumatic dart gun (Loughlin and Spraker 1989, Kiyota et al. 1992b). The SLTDR was then attached directly to the animal's back using 5-minute epoxy glue (Loughlin et al. 1987). Duration of SLTDR operation varied from 30 to more than 98 days (Table 1).

Software built into the SLTDR controlled three functions—data collection, summarization, and transmission. Once the animal had gone to sea the unit collected data on location and depth readings at 10 s intervals throughout the trip. This report contains only information on locations at sea; the foraging (dive depth and dive duration) data will be reported elsewhere. While the animal was at sea, a 32-bit data message was transmitted at 45 s intervals whenever the animal surfaced (as determined by a conductivity sensor, "saltwater switch"). These messages included summarized dive data and were used to calculate animal locations.

Service-Argos classifies the accuracy of locations as class 3 (accurate to 150 m), class 2 (accurate to 350 m), class 1 (accurate to 1 km), and class 0 (no accuracy assigned) (Service-Argos 1984). Most locations we received were class 1 and 0. Data were received from Service-Argos on floppy diskettes and formatted using software developed by Wildlife Computers. At-sea information is presented as track lines over time representing numerous locations at sea.

OSCURS

We used the computer program known as the Ocean Surface Current Simulations (OSCURS) numerical model to hind cast oceanic surface water drift in relation to fur seal movements during migration determined by SLTDR location data. OSCURS is an empirical ocean-wide surface current model that covers the Subtropic and Subarctic Pacific Region and Bering Sea from 10° to 66°N with a 90 km grid. Surface currents are calculated daily from a gridded-daily sea level atmospheric pressure database which extends back to 1946. By choosing any starting date and location within the grid (i.e., a fur seal sighting), surface water displacement trajectories may be calculated for any desired time interval (i.e., to compare the water movement with the fur seal movements).

Daily current fields are computed in three steps from empirical formulae (Ingraham and Miyahara 1988). First, wind speed and direction fields are computed from pressure gradients within the daily (0000 GMT) U.S. Fleet Numerical Meteorology and Oceanography Center (FNMOC) sea level atmospheric pressure fields after Larson (1975). Second, the wind-induced currents in the surface mixed layer (generally the upper 30-50 m) are calculated from empirical functions summarized by Huang (1979); speed is proportional to the square root of the wind speed (Witting 1909) and the angle (between 20° and 30°) to the right of the wind vector is a function of wind speed (Weber 1983). Third, the long-term mean geostrophic current (0/2,000 dbar) is added vectorially to the wind current to form the total current field. The total current velocity at the start location acting for 24 hours would then move the surface water to a new location which is the start location for the next day's calculations. Trajectories are thus formed by repeating this procedure day by day for the desired time interval (i.e., the number of days the fur seal was tracked). Speed coefficients were tuned (Ingraham and Miyahara 1989) so that the model trajectories matched the trajectories of satellite-tracked drifters (drogued at 20 m) from the Gulf of Alaska (Reed 1980).

We did not use generalized current patterns known for the study area because the OSCURS model previously indicated that near-surface winter circulation in the northern Gulf of Alaska varied considerably from year to year (Ingraham et al. 1991; Ebbesmeyer and Ingraham 1992, 1994). Instead we used current estimates derived from the model for October 1992 through January 1993.

Large scale movements through the course of each seal's migration were examined in relation to monthly summaries of OSCURS current vectors throughout the Bering Sea and North Pacific. To examine finer scale (short-term) seal movements in relation to daily surface current vectors, we chose one location per day for each of 5-15 consecutive days for selected seals using location quality and proximity to 0000 GMT as selection criteria. Daily fixes were used as starting locations for calculating surface current (surface water displacement) trajectories in OSCURS for each 24 hour period. We examined each seal's movements in relation to the daily current trajectory and the direction of surrounding current vectors.

Results

Generalized Locations and Movements

The male fur seals traversed most oceanic areas and the outer domain of the continental shelf of the Bering Sea from about 80 km north of St. Matthew Island south, and from about the 100 m isobath on the continental shelf west to Shirshov Ridge in the western Bering Sea (Fig. 1). None entered the shallow waters of the eastern Bering Sea shelf east of the Pribilof Islands although one animal (14102) was active southwest of St. Paul Island in the middle shelf domain in water about 70 m deep. Some animals moved erratically while others moved in direct lines to locations where diving occurred. Eventually all (but one) male fur seals left the Bering Sea for either the Gulf of Alaska or west toward the Kuril Islands and Japan (Fig. 2).

While in the Bering Sea, male fur seals had diving bouts, which we assumed were related to feeding, in areas associated with the outer domain of the continental slope and underwater ridges and sea mounts. Typical locations included Bowers Ridge, a large, crescent shaped ridge extending north and west from Semisopochnoi Island in the central Aleutian Island chain, and northwest of the Pribilof Islands on the continental shelf in water from 100 to 250 m deep. Relatively little time was spent over deepwater masses (>1,000 m) of the Bering Sea. Four animals spent time diving west of St. Matthew Island over the shelf and shelf break div-

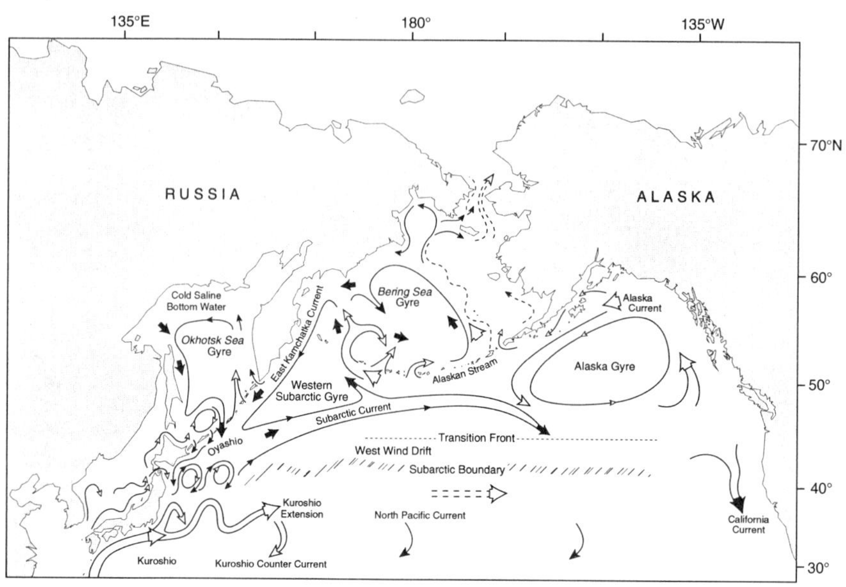

Figure 1.   Map depicting the Bering Sea and North Pacific Ocean subarctic water circulation and transition zones (modified from Ohtani 1991).

ing in water <100 m (number 14100—30 days, number 14103—20 days, number 14106—22 days, and number 14107—17 days).

One seal remained in the Bering Sea for the duration of the study period; the other seven entered the North Pacific from the Bering Sea through Aleutian Island passes from Unimak Pass west to the Commander Islands. Three used Unimak Pass (eastern Aleutian Islands); one used Samalga Pass (central Aleutian Islands); one passed through near Kiska Island (near the end of the chain); and two passed through near the Commander Islands. None had re-entered the Bering Sea by the end of the study period.

Movement of the four fur seals that migrated into the Gulf of Alaska did not follow surface current trajectories produced by OSCURS, but traveled south or southeast primarily in waters over the continental shelf where the northward current is very weak and dominated by tidal currents (Fig. 3; Stabeno and Reed 1994). Once they entered the Gulf of Alaska, the seals swam offshore and traveled in the same easterly direction as the surface water trajectories calculated by the model; such movement is consistent with the flow of the Subarctic Current into the Gulf of Alaska Gyre. Two of the seals (numbers 14101 and 14102) then moved northeast, similar to the direction of surface water movement calculated by OSCURS and into the westerly flowing Alaskan Stream during December 1992, diving and

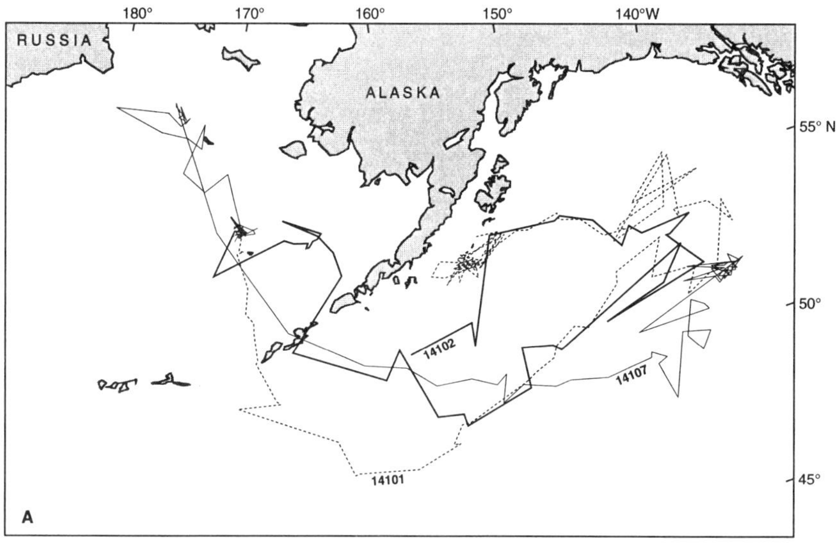

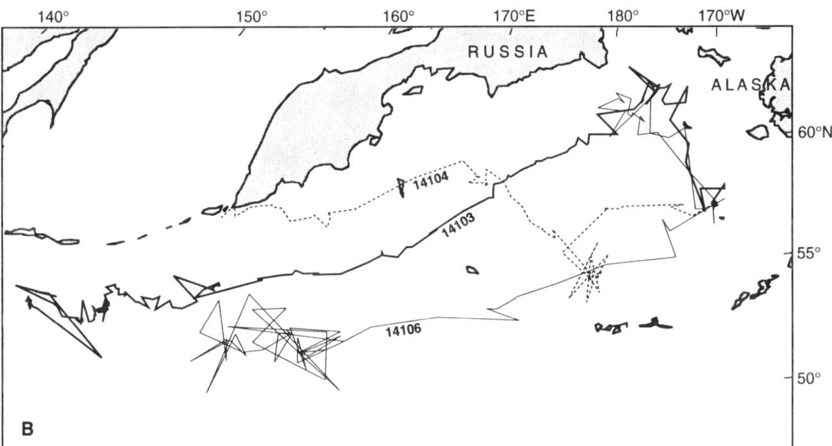

Figure 2. Movements of male northern fur seals equipped with SLTDRs in 1992. A represents three fur seals in the eastern Bering Sea and Gulf of Alaska; B represents three seals in the western Bering Sea and western Pacific Ocean (modified from Loughlin 1993).

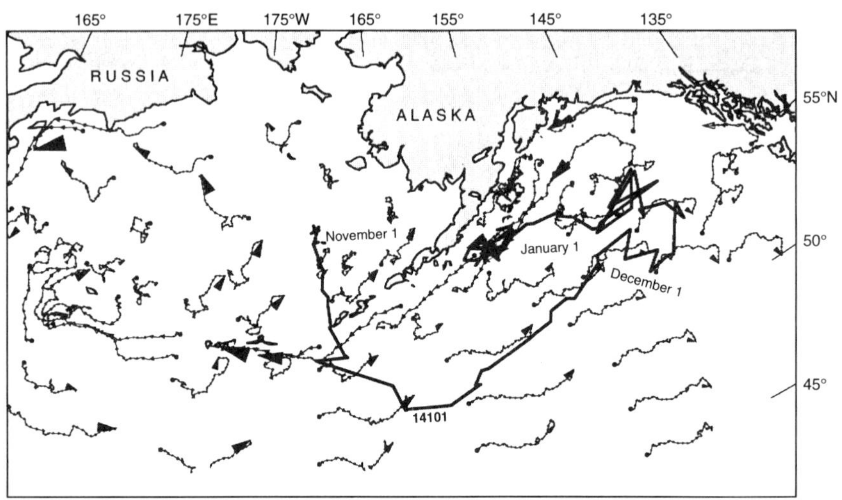

Figure 3.   Plot showing the movements of male fur seal number 14101 in the Gulf of Alaska and the movement of surface water during December 1992 as predicted by the OSCURS model in the Bering Sea and North Pacific Ocean. The thick, solid line represents the movements of fur seal number 14101 over the entire study period; locations on the first day of each month are noted for comparison to OSCURS water movement. The small solid circles represent a particle of water at the start of the month and the lines and arrows emanating from the circles show the predicted movement of the particle. Large arrows equal fast moving water and small arrows slow. The arrowhead at the end of each trajectory indicates the average direction and speed for the last ten days of the month. The figure suggests that fur seal movements at sea are similar to surface water movements as predicted by OSCURS in the area shown.

probably foraging, north of the gyre or downstream in the Alaskan Stream southwest of Kodiak Island. Location data for fur seal number 14107 ended prior to its movement into the Alaskan Stream but it did follow the same water flow as the other seals in the offshore Subarctic Current into the Gulf of Alaska Gyre. Seal number 14105 appeared to briefly follow the Subarctic Current into the western portion of the Gulf of Alaska Gyre. The last locations for seal number 14105 were received downstream of the Alaskan Stream near the Shumagin Islands.

Three male fur seals that went west (numbers 14103, 14104, and 14106) tended to follow the water flow pattern calculated by OSCURS (Fig. 4). Two of the three (numbers 14103 and 14104) left the Pribilof Islands then meandered in a northwest (number 14103—St. Matthew I.) and southwest (number 14104—Bowers Ridge) direction until offshore the Kamchatka Peninsula; they then entered the East Kamchatka Current and followed

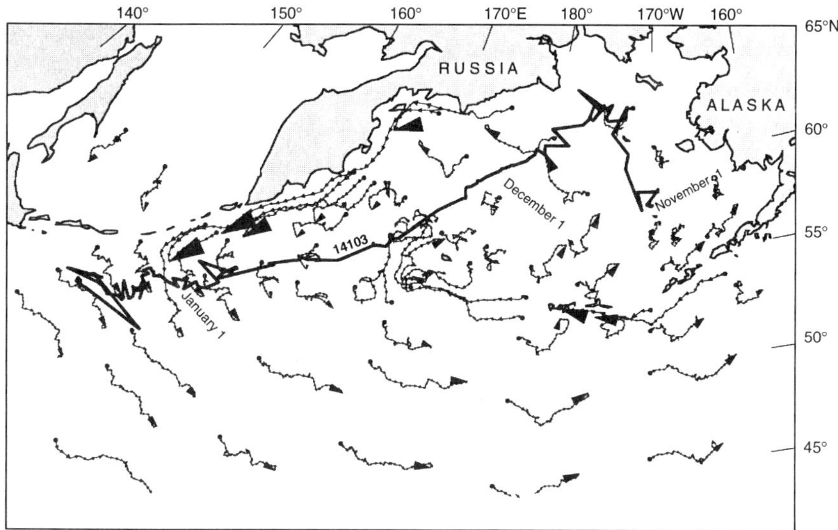

Figure 4. Plot showing the movements of male fur seal number 14103 in the western Bering Sea and western North Pacific and the movement of surface water during December 1992 as predicted by the OSCURS model in the Bering Sea and North Pacific Ocean. The thick, solid line represents the movements of fur seal number 14103 over the entire study period; locations at the first day of each month are noted for comparison to OSCURS water movement. The small solid circles represent a particle of water at the start of the month and the lines and arrows emanating from the circle show the predicted movement of the particle. Large arrows equal fast moving water and small arrows slow. The arrowheads at the end of the trajectory indicate the average direction and speed for the last ten days of the month. The figure suggests that fur seal movements at sea are similar to surface water movements as predicted by OSCURS in the area shown.

the general flow south and west out of the Bering Sea and south offshore of the Kuril Islands to northern Japan. The surface currents calculated by OSCURS show westward water movements across the Bering Sea along the northern part of the Bering Sea Gyre then south by the East Kamchatka Current. Seals entered the Western Subarctic Gyre as they left the Bering Sea where they likely fed in these productive waters. Fur seal number 14103 meandered south into the Oyashio Current and Oyashio Gyre off the Kuril Islands and remained there until the end of the study period.

Fur seal number 14106 made one trip northwest of the Pribilof Islands but then returned to the Pribilof Islands before heading southwest through the Aleutian Islands then toward the Kuril Islands. Its movements were

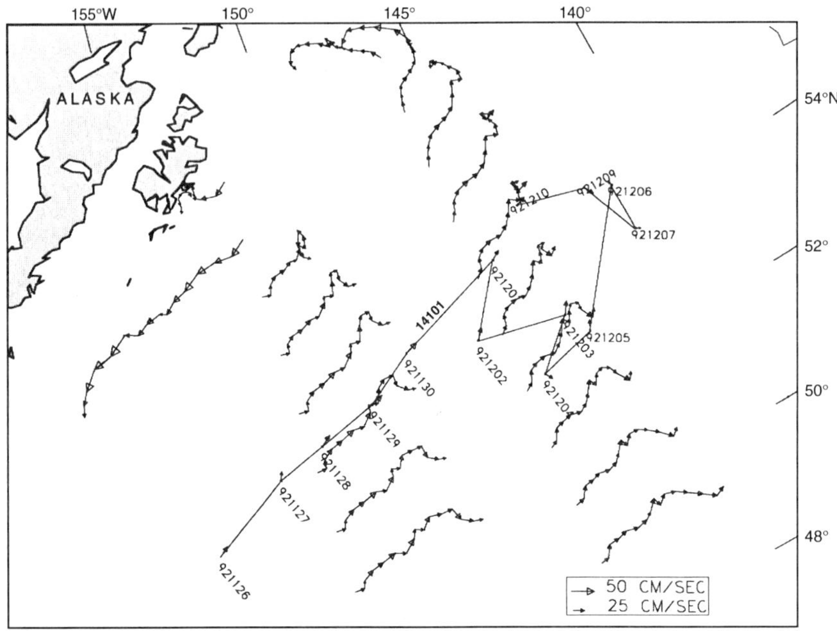

*Figure 5.   Figure depicting a fine-scale comparison between fur seal daily move-
ments and surface water movement in the Gulf of Alaska as predicted by
OSCURS. The solid line with numbers associated is the movement of fur
seal number 14101 from November 26, 1992 (921126), to December 10,
1992 (921210), in the Gulf of Alaska. The vectors and arrows (larger
arrow represents faster speed) on this line suggest the direction and speed
of a particle of water at that location as predicted by OSCURS. The other
solid lines with overlaying arrows are the fine-scale surface water movements predicted by OSCURS. The figure shows the association of the fur
seal with the general movement pattern of surface water and the location of the seal with the nearby transition zone of opposite-flowing water.
The transition zone is where high concentrations of prey are assumed to
occur.*

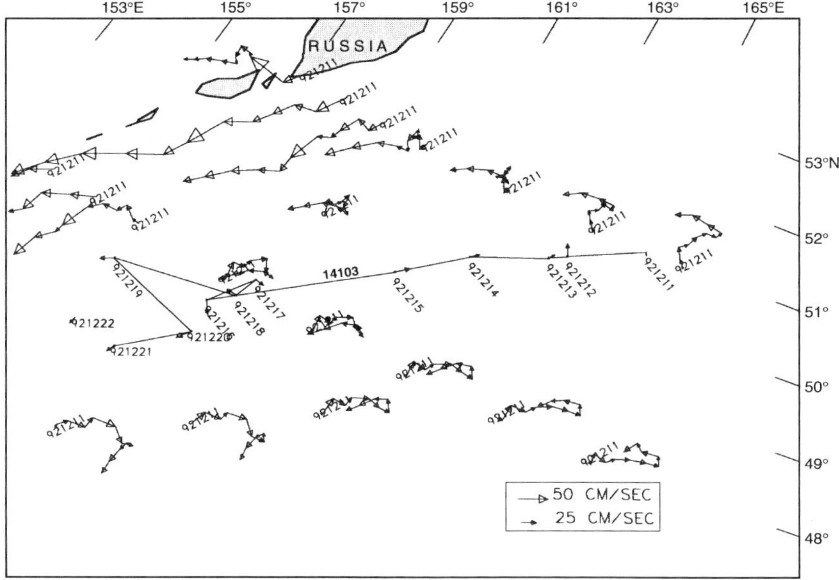

Figure 6. *Figure depicting a fine-scale comparison between fur seal daily movements and surface water movement in the western Pacific as predicted by OSCURS. The line with daily numbers associated is the movement of fur seal number 14103 from December 11, 1992 (921211), to December 22, 1992 (921222), off the Kuril Islands, Russia. The vectors and arrows (larger arrow represents faster speed) on this line suggest the direction and speed of a particle of water at that location as predicted by OSCURS. The other lines with overlaying arrows are the fine-scale surface water movements predicted by OSCURS. The figure shows the association of the fur seal with the general movement pattern of surface water and the location of the seal with the nearby transition zone of opposite-flowing water. The transition zone is where high concentrations of prey are assumed to occur.*

not as well matched to the surface water movements while in the Bering Sea as were the other two male seals that went west.

Fine-Scale Comparisons

Surface current trajectories calculated at daily locations of individual seals indicated that the movements of fur seals were influenced, but not dictated by, location and direction of surface currents. Movements of three seals that migrated into the Gulf of Alaska (numbers 14101, 14102, 14107) were consistent with daily OSCURS vectors as they followed the eastward flow in the northern portion of the Subarctic Current into the Gulf of Alaska Gyre (Fig. 5). The three seals that migrated westward showed similar movements in respect to daily OSCURS current vectors once they left the Bering Sea (Fig. 6).

Fur seals numbers 14103 and 14104 swam southward in the East Kamchatka Current then into the Western Subarctic Gyre, while fur seal number 14106 followed the direction of the Alaskan Stream extension for 2-3 days before entering the Western Subarctic Gyre from the east. While in the East Kamchatka Current or Alaskan Stream daily OSCURS vectors were consistent with fur seal movements; however, fur seals often traveled in the zone of weaker flow and eddies at the margin of these fast moving boundary currents where surface current flow, and consequently the relationship between daily directional vectors and seal movements, were less consistent.

Of the seven seals that left the Bering Sea, six arrived at similar times and remained for extended periods in one of the large cyclonic gyre systems of the North Pacific. Three of the eastward migrating seals (numbers 14101, 14102, 14107) entered the area north of 48°00′N and east of 149°00′W in the Gulf of Alaska between 30 November and 14 December 1992, remaining (presumably foraging) at the southern margin of the Alaska Gyre in the productive waters where the cold water mass from the Subarctic Current is deflected northward into the Alaska Current (Stabeno and Reed 1994). The three seals which migrated west (numbers 14103, 14104, 14106) passed south of 50°N between 11-21 December 1992 and remained in the Western Subarctic Gyre until the end of the study period. In these regions daily OSCURS vectors and fur seal locations showed more variation due to the variable current trajectories within the gyres and the localized movements of fur seals.

Discussion

Based on results from this study, we conclude that adult male northern fur seal movements at sea during migration are independent of ocean surface currents, but they likely cue on currents to facilitate migration and foraging. Our results show that seven of eight male northern fur seals that we followed swam in the same direction, or in close association with surface current trajectories calculated by the OSCURS model. Male seals that left the Bering Sea toward the west, and all seals while in the Pacific Ocean or Gulf of Alaska, had movements similar to water surface current trajectories. However, four seals that left the Bering Sea through passes in the eastern end of the Aleutian chain swam opposite to the weak northerly water flow in the southeastern Bering Sea (Overland et al. 1994, Stabeno and Reed 1994). Once in the North Pacific and Gulf of Alaska, though, their movements were similar to OSCURS' surface water trajectories.

In our fine-scale comparison, vector results showed that movements were consistent with calculated surface water trajectories in areas where these hydrographic features are associated with areas of high productivity. Four northern fur seals followed in our study went into the Gulf of Alaska along the northern margin of the Subarctic Current leading into the Alaska Gyre. This area is characterized by large aggregations of zooplankton during summer months (Brodeur et al. 1996). We postulate that the zooplankton (or similar prey) are also there in winter and provide forage to fish consumed by northern fur seals. In a similar fashion, males which migrated westward appeared to forage in association with the East Kamchatka Current while traveling along the confluence of the fast moving boundary current and the Western Subarctic Gyre.

We weren't surprised to find that migration of adult male northern fur seals coincided with surface currents. Fur seal movements are not passive or dependent on surface currents but currents could provide a mechanism for orientation and moving in the open ocean. Traveling with surface currents can result in reduced energy expenditures by seals. Probably most important is that currents and the edges of fronts and gyres are where biological productivity is concentrated and food availability for fur seals is optimal (Sinclair et al. 1994).

Studies on the fall movements of pups migrating from the Pribilof Islands showed that they moved through a number of different eastern Aleutian Island passes as they swam into the eastern and central Pacific Ocean (Ragen et al. 1995). These pups left the rookeries in November and spent, on average, about ten days swimming to the Aleutian Islands, entering the North Pacific in early November to mid-December (Ragen et al. 1995). Four males that we monitored also used these passes; however, the pups had likely migrated beyond the Aleutian Islands before arrival of the males we monitored. Adult males and pups probably feed on different size prey, dive to different depths to feed, and are probably not competing for similar prey resources.

Information collected during this study on adult male fur seals, and in earlier studies on adult females and pups (Kiyota et al. 1992a, Ragen et al. 1995) suggests that all ages and both sexes of fur seals leave the Bering Sea during winter (February-March; Bigg 1990). However, unpublished reports suggest that some male fur seals haul out onto ice in the Bering Sea during late March and April. One seal that we monitored (number 14100) remained in the Bering Sea during the time we were able to monitor its movements (through January 1993). The proportion of the male population that remains in the Bering Sea is unknown. In our study all males except the one (for which the SLTDR transmitted long enough to provide the data) left the Bering Sea before January and had not returned by the end of February.

Male fur seals that migrated into the Gulf of Alaska arrived there at about the same time (December/January) as postmolt female northern elephant seals *(Mirounga angustirostris)* and near the end of the time (February) when postbreeding elephant seals arrived there (Stewart and De-Long 1995). Male northern fur seals were in the Gulf of Alaska during

December and January, and probably into February. Elephant seals dive deeper than fur seals and feed on different prey, but it is interesting that they both utilize this area during winter months.

Acknowledgments

We express our appreciation to G. Antonelis, S. Insley and R. Ream for help in attaching the SLTDRs. The U.S. Department of the Interior, Minerals Management Service, provided financial support through Interagency Agreement IA-14411. Carolyn Kurle helped plot fur seal locations. Northern fur seals were handled under MMPA permit number 598 issued to the Alaska Fisheries Science Center. J. Baker, H. Braham, and E. Sinclair provided comments on an earlier version of the manuscript.

References

Antonelis, G.A., B.S. Stewart, and W.F. Perryman. 1990. Foraging characteristics of female northern fur seals *(Callorhinus ursinus)* and California sea lions *(Zalophus californianus)*. Canadian Journal of Zoology 68:150-158.

Bigg, M.A. 1990. Migration of northern fur seals *(Callorhinus ursinus)* off western North America. Canadian Technical Report of Fisheries and Aquatic Sciences No. 1764. 64 pp.

Brodeur, R.D., B.W. Frost, S.R. Hare, R.C. Francis, and W.J. Ingraham, Jr. 1996. Interannual variations in zooplankton biomass in the Gulf of Alaska, and covariation with California Current zooplankton biomass. CalCOFI Report 37:80-99.

DeLong, R.L., and B.S. Stewart. 1991. Diving patterns of northern elephant seal bulls. Marine Mammal Science 7:369-384.

Ebbesmeyer, C.C., and W.J. Ingraham. 1992. Shoe spill in the North Pacific. Eos, Transactions, American Geophysical Union 73:361-365.

Ebbesmeyer, C.C., and W.J. Ingraham. 1994. Pacific toy spill fuels ocean current pathways research. Eos, Transactions, American Geophysical Union 75:425-427.

Fancy, S.G., L.F. Pank, D.C. Douglas, C.H. Curby, G.W. Garner, S.C. Amstrup, and W.L. Regelin. 1988. Satellite telemetry: A new tool for wildlife research and management. U.S. Department of Interior, U.S. Fish and Wildlife Service Resource Publication 172. 54 pp.

Gentry, R.L., and G.L. Kooyman (eds.). 1986. Fur seals: Maternal strategies on land and at sea. Princeton University Press, Princeton. 291 pp.

Goebel, M.E., J.L. Bengtson, R.L. DeLong, R.L. Gentry, and T.R. Loughlin. 1991. Diving patterns and foraging locations of female northern fur seals. Fishery Bulletin, U.S. 89:171-179.

Hill, R.D., S.E. Hill, and J.L. Bengtson. 1987. An evaluation of the Argos satellite system for recovering data on diving physiology of Antarctic seals. In: Abstracts of the Seventh Biennial Conference on the Biology of Marine Mammals, Miami, Florida, p. 32.

Hills, S. 1987. New techniques for the study of Pacific walruses. In: Abstracts of the Seventh Biennial Conference on the Biology of Marine Mammals, Miami, Florida, p. 32.

Huang, N.E. 1979. On the surface drift currents in the ocean. Journal of Fluid Mechanics 91:191-208.

Ingraham, W.J., Jr., and R.K. Miyahara. 1988. Ocean surface current simulations in the North Pacific Ocean and Bering Sea (OSCURS-Numerical Model). NOAA Technical Memorandum, NMFS F/NWC-130. 67 pp.

Ingraham, W.J., Jr. and R.K. Miyahara. 1989. Tuning of OSCURS numerical model to ocean surface current measurements in the Gulf of Alaska. NOAA Technical Memorandum, NMFS F/NWC-168. 67 pp.

Ingraham, W.J., Jr., R.K. Reed, J.D. Schumacher, and S.A. Macklin. 1991. Circulation variability in the Gulf of Alaska. Eos, Transactions, American Geophysical Union 72:24.

Kajimura, H. 1984. Opportunistic feeding of the northern fur seal, *Callorhinus ursinus,* in the eastern North Pacific Ocean and eastern Bering Sea. NOAA Technical Report NMFS SSRF-779. 49 pp.

Kajimura, H., and T.R. Loughlin. 1988. Marine mammals in the oceanic food web of the eastern subarctic Pacific. Bulletin of the Ocean Research Institute, University of Tokyo No. 26(Part II):187-223.

Kiyota, M., N. Baba, T.R. Loughlin, and G.A. Antonelis. 1992a. Characteristics of winter migration of female Pribilof fur seals. In: Abstracts of the 15th Symposium on Polar Biology, December 1992, Tokyo, pp. 1-75.

Kiyota, M., T.R. Loughlin, N. Baba, M. Nakajima, and K. Kohyama. 1992b. Use of tiletamine hydrochloride-zolazepam hydrochloride mixture as an immobilizing agent for northern fur seals *(Callorhinus ursinus).* Mammal Science (Japan) 32:1-7.

Larson, S.E. 1975. A 26-year time series of monthly winds over the oceans. In: Part 1. A statistical verification of computed surface winds over the North Pacific and North Atlantic. ENVPRE-DRSCWFAC Technical Paper. U.S. Navy Environmental Prediction Research Facility, Monterey, CA, pp. 8-75.

Loughlin, T.R. 1993. Status and pelagic distribution of otariid pinnipeds in the Bering Sea during winter. Outer Continental Shelf Study, U.S. Minerals Management Service Report No. 93-0026. 58 pp. (Available from Environmental Studies Section, Minerals Management Service, Alaska OCS Region, Anchorage, AK 99508.)

Loughlin, T.R., and T. Spraker. 1989. Use of Telazol to immobilize female northern sea lions *(Eumetopias jubatus)* in Alaska. Journal of Wildlife Diseases 25:353-358.

Loughlin, T.R., J.L. Bengtson, and R.L. Merrick. 1987. Characteristics of feeding trips of female northern fur seals. Canadian Journal of Zoology 65:2079-2084.

Lowry, L.F., K.J. Frost, D.G. Calkins, G.L. Swartzman, and S. Hills. 1982. Feeding habits, food requirements, and status of Bering Sea marine mammals. North Pacific Fishery Management Council, Anchorage, AK, Document numbers 19 and 19a. 574 pp.

Merrick, R.L., T.R. Loughlin, G.A. Antonelis, and R. Hill. 1994. Use of satellite-linked telemetry to study Steller sea lion and northern fur seal foraging. Polar Research 13:105-114.

Ohtani, K. 1973. Oceanographic structure in the Bering Sea. Memoirs of the Faculty of Fisheries Hokkaido University 21:65-106.

Ohtani, K. 1991. To confirm again the characteristics of the Oyashio. Bulletin of the Hokkaido National Fisheries Research Institute No. 55:1-24.

Overland, J.E., M.C. Spillane, H.E. Hurlburt, and A.J. Wallcraft. 1994. A numerical study of the circulation of the Bering Sea basin and exchange with the North Pacific Ocean. Journal of Physical Oceanography 24:736-758.

Ragen, T.J., G.A. Antonelis, and M. Kiyota. 1995. Early migration of northern fur seal pups from St. Paul Island, Alaska. Journal of Mammalogy 76:1137-1148.

Reed, R.K. 1980. Direct measurements of recirculation in the Alaskan Stream. Journal of Physical Oceanography 10:976-978.

Service-Argos. 1984. Location and data collection system user's guide. Service-Argos, Toulouse, France. 36 pp.

Sinclair, E., T.R. Loughlin, and W. Pearcy. 1994. Prey selection by northern fur seals *(Callorhinus ursinus)* in the eastern Bering Sea. Fishery Bulletin, U.S. 92(1):144-156.

Stabeno, P.J., and R.K. Reed. 1994. Circulation in the Bering Sea Basin observed by satellite-tracked drifters: 1986-1993. Journal of Physical Oceanography 24:848-854.

Stewart, B.S., and R.L. DeLong. 1995. Double migrations of the northern elephant seal, *Mirounga angustirostris.* Journal of Mammalogy 76:196-205.

Stewart, B.S., S. Leatherwood, P.K. Yokem, and M.P. Heide-Jorgensen. 1989. Harbor seal tracking and telemetry by satellite. Marine Mammal Science 5:361-375.

Weber, J.E. 1983. Steady wind- and wave-induced currents in the open ocean. Journal of Physical Oceanography 13:524-530.

Witting, R. 1909. Zur Kenntne den vom Winde Erzeigtem Oberflashenstromes. Annalen der Hydrographis und Maritimen Meteorologia 73. 193 pp.

CHAPTER **28**

Marine Bird Populations and Carrying Capacity of the Eastern Bering Sea

George L. Hunt Jr.
University of California, Irvine, California

G. Vernon Byrd Jr.
Alaska Maritime National Wildlife Refuge, Homer, Alaska

Abstract

Marine bird populations in the Bering Sea have been monitored at selected colonies since the mid-1970s. At the Pribilof Islands, declines of black-legged *(Rissa tridactyla)* and red-legged *(R. brevirostris)* kittiwake populations occurred between the mid-1970s and the mid-1980s. Thick-billed murre *(Uria lomvia)* populations also declined at the Pribilof Islands during this period, as did common murres *(U. aalge)* at Bluff in Norton Sound. At the same time elsewhere in the Bering Sea, populations of most piscivorous seabirds remained stable, fluctuated without discernible trend, or increased. Food web dependencies differed between decreasing populations of piscivores and those that remained stable or increased. We hypothesize that between 1976 and 1985, one or more die-offs of adult kittiwakes and thick-billed murres occurred near the Pribilof Islands. In subsequent years, the remaining populations failed to increase their reproductive performance, despite markedly smaller resident populations. Therefore we hypothesize that, for kittiwakes and thick-billed murres, the carrying capacity of the waters near the Pribilof Islands has been significantly reduced. Lowered reproductive performances were associated with a decrease in the consumption of forage fish, in particular species with high fat content. Warming of surface waters, decreases in the size of the cold pool, and changes in the advective regime of the shelf-slope current could have influenced the vertical and horizontal distributions of forage fish, but data are lacking to test these hypotheses.

Introduction

Recently there has been considerable interest in using marine birds as indicators of changes in various marine ecosystems (e.g., Furness and Nettleship 1991, Montevecchi 1993). Studies that attempt to use seabirds as indicators of prey stocks (Berruti 1985, Croxall 1989, Klageges et al. 1992, Montevecchi and Myers 1992) are predicated on the assumption that variations in population size or reproductive performance of marine birds are closely linked to variations in prey populations (Cairns 1987, Hunt et al. 1991). This assumption is supported by several studies of seabird populations (e.g., Duffy 1980, Hamer et al. 1991, Ankar-Nielsen 1992) and reproductive performance (e.g., Anderson et al. 1982, Springer et al. 1986, Monaghan et al. 1989). Similar studies indicate that marine bird populations also vary with respect to regional changes in climate, but the linkages to changes in preferred prey populations have not always been evident (e.g., Aebischer et al. 1990; Ainley et al. 1996; Veit et al. 1996, 1997). Nevertheless, links between climate- or weather-driven changes in ocean conditions and marine bird foraging and reproductive ecology have been described (Dunn 1973, Birkhead 1976, Braun and Hunt 1983, Hunt et al. 1991).

In the eastern Bering Sea, results of programs to monitor breeding populations of marine birds (e.g., Byrd and Dragoo 1997) provide the basis for describing population trends and patterns in productivity over the past 20 years. Furthermore, research indicates that seabird communities in colonies and at sea vary with oceanographic regions (Hunt et al. 1981b, Springer and Roseneau 1985, Schneider et al. 1986). These data and others have been used to test hypotheses about links between variations in marine birds and various aspects of the physical (e.g., climate) and biological environment (e.g., changes in competition for prey with predatory fish and the effects of commercial fishing) (Murphy et al. 1986, 1991; Springer et al. 1986, 1996; Springer and Byrd 1989; Springer 1992, 1993; Decker et al. 1995; Hunt et al. 1996, 1997).

In this paper, we use monitoring data for black-legged *(Rissa tridactyla)* and red-legged kittiwakes *(R. brevirostris),* and thick-billed *(Uria lomvia)* and common murres *(U. aalge),* the species for which the most complete information on populations is available, to examine geographic patterns of population changes in relationship to food webs on which these piscivorous, cliff-nesting seabirds depend, or on physical processes which may influence the availability of prey.

Methods

Amount and Geographic Distribution of Data

Population data on kittiwakes and murres have been collected regularly since the mid-1970s at six colonies in the Bering Sea (Fig. 1), but the frequency and intensity of surveys has varied among sites and species (Table

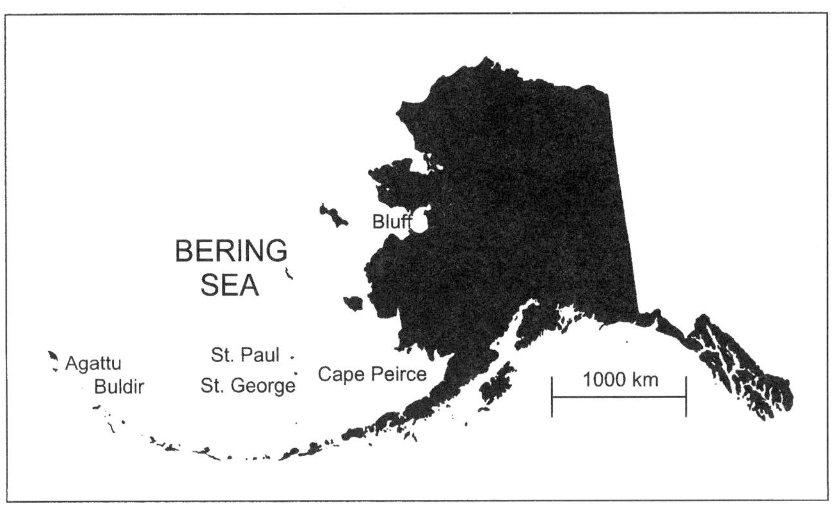

Figure 1.   Locations of seabird colonies discussed in this paper.

1). Monitoring has continued at six sites primarily because they are on the Alaska Maritime and Togiak National Wildlife Refuges. In the first few years at most sites, single counts were made of birds either in the entire colony or on a series of index plots (Sowls et al. 1978). Later, due to daily fluctuations in attendance of birds at colonies, multiple counts, usually on index plots, were made at the least variable times of day and season to improve estimates (e.g., Birkhead and Nettleship 1980, Byrd 1989, Hatch and Hatch 1988). Pelagic distribution and foraging ecology also were studied initially in the Bering Sea in the 1970s and early 1980s (e.g., Hunt et al. 1981a,c,d; Gould et al. 1982).

Geographic Settings of the Colonies

Foraging habitats for kittiwakes and murres differ among the six nesting colonies addressed herein, most likely as a function of food web structure and carbon pathways (Iverson et al. 1979; Hunt et al. 1981a; Schneider et al. 1986; Springer 1991, 1992; Springer et al. 1987, 1996) related to physical phenomena in shelf waters (Coachman 1986, Schumacher and Stabeno 1998). For example, in the western Aleutian Islands, the marine environment of Agattu and Buldir Islands is dominated by the westward flowing Alaskan Stream. This current forms the northern edge of the North Pacific gyre and enters the Bering Sea through passes between the Aleutian Islands (Stabeno et al., chapter 1, this volume). Both Agattu and Buldir support a diverse marine avifauna, but the dominant prey used by kittiwakes and murres and the marine habitats around the two islands differ

Table 1. Population trends for kittiwakes and murres at selected breeding colonies in the Bering Sea.

Location	Species[a]	Years[b] Number	Range	Overall trend[c]	Subset trends[d]
Agattu	BLKI	8	75-94	Increase ($r^2 = 0.63$, $P < 0.02$)	75-79 < 88-94 ($t = 2.746$, $P < 0.03$), no trend 88-94 ($r^2 = 0.03$)
	UNMU	7	74-94	Increase ($r^2 = 0.55$, $P < 0.06$)	74-79 < 85-94 ($t = 5.086$, $P < 0.01$), no trend 85-94 ($r^2 = 0.07$)
Buldir	BLKI	10	74-96	Increase ($r^2 = 0.87$), $P < 0.01$)	74-76 < 88-96 ($t = 12.109$, $P < 0.01$), no trend 88-96 ($r^2 = 0.14$)
	RLKI	10	74-96	Increase ($r^2 = 0.84$), $P < 0.01$)	74-76 < 88-96 ($t = 9.96$, $P < 0.01$), no trend 88-96 ($r^2 = 0.15$)
	TBMU	10	74-96	Increase ($r^2 = 0.86$), $P < 0.01$)	74-76 < 88-92 < 94-96 ($F = 155.529$, $P < 0.001$)
C. Peirce	BLKI	13	76-96	No trend ($r^2 = 0.01$)	
	COMU	13	76-96	No trend ($r^2 = 0.10$)	
Bluff	BLKI	14	79-95	Increase ($r^2 = 0.51$, $P < 0.01$)	No trend 87-95 ($r^2 = 0.07$)
	COMU[e]	8	75-82	Decline ($r^2 = 0.75$, $P < 0.01$)	
		12	79-95	No trend ($r^2 = 0.01$)	
St. Paul	BLKI	10	76-96	Decline ($r^2 = 0.63$, $P < 0.01$)	No trend 87-96 ($r^2 = 0.31$)
	RLKI	10	76-96	Decline ($r^2 = 0.75$, $P < 0.01$)	No trend 88-96 ($r^2 = 0.78$, $P > 0.10$)
	COMU	10	76-96	Decline ($r^2 = 0.53$, $P < 0.02$)	No trend 86-96 ($r^2 = 0.11$)
	TBMU	10	76-96	No trend ($r^2 = 0.12$)	76 > 82-96 ($t = 10.051$, $P < 0.01$)
St. George	BLKI	10	76-96	No trend ($r^2 = 0.24$)	Decline 76-86 ($r^2 = 0.89$, $P < 0.02$), no trend 87-96 ($r^2 = 0.64$, $P = 0.11$)
	RLKI	10	76-96	Decline ($r^2 = 0.64$), $P < 0.01$)	76-86 > 87-96 ($t = 2.086$, $P < 0.10$), no trend 87-96 ($r^2 = 0.01$)
	COMU	10	76-96	Increase ($r^2 = 0.48$), $P < 0.03$)	No trend 76-92 ($r^2 = 0.18$), increase based on 96 count
	TBMU	10	76-96	No trend ($r^2 = 0.23$)	Decline 76-88 ($r^2 = 0.94$, $P < 0.01$), apparent increase 89-96

[a] Codes: BLKI (black-legged kittiwake), RLKI (red-legged kittiwake), COMU (common murre), TBMU (thick-billed murre), UNMU (unidentified murre, includes both species).
[b] Number of years and earliest and latest year for which data are available.
[c] Trends indicated if simple linear models fit and slopes differed from zero at the 0.1 level.
[d] Trends suggested on graphs for subsets at least 4 years long were tested with regressions, and subsets were compared with t-tests or ANOVA to identify differences.
[e] Separate sets of data were analyzed for whole-colony counts (1975-1982) and plot counts (79-95).

(Springer et al. 1996). Agattu is surrounded by a broad shelf where typical prey species include sand lance *(Ammodytes hexapterus).* In contrast, Buldir Island is oceanic, and birds there depend more on prey species of oceanic origin (e.g., lanternfish [Myctophidae]). Cape Peirce and Bluff are along the mainland coast adjacent to the shallow, well-mixed water of the coastal regime. These colonies support fewer seabird species than colonies near oceanic water, and are most strongly influenced by the northward flowing Alaska Coastal Current. Prey of importance here include sand lance and saffron cod *(Eleginus gracilis).* Near the edge of the shelf in the southeastern Bering Sea, the Pribilof Islands are close to three different oceanic regimes: (1) the well-mixed waters of the coastal shelf immediately around the islands; (2) the two layer system of the middle shelf in water 50-100 m deep; and (3) the outer edge of the shelf and the shelf slope regime. For the piscivorous birds of the Pribilofs, the dominant prey is juvenile walleye pollock *(Theragra chalcogramma)* and, to a lesser extent, sand lance and lanternfish (Hunt et al. 1981a,b, 1996; Schneider and Hunt 1984; Decker et al. 1995).

Analytical Procedures

We plotted single counts or annual means with 90% confidence bounds (for years in which multiple counts were made) for each species at each site. We used quadratic regressions to fit lines to the points to illustrate how trends have changed over the past 20 years. In addition, we tested for significant changes in population size over the entire sampling period using simple linear regression. We applied these models when correlation coefficients were significant at $P = 0.10$ and subsequently tested for increasing or decreasing slopes. Subsets within the overall time-series were examined with regression, *t*-test, or ANOVA as appropriate.

Results

Population trajectories for kittiwakes and for murres have varied among colonies in the Bering Sea (Figs. 2-4). In the western Aleutians at Agattu and Buldir islands, populations of kittiwakes and murres have increased since the mid-1970s (Fig. 2). Thick-billed murres at Buldir continued to increase throughout the period, but for other species at both islands, by 1988, populations were no longer increasing (Table 1). At Cape Peirce, one of the Alaska mainland colonies, no overall trends were evident for kittiwake or murre populations over the past 20 years (Fig. 3, Table 1). In contrast, farther north at Bluff, black-legged kittiwake populations increased between 1979 to 1995 (Fig. 3). However, within that overall increase, numbers showed no trend between 1987 and 1995 (Table 1). The only decline noted at either the western Aleutian colonies or those on the Alaska mainland was the decline in common murres at Bluff between 1975 and 1982, and numbers of this species at Bluff appear to have stabilized since the late 1970s or early 1980s (Fig. 3, Table 1).

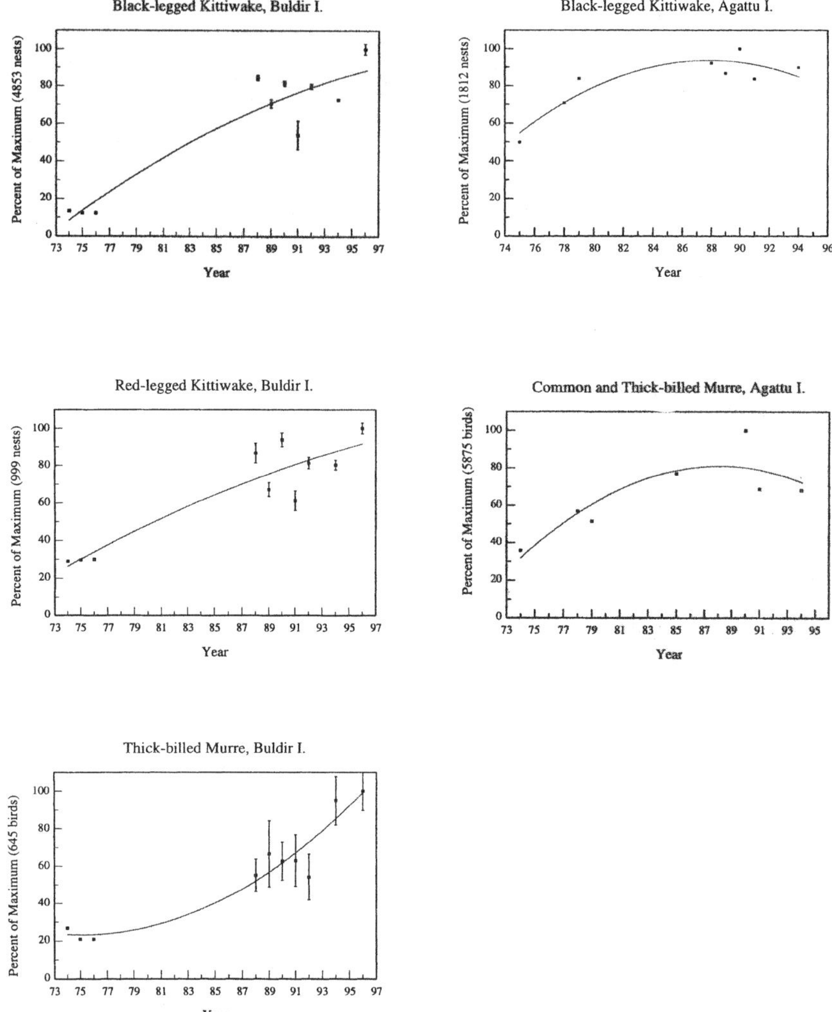

Figure 2.   Changes between 1974 and 1996 in the numbers of seabirds occurring at two colonies in the western Aleutian Islands. Error bars around points represent 90% confidence limits. Curves are quadratic regressions.

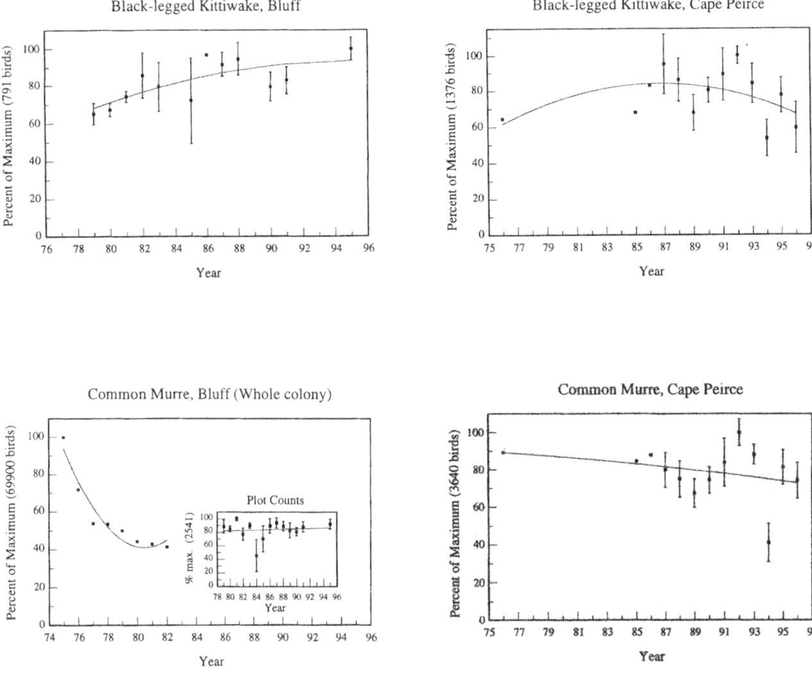

Figure 3.   Changes between 1975 and 1996 in the numbers of seabirds attending two colonies on the mainland of western Alaska. Error bars around points represent 90% confidence limits. Curves are quadratic regressions.

In 7 of 8 cases (4 species on each of 2 islands) at the Pribilof Islands, populations of kittiwakes and murres either declined over the study period as a whole, or over some portion of it (Table 1). At St. Paul Island, black-legged and red-legged kittiwake numbers have declined over the past 20 years, although populations have remained stable at reduced levels since the mid- to late 1980s (Fig. 4, Table 1). Thick-billed murre numbers were significantly higher in 1976 than they have been since, and common murres also declined (Table 1). At St. George Island, black-legged kittiwakes declined between 1976 and 1986, but numbers have stabilized or increased slightly since then, with the result that there is no overall trend in this population (Fig. 4, Table 1). Red-legged kittiwakes at St. George Island have declined over the past 20 years, with numbers between 1987 and 1996 lower than those between 1976 and 1986 (Table 1). Since 1987, red-legged kittiwake numbers on St. George Island have been stable. At St. George Island, thick-billed murre numbers showed a pattern similar to

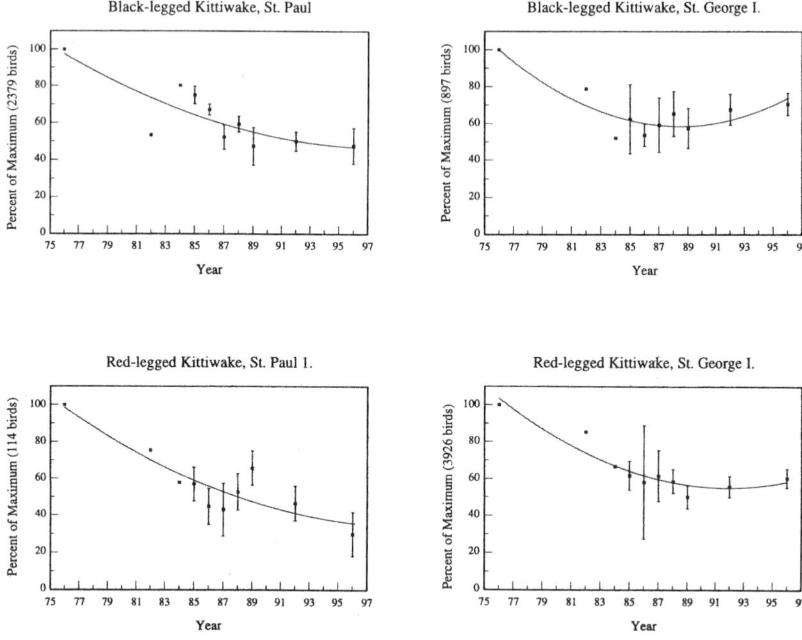

Figure 4.   Changes between 1976 and 1996 in the numbers of kittiwakes and murres attending colonies on St. George and St. Paul islands, Pribilof Islands. Error bars around points represent 90% confidence limits. Curves are quadratic regressions.

that of black-legged kittiwakes, an initial decline between 1976 and 1988, with a stable or slightly increasing population since then that has resulted in no overall population trend (Fig. 4, Table 1). The only example of a population that has increased at the Pribilof Islands over the past 20 years is the common murre at St. George Island, although prior to 1996 there was no significant trend for this population (Table 1).

Discussion

Over the past 20 years, declines in breeding populations of piscivorous seabirds at colonies in the eastern Bering Sea have occurred primarily at Bluff (common murres) and at the Pribilof Islands (kittiwakes and murres). Most of the declines occurred between the mid-1970s and the mid-1980s. Murphy et al. (1986) modeled the decline of the common murre population at Bluff and concluded that such a rapid population decline could not be accounted for by low productivity alone. They proposed that winter

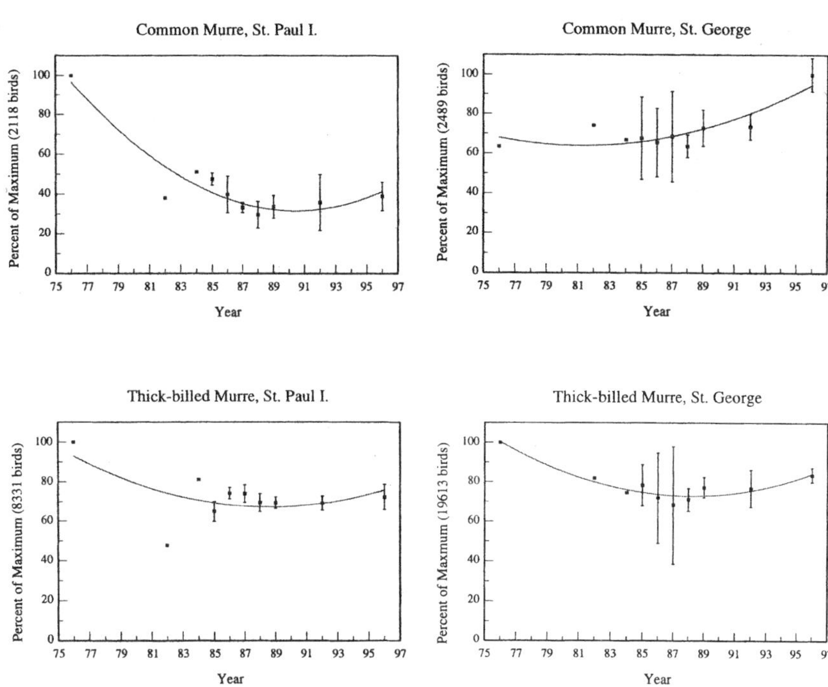

Figure 4.   (Continued.)

mortality must have played a significant role in the decline, and assumed that this mortality probably took place in the southeastern Bering sea near the ice edge where high numbers of murres spend the winter (Divoky 1978). The cause of the mortality remains unknown, although a reduction of pollock stocks by the fishery was suggested as a possible source of increased mortality of murres (Murphy et al. 1986) because the murres take juvenile pollock in winter (Divoky 1978). Another possibly critical time for murres at Bluff is spring, when water temperatures and ice cover can affect the availability of forage fish near the breeding colony.

The exact timing of declines in seabird populations at the Pribilof Islands will never be known because surveys were not conducted annually, particularly in the period of the declines. Nevertheless, by examining the population trajectories depicted by the graphs (Figs. 3, 4), it seems clear that most of the declines had ameliorated by 1985 or 1986. Whatever caused the declines must have occurred in the decade prior to the mid-1980s. For example, declines of kittiwakes at St. George may have been particularly steep between 1982 and 1984. On St. Paul Island, declines of red-legged kittiwakes were also steep during this period, although black-legged kittiwakes had their steepest decline between 1976 and 1982. Declines in both species of murres were steepest between 1976 and 1982 on St. Paul Island and between 1976 and 1987 for thick-billed murres on St. George Island. The declines of seabirds at the Pribilof Islands may have occurred after, and were of longer duration, than those at Bluff. The declines in kittiwake populations on the Pribilofs, particularly on St. George Island, almost certainly occurred after the 1970s regime shift in the North Pacific Ocean and Bering Sea (National Research Council 1996).

Population declines of kittiwakes and murres at the Pribilof Islands must have resulted from one or more of three causes: a decline in the output of young resulting in reduced recruitment, emigration to other colonies, or an increase in adult mortality. Any presumptive cause must account for these factors: a sharp decline in populations between 1976 and 1987, with the steepest portion of the drop most likely in the early 1980s; a more severe effect on kittiwakes than on murres, particularly kittiwakes on St. Paul Island; a possible difference in the timing of declines on St. Paul and St. George Islands, and an apparent lack of population declines at most other colonies in the eastern and southern Bering Sea during the same time frame.

It seems unlikely that failure to produce sufficient young was the cause of the precipitous declines of kittiwakes and murres at the Pribilof Islands. In the 1970s, production of young at the Pribilof Islands was consistently "moderate," albeit lower than has been recorded elsewhere in the Bering Sea in "good" years (Hunt et al. 1981c, 1986; Byrd and Dragoo 1997). Young fledged from 1975 to 1979 would have recruited to nesting populations in 1978 to 1984, three to five years after they fledged. These new recruits should have been sufficiently numerous to maintain stable populations in the colonies of the Pribilof Islands. However, by 1984, most of the major declines in the numbers of birds on the census plots had taken place. Thus for that period, disappearance of birds from their cliff-face nesting plots greatly exceeded recruitment. Furthermore, although murre and especially kittiwake reproductive output was very low subsequent to 1984 (Decker et al. 1995, Byrd and Dragoo 1997), these failures to produce chicks were not reflected by further dramatic declines in the populations of these species at the cliffs on St. Paul and St. George islands. Although there is a need for modeling to determine the extent to which recruitment failure might account for the seabird population declines at the Pribilof Islands, given the information available, it seems unlikely that reproductive failure alone could have been responsible for the declines in black- and red-legged kittiwakes and thick-billed murres in the late 1970s or early 1980s.

Emigration of established breeding pairs from the Pribilof Islands to other colonies is also unlikely. However, detection of emigration by black-legged kittiwakes or thick-billed murres would be difficult because there are many colonies to which they could have moved, and none has been sufficiently well monitored that immigration would have been certain to be detected. In contrast, red-legged kittiwakes are known to nest in only five or six colonies in the eastern Bering Sea, including St. George and St. Paul and Otter islands in the Pribilofs and Bogoslof and Buldir in the Aleutian Islands (Byrd 1978). The increased numbers of red-legged kittiwakes in these latter two colonies have been too low to account for the high numbers of birds that are no longer at the Pribilofs (Byrd et al. 1997). It is possible that adult birds that did not return to the colonies remained at sea and forewent breeding for several years (Hamer et al. 1991), but at least in the case of the red-legged kittiwake, it is highly unlikely that they emigrated to another colony.

A third possibility is that the mortality rate for adult kittiwakes and murres breeding at the Pribilof Islands was higher than normal between 1976 and 1984. Because declines apparently were concentrated in populations nesting in the Pribilof Islands, it would seem unlikely that excess mortality, if it indeed occurred, happened during the winter when birds from a number of colonies would be expected to be mingled. More likely, mortality events would have occurred in spring when murres and kittiwakes would have been gathering around their colonies prior to breeding, or in late September or October when birds were still in the vicinity of their colonies but were no longer being monitored. The most likely cause of this at-sea mortality would be lack of prey. Several seabird die-offs have been recorded in the Bering Sea in the 1970s and 1980s. DeGange and Rosapepe (1984) reported large numbers of dead murres in the eastern Bering Sea north of the Pribilofs in 1979, and a subsequent die-off of kittiwakes was reported in 1983 (Nysewander and Trapp 1984). These die-offs may have contributed to the population declines in the Pribilof Islands.

Several lines of evidence support the hypothesis that birds nesting at the Pribilof Islands could have experienced a shortage of food while attending their breeding colonies. Foremost is the marked reduction in the production of chicks shown by both species of kittiwakes and by thick-billed murres on the Pribilof Islands since the early 1980s (Decker et al. 1995). During the 1970s, both age-1 pollock and capelin *(Mallotus villosus)* were important components of the diets of kittiwakes and murres at the Pribilof Islands (Hunt et al. 1981a,b), and a decrease in their availability near the islands, particularly early in the season before age-0 pollock were available, could have caused increased adult mortality and/or reproductive failure. In the vicinity of the Pribilof Islands, there was a marked decline between 1979 and 1982 in the numbers of age-1 pollock taken in trawl surveys by the National Marine Fisheries Service (Hunt et al. 1997). Likewise, in the vicinity of the Pribilof Islands, the numbers of capelin declined at the same time (Hunt et al. 1996). Elsewhere, in the Gulf of Alaska, declines in the proportion of capelin in the diets of seabirds were accompanied by seabird population declines (Piatt and Anderson 1995).

Changes in the capacity of the Bering Sea to support seabird populations on the Pribilof Islands affected surface-foraging kittiwakes more severely than subsurface-foraging murres (Decker et al. 1995, Hunt et al. 1997), which can pursue prey to depths of 210 m (Croll et al. 1992). It therefore seems likely that, in addition to a decrease in the abundance of age-1 pollock within the foraging range of seabirds nesting at the Pribilof Islands, there also must have been a change in the vertical distribution of the remaining forage fish such that fewer fish were available in the upper few meters of the water column. Because red-legged kittiwakes, a species that specializes on myctophids, experienced a particularly severe decline, it is likely that the vertical distributions of fish other than pollock were also affected.

Sea surface warming that marked the Bering Sea regime shift in the late 1970s may have influenced the distribution of juvenile pollock near the Pribilof Islands (Decker et al. 1995; Hunt et al. 1996, 1997). In warm years with reduced winter ice cover, shrinking of the pool of cold water at the bottom in the middle domain may have resulted in juvenile pollock moving onto the shelf rather than being concentrated along the shelf edge, seaward of the cold bottom water (Ohtani and Azumaya 1995, Wyllie-Echeverria 1995). This shift would reduce the number of pollock close to the Pribilofs. In warm years, more of the primary production in the middle domain is captured by copepods (Walsh and McRoy 1986), which would enhance foraging prospects for juvenile pollock there. In addition, a weaker pycnocline and warmer bottom waters near the Pribilofs may have permitted juvenile pollock to move deeper in the water column at an earlier date (Olla and Davis 1990, Sogard and Olla 1993), thus reducing their availability to surface-foraging seabirds.

The possibility that predation by adult pollock depleted forage fish stocks near the Pribilof Islands cannot be rejected. However, Decker et al. (1995) argued that the limited data available on the timing of changes in forage fish stocks near the Pribilof Islands failed to reflect changes in the biomass of adult pollock there, and Hunt et al. (1997) failed to find a significant relationship between the number of age-1 pollock and the number of age-2 and older pollock in the vicinity of the Pribilof Islands or along the edge of the southeastern Bering Sea shelf.

Variations in ocean circulation affect the availability of prey in the vicinity of the Pribilof Islands. The causes of variability in the Green Belt, which occurs along the Bering Sea shelf break (Springer et al. 1996), are not well known, but must be related to the Bering Slope Current. The characteristics of the Bering Slope Current vary on time scales of months to perhaps years (Stabeno et al., chapter 1, this volume). At times it appears as a well defined, smooth flow with few eddies or meanders evident; at other times it is highly variable with numerous eddies and meanders (J. Schumacher, PMEL, 7600 Sand Point Way, N.E., Seattle, WA, Nov. 1997, pers. comm.). Transport is also variable, ranging from about 3 to 6-7 million m^3/s (Stabeno et al., chapter 1, this volume). Increased shelf-slope exchange likely occurs when the Bering Slope Current contains numerous eddies and meanders (Schumacher and Stabeno 1994; van Meurs and Stabeno, in press). Continuous flow along the 100 m isobath from the vicinity of Unimak Pass to the Pribilof Islands has been revealed in satellite tracked drifter trajectories. The strength and persistence of the flow varies significantly from year to year.

The origin of the Bering Slope Current is threefold: flow through Unimak Pass, flow up Bering Canyon, and on-shelf fluxes along the shelf break (Stabeno et al., chapter 1, this volume). The latter two sources consist of Bering Slope Current water and have high concentrations of nutrients. Flow from the vicinity of Unimak Pass would advect zooplankton and larval fish originating in that region to the Pribilof Islands. Water originating in Pribilof Canyon is often entrained in the strong current (30 cm/s) south of St. George Island (Schumacher and Stabeno 1998). Satellite-tracked drifters originating in this region often end up circling St. Paul Island (Stabeno et al., chapter 9, this volume). These on-shelf transports could affect the concentrations of zooplankton in shelf waters around the Pribilofs. Changes in zooplankton concentrations could affect the foraging opportunities, and thus the vertical distributions, of forage fish.

It is also possible that storms made foraging at the water's surface more difficult during the 1980s, compounding stress resulting from decreased abundance of prey in the water column. Surface-foraging seabirds are known to be adversely affected by stormy weather, which is alleged to have caused mass mortality of seabirds (Kazama 1968, Salt and Willard 1971, Dunn 1973) and depression of reproductive success (Braun and Hunt 1983). Likewise, subsurface-foragers also have greater difficulty foraging in rough weather (Birkhead 1976). It has been hypothesized that warming may lead to an increased storm frequency in the Bering Sea (Table 2 in U.S. GLOBEC 1996). The effects of storm frequency and the seasonal timing of storm events at the Pribilof Islands have not been investigated, but they could have considerable importance for seabird foraging and survival.

Declines in kittiwake and murre populations at the Pribilof Islands can be compared to declines in populations of Atlantic puffins *(Fratercula arctica)* and common murres on the west coast of Norway that occurred in response to collapses in prey stocks. Atlantic puffins nesting on Røst depend on migrating juvenile herring *(Clupea harengus)* for food during the summer (Myrberget 1962, Ankar-Nilssen 1992). After this stock of herring collapsed between 1957 and 1971, the more than one million puffins breeding on Røst continued to return to their colonies, but failed to breed successfully during the period when age-1 herring were absent from coastal waters (Ankar-Nilssen 1992, Wright et al. 1996). As the puffin population aged, between 1979 and 1989 the numbers of puffins returning to Røst declined at the rate of about 14% per annum, presumably from the attrition of aging individuals and the failure to produce young (Ankar-Nilssen and Røstad 1993, Wright et al. 1996). No large die-offs or beachings of dead puffins were reported. In contrast, in the north of Norway along the

Barents Sea coast, common murres depend on capelin as their major food, which are taken when the murres are at their colonies in summer, and also in winter and spring when the murres are migrating to their colonies prior to breeding (Furness and Barrett 1985, Erikstad and Vader 1989, Barrett and Furness 1990). Barents Sea capelin stocks decreased between 1975 and 1986-1987 from about 7 million t to 20,000 t (Wright et al. 1996). During the early years of the capelin decline, seabird populations appeared healthy and breeding success was high (Furness and Barrett 1985, Wright et al. 1996). However, during the winter of 1986-87, thousands of emaciated murres washed ashore along the coast of northern Norway. Counts made at the colonies in 1987 showed that between 1985-86 and 1987 the numbers of common murres at colonies had decreased by about 80% and the number of thick-billed murres by 33 to 63% (Vader et al. 1987, Wright et al. 1996).

The steep declines in the murre and kittiwake populations of the Pribilof Islands have similarities to both the crash in the common murre population in the north of Norway, and to the long-term gradual elimination of aging puffins at Røst. We favor the hypothesis that the declines in both species of kittiwakes and thick-billed murres at St. Paul and St. George Islands were caused by large die-offs of adults from these populations. The ongoing lack of reproductive output by kittiwakes and to a lesser extent by murres on the Pribilofs is similar to the situation at Røst prior to the decline in the puffins due to old age. If kittiwakes at the Pribilofs continue to have low levels of reproductive success, we can expect a further decline in their numbers as adults now attending the colonies die.

Summary

In the past two decades, kittiwake and murre populations have increased at some colonies and decreased at others. Particularly strong increases have occurred at Buldir Island, where kittiwakes and murres depend on a mesopelagic oceanic food web. In contrast, decreases in common murre populations have occurred at Bluff, an Alaska Coastal Current food web, and kittiwakes and thick-billed murres have declined on the Pribilof Islands where walleye pollock constitute a significant portion of the diets of these seabirds. At the Pribilof Islands, evidence for changes in both prey abundance and prey availability between the late 1970s and the mid-1980s supports the hypothesis that an inability to obtain sufficient food may have resulted in higher than usual mortality of adult kittiwakes and thick-billed murres in the breeding populations between 1976 and 1984. A simultaneous die-off of pre-recruits would have prevented a rebound of these seabird populations. The failure of the seabird populations on the Pribilof Islands to show enhanced reproductive performance subsequent to the reduction of breeding populations suggests that the carrying capacity of the southeastern Bering Sea declined for seabirds in the early 1980s and was reset at a new, lower, level than had existed in the mid-

1970s. Because kittiwake populations were apparently only affected at the Pribilof Islands, we suggest that the mortality must have occurred when birds would have been near their colonies.

Acknowledgments

Population data were assembled from numerous, mostly unpublished, reports. D. Dragoo (Pribilofs), L. Haggblom (C. Peirce), E. Murphy (Bluff), and J. Williams (Agattu and Buldir) helped collect and compile data and made reports available. We thank P. Hunt, D. Irons, J. Piatt, J. Schumacher, P. Stabeno, and an anonymous referee for helpful comments on previous drafts of the manuscript. J. Schumacher and P. Stabeno contributed considerably to our presentation of the physical regime. The research of G. Hunt was funded in part by grants from the National Science Foundation, Office of Polar Programs, in particular grants DPP-8308232, DPP-8521178, DPP-912283, and OPP-9321636, by the Bureau of Land Management through interagency agreement with the National Oceanic and Atmospheric Administration, Outer Continental Shelf Environmental Assessment Program (OCSEAP), and by the NOAA Coastal Ocean Program through the Southeast Bering Sea Carrying Capacity and is contribution S309. The research of V. Byrd was conducted as part of a graduate project at the University of Idaho and is contribution 852 from the College of Forestry.

References

Aebischer, N.J., J.C. Coulson, and J.M. Colebrook. 1990. Parallel long-term trends across four marine trophic levels and weather. Nature (London) 347:753-755.

Ainley, D.G., W.J. Sydeman, and J. Norton. 1996. Apex predators indicate interannual negative and positive anomalies in the California Current food web. Marine Ecology Progress Series 137:1-10.

Anderson, D.W., F. Gress, and K.F. Mais. 1982. Brown pelicans: Influence of food on reproduction. Oikos 39:23-31.

Ankar-Nilssen, T. 1992. Food supply as a determinant of reproduction and population development in Norwegian puffins. Ph.D. thesis, University of Trondheim. 150 pp.

Ankar-Nilssen, T., and O.W. Røstad. 1993. Census and monitoring of puffins, *Fratercula arctica,* on Røst, N. Norway, 1979-1988. Ornis Scandinavica 24:1-9.

Barrett, R.T., and R.W. Furness. 1990. The prey and diving depth of seabirds on Hornoy, north Norway after a decrease in Barents Sea capelin stocks. Ornis Scandinavica 21:179-218.

Berruti, A. 1985. The use of seabirds as indicators of pelagic fish stocks in the southern Benguela Current. In: L.J. Bunning (ed.), Proceedings of a symposium on birds and man. Witwatersrand Bird Club, Johannesburg, pp. 267-279.

Birkhead, T.R. 1976. Effects of sea conditions on rates at which guillemots feed chicks. British Birds 69:490-492.

Birkhead, T.R., and D.N. Nettleship. 1980. Census methods for murres, *Uria* species: A unified approach. Canadian Wildlife Service Occasional Paper 43, 25 pp.

Braun, B.M., and G.L. Hunt Jr. 1983. Brood reduction in black-legged kittiwakes. The Auk 100:469-476.

Byrd, G.V. 1978. Red-legged kittiwake colonies in the Aleutian Islands, Alaska. Condor 80:250.

Byrd, G.V. 1989. Seabirds of the Pribilof Islands, Alaska: Trends and monitoring methods. Master of Science thesis, University of Idaho. 99 pp.

Byrd, G.V., and D.E. Dragoo. 1997. Breeding success and population trends of selected seabirds in Alaska in 1996. U.S. Fish and Wildlife Service Report AMNWR 97/11. 44 pp.

Byrd, G.V., J.C. Williams, Y.B. Artukhin, and P.S. Vyatkin. 1997. Trends in populations of red-legged kittiwake, *Rissa brevirostris,* a Bering Sea endemic. Bird Conservation International 7:167-180.

Cairns, D.K. 1987. Seabirds as indicators of marine food supplies. Biological Oceanography 5:261-271.

Coachman, L.K. 1986. Circulation, water masses, and fluxes on the southeastern Bering Sea shelf. Continental Shelf Research 5:23-108.

Croll, D.A., A.J. Gaston, A.E. Burger, and D. Konnoff. 1992. Foraging behavior and physiological adaptation for diving in thick-billed murres. Ecology 73(1):344-356.

Croxall, J.P. 1989. Use of indices of predator status and performance in CCAMLR fishery management. Scientific Committee on Conservation of Marine Living Resources, pp. 353-365.

Decker, M.B., G.L. Hunt Jr., and G.V. Byrd Jr. 1995. The relationships among sea-surface temperature, the abundance of juvenile walleye pollock *(Theragra chalcogramma)* and the reproductive performance and diets of seabirds at the Pribilof Islands, southeastern Bering Sea. In: R.J. Beamish (ed.), Climate change and northern fish populations. Canadian Special Publication of Fisheries and Aquatic Sciences 121:425-437.

DeGange, A.R., and J.V. Rosapepe. 1984. A die-off of seabirds in the Bering Sea. Murrelet 65:15-19.

Divoky, G.J. 1978. The distribution, abundance, and feeding ecology of birds associated with pack ice. In: Environmental assessment of the Alaskan Continental Shelf. Final reports of principal investigators 2. BLM/NOAA/OCSEAP, pp. 167-509. (Available from NOAA, 222 W. Eighth Ave., Anchorage, AK 99513.)

Duffy, D.C. 1980. Comparative reproductive behavior and population regulation of the Peruvian Coastal Current. Ph.D. thesis, Princeton University. 105 pp.

Dunn, E.K. 1973. Changes in the fishing ability of terns associated with wind speed and sea surface conditions. Nature (London) 244:520-521.

Erikstad, K.E., and W. Vader. 1989. Capelin selection by common and Brünnich's guillemots during the prelaying season. Ornis Scandinavica 20:151-155.

Furness, R.W., and R.T. Barrett. 1985. The food requirements and ecological relationships of a seabird community in north Norway. Ornis Scandinavica 16:305-313.

Furness, R.W., and D.N. Nettleship. 1991. Introductory remarks: Seabirds as monitors of changing marine environments. ACTA XX Congressus Internationalis Ornithologici, Vol. IV, pp. 2239-2240.

Gould, P.J., D.J. Forsell, and C.J. Lensink. 1982. Pelagic distribution and abundance of seabirds in the Gulf of Alaska and the eastern Bering Sea. U.S. Fish and Wildlife Service FWS/OBS-82/48, Washington, DC. 294 pp.

Hamer, K.C., R.W. Furness, and R.W.G. Caldow. 1991. The effects of changes in food availability on the breeding ecology of great skuas, *Catharacta skua,* in Shetland. Journal of Zoology (London) 223:175-188.

Hatch, S.A., and M.A. Hatch. 1988. Colony attendance and population monitoring of black-legged kittiwakes on the Semidi Islands, Alaska. Condor 90:613-620.

Hunt, G.L., Jr., B. Burgeson, and G.A. Sanger. 1981a. Feeding ecology of seabirds of the eastern Bering Sea. In: D.W. Hood and J.A. Calder (eds.), The eastern Bering Sea Shelf: Oceanography and resources. NOAA, Washington, DC, pp. 629-647.

Hunt, G.L., Jr., M.B. Decker, and A. Kitaysky. 1996. Fluctuations in the Bering Sea ecosystem as reflected in the reproductive ecology and diets of kittiwakes on the Pribilof Islands, 1975 to 1991. In: S.P.R. Greenstreet and M.L. Tasker (eds.), Aquatic predators and their prey. Fishing News Books, Oxford, pp. 142-153.

Hunt, G.L., Jr., Z. Eppley, and W.H. Drury. 1981b. Breeding distribution and reproductive biology of marine birds in the Eastern Bering Sea. In: D.W. Hood and J.A. Calder (eds.), The eastern Bering Sea Shelf: Oceanography and resources. NOAA, Washington, DC, pp. 649-687.

Hunt, G.L., Jr., Z.A. Eppley, and D.C. Schneider. 1986. Reproductive performance of seabirds: The importance of population and colony size. The Auk 103:306-317.

Hunt, G.L., J. Piatt, and K.E. Erikstad. 1991. How do foraging seabirds sample their environment? Acta XX Congressus Internationalis Ornithologici, pp. 2272-2279.

Hunt, G.L., Jr., Z. Eppley, B. Burgeson, and R. Squibb. 1981c. Reproductive ecology, foods and foraging areas of seabirds nesting on the Pribilof Islands, 1975-1979. In: Environmental assessment of the Alaskan Continental Shelf. Final reports of principal investigators, Vol. 12 Biological Studies, BLM/NOAA/OMPA, pp. 1-250

Hunt, G.L., Jr., P.J. Gould, D.J. Forsell, and H.J. Peterson. 1981d. Pelagic distribution of marine birds in the Eastern Bering Sea. In: D.W. Hood and J.A. Calder (eds.), The eastern Bering Sea Shelf: Oceanography and resources. NOAA, Washington, DC, pp. 689-718.

Hunt, G.L., Jr., A.S. Kitaysky, M.B. Decker, D.E. Dragoo, and A.M. Springer. 1997. Changes in the distribution and size of juvenile walleye pollock, *Theragra chalcogramma,* as indicated by seabird diets at the Pribilof Islands, and by bottom trawl surveys in the eastern Bering Sea, 1975 to 1993. NOAA Technical Report Series 126:133-147.

Iverson, R.L., L.K. Coachman, R.T. Cooney, T.S. English, J.J. Goering, G.L. Hunt Jr., M.C. Macauley, C.P. McRoy, W.S. Reeburgh, and T.E. Whitledge. 1979. Ecological significance of fronts in the southeastern Bering Sea. In: R.J. Livingston (ed.), Ecological processes in coastal and marine systems. Plenum Press, New York, pp. 437-466.

Kazama, T. 1968. On the mass destruction of *Rissa tridactyla* and *Calonectris leucomelas* and their migration at Kashiwazaki, Niigata Prefecture. Tori 18:260-266.

Klageges, N.T.W., A.B. Willis, and G.B. Ross. 1992. Variability in the diet of the cape gannet at Bird Island, Algoa Bay, South Africa. In: A.L.L. Brink, K.H. Mann, and R. Hilborn (eds.), Benguela trophic functioning. South African Journal of Marine Science 12:761-771.

Monaghan, P., J.D. Uttley, M.D. Burns, C. Thaine, and J. Blackwood. 1989. The relationship between food supply, reproductive effort and breeding success in Arctic terns, *Sterna paradisaea*. Journal of Animal Ecology 58:261-274.

Montevecchi, W.A. 1993. Birds as indicators of change in marine prey stocks. In: R.W. Furness and J.J.D. Greenwood (eds.), Birds as monitors of environmental change. Chapman and Hall, London, pp. 217-266.

Montevecchi, W.A., and R.A. Myers. 1992. Monitoring fluctuations in pelagic fish availability with seabirds. Canadian Atlantic Fisheries Science Advisory Commission Research Document 92/94. Northwest Atlantic Fisheries Organization, Dartmouth, NS, Canada. 20 pp.

Murphy, E.C., A.M. Springer, and D.G. Roseneau. 1986. Population status of common guillemots, *Uria aalge,* at a colony in western Alaska: Results and simulations. Ibis 128:348-363.

Murphy, E.C., A.M. Springer, and D.G. Roseneau. 1991. High annual variability in reproductive success of kittiwakes (*Rissa tridactyla* L.) at a colony in western Alaska. Journal of Animal Ecology 60(2):515-534.

Myrberget, S. 1962. Undersølkelser over forplantningsbiologien til lunde (*Fratercula arctica* L.). Medd. St. Viltunders 2(11):1-51.

National Research Council. 1996. The Bering Sea ecosystem. National Academy Press, Washington, DC. 307 pp.

Nysewander, D.R., and J.L. Trapp. 1984. Widespread mortality of adult seabirds in Alaska, August-September 1983. Unpublished Report. U.S. Fish and Wildlife Service, 1011 E. Tudor Rd., Anchorage, AK 99503. 23 pp.

Ohtani, K., and T. Azumaya. 1995. Influence of interannual changes in ocean conditions on the abundance of walleye pollock *(Theragra chalcogramma)* in the eastern Bering Sea. In: R.J. Beamish (ed.), Climate change and northern fish populations. Canadian Special Publication of Fisheries and Aquatic Sciences 121:87-95.

Olla, B.L., and M.W. Davis. 1990. Effects of physical factors on the vertical distribution of larval walleye pollock *(Theragra chalcogramma)* under controlled laboratory conditions. Marine Ecology Progress Series 63:105-112.

Piatt, J.F., and P. Anderson. 1996. Response of common murres to the *Exxon Valdez* oil spill and long-term changes in the Gulf of Alaska marine ecosystem. In: S.D. Rice, R.B. Spies, D.A. Wolfe, and B.A. Wright (eds.), *Exxon Valdez* Oil Spill Symposium Proceedings, AFS Symposium 18:720-737.

Salt, G.W., and D.E. Willard. 1971. The hunting behavior and success of Forster's tern. Ecology 52:989-998.

Schneider, D., and G.L. Hunt Jr. 1984. A comparison of seabird diets and foraging distribution around the Pribilof Islands, Alaska. In: D. Nettleship, G. Sanger, and P. Springer (eds.), Marine birds: Their feeding ecology and commercial fisheries relationships. Canadian Wildlife Service Special Publication, pp. 86-95.

Schneider, D.C., G.L. Hunt Jr., and N.M. Harrison. 1986. Mass and energy transfer to pelagic birds in the southeastern Bering Sea. Continental Shelf Research 5:241-257.

Schumacher, J.D., and P.J. Stabeno. 1994. Ubiquitous eddies of the eastern Bering Sea and their coincidence with larval pollock. Fisheries Oceanography 3:182-190.

Schumacher, J.D., and P.J. Stabeno. 1998. The continental shelf of the Bering Sea. In: A.R. Robinson and K.H. Brink (eds.), The sea: The global coastal ocean regional studies and synthesis, Vol. XI. John Wiley and Sons, New York, pp. 869-909.

Sogard, S.M., and B.L. Olla. 1993. Effects of light, thermoclines and predator presence on vertical distribution and behavioral interactions of juvenile walleye pollock, *Theragra chalcogramma.* Journal of Experimental Marine Biology and Ecology 167:179-195.

Sowls, A.L., S.A. Hatch, and C.J. Lensink. 1978. Catalog of Alaskan seabird colonies. U.S. Fish and Wildlife Series, FWS/OBS-78/78. Washington, DC. 350 pp.

Springer, A.M. 1991. Seabird relationships to food webs and the environment: Examples from the North Pacific Ocean. In: W.A. Montevecchi and A.J. Gaston (eds.), Studies of high-latitude seabirds. 1. Behavioral, energetic, and oceanographic aspects of seabird feeding ecology. Canadian Wildlife Service, Ottawa, pp. 39-48.

Springer, A.M. 1992. A review: Walleye pollock in the North Pacific—how much difference do they really make? Fisheries Oceanography 1(1):80-96.

Springer, A.M. 1993. Report of the seabird working group. In: Is it food? Addressing marine mammal and seabird declines: Workshop summary. University of Alaska Sea Grant, AK-SG-93-01, Fairbanks, pp. 14-29.

Springer, A.M., and G.V. Byrd. 1989. Seabird dependence on walleye pollock in the southeastern Bering Sea. In: Proceedings of the International Symposium on the Biological Management of Walleye Pollock. University of Alaska Sea Grant, AK-SG-89-01, Fairbanks, pp. 667-677.

Springer, A.M., and D.G. Roseneau. 1985. Copepod-based food webs: Auklets and oceanography in the Bering Sea. Marine Ecology Progress Series 21:229-237.

Springer, A.M., J.F. Piatt, and G. Van Vliet. 1996. Seabirds as proxies of marine habitats in the western Aleutian Arc. Fisheries Oceanography 5(1):45-55.

Springer, A.M., E.C. Murphy, D.G. Roseneau, C.P. McRoy, and B.A. Cooper. 1987. The paradox of pelagic food webs in the northern Bering Sea: I. Seabird food habits. Continental Shelf Research 7:895-911.

Springer, A.M., D.G. Roseneau, D.S. Lloyd, C.P. McRoy, and E.C. Murphy. 1986. Seabird responses to fluctuating prey availability in the eastern Bering Sea. Marine Ecology Progress Series 32:1-12.

U.S. GLOBEC. 1996. U.S. Global Ocean Ecosystems Dynamics, Report on climate change and carrying capacity of the North Pacific ecosystem. U.S. GLOBEC Report 15. University of California, Berkeley. 95 pp.

Vader, W., R.T. Barrett, and K.-B. Strann. 1987. Sjøfughekking i Nord-Norge 1987, et svartår. Vår Fuglefauna 10:144-147.

van Meurs, P., and P.J. Stabeno. In press. Evidence for episodic on-shelf flow in the southeastern Bering Sea. Journal of Geophysical Research.

Veit, R.R., P. Pyle, and J.A. McGowan. 1996. Ocean warming and long-term change in pelagic bird abundance within the California current system. Marine Ecology Progress Series 139:11-18.

Veit, R.R., J.A. McGowan, D.C. Ainley, T.R. Wahls, and P. Pyle. 1997. Apex marine predator declines ninety percent in association with changing oceanic climate. Global Change Biology 3:23-28.

Walsh, J.J., and C.P. McRoy. 1986. Ecosystem analysis in the southeastern Bering Sea. Continental Shelf Research 5:259-288.

Wright, P., R.T. Barrett, S.P.R. Greenstreet, B. Olsen, and M.L. Tasker. 1996. Effect of fisheries for small fish on seabirds in the eastern Atlantic. In: G.L. Hunt Jr. and R.W. Furness (eds.), Seabird/fish interactions, with particular reference to seabirds in the North Sea. ICES Cooperative Research Report 216:44-55.

Wyllie-Echeverria, T. 1995. Sea-ice conditions and the distribution of walleye pollock *(Theragra chalcogramma)* on the Bering and Chukchi Sea shelf. In: R.J. Beamish (ed.), Climate change and northern fish populations. Canadian Special Publication of Fisheries and Aquatic Sciences 121:131-136.

CHAPTER **29**

Seabirds of the Western Bering Sea

Vyacheslav P. Shuntov
Pacific Research Institute of Fisheries and Oceanography (TINRO),
Vladivostok, Russia

Abstract

I examined the distribution, species composition, and foraging ecology of seabirds in the western Bering Sea over the past two decades. There is large variability in estimates of number of nesting seabirds in the western Bering Sea, because of differences in assessment methods. Over 4.1 million individuals were estimated to nest in the western Bering Sea in the 1960s to 1980s but more recent, precise methods provide an estimate of nearly 11 million individuals. The most abundant species include alcids, followed by gulls and tube-nosed seabirds. An estimated 2-3 million semi-aquatic birds that nest in the interior basins migrate through the Bering Sea. The number of seabirds in both the eastern and western sections of the Bering Sea appears to have increased slightly in recent years. I classified the abundance and distribution of seabirds based on their food capture method. Those species that capture prey by diving were most abundant (63% of the total) followed by those that feed on or near the surface (37%). I also examined the seasonal distribution of seabirds in the western Bering Sea. During winter, sea ice and low air temperatures limit the abundance and number of species. Species that occur there in winter include light-colored fulmars, various gull species, and some auks. Many of these species feed at the ice edge or in polynyas. As spring approaches and the ice begins to melt, the abundance and number of species entering the western Bering Sea increases dramatically. The migration route into the area varies by species and includes the shore, over the shelf in contiguous zones of deep water, and along the deep water as ice retreats. Shearwaters begin forming enormous feeding aggregations and by May increase in numbers into the millions, probably feeding on abundant krill resources developing along the retreating ice edge. A considerable number of local birds, including some immature individuals, occur at nesting sites during summer and early autumn, a period of high productivity in the western Bering Sea. Diving birds make up two-thirds of the total number of breeding birds. In autumn the dispersal of birds begins away from the coast resulting in an increase in the number of diving birds appearing off the shelf and continental slope. The long-term dynamics in the ecology and number of seabirds likely are related to ecosystem cycles on the order of 40-60 years. Associated with these cycles are changes in habitat, pulsations of faunistic complexes, and long-term tendencies for changes in the number of species.

Introduction

The Bering Sea is one of the regions of the World Ocean where large numbers of seabirds are observed throughout the year. Because they represent a large group in terms of both species and individuals, they undoubtedly play a significant role in the ecosystem, particularly where they occur in large numbers. We must, however, be cautious about attributing the influence of seabirds at certain colonies to a particular ecosystem. For example, remember that no less than half of the seabirds of the Commander Islands do not extract food in the Bering Sea but in oceanic Pacific waters (a similar situation occurs for birds of the Aleutian Islands).

In recent years, ecosystem investigations have increased particularly in regard to biological resources and the relationship and role of seabirds. Significant progress has been made in studies during seabird nesting periods, especially in the American section of the Bering Sea. Few studies on seabird ecology have occurred in the western Bering Sea, but recent studies there have provided new information on seabird distribution, nesting behavior, and the number of colonies.

Many recent publications have addressed the relationship between bird distribution and oceanographic characteristics in the Bering Sea (Schneider 1982; Kinder et al. 1983; Schneider et al. 1987; Shuntov 1993; Springer et al. 1993, 1996; Decker and Hunt 1996; Hunt et al. 1996a). These publications suggest that the continental slope is important. In the 1960s, the first large-scale complex study of the Bering Sea by TINRO and VNIRO was restricted to areas between the shelf and deepwater basin (which was then termed a life zone). It was noted that this zone was the site of constant, increased concentrations of seabirds (Shuntov 1972). Springer et al. (1996) called this zone the "Green Belt" due to the high level of primary productivity there.

The unusual role of various life zones on the continental slope in the lives of Bering Sea biota was linked to the complex dynamics of water (Kotenev 1995, Sapozhnikov 1995). Dynamic processes over the continental slope consisted of an exchange between organic matter and energy at varying depths, including enrichment of the euphotic layer by mineral biological producers. Also, the cooling of currents along the continental slope causes a series of meanders and diversely directed vortices which are characteristic of areas with complex bottom reliefs (canyons, steep slopes, or bending isobaths); these areas are important foraging areas for birds. All this complexity contributes to the accumulation of macroplankton and early life stages of nekton in localized areas. A similar situation is applicable to islands and straits where the complexity of the hydrodynamic process strengthens the ebb and flow phenomena.

It is important to emphasize that the horizontal spottiness and the vertical stratification of zooplankton and fine nekton distribution is characteristic for shelf regions. The true character of this distribution can differ at wide and narrow shelves. For example, in the wide eastern Bering Sea shallow waters of the coastal, middle, and exterior secondary fronts, which are located at the 50, 100, and 170 m isobaths, have a great effect on the distribution of birds. In the interior regions of the shelf, diving murres and shearwaters predominate, but in the exterior sections of the shelf, surface-feeders occur (i.e., fulmars, kittiwakes, and storm-petrels) (Schneider and Hunt 1982, Schneider et al. 1986). The Gulf of Anadyr is similar (Shuntov 1993); however, the warm Navarin Current there "disrupts" the cool spot in the middle of the shelf and affects the location and configuration of secondary fronts. This current has a similar effect in the Bering Strait where it meets cold, northern water and creates gradient zones where large numbers of alcids (Alcidae), especially auklets, forage. In other regions of the western Bering Sea the shelf is narrower and the water is under the influence of currents along the slope. It appears that the middle of the shelf's front is displaced toward the exterior edge of the shallow water, and along the vertical sections it is often connected with external surface fronts (Verkhunov 1995). For this reason distinct ecological groupings here are not always spatially separated.

Composition and Number of Seabirds in the Western Bering Sea

Studies during the 1960s to 1980s suggested that the number of nesting seabirds in the western Bering Sea was 4.1 million individuals (Schneider and Shuntov 1993, Shuntov and Dulepova 1995, Vyatkin 1986, Smirnov and Velizhanin 1986). Subsequent observations, particularly in the northwestern Bering Sea, indicated that the population was >7 million individuals (Table 1). Schneider and Shuntov (1993) summarized data on the number of nesting birds in the southwestern Bering Sea. Between Dezhnev Bay and Cape Olyutorsk, approximately 364,000 individual birds nested (Vyatkin 1986). Of the 7 million estimated seabirds, more than half were concentrated in the northwestern section of the Bering Sea. Alcidae were in highest abundance (61.8%), followed by gulls (Laridae, 24.8%), and tube-nosed seabirds (tubinares, 11.1%). Less than 3% were cormorants (Phalacrocoracidae), terns (Sternidae), jaegers (Stercorariidae), and phalaropes (Phalaropodidae). Six species exceeded 0.5 million individuals each, including black-legged kittiwakes *(Rissa tridactyla)*, least auklets *(Aethia pusilla)*, thick-billed murres *(Uria lomvia)*, northern fulmars *(Fulmarus glacialis)*, crested auklets *(Aethia cristatella)*, and common murres *(Uria aalge)*.

Table 1. Composition and number (thousands of individuals) of seabirds nesting in the western Bering Sea.

Species	NW Bering Sea[a]	SW Bering Sea[b]	Commander Islands	Total
Fulmar *Fulmarus glacialis*	80.0	110.0	644.0	834.0
Leach's petrel *Oceanodroma leucorhoa*	–	–	10.0	10.0
Fork-tailed storm-petrel *Oceanodroma furcata*	–	–	20.0	20.0
Pelagic cormorant *Phalacrocorax pelagicus*	56.3	33.5	7.0	96.8
Bare-faced cormorant *Phalacrocorax urile*	–	–	2.2	2.2
Red phalarope *Phalaropus fulicarius*	5.0	–	–	5.0
Red-necked phalarope *Phalaropus lobatus*	5.0	10.0	5.0	20.0
Pomarine jaeger *Stercorarius pomarinus*	2.0	–	+	2.0
Parasitic jaeger *S. parasiticus*	10.0	2.0	+	12.0
Long-tailed jaeger *S. longicaudus*	10.0	2.0	–	12.0
Slaty-backed gull *Larus schistisagus*	37.5	40.2	–	77.7
Herring gull *Larus argentatus*	2.0	–	–	2.0
Glaucous-winged gull *Larus glaucescens*	–	–	10.0	10.0
Glaucous gull *Larus hyperboreus*	2.0	–	–	2.0
Mew gull *Larus canus*	1.0	3.0	–	4.0
Black-legged kittiwake *Rissa tridactyla*	1,108.0	627.8	54.0	1,789.8
Red-legged kittiwake *Rissa brevirostris*	–	–	34.0	34.0
Black-headed gull *Larus ridibundus*	–	5.0	–	5.0
Sabine's gull *Xema sabini*	1.0	–	–	1.0

Table 1.   (Continued.)

Species	NW Bering Sea[a]	SW Bering Sea[b]	Commander Islands	Total
Common tern *Sterna hirundo*	1.0	20.0	–	21.0
Arctic tern *Sterna paradisaea*	1.0	1.0	–	2.0
Aleutian tern *Sterna camtschatica*	1.0	2.0	–	3.0
Pigeon guillemot *Cepphus columba*	21.0	2.0	3.2	26.2
Common murre *Uria aalge*	443.0	147.0	60.0	650.0
Thick-billed murre *Uria lomvia*	982.0	310.4	190.0	1,482.4
Parakeet auklet *Cyclorrhynchus psittacula*	25.0	5.0	10.0	40.0
Crested auklet *Aethia cristatella*	750.0	12.0	0.1	762.1
Whiskered auklet *Aethia pygmaea*	–	–	5.0	5.0
Least auklet *Aethia pusilla*	1,500.0	0.2	0.1	1,500.3
Marbled murrelet *Brachyramphus marmoratus*	–	0.5	–	0.5
Kittlitz's murrelet *Brachyramphus brevirostris*	5	1	–	6.0
Ancient murrelet *Synthiliboramphus antiquus*	–	–	0.1	0.1
Tufted puffin *Fratercula cirrhata*	55.0	12.5	200.0	267.5
Horned puffin *Fratercula corniculata*	40.0	1.0	5.0	46.0
Total	5,143.8	1,348.1	1,259.7	7,751.6

[a]Northwestern section of sea, north of Cape Olyutorsk.

[b]Southwestern section of sea, south of Cape Olyutorsk.

Note: This table was compiled using data from Vyatkin 1986, 1992; Artyukhin 1991; Kondrat'ev 1991, Kondrat'ev 1993; Vyatkin and Artyukhin 1994; and Shuntov 1998).

Note that these values are only estimates. For example, at one time I estimated the number of least auklets in the western Bering Sea at 1.5 million individuals. In recent "gray" literature (e.g., Konyukhov 1990, Kondrat'ev 1991), the number of least auklets ranged from 100,000 to 250,000 individuals, but in my opinion the assessment methods used often resulted in underestimates. A recent recounting of several colonies in the Okhotsk Sea and Bering Strait resulted in higher counts (Kondrat'ev et al. 1992, Zubakin et al. 1992, Springer et al. 1993, Faris et al. 1996). Thus, for Ratmanov Island (Big Diomede), different methods produced numbers from 0.7 to 4.4 million individual least auklets and from 0.3 to 2.2 million crested auklets.

The high biological productivity and optimal feeding conditions of the Bering Sea attract a large number of birds from other regions. The most numerous of these are short-tailed shearwaters *(Puffinus tenuirostris)* and sooty shearwaters *(P. griseus)* from the Southern Hemisphere. The geographic position of the Bering Sea, particularly its contact with the Arctic Ocean through Bering Strait, provides an important transit area through which the migration of northern birds is facilitated; it also provides an important wintering area for many of them. Accurately determining the number of these birds is complex considering their population dynamics, migration patterns, and redistribution between the western and eastern Bering Sea. It is particularly difficult to estimate numbers for species whose nesting ranges include the Bering Sea and contiguous regions. For example, the number of non-nesting seabirds in the western section of the Bering Sea has been conservatively (and tentatively) estimated at 3 million individuals (Table 2). In sum, the overall number of seabirds in the western Bering Sea is nearly 11 million individuals (Tables 1 and 2).

A large number of birds that nest in the interior basins migrate through the Bering Sea. Some of them (e.g., jaegers, phalaropes, and gulls) are included in Tables 1 and 2. Some loons (*Gavia* spp.) and a large number of diving ducks (Anatidae) migrate to the Bering Sea after the nesting period. A significant portion of the diving ducks (e.g., oldsquaw, *Clangula hyemalis*; king eiders, *Somateria spectabilis*; and common eiders, *S. mollissima*) remain in the Bering Sea to winter in polynyas. The number of these semi-aquatic birds which migrate through the western Bering Sea has been estimated at 2-3 million individuals.

Despite the fact that the number of seabirds in the western Bering Sea seems to be higher than recently reported, it does not change the relationship in their numbers between Russian and American waters. In American waters, the number of nesting birds has also gradually increased (Sowls et al. 1978, Lensink 1984, Schneider and Shuntov 1993, Faris et al. 1996). A recent estimate of 36 million seabirds in the eastern Bering Sea and Aleutian Islands was proposed (i.e., approximately three times more than in the western section). The number of southern shearwaters alone was estimated at approximately 12 million individuals with extreme counts from 9 to 502 million individuals (Gould et al. 1982).

Table 2.   Composition and number of non-nesting seabirds in the western Bering Sea.

Species	Numbers, thousands of individuals
Short-tailed albatross, *Diomedea albatrus*	+
Laysan albatross, *D. immutabilis*	50.0
Black-footed albatross, *D. nigripes*	+
Northern fulmar, *Fulmarus glaciailis*	200.0
Mottled petrel, *Pterodroma inexpectata*	10.0
Sooty shearwater, *Puffinus griseus*	200.0
Short-tailed shearwater, *P. tenuirostris*	1,500.0
Fork-tailed storm-petrel, *Oceanodroma furcata*	200.0
Leach's petrel, *O. leucorhoa*	+
Pelagic cormorant, *Phalacrocorax pelagicus*	4.0
Red phalarope, *Phalaropus fulicarius*	70.0
Red-necked phalarope, *P. lobatus*	150.0
Pomarine jaeger, *Stercorarius pomarinus*	20.0
Parasitic jaeger, *S. parasiticus*	40.0
Long-tailed jaeger, *S. longicaudus*	45.0
Ivory gull, *Pagophila eburnea*	5.0
Ross's gull, *Rhodostethia rosea*	25.0
Herring gull, *Larus argentatus*	25.0
Glaucous-winged gull, *L. glaucescens*	10.0
Glaucous gull, *L. hyperboreus*	5.0
Mew gull, *L. canus*	10.0
Black-legged kittiwake, *Rissa tridactyla*	200.0
Sabine's gull, *Xema sabini*	5.0
Arctic tern, *Sterna paradisaea*	50.0
Aleutian tern, *S. camtchatica*	5.0
Common tern, *S. hirundo*	5.0
Dovekie, *Alle alle*	+
Common murre, *Uria aalge*	20.0
Thick-billed murre, *U. lomvia*	350.0
Black guillemot, *Cepphus grylle*	15.0
Kittlitz's murrelet, *Brachyramphus brevirostris*	1.0
Horned puffin, *Fratercula corniculata*	2.0
Tufted puffin, *Fratercula cirrhata*	1.0
Total	3,223.0

+ Very low quantity.

To all appearances the eastern Bering Sea is a more favorable habitat for seabirds. The eastern Bering Sea has a milder climate, optimal hydrological regime, and most important, nesting on island cliffs is easier. These points optimize feeding conditions for seabirds, although according to the general level of biological productivity, the western and eastern sections are difficult to contrast. Zooplankton concentrations are approximately the same (Shuntov et al. 1993); however, Markina and Khen (1990) speculate that zooplankton biomass in the western section is 1.5 times greater than in the east. Benthic biomasses in the eastern Bering Sea are somewhat lower (Neyman 1963) but nekton and nekton-benthos there are considerably higher. In the 1980s, fish biomass in the Bering Sea was at 50-60 million t, but only one-fourth of it was in the western section (Shuntov et al. 1993, Shuntov and Dulepova 1995). But when analyzing bird feeding ecology, it is not only important to consider overall abundance of fish as prey; it is also important to consider the concentration of prey at depths accessible to the birds. These types of comparisons between the western and eastern sections of the Bering Sea are not possible at present.

Ecological Groups of Birds

The composition of birds in any region of the World Ocean depends on the structure, abundance, relative concentration, and distribution of the prey base, with the exception of biogeographical and climatic-oceanographic characteristics. Seabirds can be roughly divided into two groups: those that capture food in the surface layer (surface-feeders) and those that capture food by diving.

The predominate species (63%) which nest in the western Bering Sea capture food by diving and include cormorants and alcids. Those that capture food from the surface or in the uppermost 0.5-1.0 m (including coastal areas) comprise the remaining and include gulls, tube-nosed birds, jaegers, and phalaropes. Similar proportions of diving (67%) and non-diving (33%) nonbreeding species were observed (Table 2); among these southern shearwaters were predominant.

Non-diving birds constitute approximately 65% of the total in the western Bering Sea (around 11 million individuals). This increases to approximately 70% if one also considers birds breeding in fresh water (ducks and loons). The very large percent of diving species (number of individuals) is indirectly confirmed through high biological productivity and fish production in the Bering Sea, and by the high concentrations of macroplankton, benthos, nekton, and nekto-benthos accessible to birds.

Shuntov (1972) examined the distribution of seabirds and suggested that they fall into categories of historically neritic (coastal), neritic-oceanic (offshore), and oceanic groupings. The neritic-oceanic group also includes very specialized ice-neritic species (Shuntov 1972).

The western Bering Sea, and other water masses in the Russian Far East, is open to currents from the Pacific Ocean. Consequently, there are more oceanic bird species here, such as albatrosses (*Diomedea* spp., Fig. 1), mottled petrels *(Pterodroma inexpectata),* Leach's storm-petrels *(Oceanodroma leucorhos),* and some red-legged kittiwakes *(Rissa brevirostris).* Their distribution is restricted by deepwater basins and the continental slope where the effect of oceanic water circulating in cyclonic, large scale vortices is quite noticeable. The number of oceanic species is relatively small, particularly in the western Bering Sea which is influenced by runoff of the Kamchatka Current. Oceanic species represent approximately 1% of the total number of seabirds in the western Bering Sea.

The distribution of fork-tailed storm-petrels is similar to the oceanic type (Fig. 2). In contrast to the typical oceanic Leach's storm-petrel, the fork-tailed storm-petrel is also encountered in shallow water.

The more numerous species in the Bering Sea are the neritic or coastal groups (e.g., cormorants, *Phalacrocorax* spp.; murrelets, *Brachyramphus* spp.; pigeon guillemots, *Cepphus columba;* mew gulls, *Larus canus;* and black-headed gulls, *L. ridibundus*). These birds are no farther than several kilometers away from the coast. This group, however, is small in terms of the number of individuals in the western Bering Sea. The proportion of these neritic species to all seabirds is 1-2%, excluding semiaquatic ducks and divers.

Birds with neritic-oceanic and interzonal types of distribution predominate in the Bering Sea in terms of species and individuals. Neritic-oceanic species are more numerous at the shelf but rare in the continental slope, the deepwater basins, and within 16 km and even hundreds of kilometers from the coast. Examples of the quantitative distribution of remote-neritic types is given in Figs. 3-5. In many respects they have a similar distribution and the same number of species as slaty-backed gulls *(L. schistisagus)* and crested auklets.

In summer, when similarly large masses of birds are attracted to shelf waters, some interzonal species occur there, such as kittiwakes, tufted puffins *(Fratercula cirrhata),* and horned puffins *(Fratercula corniculata).* However, during migration and during the winter, interzonal species occur in large numbers in oceanic regions. Northern fulmars are important among the nesting birds in the Bering Sea (Fig. 6) as are sooty shearwaters among the migrants.

Quantitative Distribution of Birds: Regional and Seasonal Aspects

Winter Period

In the western Bering Sea the ecology of seabirds in winter and spring periods is nearly unstudied. At the height of winter, ice covers essentially all of the shelf making survival difficult for those birds which do not migrate south in winter. For this reason there are no prominent wintering birds there, although to a lesser extent large aggregations of birds occur in the

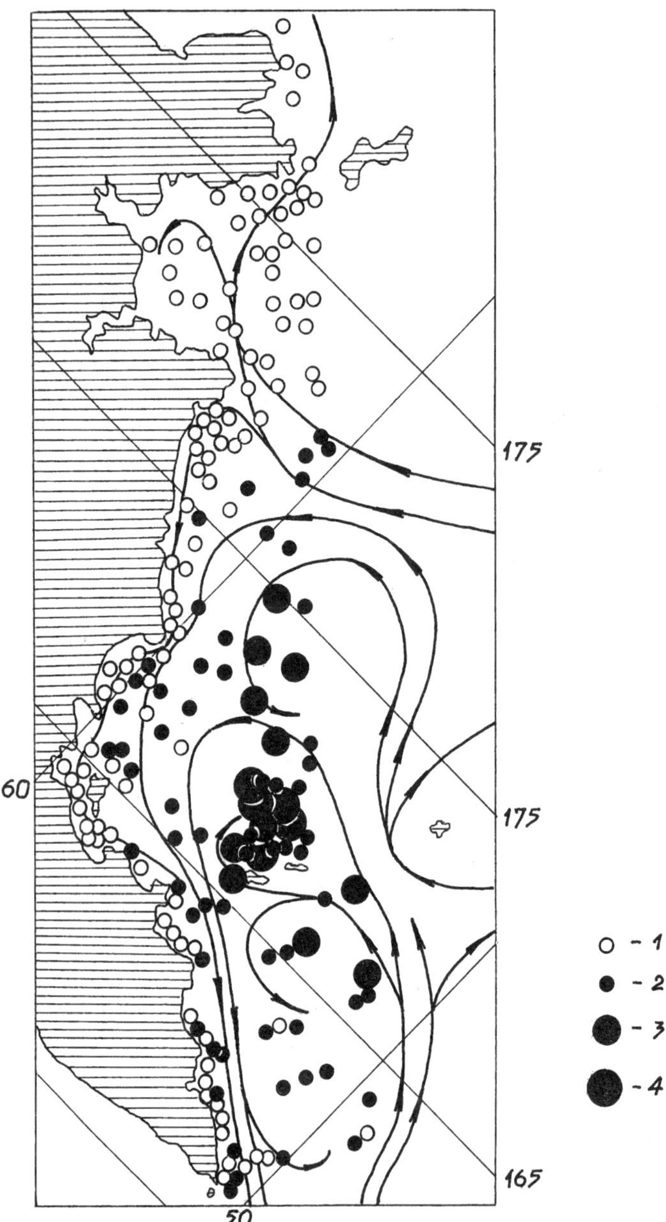

Figure 1.   The distribution of Laysan albatrosses (Diomedea immutabilis) in the western Bering Sea and contiguous ocean waters in September-November 1986. (1) none; (2) less than 0.3; (3) 0.3-1.0; (4) 1.0-2.0 individuals/km². Arrows = the generalized scheme of the currents.

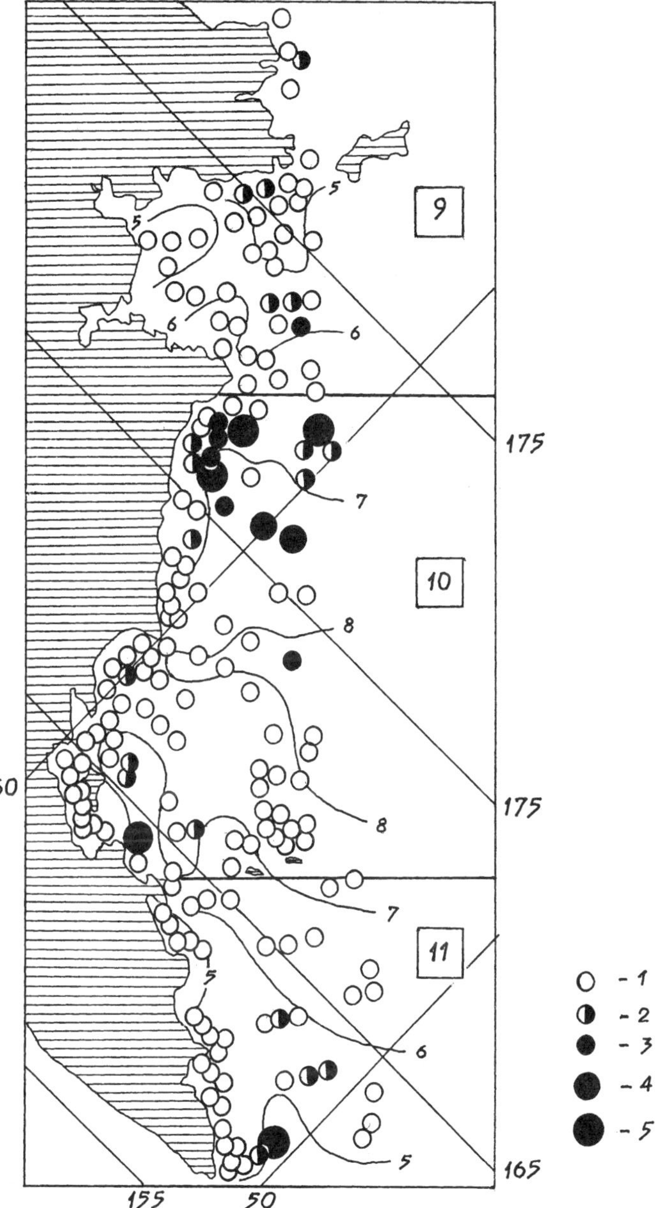

Figure 2.   The distribution of fork-tailed storm-petrels (Oceanodroma furcata) in the western Bering Sea and contiguous ocean waters in
September-November 1986. (1) none; (2) less than 0.5; (3) 0.5-
1.0; (4) 1.0-5.0; (5) 5.0-10.0 individuals/km^2. Isolines = surface
isotherms; numbers in squares = month of observation.

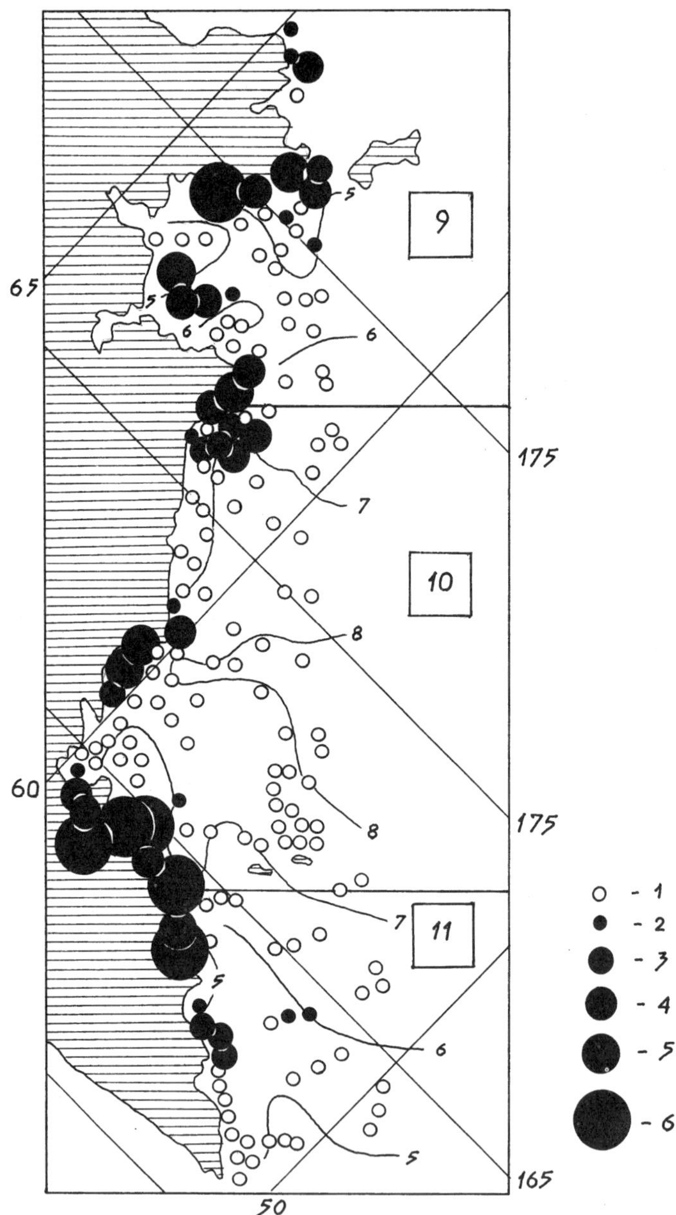

Figure 3.   The distribution of least auklets (Aethia pusilla) in the western Bering Sea and contiguous ocean waters in September-November 1986. (1) none; (2) less than 0.5; (3) 0.5-1.0; (4) 1.0-5.0; (5) 5.0-10.0; (6) more than 10 individuals/km². Isolines = surface isotherms; numbers in squares = month of observation.

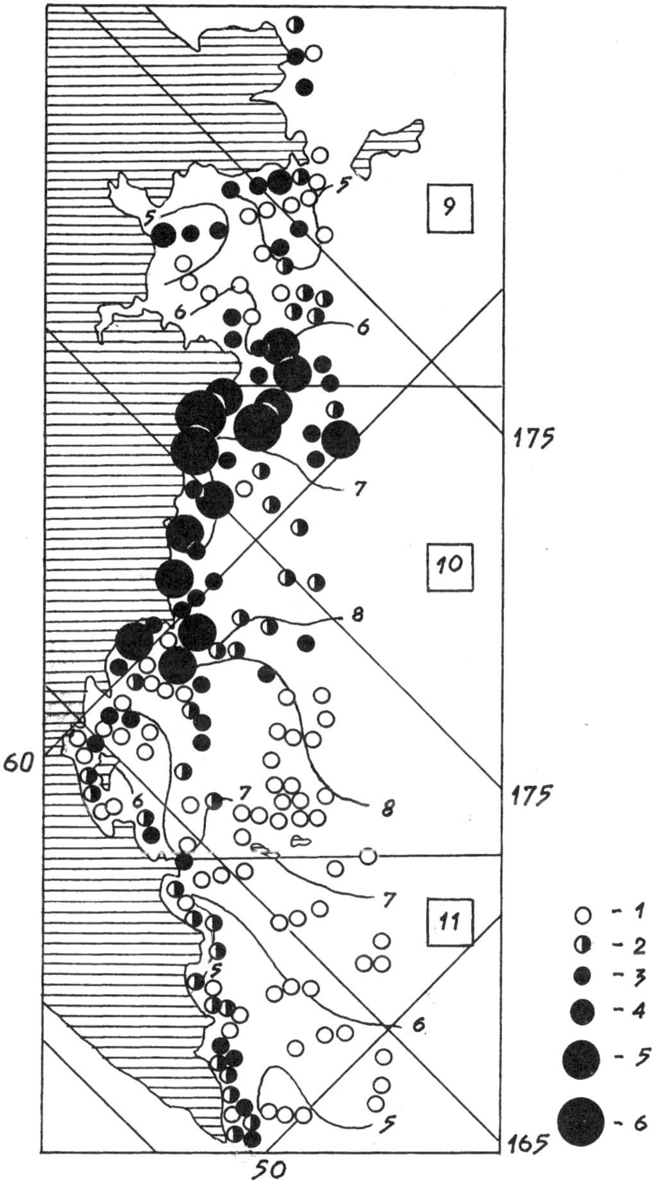

Figure 4. *The distribution of herring gulls (Larus argentatus) in the western Bering Sea and contiguous ocean waters in September-November 1986. (1) none; (2) less than 3; (3) 3-10; (4) 10-20; (5) 20-50; (6) more than 50 individuals/count (including birds near the ship). Isolines = surface isotherms; numbers in squares = month of observation.*

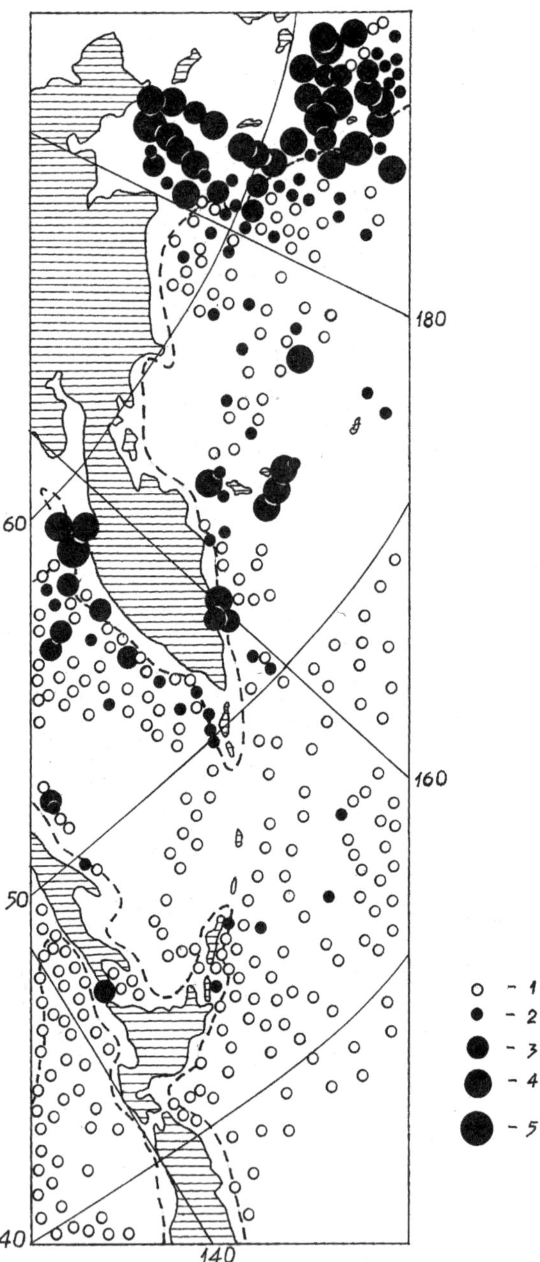

Figure 5.   The distribution of murres (Uria spp.) in the western Bering Sea and contiguous ocean waters in July to the first half of September 1960. (1) none; (2) less than 0.5; (3) 0.5-1.0; (4) 1.0-10.0; (5) more than 10 individuals/km².

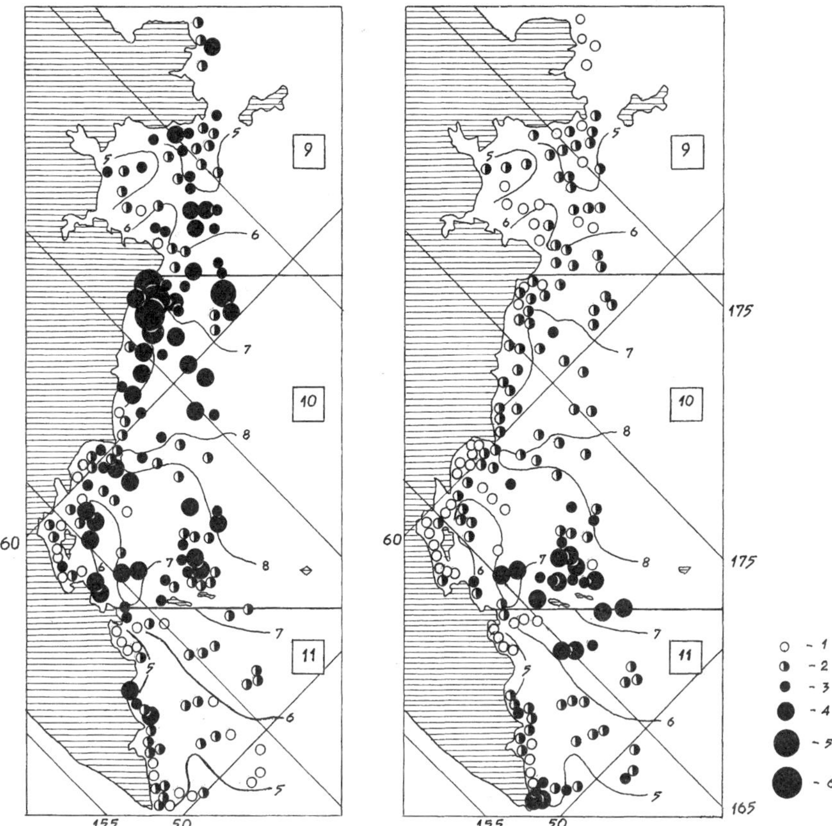

Figure 6. The distribution of light (left illustration) and dark (right illustration) morphs of northern fulmars (Fulmarus glacialis) in the western Bering Sea and contiguous ocean waters in September-November 1986. (1) none; (2) less than 0.5; (3) 0.5-1.0; (4) 1.0-5.0; (5) 5.0-10.0; (6) more than 10 individuals/km². Isolines = surface isotherms; numbers in squares = month of observation.

southeastern section of the Bering Sea near the Aleutian Islands. Very low winter air temperatures have a great effect on thermophilous and relatively thermophilous species. The magnitude of seasonal migrations of nesting birds at the Asiatic shores is greater in comparison with those birds observed in the American region. A large segment of the birds from the western region migrate to the waters of western subarctic circulation and to all appearances, to the eastern Bering Sea with the contiguous oceanic waters. In particular, counts by experts of the number of birds in winter in the western Bering Sea (including Commander Island waters) was approximately one-tenth that in other seasons.

Judging from spotty observations (Shuntov 1972, Kosygin 1985, Lobkov 1986, Trukhin and Kosygin 1987, Shuntov 1993) kittiwakes, fork-tailed petrels, Leach's storm-petrels, tufted puffin, horned puffin, all species of jaegers, phalaropes, and other species completely migrate from the western section of the Bering Sea. Seabirds that overwinter were composed of light morph fulmars, large gulls, slaty-backed and glaucous-winged gulls *(L. glaucescens)*, herring gulls *(L. argentatus)*, glaucous gulls *(L. hyperboreus)*, ivory gulls *(Pagophilia eburnea)*, Ross's gulls *(Rhodostethia rosea)*, and some alcids. Among the latter group are both species of murres, black guillemots *(Cepphus grylle)*, and Kittlitz's murrelets *(Brachyramphus brevirostris)*. Some auklets, especially least auklets, winter in the western section. Due to the sparseness of observations, there are no data regarding bird aggregations in the western section. However, concentrations of gulls and fulmars do occur near commercial fishing operations that catch walleye pollock *(Theragra chalcogramma)* in the Navarin region and the deep continental slope off Olyutorsk and Karagin bays.

All or nearly all species of seabirds avoid ice or are displaced by it. However, in polynyas and near the ice edge more favorable conditions exist for seabirds to capture prey. For ice-neritic species this habitat is preferable. There are three species of birds that overwinter in the Bering Sea: black guillemots *(Cepphus grylle)*, ivory gulls, and Ross's gulls. These gulls and a number of black guillemots descend to the ice fields and settle at the southern terminus of Kamchatka and penetrate to the Okhotsk Sea.

Polynyas sheared off by extensive ice are distinctive localities that influence winter distribution of some birds. These polynyas form even in the highest latitudes and offer a preferable habitat for some birds in winter. In the Bering Sea, such polynyas include Sireniki Polynya, near the south coast of Chukotka (Konyukhov 1990), and numerous polynyas next to St. Lawrence Island (Fay and Cade 1959). Ten bird species were noted on the Sireniki Polynya: oldsquaw *(C. hyemalis)*, common eiders *(Somateria mollissima)*, king eiders *(S. spectabilis)*, spectacled eiders *(S. fischeri)*, glaucous gulls, black guillemots, thick-billed murres, pelagic cormorants, Kittlitz's murrelets, and ivory gulls. The first two species occur in high numbers (Konyukhov 1990). The latter species and Ross's gull do not avoid the ice; however, the greatest number winter near the ice edge but not in the interior polynyas. Some birds descend into the Kuril Islands and the Okhotsk

Sea while moving along the pools of open water in the ice near the east coast of Kamchatka.

Spring Period

Ice plays an important role in the distribution of birds during spring migration which begins at the end of March and extends into April and May. In April, ice conditions are still only slightly different from winter; in Karagin Bay ice can occur into June and off Chukotka ice can be seen in May. Once spring begins, the birds cross over the ice fields and use open pools and polynyas as reference points to move in a northerly direction, toward their nesting sites.

The basic spring run of birds in the western Bering Sea occurs along the coast and over the shelf with contiguous zones of deepwater dropoffs that have a basic orientation toward the ice edge. The migration goes toward the Asiatic coasts in the northwestern Bering sea and from the southeastern Bering Sea where the ice edge forms along the contours of the shelf. The quantity of birds increases in open water of the Commander and Aleutian basins. Here, while bypassing the waters of the Aleutian-Commander Island region, the birds migrate within the wide marine range of the North Pacific.

In the western Bering Sea, migration may be later since it is related to the severity of winter conditions and the corresponding delay in the development of hydrobiological processes. Differences in the phenology of biological phenomena in the western and eastern sections of the Bering Sea are reflected in the timing and numbers of southern shearwaters (i.e., *Puffinus tenuirostris*), which fly north to molt and forage. By mid-May they are encountered in small numbers in the northwestern Bering Sea (Trukhin and Kosygin 1987). Their numbers, however, remain low even in the southwestern section at the beginning of summer (Shuntov 1972, 1998) when they form large aggregations in Asiatic regions, in oceanic waters, and in the southern half of the Okhotsk Sea. There is an earlier warming of the southeastern Bering Sea and by May shearwaters form enormous molting and foraging aggregations; they may number into the millions. Evidently, the relative stability of the feeding areas has a definite effect which must be important for the molting birds. Recent observations by me and by Hunt et al. (1996b) suggest that these shearwater feeding aggregations are focusing on krill. By spring the biomass of macroplankton in surface layers of the western Bering Sea is insignificant, resulting in no visible stable, large foraging areas. In the northwestern Bering Sea the situation is reminiscent of early summer in the southeastern section, consistent with the summer-autumn period. At the northwestern shelf and over deepwater drop-offs maximal zooplankton biomass was observed (Shuntov et al. 1993, Volkov 1996). Large masses of short-tailed shearwaters migrate here particularly in the second half of summer.

One of the important features of the spring migration in the western Bering Sea, as in other regions, is its duration during the closing stages.

The first and most powerful waves of flight consisted of mature individuals for whom breeding was imminent. In the second half of spring and in June there was a gradual gathering of immature birds toward the breeding regions. However, those less abundant species (e.g., large puffins, jaegers, and others) remained beyond the range of the Bering Sea for the entire summer.

Periods of Productivity: Summer and Early Autumn

During summer a considerable portion of the local birds, including some immature individuals, are concentrated at nest sites. Typically, in nesting locations where the continental shelf is narrow, bird colonies have a weak influence on the density of bird species 32-48 km from the colonies, and in some instances 16-24 km. In areas with a wide shelf, for example north of the Bering Sea, birds are more widely disseminated during the nesting period. This can be seen in Fig. 5 which shows the distribution of murres (*Uria* spp.). This distributional information is further supported by the information in Tables 3 and 4 and Figs. 7 and 8. These data were obtained on two expeditions in the Bering Sea from September to October 1986 and June 1991. In both instances, bird, plankton, and nekton data were collected 16-24 km beyond the coastal zone. Diving birds were the dominant (two-thirds of the total number) breeding birds in the western Bering Sea. In the southwest, surface-feeding birds (tube-nosed birds and gulls) were nearly everywhere beyond the coastal zone. There was a high proportion (but <50%) of diving birds in the Karagin region (Fig. 9).

In regions 1-8 (Fig. 10) September weather is similar to that in summer. The most numerous birds were capturing food at the surface, with the exception of the upper sections of the Gulf of Anadyr and the waters off Chukotka. Southern shearwaters, mainly short-tailed shearwater *(P. tenuirostris)*, were the dominant diving birds in Chukotka-Anadyr-Navarin waters.

In autumn with the dispersal of birds from the coast, the number and percentage of diving birds (mainly auklets, Alcidae) increased off the shelf and the continental slope. In the southwestern section, the density of birds increased 3-6 times (Table 4, Fig. 8). In deepwater regions there was a predominance of non-diving birds (tube-nosed birds) which commonly fly hundreds of kilometers from the nesting colonies.

Areas of the Bering Sea with increasing plankton biomass merge in the second half of summer and at the beginning of autumn to the north of the Bering Sea. The concentration of macroplankton there is several times higher than in the southwestern section (Table 4). This concentration gradient is the main reason for the redistribution of close-flying tube-nosed birds (particularly massive numbers of shearwaters and fork-tailed storm-petrels, *O. furcata*) from the southern regions.

Despite the high concentration of plankton in the north, the biological resources there are less than in deepwater basins, perhaps because of

Table 3.   Ratio of the density of concentration of seabirds, zooplankton, and nekton in the southwestern Bering Sea in June 1991.

Animal group	Regions				
	8	9	10	11	12
Birds, individuals/km^2	1.5	1.8	2.1	2.7	1.5
Macroplankton, mg/m^3	670	521	311	403	443
Macroplankton without *Sagitta* spp., mg/m^3	222	211	182	273	170
Nekton, t/km^2 (0-50 m)	0.9	0.05	1.6	0.3	1.3
Squids, t/km^2 (0-50 m)	0.1	+	–	+	0.3

Note: Areas as in Figs. 9 and 10.

+ Very low quantity.

Table 4.   Ratio of the density of concentration of seabirds, zooplankton, and nekton in September-October 1986 in the western Bering Sea.

Animal group	Regions											
	1	2	3	4	5	6	7	8	9	10	11	12
Birds, individuals/km^2	26.4	7.5	11.2	3.3	17.2	6.9	12.6	4.6	6.1	12.1	8.2	4.3
Macroplankton, mg/m^3	2,073	2,831	1,228	1,475	1,198	503	1,341	619	456	240	890	397
Macroplankton without *Sagitta* spp., mg/m^3	1,914	2,183	800	1,065	789	279	487	262	135	118	639	166
Nekton, t/km^2 (0-50 m)	1.93	0.91	18.0	0.37	27.0	11.5	35.0	3.72	25.0	9.18	29.7	9.4
Squids, t/km^2 (0-50 m)	–	–	–	–	0.4	+	0.7	0.2	0.1	–	0.4	0.3

Note: Areas as in Figures 9 and 10.

+ Very low quantity.

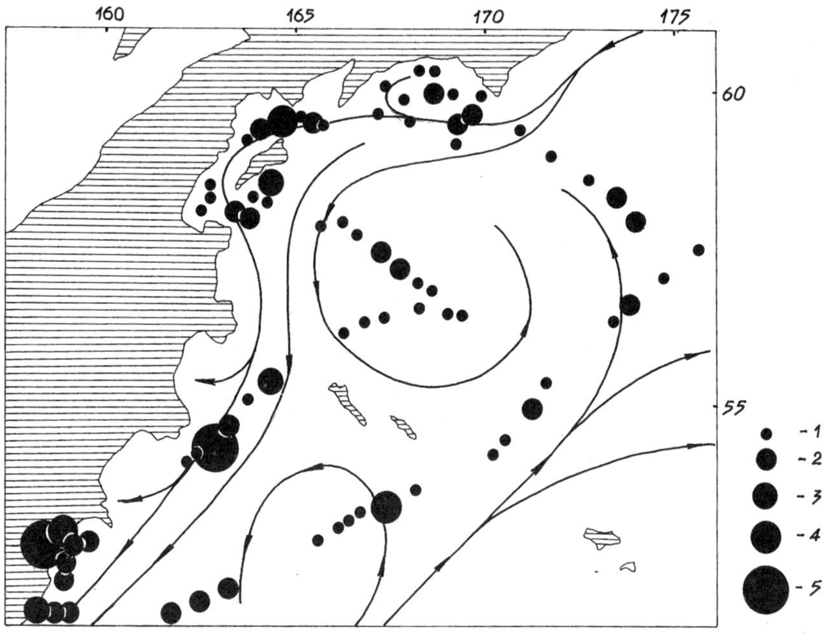

Figure 7.   *The distribution of the general density of seabirds in the southwestern Bering Sea and contiguous ocean waters in June 1991. (1) less than 2.0; (2) 2.0-5.0; (3) 5.0-10.0; (4) 10.0-20.0; (5) more than 20.0 individuals/ km². Arrows = generalized scheme of the currents.*

shallow water in the north. In deepwater areas, surface and interzonal plankton, together with mesopelagic fishes and squids, rise to the surface in dark warm periods and constitute an unlimited food source for those seabirds that have the ability to hunt in loose concentrations (e.g., albatrosses, fulmars, storm-petrels, and gadfly petrels [Procellaridae]).

As can be seen in a whole series of cases in Table 4, high densities of birds occur precisely in areas where, on average, greater concentrations of macroplankton and nekton are observed. But there are inconsistencies here also in that abundant plankton and nekton may not always be accessible to birds. On the other hand, where food resources are low, local accumulations of food may form in certain discrete areas and layers, which are accessible to foraging birds.

Figure 8 provides a quantitative distribution of birds for the western Bering Sea. In other seasons their density on the shelf and continental slope is higher than in open, deepwater basins. These high densities are likely associated with higher overall biological productivity at shallow slope areas and the presence of large secondary fronts, high gradient zones,

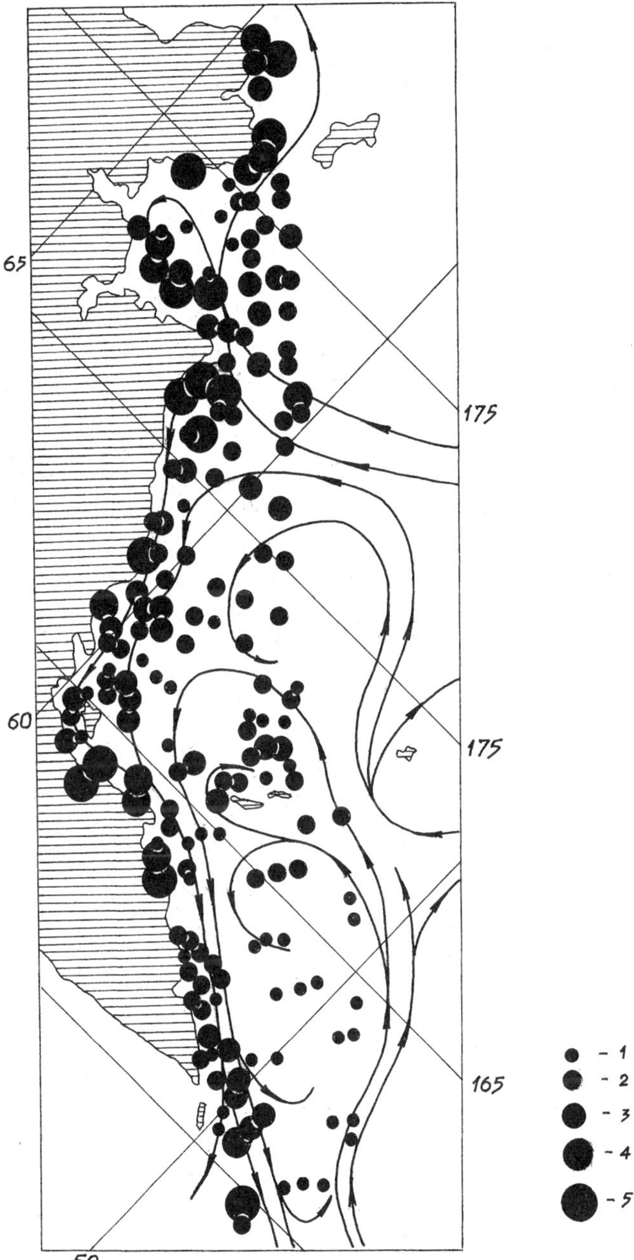

Figure 8. The distribution of the general density of seabirds in the western Bering Sea (September-November 1986) and contiguous ocean waters (November 1986). Designations are the same as in Fig. 7.

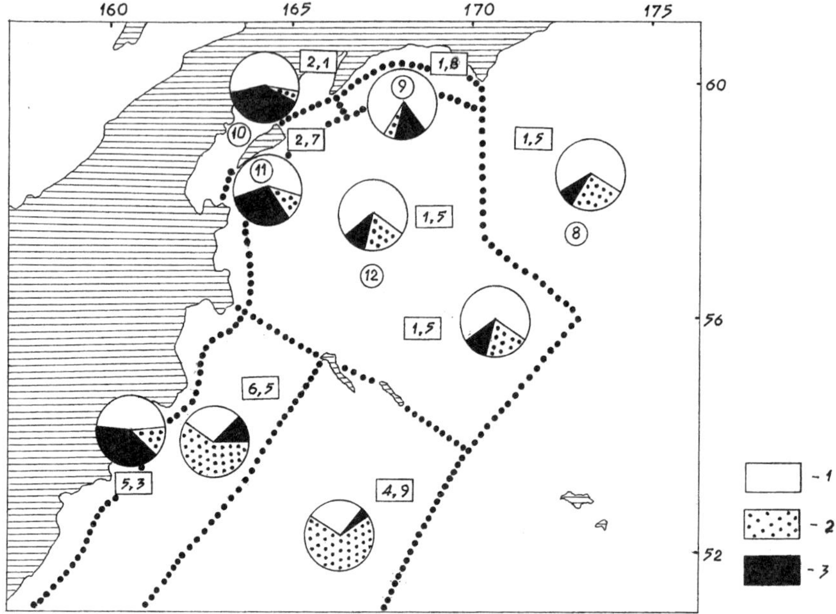

Figure 9.   *The relationship of bird groups according to the method of food capture in the southwestern Bering Sea in June 1991. (1) non-diving birds; (2) shearwaters (Puffinus spp.); (3) diving birds. Number in squares = average density of bird concentration, number/km²; numbers in small circles = the number of areas of averaged statistical information; dotted lines = limits of the areas of averaging.*

and meso-micro-circulation where there is a passive build up of plankton, larvae, and small nekton.

Late Autumn Period

The autumn migration period is drawn out. It is curious that at the end of summer (August) one can also see birds move in a counterclockwise direction and many birds begin to fly south (e.g., phalaropes, jaegers, terns, and mature shearwaters). Simultaneously in the north, the redistribution of immature shearwaters and other birds (i.e., fork-tailed storm-petrels and several gull species) continues.

The autumn migration of birds in the Bering Sea is primarily in the second half of September to the first half of November. However, in the first half of winter, with expanding ice, there is still a noticeable, gradual migration of birds in a southern direction. In waters of the western Subarctic circulation, where many birds from the Bering and Okhotsk seas winter, there is still an accumulation of birds into December.

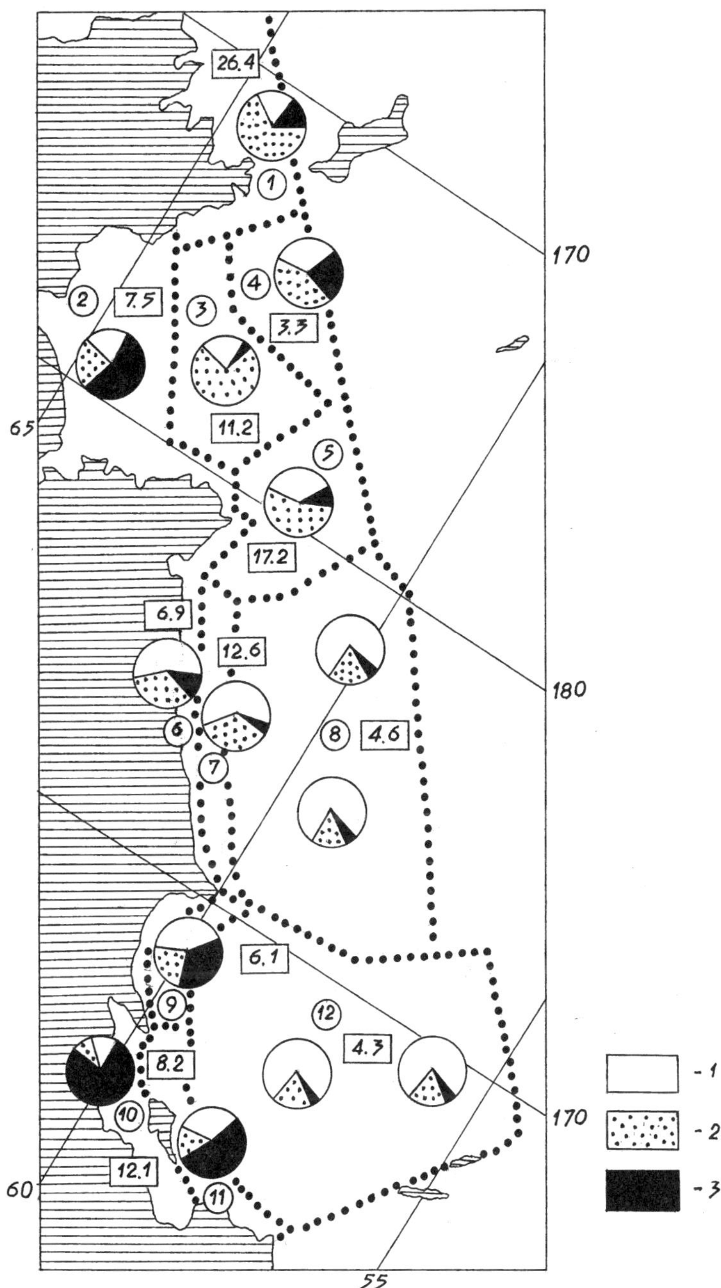

Figure 10. The relationship of bird groups according to the method of food capture in the western Bering Sea in September-October 1986. Designations the same as in Fig. 9.

Table 5. **Biomass and proportion (thousands t/%) of the basic species of fishes in the epipelagial of the Bering Sea in the 1980s and 1990s (Shuntov et al. 1997).**

Species and group	1980s	1990s
Walleye pollock, *Theragra chalcogramma*	20,000/87.9	8,000/66.4
Pacific herring, *Clupea pallasi*	700/3.1	1,500/12.4
Deepsea smelt, *Leuroglossus schmidti*	125/0.5	1,401/1.2
Northern lampfish, *Stenobrachius leucopsarus*	790/3.5	900/7.5
Salmon, *Oncorhynchus* spp.	315/1.4	470/3.9
Capelin, *Mallotus villosus*	360/1.6	500/4.1
Other fishes	465/2.0	540/4.5
Total	22,755/100	12,050/100
T/km^2	9.9	5.2

Along the edges of the western Bering Sea the migration routes are apparent in that many birds move south by following the coastline and shelf (e.g., least auklets and herring gulls; Figs. 3, 4). Some birds migrate from the northwestern section to southeastern and pre-Aleutian waters. At deepwater basins they do not follow any routes; the oceanic species migrate along the wide fronts to the south of oceanic and other waters. Tufted puffins and horned puffins from the western Bering Sea migrate south along the shelf in deepwater basins.

Long-Term Dynamics of the Ecology and Number of Birds

Although many birds are basically long-lived species, their number and distribution reflect changes in climatic, oceanographic, and biological phenomena. Examples of climatological events include alternating warm and cold epochs and periods. Within the limits of these cycles are large spans separated into cycles varying from 2 to ~10 years. Much recent attention has been given to average cycles, including those associated with the El Niño phenomenon. In my opinion the effect of this variation in climatic and oceanographic factors on the number of birds should not be exaggerated. Its effect usually is local and is related to temporal and regional parameters. More recent observation suggests that for seabirds of the North Pacific, and perhaps the entire ecosystem, a cycle of some 40-60 years should be considered (Klyashtorin and Sidorenkov 1966, Shuntov et al. 1997). Associated with these cycles are changes in habitat, pulsations of faunistic complexes, and long-term tendencies for changes in the number of fluctuating species.

The second half of a warm period in this century occurred from the early 1970s to the early 1990s, similar to periods in the 1920s and 1930s. In the middle of the 1990s the base temperature in the Far Eastern seas remained elevated; however, at the beginning of the 1990s a change began in the biota which was reminiscent of circumstances for the 1940s-1960s. For instance, a decrease in the number of pollock resulted in a decrease of fishery productivity of the Bering Sea (Table 5), but the portion of predatory chaetognaths (*Sagitta* spp.), which have little value as prey to other zooplankton (Shuntov 1986a, 1994), increased.

In my opinion this is all associated with the warm epochs of the 1970s-1980s, at least with the subsequent changes in the distribution and number of birds. A significant rise in the number of auklets in the northern Bering Sea has been reported (Konyukhov 1991, Springer et al. 1993). In the 1980s there were noticeably more common murres north of the Bering Sea which began to breed on Wrangel and Herald islands. Tufted puffins began to nest on Wrangel Island at the same time (Stishov et al. 1991, Kondrat'ev 1991). The range for some species of terns expanded to include northeast Asia in the 1970s-1980s. In the 1960s I recorded only light-colored fulmars in the Gulf of Anadyr (Shuntov 1972). In September 1980 there were approximately 5% more light-colored and more oceanographic dark-colored morphs there (Shuntov 1998).

Conversely, significant climatic, oceanographic, and ecosystem restructuring in the North Pacific will have negative consequences for seabirds. The reduction in breeding of kittiwakes (Kondrat'ev 1993, 1995; Hatch et al. 1993) was probably a consequence of such restructuring; this change originally started in American waters and extended to the Asiatic region by the 1990s. In the 1990s, a random reduction in breeding of other birds in Far Eastern seas was reported. Whether this is an expression of the long-term cycles or only the result of interannual variability must be shown by subsequent observations.

One of the effects of climatic and oceanographic epochs on bird diversity over time is the redistribution of some species of birds between the Bering Sea and the North Pacific. In the summer of the 1960s, fork-tailed storm-petrels were numerous in the Okhotsk and Bering seas (Shuntov 1972). By the 1980s their numbers in the Okhotsk Sea were several times lower. In the 1990s they became numerous again during summer in the Okhotsk Sea (Shuntov 1997). In comparing Figs. 2 and 11 one can see somewhat similar localities in the western section of the Bering Sea; during the 1980s, when the greatest concentration was observed in the same places as in the 1960s, however, the numbers were much lower. It is not known if there has been a recovery of fork-tailed storm-petrels here in the 1990s to the level of the 1960s.

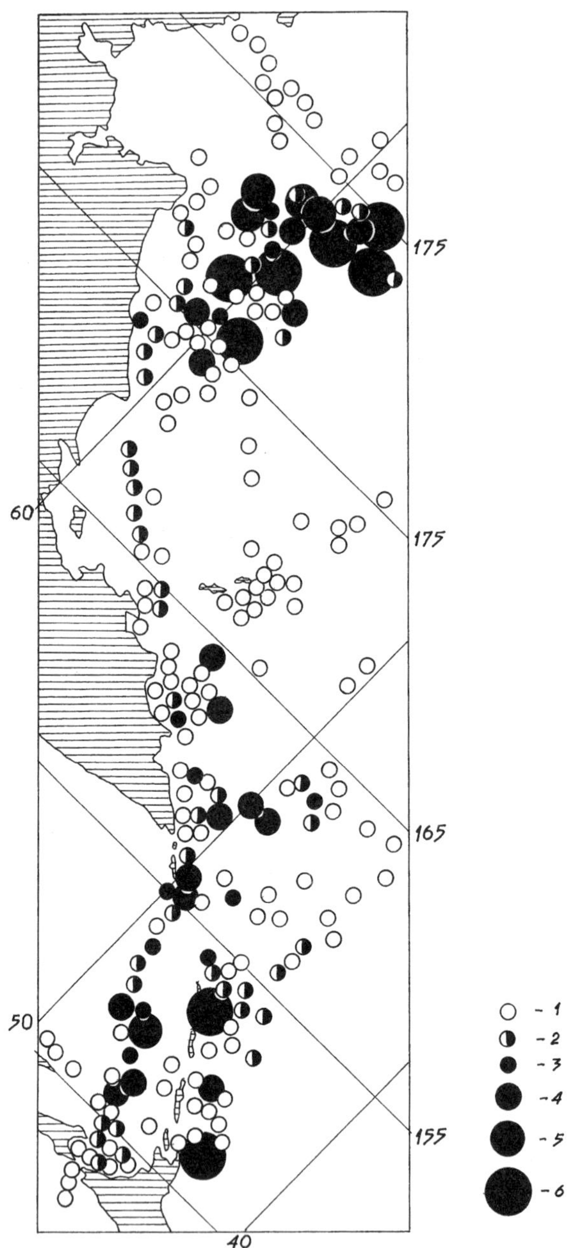

Figure 11. The distribution of fork-tailed storm-petrels (Oceanodroma furcata) in August-October 1960. (Shuntov 1998). (1) none; (2) less than 0.5; (3) 0.5-1.0; (4) 1.0-5.0; (5) 5.0-10.0; (6) more than 10.0 individuals/km².

Questions about the Role of Birds in the Bering Sea Ecosystem

When evaluating the role of birds in the marine ecosystem, consideration must be given to the complexity of their food web and the amount and diversity of prey consumed. Seabirds of the western Bering Sea were not systematically studied in this respect. However, judging from fragmentary information scattered in many publications (e.g., Ogi and Hamanaka 1982, Ogi et al. 1985), seabird feeding in the Russian section of the Bering Sea does not reflect what has been seen in the more thoroughly studied eastern section. The principal prey consists of nekton such as pollock, Pacific sand lance *(Ammodytes hexapterus),* capelin *(Mallotus villotus),* Pacific herring *(Clupea pallasi),* Atka mackerel *(Pleurogrammus monopterygius),* daubed shanny *(Lumpenus maculatus),* macroplankton (krill, copepods, amphipods), and small mesopelagic fishes and squids in deepwater regions.

Recent studies that plotted the biological balance of Bering Sea seabirds included the flow of matter and energy in the diagram (Shuntov and Dulepova 1995). The calculations, however, were done with insufficient data on the abundance of birds. One must also recognize that estimates of seabird prey consumption are complicated not only by the scarcity of data on annual rations, but also on the population dynamics of birds by season and taxonic group. Therefore, at this stage it is only possible to make rough estimates of the composition and magnitude of the annual ration. One such version is given in Table 6. The portion of nekton in the diet is assumed to be the following: For fulmars and short-tailed shearwater nekton represented 30%, storm-petrels and ducks = 10%, medium-sized alcids, kittiwakes, large gulls and albatross = 90%, cormorants = 95%, small alcids = 20%, and murres = 70%.

Using a method for estimating prey consumption by birds by season in the western Bering Sea produced an estimate of 337,000 t of which nearly half was nekton (mainly fish; Table 6). An extrapolation of these data to the eastern Bering Sea (based on the entire Bering Sea) results in approximately 2 million t (of this 0.8-1.0 million t is nekton). These figures are impressive considering the annual commercial catch from 1970 to 1990 in the Bering Sea usually fell within a range of 2.5-4.5 million t. The approximate scale of the biological resources and the production at different trophic levels are now known for the Bering Sea (Shuntov et al. 1993, Shuntov and Dulepova 1995). The mean biomass of plankton for spring was estimated at 453 million t, for summer 601 million t, for autumn 459 million t, and for annual production it was 3,810 million t. Biomass and fish production for the 1980s was estimated at 50-60 million t, of which 80% was pelagic (biomass and production are roughly 20% lower now); the biomass and production of squids was 4 and 12 million t, respectively. In the 1980s, walleye pollock consumed 189-252 million t of

Table 6.   Number of birds and their fish and invertebrate consumption in the western Bering Sea. Based on methods of Shuntov (1986b).

Species and group	Species (10^3 individuals)		Daily ration	Food consumption, (10^3/t)			Nekton consumption (10^3/t)
	Summer	Winter		Summer	Winter	Total	
Fulmar	600	200	150	18.0	5.0	23.0	8.0
Shearwaters and gadfly petrels	1,700	+	130	33.0	+	33.0	10.0
Storm-petrels	250	+	15	0.75	+	0.75	0.1
Albatrosses	50	+	500	5.0	+	5.0	4.0
Murres	2,000	200	250	100.0	8.0	108.0	76.0
Small guillemots	2,300	200	50	23.0	1.5	24.5	5.0
Average size guillemots	200	20	100	4.0	0.3	4.3	4.0
Kittiwake	1,700	20	100	34.0	0.3	34.3	30.0
Large gulls and jaegers	150	30	350	10.5	2.0	12.5	11.0
Phalaropes	90	5	400	7.2	0.3	7.5	7.0
Ducks	300	1,000	300	10.8	73.5	84.3	8.4
All birds	9,340	1,675		246.25	90.9	337.15	163.5

Note: In the calculations for the summer level of numbers we accepted 150 days for shearwaters, 120 for ducks, 200 for the remaining birds; for winter: 245 days for ducks, 165 days for the remaining birds.

+ Very low quantity.

macroplankton and small nekton which is now half that amount; pollock predators consumed 7 million t during the year. When one reflects on these values, the volume of biomass consumed by birds appears to be insignificant when considering the ecosystem as a whole (i.e., the overall role of seabirds in its biological balance is small). There is evidently a completely different picture at sites of mass aggregations of birds, particularly in shallow water regions. The Bering Strait-Navarin region, and the waters adjacent to the Commander Islands and Karagin Island, is a region of stable and long-term aggregations of large numbers of birds. In these regions many birds are concentrated both during breeding season and in other seasons. However, for concrete judgments about the role of birds in the functional biota of these regions, further specialized and complex investigations are needed.

References

Artyukhin, Y.B. 1991. The nesting avifauna of the Commander Islands and the effect of humans on their condition. Prirodnye resursy Komandorskikh ostrovov. M. Izd-vo Moskovogo Universiteta, pp. 99-137. (In Russian.)

Belizhanin, A.G. 1978. The mixing and condition of the number of colonies of seabirds in the Far East. Aktual'nye voprosy okhrany prirody na Dal'nem Vostoke. Vladivostok. DVNTs AN SSSR, pp. 154-172. (In Russian.)

Decker, M.D., and G.L. Hunt. 1996. Foraging by masses (*Uria* spp.) at tidal fronts surrounding the Pribilof Islands. Marine Ecology Progress Series 139:1-10.

Faris, R., R. Ferrero, L. Fritz, B. Mendenhall, R. Merrick, and D. Witherell. 1996. Ecosystem considerations 1997. Compiled and reviewed by the Gulf of Alaska and Bering Sea Plan Teams for the Groundfish of the Bering Sea, Aleutian Islands and Gulf of Alaska. Available from North Pacific Fishery Management Council, Anchorage, AK. 42 pp.

Fay, F.H., and T.J. Cade. 1959. An ecological analysis of the avifauna of St. Lawrence Island, Alaska. University of California Publication in Zoology 63(2):73-150.

Gould, P.J., D.J. Forsell, and C.J. Lensink. 1982. Pelagic distribution and abundance of seabirds in the Gulf of Alaska and eastern Bering Sea. U.S. Fish and Wildlife Service FWS-OBS-82/48. 294 pp.

Hatch, S.A., G.V. Byrd, D.B. Irons, and G.L. Hunt. 1993. Status and ecology of kittiwakes *(Rissa tridactyla* and *R. brevirostris)* in the North Pacific. The status, ecology and conservation of marine birds of the North Pacific. Canadian Wildlife Service Special Publication, Ottawa, pp. 140-153.

Hunt, G.L., N.M. Harrison, and J.F. Piatt. 1996a. Foraging ecology as related to the distribution of planktivorous auklets in the Bering Sea. In: The status, ecology and conservation of marine birds of the North Pacific. Canadian Wildlife Service Special Publication, Ottawa, pp. 18-26.

Hunt, G.L., K.O. Coyle, S. Hoffman, M.B. Decker, and M.V. Flint. 1996b. Foraging ecology of short-tailed shearwaters near the Pribilof Islands, Bering Sea. Marine Ecology Progress Series 141:1-11.

Kinder, H.T., G.L. Hunt, D.C. Schneider, and S.D. Schumacher. 1983. Correlation between seabirds and oceanic fronts around Pribilof Islands, Alaska. Estuarine Coastal and Shelf Science 16:309-319.

Klyashtorin, L.B., and N.S. Sidorenko. 1996. The long term climatic variation and the fluctuation of numbers of pelagic fishes of the Pacific. Izvestiya TINRO-Tsentr 119:33-54.

Kondrat'ev, A.Y. 1991. Status of the seabirds nesting in northeast U.S.S.R. ICBP Technical Publication 11:165-173.

Kondrat'ev, A.Y. 1993. Charadriiformes of northeastern Asia. Dissertatsiya v forme nauchnogo doklada. Vladivostok. DVO RAN. 47 pp. (In Russian.)

Kondrat'ev, A.Y. 1995. Summary of nesting kittiwakes on Talan Island from 1993-1995. Informatsionnyy byulleten': morskie ptitsy Beringii. Magadan. DVO RAN vyp. 3:44-45. (In Russian.)

Kondrat'ev, A.Y., V.A. Zubakin, and S.P. Kharistonov. 1992. Methods to evaluate the number of mass species of seabirds *(Aethia cristatella, Aethia pusilla)*. Pribrezhnye ekosistemy severnogo okhoto-morya O. Talan. Magadan. DVO RAN, pp. 137-152. (In Russian.)

Konyukhov, N.B. 1990. The wintering of seabirds on the Sireniki Polynya. Informatsionnye materialy izuchenie morsk. kolonial'n. ptits SSSR. Magadan. DVO RAN, pp. 36-39. (In Russian.)

Konyukhov, N.B. 1991. Some features of the biology of auklets in the colonies of the Chukotsk Peninsula. Informatsionnye materialy: Izuchenie morsk. kolonial'n. ptits SSSR. Magadan. DVO RAN, pp. 30-32. (In Russian.)

Kosygin, G.M. 1985. Material on the biology of ivory, Ross's and Sabine's gulls. Redkie i ischezayushchie ptits Dal'nego Vostoka. Vladivostok. DVNTs AN SSSR, pp. 135-138. (In Russian.)

Kotenev, B.N. 1995. Water dynamics as an important factor in the long-term variability in the bioproductivity of water and the reproduction of fishery stocks of the Bering Sea. Kompleksn. issled. ekosistemy Beringova Morya. M. Izd-vo VNIRO, pp. 7-39. (In Russian.)

Lensink, C.T. 1984. The status and conservation of seabirds in Alaska. In: Status and conservation of the world's seabirds. ICBP Technical Publication 2:13-27.

Lobkov, E.G. 1986. The nesting birds of Kamchatka. Vladivostok. DVNTs AN SSSR. 292 pp. (In Russian.)

Markina, N.P., and G.V. Khen. 1990. The basic elements of the functioning of pelagic communities of the Bering Sea. Izvestiya TINRO 111:79-93. (In Russian.)

Neyman, A.A. 1963. The quantitative distribution of benthos on the shelf and upper layers of the slope of the eastern Bering Sea. Izvestiya TINRO 50:145-205. (In Russian.)

Ogi, H., and T. Hamanaka. 1982. The feeding ecology of *Uria lomvia* in the northwestern Bering Sea region. Journal of the Yamashina Institute of Ornithology 14:270-280.

Ogi, H., H. Tanaka, and T. Tsujita. 1985. The distribution and feeding ecology of murres in the northwestern Bering Sea. Journal of the Yamashina Institute of Ornithology 17:44-56.

Sapozhnikov, V.V. 1995. A current representation about the functioning of ecosystems of the Bering Sea. Kompleksn. issled. ekosistemy Beringova Morya. M. Izd-vo VNIRO, pp. 387-392. (In Russian.)

Schneider, D.C. 1982. Fronts and seabird aggregations in the southeastern Bering Sea. Marine Ecology Progress Series 10:101-103.

Schneider, D.C., and G.L. Hunt. 1982. Carbon flux to seabirds in waters with different mixing regimes in the southeastern Bering Sea. Marine Biology 67:337-344.

Schneider, D.C., and V.P. Shuntov. 1993. The trophic organization of the marine bird community in the Bering Sea. Reviews in Fisheries Science 1(4):311-335.

Schneider, D.C., N.M. Harrison, and G.L. Hunt. 1987. Variation in the occurrence of marine birds at fronts in the Bering Sea. Estuarine, Coastal and Shelf Science 25:135-141.

Schneider, D.C., G.L. Hunt, and N.M. Harrison. 1986. Mass and energy transfer to seabirds in the southeastern Bering Sea. Continental Shelf Research 5:241-257.

Shuntov, V.P. 1972. Seabirds and the biological structure of the ocean. Vladivostok. Dal'nevostochnoe Knizhnoe. Izd. 378 pp. (In Russian.)

Shuntov, V.P. 1986a. The status of previous studies of the long-term, cyclical variation of the number of fish of the Far Eastern seas. Biologiya Morya 3:3-14. (In Russian.)

Shuntov, V.P. 1986b. Seabirds of the Okhotsk Sea. Vladivostok. DVNTs AN SSSR, pp. 6-19. (In Russian.)

Shuntov, V.P. 1993. Biological and physical determinant of marine birds distribution in the Bering Sea. The status, ecology and conservation of marine birds of the North Pacific. Canadian Wildlife Service Special Publication, Ottawa, pp. 10-17.

Shuntov, V.P. 1994. New data on the restructuring in the pelagic ecosystems of the Far Eastern sea. Vestnik DVO RAN No. 2:59-66. (In Russian.)

Shuntov, V.P. 1997. Interannual dynamics in the number and distribution of birds in the open waters of the Sakhalin-Kuril region. Izvestiya TINRO-Tsentr 122:558-570. (In Russian.)

Shuntov, V.P. 1998. Birds of the far eastern seas of Russia. Vladivostok. TINRO-Tsentr. Section I. (In Russian.)

Shuntov, V.P., and E.P. Dulepova. 1995. The current condition of the bio- and fish productivity of Bering Sea ecosystems. Kompleksnye issledovaniya ekosistemy Beringova Morya. M. Izd-vo VNIRO, pp. 358-387. (In Russian.)

Shuntov, V.P., V.I. Radchenko, E.P. Dulepova, and O.S. Temnykh. 1997. The biological resources of far eastern, Russia's economic zones: The structure of pelagic and abyssal substances and current status, tendencies of long-term dynamics. Izvestiya TINRO 122:3-15.

Shuntov, V.P., A.F. Volkov, O.S. Temnykh, and Ye.P. Dulepova. 1993. Walleye pollock in the ecosystems of far eastern seas. Vladivostok. Izd. TINRO. 426 pp.

Smirnov, G.P., and A.G. Velizhanin. 1986. The colonies of seabirds of Chukotsk National District. Byulleten' MOIP. Otd. Biologii. 91(3):29-35. (In Russian.)

Sowls, A.L., S.A. Hatch, and C.J. Lensink. 1978. Catalog of Alaskan seabird colonies. U.S. Fish and Wildlife Service FWS/OBS 78/78. p. 356.

Springer, A.M., C.P. McRoy, and M.V. Flint. 1996. The Bering Sea Green Belt: Shelf-edge processes and ecosystem production. Fisheries Oceanography 5(3-4):205-223.

Springer, A.M., A.Y. Kondrat'ev, H. Ogi, Y.V. Shibaev, and V.B. Vlient. 1993. Status, ecology and conservation of *Synthliboramphus* murrelets and auklets. In: The status, ecology and conservation of marine birds of the North Pacific. Canadian Wildlife Service Special Publication, Ottawa, pp. 187-201.

Stishov, M.S., V.I. Pridatko, and V.V. Baranyuk. 1991. Birds of Wrangel and Novosibirsk Islands. Nauka. Sibirskoe otdelenie. 254 pp. (In Russian.)

Trukhin, A.M., and G.M. Kosygin. 1987. The distribution of seabirds on the ice in the western Bering Sea and the Chukchi Sea. Rasprostranenie i biologiya morskikh ptits Dal'nego Vostoka. Vladivostok. DVO AN SSSR, pp. 6-21.

Verkhunov, A.V. 1995. The role of hydrological processes on the shelf of the Bering Sea in the formation of bioproductivity. Kompleksnye isseled. ekosis. Beringova Morya. M. Izd-vo VNIRO, pp. 52-79. (In Russian.)

Volkov, A.F. 1996. Zooplankton of the far fastern seas: Condition of the community, annual dynamics, and the significance of feeding nekton. Avtory dis. Dokt. Biol. Nauk. Vladivostok. TINRO-Tsentr. 70 pp. (In Russian.)

Vyatkin, P.S. 1986. Inventory of the colonial birds of the Kamchatka region. Morskie ptitsy Dal'nego Vostoka. Vladivostok. DVNTs AN SSSR, pp. 20-36. (In Russian.)

Vyatkin, P.S. 1992. New information about nesting fulmars on the east coast of Kamchatka. Informatsionnye materialy: izuchenie morsk. kolonial'n. ptits SSSR. Magadan. DVO RAN, pp. 29-30. (In Russian.)

Vyatkin, P.S., and Y.B. Artyukhin. 1994. Counts of the number of marine colonial birds on the Commander Islands in 1993. Informatsionnyy byulleten': Morskie ptitsy Beringii. Magadan. DVO RAN vyp. 2:40-45. (In Russian.)

Zubakin, V.A., et al., 1992. The number of seabirds on Big Diomede Islands. Informatsionnye materialy izuchenie morsk. kolonial'n. ptits SSSR. Magadan. DVO RAN, pp. 12-13. (In Russian.)

CHAPTER 30

Interdisciplinary Studies of the Bering Sea

Vera Alexander
School of Fisheries and Ocean Sciences, University of Alaska Fairbanks, Fairbanks, Alaska

The Bering Sea is a fascinating region. The name conjures up images of ice, marine mammals, severe weather, crab fisheries, Native villages, and all these do in fact describe the region. Noted as extremely productive at the upper trophic levels, the ecological basis for the Bering Sea's teeming biota has long been a puzzle. The answers are emerging slowly, and much of the progress has come about through the kind of coordinated research which we now call an "ecosystem approach." This section explores the evolution of such research and provides information on some of the significant interdisciplinary programs without trying to provide scientific details. Rather, the goals and achievements of the work are highlighted.

There have been two approaches used in interdisciplinary studies of the Bering Sea: periodic long-term cruises and comprehensive year-round ecosystem studies. Long-term cruises were done periodically over repeated cruise tracks and the data were usually seasonally incomplete and not necessarily collected annually. Examples are the joint U.S.-Russian Bering Sea and Pacific Ocean ecosystems (BERPAC) cruises of the *Akademik Korvolev* (chapter 33, this volume) and the cruises of Hokkaido University's training ship (T/S) *Oshoro Maru* (chapter 35, this volume). The former is carried out at four-year intervals, whereas the *Oshoro Maru* cruises are annual. The *Oshoro Maru* visits the Bering Sea as part of its northwestern Pacific cruise in late July of each year, with the cruise extending into August. The Russian long-term work by the Pacific Research Institute of Fisheries and Oceanography (TINRO) is another example of this approach. The second approach uses more comprehensive year-round ecosystem studies in a geographically restricted area. This approach was used in the Processes and Resources of the Bering Sea (PROBES) program (chapter 32, this volume), the Outer Continental Shelf Environment Assessment Program (OCSEAP) (chapter 36, this volume), and more recently, the Inner Shelf Transfer and Recycling (ISHTAR) program (chapter 31, this volume). These two approaches combined yield a data set which provide major insights into the dynamics of this complex marine area.

The ecosystem approach to studying the Bering Sea began early compared with other marine regions. The Bering Sea has extremely high productivity at top trophic levels and shifts in species abundance of commercial species. These factors created a "need to know" for economic reasons and was a driving force for studying this region with a comprehensive approach. PROBES was conceived to address the problem of a species abundance shift; namely to understand the system which produced a high standing stock of walleye pollock *(Theragra chalcogramma)*. By the 1970s, pollock had developed a huge biomass over the southeastern Bering Sea, an area in which the available primary productivity appeared insufficient. The "Golden Triangle" hypothesis stated that a portion of the southeastern Bering Sea shelf was a nurturing ground for pollock due to hydrographic conditions. PROBES followed the International Decade of Ocean Exploration (IDOE) pattern, but was supported by the Office of Polar Programs of the National Science Foundation. The "need to know" was not limited to programs designed for commercially important species, however. It was also relevant to the economics of oil and gas interests. The potential for oil and gas development had already spurred environmental studies (OCSEAP) which included oceanographic sampling of various kinds as well as work on birds and mammals. Nevertheless, it is safe to say that PROBES provided the major advances in ecological understanding of the southeastern Bering Sea and, by looking at new data in the context of previous physical oceanographic knowledge, advanced our understanding of the hydrographic and productivity relationships in the area. ISHTAR expanded the study of the Bering Sea shelf northward, and addressed not only regional, but also global issues relating to the connections between the Bering Sea and the Arctic Ocean. Based originally on a hypothesis which dealt with the influence of the Yukon River, attention shifted to the highly rich waters flowing north through the Anadyr Strait and through the Bering Strait. Research sponsored by OCSEAP, followed by PROBES and ISHTAR, began to demonstrate the spatial and seasonal variability in primary production in the region as well as the importance of hydrographic domains, physical transport, and ice to its biological structure and function. More recently, NOAA's Fisheries-Oceanography Coordinated Investigations (FOCI) program moved into the southeastern Bering Sea, following a successful program in Shelikof Strait, and currently this is followed up by the Southeast Bering Sea Carrying Capacity Program (SEBSCC), a focused regional ecosystem study under the Coastal Ocean Program of NOAA. A report from the latter is not included in this section since the program is in its early stages as yet, with the first major field season yet to come.

The Bering Sea has proved to be an extremely complex area and the regimes within it vary seasonally and spatially. Much more work is needed before effective management of the resources of the Bering Sea can be approached, since the vast climatic and oceanographic fluctuations and huge changes in biological components pose a challenge. The suspicion that the system is somehow in trouble is disturbing, but largely unsubstantiated, even though some marine mammal and bird populations are undergoing severe declines in numbers. A high priority must be a linking of the east and west portions of the Bering Sea, allowing effective study of biological populations which move freely across, and do not concern themselves with, national boundaries. Pollock is an example where managing the stock from one side of the Bering Sea is impossible without ecological understanding and cooperation from the other side. There continues to be a need for both approaches discussed above and it appears that the scale of the ecological work will have to expand beyond the rather circumscribed areas currently under study. Fortunately, there are programs in the planning stages which, if implemented, will allow progress. The studies described in this section provided major advances.

Interdisciplinary Bering Sea programs.

Author	Program name	Years	Funding agency
Hood, Donald W.	PROBES	1974-1982	National Science Foundation (NSF), Office of Polar Programs
Imm, Jerry, L.	OCSEAP	1974-1992	Bureau of Land Management/ Minerals Management Service; National Oceanic and Atmospheric Administration (NOAA)
McRoy, C. Peter	ISHTAR	1983-1991	NSF
Macklin, S. Allen	BS FOCI	1991-1997	NOAA, Coastal Ocean Program
Ohtani, Kiyotaka	T/S *Oshoro Maru*	1953-present	Hokkaido University
Shuntov, V.P. and Radchenko, V.I.		1983-present	TINRO
Tsyban, Alla V.	BERPAC	1977-present	U.S. Fish and Wildlife Service/ USSR State Committee for Hydrometeorology

CHAPTER **31**

Water over the Bridge: A Summing Up of the Contributions of the ISHTAR Project in the Northern Bering and Chukchi Seas

C. Peter McRoy
Institute of Marine Science, University of Alaska Fairbanks, Fairbanks, Alaska

Introduction

The Inner Shelf Transfer and Recycling (ISHTAR) Project was a multiyear, interdisciplinary project that sought to understand the processes that support the apparent high production of life in the waters between about 62° and 69°N in the Bering and Chukchi seas. All shallower than 100 m and mostly less than 50 m, this region constitutes a portion of the Inner Shelf Domain, a shelf zone expected to sustain a low overall annual production characterized by but a single spring pulse of primary production. Model predictions and the contrasting abundance of upper trophic level biota presented a challenging enigma.

The broad, shallow (less than 100 m) plateau of the northern Bering and Chukchi seas is more than 1,000 km from shelf break to shelf break, and occupies 30% of the Bering Sea and even more of the Chukchi Sea. This plateau between North America and Asia is the ancient land bridge of human migrations now flooded (Hopkins 1984). The Bering Strait in the middle of the plateau connects the Bering Sea with the Chukchi Sea and, in reality, also connects the Pacific Ocean to the arctic extension of the Atlantic Ocean (Coachman and Barnes 1961). Flow through the strait is predominantly to the north, driven by a 0.4 m difference in sea level (Overland and Roach 1987). Flow is greatest in the ice-free summer season (Coachman 1993), from mid-June through September, which determines seasonal and interannual variability and is critical to the chemical (but not physical) signature of the arctic halocline (Salmon and McRoy 1994).

The region has long been known for abundant populations of seabirds, walrus, seals, and whales that continue to sustain the subsistence lifestyle of coastal peoples (Fay 1974). Until the early 1960s, there was even a harvest of fin whales (Nasu 1974), a species characteristic of the oceanic domain and, hence, an indicator for the ocean conditions of the region. Abundance of these upper trophic level species indicates a high primary productivity that is unexpected in the inner shelf, especially where the shelf is very wide and ice-covered for most of the year (Springer and McRoy 1993).

The idea for ISHTAR was spawned by the results of earlier studies of the southeast Bering Sea, particularly the Processes and Resources of the Bering Sea Shelf (PROBES) Project (see Hood 1986, and chapter 32, this volume). The southeastern Bering Sea data for the region inside the 50 m isobath generated a cross-shelf model (McRoy et al. 1986) predicting that the carbon cycle of the inner shelf domain, in the absence of an advective nutrient source (Coachman 1986), has a single spring phytoplankton bloom fueled by nutrients from winter resupply. Concomitantly, the ecosystem in this region is limited by low primary production in the range of 50-80 g C per m^2 per year yielding a restricted food web (Walsh and McRoy 1986).

PROBES and most other shelf studies, notably the Outer Continental Shelf Environmental Assessment Program (OCSEAP), concentrated efforts on the middle and outer shelf waters of the southeastern Bering Sea so there was little supporting data for the inner shelf region. The apparent high productivity of the northern shallow shelf, as evidenced by populations of upper trophic level species, was an enigma, if not a direct contradiction, to predictions of the PROBES cross-shelf model (Springer et al. 1989, Springer and McRoy 1993). There was obvious need to reconsider the processes of the shallow shelf. Large concentrations of seabirds and marine mammals, which are characteristic of the northern Bering-Chukchi Sea region, are indicators for high productivity at the primary level in any ocean. Their presence in the Bering Strait region presented an enigma that captured the interests of the participating scientists even though these species were not targeted by the project. Hence the acronym, ISHTAR— Inner Shelf Transfer and Recycling—and the focus of the project.

The project began when the Cold War was in full progress (one of our cruises was intercepted and boarded by a Soviet military vessel) and ended in joint cruises on Soviet ships on the eve of the re-emergence of Russia. The latter event was seminal to the results of the whole program since, without such cooperation, only the models and satellite images crossed the international frontier, and our shipboard research would have been restricted to U.S. waters east of the line dividing Bering Strait.

A first analysis of historical field data from the northern Bering Sea led to a hypothesis that the reported high production was nurtured by a combination of riverine inputs from the Yukon-Kuskokwim system along the Alaska coast and supplemented by resuspension of nutrients due to the turbulence in Bering Strait (Walsh et al. 1989a). The results of the first cruise (Sambrotto et al. 1984) disproved this hypothesis and indicated that the waters west of the U.S.-U.S.S.R. frontier appeared to play a major role in the system and could not be neglected. The politics of the day presented a major roadblock to field data collection since applications for field studies on the western side of the northern Bering and Chukchi seas were routinely rejected by the Soviet government. This set the stage for the development of hydrodynamic models that then took the forefront of the project (Nihoul et al. 1990).

Scientific Program

The project was organized into ten components to test the hypothesis that interannual changes of climate forcing on water transport through Bering Strait result in two- to four-fold differences in the flux of nutrients from the Bering Sea shelf break to the Chukchi Sea (Table 1). It was hypothesized that differences also occur regionally in primary production and deposition of organic carbon, as well as energy passed up the food web to higher trophic levels. As the momentum increased, several affiliated projects used the ISHTAR platforms and data as a basis for related studies of the region (Table 2).

Field work in the project began in 1983 with a short survey cruise and ended in 1989 with a cruise to remove moored instrument arrays. Data were collected in ice-free months of 1983 to 1989. Over these years we were at sea for 441 days and occupied 2,306 oceanographic stations using six different ships. Our average was 5.2 stations per day for all ship time, port to port. In addition, we deployed 41 current meter/pressure gauge moorings, 35 with fluorometers, and six overwinter arrays. There were also two overwinter sediment trap arrays. The general hydrographic and biological data are mostly available in ISHTAR reports (Table 3).

ISHTAR investigators studied the influence of the interannual variability of physical forcing on the cycle of carbon and nutrients. The Anadyr Water in western Bering Strait originates in the southern Bering Sea and carries nutrients that sustain high seasonal production in all trophic levels. In the ice-free season, primary production is the highest reported for any portion of the world ocean. Fluctuations in transport of Anadyr Water lead to variations in carbon fixation, organic matter deposition, and mineralization on these Arctic shelves. These shelf processes contribute nutrients and organic matter to the Arctic Ocean that substantially influence biogeochemical cycles. Correlation can be made among flow through Bering Strait, climate, and deep sea ventilation in northern oceans.

Physical Transport and Hydrodynamic Models

The geography of the north Bering Sea shelf funnels three water masses into the narrow Bering Strait: Alaska Coastal Water (ACW), Bering Shelf Water (BSW), and Anadyr Water (ADW) (Coachman 1993). Ecologically, the region is essentially one of two major streams—ACW on the American side and ADW/BSW on the Russian side. The western flow (ADW/BSW) carries an oceanic fauna resulting in an east-west food web partitioning of fish, birds, and mammals. The recognition and quantitative description of this flow regime generated basic understanding of carbon and nutrient

Table 1. Core projects and investigators of ISHTAR.

Principal investigator	Institution	Project component
L.K. Coachman	U. Washington	Water Circulation and Mixing
J. Nihoul	U. Liège, Belgium	Hydrodynamic Models
C. Wirick	Brookhaven National Laboratory	Moored Instrument Arrays
C.P. McRoy	U. Alaska Fairbanks	Carbon Cycling and Productivity
T.E. Whitledge	U. Texas	Inorganic Nutrient Fields
J.J. Goering	U. Alaska Fairbanks	Pelagic Nutrient Dynamics
T.H. Blackburn	U. Aarhus, Denmark	Benthic Nutrient Mineralization
P.L. Parker	U. Texas	Stable Isotope Tracers
J.J. Walsh	U. South Florida	Ecosystem Models
C.P. McRoy	U. Alaska Fairbanks	Project Management

Table 2. Affiliated projects utilizing ISHTAR data or platforms.

Principal investigator	Institutions	Project
A.S. Naidu H.M. Feder	U. Alaska Fairbanks	Dynamics of sediments and benthos in the Chukchi Sea
R.C. Highsmith K.O. Coyle	U. Alaska Fairbanks	Ecology of benthic amphipod communities
M. Fukuchi	Natl. Institute for Polar Research, Tokyo	Particle flux variability as measured by sediment traps
A. Tsyban	Institute of Global Climate and Ecology, Academy of Sciences, Moscow	Program on long-term ecological investigations of the Bering Sea and other Pacific Ocean ecosystems (BERPAC)

Table 3.   List of ISHTAR data reports.

Report no.	Title
1	1985 Hydrographic Data STD, Nutrients, and Chlorophyll
2	1985 Current Meter and Pressure Gauge Data
3	1985 Moored Fluorometer Data
4	1986 Hydrographic Data STD, Nutrients, and Chlorophyll
6	1985-1989 Zooplankton Data
7	1986-1986 Primary Productivity Data
8	1987-1989 Primary Productivity Data
9	1987 Hydrographic Data STD, Nutrients, and Chlorophyll
10	1988 Hydrographic Data STD, Nutrients, and Chlorophyll (2 Parts)
11	1985-1989 Moored Chlorophyll *a* Fluorescence, Temperature, and Beam Attenuation Measurements
12	1986 Current Meter Data
13	1987 Current Meter Data
14	1988 Current Meter Data
15	1989 Hydrographic Data STD, Nutrients, and Chlorophyll

All reports are available at Institute of Marine Science, University of Alaska Fairbanks, Fairbanks, AK 99775.

cycling, and food web interactions. This is a major accomplishment of the project that will guide future investigations.

The three-dimensional models developed by Nihoul and colleagues (Nihoul et al. 1993) generated an architecture of the northern Bering-Chukchi system that was the foundation for subsequent studies. In this region, the geography of the ecosystem is a crucial factor in understanding the processes occurring there.

Nutrient Flux

Processes governing nutrient flux are a key to understanding the ecosystem here as elsewhere. There is a continuity within the shelf production system that extends from the southeastern Bering Sea to the Chukchi Sea (Hansell et al. 1993). Cross-shelf mixing of nitrate in winter along the shelf break in the Bering Sea determines the initial conditions for new production for the entire shelf. The location of the 90% nitrate-depletion isopleth is generally coincident with the western boundary of the ACW, reflecting the low productivity of that water. The concept of a production system extending along the shelf from the southeastern Bering Sea through Bering Strait has now been described in some detail and is known as the "Green Belt" (Springer et al. 1996).

A local nutrient source in these waters is from the mineralization of organic matter in the sediments (Henriksen et al. 1993). Nitrification rates and nutrification potentials are highest under the Anadyr system and lowest in Alaska Coastal system. Where nitrate concentration in overlying bottom water was higher than 10 µM, flux was into the sediment, and below this value flux was out of the sediment. Spatial patterns of bottom water nutrient content correlate well with sediment processes.

Primary Production

The conditions on the broad shallow northern shelf are comparable to upwelling systems of the world oceans because of the persistent supply of new nutrients to shallow water and the sustained high primary production. Indeed, for a short time in summer, possibly the global maximum rate of 14-16 g C per m^2 per day occurs (McRoy et al. 1988). Regions both north and south of Bering Strait sustain large pools of high phytoplankton biomass (Springer and McRoy 1993). The pattern of algal biomass and species clearly reflects the regional circulation (i.e., nutrient supply) and results in two contrasting production regimes that exist side by side. In ACW, production is typical of a shallow, high latitude shelf. The annual estimate is 80 g C per m^2 per day which is only about 10% of the estimate for the Anadyr system. Concomitantly, the two production systems sustain disparate food webs.

Consumers

In the pelagic realm, the real basis of the Anadyr food web is not the high diatom production of the region, though this certainly contributes to food quality, but rather it is the oceanic zooplankton that is advected into the region (Springer et al. 1989, Springer and Roseneau 1985). One result is abundant seabird colonies that are confined to the west end of St. Lawrence Island; they are absent on the east. The zooplankters feed on plankton which follows the Anadyr flow (Springer et al. 1987). The food web along the Alaska coast, characterized by small diatoms and flagellates, leads through direct herbivory to a small abundance of fish and other consumers.

While the pelagic food web is dominated by advected zooplankton, that of the benthos is directly coupled to the local primary production. Under the Anadyr waters, the settled diatoms lead to very high biomass of ampeliscid amphipods, bivalves, and other invertebrates (Grebmeier et al. 1988, 1989; Highsmith and Coyle 1990). This rich benthos is consumed by yet larger consumers, such as gray whales, seals, and walrus, that are renowned in the region, and lead to its reputation for high production

(Fay 1982, Frost and Lowry 1981). Though the benthic-related consumer biomass is large, model studies indicate that an estimated 25% of the sedimented organic matter is eventually exported from the shelf of the Chukchi Sea to the Arctic Basin (Shuert and Walsh 1993).

Organic Matter Sedimentation

Not unexpectedly, sediment conditions reflect the processes of the circulation regimes. Carbon isotope signatures in the Anadyr system reflect a marine carbon source, while in the ACW, the $\delta^{13}C$ is more negative, reflecting terrestrial carbon from riverine sources (Naidu et al. 1993). The supplemental carbon to the ACW food web is likely to be an important addition to the impoverished food web (Walsh et al. 1989b). Sediment traps in the Chirikov basin indicate that 90% of organic carbon flux, 0.5 g C per m^2 per day, can be accounted for by benthic respiration (Grebmeier and McRoy 1989).

Contributions to the Arctic Ocean

It is in this region that the Pacific (i.e., the Bering Sea) meets the Atlantic (i.e., the Chukchi Sea), and because the Pacific stands about 0.5 m higher, the flow is predominantly north across the shelves into the Arctic Ocean. This plume of Pacific water enters the Arctic Ocean as a layer of water extending from 90 to 170 m and is confined to the Canadian Basin. Using a submarine as a platform of opportunity, I was able to trace this water across the Arctic to Greenland (McRoy 1993). The Pacific water is a significant feature of the upper layers of the Arctic Ocean. The results of ISHTAR suggest that the organic production generated on the shelves and entrained in the flow makes a significant contribution to the carbon balance of the Arctic Ocean (Walsh et al. 1989b, 1997).

References

Coachman, L.K. 1986. Circulation, water masses, and fluxes on the southeastern Bering Sea shelf. Continental Shelf Research 5(1-2):23-108.

Coachman, L.K. 1993. On the flow field in the Chirikov Basin. Continental Shelf Research 13:481-508.

Coachman, L.K., and C.A. Barnes. 1961. The contribution of Bering Sea water to the Arctic Ocean. Arctic 14:147-161.

Fay, F.H. 1974. The role of ice in the ecology on marine mammals of the Bering Sea. In: D.W. Hood and E.J. Kelley (eds.), Oceanography of the Bering Sea with emphasis on renewable resources. Institute of Marine Science, University of Alaska, Fairbanks, pp. 383-399.

Fay, F.H. 1982. Ecology and biology of the Pacific walrus, *Odobenus rosmarus divergens* Iliger. U.S. Fish and Wildlife Service, Washington D.C. 279 pp.

Frost, K.J., and L.F. Lowry. 1981. Foods and trophic relations of cetaceans in the Bering Sea. In: D.W. Hood and J.A. Calder (eds.), The eastern Bering Sea shelf: Oceanography and resources, volume two. Published by the Office of Marine Pollution Assessment, National Oceanic and Atmospheric Administration and Bureau of Land Management, pp. 825-836. (Distributed by the University of Washington Press, Seattle, WA 98105.)

Grebmeier, J.M., and C.P. McRoy. 1989. Pelagic-benthic coupling on the shelf of the northern Bering and Chukchi seas. III. Benthic food supply and carbon cycling. Marine Ecology Progress Series 53:79-91.

Grebmeier, J.M., H.M. Feder, and C.P. McRoy. 1989. Pelagic-benthic coupling on the shelf of the northern Bering and Chukchi Seas. II. Benthic community structure. Marine Ecology Progress Series 51(3):253-268.

Grebmeier, J.M., C.P. McRoy, and H.M. Feder. 1988. Pelagic-benthic coupling on the shelf of the northern Bering and Chukchi Seas. I. Food supply source and benthic biomass. Marine Ecology Progress Series 48(1):57-67.

Hansell, D.A., T.E. Whitledge, and J.J. Goering. 1993. Patterns of nitrate utilization and new production over the Bering/Chukchi shelf. Continental Shelf Research 13:601-627.

Henriksen, K., T.H. Blackburn, B.A. Lomstein, and C.P. McRoy. 1993. Rates of nitrification, distribution of nitrifying bacteria and inorganic N fluxes in northern Bering-Chukchi shelf sediments. Continental Shelf Research 13(5-6):629-651.

Highsmith, R.C., and K.O. Coyle. 1990. High productivity of northern Bering Sea benthic amphipods. Nature 344:862-864.

Hood, D.W. (ed.) 1986. Processes and resources of the Bering Sea shelf (PROBES). Continental Shelf Research 5(1-2):1-288.

Hopkins, D.M. 1984. Sea-level history in Beringia during the past 250,000 years. In: V.L. Kontrimavichus (ed.), Beringia in the Cenozoic Era. Amerind Publishing Co., New Delhi, pp. 3-29.

McRoy, C.P. 1993. Pacific waters in the Arctic Ocean: Chemical tracers of the plume. Unpublished report to the U.S. Navy, Institute of Marine Science, University of Alaska Fairbanks.

McRoy, C.P., D.A. Hansell, A.M. Springer, J.J. Walsh, and T.E. Whitledge. 1988. Global maximum of primary production in the north Bering Sea. Eos 68:L1727.

McRoy, C.P., D.W. Hood, L.K. Coachman, J.J. Walsh, and J.J. Goering. 1986. Processes and resources of the Bering Sea shelf (PROBES): The development and accomplishments of the project. Continental Shelf Research 5(1-2):5-21.

Naidu, A.S., R.S. Scalan, H.M. Feder, J.J. Goering, M.J. Hameedi, P.L. Parker, E.W. Behrens, M.E. Caughey, and S.C. Jewett. 1993. Stable organic carbon isotopes in sediments of the North Bering-South Chukchi Sea, Alaska-Soviet Arctic Shelf. Continental Shelf Research 13:669-691.

Nasu, K. 1974. Movement of baleen whales in relation to hydrographic conditions in the northern part of the North Pacific Ocean and Bering Sea. In: D.W. Hood and E.J. Kelley (eds.), Oceanography of the Bering Sea with emphasis on renewable resources. Institute of Marine Science, University of Alaska, Fairbanks, pp. 345-361.

Nihoul, J.C., E. Deleersnijder, and S. Djenidi. 1990. Mathematical model of the northern Bering Sea. Liége, Belgium, IRMA (Institute de recherches marines et d'interactions aire-mer). 138 pp.

Nihoul, J.C.J., P. Adam, P. Brasseur, E. Deleersnijder, S. Djenidi, and J. Haus. 1993. Three-dimensional general circulation model of the northern Bering Sea's summer ecohydrodynamics. Continental Shelf Research 13:509-542.

Overland, J.E., and A.T. Roach. 1987. Northward flow in the Bering and Chukchi seas. Journal of Geophysical Research 92:7097-7105.

Salmon, D.S., and C.P. McRoy. 1994. Nutrient-based tracers in the western Arctic: A new lower halocline water defined. In: O.M. Johannesseen, R.D. Muench, and J.E. Overland (eds.), The polar oceans and their role in shaping the global environment. Geophysical Monograph 85. American Geophysical Union, Washington D.C., pp. 47-61.

Sambrotto, R.N., J.J. Goering, and C.P. McRoy. 1984. Large yearly production of phytoplankton in the western Bering Strait. Science 255:1147-1150.

Shuert, P.G., and J.J. Walsh. 1993. A coupled physical-biological model of the Bering-Chukchi seas. Continental Shelf Research 13(5-6):543-573.

Springer, A.M., and C.P. McRoy. 1993. The paradox of pelagic food webs in the Northern Bering Sea—III. Patterns of primary productivity. Continental Shelf Research 13:575-599.

Springer, A.M., and D.G. Roseneau. 1985. Copepod-based food webs: Auklets and oceanography in the Bering Sea. Marine Ecology Progress Series 21:229-237.

Springer, A.M., C.P. McRoy, and M.V. Flint. 1996. The Bering Sea Green Belt. Fisheries Oceanography 5:205-223.

Springer, A., C.P. McRoy, and K.R. Turco. 1989. The paradox of pelagic food webs in the northern Bering Sea—II. Zooplankton communities. Continental Shelf Research 9(4):359-386.

Springer, A., E.C. Murphy, D.G. Roseneau, C.P. McRoy, and B.A. Cooper. 1987. The paradox of pelagic food webs in the northern Bering Sea—I. Seabird food habits. Continental Shelf Research 7:895-911.

Walsh, J.J., and C.P. McRoy. 1986. Ecosystem analysis in the southeastern Bering Sea. Continental Shelf Research 5:259-288.

Walsh, J.J., D.A. Dieterle, F.E. Müller-Karger, K. Aagaard, A.T. Roach, T.E. Whitledge, and D. Stockwell. 1997. CO_2 cycling in the coastal ocean. II. Seasonal organic loading of the Arctic Ocean from source waters in the Bering Sea. Continental Shelf Research 17(1):1-36.

Walsh, J.J., C.P. McRoy, T.H. Blackburn, L.K. Coachman, J.J. Goering, K. Henriksen, P. Andersen, J.J. Nihoul, P.L. Parker, A.M. Springer, R.D. Tripp, T.E. Whitledge, and C.D. Wirick. 1989a. The role of Bering Strait in the carbon/nitrogen fluxes of polar marine ecosystems. In: L. Rey and V. Alexander (eds.), Proceedings of the Sixth Conference of the Comité Arctique International 13-15 May 1985, pp. 90-120. E.J. Brill, Leiden.

Walsh, J.J., C.P. McRoy, L.K. Coachman, J.J. Goering, J.J. Nihoul, T.E. Whitledge, T.H. Blackburn, P.L. Parker, C.D. Wirick, P.G. Shuert, J.M. Grebmeier, A.M. Springer, R.D. Tripp, D.A. Hansell, S. Djenidi, E. Deleersnijder, K. Henriksen, B.A. Lund, P. Andersen, F.E. Müller-Karger, and K. Dean. 1989b. Carbon and nitrogen cycling within the Bering/Chukchi seas: Source regions for organic matter effecting AOU demands of the Arctic Ocean. Progress in Oceanography 22(4):277-359.

CHAPTER 32

PROBES: Processes and Resources of the Eastern Bering Sea Shelf

Donald W. Hood
Professor Emeritus, University of Alaska Fairbanks, Fairbanks, Alaska

Introduction

The stimulus for Processes and Resources of the Eastern Bering Sea Shelf (PROBES) was a mutual U.S.-Japan scientific concern about the ability of the Bering Sea to support the huge annual catch, on a sustained basis, of groundfish, pelagic fish, crabs, and other marine resources. The question was large enough to demand the attention of oceanographic disciplines from both countries as well as from others. After several conferences, it was decided to emphasize the "Golden Triangle" (the area encompassed within longitudes 165° and 170°W and latitude 54°N) and the adjacent southeastern Bering Sea shelf in an attempt to understand the bioproductivity of this region (Hood and Kelley 1974, Hood and Takenouti 1975). In the beginning, it appeared most expedient, for logistic and funding reasons, to have each country pursue its part of the program independently.

PROBES (1974-1982) was designed as an interdisciplinary, multi-institutional, largely U.S. effort to track the events in a pelagic ecosystem that lead to high production in the higher trophic levels of the southeastern Bering Sea shelf. Its development is discussed in greater detail in the Continental Shelf Research volume dedicated to the PROBES effort (Hood 1986, McRoy et al. 1986).

Development of Hypothesis

Based on conventional trophic dynamics, it appeared that primary productivity was too low to support the higher trophic levels of crab, fish, birds, and mammals known to exist in the region. We reasoned that a combination of oceanographic factors, particularly the wide (500 km) shallow shelf, its inferred circulation pattern, and the timing of crucial events in the production cycle, both primary and secondary, resulted in a highly efficient energy transfer among the trophic levels.

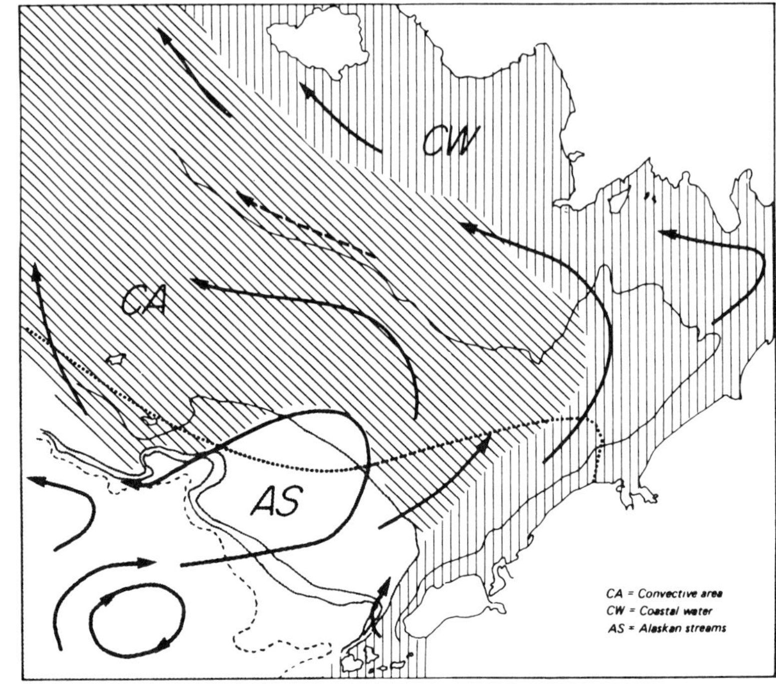

*Figure 1.   Early concepts of water masses and circulation on the southeast-
ern shelf of the Bering Sea (Takenouti and Ohtani 1974, Muench
1976). Dotted line is winter sea ice.*

Walleye pollock *(Theragra chalcogramma)* was selected as the biological tracer to check this energy transfer because it is a major fishery species and thus, considerable data were available on it. Also, during their
early life history, pollock are not very different trophically from zooplankton or the larva of fish and crabs. These plankton must have food
readily available upon hatching for their survival (Lasker 1975). Food for
these early life stages is largely provided by species of phytoplankton and
zooplankton which are dependent upon ocean conditions such as available nutrients, light, sea surface stability, circulation, salinity, temperature, growth factors, and the presence of an appropriate seed population
of phytoplankton and zooplankton. It was necessary to check as many
factors as possible to determine the appropriate oceanic conditions.

The Bering Sea has two major geomorphological features—the deep
ocean basin and the continental shelf—each occupying similar surface
areas. The sea is bounded on the south by the Aleutian Arc which permits
free exchange with Pacific Ocean water. Oceanographically, it is an exten-

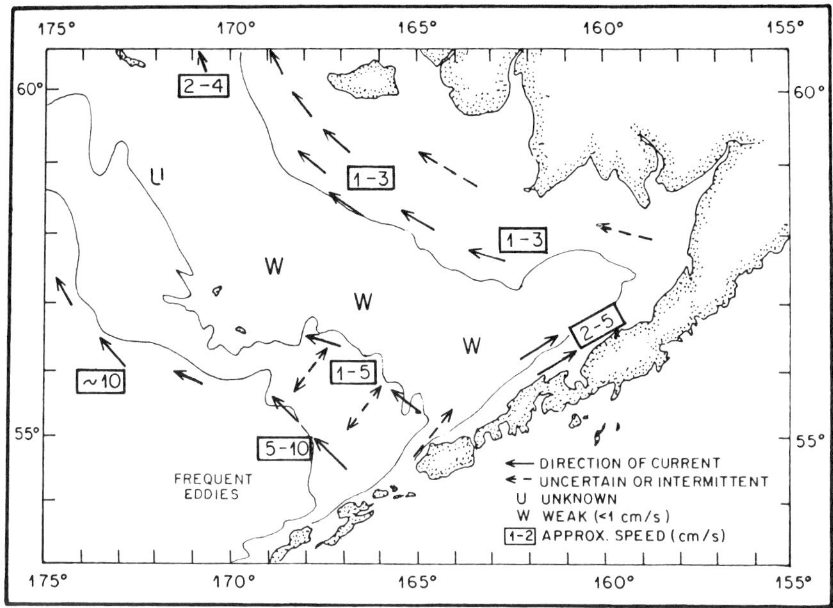

Figure 2. *Estimated longer-term circulation. The dashed arrows in the northern coastal regime suggest probable seasonal variability, while those of Unimak Island and the outer shelf domain are mostly subtidal variations, much of which are of 2 to 10 days' duration. Flow over the shelf is mostly tidal, so that the instantaneous flow is quite different from this depiction; however, it is this flow which affects the net advective transport of properties (adapted from Kinder and Schumacher 1981).*

sion of the North Pacific Ocean (Hood 1983). The pre-PROBES model of the physical circulation of the eastern Bering Sea shelf showed the Alaska Stream flowing onto the shelf creating coastal and convective area currents of great magnitude. This led to a hypothesis in which the Alaska Stream was visualized as a "river in the sea" over the shelf, beginning near Unimak Pass and flowing toward the Bering Strait (Fig. 1). Using this model, we could study the sequential events, both in space and time, that lead to efficient transfer to higher trophic levels moving downstream from the source. However, the results of several investigations occurring simultaneously with our first field season showed this characterization to be in error (Fig. 2.; Kinder and Schumacher 1981). In contrast to our concept of spatial separation of pollock eggs, larvae, and post-larvae in an upstream-downstream manner, all early stages of the fishes were occurring together.

A revised hypothesis was developed for PROBES based on a better understanding of the physical oceanography of the shelf region (Coachman

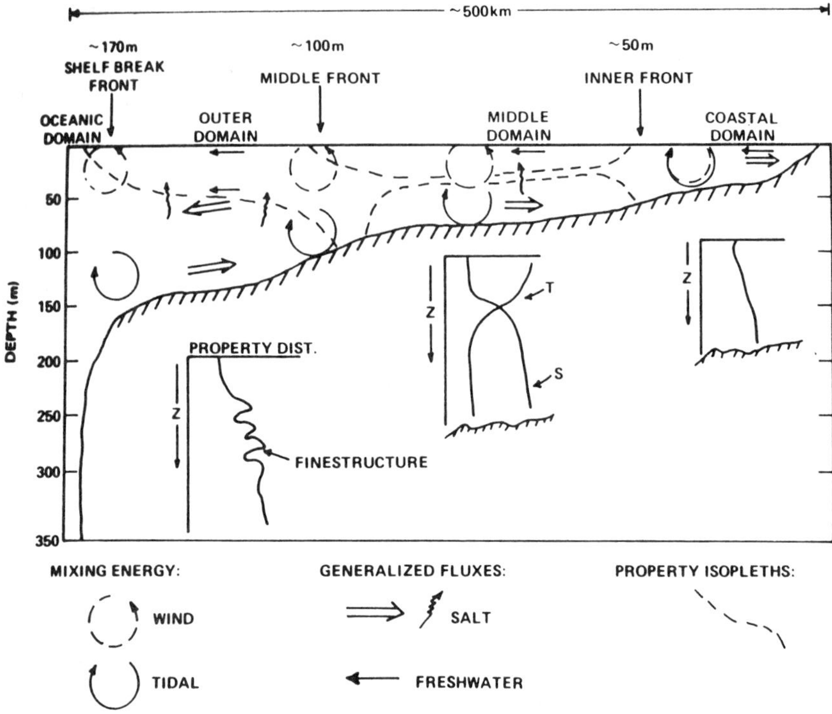

*Figure 3.   The cross-shelf advection/diffusion model developed by PROBES investi-
gators (from Coachman et al. 1980).*

1986). We developed a cross-shelf advection-diffusion model (Fig. 3; Coachman et al. 1980), proposing that this conceptual framework could explain
primary and secondary production processes and patterns. The enormous
breadth of the shelf (the widest in the world outside of the Arctic) has the
effect of laterally stretching out the hydrographic domains and cross-shelf-
active processes, exposing them to better observation (Coachman and
Charnell 1979). This allowed the advection-diffusion model to be easily
tested.

The waters over the shelf are highly structured, consisting of discrete
domains divided by oceanographic fronts (Coachman 1986). Fronts occur
where there is a change in the lateral flux rates due to alteration in water
mixing energies enhanced by the broad, shallow shelf. Two water masses
are formed on the shelf. These are the Alaskan Coastal waters which occupy the domain inshore of the 50 m isobath. This water is a mixture of land
runoff with saline basin water and tends to be vertically homogeneous
due to wind and tidal mixing. It is separated from the middle shelf domain
by the inner front that occurs at the 50 m isobath. This domain is made up

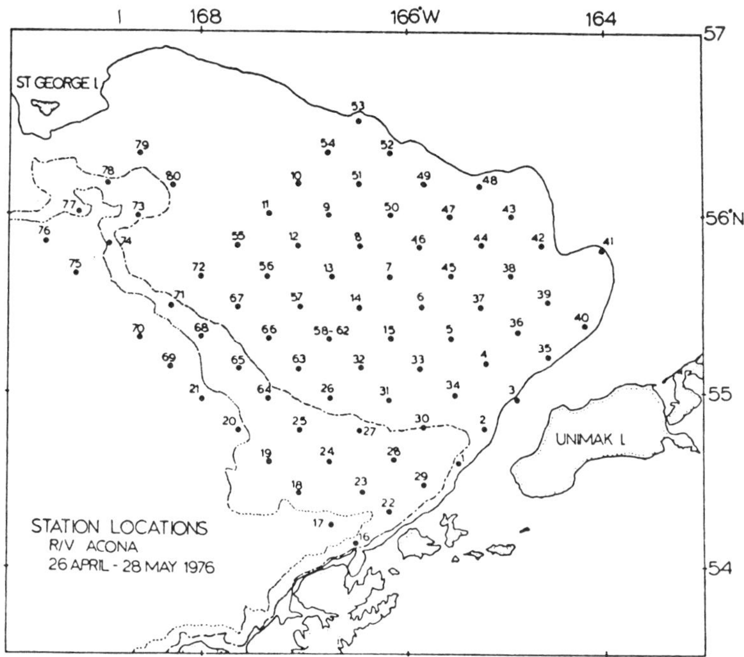

Figure 4.   The initial PROBES station grid based on the first hypothesis concerning early life history of pollock.

of saline water that diffuses across the middle front and fresher water moving seaward from the coastal domain. In the middle domain, the lower layer becomes isolated from seasonal heating, resulting in a two-layer system. The middle domain is separated from the outer domain by a front that occurs at the 100 m isobath. The outer domain does not contain an identifiable water mass but is a zone of lateral interaction between central domain shelf water and the Bering Sea basin water. The water column can be vertically stratified in which the middle layer, with little or no mixing energy, is characterized by a fine structuring of properties. Finally, the outer edge of the shelf is characterized by the shelf-break front in the upper 50 m over the 150 to 200 m isobaths and separates the central shelf from the Bering Sea water. This description of the physical system of the shelf water regime was one of the most important PROBES contributions. Instead of a station pattern that sampled the whole shelf as shown in Fig. 4, the station pattern shown in Fig. 5 was adopted. This pattern was reduced to a single main transect of 23 stations. The single main transect was designated with the letter "A" and now offers a well documented record of oceanographic conditions on a broad northern shelf. Once the physical

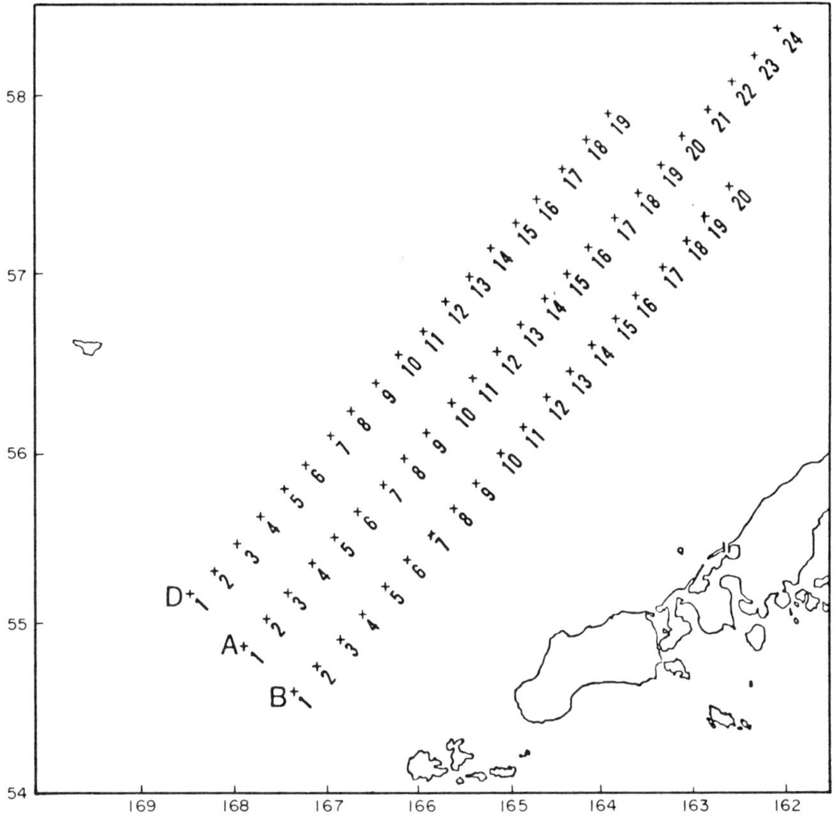

*Figure 5.   The revised PROBES station plan. The "A" line crosses all fronts and do-
mains and was the primary sampling transect.*

regime was understood, we applied its consequence to structural and functional aspects of the shelf ecosystem.

We assumed that the sources and supply of nutrients to the shelf
ecosystem were limited to the deeper waters of the outer and middle shelf
domains below the seasonal surface mixed layer. In winter, high nutrients
in all domains were expected, but with the onset of spring and the setup
of the middle and inner fronts, nutrients would be rapidly depleted by
phytoplankton in the photic zone. The spring bloom of phytoplankton
was expected to occur first in the inner and middle shelf, then to move
seaward as light increased and the water column stabilized (Hood 1981,
Codispoti et al. 1986, Sambrotto et al. 1986, Whitledge et al. 1986). Because the nutrient rich waters can be mixed upward, the primary produc-

tivity of the shelf break front, the middle front, and domain has a longer bloom period for primary producing phytoplankton. The coastal domain, limited to wintertime regeneration processes, would only have a single annual bloom.

The results of early PROBES work established the existence of two distinct zooplankton communities—a shelf group dominating the waters shoreward of the middle front and an oceanic group existing seaward of the front (Cooney and Coyle 1982). The shelf group consisted of small animals (e.g., *Pseudocalanus* spp. and *Acartia* spp.) that are year-round residents which apparently reproduce and develop large populations following the spring bloom. These animals are ineffective grazers of larger phytoplankton such as acentric diatoms. The oceanic group is composed of large herbivores (e.g., euphausiids and the large copepods, *Calanus* spp. and *Eucalanus* spp.) that find winter refuge in the deep waters beyond the shelf. In the spring, they migrate to the outer shelf but are restricted by the middle front. These animals graze on large phytoplankton cells and do not lag behind the phytoplankton development. This creates two regimes in the energy food web for the middle and outer shelf systems—one rich in detritus for bottom feeders, e.g., crabs and clams, and one where most of the production goes into pelagic herbivores, micronekton, pelagic fishes, seabirds, and mammals.

On the outer shelf, where in warm or cold years the first feeding larvae survive, primary production supports an essentially pelagic food web leading to a major fishery of adult pollock. Based on copepod ingestion (Dagg et al. 1982), respiration, and growth rates (Vidal and Smith 1986), the small zooplankton prey of larval pollock and the larger, ontogenetic migrator together remove 68 g C per m^2 per year. The rest of the outer shelf primary production evidently sinks out of the water column as phytodetritus to be consumed on the bottom or to be advected seaward. In the middle shelf, the food web is primarily benthic because of the larger amount of algae production that sinks to the sea floor; only 26 g C per m^2 per year is taken annually by zooplankton. The macrobenthos reaches its greatest abundance in the middle shelf (Haflinger 1981) and numerous benthic predators congregate there. Walrus *(Odobenus rosmarus)* are abundant and there were intensive fisheries of the yellowfin sole, *Limanda aspera* and the king crab, *Paralithodes camtschaticus* (Pereyra et al. 1976). Macrobenthic infaunal biomass at mid-shelf varies from <4 to >24 g C/m^2, tenfold that of the outer shelf. A carbon budget, and therefore an energy flow indicator, of these two systems is given in Figs. 6 and 7 (Walsh and McRoy 1986). These data are supported by the available information for marine mammals (Nasu 1974, Fay 1982), and for fish and crabs (Pereyra et al. 1976). Also, the seabird distribution would fit the food web separation scheme (Schneider et al. 1986). The PROBES "A" line, which proved so valuable in the above considerations, was now used to determine interseasonal and interannual changes, effect of weather on the mixing zone (Sambrotto et al. 1986), comparison of methods of primary production measurements

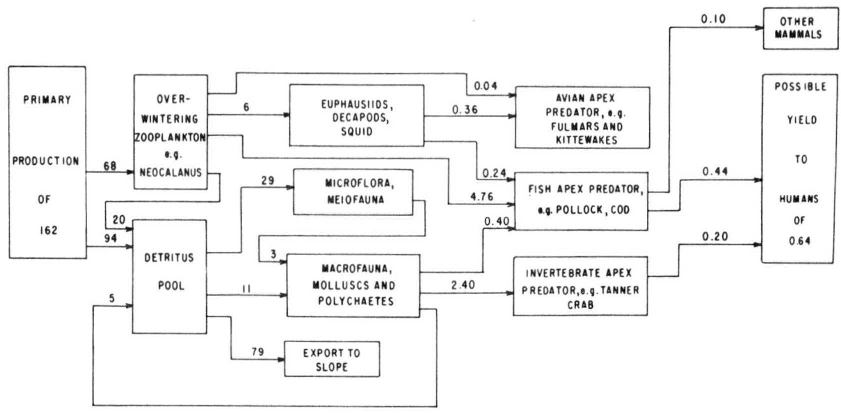

Figure 6.   An annual carbon budget (g C per m² per year) for the outer Bering Sea shelf.

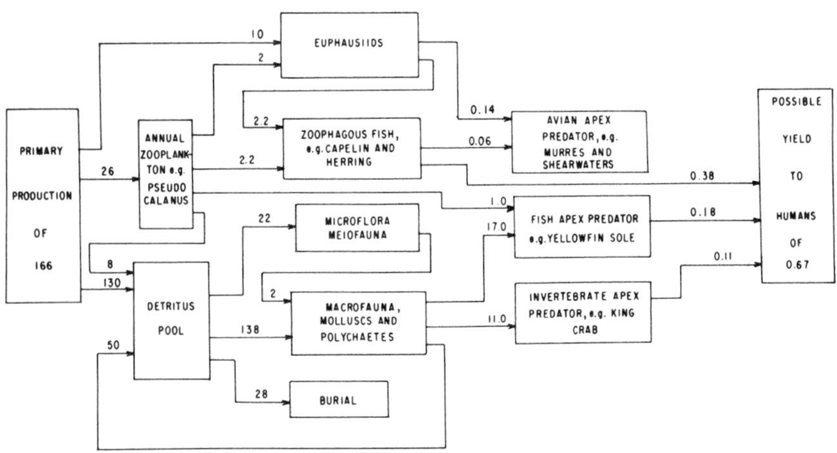

Figure 7.   An annual carbon budget (g C per m² per year) for the middle Bering Sea shelf.

(Hood 1981, Codispoti et al. 1986), variations in the zooplankton distributions (Smith and Vidal 1986), and other related matters.

The Field Program

Early in the field effort (1976, 1977, and part of 1978), the R/V *Acona* carried out most of the field observations for PROBES on the southeastern Bering Sea shelf. This 28-m vessel fought hard to carry out its functions at sea and made notable early observations (Hirano and Nishiyama 1981, Hood 1981, Niebauer et al. 1981, Hood and Codispoti 1984). Our seagoing capability was greatly improved in 1978 by the addition of the R/V *T.G. Thompson,* a 68-m vessel (AGOR 9) operated by the University of Washington. This larger ship allowed PROBES the unique opportunity to place the field work for various interdisciplinary programs on the same vessel at the same time. This proved to be most beneficial for each component of the total effort. In 1979, the R/V *Alpha Helix* replaced the retired R/V *Acona* until the end of the project in 1982.

In total, there were 2,727 stations occupied during PROBES. A standard station consisted of the following measurements: depth profiles of salinity, temperature, sigma-*t*, dissolved oxygen, nitrate, nitrite, ammonium, phosphate, silicate, total CO_2, alkalinity, and pH; also, primary production, nitrogen uptake, chlorophyll *a*, partial pressure of surface CO_2 and total CO_2, and zooplankton species and numbers. These data are listed in McRoy et al. (1986) and are available through the Rasmuson Library at the University of Alaska Fairbanks.

The project was organized according to the outline given in Fig. 7. There were 20 principal investigators and a total of 167 people contributed to the overall success of the project. Results have been published widely in the literature, as well as in reports, books, and theses. A special issue of *Continental Shelf Research* (Vol. 5, Nos. 1/2, 1986) was devoted to PROBES. Since there was no list of publications kept by the PROBES investigators, it is now difficult to evaluate the true contributions to the literature of this project, but many years after the official end of the project some outstanding papers were published from data obtained from the field work of PROBES (Hansell et al. 1989; Walsh et al. 1989a, 1989b; Sambrotto et al. 1993; Walsh and Dieterle 1994).

Accomplishments of PROBES

It has now been 15 years since the major field work of PROBES was completed. There has been time to reflect on the true accomplishments of this project and relate them to other similar efforts made to understand ocean phenomena. The following is a list of some of the contributions made by the PROBES program.

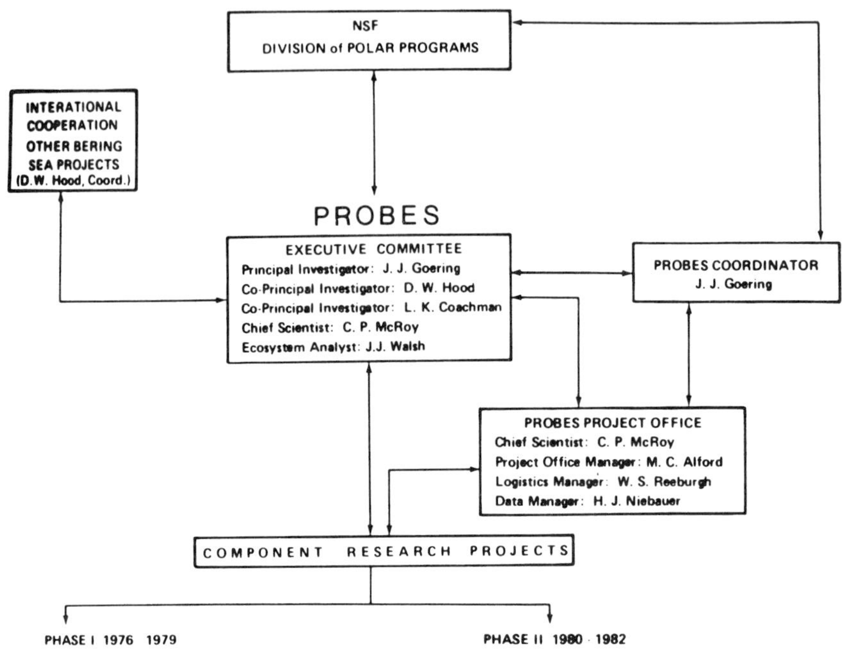

Figure 8.   The organizational structure for Phase I and Phase II of PROBES.

1. PROBES was an unusually well planned program. Several years of effort were spent in determining how to answer the question of whether the biological productivity of the Bering Sea could support the annual catch on a sustained basis. After several symposia (Hood and Kelley 1974, Hood and Takenouti 1975), books (Hood and Calder 1981), and workshops (Kelley and Hood 1974), a final working plan evolved sponsored by the National Science Foundation's Office of Polar Programs (Fig. 8).

2. The establishment of the circulation, water masses, and fluxes of the southeastern Bering Sea shelf (Kinder 1981, Coachman 1986) made possible the establishment of the PROBES "A" station line. This allowed a concentrated effort focusing on chemical-biological changes that occurred over the spring bloom period (Hood and Codispoti 1984, Codispoti et al. 1986, Sambrotto et al. 1986, Whitledge et al. 1986).

3. The cross-shelf advection/diffusion model of Coachman et al. (1980) gave the PROBES investigators the rationale to establish the inner, middle, and outer domains for the southeastern Bering Sea shelf. This explains why the pelagic and benthic food chains are located near the

physical fronts on the shelf (Pereyera et al. 1976, Cooney and Coyle 1982, Fay 1982).

4. The sampling of the PROBES "A" line from April until July on a continuous basis allowed for a greater understanding of interannual and interseasonal differences in primary and secondary production for the region. These measurements were used by numerous researchers (Sambrotto et al. 1986, Smith and Vidal 1986, Vidal and Smith 1986).

5. The inorganic carbon system (pCO_2, total CO_2) in this nitrogen (ammonium, nitrite, and nitrate) limited environment showed that when nitrogen is limiting, the organic matter synthesized increases its C/N ratio substantially over that found by Redfield et al. (1963). This ratio is important to efforts to account for the ocean's role in the global carbon cycle, and ultimately, to predict the effect on anthropogenic CO_2 emissions (Sambrotto et al. 1993).

6. PROBES developed a new approach in oceanographic research that proved most useful in solving problems on the shelf environment. This was the use of a single vessel with several disciplines participating in a single experiment. The R/V *T.G. Thompson* had the size and sea-keeping capability to support the scientists needed to make a complete station. This meant that physical, meteorological, chemical, and biological data were all available from the same water samples. Because of this uniqueness of sample data, several disciplines outside the PROBES venue were able to join these efforts to help answer their own specific questions (Schneider and Hunt 1982, Incze 1983, Schneider et al. 1986).

PROBES was managed by a five member executive committee (Fig. 8) which had authority to shift emphasis (and funding) among components or to add new ones to the program. Annually, the committee reported accomplishments and plans for future research to the Office of Polar Programs. This mechanism of management kept the program focused on specific ecosystem-related questions and allowed for answering these questions from different perspectives.

PROBES occurred at a point in oceanographic research history when different disciplines in the field were attempting to get together with their research interests to answer specific environmental questions. The barriers between disciplines broke down and biologists were talking to physicists, chemists to biologists, etc., in a very productive way. This was facilitated by the executive committee and also extended itself into the personnel's attitude about the project. I give credit to Dr. Larry K. Coachman, now deceased, for being a great stimulus and one of the few physical oceanographers who considered biological problems intellectually challenging.

Conclusion

PROBES was successful in combining various people, institutions, and disciplines to carry out a well planned scientific experiment. We were not constrained to continue a component to its completion unless it was contributing to the total objective. The first two years of PROBES were somewhat chaotic since the circulation of the shelf waters had not been established and because, based on the new information on circulation, the model of "rivers in the sea" had to be abandoned for the "lake in the sea" concept. The flexibility in the project allowed for a quick retreat and new plans were quickly made to better approach the new observations.

The program developed what amounted to a benchmark for research on the continental shelves of the world. The manner in which it was carried out and the results obtained by it serve as a pattern for scientific endeavor on the oceanic environment.

References

Coachman, L.K. 1986. Circulation, water masses, and fluxes on the southeastern Bering Sea shelf. Continental Shelf Research 5:23-108.

Coachman, L.K., and R.L. Charnell. 1979. On lateral water mass interaction: A case study, Bristol Bay, Alaska. Journal of Physical Oceanography 6:278-297.

Coachman, L.K., T.H. Kinder, J.D. Schumacher, and R.B. Tripp. 1980. Frontal systems of the southeastern Bering Sea shelf. In: T. Carstens and T. McClimans (eds.), Stratified flows, second IAHR symposium, Trondheim, June 1980, Vol. 2. TAPIR, pp. 917-933.

Codispoti, L.A., G.E. Friederich, and D.W. Hood. 1986. Variability in the inorganic carbon system over the southeastern Bering Sea shelf during spring 1980 and spring-summer 1981. Continental Shelf Research 5:133-160.

Cooney, R.T., and K.O. Coyle. 1982. Trophic implications of cross-shelf copepod distributions in the southeastern Bering Sea. Marine Biology 70:187-196.

Dagg, M.J., J. Vidal, T.E. Whitledge, R.L. Iverson, and J.J. Goering. 1982. The feeding, respiration, and excretion of zooplankton in the Bering Sea during a spring bloom. Deep-Sea Research 29:45-63.

Fay, F.H. 1982. Ecology and biology of the Pacific walrus, *Odobenus rosmarus divergens* Illiger. North American Fauna 74, U.S. Fish and Wildlife Service. 279 pp.

Haflinger, K. 1981. A survey of benthic infaunal communities of the southeastern Bering Sea shelf. In: D.W. Hood and J.A. Calder (eds.), The eastern Bering Sea shelf: Oceanography and resources, Vol. 2. Office of Marine Pollution Assessment, NOAA, pp. 1091-1103. (Distrib. by University of Washington Press, Seattle.)

Hansell, D.A., J.J. Goering, J.J. Walsh, C.P. McRoy, L.K. Coachman, and T.E. Whitledge. 1989. Summer phytoplankton production and transport along the shelf break in the Bering Sea. Continental Shelf Research 9:1085-1104.

Hirano, K., and T. Nishiyama. 1981. Species composition and zooplankton by depth at stations 15, 16 and 31 in southeast Bering Sea (*Acona* Cruise 278), 30 May-8 June 1979. PROBES Data Report PDR81-001. Institute of Marine Science, University of Alaska, Fairbanks. 10 pp.

Hood, D.W. 1981. Preliminary observations of the carbon budget of the eastern Bering Sea shelf. In: D.W. Hood and J.A. Calder (eds.), The Bering Sea shelf: Oceanography and resources, Vol. 1. Office of Marine Pollution Assessment, NOAA, pp. 347-358. (Distrib. by University of Washington Press, Seattle.)

Hood, D.W. 1983. The Bering Sea. In: B.H. Ketchum (ed.), Estuaries and enclosed seas, Ecosystems of the world, Vol. 26. Elsevier, pp. 337-373.

Hood, D.W. (ed.). 1986. Processes and resources of the Bering Sea shelf (PROBES). Continental Shelf Research (Spec. Issue) 5:1-288.

Hood, D.W., and J.A. Calder (eds.). 1981. The eastern Bering Sea shelf: Oceanography and resources, Vol. 1 and 2. Office of Marine Pollution Assessment, NOAA and BLM. 1,339 pp. (Distrib. by University of Washington Press, Seattle.)

Hood, D.W., and L.A. Codispoti. 1984. The effect of primary production on the carbon dioxide components of the Bering Sea shelf. In: J.H. McBeath (ed.), The potential effects of carbon dioxide–induced climatic changes in Alaska. Miscellaneous Publication 83-1, School of Agriculture and Land Resource Management, University of Alaska, Fairbanks, pp. 33-39.

Hood, D.W., and E.J. Kelly (eds.). 1974. Oceanography of the Bering Sea with emphasis on renewable resources. Occasional Publication No. 2, Institute of Marine Science, University of Alaska, Fairbanks. 623 pp.

Hood, D.W., and Y. Takenouti (eds.). 1975. Bering Sea oceanography: An update 1972-1974. Rep. 75-2, Institute of Marine Science, University of Alaska, Fairbanks. 292 pp.

Incze, L.S. 1983. Larval life history of Tanner crabs, *Chionoecetes bairdi* and *C. opilio,* in the southeastern Bering Sea and relationships to regional oceanography. Ph.D. thesis, University of Washington, Seattle. 192 pp.

Kelley, E.J., and D.W. Hood. 1974. PROBES: A prospectus on processes and resources of the Bering Sea shelf 1975-1985. University of Alaska Sea Grant, AK-SG-73-10, Fairbanks. 71 pp.

Kinder, T.H. 1981. A perspective of physical oceanography in the Bering Sea, 1979. In: D.W. Hood and J.A. Calder (eds.), The Bering Sea shelf: Oceanography and resources, Vol. 1. Office of Marine Pollution Assessment, NOAA, pp. 5-13. (Distrib. by University of Washington Press, Seattle.)

Kinder, T.H., and J.D. Schumacher. 1981. Circulation over the continental shelf of the southeastern Bering Sea. In: D.W. Hood and J.A. Calder (eds.), The Bering Sea shelf: Oceanography and resources, Vol. 1. Office of Marine Pollution Assessment, NOAA, pp. 53-75. (Distrib. by University of Washington Press, Seattle.)

Lasker, R. 1975. Field criteria for survival of anchovy larvae: The relation between inshore chlorophyll maximum layers and successful first feeding. Fisheries Bulletin 73:453-462.

McRoy, C.P., D.W. Hood, L.K. Coachman, J.J. Walsh, and J.J. Goering. 1986. Processes and resources of the Bering Sea shelf (PROBES): The development and accomplishments of the project. Continental Shelf Research 5:5-21.

Muench, R.D. 1976. A note on eastern Bering Sea shelf hydrographic structure, August 1974. Deep-Sea Research 23:245-247.

Nasu, K. 1974. Movement of baleen whales in relation to hydrographic conditions in the northern part of the North Pacific Ocean and the Bering Sea. In: D.W. Hood and E.J. Kelley (eds.), Oceanography of the Bering Sea. Occasional Publication No. 2, Institute of Marine Science, University of Alaska, Fairbanks, pp. 345-361.

Niebauer, H.J., C.P. McRoy, and J.J. Goering. 1981. May 1976, May-June, September-October 1977, R/V *Acona* cruises 227, 242, 243, 250 and 251, hydrographic (bottle) data. PROBES Data Report PDR81-003, Institute of Marine Science, University of Alaska, Fairbanks. 205 pp.

Pereyra, W.T., J.E. Reeves, and R.G. Bakkala. 1976. Demersal fish and shellfish resources of the eastern Bering Sea in the baseline year 1975. Northwest Fisheries Science Center, National Marine Fisheries Service. 619 pp.

Redfield, A.C., B.H. Ketchum, and F.A. Richards. 1963. The influence of organisms on the composition of seawater. In: M.N. Hill (ed.), The sea, Vol. 2. Wiley, pp. 26-77.

Sambrotto, R.N., H.J. Niebauer, J.J. Goering, and R.L. Iverson. 1986. Relationships among vertical mixing, nitrate uptake, and phytoplankton growth during the spring bloom in the southeast Bering Sea middle shelf. Continental Shelf Research 5:161-198.

Sambrotto, R.N., G. Savidge, C. Robinson, P. Boyd, T. Takahashi, D.M. Karl, C. Langdon, D. Chipman, J. Marra, and L. Codispoti. 1993. Elevated consumption of carbon relative to nitrogen in the surface ocean. Nature 363:248-250.

Schneider, D., and G.L. Hunt. 1982. Carbon flux to seabirds in waters with different mixing regimes in the southeastern Bering Sea. Marine Biology 67:337-344.

Schneider, D.C., G.L. Hunt Jr., and N.M. Harrison. 1986. Mass and energy transfer to seabirds in the southeast Bering Sea. Continental Shelf Research 5:241-257.

Smith, S.L., and J. Vidal. 1986. Variations in the distribution, abundance, and development of copepods in the southeastern Bering Sea in 1980 and 1981. Continental Shelf Research 5:215-239.

Takenouti, A.Y., and K. Ohtani. 1974. Currents and water masses in the Bering Sea: A review of Japanese work. In: D.W. Hood and E.J. Kelley (eds.), Oceanography of the Bering Sea. Occasional Publication No. 2, Institute of Marine Science, University of Alaska, Fairbanks, pp. 39-57.

Vidal, J., and S.L. Smith. 1986. Biomass, growth and development of populations of herbivorous zooplankton in the southeastern Bering Sea during spring. Deep-Sea Research 33:523-556.

Walsh, J.J., and D.A. Dieterle. 1994. CO_2 cycling in the coastal ocean. I. A numerical analysis of the southeastern Bering Sea with applications to the Chukchi Sea and the northern Gulf of Mexico. Progress in Oceanography 34:335-392.

Walsh, J.J., and C.P. McRoy. 1986. Ecosystem analysis of the southeastern Bering Sea. Continental Shelf Research 5:259-288.

Walsh, J.J., C.P. McRoy, T.H. Blackburn, L.K. Coachman, J.J. Goering, K. Henriksen, P. Andersen, J.J. Nihoul, P.L. Parker, A.M. Springer, R.D. Tripp, T.E. Whitledge, and C.D. Wirick. 1989a. The role of Bering Strait in the carbon/nitrogen fluxes of polar marine ecosystems. In: L. Rey and V. Alexander (eds.), Proceedings of the Sixth Conference of the Comité Arctique International 13-15 May 1985, pp. 90-120. E.J. Brill, Leiden.

Walsh, J.J., C.P. McRoy, L.K. Coachman, J.J. Goering, J.J. Nihoul, T.E. Whitledge, T.H. Blackburn, P.L. Parker, C.D. Wirick, P.G. Shuert, J.M. Grebmeier, A.M. Springer, R.D. Tripp, D.A. Hansell, S. Djenidi, E. Deleersnijder, K. Henriksen, B.A. Lund, P. Andersen, F.E. Müller-Karger, and K. Dean. 1989b. Carbon and nitrogen cycling within the Bering/Chukchi Seas: Source regions for organic matter effecting AOU demands of the Arctic Ocean. Progress in Oceanography 22:277-359.

Whitledge, T.E., W.S. Reeburgh, and J.J. Walsh. 1986. Seasonal inorganic nitrogen distributions and dynamics in the southeastern Bering Sea. Continental Shelf Research 5:109-132.

CHAPTER **33**

The BERPAC Project: Development and Overview of Ecological Investigations in the Bering and Chukchi Seas

Alla V. Tsyban
Institute of Global Climate and Ecology, Moscow, Russia

Development of the BERPAC Project— Its Goals and Objectives

The scientific concept of the BERPAC project was developed gradually based on earlier Bering Sea studies (Ratmanov 1937a, 1937b; Natarov 1963; Neiman 1963; Arsenev 1967; Hood and Kelley 1974; Hood and Calder 1981; McRoy et al. 1986). Intensive joint investigations (experimental and analytical) by Russian and American scientists in the Bering and Chukchi seas led to new hypotheses and revisions of the project's scientific direction. This resulted in the creation of a new scientific program—the BERPAC project, originally called the Long-Term Ecological Research of Marine Ecosystems in the Arctic and Pacific Oceans, then shortened to BERPAC (O'Connor et al. 1992).

The goal of the BERPAC project was to study the status and dynamics of arctic marine ecosystems in light of anthropogenic impacts and possible climate change. To achieve this goal, the following basic objectives were formulated:

1. Investigation of oceanographic and hydrochemical processes in the Bering and Chukchi sea ecosystems.

2. Study the biological processes occurring in the pelagic and benthic environments.

3. Study of the biogeochemical cycles of contaminants and assessment of the ecological consequences of pollution in the Bering and Chukchi seas.

4. Study of the processes determining the assimilative capacity of arctic marine ecosystems in respect to contaminants.

5. Assessment of the ecological consequences of other anthropogenic impacts, including possible climate changes, in the Bering and Chukchi seas.

The BERPAC project has achieved certain advantages over other international scientific projects aimed at studying marine ecosystems. The most important of these are:

1. Long-term (about 20 years) series of observations.

2. Constant recurrence of investigations in the same regions.

3. An interdisciplinary approach.

4. Conduct of joint expeditions, at regular intervals, and discussion of scientific results at many bilateral seminars and symposia in Russia and the United States.

5. Systematic publication of monographs generalizing the data of bilateral investigations, both field research and laboratory experiments (Izrael and Tsyban 1983, 1990, 1992; Roscigno 1990; Nagel 1992).

Specialists at many scientific institutions in the USSR/Russia and United States have participated in activities within the framework of the BERPAC project over the past 20 years (Table 1).

BERPAC Project Expeditions

The joint Soviet/Russian-American investigations were started in the Bering Sea in August 1977. Six integrated ecological expeditions have been carried out in the Bering Sea (Figs. 1 and 2), Chukchi Sea (Fig. 2), eastern Siberian Sea, and in the North Pacific Ocean (Fig. 3). The total duration of the cruises was 375 days and 1,870 standard oceanographic and ecological stations were occupied onboard Russian research vessels of 4,500-7,500 t displacement. Approximately 50-60 scientists and technicians participated in the field work. They represented various oceanographic disciplines including meteorology, hydrochemistry, hydrology, hydrobiology, microbiology, ecology, and geochemistry. This interdisciplinary approach made possible a comprehensive study of the marine ecosystem using modern oceanographic methods. Each subsequent expedition was expanded and compared to prior ones, and often included new scientific lines of investigation. It is significant that many processes in different expeditions were studied by the same specialists since the basic composition of the participants remained unchanged.

During 20 years of the BERPAC project, 19 scientific meetings and symposia were held. The programs, results of joint expeditions, and plans for the preparation of joint monographs were discussed.

Table 1. Specialists at many scientific institutions have participated in the BERPAC project.

USSR/Russia	United States
Goskomgidromet of the USSR/ Rosgidromet	U.S. Department of the Interior, Fish and Wildlife Service
USSR Academy of Sciences/ Russian Academy of Sciences	Louis Calder Conservation and Ecology Study Center, Fordham University
Academy of Sciences of the Ukrainian SSR/ Ukrainian Academy of Sciences	Patuxent Wildlife Research Center, U.S. Fish and Wildlife Service
Academy of Sciences of the Byelorussian SSR/Byelorussian Academy of Sciences	Institute of Marine Science, University of Alaska Fairbanks
Academy of Sciences of the Estonian SSR/ Estonian Academy of Sciences	Institute of Marine Science, Florida Institute of Oceanography, St. Petersburg
Academy of Sciences of the Uzbek SSR/ Uzbek Academy of Sciences	University of New England, Biddeford, Maine
Moscow State University	Marine Science Institute, University of Texas
Board of Hydrometeorology of Lithuania Goskomgidromet of USSR/ Lithuanian Academy of Sciences	Great Lakes Science Center, Department of the Interior, National Biological Service
	University of Tennessee

Scientific Results

To assess the ecological status of the Bering and Chukchi seas, the BERPAC project focused on three interconnected areas of study. The first included processes that determine the balance between the formation of organic matter and its breakdown within an ecosystem, as well as the biomass of various trophic groups. The second involved studying circulation and levels of hazardous contaminants in the marine environment. The third involved natural processes aimed at the removal and destruction of contaminants and, ultimately, determining their fate within marine ecosystems.

Dynamics of Some Biological Processes in the Bering and Chukchi Seas

Current investigations show that the Bering and southern Chukchi seas are some of the most productive, not only in the Arctic but in the World

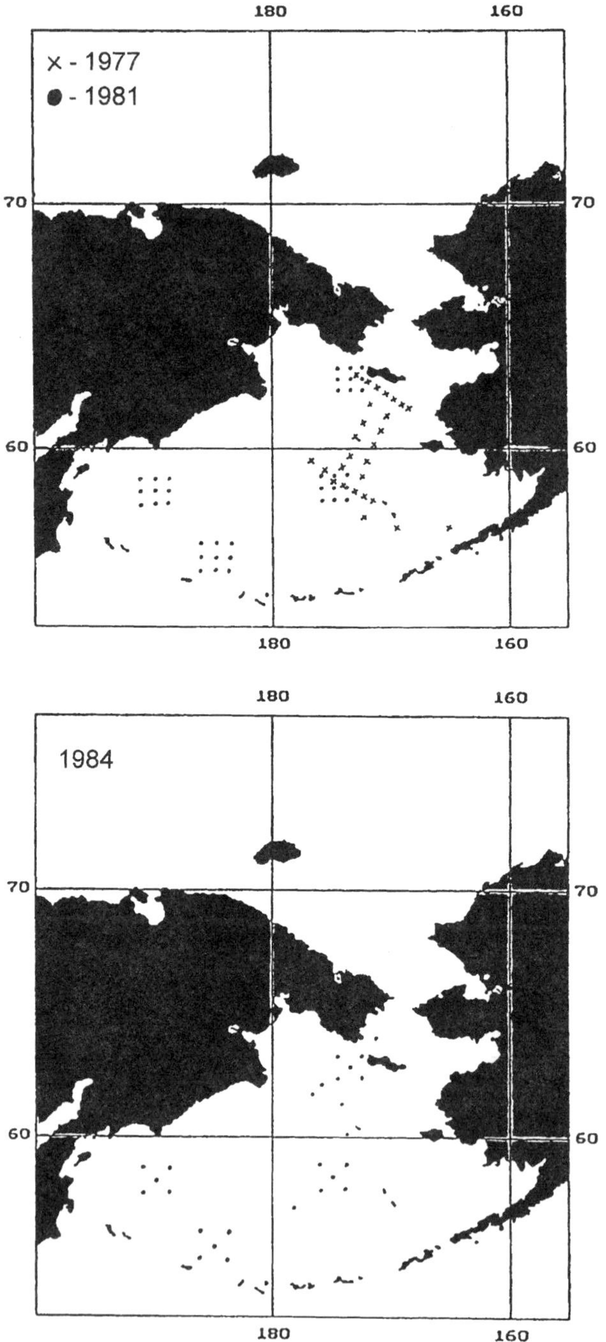

Figure 1.   Stations in the Bering Sea—1977, 1981, and 1984.

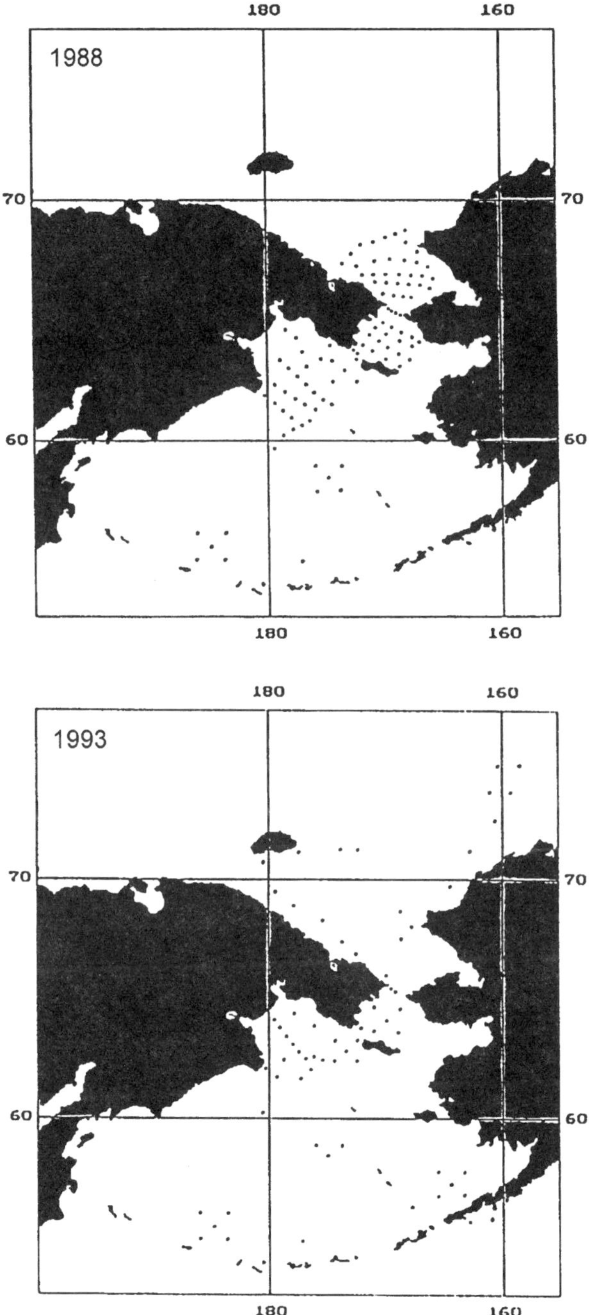

Figure 2. Stations in the Bering and Chukchi seas—
1988 and 1993.

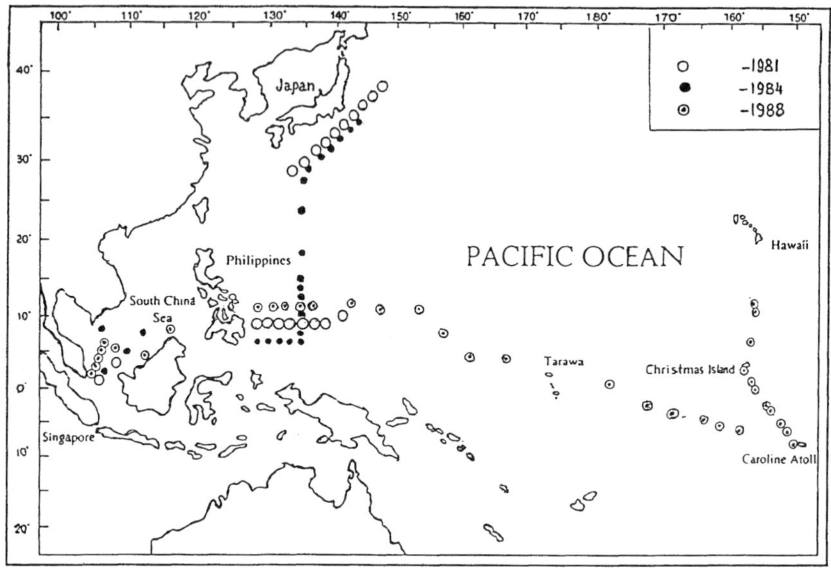

Figure 3.   Stations in the North Pacific Ocean—1981, 1984, and 1988.

Ocean as a whole. For example, according to estimates by Russian and American scientists, the annual primary production in the Bering Sea reaches the order of 2.0×10^8 t C, which accounts for 1.5% of the total volume of organic matter fixed through photosynthesis in the ocean (Tsyban et al. 1990, Izrael and Tsyban 1992).

Long-term investigations of the Bering Sea carried out in the framework of the BERPAC program allowed estimation of the levels and dynamics of production/destruction processes, and to characterize the quantitative characteristics of plankton communities (Table 2) (McLaughlin et al. 1977; Izrael and Tsyban 1989, 1990, 1992; Roscigno 1990; Nagel 1992; BERPAC-93 1993). These studies also enhanced our understanding of the carbon biochemical cycle. When the data were analyzed, it became obvious that the interannual variability of production proved to be rather distinct, especially in the southern part of the sea, which is subject to the influence of Pacific waters. In 1991-1993, the average phytoplankton production for the entire Bering Sea varied from 0.5 to 2.2 g C per m² per day. Maximum values of primary production were found in the northern Bering Sea (near St. Lawrence Island). In August 1988, primary production in the northern Bering Sea reached 15 g C per m² per day, whereas the highest values in the southern Chukchi Sea amounted to 4-5 g C per m² per day.

In 1981-1993, the rate of organic matter breakdown by bacteria in the Bering Sea varied from 0.7 to 4.8 g C per m² per day in the 0-45 m depth

Table 2. Elements of the biotic balance of the Bering Sea ecosystems from the results of long-term investigations.

Indices	Month and year of investigation	South	East	North	West	Chirikov Basin
Primary production (P_p), g C per m^2 per day	06.1981	0.32	0.53	0.87	0.49	–
	07.1984	0.62	0.43	1.95	0.56	–
	08.1988	1.83	0.92	7.87	–	1.03
	08.1993	3.09	2.55	1.87	–	2.68
Bacterial destruction (D), g C per m^2 per day	06.1981	1.30	0.73	0.86	0.64	–
	07.1984	1.70	1.60	1.44	2.20	–
	08.1988	4.80	3.60	1.70	–	
	08.1993	4.48	2.28	5.32	–	2.57
Biosedimentation (F), g C per m^2 per day	06.1981	0.35	0.18	0.76	0.46	–
	07.1984	0.54	0.44	0.55	0.56	–
	08.1988	1.02	0.75	0.29	–	0.42
	08.1993	0.70	0.64	0.68	–	0.52
Bacterioplankton biomass (B_b), g C per m^2 per day	06.1981	0.52	0.22	0.45	0.24	–
	07.1984	0.10	0.16	0.12	0.14	–
	08.1988	0.70	1.00	0.77	–	
	08.1993	1.42	1.69	1.48	–	0.83
Phytoplankton biomass (B_p), g C per m^2 per day	06.1981	0.80	1.50	1.00	2.20	–
	07.1984	1.60	1.00	2.10	0.95	–
	08.1988	0.32	1.40	1.64	–	3.03
	08.1993	*	*	*	–	*
Zooplankton biomass (B_z), g C per m^2 per day	06.1981	3.70	5.50	1.10	4.00	–
	07.1984	3.50	3.50	1.10	4.40	–
	08.1988	1.80	1.90	1.40	–	1.00
	08.1993	1.67	0.84	2.29	–	3.78
P_p/B_p	06.1981	0.40	0.35	0.87	0.22	–
	07.1984	3.50	3.50	1.10	4.40	–
	08.1988	0.50	1.20	0.80	–	0.34
	08.1993				–	
P_p/D	06.1981	0.25	0.73	1.01	0.77	–
	07.1984	0.36	0.27	1.35	0.25	–
	08.1988	0.02	0.26	4.63	–	
	08.1993	0.69	1.12	0.35	–	1.04

NOTE: The complete results of the statistical analysis of the quantitative estimates of the parameters are not presented since the work is in review.

* Under investigation.

range (Izrael and Tsyban 1992, BERPAC-93 1993). This variability also occurred interannually and interseasonally by region. For example, the lowest rate of organic matter breakdown was observed in 1984 in the southern and central portions of the Bering Sea. The highest values were found in the same regions in 1988 and 1993 (although in 1993 levels in the southern and central Bering Sea proved to be 1.8-2 times lower than in 1988). Note that the daily organic matter breakdown by bacteria in the southern and central Bering Sea exceeded primary production values by 1.5 and 3 times, respectively.

Studies in 1981-1993 showed that phytoplankton biomass averaged 0.9-1.4 g C/m^2 in all the areas studied in the Bering Sea. Maximum values of microalgae biomass (1.0-2.1 g C/m^2) and the lowest values (0.3-0.4 g C/m^2) were found in the spring-summer session in the northern and in the southern-western sections, respectively. The wide variations of phytoplankton biomass were caused by differences in primary productivity and by differing patterns of succession in the plankton communities.

The distribution of microflora (which was obtained for this region for the first time) has shown a considerable zonality typical for water bodies with a complicated water mass structure. The total number of bacteria in the Bering Sea varied over a wide range—from 0.12 to 3.3×10^6 cells/ml—the average biomass of microorganisms in different study areas of the Bering Sea ranged from 0.1 to 1.0 g C/m^2. The highest density of bacteria was observed in the northern portion, and in the absence of a clearly defined water mass layer, the vertical distribution of bacteria was rather uniform. A substantially lower level of microbial production was found in the southern region under the influence of Pacific waters, as well as a more complex vertical structure in the bacterioplankton.

In the Chukchi Sea, the total number of microorganisms varied from 0.31 to 2.0 million cells/ml. The most developed microbiocenoses were found in the offshore waters of the Chukot Peninsula and Alaska.

Zooplankton biomass varied from 0.8 to 5.5 g C/m^2 in the 100 m surface layer over the deep Bering Sea basin in spring and summer. The highest values were found in surface water in the Chirikov Basin (see Table 2).

Further investigations are necessary to clarify the temporal scale of variability of the biological processes occurring in the ecosystems of the Bering and Chukchi seas. This work is important in the study of the ecological consequences of global warming in the arctic seas and dictates the necessity for a comprehensive study of the biological processes constituting the carbon cycle of these seas.

Chemical Pollution of Some Bering Sea Ecosystems

The study of biogeochemical cycles of pollutants is an important objective of the BERPAC program. It is aimed at the investigation of the chemical regime and assessment of the assimilative capacity of its ecosystem in respect to anthropogenic pollution.

Over the past 16 years, the distribution of organic pollutants in the Bering Sea has increased annually. The concentrations of hexachlorocyclo-hexanes (HCHs) in water samples exceeded the concentrations of other chlorinated hydrocarbons such as polychlorinated biphenyls (PCBs) and DDTs (Izrael and Tsyban 1992, BERPAC-93 1993, Bidleman et al. 1995).

In 1993, the mean concentrations of α- and γ-HCH in the Bering Sea were 2.00 ng/L and 0.16 ng/L. Data on the source of isomeric HCH composition (a considerable amount of α-isomer over γ-isomer) show that HCHs are transported to these regions in the atmosphere and come from southeast Asia where hexachlorane is used. Hexachlorane is 85% saturated with the α-isomer.

Recent investigations (Bidleman et al. 1995) show that the atmospheric concentrations of HCH isomers have decreased considerably in the last few years; however, the α-HCH content of seawater has remained at the same level over the last 5 years, while the γ-HCH has decreased four-fold (Hinckley et al. 1992, Bidleman et al. 1995). This suggests that the Bering and Chukchi seas are losing their function as an HCH sink and are becoming a new source of HCHs for the arctic atmosphere.

Pollution of the Bering Sea ecosystem by PCBs causes serious concern. Studies in 1984 show that PCB concentrations varied in surface waters from 0.5 to 0.8 ng/L. However, in 1988 their concentrations were lower and varied from 0.2 to 0.6 ng/L, but in 1993 they increased again to 0.6-1.1 ng/L (Izrael and Tsyban 1990, Chernyak et al. 1992, BERPAC-93 1993). It is worth noting that chlorinated hydrocarbons have been discovered in Chukchi Sea ice: 3.4 ng/L HCH, 0.016 ng/L DDTs, and 0.9 ng/L PCBs.

The long residence time of chlorinated hydrocarbons (several decades) and their resistance to microbial degradation (especially in low temperature conditions; Izrael and Tsyban 1989), determines their active circulation along food webs and accumulation in marine organisms, including commercial species. For example, in 1984 the PCB content of particulate matter in the Bering Sea ranged from 2.7 to 5.4 ng/L dry weight; whereas in 1988 and 1993 the PCB content ranged from 2.3 to 7.0 ng/L dry weight. The coefficients of PCB accumulation in particulate matter, plankton, and neuston samples were 10^2-10^5.

Systematic investigations of the biogeochemical cycle of benzo[*a*]-pyrene—an indicator of carcinogenic polycyclic aromatic hydrocarbons (PAHs)—began in 1977 in the Bering Sea and a little later in the Chukchi Sea. Benzo[*a*]pyrene was found in all components of the marine ecosystem even though concentrations abruptly decreased in the last few years (in 1981, the most frequently encountered benzo[*a*]pyrene concentrations amounted to 20-40 ng/L; recently they decreased to 3-5 ng/L). However, the coefficients of benzo[*a*]pyrene accumulation in the particulate matter and biota remain rather high: 10^2-10^3 (Tsyban et al. 1986a; Izrael and Tsyban 1990, 1992; BERPAC-93 1993).

The distribution of heavy metals (copper, cadmium, manganese, zinc, and lead) in the water, bottom sediments, and biota of the Bering and

Chukchi seas shows much lower concentrations (up to 2.05 µg of copper, up to 3.69 µg of cadmium, etc.) than, for example, in the Black and Baltic seas (Izrael and Tsyban 1992).

The deleterious effects of chemical pollution in the Bering and Chukchi seas can be determined by monitoring the number and distribution of certain indicator microorganisms, i.e., the heterotrophic microorganisms that are able to destroy organic toxic pollutants at the expense of changes in the genotype. The number of indicator microorganisms present reflects the mutation dynamics in the microbial populations and hence, ecosystem variability. The number of these forms is not yet high in the Bering and Chukchi seas; however, their distribution has increased from year to year and they are now found almost everywhere (Tsyban 1974, 1980; Izrael and Tsyban 1989; Tsyban et al. 1992d)

Long-term (1981-1993) research in a region termed the "polygon north" (Fig. 4) regarding the number and vertical distribution of the PCB-destroying microflora shows that the number of these indicator bacteria has increased by 1-2 orders of magnitude, and their distribution has widened. These findings are indicative of the gradual pollution of the sea by chlorinated hydrocarbons.

Figures 5 and 6 show HCH concentrations as well as the amount of microflora that break down HCH within the same regions of the Chukchi Sea. It appears that indicator microflora distribution resembles that of HCH isomers, which is also indicative of changes in plankton communities. Undoubtedly, this is one of the negative consequences of chlorinated hydrocarbon pollution in some regions in the Bering and Chukchi seas. (Earlier studies on toxic and genotoxic properties of indicator microflora are important here [Tsyban et al. 1992a]. Indicator microflora play a leading role in the breakdown and transformation of organic pollutants in the ocean [Tsyban 1974, Seki 1982, Tsyban et al. 1992c]).

Role of Biological Processes in the Transport and Elimination of Pollutants

Studies during the last decades established the leading role of marine microorganisms in the breakdown and transformation of different organic substances and contaminants in the ocean. Bacterial degradation of organic pollutants is of particular importance in high-latitude regions of the World Ocean where other natural processes, such as chemical and photochemical oxidation, are only moderate in low temperatures and limited insolation.

In the Bering and Chukchi seas, assessment of microbial degradation of aromatic and chlorinated hydrocarbons has been carried out in the last few decades. The results of the investigations have been published in a number of monographs and articles (Izrael and Tsyban 1989, 1990, 1992; Izrael et al. 1990). Note the following points:

1.   In the Bering and Chukchi seas, from 8 to 65% of benzo[*a*]pyrene is destroyed at the expense of microbial degradation. The highest activity

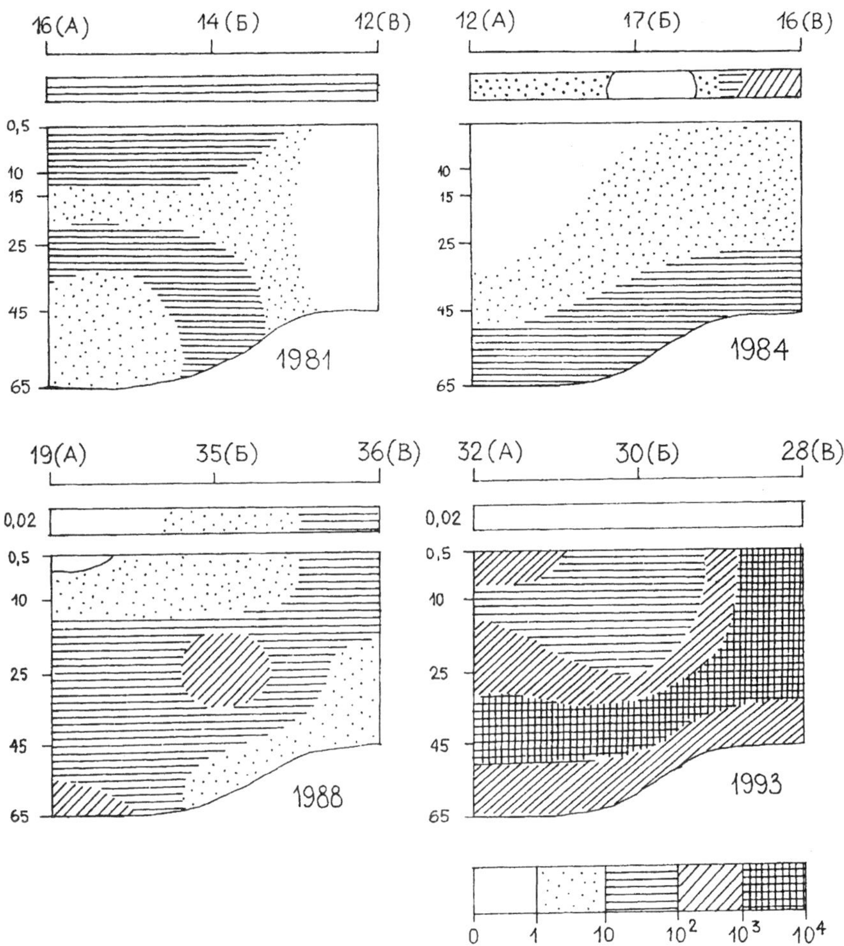

Figure 4. Vertical distribution of PCB-transforming bacteria at three stations of Polygon North in 1981, 1984, 1988, and 1994.

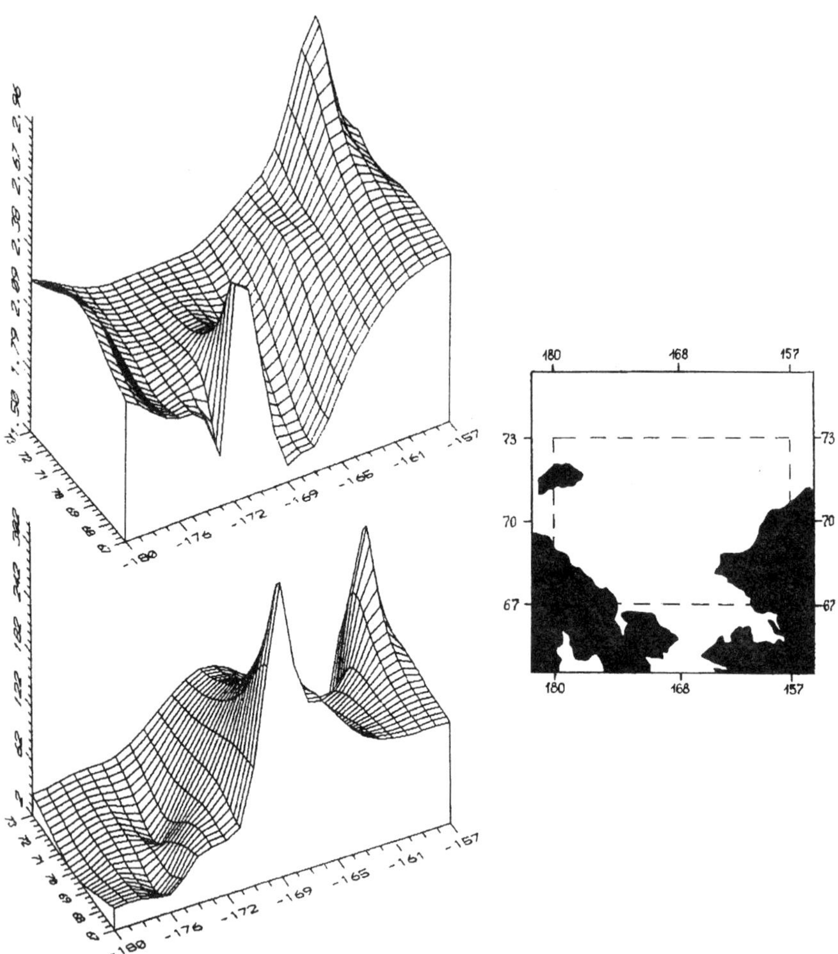

Figure 5. Concentration of α-HCH (ng/L) (top) and number of HCH-oxidizing microorganisms (cells/ml) (bottom) in the Chukchi Sea (1993).

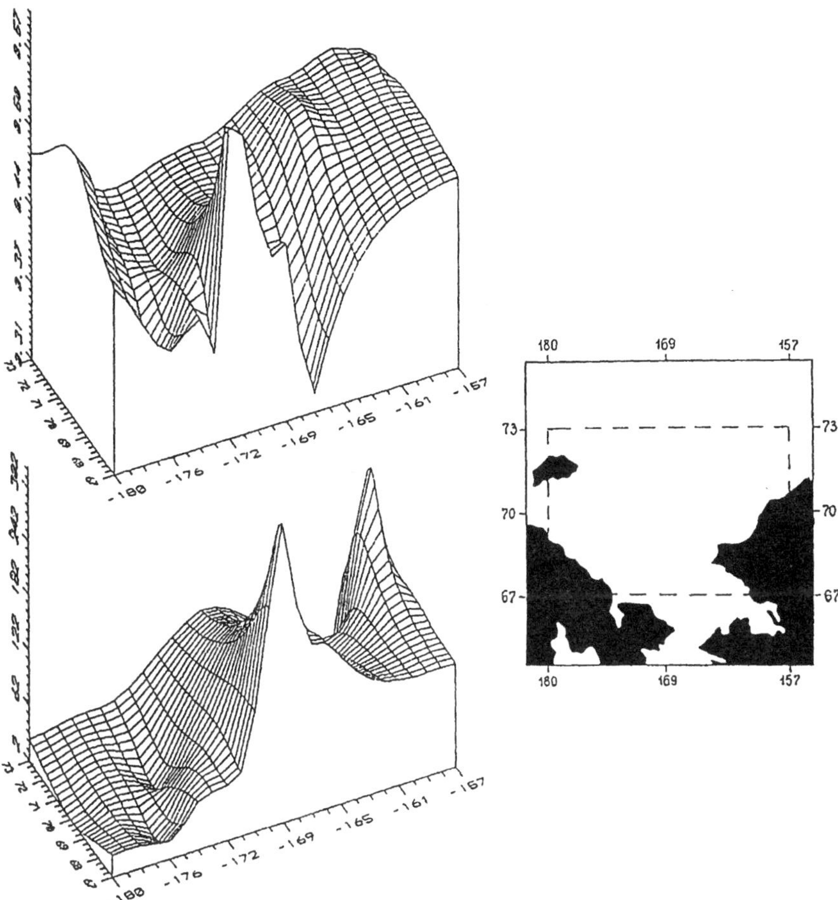

Figure 6.   Concentration of γ-HCH (ng/L) (top) and number of HCH-oxidizing micro-organisms (cells/ml) (bottom) in the Chukchi Sea (1993).

of microbial populations was in the northern Bering Sea (about 70%) and in the southern Chukchi Sea (about 80%). The results of long-term monitoring of benzo[*a*]pyrene biodegradation (and the quantitative assessment of its fluxes) shows that in summer bacterioplankton in the 0-45 m layer of the northern Bering Sea are able to transform from 20 to 90 µg benzo[*a*]pyrene under 1 m²/day. In the southeastern East Siberian Sea, heterotrophic microflora are able to transform benzo[*a*]pyrene at a level of 47-58%.

2.   Contrary to the widespread idea of PCB stability in northern water, extensive samples on the decomposition of a whole group of PCB coherents by marine microorganisms were obtained during the BERPAC expeditions. Only low-chlorinated coherents, from mono- to tetrachlorobiphenyls, which account for about 18% of the sum total of all the PCB coherents, are subject to pronounced microbial transformation. For example, in experimental situations, 95-100% of dichlorobiphenyls are destroyed in 20 days, while the concentration of some hexachlorobiphenyls only decreased by 7%. High-chlorinated PCB components containing more than 6 atoms of chlorine in molecules proved resistant to microbial degradation in low-temperature conditions (Tsyban et al. 1992b).

3.   In 1993, the microbial degradation of α- and γ-HCH in the Bering and Chukchi seas was studied for the first time. In contrast to polychlorinated biphenyls, these compounds are subjected to more active microbial degradation. For example, in the northern Bering Sea at 62°N, the microflora on the surface were able to transform up to 40% of the entire α-HCH input for five days (the initial concentration was 40 ng/L).

Thus, a substantial portion (from 40-85%) of recalcitrant organic pollutant is not subject to microbial transformation in arctic seas, and accumulates actively in marine organisms and bottom sediments. At the same time, microorganisms are not the only component of the marine biota with the ability to destroy organic pollutants. Protozoa, zooplankton, bacteria, and macrophytes have some ability for assimilating organic pollutants, such as petroleum.

One of the most important processes for purification of the seas is the accumulation of toxicants on particulate organic matter and the deposition of both under gravity (biosedimentation) (Elder and Fowler 1977, Izrael and Tsyban 1989). Biogenic particles ranging in size from several to hundreds of millimeters remove from the photic layer a great amount of pollutants and facilitate their redistribution with depth, elimination from the water column, and burial in bottom sediments.

In the Bering and Chukchi seas as a whole, researchers conclude that high biosedimentation intensity is associated with high biological productivity. Investigations carried out during a number of years in different seasons show that periods of high biosedimentation intensity are extended and cover a large fraction of the growing period. This stable flux ensures the existence of rich bottom fauna in shallow regions of the Bering and Chukchi seas (Glebov et al. 1992).

Table 3.   Biosedimentation flux of PCBs from the upper 40 m layer of the Bering and Chukchi Seas in August 1993.

Study area	Avg flux, g dry weight per m² per day	Flux of PCBs, ng per m² per day
Bering Sea	2.4	1200
Bristol Bay	2.6	1800
Southern Sea	3.5	1750
Polygon South	3.2	1600
Polygon East	3.2	1600
Polygon Navarin	3.5	1725
Gulf of Anadyr	3.4	1700
Chirikov Basin	2.6	1300
Chukchi Sea		
Western Sea	3.0	1475
Southern Sea	3.5	1722

Studies to estimate the biosedimentation fluxes of PCBs from the productive layers of the Bering and Chukchi seas were conducted using data from studies on the content of organic pollutants in particulate matter (Table 3). Results showed that up to 20% of organic carbon accumulates in the bottom sediments in the Chirikov Basin (Walsh et al. 1989). The rest of the organic matter decomposes to mineral compounds or returns to the water in the form of transformed organic compounds. The regularities of accumulation and decomposition of organic matter in the benthic communities can also be extrapolated in the first approximation to organic pollutants. In this case, PCB deposition in the Gulf of Anadyr can amount to 390-420 ng per m² per day, in the Chirikov Basin to 260 ng per m² per day, and in the Chukchi Sea to 240-350 ng per m² per day.

From the above discussion it follows that due to the process of biodegradation and biogenic sedimentation, organic pollutants (5-65%) are destroyed, transformed, and removed from the photic layers of the Bering and Chukchi seas and deposited in the bottom sediments. However, they can return to the water and food webs. These problems and the basic scientific results of activities in the framework of the BERPAC project are published in Russia and in the United States in a large number of scientific journals, reviews, and in eight monographs (McLaughlin et al. 1977; Tsyban et al. 1986a, 1986b; Izrael and Tsyban 1989, 1990, 1992; Izrael et al. 1988; Grebmeier and McRoy 1989; Roscigno 1990; Nagel 1992; BERPAC-93 1993; Grebmeier 1993; Springer and McRoy 1993; Shuert and Walsh 1993;

Bidleman et al. 1995; Jantunen and Bidleman 1995; Chernyak et al. 1996; and other publications).

Studies in the Bering and Chukchi seas raise many questions and challenges. At the XIX BERPAC Symposium held in Moscow in September 1996, the perspectives for further development of joint work in the framework of the BERPAC project were outlined. This included preparation and publication of a joint monograph, as well as planning for new experiments and expeditions.

Acknowledgments

Many people, institutions, and agencies have devoted their time and efforts to the BERPAC project. It is with great pleasure that the Russian scientists participating in the project express their deep gratitude to the American scientists and representatives of the American administration whose professional skills and ready help have contributed to the development of the BERPAC project. Russian participants of BERPAC express their sincere appreciation to the U.S. project leader, H.J. O'Connor, whose productive term in this position has contributed to the strengthening of the BERPAC project.

References

Arsenev, V.S. 1967. Currents and water masses in the Bering Sea. Nauka Publishing House, Moscow. 137 pp.

BERPAC-93. 1993. Report on the preliminary results of the fourth Russian-American expedition BERPAC-93 (58 cruise, R/V *Okean*), IGCE. 127 pp. (In Russian.)

Bidleman, T.F., L.M. Jantunen, R.L. Falconer, L.A. Barrie, and P. Fellin. 1995. Decline of hexachlorocyclohexanes in the arctic atmosphere and reversal of the air-water gas exchange. Geophysical Research Letters 22(3):219-222.

Chernyak, S.M., C.P. Rice, and L.L. McConnell. 1996. Evidence of currently used pesticides in air, ice, fog, seawater and surface microlayer in the Bering and Chukchi seas. Marine Pollution Bulletin 32(5):410-419.

Chernyak, S.M., Vronskaya, T.P. Kolobova. 1992. Migratory and cumulative peculiarities in the biogeochemical cycling of chlorinated biocarbons. In: P.A. Nagel (ed.), Results of the third joint U.S.-U.S.S.R. Bering and Chukchi seas expedition (BERPAC), Summer 1988. U.S. Fish and Wildlife Service, Washington DC, pp. 279-284.

Elder, D.L. and S.W. Fowler. 1977. Polychlorinated biphenyls: Penetration into the deep ocean by zooplankton fecal pellet transport. Science 197:459-561.

Glebov, B.V., V.I. Medinetz, and V.G. Solovev. 1992. Biogenic sedimentation. In: A.Y. Izrael and A.V. Tsyban (eds.), Ecosystem investigation of the Bering and Chukchi seas. Gidrometeoizdat, St. Petersburg, pp. 421-431. (In Russian.)

Grebmeier, J.M. 1993. Benthic carbon cycling in the Bering and Chukchi seas. Continental Shelf Research 13:653-668.

Grebmeier, J.M., and C.P. McRoy. 1989. Pelagic-benthic coupling on the shelf of the northern Bering and Chukchi seas. III. Benthic food supply and carbon cycling. Marine Ecology Progress Series 53:79-91.

Hinckley, D.A., T.F. Bidleman, and C.P. Rice. 1992. Long-range transport of atmosphere organochlorine pollutants and air-sea exchange of hexachlorocyclohexane. In: P.A. Nagel (ed.), Results of the third joint US-USSR Bering and Chukchi seas expedition (BERPAC), Summer 1988. U.S. Fish and Wildlife Service, Washington DC, pp. 267-278.

Hood, D.W. and J.A. Calder (eds.). 1981. The eastern Bering Sea shelf: Oceanography and resources, Vol. I and II. Published by the Office of Marine Pollution Assessment, National Oceanic and Atmospheric Administration and Bureau of Land Management. Distributed by the University of Washington Press, Seattle, WA 98105. 1,339 pp.

Hood, D.W. and E.J. Kelley (eds.). 1974. Oceanography of the Bering Sea with emphasis on renewable resources. Occasional Publication No. 2, Institute of Marine Science, University of Alaska, Fairbanks. 623 pp.

Izrael, Y.A., and A.V. Tsyban (eds.). 1983. Investigation of the Bering Sea ecosystem. Gidrometeoizdat, Leningrad. 157 pp. (In Russian.)

Izrael, Y.A., and A.V. Tsyban. 1989. Anthropogenic ecology of the ocean. Gidrometeoizdat, Leningrad. 528 pp. (In Russian.)

Izrael, Y.A., and A.V. Tsyban (eds.). 1990. Investigation of the Bering Sea ecosystem. Gidrometeoizdat, Leningrad. 344 pp. (In Russian.)

Izrael, Y.A., and A.V. Tsyban (eds.). 1992. Investigation of the ecosystems of the Bering and Chukchi seas. Gidrometeoizdat, St. Petersburg, 656 pp. (In Russian.)

Izrael, Y.A., A.V. Tsyban, G.V. Panov, and S.M. Chernyak. 1990. Microbial transformation of polychlorinated biphenyls in polar marine regions. DAN SSR 310(N2):502-505. (In Russian.)

Izrael, Y.A., A.V. Tsyban, G.V. Panov, M.N. Korsak, V.M. Kudravtsev, Y.L. Volodkovich, and S.M. Chernyak. 1988. Comprehensive analysis of the Bering Sea ecosystem. In: Proceedings of the fifth Soviet-American Symposium, Washington DC, December 1986. Gidrometeoizdat, Leningrad, pp. 77-111. (In Russian.)

Jantunen, L. M., and T. F. Bidleman. 1995. Reversal of the air-water gas exchange direction of hexachlorocyclohexanes in the Bering and Chukchi seas: 1993 versus 1988. Environmental Science and Technology 29(4):1081-1089.

McLaughlin, J.J.A., A.Y. Izrael, R.E. Putz, A.V. Tsyban, et al. (eds.). 1977. Joint USA-USSR ecosystem investigation of the Bering Sea, July-August, 1977. Project 02.05-41 Biosphere Reserves "Bering Sea Studies." Library of Congress Catalog Number 82-084513. 271 pp.

McRoy, C.P., D.W. Hood, L.K. Coachman, J.J. Walsh, and J.J. Goering. 1986. Processes and resources of the Bering Sea shelf (PROBES): The development and accomplishments of the project. Continental Shelf Research 5:5-21.

Nagel, P.A. (ed.) 1992. Results of the third joint US-USSR Bering and Chukchi seas expedition (BERPAC), Summer 1988. U.S. Fish and Wildlife Service, Washington, DC.

Natarov, U.V. 1963. Water masses and currents at the Bering Sea. Trudy VNIRO 48:111-133. (In Russian.)

Neiman, A.A. 1963. Quantitative benthic distribution on the shelf and upper slope horizons in the eastern part of the Bering Sea. VNIRO 48:145-206.

O'Connor, H.J., Y.A. Izrael, A.V. Tsyban, T.E. Whitledge, C.P. McRoy, and L.K. Coachman. 1992. Program on long-term ecological investigations of the Bering Sea and other Pacific Ocean ecosystems (BERPAC). In: P.A. Nagel (ed.), Results of the third joint US-USSR Bering and Chukchi seas expedition (BERPAC), Summer 1988. U.S. Fish and Wildlife Service, Washington DC, pp. 3-6.

Ratmanov, G.E. 1937a. On the hydrology of the Bering and Chukchi seas. In: Investigations of the USSR seas. Gidrometeoizdat, N25, pp. 110-118. (In Russian.)

Ratmanov, G.E. 1937b. On the problem of water exchange through the Bering Strait. In: Investigations of the USSR seas. Gidrometeoizdat, N25, pp. 120-137. (In Russian.)

Roscigno, P.E. (ed.) 1990. Results of the second joint US-USSR Bering Sea expedition, Summer 1984. U.S. Fish and Wildlife Service, Biological Report 90(13). 317 pp.

Seki, H. 1982. Organic materials in aquatic ecosystems. CRC Press, Inc., Boca Raton.

Shuert, P.G. and J.J. Walsh. 1993. A coupled physical/biological model of the Bering-Chukchi seas. Continental Shelf Research 13:543-573.

Springer, A.M., and C.P. McRoy. 1993. The paradox of pelagic food webs in the northern Bering Sea. III. Patterns of primary production. Continental Shelf Research 13:575-599.

Tsyban, A.V. 1974. Bacterioneuston and problem of degradation in surface films of organic substances released into sea. Progress in Water Technology 7(N3/4).

Tsyban, A.V. 1980. Scientific aspects of the organization of biological monitoring of the Baltic Sea ecosystem state. In: Integrated global monitoring of environmental pollution. Proceedings of the International Symposium, Riga, December 12-15, 1978. Gidrometeoizdat, Leningrad, pp. 332-340. (In Russian.)

Tsyban, A.V., V.A. Ivanitsa, G.V. Khudchenko, and G.V. Panov. 1992a. Characteristic of the predominant microflora in the open Bering Sea. In: Ecosystem investigation of the Bering and Chukchi seas. Gidrometeoizdat, Leningrad. (In Russian.)

Tsyban, A.V., M.N. Korsak, Y.L. Volodkovich, and J. McLaughlin. 1986a. Ecological investigations in the Bering Sea. Integrated Global Ocean Monitoring. Proceedings of the first International Symposium, Tallin, USSR, October 2-10, 1983. Gidrometeoizdat.

Tsyban, A.V., C.P. Rice, G.V. Panov, and S.M. Chernyak. 1992b. General conclusion. In: Ecosystem investigation of the Bering and Chukchi seas. Gidrometeoizdat, Leningrad, pp. 638-641. (In Russian.)

Tsyban, A.V., M.V. Venttsel, G.V. Panov, Y.L. Volodkovich, M.N. Korsak, and N.V. Kov-
aleva. 1986b. Long-term ecological investigations in impact and background
regions of the World Ocean. In: Proceedings of the Third International Symposium, Integrated Global Monitoring of the Biosphere State, Tashkent, October
14-19, 1985. Gidrometeoizdat, Leningrad, pp. 45-60. (In Russian.)

Tsyban et al. 1990. World Ocean and coastal zones. In: Climate change: The IPCC
impacts assessment. Report of WGII. Australian Government Publishing Service, Canberra, Australia, pp. 6-1-6-27.

Tsyban et al. 1992c. Microorganisms and microbiological process. In: Ecosystem
investigation of the Bering and Chukchi seas. Gidrometeoizdat, Leningrad,
pp. 93-212. (In Russian.)

Tsyban et al. 1992d. Heterotrophic saprophytic microflora. In: Ecosystem investigation of the Bering and Chukchi seas. Gidrometeoizdat, Leningrad, pp. 143-
171. (In Russian.)

Walsh, J.J., C.P. McRoy, L.K. Coachman, J.J. Goering, J.J. Nihoul, T.E. Whitledge, T.H.
Blackburn, P.L. Parker, C.D. Wirick, P.G. Shuert, J.M. Grebmeier, A.M. Springer,
R.D. Tripp, D.A. Hansell, S. Djenidi, E. Deleersnijder, K. Henriksen, B.A. Lund, P.
Anderson, F.E. Müller-Karger, and K. Dean. 1989. Carbon and nitrogen cycling
within the Bering/Chukchi seas: Source regions for organic matter effecting
AOU demands of the Arctic Ocean. Progress in Oceanography 22:277-359.

CHAPTER 34

Bering Sea FOCI

S. Allen Macklin
Pacific Marine Environmental Laboratory, Seattle, Washington

Introduction

Bering Sea Fisheries-Oceanography Coordinated Investigations (BS FOCI) was a coastal fisheries ecosystems project of NOAA's Coastal Ocean Program (Boyles and Scavia 1993, Wenzel and Scavia 1993) from 1991 through 1997. BS FOCI was established to seek understanding of the factors that control abundance of fish populations, with a specific goal of reducing uncertainty in resource management decisions through ecological research on walleye pollock *(Theragra chalcogramma)* recruitment and stock structure in the Bering Sea. Using a competitive process, BS FOCI awarded research components to academic scientists from several universities and to scientists from NOAA's Alaska Fisheries Science Center and Pacific Marine Environmental Laboratory. BS FOCI developed collaborations with Russian, Korean, and Japanese scientists. The National Research Council (1994) reviewed accomplishments and plans of BS FOCI.

For the past two decades, pollock has been the most abundant component of the Bering Sea fishery in the United States. Historically a non-U.S. endeavor, the Bering Sea groundfish fishery experienced significant domestic growth during the 1980s. The groundfish catch by American fishermen in the U.S. Exclusive Economic Zone (EEZ) increased from 42,000 t in 1981 to over 2 million t in 1991 when the ex-vessel (unprocessed fish) value was more than $350 million. Eighty percent of the catch by weight was pollock. Prior to 1980, most of the pollock resources in the North Pacific were taken within U.S. and Soviet waters. As both countries tightened fishing regulations, foreign fisheries were forced to exploit pollock in international waters of the central basin of the Bering Sea ("donut hole," Fig. 1). The 1988 catch in the "donut hole" was over 1 million t. This fishery was unregulated and the impact of extensive exploitation on U.S. zone fisheries was unknown. During the early 1990s the "donut hole" fishery experienced decreasing yields and was closed by international moratorium.

BS FOCI built on knowledge established by studies in the early 1970s addressing the international fisheries, and, from the mid-1970s, focused

PMEL contribution 1773, FOCI contribution B287.

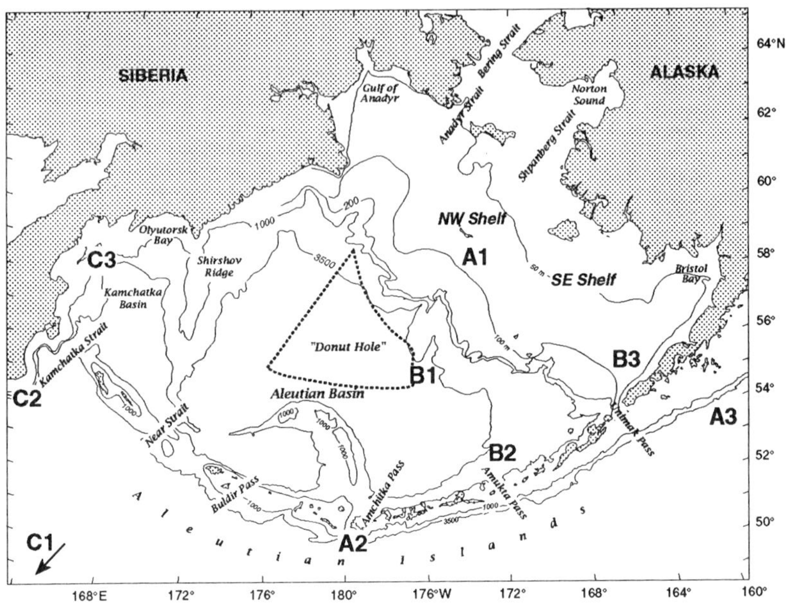

Figure 1.   The Bering Sea showing the international area known as the "donut hole" and sampling locations (A1, A2, etc.) for genetic testing of pollock.

on programs assessing shelf resources, ecology, and sea-ice dynamics. These latter programs were the Outer Continental Shelf Environmental Assessment Program (OCSEAP; Hood and Calder 1981), Processes and Resources of the Bering Sea Shelf (PROBES; Hood 1986), Inner Shelf Transfer and Recycling (ISHTAR; Coachman and Hansell 1993), and Bering Sea Marginal Ice Zone Experiment (MIZEX; Muench 1983). Information on pollock and the biophysical processes that affect the species in Shelikof Strait, Gulf of Alaska (Kendall et al. 1996), and associated studies in the Bering Sea (Schumacher and Kendall 1995), provided guidance for formulation of hypotheses for BS FOCI.

The collective knowledge was more extensive over the continental shelf of the eastern Bering Sea than over the basin and slope. Previous research indicated that the basin and the shelf are biologically separate domains with different likelihood of pollock survival. The scanty knowledge of stock structure and basin and slope dynamics, coupled with recommendations from an international symposium on pollock (Aron and Balsiger 1989), provided two research objectives for BS FOCI: (1) determine stock structure of pollock in the Bering Sea and its relationship to physical features, and (2) understand recruitment processes in the east-

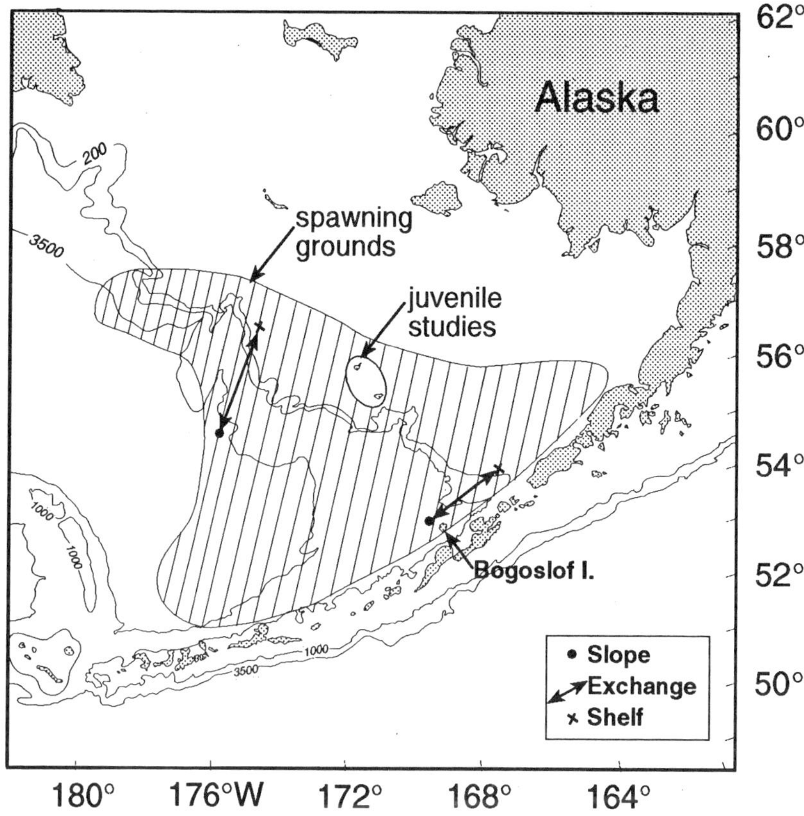

Figure 2.   The southeastern Bering Sea shelf became the focus of BS FOCI's recruitment studies.

ern Bering Sea, especially as they pertain to basin and slope waters. During the last three years of the project, the emphasis of recruitment studies was shifted to the southeastern shelf (Fig. 2). A component examining juvenile pollock near the Pribilof Islands during the late summer and fall of their birth year was added in 1994. Table 1 lists research components and investigators for BS FOCI. A technical advisory committee (Table 2) and an executive committee provided guidance.

Stock Structure Studies

International boundaries and the complex nature of aggregations of pollock in the Bering Sea make allocation of resources based on an understanding of population dynamics a difficult management problem. The

Table 1.   Bering Sea FOCI research components and principal investigators.

Component	Principal Investigator
Modeling Bering Sea circulation	James E. Overland NOAA/Pacific Marine Environmental Laboratory
A retrospective analysis of GEOSAT altimeter data in the Bering Sea	Robert Leben University of Colorado
The use of molecular techniques to dissect the genetic architecture of pollock populations: analysis of mitochondrial and nuclear genes by the polymerase chain reaction	Dennis A. Powers Stanford University
Chemical tracers of pollock origin	Brenda L. Norcross University of Alaska Fairbanks
Ichthyoplankton collections in the eastern Bering Sea	Arthur W. Kendall Jr. NOAA/Alaska Fisheries Science Center
Growth, transport, and mortality studies of larval walleye pollock	Robert C. Francis University of Washington
Larval feeding mechanisms of walleye pollock	Kevin M. Bailey NOAA/Alaska Fisheries Science Center
Feeding ecology of larval walleye pollock in oceanic and neritic domains of the Bering Sea: effects of variation in prey	Lewis J. Haldorson University of Alaska Fairbanks
Role of protozoa in the diet of larval pollock in the Bering Sea	Evelyn J. Lessard University of Washington
Microcrustacean production potential during the onset of feeding by walleye pollock larvae in the Aleutian Basin and Bering Sea shelf	A.J. Paul University of Alaska Fairbanks
Nutritional condition of larval walleye pollock	Gail H. Theilacker NOAA/Alaska Fisheries Science Center
Impact of invertebrate predators on walleye pollock larvae	Richard D. Brodeur NOAA/Alaska Fisheries Science Center

Table 1.   (Continued.)

Component	Principal Investigator
Modeling the upper ocean production dynamics of plankton and larval pollock in the Bering Sea	Stephen Bollens San Francisco State University
Phytoplankton dynamics from moored optical instruments in the southeastern Bering Sea	J. Ronald V. Zaneveld Oregon State University
Determining chlorophyll concentration from ocean color measurement in the Bering Sea	John Cullen and Richard Davis Dalhousie University
Food chain dynamics: spring bloom hypothesis and larval survival	Jeffrey M. Napp NOAA/Alaska Fisheries Science Center
Basin vertical processes: mixed-layer dynamics from the *Peggy Bering Sea* mooring	Edward D. Cokelet NOAA/Pacific Marine Environmental Laboratory
Circulation of the eastern Bering Sea: Aleutian North Slope Current	Ronald K. Reed and Phyllis J. Stabeno NOAA/Pacific Marine Environmental Laboratory
Shelf-slope exchange along the eastern Bering Sea shelf break	James D. Schumacher and Phyllis J. Stabeno NOAA/Pacific Marine Environmental Laboratory
Monitoring of mesoscale ocean processes by synthetic aperture radar (SAR)	Antony K. Liu NASA/Goddard
Juvenile pollock studies	Richard D. Brodeur NOAA/Alaska Fisheries Science Center

Table 2.   Bering Sea FOCI Technical Advisory Committee

Dr. Michael Dagg Louisiana University Marine Consortium Chauvin, LA 70344	Dr. Thomas Royer Old Dominion University Norfolk, VA 23505
Mr. Bart Eaton Trident Seafoods Corporation Seattle, WA 98107	Dr. Al Tyler University of Alaska Fairbanks Fairbanks, AK 99775-1080
Dr. Eileen Hofmann Old Dominion University Norfolk, VA 23505	Dr. Warren Wooster University of Washington Seattle, WA 98195-5682
Dr. William Leggett Queens University Kingston, Ontario Canada K7L 3N6	

importance of pollock in the ecosystem (e.g., Springer 1992), as well as the relationships and interchange among stocks, are largely unknown. Although many of the characteristics of pollock early life history are common to all populations of the species, in the Bering Sea the population structure and early life history pattern are more complex than in the Gulf of Alaska. For BS FOCI, linkages between off-shelf (Aleutian Basin) and on-shelf stocks of the eastern Bering Sea pollock resource needed to be understood. There appeared to be several distinct spawning aggregations, e.g., groups of fish with distinct reproductive characteristics behaving as cohesive units (Hinckley 1987). Indirect evidence on the reproductive biology and growth of Bering Sea pollock and the physical oceanography of the region indicated that larvae spawned over the southern deep basin could be transported onto the shelf. What happens along the central and western part of the extensive basin-shelf border was largely unknown. A case was made for a counterclockwise movement of pollock with life stage: eggs are predominantly spawned along the southern and eastern slopes and basins, hatch into larvae along the eastern and northeastern slopes, and develop into juveniles that may be found predominantly on the northwestern and western slopes and shelves.

To determine stock structure of Bering Sea pollock and its relationship to physical features, BS FOCI field and modeling studies investigated circulation throughout the deep basin. Satellite-tracked drifters were useful in monitoring circulation, and a statistical treatment of their composite tracks (Fig. 3) lent credence to the hypothesis of counterclockwise movement of pollock with age around the Bering Sea. A related study on pollock used mitochondrial DNA to establish genetic "fingerprints" to eval-

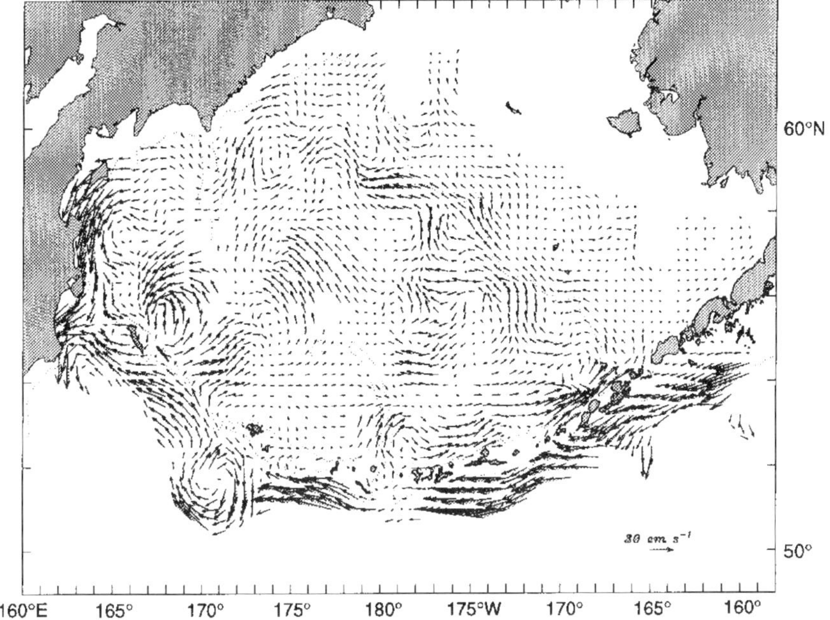

Figure 3. Mean current velocities of the Bering Sea interpolated to 0.5° longitude × 0.25° latitude from trajectories of 86 satellite-tracked drifters released from 1986 through 1993. A minimum of three independent velocities were required to determine mean velocity in each grid cell. (Updated from Stabeno and Reed 1994.)

uate stock structure. Preliminary conclusions showed that eastern and western Bering Sea pollock populations were separated by significant genetic distance (Fig. 4; Bailey et al. 1997; Bailey et al., chapter 26, this volume). Genetic characteristics (Mulligan et al. 1992) and length-at-age and fecundity relationships (Hinckley 1987) suggested that several spawning stocks exist in the eastern Bering Sea. Studies analyzing chemical deposition in the otoliths as a record of the environment experienced by the larvae at various times during their development indicated the possibility of discriminating among juvenile pollock of various geographic and genetic origins (Severin et al. 1995).

Recruitment Studies

To understand recruitment processes of eastern Bering Sea pollock, BS FOCI focused on their early life history. Investigators addressed differences between survival potential of eggs and larvae over the deep waters

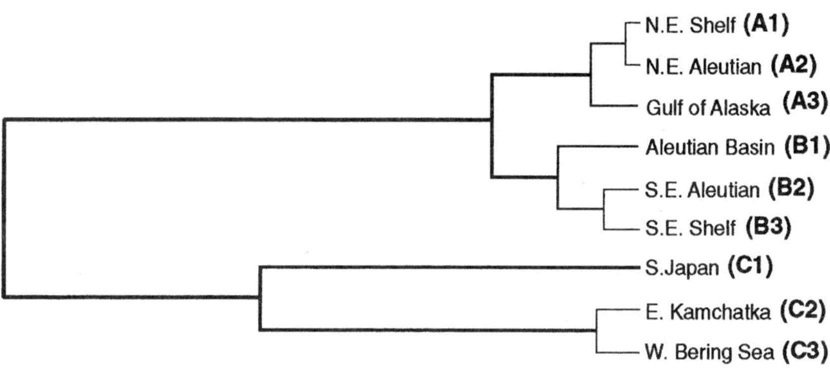

Figure 4.   Dendrogram showing genetic distance between North Pacific pollock based on 164 samples using mtDNA haplotypes. For sampling locations, see Fig. 1.

to that over the adjacent shelf. Spawning begins earlier in the year in some parts of the Bering Sea than it does in Shelikof Strait, in the Gulf of Alaska, and apparently different groups of fish spawn at different times and places. Efforts focused first on the population that spawns in February over the southeastern slope and that supported a substantial fishery in the late 1980s and 1990s. There were indications that some of the eggs and larvae were much deeper in the water column (400 m) than in Shelikof Strait. Also, feeding conditions did not seem to be adequate for optimal growth in this area. Prey concentrations were low, dominated by species that were not preferentially ingested by pollock (Fig. 5; Hillgruber et al. 1995). An alternative food source, protozoa, was available in quantities sufficient to supplement the low levels of desirable prey (Howell-Kübler et al. 1996). Microzooplankton samples collected over the slope during 1992 and 1993 (Paul et al. 1996) were compared with samples collected over the shelf during 1994 and 1995 (J. Napp, NOAA/AFSC, 7600 Sand Point Way N.E., Seattle, WA 89115, pers. comm., August 1996). Results suggested that the number of prey available to pollock larvae are not vastly different in these regions; however, the ratio of desirable to non-desirable prey is higher over the shelf. Larvae were associated with eddies in this area (Schumacher and Stabeno 1994), as in Shelikof Strait. Later, BS FOCI focused on the spawning (April-June) that occurs over the continental shelf of the southeastern Bering Sea.

Basin Circulation and Mesoscale Features

Prior to BS FOCI research, many schematics existed of circulation in the Bering Sea, and wind stress was considered to provide the primary forcing (Hughes et al. 1974). Results from BS FOCI have refined our knowledge of

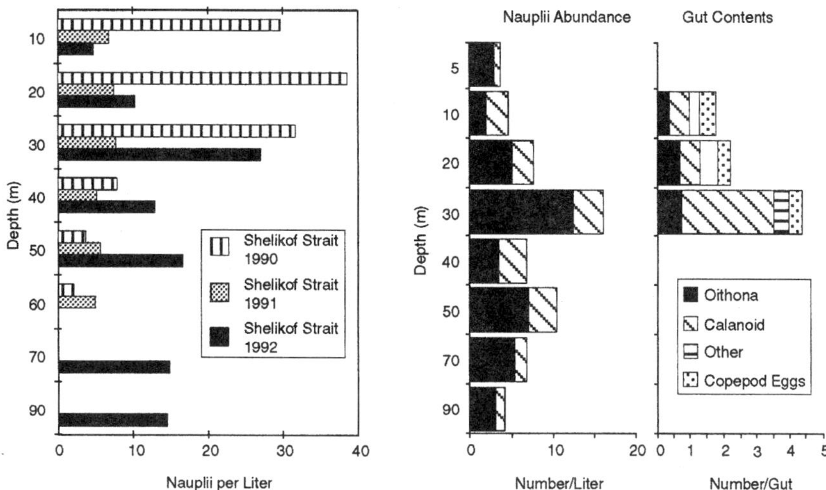

Figure 5. (Left) Prey concentrations of Shelikof Strait and the Bering Sea slope, and (right) prey availability and ingestion for pollock on the outer Bering Sea slope during spring 1992.

circulation and meteorological forcing over the basin from observations (Stabeno and Reed 1994, Cokelet et al. 1996) and model studies (Overland et al. 1994). A cyclonic gyre dominates circulation over the basin with a western boundary current (Kamchatka Current) along the Asian side of the basin (Reed et al. 1993). This gyre is mainly an extension of the Alaskan Stream, and the majority of volume transport enters through Near Strait (~10 × 10⁶ m³/s) and exits by the Kamchatka Current (Stabeno and Reed 1994). When instabilities in the Alaskan Stream inhibit flow into the Bering Sea through Near Strait (Stabeno and Reed 1992), transport in the Kamchatka Current can be reduced by half. Such conditions existed from 1990 to 1991, then returned to normal in late 1991 (Reed and Stabeno 1993). A climatology of wind forcing showed that eastward and northward-propagating storm systems dominate surface stress at short periods (<1 month) and serve principally to mix the upper ocean (Bond et al. 1994). At longer periods (>1 month), estimated wind-driven transport accounts for about half the observed transport within the Kamchatka Current. Interannual variations in the transports are about a quarter of the mean. A flux (~3.0 × 10⁶ m³/s) of Alaskan Stream waters through the eastern passes (Amchitka and Amukta passes) affects regional water properties and circulation (Reed and Stabeno 1994, Reed et al. 1994, Schumacher and Stabeno 1994, Reed 1995). These waters flow northeastward as the Aleutian North Slope Current before turning northwestward to flow along the slope (Schumacher and Reed 1992). They carry a subsurface temperature maximum that can be traced hundreds of kilometers (Cokelet and Stabeno 1997). The southeastern basin waters are also rich in eddies, some of which are formed by flow through Amukta Pass (Schumacher and Stabeno 1994).

In the Bering Sea and Shelikof Strait, eddies play a role in coupled biophysical processes. In the eastern Bering Sea, the presence of relatively small eddies (diameter <50 km) was documented by Schumacher and Stabeno (1994). These eddies are formed in regions of high current shear in open waters, or by interaction of inflowing Alaskan Stream water with topography of passes in the eastern Aleutian Island chain. Eddy formation during periods of inflow through Amukta Pass occurred in a numerical model of the region (Schumacher and Stabeno 1994). Since 1986, satellite-tracked buoys have been deployed in the Bering Sea to support studies of pollock and their environment. In four of these years, five regions of relatively high abundance of pollock larvae were found and buoys were deployed in them. In all but one case, the trajectories of the buoys indicated eddies. Buoys that were not deployed in a patch of abundant pollock larvae did not indicate eddies. This association of pollock larvae and eddies may have significantly affected larval survival. Research on pollock in Shelikof Strait indicates that larval survival may be aided by their retention within eddies (Canino et al. 1991, Bograd et al. 1994).

Influence of Sea Ice

Investigation of the seasonal ice pack in the eastern Bering Sea (Wyllie-Echeverria 1995) showed a relationship between southward extension of the ice and the distribution of pollock over the shelf. The extent of sea ice determines the size of the cold pool—a vast, permanent feature of the shelf. Pollock surveys rarely find age-0 juveniles within the cold pool. During years when the cold pool is extensive, it is hypothesized that young pollock are forced to cohabit the same shelf regions as cannibalistic adult pollock. This was the case in 1976 (Fig. 6). Increased predation by cannibalism may significantly limit recruitment.

Upper Ocean Phytoplankton Survey

For its last field experiment in spring 1996, BS FOCI conducted a survey of sea surface chlorophyll over the southeastern Bering Sea using a research aircraft and ship. The project documented the spatial and temporal evolution of chlorophyll concentration during the spring bloom in the Bering Sea, and how this concentration related to the ocean's physical characteristics. Chlorophyll concentration was estimated from the aircraft using remote color-sensing instrumentation designed to detect upwelling radiance from the ocean; the ocean's physical characteristics were assessed using air-expendable bathythermographs and observations by moored buoys and by ship. Preliminary results showed higher chlorophyll concentrations in the northern portion of the domain where sea ice was melting, and a prominent 100×200 km patch southeast of the Pribilof Islands with concentrations an order of magnitude greater than background lev-

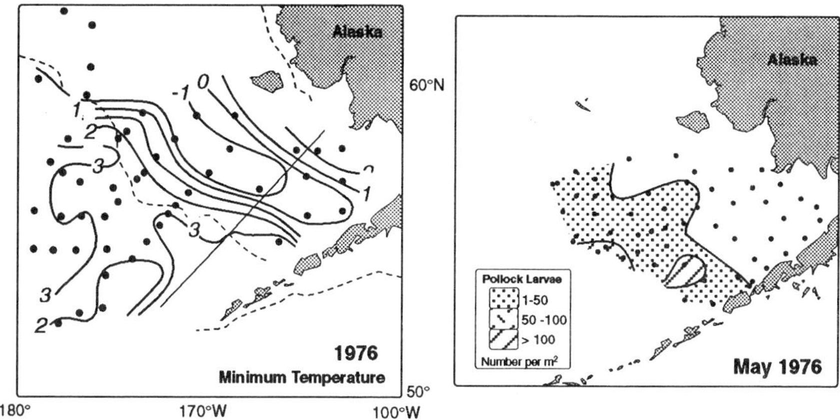

Figure 6. (Left) The location of the cold (<2°C) pool, and (right) the distribution of larvae in 1976 suggest that young pollock avoid the cold pool.

els (Fig. 7; Davis et al. 1997b). This patch was in a region of the shelf where pollock spawning occurs.

Models and New Technology

BS FOCI used models to conceptualize program strategies, to expand the time-space domain of field observations, to guide program direction through hypothesis generation, and to understand features of biophysical processes. A numerical model of circulation of the Bering Sea basin and exchange with the North Pacific Ocean demonstrated that flow instabilities contribute to substantial interannual variability in circulation (Overland et al. 1994). A coupled, one-dimensional, biophysical model investigated production dynamics of the pelagic ecosystem with respect to growth of larval pollock. The approach used field observations to determine rates and appropriate species composition for several of the distinct physical (Coachman 1986) and biological (Cooney and Coyle 1982, Smith and Vidal 1986) domains in the eastern Bering Sea. The model included stage-structured dynamics of copepod populations (*Calanus* spp. and *Neocalanus* spp.) and larval pollock feeding and growth (Francis et al., chapter 20, this volume; Bollens et al., in preparation). The temporal behavior of the mixed layer was obtained from observations from *Peggy Bering Sea,* a moored biophysical platform described below (Schumacher and Kendall 1995). Model results showed that the species composition of zooplankton has a strong influence on growth of larval pollock, and the presence of protozoan prey becomes important when young copepods are scarce (i.e., during early spring over the slope). Results from the model also suggested that variability of the mixed-layer depth has significant

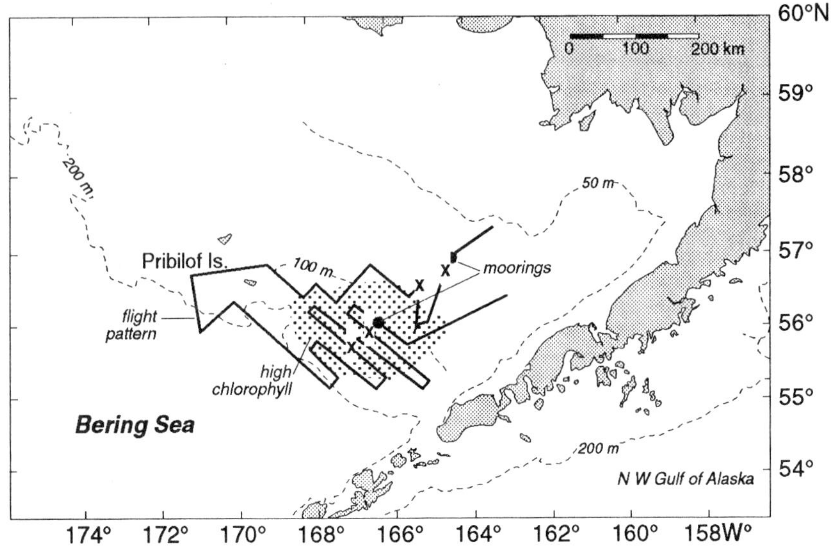

Figure 7.   *A patch of water containing relatively high amounts of chlorophyll as detected by aircraft survey was in an area where pollock spawn.*

impact on larval growth by affecting lower trophic level production. These results led to the addition of a field research component that developed a new method for live-staining protists (Lessard et al. 1996) and showed that microprotozoan biomass levels are sufficient to augment prey levels and support estimated metabolic needs of first-feeding larval pollock (Howell-Kübler et al. 1996).

Because there were no continuous time series of biophysical conditions from pre-spring bloom conditions through the summer, BS FOCI deployed a mooring, *Peggy Bering Sea* (Fig. 8), over the outer slope (~2,200 m). The platform was similar to moorings deployed in the tropical Pacific Ocean (McPhaden et al. 1991), but modified to withstand the arctic marine climate of the Bering Sea. Included in its suite of measurements were winds, insolation, air temperature, humidity, salinity and temperature at ten depths, currents from acoustic Doppler current profiler and acoustic current meters, acoustic backscatter, and chlorophyll absorption (Davis et al. 1997a). Some of the observations were telemetered real-time via satellite. When a significant alteration of the diel migration occurred, the availability of real-time data permitted direction of field sampling to provide in situ measurements by ship. Results elucidated characteristics of eddies, including their density and velocity structure. Also, the observations provided time series of mixed layer depth to a coupled, one-dimensional, biophysical model.

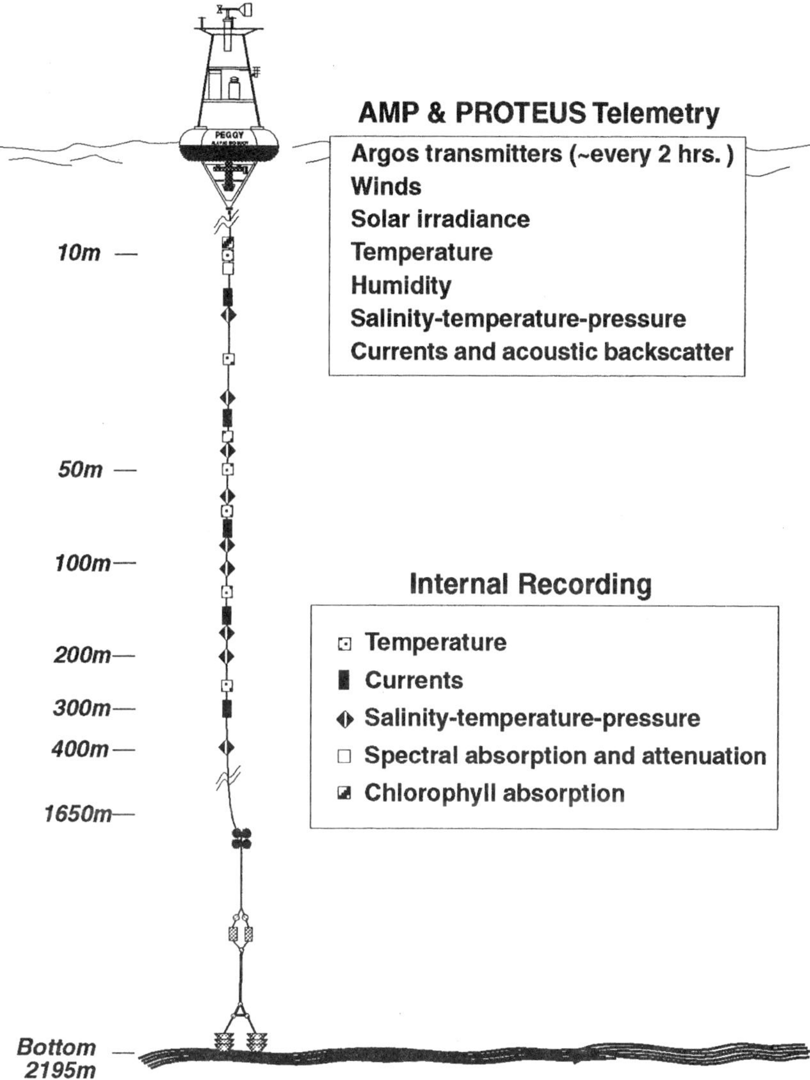

Figure 8.   Peggy Bering Sea, a biophysical mooring, was deployed during spring and summer in the same location over the eastern Bering Sea basin for three consecutive years.

The mooring design later was modified for shallow water use, and two moorings were deployed during early March 1995—one over the outer and one over the middle southeastern Bering Sea shelf. They were inundated by sea ice, providing unique time series of water column temperature, salinity, estimated chlorophyll *a,* and phytoplankton fluorescence during ice advance and retreat. Contrary to earlier findings, chlorophyll *a* concentration increased prior to formation of a well-defined, two-layer system, and events of strong advection, rather than vertical fluxes, dominated changes in water column structure and properties (Stabeno et al. 1998). This was especially apparent at the weakly advective middle shelf location. During winter such advective events can replenish nutrients that support the spring bloom over the middle shelf.

Real-Time Detection of Eddies

Results from the slope waters of the eastern Bering Sea indicated that the highest abundances of pollock larvae often reside in eddies. To examine the nature of biophysical processes present in these features and determine their influence upon survival required in situ observations. Finding a reliable method to locate an eddy for field studies was a challenge. Although infrared imagery has proved useful, cloud cover and generally weak sea surface temperature gradients limit this approach. High resolution synthetic aperture radar (SAR) eliminates both of these constraints. Mesoscale features are imaged by SAR through several possible mechanisms that are not well understood, including current-induced wave refraction. SAR images were examined for mesoscale features in Alaskan coastal waters (Liu et al. 1994), and BS FOCI maintained an image library. Acoustic backscatter signals also were used to identify and characterize mesoscale biophysical features in the ocean (Aoki and Inagaki 1992, Brodeur et al. 1997), thereby permitting real-time studies of these features. Real-time, remote tracking of eddies enabled BS FOCI to guide sampling by a Japanese research vessel during summer 1993.

Juvenile Studies

A BS FOCI research component established in 1994 compared habitats of juvenile pollock around the Pribilof Islands. A structural front resulting from tidal mixing exists around the Pribilof Islands in the eastern Bering Sea. This feature separates weakly stratified nearshore water from the strongly stratified middle shelf water farther offshore. An enhanced vertical flux of nutrients at the structural front results in high abundance of phytoplankton, zooplankton, and micronekton. The unique biophysical conditions associated with the structural front may provide a more suitable habitat for age-0 pollock than elsewhere on the shelf. BS FOCI tested this hypothesis using observations from late summer and fall during oceanographically contrasting years, 1994 and 1995.

Comparisons were made of abundance, size composition, feeding, growth, and condition of pollock from inshore, at, and seaward of the structural front, and also from the vicinity of the middle shelf front (Napp et al. 1995, Brodeur et al. 1997). The frontal region occurred 12-20 km offshore during both years, but the width of the front during 1995 was about twice that in 1994. The thermocline was much deeper (~40 m) in 1994 than in 1995 (20-25 m). The isothermal (<8°C) inner domain had low concentrations of chlorophyll and zooplankton. The highest chlorophyll concentrations occurred seaward of the front and decreased thereafter with increasing distance from shore. Small zooplankton (mainly copepods) occurred at highest density in the frontal region. Large zooplankton (mainly euphausiids) were most abundant in the stratified offshore waters below the thermocline. Age-0 pollock dominated the midwater fish catches during both years (>84% by weight, >99% by number), although large medusae dominated the biomass of the overall catch. Age-0 pollock densities averaged about three times higher in 1994 than in 1995. The length distributions of age-0 pollock were significantly smaller nearshore and at the front in 1994 than in 1995. Pollock densities generally decreased, but their mean size increased, with increasing distance from shore.

Catches in summer 1995 were highest in the central shelf region and near the Pribilof Islands, but were low in the northwest part of the region. The size of juvenile pollock caught generally increased with increasing latitude. Several species of jellyfish were the only other dominant taxa collected with age-0 pollock in these trawls, and there was some indication of feeding by these large medusae on small juvenile pollock. During September, acoustical surveys, midwater and bottom trawls, and surveys using a remotely operated vehicle with video cameras detected abundant juvenile pollock in midwater deployments. The juveniles were often associated with large medusae during daytime, then found in large aggregations close to the surface at night (Brodeur 1998). Special sampling was done in areas where satellite-tagged fur seals were located and where seabirds were observed to be foraging.

These results suggested that fronts associated with tidal mixing near the Pribilof Islands provide a suitable habitat for juvenile pollock. Because pollock were the dominant pelagic fish species in the Bering Sea ecosystem at the time of BS FOCI, juvenile pollock undoubtedly played a pivotal role in the food chain, serving as both predator and prey. Much of the biological energy of the Bering Sea passed through the juvenile pollock stage.

Summary

Bering Sea FOCI was a coastal fisheries ecosystem project of NOAA's Coastal Ocean Program. The project enlisted the research efforts of about twenty scientists of various disciplines from seven universities and three government laboratories. In its 7-year lifetime, BS FOCI researched two central themes concerning walleye pollock of the Bering Sea: stock structure and recruitment processes. Based on observations and models of circulation, there is probably counter-clockwise transport of early-life stages of pollock around the basin. Major spawning occurs over the slope and shelf in the southeastern and western regions, and the larvae are slowly advected downstream. A significant population of age-0 juveniles from the southeastern shelf spawn are known to inhabit the waters of the Pribilof Islands in the fall. Despite the potential for far-ranging transport of early-life stages, populations remain distinct. Genetic tests have shown that stocks of the western and eastern Bering Sea are separate and discernible. Recruitment studies in the southeastern Bering Sea determined the importance of eddies, sea ice, and shelf-slope exchange on the well-being of pollock, and suggested that conditions in the basin are not conducive to survival. Studies of juvenile pollock have uncovered important relationships between shelf regimes and foraging behavior in the marine ecosystem of the Pribilof Islands.

New programs such as NOAA Coastal Ocean Program's Southeast Bering Sea Carrying Capacity and NOAA's Arctic Research Initiative, as well as ongoing research by FOCI, will build on these concepts developed by BS FOCI, and will continue to advance our understanding of the complex, productive ecosystem of the Bering Sea.

References

Aoki, I., and T. Inagaki. 1992. Acoustic observation of fish schools and scattering layers in a Kuroshio warm-core ring and its environs. Fisheries Oceanography 1:137-142.

Aron, W., and J. Balsiger. 1989. Proceedings of the international scientific symposium on Bering Sea fisheries. NOAA Technical Memo NMFS F/NBC-163, 424 pp. (Available from Alaska Fisheries Science Center, 7600 Sand Point Way N.E., Seattle, WA 98115.)

Bailey, K.M., P.J. Stabeno, and D.A. Powers. 1997. The role of larval retention and transport features in mortality and gene flow of walleye pollock, *Theragra chalcogramma*. Journal of Fish Biology 51:135-154.

Bograd, S.J., P.J. Stabeno, and J.D. Schumacher. 1994. A census of mesoscale eddies in Shelikof Strait, Alaska, during 1989. Journal of Geophysical Research 99:18243-18254.

Bollens, S.M., M.C. Landsteiner, and B.W. Frost. In prep. A model of the upper ocean production dynamics of plankton and larval pollock in the southeastern Bering Sea. Marine Ecology Progress Series. (Available from S. Bollens, Department of Biology, San Francisco State University, San Francisco, CA 94132.)

Bond, N.A., J.E. Overland, and P. Turet. 1994. Spatial and temporal characteristics of the wind forcing of the Bering Sea. Journal of Climate 7:1119-1130.

Boyles, R., Jr., and D. Scavia. 1993. Innovative coastal ocean science at NOAA. Sea Technology 34:10-16.

Brodeur, R.D. 1998. In situ observations of the association between juvenile fishes and scyphomedusae in the Bering Sea. Marine Ecology Progress Series. In press. (Available from R. Brodeur, NMFS/AFSC, 7600 Sand Point Way N.E., Seattle, WA 98115.)

Brodeur, R.D., M.T. Wilson, J.M. Napp, P.J. Stabeno, and S. Salo. 1997. Distribution of juvenile walleye pollock relative to frontal structure near the Pribilof Islands, Bering Sea. In: Forage fishes in marine ecosystems: Proceedings of the International Symposium on the Role of Forage Fishes in Marine Ecosystems. University of Alaska Sea Grant, AK-SG-97-01, Fairbanks, pp. 573-589.

Canino, M.F., K.M. Bailey, and L.S. Incze. 1991. Temporal and geographic differences in feeding and nutritional condition of walleye pollock larvae *Theragra chalcogramma* in Shelikof Strait, Gulf of Alaska. Marine Ecology Progress Series. 79:27-35.

Coachman, L.K. 1986. Circulation, water masses and fluxes on the southeastern Bering Sea shelf. Continental Shelf Research 5:23-108.

Coachman, L.K., and D.A. Hansell (eds.). 1993. ISHTAR: Inner shelf transfer and recycling in the Bering and Chukchi seas. Continental Shelf Research 13:473-704.

Cokelet, E.D., and P.J. Stabeno. 1997. Mooring observations of the thermal structure, salinity and currents in the SE Bering Sea basin. Journal of Geophysical Research 102(c10):22947-22964.

Cokelet, E.D., M.L. Schall, and D. Dougherty. 1996. ADCP-referenced geostrophic circulation in the Bering Sea basin. Journal of Physical Oceanography 26:1113-1128.

Cooney, R.T., and K.O. Coyle. 1982. Trophic implications of cross-shelf copepod distributions in the southeastern Bering Sea. Marine Biology 70:187-196.

Davis, R.F., C.C. Moore, J.R.V. Zaneveld, and J.M. Napp. 1997a. Reducing the effects of fouling on chlorophyll estimates derived from long-term deployments of optical instruments. Journal of Geophysical Research 102:5851-5856.

Davis, R.F., G. Lazin, J. Bartlett, A. Ciotti, and P. Stabeno. 1997b. Remote sensing of a pigment patch in the southeastern Bering Sea. Ocean Optics XIII, Proceedings of the Society for Photo-Optical Instrumentation Engineers, P.O. Box 10, Bellingham, WA 98227-1101, pp. 654-657.

Hillgruber, N., L.J. Haldorson, and A.J. Paul. 1995. Feeding selectivity of larval walleye pollock *Theragra chalcogramma* in the oceanic domain of the Bering Sea. Marine Ecology 120:1-10.

Hinckley, S. 1987. The reproductive biology of walleye pollock, *Theragra chalcogramma,* in the Bering Sea, with reference to spawning stock structure. Fisheries Bulletin, U.S. 85:481-498.

Hood, D.W. (ed.). 1986. Processes and resources of the Bering Sea shelf (PROBES). Continental Shelf Research 5:1-286.

Hood, D.W. and J.A. Calder (eds.). 1981. The eastern Bering Sea shelf: Oceanography and resources, Vol. I and II. Published by the Office of Marine Pollution Assessment, NOAA and BLM. 1,339 pp. (Distributed by the University of Washington Press, Seattle, WA 98105.)

Howell-Kübler, A.N., E.J. Lessard, and J.M. Napp. 1996. Springtime microprotozoan abundance and biomass in the southeastern Bering Sea and Shelikof Strait, Alaska. Journal of Plankton Research 18:731-745.

Hughes, F.W., L.K. Coachman, and K. Aagaard. 1974. Circulation, transport, and water exchange in the western Bering Sea. In: D.W. Hood and E.J. Kelley (eds.), Oceanography of the Bering Sea. Occasional Publication 2, Institute of Marine Science, University of Alaska, Fairbanks, pp. 59-98.

Kendall, A.W., Jr., R.I. Perry, and S. Kim. 1996. Fisheries oceanography of walleye pollock in Shelikof Strait, Alaska. Fisheries Oceanography 5(Suppl. 1). 203 pp.

Lessard, E.J., M.P. Martin, and D.J.S. Montagnes. 1996. A new method for live-staining protists with DAPI for use as a tracer of ingestion by walleye pollock *(Theragra chalcogramma)* larvae. Marine Ecology Progress Series 204:43-57.

Liu, A.K., C.Y. Peng, and J.D. Schumacher. 1994. Wave-current interaction study in the Gulf of Alaska for detection of eddies by SAR. Journal of Geophysical Research 99:10075-10085.

McPhaden, M.J., H.B. Milburn, A.I. Nakamura, and A.J. Shepherd. 1991. PROTEUS: Profile telemetry of upper ocean currents. Sea Technology Magazine, February, 10-19.

Muench, R.D. (ed.). 1983. Marginal ice zones. Journal of Geophysical Research 88:2713-2966.

Mulligan, T.J., R.W. Chapman, and B.L. Brown. 1992. Mitochondrial DNA analysis of walleye pollock, *Theragra chalcogramma,* from the eastern Bering Sea and Shelikof Strait, Gulf of Alaska. Canadian Journal of Fisheries and Aquatic Sciences 49:319-326.

Napp, J.M., R.D. Brodeur, D. Demer, and R. Hewitt. 1995. Acoustic determination of plankton and nekton concentrations at tidally-generated fronts in the eastern Bering Sea. International Symposium on Fisheries and Plankton Acoustics, Aberdeen, Scotland, June. (Available from ICES Publications, Charlottenlund, 2920 Denmark.)

National Research Council. 1994. A Review of the accomplishments and plans of the NOAA Coastal Ocean Program. National Academy Press, Washington, DC. 115 pp.

Overland, J.E., M.C. Spillane, H.E. Hurlburt, and A.J. Wallcraft. 1994. A numerical study of the circulation of the Bering Sea basin and exchange with the North Pacific Ocean. Journal of Physical Oceanography 24:736-758.

Paul, A.J., J.M. Paul, and K.O. Coyle. 1996. Abundance and taxa composition of copepod nauplii over southeastern Bering Sea deep water spawning grounds of walleye pollock, *Theragra chalcogramma.* Crustaceana 69:494-508.

Reed, R.K. 1995. On the variable subsurface environment of fish stocks in the Bering Sea. Fisheries Oceanography 4:317-323.

Reed, R.K., and P.J. Stabeno. 1993. The recent return of the Alaskan Stream to Near Strait. Journal of Marine Research 51:515-527.

Reed, R.K., and P.J. Stabeno. 1994. Flow along and across the Aleutian ridge. Journal of Marine Research 52:639-648.

Reed, R.K., G.V. Khen, P.J. Stabeno, and A.V. Verkhunov. 1993. Water properties and flow over the deep Bering Sea basin, summer 1991. Deep-Sea Research 40:2325-2334.

Reed, R.K., J.D. Schumacher, C.H. Pease, P.J. Stabeno, A.W. Kendall Jr., and P. Dell'Arciprete. 1994. Bering Sea circulation explored. Eos, Transactions of the American Geophysical Union 75:342.

Schumacher, J.D., and A.W. Kendall Jr. 1995. An example of fisheries oceanography: Walleye pollock in Alaskan waters. U.S. National Report 1991-1994, 1153-1163. (Available from Amer. Geophys. Union, 2000 Florida Ave. N.W., Washington, DC 20009.)

Schumacher, J.D., and R.K. Reed. 1992. Characteristics of currents over the continental slope of the eastern Bering Sea. Journal of Geophysical Research 97:9423-9433.

Schumacher, J.D, and P.J. Stabeno. 1994. Ubiquitous eddies of the eastern Bering Sea and their coincidence with concentrations of larval pollock. Fisheries Oceanography 3:182-190.

Severin, K.P., J. Carroll, and B.L. Norcross. 1995. Electron microprobe analysis of juvenile walleye pollock, *Theragra chalcogramma*, otoliths from Alaska: A pilot stock separation study. Environmental Biology of Fishes 43:269-283.

Smith, S.L., and J. Vidal. 1986. Variations in the distribution, abundance, and development of copepods in the southeastern Bering Sea in 1980 and 1981. Continental Shelf Research 5:215-239.

Springer, A.M. 1992. A review: Walleye pollock in the North Pacific—how much difference do they really make? Fisheries Oceanography 1:80-96.

Stabeno, P.J., and R.K. Reed. 1992. A major circulation anomaly in the western Bering Sea. Geophysical Research Letters 19:1671-1674.

Stabeno, P.J., and R.K. Reed. 1994. Circulation in the Bering Sea basin observed by satellite-tracked drifters: 1986-1993. Journal of Physical Oceanography 24:848-854.

Stabeno, P.J., J.D. Schumacher, R.F. Davis, and J.M. Napp. 1998. Under-ice observations of water column temperature, salinity and spring phytoplankton dynamics: Eastern Bering Sea shelf. Journal of Marine Research 56:239-255.

Wenzel, L., and D. Scavia. 1993. NOAA's Coastal Ocean Program. Oceanus 36:85-92.

Wyllie-Echeverria, T. 1995. Sea-ice conditions and the distribution of walleye pollock *(Theragra chalcogramma)* on the Bering and Chukchi sea shelf. Canadian Special Publication of Fisheries and Aquatic Sciences 121:131-136.

CHAPTER 35

Summary of *Oshoro Maru* Cruises in the Eastern Bering Sea

Kiyotaka Ohtani
Laboratory of Physical Oceanography, Hokkaido University, Hokkaido, Japan

The T/S *Oshoro Maru* has made a research and training cruise to the North Pacific Ocean and Bering Sea every summer since 1953. Oceanographic observations, biological sampling of plankton and fishes, and other measurements have been carried out by the research staff, captain, and officers with the assistance of the crew and of cadets from the University of Hokkaido and the University of Nagasaki as part of their on-board fisheries training.

Data records of each cruise were issued by the Faculty of Fisheries, Hokkaido University, in the *Data Record of Oceanographic Observations and Exploratory Fishing,* No. 1 (1957) to No. 40 (1997). Routine observations included temperature, salinity, nutrients, water transparency and color, as well as results from dynamic computations. Other observations included radioactivity of seawater, turbidity, chemical substances, and chlorophyll *a*.

Plankton sampling using a Norpac net was carried out at each station. Fish larva nets were towed every evening one hour after sunset. Horizontal tows using MTD nets, Isaac-Kidd nets, and other nets were occasionally made at selected stations. Biological data on fishes caught by drift gillnet and trawl fishing were published in the *Data Record of Oceanographic Observations and Exploratory Fishing.* The biological characteristics of fishes caught by drift gillnet research were measured for species, fork length, body weight, gonad weight, sex, and age. Results of visual observations of seabirds and marine mammals were also recorded.

Information for each cruise is summarized in Table 1. The number of observations taken in the Bering Sea totals 2,480 points. These observations account for 62% of all observations of the summer cruises over the past 44 years. Three generations of *Oshoro Maru* vessels have been in the Bering Sea for a total of 1,047 days of observation and sampling. They carried 309 researchers, 131 guest scientists from the United States (U.S. Fish and Wildlife Service, NOAA National Marine Fisheries Service, Point

Table 1. Summary of *Oshoro Maru* cruises in the Bering Sea, 1953-1996.

Year	Month, day period	Number of stations Bering Strait (points)	Total (points)	Sampling gear	Research area	Research staff
1953	5,17-20	6	21	N	Bs	4
1954	6,2-16	11	28	N	Bs,Ss	3
1955	7,2-20	27	37	G	B,Ss	10
1956	7,23-8,13	40	50	G,T	B,Sn	10
1957	7,7-13	6	376	G	B	7
1958	6,8-18.23-27	17	22	G	B,Sn	5
1959	6,15-18.25-7,14	29	67	G	B	8
1960	6,16-22.28-7,14	25	64	G	B,S	8
1961	6,28-7,15	21	63	G	Be	4
1962	6,7-6,21	21	62	G	B	7
1963	6,13-7,19	75	106	G,T	B,Ss	4
1964	6,12-7,7.7,21-8,9	89	127	G,T	B,S,C	5
1965	6,3-27.7,7-8,1	96	114	G,T	B,Ss	7
1966	6,13-30.7,10-8,2	81	81	G,T	Be,S	8
1967	6,11-7,2.28-8,17	93	97	G,T	Be,S	11
1968	6,11-7,4.26-8,9	82	83	G,T	Be,S	12(3)
1969	6,13-8,3	97	99	G,T	BeS	12(1)
1970	6,23-7,3.8,4-15	45	63	G,T	Be,S	11(2)
1971	6,12-7,8.20-23	59	73	G	Bw,S	4
1972	6,11-8,10	128	141	G,T	Be,S,C	4(2)
1973	6,11-7,23	116	117	G	Be,S	3(3)
1974	6,13-7,22	142	143	G,T	Be,S	6(4)
1975	6,13-7,5	56	68	G,L	Be,S	5(1)
1976	6,14-29.7,8-26	84	89	G	Be,S	1
1977	6,12-29.7,12-24	68	71	G	Be,S	0(3)
1978	6,12-30	51	74	G	Be,S	4(6)
1979	7,10-26	40	84	G	Be,S	4(5)
1980	6,22-7,1	32	72	N	Be,S	5(7)
1981	6,22-7,1	40	83	N	Be,S	0(7)
1982	6,23-7,7	31	72	N	Ss	9(8)
1983	7,18-8,3	45	78	Gs,L	Ss	3(11)
1984	6,17-25	30	108	N	S	5(15)
1985	6,18-25	30	82	N	Ss	2(5)
1986	6,20-7,10.23-8,5	114	141	Gf,L	Be,S	8(4)
1987	6,11-7,6	37	92	G,L	Ss	14(5)
1988	6,23-7,2	20	145	G	Ss	9(3)
1989	6,22-26	21	90	N	Be,Sn	11(2)
1990	6,19-24.7,24-8,5	60	133	N	S,C	20(5)
1991	6,19-22	76	148	T	S,C	11(5)
1992	7,16-8,6	90	128	T	S,C	10(2)
1993	6,22-24.7,19-8,1	51	121	T	Sn	9(1)
1994	6,20-21.7,16-8,2	61	157	T	Be,S	6(4)
1995	7,19-8,5	66	147	Tb,Tm	S	5(9)
1996	7,21-30	71	130	T,Tb,L	S	15(8)
TOTALS: 1,047		2,480	4,008			309+(131)

Sampling gear: G=salmon gillnet fishing, Gs=small size gillnet, Gf=fine mesh gillnet for sampling juvenile walleye pollock, T=otter trawl fishing, Tb=beam trawl, Tm=mid-layer trawl, L=longline fishing for tagging salmon. N=no experimental fishing.

Table 1.   (Continued.) Summary of *Oshoro Maru* cruises in the Bering Sea, 1953-1996.

Year	Month, day period	Cadets	Captain	Remarks	Date Rec. O.O.E.F.[a]
1953	5,17-20	4,008	Mishima	*Oshoro Maru* II	1
1954	6,2-16	14	Mishima	Nansen bottles with reversing thermometers	1
1955	7,2-20	32	Fujii	Silver titration	1
1956	7,23-8,13	27	Fujii		2
1957	7,7-13	28	Fujii		2
1958	6,8-18.23-27	37	Mishima		3
1959	6,15-18.25-7,14	23	Fujii		4
1960	6,16-22.28-7,14	23	Fujii		5
1961	6,28-7,15	20	Fujii		6
1962	6,7-6,21	22	Fujii		7
1963	6,13-7,19	19	Fujii	*Oshoro Maru* III	8
1964	6,12-7,7.21-8,9	7	Fujii		9
1965	6,3-27.7,7-8,1	14	Fujii		10
1966	6,13-30.7,10-8,2	12	Fujii	T.S.-E2 Salinometer	11
1967	6,11-7,2.28-8,17	17	Fujii		12
1968	6,11-7,4.26-8,9	9	Fujii		13
1969	6,13-8,3	15	Fujii		14
1970	6,23-7,3.8,4-15	17	Fujii		15
1971	6,12-7,8.20-23	14	Fujii	Auto Lab Salinometer	16
1972	6,11-8,10	15	Fujii		17
1973	6,11-7,23	10	Fujii		18
1974	6,13-7,22	10	Fujii		19
1975	6,13-7,5	11	Fujii		19
1976	6,14-29.7,8-26	14	Fujii		20
1977	6,12-29.7,12-24	10	Fujii		21
1978	6,12-30	15	Fujii		22
1979	7,10-26	14	Fujii		23
1980	6,22-7,1	13(3)	Fujii		24
1981	6,22-7,1	14(3)	Fujii		25
1982	6,23-7,7	17(7)	Fujii		26
1983	7,18-8,3	8(2)	Fujii		27
1984	6,17-25	12(5)	Masuda	*Oshoro Maru* IV	28
1985	6,18-25	16(6)	Masuda	CTD[b] Neil Brown Mark III-B	29
1986	6,20-7,10.23-8,5	13(3)	Masuda		30
1987	6,11-7,6	13(8)	Masuda		31
1988	6,23-7,2	21(7)	Masuda		32
1989	6,22-26	25(13)	Masuda		33
1990	6,19-24.7,24-8,5	15(9)	Masuda		34
1991	6,19-22	16(6)	Masuda		35
1992	7,16-8,6	23(12)	Masuda		36
1993	6,22-24.7,19-8,1	10	Anma		37
1994	6,20-21.7,16-8,2	4	Anma		38
1995	7,19-8,5	22(8)	Anma		39
1996	7,21-30	21(10)	Anma		40
	TOTALS:	730(102)			

Research areas: B=Bering basin, Be=eastern basin, Bs=southern basin, S=shelf, Sn=northern shelf, Ss=southern shelf, C=Chukchi Sea.

Research staff: Numbers in parentheses indicate guest scientists.

[a] Oceanographic Observations and Exploratory Fishing

[b] Conductivity, Temperature, and Depth meter

Reyes Bird Observatory, University of Alaska, University of Washington, Oregon State University, Woods Hole Oceanographic Institution), Japanese National Institute of Polar Research, and 730 cadets.

We wish to recognize the efforts of Captains T. Fujii, S. Mishima, K. Masuda, and G. Anma, as well as everyone who participated in the Bering Sea cruises.

CHAPTER **36**

United States Oil and Gas Exploration in the Bering Sea

Cleveland J. Cowles, Jerry L. Imm, and Jeff Walker
U.S. Minerals Management Service, Alaska Outer Continental Shelf Region, Anchorage, Alaska

Abstract

The Bering Sea Outer Continental Shelf (OCS) held great interest for the oil companies as OCS leasing accelerated in the United States following the 1973 oil embargo. The U.S. Department of the Interior's Minerals Management Service (MMS) has held four OCS lease sales in the Bering Sea. More than $100 million has been spent on environmental studies to support OCS lease-sale proposals in the Bering Sea region. Major issues and concerns associated with OCS exploration activities have been identified and mitigated. Twenty-four exploratory wells have been drilled without finding commercial quantities of oil or gas. Except for the North Aleutian Basin, where there has been no exploratory drilling, there is limited interest in future near-term leasing and exploration opportunities in the Bering Sea OCS. The MMS environmental studies effort has been significantly reduced in the Bering Sea, with the exception of completion of certain studies and monitoring of protected species as part of other ongoing studies.

Introduction

In 1974, President Richard M. Nixon initiated "Project Energy Independence" as a response to the Arab Oil Embargo of 1973. Its purpose was to allow the United States to be less dependent on non-U.S. sources of oil and oil products. As part of the project, the U.S. Department of the Interior (USDOI) began to accelerate leasing of Outer Continental Shelf (OCS) lands around the United States. The Alaska OCS was considered important to the national effort. The Bering Sea OCS held great interest for the oil companies and was believed to be on a par with other areas of the Alaska OCS, such as the Gulf of Alaska and the Chukchi and Beaufort seas. In 1973, a decision was made to begin a massive environmental study program in order to comply with the U.S. National Environmental Policy Act (NEPA) and other legislation. The goal of the program was to provide information

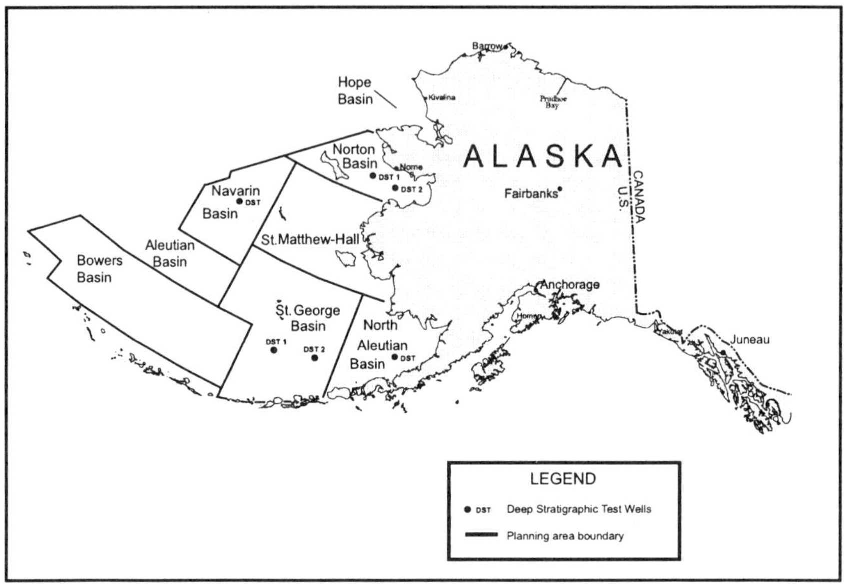

Figure 1.   *Planning area boundaries and deep stratigraphic test wells, Bering Sea, Alaska. Note: The maritime boundaries and limits shown, as well as the divisions between the planning areas, are for initial planning purposes only and do not prejudice or affect U.S. jurisdiction.*

and data sufficient to permit environmental assessments to be made to predict the potential effects of activities associated with the exploration, development, and production of oil and gas on OCS lands.

Oil and gas exploration and development in the U.S. portion of the Bering Sea is managed by the U.S. Minerals Management Service (MMS) and the state of Alaska. The MMS manages oil and gas activities on the federal OCS. The OCS comprises those lands located generally 3 miles offshore of the coast and seaward. The state administers these activities on submerged lands located from the coast to the state/federal boundary. To date, there has been no oil and gas exploratory drilling or development activity on submerged lands in the offshore Bering Sea area. This paper focuses on activities on the federal OCS in the Bering Sea.

Leasing History

The MMS lease-sale program divides the state into several planning areas. There were five original planning areas in the Bering Sea (Fig. 1)—the St. George, Navarin, Bowers, North Aleutian, and Norton basins. Four OCS lease sales were held in four of these planning areas. The sale results are

Table 1.   Alaska OCS region oil and gas lease sales.

Planning area	Sale number	Sale date	No. leases issued
Norton Basin	Sale 57	March 1983	59
St. George Basin	Sale 70	April 1983	96
Navarin Basin	Sale 83	April 1984	163
North Aleutian Basin[a]	Sale 92	October 1988	23

[a] Twenty-three oil and gas leases were issued as a result of Sale 92. Following lease issuance, U.S. Congressional appropriations included moratoriums which provided that no funds be expended by the USDOI for the approval or permitting of any drilling or other exploration activity on lands within this planning area. Lessees sued the government to buy back the leases. In 1995, MMS announced the settlement of a portion of the lawsuit. As part of the settlement agreement, companies relinquished all of the Alaska OCS leases issued in Sale 92. All Bering Sea leases are now terminated

summarized in Table 1. Several other sales were proposed but were canceled or dropped from the proposed sale schedule for various reasons.

These frontier areas had high resource potential and tremendous industry interest. Because of the frontier nature of these sale areas, this was the first time the MMS offered leases with 10-year (compared to more traditional 5-year) primary lease terms to provide extra time and opportunity for industry to properly plan and conduct operations in these biologically sensitive areas.

Leases were 9-square-mile tracts, comprising about 2,304 hectares each. Approximately 779,651 hectares were leased, for a total high bonus bid value of $1.479 billion.

Environmental Issues and Mitigation

The Bering Sea is characterized by unique arctic and subarctic environmental conditions, pristine air and water quality, diverse biological resources, and multiple use including world-class commercial fisheries and Native subsistence-harvest activities. Among the challenges MMS faced in developing a comprehensive leasing program was the identification of major issues and concerns, and development of mitigating measures that would protect the environment and other users. This was accomplished in large part through the preparation of individual lease-sale environmental impact statements (EIS), which provided multiple opportunities for public testimony and written comments. The MMS was able to identify public concerns and issues associated with oil and gas exploration and development and related mitigation. As the program matured, a set of broad subjects evolved to direct study efforts. These included:

1.   Contaminant sources and effects—Studies to determine the pre-development distribution and concentration of potential contaminants

in the environment that are commonly associated with oil and gas development.

2. Pollutant transport—Studies to simulate hypothetical oil spill transport in open and ice-covered waters by means of circulation risk models.

3. Oil spill fate and effects—A vital portion of the program was to determine the fate and weathering of spilled oil, and the effects that a spill could have upon marine biota.

4. Ecosystems—Several study efforts addressed key Bering Sea ecosystems, such as the North Aleutian Basin, Yukon River Delta, and along the nearshore areas of the Alaska Peninsula.

5. Living resources—This program funded a broad array of investigations into the life history, food habits, abundance, and distribution of several vertebrate and invertebrate species important to the Bering Sea ecosystem and that are of commercial importance.

6. Endangered species—Bowhead *(Balaena mysticetus)* and gray whales *(Eschrichtius robustus)* were the focus of these investigations.

7. Environmental monitoring—If oil and gas activities had occurred, an environmental monitoring plan was developed for the Bering Sea.

8. Environmental geology—Early in the program, there were numerous geologic, geophysical, and chemical oceanographic studies to obtain information related to hazards to oil and gas activities and associated structures. Other studies were performed to gather information on seismicity processes, ice gouging, suspended particulate matter, heavy metal content, carbon budgets, and dissolved hydrocarbons.

9. Social and economic—These studies examined the effects of oil and gas development on the culture of Native communities near oil and gas related activities.

Major operational interactions associated with Bering Sea exploration varied depending on the lease-sale location. More northerly and remote sales in the Norton or Navarin basins were typified by resolving issues concerning sea ice and how operations should be conducted in such physical conditions. In more southerly areas such as the St. George or North Aleutian basins, concerns regarding interactions with commercial fishing and other socioeconomic components were particularly important.

In some cases due to the known high value placed on diverse assemblages of biota in nearshore or key habitats (e.g., Unimak Pass), tract-selection processes that occur prior to identification of a proposed sale area eliminated many potential conflicts. For example, the tract selection for proposed North Aleutian Lease Sale 75, situated to the north of the Alaska Peninsula in Bristol Bay, excluded a band of potential tracts between roughly 3 and 12 miles offshore to minimize potential effects on migratory species, coastal habitats, and human uses in those areas. This tract selection resulted in part from application of a simple, computer-based geographic information system using available aggregated resource information in conjunction with environmental protection goals (Cowles 1981). In many cases, the scale of potential interactions can be dealt with in a predictive sense and the results of such predictions applied either in an EIS analysis, or by the public to evaluate the importance of perceived or anticipated effects (e.g., on local resources or sensitive habitats). A variety of quantitative models for determining scale of effects, recovery times, or risk assessment have been developed for Bering Sea resources and are described elsewhere (e.g., Cowles and Imm 1988, Cowles 1989).

Of particular importance, perhaps most important, is the federal government's consistent encouragement for industry to work directly with affected communities or other industry in explaining proposed programs and working directly with such parties to plan and avoid conflicts. This approach continues to yield optimal achievement of human uses of marine and subsea resource potential in other areas, as it did for specific Bering Sea oil and gas explorations.

Certainly the most prominent concern (but not necessarily the most probable of all risks) of oil and gas activity is the specter of the oil spill and associated secondary issues. Although the exposure of resources or environments to oil-spill risks varies greatly depending on the location of an exploration area, oil spills have been and remain the paramount issue in terms of public concern. Other issues such as the potential effects of noise on endangered or protected species, effects of the disposal of muds and cuttings, effects of industrial activities on Native subsistence-harvest success, effects on local economies or social systems, and effects on commercial fisheries/fishing are of particular concern in Bering Sea locales. A particularly difficult issue is the concept of the cumulative effect, e.g., how do the effects of offshore oil and gas exploration compare to or interact with environmental effects associated with commercial fishing? The "local" significance of these major issues is intensified in consideration of the Bering Sea's dynamic and often extreme character. Not only is this vast region characterized by a general paucity of quantitative information, it is also noted for its extensive seasonal sea-ice incursions and a notorious frequency of intensive storms; and at least one of the proposed lease-sale areas (Navarin Basin) was the most remote area proposed for exploration in any U.S. waters. Lastly, if exploration and development did proceed to production, how would chronic or other long-term effects be monitored?

Existing laws and regulations mitigated many of these concerns and issues; operational discharges were subject to permitting under the Clean Water Act, and endangered species were protected under the Endangered

Species Act and Marine Mammal Protection Act. Where existing regulations did not provide adequate protection, the MMS developed additional mitigation. Some of the more common types of mitigation that have been used in OCS lease sales throughout Alaska and nationwide include: requirements for lessees to conduct biological or archaeological surveys when directed by the MMS; requirements for lessees to provide training programs to workers about environmental, social, and cultural conditions in the areas of activity; establishment of a Biological Task Force (an MMS advisory group on biological issues); and identification of critical or sensitive habitat for oil-spill-response planning.

Some of the additional mitigating measures that were developed in response to specific concerns for the Bering Sea were:

1. In the Norton Basin Sale 57 area, exploratory drilling during periods of broken and pack ice was prohibited until the lessee demonstrated the ability to clean up, contain, and dispose of spilled oil under such conditions. This measure would set a precedent for seasonal drilling restrictions for future lease-sale proposals in the Beaufort Sea.

2. In the North Aleutian Basin Sale 92 area, lessees were required to coordinate proposed activities with potentially affected commercial fishing organizations to minimize conflicts and to document plans for coordination and conflict resolution. This measure would also establish a precedent for similar measures in other high-use commercial fishing areas in Alaska, including Cook Inlet, and for bowhead whale subsistence-use areas in the Beaufort Sea.

3. In the Navarin Basin Sale 83 area, lessees were required to conduct monitoring programs for endangered gray whales during certain times of the year. Similar provisions for monitoring endangered species in association with oil and gas activities have since been incorporated into the Marine Mammal Protection Act and its implementing regulations to allow for incidental take of endangered species by harassment.

Significant changes in laws and U.S. policies have strengthened environmental protection since leasing began in the Bering Sea in the early 1980s. For example, the Oil Pollution Act of 1990 established new standards for oil-spill prevention, response capabilities, and limits for financial responsibility for tankers and other oil and gas activities. The Coastal Zone Management Act, under which coastal states have the opportunity to determine if federal actions are consistent with state coastal management plans, was amended to specifically provide for state concurrence with lease-sale proposals, giving states a more meaningful role in the OCS lease-sale process.

Use of Environmental Information in Decision Making

In consideration of the ambitious U.S. leasing schedules for the Bering Sea in the 1980s, a formidable amount of multidisciplinary information was needed to complete EISs on the Bering Sea natural and human environment and to support lease-sale and post-lease decisions. During and somewhat prior to that period, MMS alone spent more than $100 million on environmental studies applicable to Bering Sea pre-lease environmental assessment and post-lease monitoring-decision processes. As an example, MMS attributed 171 of the environmental and 55 of the socioeconomic studies that it sponsored as applicable to leasing decisions in the St. George Basin Planning Area, situated between Unimak Pass and the Pribilof Islands.

The process of synthesis is critically important considering the numerical scope and disciplinary diversity of the above research effort. This vital component of integrating and summarizing numerous study reports was accomplished by a series of "synthesis," "information transfer," and "information update" meetings, as well as production of hard-cover volumes such as Hood and Calder (1981). Thus, the potentially ephemeral and/or specialized literature of individual reports was made more useful, accessible, and available to a wider public. The most recent synthesis reports for specific Bering Sea planning areas are Jarvela (1984) for the Navarin Basin, Hameedi (1982) for the St. George Basin, Thorsteinson (1984) and Jarvela and Thorsteinson (1989) for the North Aleutian Basin, and Truett (1985) for the Norton Basin. Reports from MMS information transfer meetings such as Minerals Management Service (1990a) also reviewed studies that were ongoing in the Bering Sea. Additional synthesis efforts for key topics of concern such as the bowhead whale (Burns et al. 1993), effects of noise on Bering Sea pinnipeds (Johnson et al. 1989), the Yukon Delta (Thorsteinson et al. 1989), forage fish (Minerals Management Service 1987), and effects of OCS mining (Minerals Management Service 1989, 1990b) are also available.

Dissemination of information goes hand in hand with the synthesis process, and MMS-sponsored researchers have been encouraged and required to publish their results. For Bering Sea physical oceanography, of 200 reports and publications known to have been published since 1981, about 75 were sponsored by MMS; and more than 35 of them integrated physical and biological research (R. Prentki, MMS, 949 E. 36th Ave., Anchorage, AK 99508-4302, pers. comm., October 1996).

Federal law is the framework for MMS studies management and its process for moving environmental information into decision making. The OCS Lands Act as Amended (OCSLAA) (43 U.S.C. 1331-1356 [1994]) provides statutory authorization for the expenditure of funds to provide information needed for the prediction, assessment, and management of effects on the human, marine, and coastal environments of the OCS and nearshore areas that may be affected by OCS oil and gas activities. These funds are administered by the MMS Environmental Studies Program (ESP) with the intent to achieve objectives consistent with the OCSLAA. These objectives are the following: (1) provide information on the status of the environment upon which the prediction of impacts on the human, marine, and coastal environment may be based, (2) ensure that information already available or being collected is in a form that can be used in the decision-making process, and (3) provide a basis for future monitoring of OCS operations. Regarding the latter objective, a legal basis is also found in Section 20b of the OCSLAA (43 U.S.C. 1346), which states:

> ... the Secretary [of the Interior] shall ... monitor the human, marine, and coastal environments ... in a manner designed to provide time-series and data-trend information which can be used for comparison with any previously collected data for the purpose of identifying any significant changes in the quality and productivity of such environments ....

The following concepts are important in understanding how environmental information is obtained and brought to bear in a decision process: first, the greater the uncertainty surrounding a decision, the greater the need for additional information; second, the greater the information needs, the greater the likely cost in obtaining the information; and third, the more information obtained, the lower the expected cost of not having it.

Thus, the first part of applying new information to decisions is the subordinate decision process of selecting topics to study and prioritizing those topics to assure that expenditures are made wisely (i.e., to maximize reduction of uncertainty and minimize the costs of unresolved issues). When the MMS was developing its Bering Sea information base, criteria were applied to proposed studies to determine their ranking for potential funding. Program-wide criteria for evaluating proposed studies included an assessment of each proposed acquisition in terms of its mandate for study (e.g., critical, essential, useful) and timing (i.e., cannot be deferred versus can be deferred). Other criteria applied included peer assessment of scientific design quality, availability of other applicable information, and cost effectiveness of methods. The information used to make these judgments came from a variety of sources including internal and external government components and the MMS Scientific Advisory Committee, and public comment on draft regional study plans.

The Alaska OCS offshore oil and gas risk-assessment analyses have been structured (and defined by the courts) in relation to three main phases—pre-lease, exploration, and development. Because the pre-lease analyses have been required to address large geographic areas within the Bering Sea, without knowledge of exactly where exploration would occur in those areas, uncertainty is great. Instead of small surveys, large surveys were required to obtain descriptive information. Instead of small, tightly bounded models, large models with some unbounded dimensions were required. Instead of a localized focus on ecological processes, Bering Sea–wide process understanding was sought. Thus, MMS devoted the majority of its Bering Sea ESP funding and personnel resources toward broad, pre-lease information to reduce geographically broad uncertainty.

Although we do not have sufficient documentation of study expenditures in relation to a decision phase, there is no doubt that we have spent an order of magnitude more for Bering Sea pre-lease information than for post-lease. Direct examples of this process of moving information into decisions include Final EISs such as those issued for Norton Sound Lease Sale 57 (Minerals Management Service 1982) and North Aleutian Basin Lease Sale 92 (Minerals Management Service 1985). The former includes more than 300 pages and the latter more than 400 pages of discussion of the proposed lease sale, alternatives to the proposal, mitigating measures, and environmental impact predictions related to each alternative. Much of the information cited in these documents is directly referenced to Bering Sea environmental study results. A critical component of the impact assessments for various human, marine, and coastal resources is the use of oil-spill-trajectory models to predict the probabilities of oil spills intersecting shoreline segments or offshore resource areas. These models are based on the MMS-sponsored physical oceanography and ocean-circulation studies conducted in or near the proposed sale areas. For more information on recent ocean circulation and oil-spill modeling related to offshore oil and gas exploration, see Minerals Management Service (1991) or contact the MMS, Alaska OCS Region, in Anchorage, Alaska.

As the information regarding location and timing of exploration and development improved, the scale of desired environmental information and the related uncertainties decreased. During post-lease phases, because exploration locations and issues are more clearly defined, much more focused efforts such as monitoring studies were implemented. For example, Houghton et al. (1987) and Brueggeman (1987) provided research designs for monitoring Bering Sea hydrocarbon pollution and endangered whales, respectively. The former may be of particular interest to parties currently designing studies of other potential Bering Sea contaminants. With the termination of Bering Sea post-lease activity, as well as relinquishment of leases (see Exploration History, below), the MMS environmental studies mandate for the Bering Sea was significantly lessened. Thus, the MMS environmental studies effort has been significantly reduced in the Bering Sea, with the exception of completion of certain physical oceanographic studies, social indicator monitoring, and monitoring of protected species as part of ongoing region-wide studies. Several of the ongoing studies, although managed by entities other than MMS and cited elsewhere in this book, were cooperatively designed and initiated by MMS in the early- to mid-1980s, including seabird colony monitoring and the Alaska Marine Mammal Tissue Archival Program (AMMTAP). The latter has served as a model for national monitoring programs.

Table 2.   Summary of deep stratigraphic test wells drilled in the Bering Sea.

Sale number and planning area	Well location and G&G permit	Operator	Drilling unit	Depth (feet)	Year
Sale 70 St. George Basin	76-11 3-1 Blk 459	ARCO	Ocean Ranger	13,771	1976
Sale 70 St. George Basin	82-19 3-1 Blk 390	ARCO	Sedco 708	14,626	1982
Sale 57 Norton Basin	80-7 3-1 Blk 197	ARCO	Dan Prince	14,683	1980
Sale 57 Norton Basin	82-16 3-2 Blk 273	ARCO	Key Singapore	14,889	1982
Sale 92 N. Aleutian Basin	82-18 4-7 Blk 666	ARCO	Sedco 708	17,155	1982-83
Sale 83 Navarin Basin	82-17 1-8 Blk 801	ARCO	Sedco 708	16,400	1983

Exploration History

Exploration in the Bering Sea has taken place in three stages—collection of geophysical deep seismic data, drilling of deep stratigraphic test (DST) wells, and exploratory drilling.

Geophysical deep seismic data collection typically takes place prior to a lease sale to gather information about the geologic potential of a planning area in preparation for a competitive lease sale. Over 300,000 line miles of deep seismic data have been collected in the Bering Sea OCS under MMS permit.

The DST wells are also conducted prior to a lease sale to improve and refine industry's knowledge of the oil and gas potential of a region. The DST wells are generally drilled off of a geologic structure to the basement and are logged and tested more extensively compared to exploratory wells. Six DSTs have been drilled in the Bering Sea (see Fig. 1 and Table 2).

Twenty-four exploratory wells have been drilled in the Bering Sea (see Table 3). There were no reported discoveries of oil or gas, and all of the wells were permanently plugged and abandoned. The absence of a discovery of economically commercial quantities of hydrocarbons has generally discouraged industry interest in additional leasing and exploration in previously drilled planning areas. Industry has continued to express interest in drilling in the North Aleutian Shelf Planning Area, which is still considered

Table 3. Summary of exploratory wells drilled in the Bering Sea.

Sale number	Lease number	Operator	Drilling unit	Spud date and (abandonment date)	Depth (feet)
Sale 57	Y-0414	Exxon	Rowan Middle Town	6/19/84 (7/23/84)	3,636
	Y-0398	Exxon	Key Hawaii	7/2/85 (7/23/85)	6,913
	Y-0407	Exxon	Key Hawaii	7/24/85 (8/11/85)	7,867
	Y-0425	Exxon	Key Hawaii	8/13/85 (8/22/85)	6,093
	Y-0430	Exxon	Rowan Middle Town	7/25/84 (8/16/84)	4,951
	Y-0436	ARCO	Key Hawaii	6/25/84 (8/19/84)	10,950
Sale 70	Y-0530	Exxon	Doo Sung	6/29/84 (9/4/84)	8,800
	Y-0519	Chevron	Sedco 712	7/20/84 (9/25/84)	11,595
	Y-0537	ARCO	Sedco 708	8/4/84 (10/30/84)	12,456
	Y-0527	Exxon	Doo Sung	9/13/84 (11/19/84)	12,433
	Y-0466	Mobil	Sedco 712	9/29/84 (11/3/84)	8,085
	Y-0511	ARCO	Sedco 708	11/7/84 (12/8/84)	10,862
	Y-0454	Shell	Ocean Odyssey	11/20/84 (1/25/85)	10,277
	Y-0477	Gulf	Doo Sung	11/27/84 (1/23/85)	9,592
	Y-0463	Shell	Ocean Odyssey	1/26/85 (3/26/85)	8,510
	Y-0511-1A	ARCO	Sedco 708	12/17/84 (2/14/85)	13,937
Sale 83	Y-0560	AMOCO	Sedco 708	8/22/85 (10/8/85)	9,085
	Y-0583	Exxon	Doo Sung	8/31/85 (10/12/85)	11,570
	Y-0586	ARCO	Sedco 712	8/5/85 (10/23/85)	13,741
	Y-0599	Exxon	Doo Sung	6/14/85 (8/30/85)	11,536
	Y-0639	AMOCO	Sedco 708	6/19/85 (8/20/85)	10,045
	Y-0673	AMOCO	Ocean Odyssey	9/1/85 (10/21/85)	7,962
	Y-0707	AMOCO	Ocean Odyssey	6/7/85 (8/27/85)	11,030
	Y-0719	AMOCO	Sedco 708	10/12/85 (11/25/85)	8,708

a frontier area. During the development of the MMS 5-year oil and gas leasing program for 1997-2002, industry indicated a small interest in additional leasing in the Norton Basin Planning Area. However, under the final 5-year program, no lease sales in the Bering Sea planning areas are proposed for the 1997-2002 period (Minerals Management Service 1996).

The MMS issued an updated National Assessment of Oil and Gas Resources on the OCS in May 1996 (Sherwood et al. 1996). The assessment is based on complex modeling of geologic, engineering, and economic variables and probabilities. Under anticipated economic conditions, the Bering Sea planning areas were assigned small to negligible estimates for economically recoverable oil or gas resources.

Conclusions and Findings

The MMS has held four OCS lease sales in the Bering Sea. More than $100 million has been spent on environmental studies to support OCS lease-sale proposals in the Bering Sea region. Major issues and concerns associated with OCS exploration activities have been identified and mitigated. Twenty-four exploratory wells have been drilled without finding commercial quantities of oil or gas. Except for the North Aleutian Basin, where there has been no exploratory drilling, there is limited interest in future near-term leasing and exploration opportunities in other OCS Bering Sea planning areas.

References

Burns, J.J., J.J. Montague, and C.J. Cowles (eds.). 1993. The bowhead whale. Society for Marine Mammalogy Special Publication Number 2. Lawrence, KS. 787 pp.

Brueggeman, J.J. 1987. Monitoring the winter presence of bowhead whales in the Navarin Basin through association with sea ice. OCS Study MMS 87-0028. MMS, Alaska OCS Region, Anchorage, AK. 186 pp.

Cowles, C.J. 1981. Application of a simple algorithm in portrayal of environmental sensitivity to effects of offshore oil and gas development. In: Proceedings of the 32nd Alaska Science Conference, pp. 168-169. (Abstract.)

Cowles, C.J. 1989. Biological models as predictive tools for assessment of potential effects of Alaska outer continental shelf oil and gas exploration. In: Oceans '89: An international conference addressing methods for understanding the global ocean. Vol. 1, Fisheries, global ocean studies, marine policy and education, oceanographic studies. Marine Technology Society and Institute of Electrical and Electronic Engineers. IEEE Publication No. 89CH2780-5, pp. 307-310.

Cowles, C.J., and J.L. Imm. 1988. Review of studies of man-induced noise on marine mammals of the Bering, Chukchi, and Beaufort seas and how the results have been applied to federal offshore oil and gas management decisions. In: Port and ocean engineering under arctic conditions. Vol. II, Symposium on noise and marine mammals. Geophysical Institute, University of Alaska Fairbanks, pp. 1-8.

Hameedi, M.J. (ed.). 1982. Proceedings of a synthesis meeting: The St. George Basin environment and possible consequences of planned offshore oil and gas development, Anchorage, AK, April 28-30, 1981. NOAA and BLM, Juneau, AK. 162 pp.

Hood, D.W., and J.A. Calder (eds.). 1981. The eastern Bering Sea shelf: Oceanography and resources, Vol. I and II. Published by the Office of Marine Pollution Assessment, NOAA and BLM. 1,339 pp. (Distributed by the University of Washington Press, Seattle, WA 98105.)

Houghton, J.P., W.M. Blaylock, J.E. Zeh, and D.A. Segar. 1987. Bering Sea monitoring program: Proceedings of a workshop and sampling design recommendations, Anchorage, AK, January 1987. OCS Study MMS 87-0048. MMS, Alaska OCS Region, Anchorage, AK. 128 pp.

Jarvela, L.E. (ed.). 1984. The Navarin Basin: Environmental and possible consequences of planned offshore oil and gas development. NOAA and MMS, Juneau, AK. 157 pp.

Jarvela, L.E., and L.K. Thorsteinson (eds.). 1989. Proceedings of the Gulf of Alaska, Cook Inlet, and North Aleutian Basin information update meeting, Anchorage, AK, February 7-8, 1989. OCS Study MMS 89-0041. NOAA and MMS, Alaska OCS Region, Anchorage, AK. 179 pp.

Johnson, S.R., J.J. Burns, C.I. Malme, and R.A. Davis. 1989. Synthesis of information on the effects of noise and disturbance on major haulout concentrations of Bering Sea pinnipeds. OCS Study MMS 88-0092. MMS, Alaska OCS Region, Anchorage, AK. 267 pp.

Minerals Management Service. 1982. Norton Sound Oil and Gas Lease Sale 57 Final EIS. MMS, Alaska OCS Office, Anchorage, AK.

Minerals Management Service. 1985. North Aleutian Basin Oil and Gas Lease Sale 92 Final EIS, Vol. 1 and 2. OCS EIS/EA MMS 85-0052, MMS, Alaska OCS Region, Anchorage, AK.

Minerals Management Service. 1987. Forage fishes of the southeastern Bering Sea: Proceedings of a conference, November 4-5, 1986. OCS Study MMS 87-0017, MMS, Alaska OCS Region. 90 pp.

Minerals Management Service. 1989. Mercury in the marine environment: Workshop proceedings. OCS Study MMS 89-0049, MMS, Alaska OCS Region. 199 pp.

Minerals Management Service. 1990a. Proceedings of Alaska OCS Region third information transfer meeting, Anchorage, AK, January 30-February 1, 1990. OCS Study MMS 90-0041, MMS, Alaska OCS Region. 220 pp.

Minerals Management Service. 1990b. Design of baseline and monitoring studies for the OCS mining program, Norton Sound, Alaska: Workshop proceedings. OCS Study MMS 90-0059, MMS, Alaska OCS Region, Anchorage, AK. 130 pp.

Minerals Management Service. 1991. Offshore oil-spill movement and risk assessment workshop. Interim Report. OCS Study MMS 91-0007, MMS, Alaska OCS Region, Anchorage, AK. 27 pp. plus appendices.

Minerals Management Service. 1996. Proposed OCS Oil and Gas Leasing Program 1997 to 2002. Decision Document, February 1996. MMS, Washington, DC.

Sherwood, K.W., J.D. Craig, and L.W. Cooke. 1996. Endowments of undiscovered conventionally recoverable and economically recoverable oil and gas in the Alaska federal offshore as of January 1995. OCS Report MMS 96-0033. MMS, Alaska OCS Region, Anchorage, AK. 17 pp.

Thorsteinson, L.K. (ed.). 1984. Proceedings of a synthesis meeting: The North Aleutian shelf environment and possible consequences of offshore oil and gas development, Anchorage, AK, March 9-11, 1982. NOAA, Outer Continental Shelf Environmental Assessment Program (OCSEAP), and MMS, Alaska OCS Office, Juneau, AK. 159 pp.

Thorsteinson, L.K., P.R. Becker, and D.A. Hale. 1989. The Yukon Delta: A synthesis of information. OCS Study MMS 89-0081. NOAA, OCSEAP, and MMS, Alaska OCS Region. 89 pp.

Truett, J.C. (ed.). 1985. Proceedings of a synthesis meeting: The Norton Basin environment and possible consequences of offshore oil and gas development, Denali National Park, AK, June 5-7, 1984. OCS Study MMS 85-0081. NOAA, OCSEAP, and MMS, Alaska OCS Region, Anchorage, AK.

CHAPTER 37

Summary of TINRO Ecosystem Investigations in the Bering Sea

Vyacheslav P. Shuntov and Vladimir I. Radchenko
*Pacific Research Institute of Fisheries and Oceanography (TINRO),
Vladivostok, Russia*

Introduction

Investigations in the Bering Sea were carried out by the Pacific Research Institute of Fisheries and Oceanography (TINRO) from an ecological point of view. The composition, structure, interannual dynamics, and function of the pelagic and demersal communities were studied on a macroecosystem scale. This work was primarily undertaken by the Laboratory of Applied Biocenology (i.e., ecology) and the Laboratory of Research of Plankton of Far-Eastern Seas. Ecosystem investigations were included as part of fishery research and were supported by a research vessel. The mission of the first expedition in 1983 was to assess demersal fishes in Karagin Bay. Only demersal surveys were conducted until 1985, with pelagic surveys starting after 1986. Eleven expeditions have been completed.

The Program

Expeditions by TINRO involved a wide range of meteorological, hydrological, hydrochemical, trawl-acoustic, plankton, and icthyoplankton investigations in the Russian far-eastern economic zone and in some international waters. The distribution, migration, feeding, and physiology of nektonic organisms were studied. Nekton was assessed mainly in the epipelagic (0-200 m) and pelagic (0-1,000 m) layers. Data on primary production, bacteria, and protozoa were collected, examined, and processed onboard the research vessel.

The original database contains information on species composition, biomass, and resources of zooplankton and nekton from more than 2,000 oceanographic, plankton, and trawl stations. This database includes the results of bioanalysis of more than 65,000 fishes and 620,000 measurements of fishes. The feeding habits of tens of thousands of specimens of more than 30 nekton species were studied to develop community trophic structure; however, all of these data did not suffice to describe the complex ecosystem cycle.

To further enhance our understanding of the Bering Sea ecosystem, a thorough review of the literature on the eastern Bering Sea was conducted. That review, when combined with some parameters from the western Bering Sea, allowed for the derivation of a model for energy flow throughout the Bering Sea. This model was presented at the PICES Scientific Workshop in Seattle, 1991.

Results

The most substantial data about community structures of the Bering Sea were collected in the 1980s during the historical peak of fishery production in this region. Compared to the Okhotsk Sea during the same period, the Bering Sea ecosystem functioned more effectively, characterized by the output of high trophic level production. Biomass and production per unit area of the second trophic level organisms were higher in the Okhotsk Sea, but top trophic level organisms were more abundant in the Bering Sea (Table 1). Results of research on the Bering Sea pelagic communities were summarized by Shuntov et al. (1993).

Beginning in the 1990s, substantial reorganization of pelagic communities of the far-eastern seas occurred and was likely related to large changes in global climate and oceanographic conditions in the northern Pacific. Changes in composition and biomass were recorded in both plankton and nekton communities (Tables 2-4). Results of these observations were presented at the PICES Second Annual Meeting in Seattle in 1993. Information gained by TINRO scientists about the dynamics and status of pelagic communities is being prepared by V.I. Radchenko.

Bering Sea nekton is made up of 400 fish species and more than twenty squid species. TINRO scientists have separated the nekton communities of the Bering Sea into three groups: (1) the communities of the shallow Gulf of Anadyr and Karagin Bay are classified as coastal; (2) the communities of Navarin, Koryak, and Olyutorsk bays are classified as neritic; and (3) the epipelagic and mesopelagic communities of deepwater basins are classified as offshore. The epipelagic nektonic communities appear to function in dynamic environmental conditions resulting in periodic changes. Thus, epipelagic communities are characterized by a small number of species, which are further characterized by substantial fluctuations in abundance and biomass. Periods of fluctuation coincide with the phases of climatic-oceanographic change.

The dynamics of environmental conditions of the Bering Sea epipelagic zone are determined by periodic revision of current patterns under the influence of changes in atmospheric circulation patterns (Radchenko 1993). During the 1970s and 1980s (and possibly in the 1930s and 1940s), the greater part of Pacific Ocean water was flowing with the Alaskan Stream into the Bering Sea. The general scheme of water circulation had been characterized by cyclonic patterns before the 1990s. During the 1950s, 1960s, and 1990s, significant segments of Pacific Ocean waters flowed

Table 1. Biomass and production of the main groups of organisms in the Bering and Okhotsk seas in the 1980s. Units in g/m^2.

	Biomass		Production	
Groups	Bering Sea	Okhotsk Sea	Bering Sea	Okhotsk Sea
Zooplankton (0-200 m)				
Nonpredatory				
Summer	136	205	548	754
Winter	74	131	148	217
Predatory				
Summer	37	39	122	133
Winter	56	60	61	47
Zooplankton (whole year)				
Nonpredatory	–	–	1,343	1,678
Predatory	–	–	313	320
Zoobenthos				
Nonpredatory	83	125	113	199
Predatory	5.4	12.7	7.5	15.0
Pelagic fishes	16.8	15.8	8.4	7.9
Demersal fishes	5.0	3.2	1.5	0.8
Total nekton	18.5	16.4	13.8	9.8
Total nektobenthos	6.0	4.1	2.1	1.1
Mammals and birds	0.6	0.1	–	–

Table 2. Biomass ($\times$ 1,000 t) of zooplankton in the southwestern Bering Sea in summer.

Date	Small plankton (<1.5 m)	Medium plankton (1.5-3.5 mm)	Large plankton (>3.5 mm)
1989	31,074	5,481	76,973
1991	7,301	1,909	33,485
1993	8,098	3,072	67,127

Table 3.   Interannual changes in biomass of different macroplankton groups in the southwestern Bering Sea.

Date	Euphausiids	Hyperiids	Copepods	Chaetognaths	Other
1989	4,665	645	56,735	14,114	802
1991	2,891	457	6,985	20,777	2,374
1993	6,131	1,000	18,281	39,201	2,514

Table 4.   Concentrations (t/km^2) of nekton and jellyfish in the upper epipelagic (0-50 m) in the southwestern Bering Sea in June-July 1991 and 1993.

Group	1991	1993
Fishes	0.98	0.47
Squids	0.21	0.25
Total nekton	1.19	0.72
Jellyfish	0.02	0.55

into the Bering Sea with the western subarctic gyre currents. This phenomenon determines direct Pacific water inflow from Near Strait to Kamchatka Strait through the southern region of the Kamchatka Basin. Eastward dislocation of the cyclonic Bering Sea gyre occurs from the Kamchatka Basin to the western part of the Aleutian Basin. The Central Bering Sea Current (CBSC) shifts to the same direction, to the eastern Bering Sea slope and outer shelf. These changes in the current pattern are unfavorable for the survival of walleye pollock *(Theragra chalcogramma)* roe and larvae. An eastern shift of the CBSC will increase flow intensity of the pollock larvae and fry into the Gulf of Anadyr where environmental conditions are more severe than in the Navarin region.

Only a few species make up a significant proportion of total biomass of epipelagic nekton. Pollock predominated in all epipelagic communities with the exception of coastal communities during the summer. During the second half of the 1980s, pollock biomass was assessed at 20-25 million t and annual production at 8.5-10.5 million t in the Bering Sea (Shuntov et al. 1993). After 1988, pollock abundance in the Bering Sea pelagic community decreased as a result of the influence of density factors and of changed environmental conditions. Presently, pollock biomass is assessed at 6 million t (annual production at about 3 million t).

Table 5. **Changes in biomass (million t) and consumption (million t) for major fish species and groups in the Bering Sea during the 1980s and 1990s.**

	1980s		1990s	
Species and groups	Biomass	Consumption	Biomass	Consumption
Walleye pollock	22.5	220	6	66
Mesopelagic fishes[a]	9.3	46	10.5	51.9
Pelagic squid	0.6	5.9	0.65	6.4
Herring	0.7	10.6	0.9	12.9
Capelin	0.36	2	0.5	2.8
Smooth lumpsucker	0.25	4	0.25	4
Pink salmon[b]	0.09	0.1	0.1	0.11
Chum salmon	0.1	0.56	0.126	0.71
Sockeye salmon	0.09	0.51	0.1	0.57
Chinook salmon	0.03	0.14	0.03	0.14
Coho salmon	0.005	0.02	0.003	0.01
Dolly Varden	+	+	0.08	0.01

[a] Average part of total grouping biomass, which consists of species migrated in epipelagic zone at night.

[b] Biomass for salmonids are adapted to warm season.

Light-ray lanternfish *(Stenobrachius leucopsarus)* predominate in the mesopelagic community under stable environmental conditions. Biomass of mesopelagic fishes is assessed at 15-20 million t in the Bering Sea, of which 9.1-10.5 million t rise every night in the epipelagial during diurnal vertical migrations. During the second part of the 1980s, annual consumption of zooplankton by nekton in the epipelagic zone came to 265 million t, from which more than 90% were nonpredatory plankton. Euphausiids made up 36.6% of annual planktonic diets of fish and squid, 48.0% were copepods, 6.6% were amphipods, 1.2% were pteropods, 2.3% were jellyfish, 2.2% were sagittas, and 3.1% were others.

Annual consumption of zooplankton resources in the 1990s by nekton have decreased to 158 million t in relation to the decrease of pollock biomass in the Bering Sea. Surplus nonpredatory zooplankton production creates favorable conditions for an increase of biomass and productivity of predatory zooplankton, especially chaetognaths (i.e., *Sagitta* spp.), resulting in an increase of short-lived nektonic species abundance and in Dolly Varden *(Salvelinus malma)* migration from the coastal zone to the epipelagic zone of the deepwater basins (Table 5). However, the total increase in biomass of this species is not significant when compared to the pollock biomass decrease.

Decline of nekton biomass has continued in the Bering Sea epipelagic zone. We believe that there will be changes in the overall biota in this zone in the future. Among these changes could be the stabilization of biomass of predatory plankton due to consumption by pelagic fishes (especially herring, *Clupea pallasi*). The abundance of herring in the Korf-Karagin region is increasing in the western Bering Sea, where herring biomass is now at 400,000 t. In December 1993, the abundance of herring fingerlings was assessed at 120 million individuals. This is only half the abundance of pollock fingerlings. In the late 1980s, the ratio of herring to pollock fingerlings was 1:25 to 1:100. In the future, an increase of herring abundance to the level of the 1960s is possible.

The trophic structure of epipelagic nekton in the Bering Sea is characterized by the existence of a one-species "block" (pollock) and two multi-species "blocks" (pelagic squid and mesopelagic fishes) through which the organic matter of plankton pass to higher trophic levels. Resources of these "blocks" are exploited by euryphagous fishes. The value of their consumption represents a significant part of the biomass of species which are included in "blocks." Capelin *(Mallotus villosus)* are consumed by carnivorous fish in significant proportion to its relatively low biomass; however, the contribution of capelin in total biomass to ecosystem stability in the Bering Sea pelagic zone is low.

Currently, planktonic and nektonic communities of the Bering Sea epipelagic zone are in a transitional state. The situation is characterized by an increase in the proportion (total biomass) of predatory plankton and production of planktonic communities. It is also characterized by the abundance of short-lived fishes. These processes compensate for the decrease in nonpredatory zooplankton consumption by pollock. Fish productivity of the Bering Sea pelagic zone as a whole would have been comparatively low in this period, at about 2 t/km^2. Such a low level of fish productivity would persist pending an increase in herring biomass, which is expected in the Bering Sea in the latter half of the 1990s.

References

Radchenko, V.I. 1993. Long-term variability in the Bering Sea surface geostrophic circulation and its possible influence on the pelagic fish community. In: Abstracts of PICES Second Annual Meeting, Seattle, WA, pp. 29-30.

Shuntov, V.P., A.F. Volkov, O.S. Temnykh, and Ye.P. Dulepova. 1993. Walleye pollock in the ecosystems of the far-eastern seas. TINRO Monograph, Vladivostok, 425 pp. (In Russian.)

CHAPTER **38**

Summary, Conclusions, and Recommendations

Alan M. Springer
Institute of Marine Science, University of Alaska Fairbanks, Fairbanks, Alaska

Much of the interest in dynamics in the Bering Sea, now and in the past, has been spurred by concerns over the stability and sustainability of its vast living resources. Particularly prominent today are depressed populations of several species of marine mammals, notably great whales, Steller sea lions, fur seals, harbor seals, and sea otters, and of additional species of considerable economic importance, such as king crabs, shrimp, and Pacific Ocean perch. The reason for the collapse of whales, shrimp, and Pacific Ocean perch is known—they were killed by commercial fisheries. The recent decline of sea otters in the Aleutian Islands is thought to have been caused by increased predation. The reasons for diminished populations of other species are not known, or at least not agreed upon, and have been the stuff of extensive, often rancorous debate. No less dramatic, however, have been spectacular increases of certain fishes, such as flatfishes, walleye pollock that grew in abundance by nearly an order of magnitude between the 1960s and 1980s, and Pacific salmon that provided record harvests across the northeastern North Pacific for many years in the 1980s and 1990s before collapsing in some regions, notably the Bering Sea, in 1997 and 1998.

This volume is the most recent in a growing series of publications devoted to the Bering Sea and is basically a report on certain advances that have been made in our understanding of the ecosystem—what it is and why it behaves as it does—primarily in regard to issues of interest to PICES. Thus, many of the papers presented here, as the general title indicates, address dynamics—dynamics of physical processes like meteorology and ocean circulation, and of ecosystem processes that are important to biomass yield at higher trophic levels in pelagic food webs. Others provide more descriptive information that will be useful in future dynamic contexts. My brief summary touches on some of the highlights of the foregoing chapters, but each must be read to fully appreciate the state of knowledge revealed in their pages.

Physical Dynamics

Studies of meteorology have broadened in recent years with (1) the recognition of teleconnections between the El Niño Southern Oscillation (ENSO) in the equatorial Pacific and meteorological variability over the North Pacific and Western Arctic and (2) a heightened awareness of the importance of meteorology to physical oceanographic conditions and the biological realm (Trenberth and Hurrell 1994; Roach et al. 1995; Mantua et al. 1997; Klyashtorin 1997; Springer 1998; Francis et al. 1998; Niebauer, in press; Niebauer et al., chapter 2; Schumacher and Alexander, chapter 6; Wyllie-Echeverria and Ohtani, chapter 21, this volume). Of particular interest are relationships between ENSO; fluctuations in the strength and position of the Aleutian Low pressure system between quasi-stable states with periods of decadal-scale proportions, otherwise known as regimes or the Pacific interDecadal Oscillation (PDO); oscillations in the Earth Rotation Velocity Index and the Atmospheric Circulation Index; and physical oceanographic variability in the subarctic North Pacific and Bering Sea. One significant example described here is the relationship between sea ice in the Bering Sea and the PDO (Niebauer et al., chapter 2; Niebauer, in press).

Good spatial descriptions of the physical seascape of the Bering Sea have existed for some time. The hydrographic domains across the eastern and northern shelf (inner, middle, outer) were elucidated in the 1970s and 1980s (Stabeno et al., chapter 1; McRoy, chapter 31; Hood, chapter 32) and are similar to domains in the western Bering Sea, with the exception that domain boundaries on the eastern shelf are comparatively stationary, whereas those on the narrow western shelf are dynamic and depend on the position of the Kamchatka Current (Khen, chapter 7). Another striking difference is that on the eastern shelf, the shallow inner domain is well mixed, a hallmark characteristic, whereas the inner domain of the western shelf is not.

Progress has been made in confirming earlier views about the general circulation of the Bering Sea and in discovering additional details about important broadscale and mesoscale processes and variability. Most attention has focused on volume transport and the trajectory of Alaskan Stream inflow through Aleutian Island passes, the generation and propagation of eddies over the basin and along the shelf edge, and characteristics of flow fields along the edge of the shelf and around the Pribilof Islands (Stabeno et al., chapter 1; Kowalik, chapter 4; Reed and Stabeno, chapter 8; Sapozhnikov, chapter 11). As a result, we now have a much improved understanding of dynamics. For example, transport variability in the Bering Sea gyre is >50% due to variations in wind-driven transport and inflow of Alaskan Stream waters through Aleutian Island passes, principally Near Strait (Stabeno et al., chapter 1), and transport of water out of the Bering Sea to the Arctic Ocean has apparently declined by 25-30% in the past 50 years (Roach et al. 1995; T. Weingartner, unpubl. data).

In contrast, little attention has been paid recently to the Alaska Coastal Current on the eastern shelf and it is hardly mentioned in any of the papers in this book. It originates on the southeastern shelf, contributes about 30% of the volume transport through Bering Strait, and can be traced well into the western Beaufort Sea (Coachman et al. 1975, Coachman 1993).

Chemical Dynamics

Chemical dynamics in the Bering Sea are forced by both physics and biology, and general principles of the dynamics of plant nutrients in particular have existed for some time (Whitledge and Luchin, chapter 10). Bottom waters in the basin are supplied by deep inflow from the North Pacific, are perhaps the oldest in the world, and contain the highest concentrations of naturally occurring macronutrients (e.g., nitrate, phosphate, silicate) in the world ocean (Coachman et al., chapter 13). Residence time for deep water displaced upward by bottom water flowing in from the North Pacific is 250-300 years. Over the shelf, nutrient concentrations are governed by rates of biological uptake and remineralization, by physical structural fronts that retard cross-shelf flow of nutrient-laden basin waters, and by countervailing physical features that distribute basin waters to the shelf. Interannual variability in nutrient flux over the shelf is related to the frequency of storms that force onwelling of basin waters laterally to impoverished middle and inner regions, as well as vertically to the euphotic zone by mixing of the water column. Such dynamics in turn have a large effect on the annual amount of primary production.

Currents also supply nutrients to regions that are otherwise limited. The most extreme example is on the northern shelf where the flow of Anadyr Water, which originates at depth along the shelf edge in the northwestern Bering Sea and is thus highly enriched in nutrients, creates a chemostat-like environment that fuels levels of primary production of world record proportions (McRoy et al. 1987, Springer and McRoy 1993). In the western Bering Sea, the strong Kamchatka Current likely plays a similar role in delivering nutrients to the narrow shelf there.

Eddies may also be important to nutrient budgets in the Bering Sea and thus to primary production. Sapozhnikov et al. (chapter 17) believe eddies off the Kamchatka Peninsula and in the Aleutian Basin have an important role in nutrient distributions (horizontal and vertical) and thus also in primary productivity. They cite as evidence depleted nitrate and supersaturated oxygen in eddies. This conforms to evidence of enhanced primary production in eddies along the shelf edge derived from remote sensing imagery (Eslinger 1994). Indeed, eddies may have further biological significance as described by Schumacher and Stabeno (1994) in the context of pollock survival and production.

Biological Dynamics

Phytoplankton

The oceanic domain over the basin, the largest domain of the Bering Sea, is characterized as high nutrient/low chlorophyll (HNLC); as in much of the open North Pacific, surface waters always have concentrations of macronutrients, particularly nitrogen, sufficient for phytoplankton growth, yet phytoplankton biomass is consistently low. The condition has been explained in terms of iron limitation of phytoplankton growth, phytoplankton uptake kinetics and competition for several forms of nitrogenous nutrients (nitrate, ammonia, urea), and grazing (Frost 1987, Martin and Fitzwater 1988, Miller 1993, Banse 1995). Here, Shiomoto (chapter 15) emphasizes phytoplankton size structure in the basin and controls that have to do with dynamics and standing stocks of nutrients, particularly nitrate, ammonia, silicate, and iron. Picoplankton (< 2 μm) numerically dominate the floral community in the basin, although nano- and microplankton can contribute substantially to biomass and production. Shiomoto subscribes to the model of ammonium inhibition of nitrate uptake as the explanation for the small size of phytoplankton and the HN condition. He believes that silicate might further limit the size of plankton, since diatoms, which are generally larger, need it, and he discounts iron as the reason for low primary production. But he presents no measurements of iron, which apparently do not exist, so this remains an unanswered question.

There are related questions about the actual level of primary production in the basin. Sapozhnikov et al. (chapter 17) contend, as does conventional wisdom, that it is low compared to the shelf. But Maita et al. (chapter 16) believe that primary production in the basin is much higher than previously thought, up to 250 g C per m^2 per yr compared to earlier estimates that average just 60 g C per m^2 per yr (Springer et al. 1996). They ascribe the high production to the occurrence of three blooms each year and a low grazing stress by mesozooplankton, primarily calanoid copepods.

An extension of this issue is the question of what accounts for the Green Belt along the shelf edge (Springer et al. 1996), where primary production is higher than over either the shelf or the basin, including even the higher estimates of Maita. Various possibilities have been considered, such as upwelling, stabilization of the water column by the shelf break front, and one that invokes nutrient dynamics. Whereas the basin is a HNLC region where primary production clearly is not limited by nitrate, the euphotic zone of the shelf is depleted of nitrate during most of the productive season and clearly is not iron limited. Thus, at the interface of basin and shelf waters at the shelf edge, each may contribute essential nutrients—nitrate from basin waters and iron from shelf waters—forming an "Iron Curtain" that leads to the Green Belt phenomenon (P. McRoy, pers. comm.).

Over the shelf, such things are better known and have to do primarily with a deterministic production regime that depends on light intensity, water column stratification, and nutrient supply. The annual production cycle typically begins with an ice edge bloom in the low-salinity buoyant layer near the retreating ice front (McRoy and Goering 1974, Niebauer and Alexander 1985). Net production can be intense, up to at least 13 g C per m^2 per day, but growth is generally short-lived because of rapid nutrient depletion in the thin stratified layer or because winds break it down (Niebauer et al. 1995). Essentially all of the production at the ice edge is lost to pelagic food webs because there are no grazers present at that time (Coyle and Cooney 1988).

The ice edge bloom is succeeded in most years by a second, more prolific bloom following deeper thermal stratification of the water column that persists until nutrients are stripped from the euphotic zone. Sukhanova et al. (chapter 22) point out the magnitude of interannual variability in production over the middle shelf that is driven by storms, as previously described by Sambrotto et al. (1986). Storms can effectively resupply nutrients to the euphotic zone through onwelling of deep water laterally across the shelf and vertical mixing of the water column. Storms account for 10-50% of total production, an important proportion of which is new (nitrate) production, and are thus the most important factor influencing the annual production budget.

Production over the eastern shelf is not especially high, with the bulk occurring during the spring bloom and little thereafter (Walsh and McRoy 1986). On the western shelf, however, Sapozhnikov et al. (chapter 17) estimate that post-bloom production amounts to 4-6 g C per m^2 per day, rates typical of growth during the bloom. They also note that it is fueled by ammonia and urea, i.e., recycled nitrogen, and thus does not contribute to export production.

So, which plants do all this growing? Sukhanova et al. (chapter 22) count 266 species among the planktonic flora of the Bering Sea. They say the high taxonomic diversity exists because of the great latitudinal expanse of the Bering Sea and its position sandwiched between the Arctic Ocean and temperate North Pacific. Patterns in distribution are related to both season and location (i.e., habitats): for example, picoplankton (diameter $\leq$2-2.5 μm) are more abundant by approximately an order of magnitude at most times over the outer shelf and basin than over the middle shelf and coastal zone. But they are small and contribute generally only a small to moderate amount (0.2-16%) to total biomass. Picoplankton might contribute more to standing stocks of phytoplankton if they were not grazed so efficiently by microheterotrophic protozoa. They are small and those that are not grazed sink very slowly so they contribute little to benthic production. Nevertheless, because they are grazed efficiently, their annual production is much higher than their biomass indicates, as might their contribution to the transfer of carbon and energy to higher trophic levels in the pelagic food web.

Issues of taxonomic composition of the phytoplankton community are major ones, since not all of the many species are of equal importance in the scheme of food webs. During the PROBES studies in the 1970s it was proposed that as much as half of the annual primary production on the southeastern shelf was contributed by *Phaeocystis pouchetti*, a colonial haptophyte that is not efficiently grazed by herbivorous zooplankton (Kokur 1982). The other half was by diatoms that are among the favorite foods of grazers and were consumed almost completely. Temporal variability in production of various size classes of phytoplankton, or in the ratio of *Phaeocystis* to diatoms, therefore, could be expected to affect production at higher trophic levels. Data presented by Sugimoto and Tadokoro (1997) reveal considerable interannual and decadal variability in summer phytoplankton standing stocks in the Bering Sea, as well as intriguing patterns in the relationship of phytoplankton to zooplankton abundance that hint at changes in trophic dynamics of likely significance to food web structure and efficiency (Springer 1998).

Zooplankton

Unfortunately, there is not a chapter devoted to zooplankton, in spite of the crucial roles they play in marine food webs and the dynamic nature of their life histories and abundances at many time scales. It is not possible here to adequately cover such an important field but only to mention certain known or suspected facts about them.

Over the shelf, each of the domains has characteristic zooplankton communities and differential grazing stress is responsible for contrasting pathways of energy flow through regional food webs. The large calanoid copepods *Neocalanus cristatus*, *N. plumchrus*, *Eucalanus bungii*, and *Metridia pacifica* are efficient grazers but are restricted to the outer and oceanic domains because of physical fronts (Cooney 1981, Smith and Vidal 1984). Herbivorous zooplankton in the middle and inner domains are less efficient at harvesting phytoplankton of the spring bloom and a greater proportion of the production in these domains goes to the bottom than in the outer and oceanic domains (Cooney and Coyle 1982). Thus, benthic communities are much more developed in the middle domain than in the outer and oceanic domains, where pelagic food webs predominate (Walsh and McRoy 1986).

Zooplankton growth rates and biomass vary between years and over longer intervals (Vidal and Smith 1986, Sugimoto and Tadokoro 1997) as does the ratio of herbivorous to predatory species (Shuntov et al. 1996). The biomass of jellyfish on the eastern shelf has averaged an order of magnitude greater in the 1990s than in the 1980s (R. Brodeur, unpubl. data). Variability such as this is sufficiently great that it could be expected to influence production at other trophic levels.

Zooplankton biomass on the order of 2×10^{12} g C per yr is advected annually from the Bering Sea to the Chukchi Sea, with roughly 70% of oceanic origin and the remainder from the shelf (Springer et al. 1989). Euphausiids and oceanic copepods transported in the flow support immense numbers of planktivorous seabirds on the northern shelf (Springer and Roseneau 1985, Springer et al. 1987). The flux of zooplankton biomass is proportional to volume transport, as well as to seasonal standing stocks and annual production at the shelf edge, such that the long-term decline in transport through Bering Strait could be of significance to regional predators.

Fishes

Fishes have extremely high profiles because of their economic and ecosystem importance. They have been the focus of numerous major oceanographic research programs in the Bering Sea, beginning in the 1950s when Japan and Russia began exploring fishing grounds (e.g., Hokkaido University 1957-1970, Moiseev 1963-1970) and continuing with Processes and Resources of the Bering Sea Shelf (PROBES) in the 1970s and Fisheries Oceanography Coordinated Investigations (FOCI) and Southeast Bering Sea Carrying Capacity (SEBSCC) in the 1990s. As a result, the fisheries literature is large. A short list of recent examples with information on the Bering Sea includes the annual groundfish report of the National Marine Fisheries Service (e.g., NPFMC 1996), proceedings volumes of symposiums (e.g., Alaska Sea Grant 1989, 1997; Beamish 1995; Brodeur et al. 1996), and many journal articles cited here and in other papers in this volume.

One thing about fish which is very well known is that many populations in the North Pacific and Bering Sea are highly dynamic. They fluctuate in abundance because of commercial fishing pressure, in relation to weather and climate, and perhaps for reasons that involve some combination of fishing, climate change, and predator/prey interactions, including cannibalism (Wakabayashi et al., cited in Bakkala 1981; Wespestad and Fried 1983; Francis and Hare 1994; Quinn and Niebauer 1995; Ito and Ianelli 1996; Anderson et al. 1997; Wespestad et al., in press).

Another well-known fact is that fishes are critical prey of many species of seabirds, marine mammals, and other fishes (Hunt et al. 1981, Springer et al. 1986, Perez and Bigg 1986, Livingston 1993, Merrick and Calkins 1996) including the very species that have experienced the greatest declines in the Bering Sea and Gulf of Alaska in the past two decades (Steller sea lions, fur seals, harbor seals, red-legged kittiwakes, and black-legged kittiwakes). It is believed that a lack of one or more species of forage fish is the cause of the declines (Alaska Sea Grant 1993).

In this volume, two papers show how water temperature is important to fish distributions on the continental shelf—certain species prefer colder temperatures, while others prefer warmer temperatures (Wyllie-Echeverria and Ohtani, chapter 21; Brodeur et al., chapter 24). Cold shelf bottom water temperature in winter drives most species to deeper, warmer waters at the shelf edge, increasing predator-prey interactions (Favorite and Laevastu 1981). Interannual and decadal variability in air temperatures, water temperatures, and sea ice extent alter patterns of distribution and abundance of several species of forage fishes that in turn likely change important predator-prey relationships.

Issues of trophic interactions among fishes are particularly prominent, since fishes are by far the greatest predators of each other (Livingston
1993). One potentially significant relationship not discussed here is that
of pollock cannibalism, which is thought to be important to recruitment
dynamics and may be proportional to wind-driven transport of surface
waters on the southeastern shelf (Wespestad et al., in press). According to
this model, interannual and longer-term variability in winds over the Bering
Sea could be a significant factor affecting production dynamics of pollock.

An important variation on the theme of dynamics concerns stock structure and discreteness of abundant, widespread species. Bailey et al. (chapter 26) reviewed questions of pollock stock structure, migrations, spawning,
harvesting, and management. They point out that years of strong recruitment in the Bering Sea were strong in all regions (stocks) and that intermittent recruitment events are important to fluctuations in stock biomass.
The issue of panmictic or discrete stocks is critical to appropriate management discussions, since there is growing evidence of a rich pattern of
pollock stock structure.

The chapter by Sinclair et al. (chapter 23) helps to fill the major gap in
our knowledge of fishes in the basin of the Bering Sea, where biomass
rivals that on the shelf. The biomass of the lanternfish *Stenobrachius
leucopsarus* alone is estimated to be in the order of 10×10^6 t in just the
western portion of the basin, making it perhaps the most abundant species of fish in the Bering Sea. Distribution and abundance of fishes vary
between years in the basin, as they do on the shelf, with little known
about driving forces in either region.

Mito et al. (chapter 25) review diets of numerous species and size
classes of fishes. The information will be valuable in describing interspecific relationships, predator/prey interactions, and food web structure, and therefore to models of individual species and the ecosystem.

Benthos

Nothing was written about dynamics of benthic organisms or communities and the significance of variability in populations of several species to
an understanding of the larger ecosystem of the Bering Sea. Two conspicuous examples are pink shrimp and red king crabs. Forty years ago, the
pink shrimp stock of the eastern shelf region northwest of the Pribilof
Islands was removed by an intense commercial fishery, and it has not
recovered. The stock was not insignificant—it was distinguished by "... its
huge area covering up to 120 miles in length and 20-30 miles in width,"
with winter test fishery catches reaching "... 100 q/hr [=10,000 k/h], which
is probably a world record with respect to absolute size" (Ivanov 1970).
More recently, the collapse of the red king crab stock on the southeastern
shelf in the 1980s was a major economic loss to many people. The cause
is not agreed upon, and the failure of crabs to recover has not been explained.

Another clear signal from the bottom of the sea of ecosystem dynamics has been the prolonged increase of several species of flatfishes resulting from strong to exceptional recruitment of year classes spawned in the 1980s (NPFMC 1996). The contrasting decline of one species, Greenland turbot, adds additional uncertainty about the processes at work.

Seabirds

The most extreme case of population dynamics of seabirds in the Bering Sea was that of the flightless cormorant, at one time abundant in the Commander Islands but exterminated by Russian explorers by the early 1800s (Stejneger 1885). However, numerous other nesting species were severely depleted in the two centuries following the discovery of the Aleutian Islands and the introduction of foxes and other terrestrial mammals to them (Murie 1959, Bailey and Kaiser 1993). And one migrant species, the short-tailed albatross, which was formerly very abundant in the Bering Sea, was decimated on its nesting grounds off Japan and has become an extreme economic concern to the fishing industry because of restrictions on take imposed by the U.S. Endangered Species Act.

Detailed information on dynamics of seabirds in the Bering Sea does not begin until the 1970s, when colonies were first systematically censused. Populations monitored since then have experienced a variety of trends in abundance and productivity (e.g., Byrd et al. 1998). The chief concern has been dramatic declines of piscivorous red-legged and black-legged kittiwakes on the Pribilof Islands beginning in the mid-1970s in association with low productivity. This is particularly troubling in the case of red-legged kittiwakes due to their restricted numbers and breeding distribution: over 80% of the world population nests on St. George Island, where the decline was about 40% (Kildaw 1998). Other primarily piscivorous species on the Pribilof Islands, such as murres, the most abundant seabirds there, variously increased or decreased. Common murres on St. George Island were stable from the 1970s to the early 1990s and have increased by about 60% since then, while on St. Paul Island they declined by some 70% through the mid-1980s but have recovered by over 30% since. Thick-billed murres declined by about 30% on both islands through the mid-1980s, and may have increased by about 20% in the 1990s on St. George.

In contrast, these same four species all increased in the 1970s and 1980s in the western Aleutian Islands, indicating that conditions in that region of the Bering Sea were markedly different than on the eastern shelf near the Pribilof Islands. Hunt and Byrd (chapter 28) make this point, that seabird declines in the Bering Sea were unique to the Pribilofs, and they contend that the declines were precipitated by one or more large die-offs near the colonies. They further believe that a decline in carrying capacity of the Bering Sea has been limiting the populations by depressing reproductive output. In particular they cite a lack in diets of prey species with a high fat content, notably capelin.

Unfortunately, as the authors point out, there is no direct evidence of die-offs of these seabirds of a magnitude that could cause such large declines in abundance. For the population of red-legged kittiwakes on the Pribilofs to have declined by half, some 75,000 birds would have to have died. Likewise, a decline of 30% in numbers of thick-billed murres due to die-offs would have required 500,000 deaths. These are very large numbers. Furthermore, it would be surprising if such a universal prey failure occurred: the four afflicted species exploit a wide range of prey, including myctophids and squid of the oceanic domain and pollock, capelin, sand lance, and euphausiids of shelf domains. And, although kittiwakes are restricted to foraging in surface waters, murres are capable of diving to great depths. Still, the declines were abrupt and it is difficult to explain them in terms other than high mortality of grown birds (adults and juveniles). Declining productivity alone appears to be inadequate, although as Hunt and Byrd note, the appropriate population modeling has not been done.

Another factor, which was discounted, is emigration. Seabirds generally are considered to be strongly philopatric, with low rates of emigration and immigration. While this is true of adults, Divoky (1998) has shown that juvenile black guillemots, a species closely related to murres, are much more adventuresome. At Buldir Island in the western Aleutians, thick-billed murres increased at a rate of 15% per year from 1991 to 1994 and black-legged kittiwakes increased by 23% per year from 1975 to 1985. These population growth rates apparently exceed those that can reasonably be accounted for by productivity alone (Nur and Ainley 1992). Therefore, the increases at Buldir must have occurred in part from immigration. The actual numerical increase of birds at Buldir does not equal the decrease at the Pribilof Islands, which allows for some mortality or immigration to other colonies in the Bering Sea where populations are not closely monitored.

Information on planktivorous species, the auklets, is inadequate to confidently conclude anything about trends in population abundance or productivity (Hunt and Byrd, chapter 28). This is unfortunate, since these birds are supported by the same species of mesozooplankton that are critical links in the transfer of carbon and energy from phytoplankton to most higher trophic levels in pelagic food webs, namely *Neocalanus cristatus, N. plumchrus, Calanus marshallae*, and various species of euphausiids. Elements of auklet breeding biology are very sensitive to fluctuations in the prey availability (Roseneau et al. 1985, Springer et al. 1986) and as a group they could provide an inexpensive, precise index of the abundance of various species of zooplankton over a broad area of the Bering Sea.

For the western Bering Sea, Shuntov (chapter 29) describes seasonal abundances of seabirds and shows how numbers change. He believes that fluctuations in abundance should be considered in time scales of 40-60 years, which correspond to broad changes in faunistic complexes. Shorter-term climatic variability, including the scale of El Niño, is of little consequence by comparison. Yet, when time scales of change of murres and kittiwakes in the western Aleutian Islands and on the Pribilof Islands are examined, it would seem that certain shorter periods are highly relevant to dynamics. There, abundances have changed significantly (up and down) over periods of a decade or less. Indeed, if the hypothesis of Hunt and Byrd (chapter 28) is correct, interannual time scales may be quite significant.

Marine Mammals

Dynamics of marine mammal populations in the Bering Sea are considered in just one chapter. Loughlin et al. (chapter 27) describe movements of adult male fur seals out of Bering Sea in winter. In the North Pacific they generally follow the direction of surface circulation, which in the eastern subarctic takes them around the perimeter of the Alaska Gyre. The authors speculate that prey abundance is higher around the edge than in the center of the gyre, a view that is supported by a variety of other evidence (Springer et al., in press).

Population dynamics of marine mammals in the Bering Sea and greater North Pacific have been extreme in the past two centuries, mostly, or perhaps entirely as some would argue, due to the depravities of people. Steller sea cows were driven to extinction in the same period as was the flightless cormorant (Stejneger 1887). In the 1800s, bowhead and right whales were depleted by commercial whalers (Bockstoce and Botkin 1983) to the extent that bowheads have not yet reestablished a resident population in the Bering Sea and right whales remain one of the most endangered cetaceans in the world (Brownell et al. 1986). Polar bears were eliminated as a resident species in the Bering Sea when the last one was killed on St. Matthew Island in the 1890s (Hanna 1917). Others, notably fur seals, sea lions, walruses, and sea otters, have suffered intervals of heavy persecution with large swings in abundance as populations collapsed and then recovered (Kenyon 1962, 1969; Lander and Kajimura 1982; Fay et al. 1989). On the Pribilof Islands, walruses and vast numbers of sea otters were extirpated by humans and sea lions were greatly depleted. Since the end of World War II, other species of great whales—fin whales, sperm whales, and humpback whales—were nearly eliminated as well and are still at very low numbers.

Thus, the history of marine mammals in the Bering Sea is a sad one. And in some respects it has not improved in recent years. In the past two decades Steller sea lions have declined by approximately 80%, harbor seals appear to have declined by an undetermined but significant amount, fur seals are perhaps half to a third of their former abundance, and sea otter populations in the Aleutian Islands have collapsed (Lander and Kajimura 1982; York and Kozloff 1987; NMML 1994; Estes et al. 1998; NMFS, unpubl. data). The reasons for most of these recent declines are not known, and this has caused high anxiety over potential effects of commercial fisheries, climate change, and historical changes in the ecosystem, e.g., the wholesale slaughter of whales. The classification of the bulk of the sea lion population as endangered in 1997 highlights the problem.

All is not doom and gloom, however. Bowhead whales have been recovering at a rate of 2% to 3% per year and the population numbered around 8,000 in 1993 (Zeh et al. 1995). Although none yet summer in the Bering Sea, as they did before exploitation, their summer range has apparently been expanding southward in the eastern Chukchi Sea from the Beaufort Sea where they found refuge from whalers in the nineteenth century (R. Suydam, pers. comm.). Gray whales have recovered (Buckland et al. 1993). Right whales have been sighted repeatedly in the southeastern Bering Sea during the past three summers, albeit in very small numbers (e.g., Goddard and Rugh 1998). Walruses have been abundant and seem to have been in equilibrium with carrying capacity since recovering from the most recent round of excessive commercial harvests from the 1930s to 1950s (Fay 1997). The status of populations of pagophilic seals—spotted, ribbon, ringed, and bearded—is not known, but at least there is no evidence of calamitous declines in numbers (L. Lowry, pers. comm.). It is not known if the decrease in sea ice extent in the Bering Sea during the past meteorological regime affected them.

Humankind

The book also stops short of considering dynamics of human populations in the Bering Sea, in spite of the fact that our heavy hand is known to have been responsible for many extreme changes and is suspected of causing or contributing to others. The invasions of explorers and pillagers from many nations since the 1700s left deep wounds, some of which have not yet healed. Commercial fisheries continue to alter habitats and populations in ways that are of concern to many people who find their livings and cultural identities in the sea or who wonder about unseen changes that may be accumulating in the ecosystem. Reassurances from fisheries managers have not allayed these fears.

Modeling

Francis et al. (chapter 20) tackle issues of ecosystem modeling and management. They make a strong case for paying close attention to scales of time and space when attempting to define ecosystems and understand how they work and why they vary. To do this, one must recognize patterns and use them as tools to identify scales—"... once patterns are detected and described, we can seek to discover the determinants of pattern, and the mechanisms that generate and maintain those patterns." They remind us that ecology has four dimensions—that time defines rates, and processes lose significance without reference to rates. Time has many scales and complex behavior will arise from the interaction of differently scaled processes.

For the Bering Sea, a final management strategy could range from single actions to a mix of actions combined with experimental manipulations (Francis et al., chapter 20). But this requires agreement that there is a problem with the current approach—if it isn't broken, don't fix it—and the resolve to try something new. It is likely that such an approach would lead to important new knowledge about individual species and the larger system, but it is less certain, as some argue, that it would be worth the economic disruption, political turmoil, or other probable consequences, especially if there is nothing inherently wrong with the status quo. And this comes back to our sense of pattern and scale—we are just beginning to know the magnitude of fluctuations in biomass over appropriate time and space scales. The work of Baumgartner et al. (1992) showed how greatly species vary naturally, i.e., not influenced by humans, yet we have limited knowledge about such things for the Bering Sea. We know only that in recent years of intense human involvement in the affairs of the Bering Sea ecosystem, there have been many big changes, some of which are disturbing.

Conclusions

The physical environment of the Bering Sea is intimately connected to the meteorology of the greater North Pacific Ocean and Western Arctic, which in turn has strong teleconnections with the El Niño Southern Oscillation over the equatorial Pacific. Interactions between the Aleutian Low and the Siberian High pressure systems complicate an understanding of ocean responses in the Bering Sea to atmospheric forcing.

Nevertheless, there are clear indications that Bering Sea physics do adjust to state changes in atmospheric circulation over the North Pacific, e.g., regimes of sea ice that conform to meteorological regimes of the North Pacific. And just as the physical environment shifts between states, or more appropriately because of the shifts, the biosphere of the Bering Sea appears to shift between states as well. A strong case has been made in this regard for patterns in production of Pacific salmon and groundfish. Can this model be applied to other species and to the ecosystem as a whole? There is a variety of evidence that it can be (Francis et al. 1998, Springer 1998). However, past performance is no guarantee of future returns, and as the global climate continues to warm, predictions based on historical patterns may not be possible.

To improve on these models for the Bering Sea, however, a much broader net will need to be cast. For example, in most respects the basin of the Bering Sea continues to receive little scientific attention. The principal exception is that of circulation, which we now have a much improved understanding of and know to be much more dynamic than previously recognized. Circulation is highly variable between years and decades, and eddies are prominent features. They are thought to affect hydrography and nutrient distributions and deepwater renewal, primary production and phytoplankton distributions, and in the shelf edge region might be important to pollock growth and survival. The formation of bottom water over the basin and the advection of bottom water into the basin from the North Pacific are important to oceanic circulation and nutrient budgets and may influence carbon and contaminant budgets as well.

Primary production over the basin may be higher than earlier estimates indicated. Higher production and an efficient biological pump would enhance the role of the basin in large-scale carbon budgets. Greater production over the basin would also help to balance carbon budgets for production at higher trophic levels there, e.g., by the apparently huge biomass of mesopelagic fishes and squids, as well as for walleye pollock that apparently must utilize shelf and basin regions to maintain a biomass as great as that to which it grew in the last decade.

Other questions remain about production in the basin. What limits primary production there and what accounts for the HNLC condition? Or, conversely, what accounts for the recent estimates of high production? What is the annual new (nitrate) production available for export? What are the roles of the several size classes of plankton, from picoplankton to mesoplankton, in food web dynamics? Likewise, we know very little about fishes and squids in the basin, primarily because there are relatively few commercially important species harvested there. If the estimated biomass of mesopelagic fishes in the western basin is applied to the entire basin, the resulting value would surpass that of any shelf species, including pollock. Does this apparently huge biomass of myctophids constrain secondary production and thus enhance primary productivity?

We still have only partial pictures of the overall distributions of many fishes, and much less knowledge of variability in distribution between years of contrasting physical conditions. The point was well made by Brodeur et al. (chapter 24) in their analysis of forage fish distributions between 1986, a cold year, and 1987, a warm year. The survey coverage they drew upon was not adequate to establish whether distribution, biomass, or a combination of the two was responsible for several of the interannual differences observed in the data. The reason for this is that standard surveys cover the region normally occupied by the bulk of the commercially important species and age classes, and excludes typically marginal habitat that is important in all years to some forage species (capelin, herring) and in some years to many or all age classes of others (pollock). A combination of economics and politics is responsible—broad-scale surveys are very expensive and the political climate between Russia and the United States has restricted the geographical coverage to territorial waters in most years. Even in the case of the broad Russian surveys, which were very revealing, important portions of the eastern inner shelf were omitted and thus species such as herring, capelin, and pollock still were not adequately sampled. This is not a minor detail, because we know that fishes do move around and fish abundance in particular areas is critical to the success of predator populations. Yet in most cases we still do not know whether in years of low regional fish abundance it is because of diminished biomass, a change in distribution, or some combination.

The lack of comprehensive survey coverage also hampers our ability to answer questions posed by Bailey et al. (chapter 26) as revealed by their figures 4-6. In several cases an unknown, but possibly substantial, portion of the target group was outside the survey boundaries. It would seem that in the case of a species like pollock, which alone generates more than 1×10^9 \$U.S. annually for the economies of the United States, Japan, Russia, Norway, and other nations, there could be proportional expenditures on such basic research elements as distribution and abundance and their relationship to a dynamic physical and biological environment.

We have made equally little progress toward an understanding of causes of population dynamics of species at higher trophic levels that may be initiated by fluctuations in abundance of species at the base of the food web or at intermediate levels. How do changes in the production base (phytoplankton and zooplankton community structure and production) influence rates and quantities of material and energy transferred to higher trophic levels? What are the magnitudes of fluctuations in these basic elements between years, between decades, and across space? Phytoplankton and zooplankton are *sine qua non* for most denizens of the oceans. Likewise, the biology of species at middle trophic levels is virtually unknown. These include the majority of forage species that are critical prey of many fishes, seabirds, and marine mammals, e.g., euphausiids, sand lance, capelin, herring, and myctophids, as well as species such as jellyfish that may play an important role in dynamics at higher and lower trophic levels.

There are emerging issues of inter-ecosystem interactions in the Bering Sea of potentially great significance that are not discussed here. One case is the dynamic benthic ecology on the eastern shelf and its relationship to pelagic production processes and variability. The rapid growth of most flatfish populations and continued low production of red king crabs and shrimp are conspicuous signs of broad state changes of importance to food web production and commercial interests. What is their connection, if any, to pelagic production processes and food web dynamics?

Another case in point is the collapse of sea otters in the Aleutian Islands (Estes et al. 1998). Sea otters are members of the kelp forest ecosystem, yet might interact at times in critical ways with members of the pelagic ecosystem. Changes in predator-prey relationships at numerous trophic levels form connections between the two ecosystems. As an extension of this notion, one can ask if there are emergent properties of the Bering Sea ecosystem as suggested by Francis et al. (chapter 20), such as inherent stability and resistance to external perturbations (e.g., climate shifts) by virtue of diverse and abundant long-lived species at high trophic levels. Was that compromised by the depletions of certain whales and fishes?

The effect of top-down interactions, originating with marine mammals, in structuring ecosystems in the Bering Sea is conspicuous in some cases (e.g., sea otter effects on kelp forests, and walrus and gray whale effects on benthic communities), but here and elsewhere in the open ocean it is poorly understood (Bowen 1997). If, for instance, killer whales are responsible for the collapse of sea otter populations in the Aleutian Islands and the cascade of effects in the nearshore ecosystem, what might have been their role in the earlier declines of sea lions and harbor seals, or their role in the recovery, or lack thereof, of these species? Killer whales are the apex predators in the Bering Sea and should not be ignored as a possible factor in dynamics of prey populations. What was the effect on trophic linkages in Bering Sea food webs of the rapid removal of most of the great whales in the 1950s to 1970s? Further development of multispecies models for the Bering Sea, e.g., Trites et al. (1999), may help answer such questions.

Much complementary information on features of the Bering Sea ecosystem obtained by scientists from several countries has not been integrated into a single body of knowledge. The various data sets cover a large amount of geography and a growing length of time. Understandably, much has been reported in national outlets in native languages and is thus not widely utilizable to scientists outside the individual countries. One of the notable examples is trophic dependencies of groundfishes: the paper by Mito et al. (chapter 25) summarizes much of the Japanese data and refers it in some degree to U.S. data but hardly to Russian data. One message from this overview is that there remains a need to integrate disparate data sets to see what knowledge might emerge on spatial and temporal variability in components of the ecosystem that are particularly important to present scientific, economic, and conservation concerns. One must bear in mind, however, that such analyses might show only the futility in attempting comparisons between data sets for a variety of reasons. If so there might yet be a silver lining to the exercise—it might provide an avenue for cooperation in designing sampling strategies and methods of data reduction for future studies that would allow comparisons to be made in a dynamic context, rather than only the descriptive ways now possible. A nice example where this has begun is presented in the paper by Sinclair et al. (chapter 23) in the compilation of Japanese, Russian, and U.S. data on mesopelagic fishes and squids.

Recommendations

- Do not think of the Bering Sea as pristine, or of the same ecological maturity as it was before being so greatly disturbed by people in the past two centuries. The history of excessive exploitation must be borne in mind when developing ecosystem models and when planning long-range conservation and management strategies.

- Keep a broad view of spatial scales when thinking about the Bering Sea. It is characterized by a variety of habitats over the shelf and basin, each with particular dimensions. At the same time, the Bering Sea is but a small embayment of the North Pacific and tractable as a study area at this scale given innovative approaches.

- Improve our limited knowledge of dynamics of important food web organisms, particularly phytoplankton and zooplankton, the base of the food web that determines in large measure biomass production at higher trophic levels, and forage species at intermediate levels that are critical prey of many fishes, seabirds, and marine mammals.

- Give the basin a little respect—it is an integral part of the larger ecosystem but is virtually ignored.

- Ensure that, in the continuing search for clues to ecosystem processes over interannual time scales, the information will be useful in identifying patterns over decadal and longer time scales.

- Expand efforts to meld data on the Bering Sea held by Pacific Rim nations and to pool their financial and intellectual resources to more efficiently devise and undertake research of interest and value to all. This should include a much greater emphasis on translations of national literature so it can be entered into the collective knowledge pool.

- Seek greater financial and in-kind participation by the commercial fishing industry in Bering Sea research.

- Implement a multinational, long-term monitoring strategy for key ecosystem components. This could include, for example, expanding the boundaries of annual groundfish surveys; improving sampling of noncommercial forage species such as capelin, sand lance, myctophids, and squids; tracking primary and secondary production and the composition of phytoplankton and zooplankton assemblages across domains over time; initiating a dedicated whale censusing program; and making better use of existing databases to investigate change in distributions and abundance of species in both time and space.

- The biodiversity of the Bering Sea is by many measures much reduced from former times—two species are extinct and a third nearly so, others have been locally or regionally extirpated, and still others exist at diminished abundances. There are many fewer examples of population increases, although these are dramatic. Thus, in a comparative sense the Bering Sea does not "team with wildlife" as in the past. Many of the changes were caused by people directly, while others may have been exacerbated by our actions. To the extent that changes in exploitation rates and practices might improve this situation, every opportunity should be taken to test innovative management approaches that could benefit a broad spectrum of inhabitants and users.

Acknowledgments

I thank T. Loughlin, P. McRoy, and G. Van Vliet for comments that improved this manuscript. Support for the author was provided by the Institute of Marine Science, University of Alaska Fairbanks, and by the North Pacific Universities Marine Mammal Research Consortium. This is contribution No. 2542, Institute of Marine Science, University of Alaska Fairbanks, Fairbanks, AK 99775.

References

Alaska Sea Grant. 1989. Proceedings of the International Symposium on the Biology and Management of Walleye Pollock. University of Alaska Sea Grant, AK-SG-89-01, Fairbanks. 789 pp.

Alaska Sea Grant. 1993. Is it food? Addressing marine mammal and seabird declines: Workshop summary. University of Alaska Sea Grant, AK-SG-93-01, Fairbanks. 59 pp.

Alaska Sea Grant. 1997. Forage fishes in marine ecosystems: Proceedings of the International Symposium on the Role of Forage Fishes in Marine Ecosystems. University of Alaska Sea Grant, AK-SG-97-01, Fairbanks. 816 pp.

Anderson, P.J., J.E. Blackburn, and B.A. Johnson. 1997. Declines of forage species in the Gulf of Alaska, 1972-1995, as an indicator of regime shift. In: Forage fishes in marine ecosystems: Proceedings of the International Symposium on the Role of Forage Fishes in Marine Ecosystems. University of Alaska Sea Grant, AK-SG-97-01, Fairbanks, pp. 531-543.

Bailey, E.P., and G.W. Kaiser. 1993. Impacts of introduced predators on nesting seabirds in the northeast Pacific. In: K. Vermeer, K.T. Briggs, K.H. Morgan, and D. Siegel-Causey (eds.), The status, ecology, and conservation of marine birds of the North Pacific. Canadian Wildlife Service, Ottawa, pp. 218-226.

Bakkala, R.G. 1981. Population characteristics and ecology of yellowfin sole. In: D.W. Hood and J.A. Calder (eds.), The eastern Bering Sea shelf: Oceanography and resources, Vol. I. University of Washington Press, Seattle, pp. 553-574.

Banse, K. 1995. Zooplankton: Pivotal role in the control of ocean production. ICES Journal of Marine Science 52:265-277.

Baumgartner, T.R., A. Soutar, and V. Ferreira-Bartrina. 1992. Reconstruction of the history of Pacific sardine–northern anchovy populations over the past two millennia from sediments of the Santa Barbara basin, California. CALCOFI Report 33:24-40.

Beamish, R.J. (ed.). 1995. Climate change and northern fish populations. Canadian Special Publications of Fisheries and Aquatic Sciences 121:1-739.

Bockstoce, J.R., and D.B. Botkin. 1983. The historical status and reduction of the western arctic bowhead whale (*Balaena mysticetus*) population by the pelagic whaling industry, 1848-1914. Reports of the International Whaling Commission Special Issue 5:107-141.

Bowen, W.D. 1997. Role of marine mammals in aquatic ecosystems. Marine Ecology Progress Series 158:267-274.

Brodeur, R.D., P.A. Livingston, T.R. Loughlin, and A.B. Hollowed (eds.). 1996. Ecology of juvenile walleye pollock, *Theragra chalcogramma.* NOAA Technical Report NMFS 126:1-227.

Brownell, R.L., Jr., P.B. Best, and J.H. Prescott. 1986. Right whales: Past and present status. Reports of the International Whaling Commission Special Issue 10:27-33.

Buckland, S.T., J.M. Breiwick, K.L. Cattanach, and J.L. Laake. 1993. Estimated population size of the California gray whale. Marine Mammal Science 9:235-249.

Byrd, G.V., D.E. Dragoo, and D.B. Irons. 1998. Breeding status and population trends of seabirds in Alaska in 1997. U.S. Fish and Wildlife Service, Homer, AK. 59 pp.

Coachman, L.K. 1993. On the flow field in the Chirikov Basin. Continental Shelf Research 13:481-508.

Coachman, L.K., K. Aagaard, and R.B. Tripp. 1975. Bering Strait: Regional physical oceanography. University of Washington Press, Seattle. 172 pp.

Cooney, R.T. 1981. Bering Sea zooplankton and micronekton communities with emphasis on annual production. In: D.W. Hood and J.A. Calder (eds.), The eastern Bering Sea shelf: Oceanography and resources, Vol. I. University of Washington Press, Seattle, pp. 947-974.

Cooney, R.T., and K.O. Coyle. 1982. Trophic implications of cross-shelf copepod distributions in the southeastern Bering Sea. Marine Biology 70:187-196.

Coyle, K.O., and R.T. Cooney. 1988. Estimating carbon flux to pelagic grazers in the ice-edge zone of the eastern Bering Sea. Marine Biology 98:299-306.

Divoky, G.J. 1998. Factors affecting the growth of a black guillemot colony in northern Alaska. Ph.D. thesis, University of Alaska Fairbanks. 144 pp.

Eslinger, D.L. 1994. Satellite and optical drifter observations of mesoscale eddies in the Bering Sea. In: M.O. Jeffries and K.G. Dean (eds.), 3rd Circumpolar Symposium on Remote Sensing and the Environment, Programs and Abstracts. University of Alaska Fairbanks, p. 21.

Estes, J.A., M.T. Tinker, T.M. Williams, and D.F. Doak. 1998. Killer whale predation on sea otters linking oceanic and nearshore ecosystems. Science 282:473-476.

Favorite, F., and T. Laevastu. 1981. Finfish and the environment. In: D.W. Hood and J.A. Calder (eds.), The eastern Bering Sea shelf: Oceanography and resources, Vol. I. University of Washington Press, Seattle, pp. 597-610.

Fay, F.H. 1997. Status of the Pacific walrus population, 1950-1989. Marine Mammal Science 13:537-565.

Fay, F.H., B.P. Kelly, and J.L. Sease. 1989. Managing the exploitation of Pacific walruses: A tragedy of delayed response and poor communication. Marine Mammal Science 5:1-16.

Francis, R.C., and S.R. Hare. 1994. Decadal-scale regime shifts in the large marine ecosystems of the Northeast Pacific: A case for historical science. Fisheries Oceanography 3:279-291.

Francis, R.C., S.R. Hare, A.B. Hollowed, and W.S. Wooster. 1998. Effects of interdecadal climate variability on the oceanic ecosystems of the NE Pacific. Fisheries Oceanography 7:1-21.

Frost, B.W. 1987. Grazing control of phytoplankton stock in the open subarctic Pacific Ocean: A model assessing the role of mesozooplankton, particularly the large calanoid copepods *Neocalanus* spp. Marine Ecology Progress Series 39:49-68.

Goddard, P., and D.J. Rugh. 1998. A group of right whales seen in the Bering Sea in July 1996. Marine Mammal Science 14:344-349.

Hanna, G.D. 1917. The summer birds of the St. Matthew Island Bird Reservation. Auk 34:403-410.

Hokkaido University. 1957-1970. Data records of oceanographic observations in exploratory fisheries, Hokkaido University No. 1-4, Hakodata, Hokkaido, Japan.

Hunt, G.L., Jr., B. Burgesson, and G.A. Sanger. 1981. Feeding ecology of seabirds of the eastern Bering Sea. In: D.W. Hood and J.A. Calder (eds.), The eastern Bering Sea shelf: Oceanography and resources, Vol. I. University of Washington Press, Seattle, pp. 629-648.

Ito, D.H., and J.N. Ianelli. 1996. Pacific Ocean perch. In: Stock assessment and fishery evaluation report for the groundfish resources of the Bering Sea/Aleutian Islands region. North Pacific Fishery Management Council, Anchorage, AK, pp. 332-359.

Ivanov, B.G. 1970. Distribution of the deep-sea prawn (*Pandalus borealis* Kr.) in the Bering Sea and Gulf of Alaska. In: P.A. Moiseev (ed.), Soviet fisheries investigations in the northeastern Pacific. Pishchevaya Promyshlennost, Moscow, pp. 125-142.

Kenyon, K.W. 1962. History of the Steller sea lion and the Pribilof Islands, Alaska. Journal of Mammalogy 43:68-75.

Kenyon, K.W. 1969. The sea otter in the eastern Pacific Ocean. North American Fauna 68:1-352.

Kildaw, S.D. 1998. Population status and patterns of distribution and productivity of kittiwakes on St. George Island, Alaska. Ph.D. thesis, University of Alaska Fairbanks. 152 pp.

Klyashtorin, L.B. 1997. Global climate cycles and Pacific forage fish stock fluctuations. In: Forage fishes in marine ecosystems: Proceedings of the International Symposium on the Role of Forage Fishes in Marine Ecosystems. University of Alaska Sea Grant, AK-SG-97-01, Fairbanks, pp. 545-557.

Kokur, C. 1982. Phytoplankton distribution in southeastern Bering Sea shelf waters during spring. M.Sc. thesis, Florida State University, Tallahassee, FL. 75 pp.

Lander, R.H., and H. Kajimura. 1982. Status of northern fur seals. FAO Fisheries Series 5:319-345.

Livingston, P.A. 1993. Importance of predation by groundfish, marine mammals and birds on walleye pollock *Theragra chalcogramma* and Pacific herring *Clupea pallasi* in the eastern Bering Sea. Marine Ecology Progress Series 102:205-215.

Mantua, N.J., S.R. Hare, Y. Zhang, J.M. Wallace, and R.C. Francis. 1997. A Pacific interdecadal climate oscillation with impacts on salmon production. Journal of the American Meteorological Society 78:1069-1079.

Martin, J.H., and S. Fitzwater. 1988. Iron deficiency limits phytoplankton growth in the northeast Pacific subarctic. Nature 331:341-343.

McRoy, C.P., and J.J. Goering. 1974. The influence of ice on the primary productivity of the Bering Sea. In: D.W. Hood and E.J. Kelley (eds.), Oceanography of the Bering Sea with emphasis on renewable resources. Occasional Publication No. 2, Institute of Marine Science, University of Alaska, Fairbanks, pp. 403-421.

McRoy, C.P., D.A. Hansell, A.M. Springer, J.J. Walsh, and T.E. Whitledge. 1987. Global maximum of primary production in the north Bering Sea. Eos 68:L1727.

Merrick, R.L., and D.G. Calkins. 1996. Importance of juvenile walleye pollock, *Theragra chalcogramma*, in the diet of Gulf of Alaska Steller sea lions, *Eumetopias jubatus*. NOAA Technical Report NMFS 126:153-166.

Miller, C.B. 1993. Pelagic production processes in the subarctic Pacific. Progress in Oceanography 32:1-15.

Moiseev, P.A. (ed.). 1963-1970. Soviet fisheries investigations in the northeastern Pacific, Part 1-5. All-Union Research Institute of Marine Fisheries Oceanography and Pacific Scientific Research Institute of Fisheries and Oceanography.

Murie, O.J. 1959. Fauna of the Aleutian Islands and Alaska Peninsula. North American Fauna 61:1-406.

Niebauer, H.J. In press. Variability in Bering Sea ice cover as affected by a "regime shift" in the north Pacific in the period 1947-96. Journal of Geophysical Research.

Niebauer, H.J., and V. Alexander. 1985. Oceanographic frontal structure and biological production at an ice edge. Continental Shelf Research 4:367-388.

Niebauer, H.J., V. Alexander, and S.M. Henrichs. 1995. A time-series study of the spring bloom at the Bering Sea ice edge. I: Physical processes, chlorophyll and nutrient chemistry. Continental Shelf Research 15:1859-1878.

NMML. 1994. Status review of Steller sea lions (*Eumetopias jubatus*). National Marine Mammal Laboratory, National Marine Fisheries Service, Seattle, Washington. 40 pp.

NPFMC. 1996. Stock assessment and fishery evaluation report for the groundfish resources of the Bering Sea/Aleutian Islands regions. Compiled by the Plan Team for the groundfish fisheries of the Bering Sea and Aleutian Islands. North Pacific Fishery Management Council, Anchorage, AK. 384 pp.

Nur, N., and D.G. Ainley. 1992. Comprehensive review and critical synthesis of the literature on recovery of marine bird populations from environmental perturbations. Point Reyes Bird Observatory, Stinson Beach, CA. 38 pp.

Perez, M.A., and M.A. Bigg. 1986. Diet of northern fur seals, *Callorhinus ursinus*, off western North America. Fishery Bulletin 84:957-971.

Quinn, T.J., II, and H.J. Niebauer. 1995. Relation of eastern Bering Sea walleye pollock *(Theragra chalcogramma)* recruitment to environmental and oceanographic variables. Canadian Special Publications in Fisheries and Aquatic Sciences 121:497-507.

Roach, A.T., K. Aagaard, C.H. Pease, S.A. Salo, T. Weingartner, V. Pavlov, and M. Kulakov. 1995. Direct measurement of transport of water properties through the Bering Strait. Journal of Geophysical Research 100:18443-18457.

Roseneau, D.G., M.I. Springer, E.C. Murphy, and A.M. Springer. 1985. Population and trophics studies of seabirds in the northern Bering and eastern Chukchi seas, 1981. Outer Continental Shelf Environmental Assessment Program Final Reports 30:1-58.

Sambrotto, R.N., H.J. Niebauer, J.J. Goering, and R.L. Iverson. 1986. Relationships among vertical mixing, nitrate uptake, and phytoplankton growth during the spring bloom in the southeast Bering Sea middle shelf. Continental Shelf Research 5:161-198.

Schumacher, J.D., and P.J. Stabeno. 1994. Ubiquitous eddies of the eastern Bering Sea and their coincidence with concentrations of larval pollock. Fisheries Oceanography 3:182-190.

Shuntov, V.P., E.P. Dulepova, V.I. Radchenko, and V.V. Lapko. 1996. New data about communities of plankton and nekton of the far-eastern seas in connection with climate-oceanological reorganization. Fisheries Oceanography 5:38-44.

Smith, S. L., and J. Vidal. 1984. Spatial and temporal effects of salinity, temperature and chlorophyll on the communities of zooplankton in the southeastern Bering Sea. Journal of Marine Research 42:221-257.

Springer, A.M. 1998. Is it all climate change? Why marine bird and mammal populations fluctuate in the North Pacific. In: G. Holloway, P. Muller, and D. Henderson (eds.), Biotic impacts of extratropical climate variability in the Pacific. Proceedings 'Aha Huliko'a Hawaiian Winter Workshop, University of Hawaii, Honolulu, pp. 109-119.

Springer, A.M., and C.P. McRoy. 1993. The paradox of pelagic food webs in the northern Bering Sea. III: Patterns of primary production. Continental Shelf Research 13:575-599.

Springer, A.M., and D.G. Roseneau. 1985. Copepod-based food webs: Auklets and oceanography in the Bering Sea. Marine Ecology Progress Series 21:229-237.

Springer, A.M., C.P. McRoy, and M.V. Flint. 1996. The Bering Sea Green Belt: Shelf edge processes and ecosystem production. Fisheries Oceanography 5:205-223.

Springer, A.M., C.P. McRoy, and K.R. Turco. 1989. The paradox of pelagic food webs in the northern Bering Sea. II: Zooplankton communities. Continental Shelf Research 9:359-386.

Springer, A.M., E.C. Murphy, D.G. Roseneau, C.P. McRoy, and B.A. Cooper. 1987. The paradox of pelagic food webs in the northern Bering Sea. I: Seabird food habits. Continental Shelf Research 7:895-911.

Springer, A.M., D.G. Roseneau, D.S. Lloyd, C P. McRoy, and E.C. Murphy. 1986. Seabird responses to fluctuating prey abundance in the eastern Bering Sea. Marine Ecology Progress Series 32:1-12.

Springer, A.M., J.F. Piatt, V.P. Shuntov, G.B. Van Vliet, V.L. Vladimirov, A.E. Kuzin, and A.S. Perlov. In press. Marine birds and mammals of the Pacific subarctic gyres. Progress in Oceanography.

Stejneger, L. 1885. Results of ornithological explorations in the Commander Islands and in Kamtschatka. Bulletin of the United States National Museum 29:1-382.

Stejneger, L. 1887. How the great northern sea cow *(Rytina)* became extinct. American Naturalist 21:1047-1054.

Sugimoto, T., and K. Tadokoro. 1997. Interannual-interdecadal variations in zooplankton biomass, chlorophyll concentration and physical environment in the subarctic Pacific and Bering Sea. Fisheries Oceanography 6:74-93.

Trenberth, K.E., and J.W. Hurrell. 1994. Decadal atmospheric-ocean variations in the North Pacific. Climate Dynamics 9:303-319.

Trites, A.W., P.A. Livingston, M.C. Vasconcellos, S. Mackinson, A.M. Springer, and D. Pauly. 1999. Ecosystem considerations and the limitations of ecosystem models in fisheries management: Insights from the Bering Sea. In: Ecosystem approaches for fisheries management. University of Alaska Sea Grant, AK-SG-99-01, Fairbanks, pp. 609-619.

Vidal, J., and S.L. Smith. 1986. Biomass, growth, and development of populations of herbivorous zooplankton in the southeastern Bering Sea during summer. Deep-Sea Research 33:523-556.

Walsh, J.J., and C.P. McRoy. 1986. Ecosystem analysis in the southeastern Bering Sea. Continental Shelf Research 5:259-288.

Wespestad, V.G., and S.M. Fried. 1983. Review of the biology and abundance trends of Pacific herring *(Clupea harengus pallasi)*. In: W.S. Wooster (ed.), From year to year. University of Washington Sea Grant, Seattle, pp. 17-29.

Wespestad, V.G., L.W. Fritz, W.J. Ingraham, and B.A. Megrey. In press. On relationships between cannibalism, climate variability, physical transport and recruitment success of Bering Sea walleye pollock, *Theragra chalcogramma*. ICES Journal of Marine Science.

York, A.E., and P. Kozloff. 1987. On the estimation of numbers of northern fur seal, *Callorhinus ursinus*, pups born on St. Paul Island, 1980-86. Fishery Bulletin 85:367-375.

Zeh, J.E., J.C. George, and R. Suydam. 1995. Population size and rate of increase, 1978-1993, of bowhead whales, *Balaena mysticetus*. Reports of the International Whaling Commission 45:339-344.

APPENDIX

Principal Scientific Questions on Bering Sea Ecosystem Function and Productivity

Introduction

The Science Board of the North Pacific Marine Science Organization (PICES) created a number of working groups to assist them in their efforts to coordinate collaborative research efforts in the North Pacific Ocean and adjoining seas. Working Group 5–Bering Sea was created as a temporary working group to review the present knowledge of atmospheric and oceanic circulation in the Bering Sea, to review present knowledge of the Bering Sea ecosystem and its responses to environmental variability, to develop a plan for a symposium on the Bering Sea ecosystem, and to identify major gaps in present knowledge. The working group was chaired by Dr. Al Tyler of the University of Alaska Fairbanks. Members included representatives from the United States, Canada, Japan, China, and Russia. This summary represents an edited version of working group reports submitted to PICES by Dr. Tyler.

The working group proposed a series of research questions that it considered important for understanding the environmental and ecosystem functions of the Bering Sea. The purpose of the questions was to provide a focus for cooperative international research in the region. As research progresses, it may be desirable to add additional questions, or to refine the existing questions.

The working group agreed on three points which provided a common perspective for framing the scientific questions. The first point of consensus was that the abundance of animals in the Bering Sea fluctuated widely. The work should further understanding of the nature and reasons for such fluctuations. The second point of consensus was to integrate studies of the physical and biotic environment to address biological productivity. The third point was that comparative studies of the Bering Sea and other boreal ecosystems could provide additional insights to high biological productivity in the Bering Sea.

In framing the questions, the working group noted that the identification of ecosystem emergent properties that contribute significantly to Bering Sea productivity as a whole would develop only upon taking a broad view of ecological interactions. The complexity of the system has to do with the number and intricacy of vital links that contribute significantly to ecosystem structure.

To stimulate international cooperative research in the Bering Sea, the working group posed four categories of questioning: the first deals with broad decadal change and accumulation in the basins, the second with the physics of water mass exchange, the third with sea-ice dynamics, and the fourth with prey-predator relationships. These broad areas can serve to focus research on special environmental features and high biological productivity in the Bering Sea by member governments of PICES.

Decadal Scale Changes

Decadal and possibly century scale atmospheric and oceanic variability has been observed in the surface pressure distribution, in ocean conditions such as sea surface temperature (SST), in sea level, and in sea ice extent. What are the mechanisms whereby these frequencies are determined for the North Pacific? To what extent can the variations be predicted; for example, when will a warm era begin and end? How would the projected global warming affect variability of this scale?

Variability in atmospheric and oceanic condition appears to have a strong multi-decadal component which is reflected in the abundance of marine organisms including stocks of fish, shellfish, and piscivores. The physical environment appears to alternate between warm and cold eras. It is not known whether the biological subsystem, in response, changes continuously or whether the system varies between two or more quasi-stable states. An example could be an alteration in dominance of shelf production by benthic versus pelagic components of the ecosystem. In eastern boundary current ecosystems, sardine and anchovy dominance seem to alternate. In the Bering Sea, would pollock dominance be replaced by that of some other species? Climate shifts are mediated via biological processes related to survival and productivity—processes that are influenced directly by ocean physics. Mechanisms of biological response to climate change should be examined in laboratory and research at sea.

Water Mass Interchange and Formation of Deep Water

Interchange between the North Pacific Ocean and the Bering Sea

What are the special features of water interchange between the North Pacific Ocean and the Bering Sea through the two deep Kamchatka and Near Straits, and the shoaler straits to the east? It is obvious that North Pacific water plays an important role in the formation of Bering Sea thermohaline structure, especially to the circulation of deepwater basins. The North Pacific water masses influence the thermal regime (important to the biological production) of the Bering Sea and impact the main features and variability (annual and longer term) of Bering Sea currents.

Deep Basin Circulation

Do lateral and vertical input processes cause the deep Bering Sea to be a significant repository for global deep water? The deep Bering Sea is the terminus of the deep circulation in the Atlantic and Pacific Oceans as evidenced by the gradients of dissolved substances. This sink of the deep ocean flow therefore receives lateral input from other deep ocean basins and also the vertical flux of materials in the open Bering Sea. The pronounced oxygen minimum and enhanced concentrations of dissolved constituents probably result from a combination of changes in physical, chemical, and biological fluxes to the deep ocean basin. Perhaps the best location to study questions concerning anthropogenic materials is the deep ocean. Both lateral and vertical input processes could be concentrating contaminants such as chlorinated organics, synthetic organics, and radionuclides in the bottom waters. The deep Bering Sea may be an important region to investigate the possible results of changes in global ocean circulation.

Interaction between Deep Basin and Shelf Waters— Biological and Physical

How do the differing biological regimes and transport between the Aleutian basin and the eastern and western shelves contribute to the overall production of the Bering Sea? The Bering Sea shelf has been studied intensely through several programs but the extent of interaction of shelf and deep basin waters is not well understood. More attention should be given to the physical transport of waters, exchange of water across the shelf break through submarine canyons and valleys, shelf break upwelling driven by winds or bathymetrically induced by flows, and the role of eddies. Significant increases in primary production may result from the enrichment of nutrients through cross-shelf processes. The herbivores of the deep Aleutian basin generally are large and spawn over a February-to-April period, primarily independent of local phytoplankton bloom conditions. Herbivores on the shelf are small and closely linked to the blooms during May. It appears that pollock make use of both shallow and deep regions in their life cycle. How does the year-to-year variability in transport onto the shelf contribute to nutrient supply and larval survival? How does the year-to-year variability in storminess, ice cover, and air temperature regulate growth of herbivores on the shelf?

Investigation of the most intensive Bering Sea current, the Kamchatka current, and its relationship to water exchange with the North Pacific Ocean is recommended. The Kamchatka current influences water mass interchange in two modes: meandering and nonmeandering. In the meandering mode the current moves away from the coast and does not influence coastal waters. In the nonmeandering mode it has strong influence on the onshore regime and prevents the large formations of river plumes in the Bering Sea that change vertical stratification due to weak freshwater flux from the coast. Large mesoscale eddies enhance vertical motions and change biological conditions at coastal waters. What are the relationships between the two modes, their frequencies, and other components of water balance as they relate to biological productivity?

Ice Dynamics

Influence of Ice Dynamics on Productivity

What are the effects of year-to-year variations in the maximum southerly extent of sea ice? Specifically, we need to know the effects on primary production, including total production during the spring bloom, timing of the spring bloom, and its duration and intensity. Timing of the spring bloom as determined by sea ice is important to the allocation of the products between the benthic versus pelagic sectors.

What are the important effects of ice dynamics on reproductive biology of animals?

Melting sea ice triggers an early and intense spring bloom, which may be important in satisfying the needs of higher trophic levels. (Note that the APPRISE Project clearly showed the importance of timing of the spring bloom and its intensity in the allocation of the products.) The extremely low water temperatures associated with the marginal ice zone may also inhibit spawning of the shelf species that may not be tolerant of such cold-water regimes (for example, pollock).

What are the effects of ice dynamics on zooplankton grazing?

Grazing at the ice edge appears to be depressed due to low temperatures and lack of overwintering animals. Copepod grazing has been found to be low and depressed, whereas euphausiid grazing dominates. Thus, much of the primary production may sink to the bottom.

Role of the Ice Edge in North-South Retreat

All work on the ice edge system to date has been done either near the shelf break or near the stationary fronts. Farther north over the eastern portion of the Bering Sea shelf, nutrient concentrations are lower and probably limited to the amount regenerated in situ or in the sediments. Here, in shallow waters, the residue of the brief, ice-edge spring bloom may be quantitatively important to primary production when it reaches the sediment surface. This question can be addressed by following the retreating ice (along with the ice edge bloom) northward over the shelf. Remote sensing with the SeaWIF sensor can determine the position and extent of the bloom with time. The distribution and activity of phytoplankton within the water column and over the sediment can be evaluated during the ship's return southward.

Polynyas

The Bering Sea has a number of recurrent polynyas found along the southward-facing coasts of islands as well as land masses. These polynyas are recognized as very important for marine mammals and birds, at a minimum providing access to water in winter, and most likely in other ways too. The larger ones, such as the polynya in the Gulf of Anadyr and the St. Lawrence Island polynya, play an important physical role. As major sites of ice formation and export, they produce saline water which is carried south and northward. These brine waters ultimately form deep water in the North Pacific and Bering Basin as well as move through the Bering Strait to contribute significantly to the halocline of the Arctic Ocean.

While polynyas are assumed to have distinct biological regimes, this has not as yet been examined for polynyas in the Bering Sea. These wind-formed coastal polynyas differ from those found elsewhere, with the possible exception of polynyas in the Sea of Okhotsk. Western Bering Sea polynyas are found in a region which is characterized by high nutrient import and high summer primary production. The interaction of the winter-early spring regime with the subsequent summer regime needs to be examined in relation to the biological consequences.

Biology of Predator-Prey Relationships
Natural Predator-Prey Links

What are the seasonal changes in predator-prey relationships?

Our present understanding of predator-prey relationships in the Bering Sea is limited to when the area is ice free. During the ice season, some top-level predatory species migrate south to warmer waters, e.g., gray whales and northern fur seals, while others move into the area (pagophilic pinnipeds). The seasonal prey-predator links have to be identified. Impact of seasonal predators on prey abundance and recruitment are unknown and should be a subject of priority coordinated research.

What are the key nodal species in the food web?

From empirical directions and food-web theory it is evident that a small number of species at an intermediate position in the food web seem to have an inordinate influence on ecosystem productivity. In the Bering Sea we need to identify the critical food-web nodes, and explore their extent and prevalence. Is there evidence for seasonal change in the nodes over annual production cycle? Do nodal species have properties in common other than their position in the food web? What are the causes and consequences of interannual variation in nodal species in food webs?

Surveys and data analysis

Surveys and data analysis are needed on noncommercially exploited fishes, cephalopods, and other macro-invertebrates. The abundances and distributions of noncommercially exploited fishes and cephalopods in the Bering Sea are not known. Some of these species, particularly cephalopods and mesopelagic fishes, may be important prey and crucial links to the survival of many predatory species. The status and trend of these noncommercially important species should be determined by assessment surveys conducted in conjunction with those for commercially important fish species.

Migratory animals

Long-distance, migratory animals such as whales and seabirds enter the Bering Sea from neighboring and remote ecosystems and use habits seasonally. Some species use the region for reproduction, others use it as a winter area. What role do these highly mobile animals play in the transfer of energy and materials between regions?

Commercial Fishing as a Predator—Effect of Commercial Fisheries on High-Level Trophic Predators, Particularly Marine Mammals and Seabirds

How does fishing change the food web?

Commercial fisheries change the age structure and reduce the abundance of targeted species, even when managed exactly according to recommended, single-species exploitation rates. Secondary effects of this exploitation may occur through predator-prey interactions and may or may not have changed the availability of commercial and noncommercial prey for marine mammals and birds. As an example of cost-effective research, the working group proposes that historical survey data and food habits information be analyzed for evidence of changes in prey and its utilization.

Comparative ecosystem studies

The Bering Sea marine mammal and seabird fauna has generally declined since the late 1960s, coincident with the increase in commercial groundfish fisheries. Conversely, pinniped populations in the California current ecosystem have significantly increased during the same period, as have commercial fisheries. A comparative approach to the study of each system will identify similarities and differences between the two systems and ultimately the mechanisms that determine marine mammal and seabird abundance and trends.

INDEX